POWER PLANTS AND POWER SYSTEMS CONTROL 2003

A Proceedings volume from the 5th IFAC Symposium,
Seoul, South Korea, 15 – 19 September 2003

Edited by

K.Y. LEE
Department of Electrical Engineering,
The Pennsylvania State University, PA, USA

and

M.C. SHIN
Information and Computer Engineering,
Sungkyunkwan University, Suwon-city, Korea

(In three volumes)

Volume 3

Published for the

INTERNATIONAL FEDERATION OF AUTOMATIC CONTROL

by

ELSEVIER LTD

ELSEVIER Ltd
The Boulevard, Langford Lane
Kidlington, Oxford OX5 1GB, UK

Elsevier Internet Homepage
http://www.elsevier.com

Consult the Elsevier Homepage for full catalogue information on all books, journals and electronic products and services.

IFAC Publications Internet Homepage
http://www.elsevier.com/locate/ifac

Consult the IFAC Publications Homepage for full details on the preparation of IFAC meeting papers, published/forthcoming IFAC books, and information about the IFAC Journals and affiliated journals.

First edition 2004

Library of Congress Cataloging in Publication Data

A catalogue record for this book is available from the Library of Congress

British Library Cataloguing in Publication Data

A catalogue record for this book is available from the British Library

ISBN 0-08-044210 2
ISSN 1474-6670

Printed in Great Britain

To Contact the Publisher

Elsevier welcomes enquiries concerning publishing proposals: books, journal special issues, conference proceedings, etc. All formats and media can be considered. Should you have a publishing proposal you wish to discuss, please contact, without obligation, the publisher responsible for Elsevier's industrial and control engineering publishing programme:

Christopher Greenwell
Publishing Editor
Elsevier Ltd
The Boulevard, Langford Lane
Kidlington, Oxford
OX5 1GB, UK

Phone: +44 1865 843230
Fax: +44 1865 843920
E.mail: c.greenwell@elsevier.com

General enquiries, including placing orders, should be directed to Elsevier's Regional Sales Offices – please access the Elsevier homepage for full contact details (homepage details at the top of this page).

5th IFAC SYMPOSIUM ON POWER PLANTS AND POWER SYSTEMS CONTROL 2003

Sponsored by
International Federation of Automatic Control (IFAC)
Technical Committee on Power Plants and Power Systems Control

Organized by
KIEE (The Korean Institute of Electrical Engineers)
on behalf of ICASE, NMO of IFAC

Co-sponsored by
CIGRE
IEEE Power Engineering Society

Supported by
KEPCO (Korea Electric Power Corporation)
Korea Southern Power Co., Ltd
Korea Western Power Co., Ltd
Korea Hydro & Nuclear Power Co., Ltd
Korea East-West Power Co., Ltd
Korea Midland Power Co., Ltd
Korea South-East Power Co., Ltd
Korea Power Exchange
KOPEC (Korea Power Engineering Corporation, Inc.)
KERI (Korea Electrotechnology Research Institute)
KEPRI (Korea Electric Power Research Institute)
EESRI (Electrical Engineering & Science Research Institute)
Korea Electrical Contractors Association

International Programme Committee (IPC)

National Organizing Committee (NOC)

Chair: Dr. Rimtaig Lee
Editors: Prof. Kwang Y. Lee
Prof. Myung-Chul Shin
Vice-Chair: Prof. Young-Hyun Moon
Secretary: Prof. Kyung-Bin Song

Jae-Seok Choi
Gilsoo Jang
Jae Won Jang
Heui Tai Kang
Baik Kim
Chul-Hwan Kim
Jae-Chul Kim
Jae-Eon Kim
Jin-O Kim
Jung-Hoon Kim
Kwang-Ho Kim
Byung-Ha Lee
Kwang-Ho Lee
Seung-Jae Lee
Young-Hwan Moon
Yong-Taek Oh
June Ho Park
Dae-Seok Rho
Kyoung-Soo Ro
Hyo-Seong Song
Kyung-Bin Song

Steering Committee

Chair: Prof. Hee-Ro Kwak
Co-Chairs: Dr. Tay-Ho Jung, Mr. Kap-Koo Yoon

Jeong-Shik Ahn
Young-Sik Baek
Jong-Woong Choe
Jin-Boo Choo
Yong-Chul Chung
KabJu Hwang
Hoyong Kim
Il-Nam Kim
In-Sup Kim
Kem-Joong Kim
Tae-Won Kwon
InKyu Lee
Jin-Moon Lee
Seung-Chul Lee
Tae-Kyoo Oh
Dong-Wook Park
Jin Bae Park
Sang Doug Park
Jang Sun Seo
Hee Seung Shin
Joong-Rin Shin

Sponsoring Committee:
Chair: Mr. In-Seok Koh
Co-Chairs: Mr. In Soon An, Mr. Young-Ki Baek,
Members:
Mr. Yang Uoo Choe
Dr. Rimtaig Lee
Mr. Yong-Oh Lee
Mr. Young Cheol Lee
Dr. Pyung-Sam Ji
Dr. Tay-Ho Jung
Mr. Chang-Joon Kim
Mr. Chun Chae Kim
Mr. Young-Joon Kim
Mr. Yung-Chul Kim
Dr. Young-Han Kwun
Mr. Haeng-Soon Yoon
Dr. Meng-Hyun Yoon

Secretary: Prof. Jung-Hoon Kim

Local Technical Program Committee:

Chair: Prof. Jong-Keun Park
Co-Chairs: Dr. Young-Chang Kim, Prof. Sae-Hyuk Kwon, Dr. Hong-Woo Rhew

Members
Sang-Hoon Chang
Bong-Hee Kim
Kyu-Ho Kim
Moon-Duk Kim
Se-Ho Kim
Seog-Joo Kim
Byong-Jun Lee
Hahk Sung Lee
Heung-Jae Lee
Jong-Beom Lee
Ick-Hun Lim
Seung-Il Moon

PREFACE

This Proceedings volume of the IFAC Symposium on Power Plants and Power Systems Control 2003 held in Seoul, South Korea contains 203 papers presented in 32 oral sessions and a poster session. In addition to these contributed papers, plenary speeches, papers in three panel sessions and a summary of a tutorial session were presented. All of the papers are original and were provided by specialists from 26 different countries.

The papers duly presented at the Symposium are archived in this Proceedings volume published by Elsevier Ltd., Oxford, UK. Upon recommendation, some of the papers will also be published in the IFAC Journals after going through a formal review process.

All this is made possible by the efforts of the contributors and we would like to take this opportunity to thank all delegates for participating. We would also like to thank the organizers, sponsors, co-sponsors and the industrial partners who supported the meeting financially.

The Editors:
Kwang Y. Lee, The Pennsylvania State University, USA
Myong-Chul Shin, Sungkyunkwan University, Korea

CONTENTS

VOLUME 1

PLENARY PAPER

PANEL PAPERS I

PANEL PAPERS II

PANEL PAPERS III

TUTORIAL PAPER

ORAL PAPERS

Power Plant Operation and Control I

Power System Modeling I

Power System Monitoring and Diagnosis

Distribution System Operation

Electricity Pricing/Bidding Schemes

Power System Control I

Distributed Generation I

Power Plant Operation and Control II

Power Plant Operation and Control III

Power Plant Operation and Control IV

Flexible Transmission Systems

Information Management

Power Plant Operation and Control V

Power System Control II

Distributed Generation II

Power System Protection

VOLUME 2

ORAL PAPERS (continued)

Operation of Competitive Electricity Markets I

Power Plant Modeling I

Transmission/Distribution Apparatus

Power System Stability I

Unit Commitment

Simulators/Simulation Methods

Power System Restoration

Voltage and Reactive Power Control

Power System Stabilizer

Power Plant Operation and Control VI

Power System Analysis I

Power System Modeling II

Operation of Competitive Electricity Markets II

Power Plant Modeling II

Power System Analysis II

Power System Stability

POSTER PAPERS

VOLUME 3

POSTER PAPERS (continued)

ELSEVIER
IFAC
PUBLICATIONS
www.elsevier.com/locate/ifac

LOSS MINIMIZATION BY DGs INSTALLATION AND OPERATION IN UNBALANCED DISTRIBUTION SYSTEMS

K. H. Kim[a], Y. J. Lee[b], S. B. Rhee[b], S. K. Lee[c], S. K. You[b]

[a] *Dept. of Electrical Engineering, Ansan College of Technology, Ansan, Korea*
[b] *Dept. of Electrical Engineering, Hanyang University, Ansan, Korea*
[c] *Dept. of Electrical Engineering, National Wonju College, Wonju, Korea*

Abstract: This paper presents the scheme for load model based dispersed generation system (DGs) installation and operation in unbalanced distribution systems. Groups of each individual load model consist of residential, industrial, commercial, official and agricultural load. The problem formulation considers an objective to reduce power loss and the constraints such as the number or total capacity of DGs and the deviation of the bus voltage. The main idea of solving fuzzy nonlinear goal programming is to transform the original objective function and constraints into the equivalent multiple objective functions with fuzzy sets to evaluate their imprecise nature for the criterion of power loss minimization, the number or total capacity of DGs and the bus voltage deviation, and then solve the problem using genetic algorithms. The method proposed was applied to IEEE 13 bus test systems to demonstrate its effectiveness. *Copyright © 2003 IFAC*

Keywords: DGs Installation, Fuzzy Goal Programming, GA, Load Model, Compromised Solution, Unbalanced Distribution Systems

1. INTRODUCTION

Dispersed generators system (DGs) can reduce distribution losses and replace large-scale generators if they are placed appropriately in the distribution systems. DGs are closer to customers so that transmission and distribution cost are avoided or reduced and the latest technologies are perfected and are widely practiced (Dugan and Thomas, 1984; Rizy and Jewell, 1985).

The second order algorithms are used to identify optimal locations of distributed resources in a network to minimize losses, line loadings, and reactive power requirement (Rau and Wan, 1994). Further, the impacts of dispersed generation at the distribution level were performed with an emphasis on resistive losses and capacity savings and proposed the algorithm to determine the near optimal, with respect to system losses, placement of these units on the power grid (Griffin, *et al.*, 2000). Genetic algorithms are introduced to establish the optimal distributed generation allocation on distribution network considering all the technical constraints, like feeder capacity limits, feeder voltage profile and three-phase short circuit current in the network nodes (Celli and Pilo, 2001). Tabu search algorithms are applied for finding the optimal allocation of DGs from a viewpoint of losses minimization (Nara, *et al.*, 2001). Also, the authors proposed a Fuzzy-GA method to resolve DGs installation in balanced distribution systems (Kim, *et al.*, 2002).

This paper presents the scheme for DGs installation and operation based on the voltage characteristics of each individual load component in unbalanced distribution systems. The problem was formulated with an objective to reduce power loss of unbalanced distribution systems and the constraints such as the number or total capacity of dispersed generators and the deviation of the bus voltage. These objective function and constraints are transformed into multiple objective functions and modeled with fuzzy sets to evaluate their imprecise nature for the criterion of power loss minimization, the number or total capacity of DGs and the bus voltage deviation, and then solve the problem using genetic algorithms. The method proposed was applied to IEEE 13 bus test systems to demonstrate its effectiveness.

2. THREE-PHASE POWER FLOW

2.1 Load Model

Each individual load component was expressed by the fourth order equation about voltage variation and was represented by injection currents. Each injected current and load model are detailed in Final Report for Consolidated Edison Company Research and Development Project (1989). The active and reactive powers in terms of the incremental change of the voltage were showed as follows:

$$
\begin{aligned}
P_1 &= P_0\left(a_p + b_p\Delta V + c_p\Delta V^2 + d_p\Delta V^3 + e_p\Delta V^4\right) \\
Q_1 &= Q_0\left(a_q + b_q\Delta V + c_q\Delta V^2 + d_q\Delta V^3 + e_q\Delta V^4\right)
\end{aligned}
\qquad (1)
$$

where

P_0, Q_0 : active and reactive power in base voltage

P_1, Q_1 : active and reactive power in actual voltage

$a_p \cdots e_p$: constant of P for voltage in individual load

$a_q \cdots e_q$: constant of Q for voltage in individual load

Groups of each individual load model consist of residential, industrial, commercial, official and agricultural load. Each load includes kitchen facilities, incandescent lightings, fluorescent lightings, televisions, cooling facilities, heating facilities, cooling fans, refrigerators, pumps, the various size of induction motors and others.

2.2 Flowchart of Three-Phase Power Flow

The Z_{Bus} Gauss approach has been selected for a three-phase power flow (Chen, *et al.*, 1991). The Z_{Bus} Gauss method uses the sparse bifactored Y_{Bus} matrix and equivalent current injections to solve network equations.

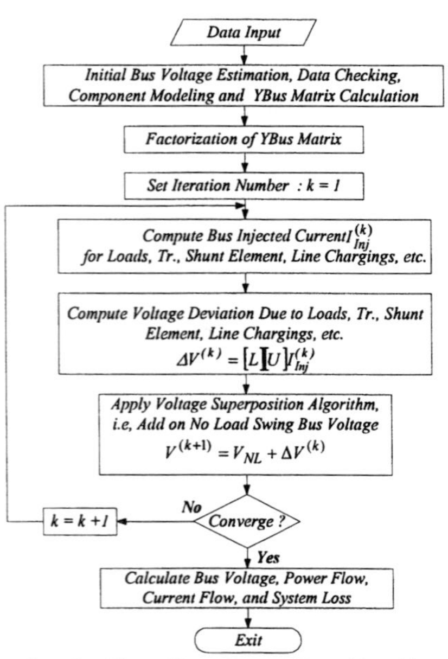

Fig. I. Flowchart for Three-Phase Power Flow Algorithm

The convergence behavior of the Z_{Bus} method is highly dependent upon the number of voltage specified buses in the system. The Z_{Bus} method meets the requirements for both rapid convergence rate and minimum memory usage. Therefore, if the only voltage specified bus in the system is the substation bus, the distribution system is well suited for the Z_{Bus} method. The flowchart of three-phase power flow algorithm is shown in Fig. I.

3. PROBLEM FORMULATION

The loads are assumed to vary in a conformal way and the load variations are supposed to be in discrete levels. Under these assumptions, the so-called load duration curve can be approximated by a piecewise linear function (Grainger and Lee, 1981).

3.1 Objective Function

The objective function in the dispersed generators installation problem is formulated by summing up the power loss multiplied by time duration of each load level (Baran and Wu, 1989).

$$
Min.\ objective\ func. = \sum_{i=0}^{nt} T^i P_{loss}^i \qquad (2)
$$

where P_{loss}^i is the power loss for load level *i*. T^i means a time duration. *nt* is the number of load levels respectively.

3.2 Constrains

Power Flow Equations

$$
F^i(x^i, P_{G,k}^i) = 0 \qquad (3)
$$

where $P_{G,k}^i$ is the dispersed generation control variable at bus *k* and *x* means the state variable such as voltage magnitude.

Operational constraints on bus voltages

$$
V_{min}^i \le V^i \le V_{max}^i \qquad (4)
$$

where *V* is voltage magnitude, min and max indicate the lower and upper limit of variables.

Constraints on the number or total capacity of DGs

$$
P_{G,k,min}^i \le P_{G,k}^i \le P_{G,k,max}^i \qquad (5)
$$

where $P_{G,k}^i$ is the number or total capacity of DGs at bus *k*.

4. SOLUTION ALGORITHMS FOR DGs INSTALLATION

4.1 Fuzzy Membership of Objective Function and Constraints

The objective function and constraints of section 3 are transformed into multiple objective functions and modeled with fuzzy sets to evaluate their imprecise nature. A fuzzy set is generally represented by a membership function. The higher the value of the membership function implies a greater satisfaction with the solution. As the value of the membership function increase, the optimal solution will reach the crisp limits of the variables (Kim, *et al.*, 2002; Gen, *et al.*, 1997).

Minimize the Power Loss

$$Max.\ \ f_1(X) = \sum_{i=0}^{nt} T^i \left| P_{loss}^{i,0} - P_{loss}^i \right| \qquad (6)$$

f_1 represents power loss deviation. $P_{loss}^{i,0}$ and P_{loss}^i are the power loss at load level i before and after DGs installation.

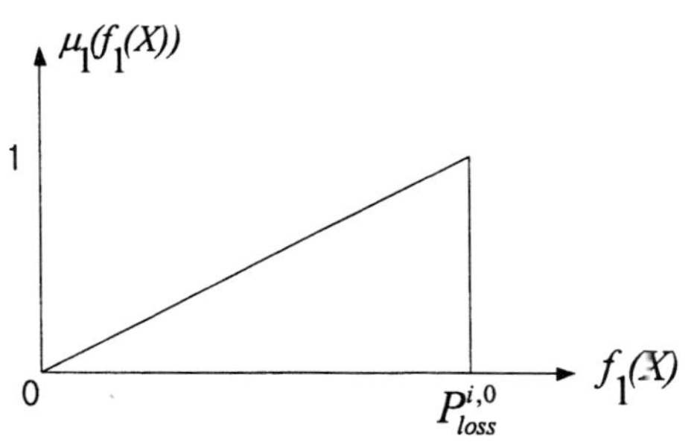

Fig. II. The membership function for the minimization of power loss

Minimize the number or total capacity *of DGs installation*

$$Min.\ \ f_2(X) = \max\left[\left|\Delta P_{G,k}^i\right|\right]\ \ k = 1,2,\cdots,N_G \qquad (7)$$

where $\Delta P_{G,k}^i = P_{G,k}^i - P_{G,k}^{i,min}$, f_2 represents the maximal deviation of the number or total capacity of DGs. N_G is the total number of DGs. i means load level. In Fig. III $\Delta P_{G,k,\min}^i$ is minimum available number or total capacity of DGs and $\Delta P_{G,k,\max}^i$ is maximum available number or total capacity of DGs.

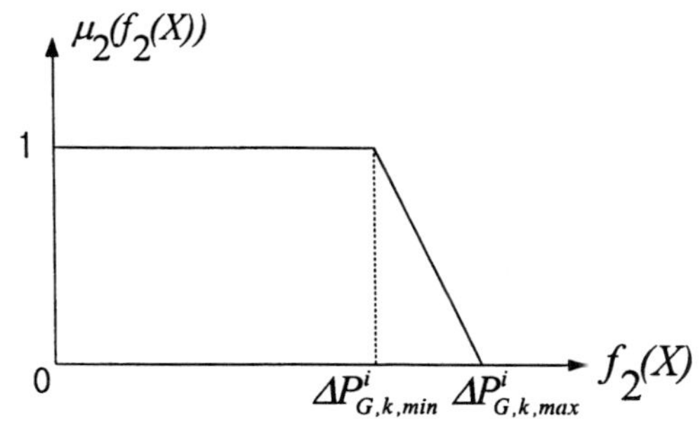

Fig. III. The membership function for the minimization of the number or total capacity of DGs

Minimize the deviation of the bus voltage

$$Min.\ \ f_3(X) = \max\left[\left|\Delta V_j\right|\right]\ \ j = 1,2,\cdots,N_B \qquad (8)$$

where $\Delta V_j = 1.0 - V_j$, f_3 represents the maximal deviation of the bus voltage. N_B is the total number of the buses. V_j denotes the voltage on bus j. In Fig. IV $\Delta V_{j,min}$ is minimum deviation of voltage at bus j and $\Delta V_{j,max}$ is maximum deviation of voltage at bus j.

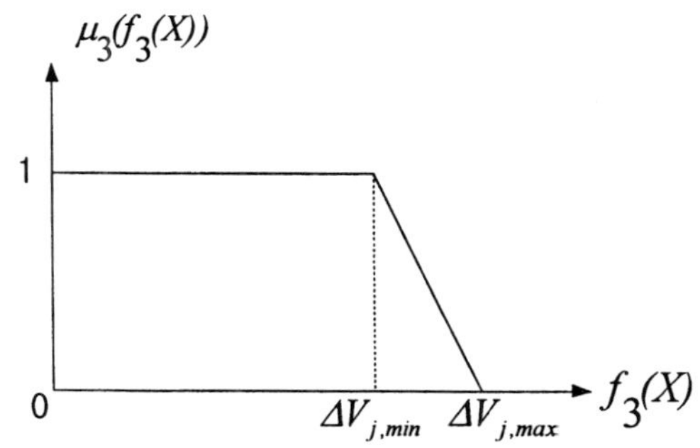

Fig. IV. The membership function for the minimization of the deviation of the bus voltage

Three objective functions (6)~(8) can be modified by goal programming as follow:

$$\begin{aligned} & Max.\ \ \sum_{k=1}^{3} \omega_k \mu_k(f_k(x)) \\ & \text{s.t.} \\ & 0 \le \mu_k(f_k(x)) \le 1 \quad k = 1,2,3 \end{aligned} \qquad (9)$$

We try to get the compromised solution of multiple objectives and imprecise information using genetic algorithms.

4.2 Genetic Algorithms

Genetic algorithms (GA) are very powerful search method based on the mechanics of natural selection and natural genetics (Michalewicz, 1992). GA searches from a population of points in the search space simultaneously. GA does not need derivatives or other auxiliary knowledge. The Algorithm procedure is as follows:

Step 1. Representation and Initialization
Initialize N subpopulation assigned to each load level. The genes of each individual in N subpopulation are randomly generated within the maximum and minimum limits of each variable

Step 2. Fitness Evaluation and Penalty Function
Evaluate the fitness after distribution load flow using each gene of individual in N subpopulation.

$$Fitness\ \ func. = \sum_{k=1}^{3} \omega_k \mu_k(f_k(X)) \qquad (10)$$

Step 3. Criterion of convergence
If it is maximum generation, stop. Otherwise, go to *step 4*.

Step 4. Manipulate Genes of Individual
In order to generate a new and improved population

from old one, GA uses genetic operators such as reproduction, whole arithmetical crossover, non-uniform mutation. After the manipulation of genes, go to *step 2*. Each operator is detailed in the book of Michalewicz (1992).

5. NUMERICAL RESULTS

The proposed scheme for DGs installation and operation was implemented by Visual C++ language on a Pentium-III computer to prove the efficiency of the algorithms. The test system is IEEE 13 bus unbalanced distribution systems, as shown in Fig. V. A system load data is presented in Table 1. The other data such as line configuration and transformer data can be found in IEEE Distribution Planning Working Group Report (1991). Three load levels and load duration time data are given in Table 2. To investigate the scheme for the optimal DGs installation and operation based on the voltage characteristics of each individual load component, the simulation of different cases has been implemented as follows:

Case 1: Load model in bus is considered. Groups of each individual load model consist of residential, industrial, commercial, official and agricultural load. Each load includes kitchen facilities, incandescent lightings, fluorescent lightings, televisions, cooling facilities, heating facilities, cooling fans, refrigerators, pumps, the various sizes of induction motors and others. The type of DGs can be controlled with load variation.
Case 2: All loads are PQ constant. The type of DGs can be controlled with load variation.
Case 3: Load model in bus is considered. The type of DGs cannot be controlled with load variation.
Case 4: All loads are PQ constant. The type of DGs cannot be controlled with load variation.

The simulation constants are as follows:

Available capacity of DGs[kW]: 30, 60, 90, 120, 150
$\Delta P^{i}_{G,k,\min}$: 1EA(Number), 30[kW](Total capacity)
$\Delta P^{i}_{G,k,\max}$: 6EA(Number), 900[kW](Total capacity)
$\Delta V_{j,min}$: 0.07
$\Delta V_{j,max}$: 0.15
Time period [h]: 8(9:00 ~ 17:00)
Test years: 5

Table 3 shows the locations and capacity of DGs installation. In number of available installation of DGs, optimal locations are bus 611, 652, 675 and 692. Optimal capacities are different according to load levels. Especially, in case of Case 1 and Case 2 DGs capacities are very different because of various industrial loads such as induction motors. It is seen that DGs is well installed and operated according to load model and load variation. It can be found that DGs installation based on load model is very important.

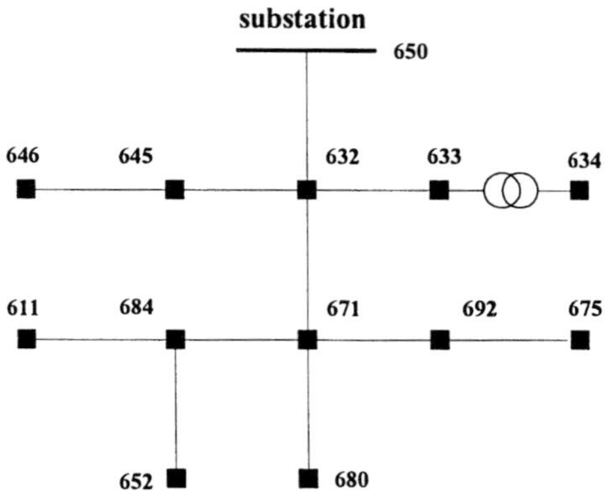

Fig. V. IEEE 13 bus test systems

Table 1 Load model ID and Bus load data

Node	Load Model ID	Ph-1 kW	Ph-1 kVAR	Ph-2 kW	Ph-2 kVAR	Ph-3 kW	Ph-3 kVAR
611	2	0	0	0	0	170	80
634	3	160	116	120	87	120	87
645	5	0	0	170	125	0	0
646	4	0	0	230	132	0	0
652	2	128	86	0	0	0	0
671	3	385	220	385	220	385	220
675	3	485	190	68	60	290	212
692	4	0	0	0	0	170	151

ID = 1: PQ constant, 2: residential, 3: industrial, 4: commercial, 5: official, 6: agricultural

Table 2 Load level and load duration time

Load level (%)	100	90	70
Time interval (hours)	3650	7300	3650

Table 3 Optimal DGs locations and capacities

Optimal location	Case 1[kW]			Case 2 [kW]		
	100%	90%	70%	100%	90%	70%
611	90	90	90	90	60	0
652	120	120	90	150	150	120
675	150	150	150	120	120	90
692	90	60		150	150	150
Total	450	420	330	510	480	360

Table 4 Real power losses and voltage profiles

	Load level [%]	Case 1 Without DGs	Case 1 With DGs	Case 2 Without DGs	Case 2 With DGs
Real power losses [kW]	100	434.229	290.175	492.780	302.080
	90	698.028	448.521	782.118	473.434
	70	208.431	128.344	227.669	142.934
Total losses [kW]		1,340.689	867.041	1,502.568	918.450
V_{min} [p.u.]	100	0.8999	0.9411	0.8884	0.9364
	90	0.9105	0.9510	0.9012	0.9470
	70	0.9311	0.9619	0.9255	0.9595
V_{ave} [p.u.]	100	0.9434	0.9566	0.9388	0.9558
	90	0.9494	0.9624	0.9456	0.9611
	70	0.9610	0.9721	0.9587	0.9697

In Table 4, the comparison of losses without/with DGs installation is shown. The system losses are reduced in each load level of Case 1 and 2. The total losses for the system without any DGs installation in

Case 1 were found to be 1,340,689 [kW]. After DGs installation, the total losses were 867,041 [kW]. The saving losses can be seen 473,648 [kW] during 5 years. The voltage profiles are improved significantly.

In Case 3 and 4, the test results are presented for type of DGs that cannot be controlled with load variation. The optimal locations and capacities are shown in Table 5 and Table 6. In comparison with Case 1 and 2, the type of DGs that can be controlled with load variation is very efficient.

Table 5 Optimal DGs locations and capacities

Optimal location	Case 3[kW]			Case 4 [kW]		
	100%	90%	70%	100%	90%	70%
611	120	120	120	120	120	120
652	150	150	150	150	150	120
671	60					
675	150	120	90	150	150	120
680	60			60	60	
684				60		
692	30	30				
Total	570	420	360	570	480	360

Table 6 Real power losses and voltage profiles

	Load level [%]	Case 3		Case 4	
		Without DGs	With DGs	Without DGs	With DGs
Real power losses [kW]	100	434.229	320.705	492.780	360.057
	90	698.028	548.999	782.118	581.339
	70	208.431	160.405	227.669	171.670
Total losses [kW]		1,340.68	1,030.11	1,502.56	1,113.06
V_{min} [p.u.]	100	0.8999	0.9396	0.8884	0.9348
	90	0.9105	0.9482	0.9012	0.9444
	70	0.9311	0.9609	0.9255	0.9588
V_{ave} [p.u.]	100	0.9434	0.9480	0.9383	0.9445
	90	0.9494	0.9526	0.9456	0.9503
	70	0.9610	0.9638	0.9587	0.9621

6. CONCLUSION

This paper presents the scheme for DGs installation and operation based on the voltage characteristics of each individual load component in unbalanced distribution systems. Groups of each individual load model consist of residential, industrial, commercial, official and agricultural load.

The objective function and constraints are transformed into multiple objective functions and modeled with fuzzy sets to evaluate their imprecise nature for the criterion of power loss minimization, the number or total capacity of DGs and the bus voltage deviation. Moreover, we tried to get the compromised solution of multiple objectives and imprecise information using goal programming and genetic algorithms. Also, it is seen that DGs installation based on load model is very important.

REFERENCES

Final Report for Consolidated Edison Company Research and Development Project 92087 (1989). Generalized Distribution Analysis System.

Michalewicz, Z. (1992). *Genetic algorithms + Data structures = Evolution Programs*. Second Edition, Springer Verlag

Baran, M. and F. Wu (1989). Optimal capacitor placement on radial distribution system. IEEE Trans. on Power Delivery, **Vol. 4**, **No. 1**, pp.725-734

Celli, G. and F. Pilo (2001). Optimal distributed generation allocation in MV distribution networks. 22nd IEEE Power Engineering Society International Conference, pp.81-86,

Chen, T. H., M. S. Chen, K. J. Hwang, P. Kotas and E. A. Chebli (1991). Distribution System Power Flow Analysis – A Rigid Approach. IEEE Trans. on Power Delivery, **Vol. 6, No. 3**, pp.1146-1152

Dugan, R. C. and S. A. Thomas (1984). Integrating Dispersed Storage and Generation (DGS) with An Automated Distributed System. IEEE Trans., **PAS**, pp.1142-1146

Gen, M., K. Ida and J. Kim (1997). Fuzzy goal programming using genetic algorithm. IEEE International Conference, pp. 413 –418

Grainger, J. J. and S. H. Lee (1981). Optimum size and location of shunt capacitors for reduction of losses on distribution feeders. IEEE Trans. on Power Apparatus and Systems, **Vol. PAS-100**, pp.1105-1116

Griffin, T., K. Tomsovic, D. Secrest and A. Law (2000). Placement of dispersed generations systems for reduced losses. *Proceedings of the 33rd Annual Hawaii International Conference*, pp.1446 –1454,

IEEE Distribution Planning Working Group Report (1991). Radial Distribution Test Feeders. IEEE Trans. on Power System, **Vol. 6, No. 3**, pp.975-985

Kim, K. H., Y. J. Lee, S. B. Rhee, S. K. Lee and S. K. You (2002). "Dispersed Generator Placement using Fuzzy-GA in Distribution. IEEE Power Engineering Society Summer Meeting, pp.1148-1153

Nara, K., Y. Hayashi, K. Ikeda and T. Ashizawa (2001). Application of tabu search to optimal placement of distributed generators. IEEE Power Engineering Society Winter Meeting, **Vol. 2**, pp.918 –923,

Rau, N. S. and Y. H. Wan (1994). Optimum location of resources in distributed planning. IEEE Trans. on Power Systems., **Vol. 9**, pp.2014 –2020

Rizy, D. T. and W. T. Jewell (1985). Operation and design Considerations for electric Distribution Systems with Dispersed Storage and Generation (DGS). IEEE Trans., **PAS**, pp.2864-2871

ELSEVIER
IFAC
PUBLICATIONS
www.elsevier.com/locate/ifac

NETWORK RECONFIGURATION USING CHAOS SEARCH METHOD IN UNBALANCED DISTRIBUTION SYSTEMS

Sang-Bong Rhee [a*], Yu-Jeong Lee [a], Kyu-Ho Kim [b], Seok-Ku You [a]

[a] *Department of Electrical Engineering, Hanyang University, Korea*
[b] *Department of Electrical Engineering, Ansan College of Tech., Korea*

Abstract: In this paper, we applied a chaos optimization method to reduce computational effort and enhance optimality of the solution in feeder reconfiguration problem. Chaos method, in optimization problem, searches the global optimal solution on the regularity of chaotic motions and more easily escapes from local or near optimal solution than stochastic optimization algorithms. The chaos optimization method applied to the IEEE 13 unbalanced test feeder systems, and the test results indicate that it is able to determine appropriate switching options for global optimum configuration with less computation efforts. *Copyright © 2003 IFAC*

Keywords: Distribution feeder configuration, Three-Phase load flow, Chaos theory, Optimal search techniques, Global optimization

1. INTRODUCTION

Network reconfiguration is a very effective and important operation to reduce distribution resistive line losses and improve system security performance (voltage profile). But reconfiguration problem is a very complicated non-linear optimization problem. There are many works and researches have been proposed for optimal feeder reconfiguration. The reconfiguration problem is formulated using heuristics rules, seeking to reduce the computation burden (Civanlar, *et al.*, 1998; Goswami & Basu, 1992). Other works (Baran & Wu, 1989; Jasmon, *et al.*, 1991) try to achieve better and faster solutions, incorporating both optimization methods and new load flow algorithms. But these methods are not, however, designed to solve unbalanced three phases distribution systems.

In the last few years, new technologies based on artificial intelligence and advanced heuristic methods have been applied to this problem, trying to deal this combinatorial optimization problem more efficiently. In 1998, Li and Jiang proposed the utilization of the chaos search algorithm (CSA) for guaranteeing the optimality of the solution in optimization problems (Li & Jiang, 1998). This is the algorithm, which used chaotic movement for new search surface of the solutions. CSA is used in this paper just for obtaining the solution sets, which satisfied the optimality of the optimization problems. CSA is not like some stochastic optimization algorithms that escape from local minima by accepting some bad solutions according to certain probability. It searches on the regularity of chaotic motions. CSA easily escape from local minima than stochastic optimization algorithms such as GA, SA, and tabu search.

In this paper we proposed the application of the CSA to the network reconfiguration problem in IEEE 13 unbalanced distribution systems. In the case studies the proposed method was significantly more efficient than those optimization algorithms currently in use.

2. CHAOTIC SEARCH ALGORITHM (CSA)

Chaos often exists in non-linear systems. It is the highly unstable motion of deterministic systems in finite phase space. High instability means that the neighbour traces depart exponentially with time. Because of the instability, the long-time motion of systems shows some confusion and typical stochastic properties. But, Chaos is not really chaotic it has exquisite structure. Chaos has some good properties such as stochastic property and regularity. The Chaotic movement can go through every state in the certain area according to its own regularity, and every state is obtained only once. Taking advantage of these chaos properties, a new chaos search algorithm (CSA) is proposed by Li and Jiang for global optimization problems (Li & Jiang, 1998).

2.1 Chaotic Behaviour for Optimization

The long-time behavior of a chaos system shows a kind of stochastic property. These systems are very sensitive to their initial conditions. Very small differences in the initial conditions will cause large differences in the long-time behavior of systems.

Suppose the well-known logistic mapping

$$x_{n+1} = f(\mu, x_n) = \mu x_n (1 - x_n) \tag{1}$$

where μ is a control parameter and $0 \le x_0 \le 1$, $0 \le \mu \le 4$. It is easy to find that (1) is a deterministic dynamic system without any stochastic disturbance. It seems that long-time behavior can be predicted. But that is not true. The behavior of (1) is greatly changed with the variation of μ. Test results are shown in Fig. Ⅰ.

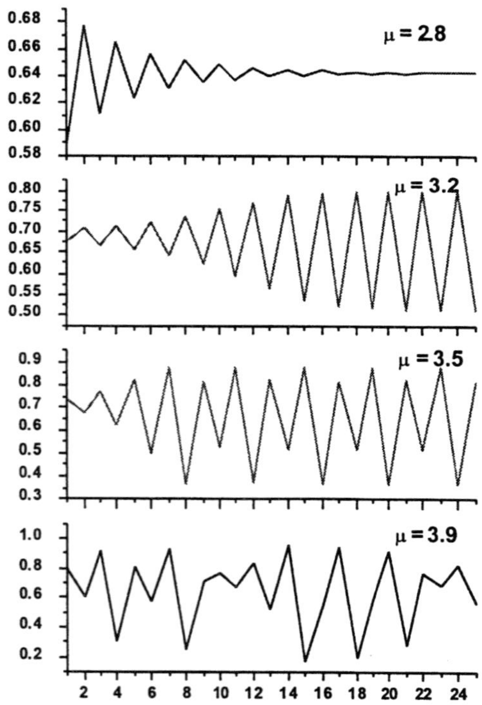

Fig. I. The Output of equation (1) According to Various μ

When $\mu \ge 3.5$, (1) becomes chaotic, and the (1) is changed to

$$x_{n+1} = 4 x_n (1 - x_n) \tag{2}$$

2.2 Chaotic Search Algorithm (CSA)

General optimization problems can be described as (3).

$$\begin{aligned} &\min \quad f(x_i) \quad i = 1, \cdots, n \\ &s.t. \quad a_i \le x_i \le b_i \end{aligned} \tag{3}$$

Then chaotic search algorithm can be used to optimize these problems. The main procedures of the algorithm are shown as follows.

Step 1. Generate chaos variables
Give i initial values, which have very small difference to $x(n)$ of (2), then i chaotic states can be obtained by (2).

Step 2. First carrier wave
By the carrier wave method, change i optimization variables to chaos variables. Furthermore, amplify the ergodic areas of the i chaos variables to the variance ranges of optimization variables by (4).

$$x_i'(n+1) = a_i + (b_i - a_i)\, x_i(n+1) \tag{4}$$

where $x_i(n+1)$ are i chaotic states generated by (2), a_i and b_i are bounds of (3). $x_i'(n+1)$ are i chaotic variables of the optimization problem.

Step 3. Rough search
Let $x_i^* = x_{i0}$, and calculate the initial value of the objective function f, set $f^* = f$. Do $n = n+1$, $x_i = x_i'(n+1)$ and calculate the value of objective function f. If $f \le f^*$ then $f^* = f$, $x_i^* = x_i'$. Else if then give up the solution. Loop until f^* does not improve after k searches.

Step 4. Second carrier
Do the second carrier wave by (5)

$$x_i''(n+1) = x_i^* + \alpha x_i(n+1) \tag{5}$$

where x_i^* is the best solution up to now, α is a constant, and $\alpha x_i(n+1)$ generates i chaotic states with small ergodic ranges around x_i^*.

Step 5. Fine search
Do $n = n+1$. Let $x_i = x_i''(n+1)$, and calculate the objective function f. If $f \le f^*$ then $f^* = f$, $x_i^* = x''$. Else if give up the solution. Loop until the stop criterion is satisfied.

Step 6. End
Stop the search process and put out x^*, f^* as the best solution.

3. FEEDER RECONFIGURATION

Feeder reconfiguration problem, which is minimized the line losses under the several constraints, can be mathematically formulated as follows:

$$Minimize \quad f_c = \sum_{i=1}^{LN} Loss_i \tag{6}$$

where LN is total number of branches. $Loss_i$ means real power loss in branch i.

Subject to:
a) The electrical constraints have to be satisfied.
b) No feeder section can be left out-service.
c) Radial network structure must be retained.

4. THREE PHASE LOAD FLOW FOR UNBALANCED DISTRIBUTION SYSTEMS

Distribution load flow is a very important tool for the analysis of distribution systems and is used in network reconfiguration as well as operating and planning the systems. But distribution systems have some of the inherent features i.e., a radial network structure, an unbalanced load and unbalanced operation, an extremely large number of branches /nodes, and a wide range of resistance and reactance. These features cause the Gauss-Seidel and Newton-Raphson methods to lack either computer economy or robustness in distribution systems applications.

Ghosh and Das proposed simple and easy ideas on how to deal the distribution networks. Ghosh and Das's method involves only the evaluation of a simple algebraic expression. Time-consuming procedures, such as LU factorization and Jacobian matrix construction are not needed.

4.1 Load Flow Methodology

To illustrate the method, consider the radial network in Fig. II. that is assumed balanced and presented by their equivalent single-line diagrams. The branch number sending-end and receiving-end node of this feeder are given in Table 1.

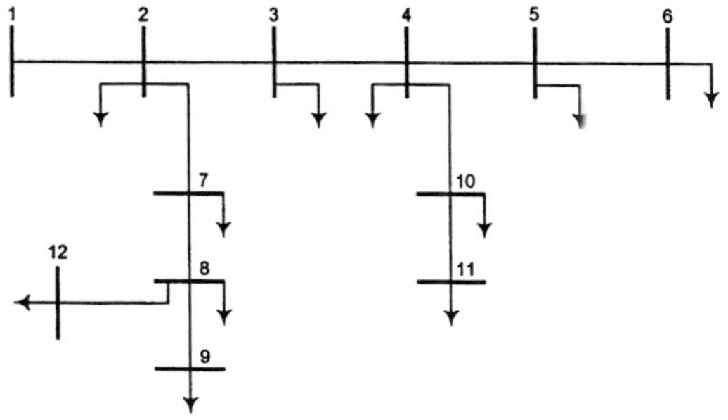

Fig. II. Single-Line Diagram of Radial Distribution Network

The branch 1, sending and receiving node are 1 and 2, the receiving-end node voltage can be written as

$$V[2] = V[1] - I[1]\,Z[1] \tag{7}$$

Similarly for branch 2,

$$V[3] = V[2] - I[2]\,Z[2] \tag{8}$$

TABLE 1

B[JJ]	IS[JJ]	IR[JJ]	NODES BEYOND BRANCH JJ	N[JJ]
1	1	2	2, 3, 4, 7, 8, 5, 10, 9,12, 6, 11	11
2	2	3	3, 4, 5, 10, 6, 11	6
3	3	4	4, 5, 10, 6, 11	5
4	4	5	5, 6	2
5	5	6	6	1
6	2	7	7, 8, 9, 12	4
7	7	8	8, 9, 12	3
8	8	9	9	1
9	4	10	10, 11	2
10	10	11	11	1
11	8	12	12	1

B[JJ] : BRANCH NUMBER, IS[JJ] : SENDING-END NODE
IR[JJ] : RECEIVING-END NODE
N[JJ] : TOTAL NUMBER OF NODES BEYOND BRANCH B[JJ]

As the substation voltage $V[1]$ is known, so if $I[1]$ is known, i.e. current of branch 1, it is easy to calculate $V[2]$ from (7). Once $V[2]$ is known, it is easy to calculate $V[3]$ from (8), if the current through branch 2 is known. Similarly, voltages of nodes 4, 5, ..., NB can easily be calculated if all the branch currents are known. Therefore, a generalized equation of receiving-end voltage, sending-end voltage, branch current and branch impedance is

$$V[IR(jj)] = V[IS(jj)] - I[jj]\,Z[jj] \tag{9}$$

where jj, $IR(jj)$, and $IS(jj)$ is the branch number, receiving node and sending node respectively. Equation (9) can be evaluated for jj=1, 2, ..., $LN(LN=NB$ -1 = number of branches). Current through branch 1 is equal to the sum of the load currents of all the nodes beyond branch 1 plus the sum of the charging currents of all the nodes beyond branch 1, i.e.

$$I[1] = \sum_{i=2}^{LN} IL[i] + \sum_{i=2}^{LN} IC[i] \tag{10}$$

Therefore, if it is possible to identify the nodes beyond all the branches, it is possible to compute all the branch currents. Identification of the nodes beyond all the branches is realized through an algorithm as explained in (Ghosh & Das, 1999). The load current of node i is

$$IL[i] = \frac{PL[i] - jQL[i]}{V^*[i]} \qquad i = 2, 3, \ldots, NB \tag{11}$$

The charging current at node i is

$$IC[i] = y_0[i]\,V[i] \qquad i = 2, 3, ..., NB \tag{12}$$

Load currents and charging currents are computed iteratively. Initially, a flat voltage of all nodes is assumed and load currents and charging currents of all the loads are computed using (11) and (12). The real and reactive power loss of branch *jj* are given (13) and (14).

$$Loss\ P[jj] = |I[jj]|^2\ R[jj] \tag{13}$$

$$Loss\ Q[jj] = |I[jj]|^2\ X[jj] \tag{14}$$

4.2 Three-Phase Model

Fig. Ⅲ. shows the phase and neutral lines between node i and j.

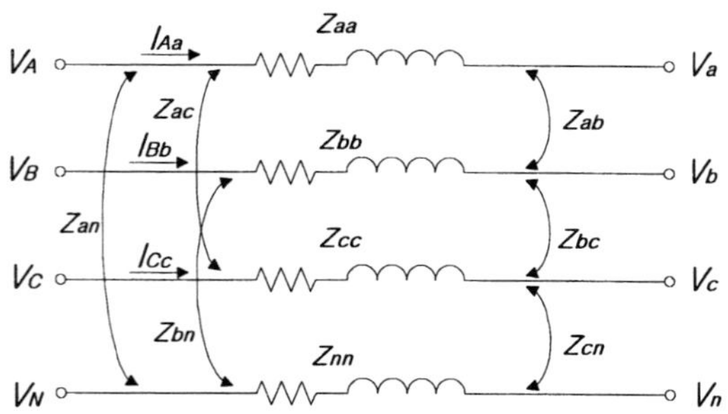

Fig. Ⅲ. Three-Phase Line

After Kron's reduction applied the equivalent series impedance matrix is modified 3x3 matrix.

$$[Z_{abc}] = \begin{bmatrix} Z_{aa-n} & Z_{ab-n} & Z_{ac-n} \\ Z_{ba-n} & Z_{bb-n} & Z_{bc-n} \\ Z_{ca-n} & Z_{cb-n} & Z_{cc-n} \end{bmatrix} \tag{15}$$

The (9) can be expanded to a three-phase line as (16).

$$[V_{abc}] = [V_{ABC}] - [Z_{abc}] \cdot [I_{ABC}]$$

$$\begin{bmatrix} V_a \\ V_b \\ V_c \end{bmatrix} = \begin{bmatrix} V_A \\ V_B \\ V_C \end{bmatrix} - \begin{bmatrix} Z_{aa-n} & Z_{ab-n} & Z_{ac-n} \\ Z_{ba-n} & Z_{bb-n} & Z_{bc-n} \\ Z_{ca-n} & Z_{cb-n} & Z_{cc-n} \end{bmatrix} \cdot \begin{bmatrix} I_{Aa} \\ I_{Ba} \\ I_{Ca} \end{bmatrix} \tag{16}$$

The report (ECRC, 1991) explained in detail the rest line components, transformer, real and reactive load, and shunt capacitor.

5. CSA FOR RECONFIGURATION

The network reconfiguration for loss minimization is performed by opening and (or) closing the sectionalizing and tie-switches. And the radial of the network is retained and at the same time power loss is minimized. CSA has advantages to solve problems with non-linear, non-derivative, discrete objective functions and constraints. In this paper, the CSA approach is employed to obtain the feasible combination of the switches and determine the switching status of the individual switches.

At first, create meshed network by closed the tie-switch and assign the switch of the each loop to chaos variable. Then generate chaos value that is open switch number of the created loop. The infeasible solutions are omitted from the total combinations by checking the radial structure. The three-phase load flow is calculated for each solution of the switching configuration and the total loss is estimated. Three-phase load flow calculation is separated from the optimization process. That is, once every switch state are generated from CSA, run load flow to update the objective value. Because the every stage is needed load flow analysis, the fast load flow method, which is discussed in section 4, is used in this paper.

A flow diagram of CSA method for reconfiguration is shown in Fig. Ⅳ..

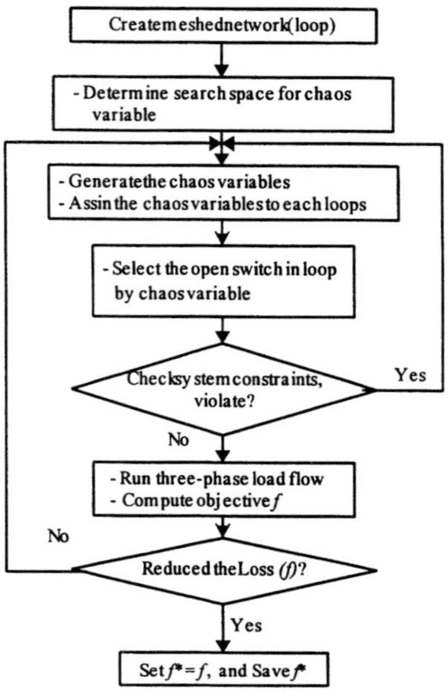

Fig. Ⅳ CSA Algorithm for Reconfiguration

6. NUMERICAL RESULTS

To demonstrate the usefulness of CSA for three-phase distribution system, the IEEE 13 node test feeder, as shown in Fig. Ⅴ., is used.

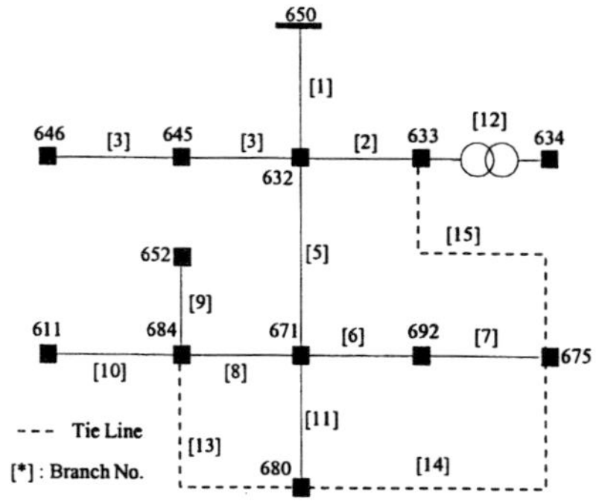

Fig. Ⅴ IEEE 13 Node Test Feeder

The tie-line data is presented in Table 2. The other data can be found in (IEEE, 1991). Among the other data the load model is regard as constant PQ load and multiplied by 1.3. Also the regulator is neglected.

TABLE 2 TIE-LINE SEGMENT DATA

BR. NO	NODE A	NODE B	PHASE	LENGTH (FT)	CONFIG
[13]	684	680	ABCN	1000	606
[14]	680	675	ABCN	1000	606
[15]	675	633	ABCN	1000	606

The base system (before configuration) voltage profile is listed in Table 3.

TABLE 3 VOLTAGE PROFILE (BEFORE CONFIGURATION)

NODE	PHASE A	PHASE B	PHASE C
650	1.0000	1.0000	1.0000
632	0.9783	0.9846	0.9609
671	0.9648	0.9897	0.9179
680	0.9648	0.9897	0.9179
645	-	0.9743	0.9634
633	0.9764	0.9834	0.9592
646	-	0.9708	0.9643
634	0.9670	0.9762	0.9521
684	0.9635	-	0.9162
611	-	-	0.9144
652	0.9588	-	-
692	0.9648	0.9897	0.9179
675	0.9602	0.9912	0.9160
P LOSS	66.651 [KW]	40.385 [KW]	84.863 [KW]
Q LOSS	183.671 [KVAR]	111.898 [KVAR]	251.006[KVAR]
TOTAL	P : 191.8987 [KW]	Q :546.5746 [KVAR]	

TABLE 4 THE SOLUTION OF EACH SEARCH STEP BY CSA

STEP	POWER LOSS CHANGE [KW]	SOLUTIONS [BRANCH IN-OUT]
1	191.89	[13]-[8], -[14], [15]-[7]
2	170.34	[13]-[11], [6]-[14], -[15]
3	162.87	[13]-[8], -[14], [15]-[5]
4	160.02	[13]-[11], -[14], [15]-[5]
5	E/S**	

The CSA method is tested on test system using C language. The test result obtained by the proposed algorithm is given in Table 3. The real power losses are also listed. The E/S means the end of search by CSA. The overall loss minimum configurations for the test system is obtained which give the optimal or near optimal solutions for loss minimization.

The minimum voltage magnitudes on node at each search step are listed in Table 4. The voltage profile of the test system is considerably improved by proper reconfiguration. For instance, the minimum voltage magnitude at each phase 0.9588, 0.9708, and 0.9144 [p.u] respectively, before reconfiguration and is improved to phase A: 0.9628, phase B: 0.9734, and phase C: 0.9234 [p.u] after reconfiguration.

The real power losses are improved by proper network reconfiguration as shown in Table 5. The total real power loss before/after reconfiguration is 190.89 [kW] and 160.139.02 [kW] in test system. This represents near 16 % reduction in real power loss.

TABLE 5 VOLTAGE PROFILE (AFTER CONFIGURATION)

NODE	PHASE A	PHASE B	PHASE C
650	1.0000	1.0000	1.0000
632	0.9847	0.9889	0.9733
671	0.9753	0.9934	0.9431
680	0.9753	0.9934	0.9431
645	-	0.9810	0.9752
633	0.9833	0.9881	0.9720
646	-	0.9784	0.9758
634	0.9765	0.9829	0.9670
684	0.9742	-	0.9419
611	-	-	0.9407
652	0.9707	-	-
692	0.9753	0.9934	0.9431
675	0.9719	0.9946	0.9419
P LOSS	55.913 [KW]	33.979 [KW]	70.123 [KW]
Q LOSS	154.042 [KVAR]	94.007 [KVAR]	207.396[KVAR]
TOTAL	P : 160.0152 [KW]	Q : 455.4437 [KVAR]	

7. CONCLUSIONS

In this paper, a new feeder reconfiguration method based on the chaotic search algorithm for three-phase radial distribution systems is proposed. By performing the computer simulation for the IEEE 13 node test systems, the CSA algorithm can give optimal solution within a reasonable computing effort in the area of network reconfiguration. Therefore, the CSA algorithm can be used for real time online implementation and other optimization problems analysis tool in the power systems.

REFERENCES

Baran, M. E. & Wu, F. F. (1989). Network Reconfiguration in Distribution Systems for Loss Reduction and Load Balancing. *IEEE Trans. on PWRD*, vol. 4, pp. 1401-1407.

Civanlar, S., Grainger, J. J., Yin, H., & Lee S. H. (1988). Distribution Feeder Reconfiguration for Loss Reduction. *IEEE Trans. on PWRD*, vol. 3, pp. 1217-1223.

ECRC. (1991) Final Report for Consolidated Edison Company Research and Development Project 92087, "Generalized Distribution Analysis System", June, 1991.

Ghosh, S. & Das, D. (1999). Method for load-flow solution of radial distribution networks. *IEE Proc.-Gener. Trans. Distrib.*, vol. 146, No. 6, pp. 641-648.

Goswami, S. K., & Basu, S. K. (1992). A New Algorithm for the Reconfiguration of Distribution Feeder for Loss Minimization. *IEEE Trans. on PWRD,* vol. 4, pp. 1484-1490.

IEEE. (1991). IEEE Distribution Planning Working Group Report, "Radial Distribution Test Feeders," *IEEE Trans, on Power Systems,* vol. 6, No. 3, pp. 975-985, 1991.

Jasmon, G. B., Callistus, L. H., & Lee, C. (1991). A Modified Technique for Minimization of Distribution System Losses. *Electrical Power System Research*, vol. 20, pp. 81-88.

Li, B. & Jiang W. (1998). Optimizing complex Functions by Chaos search. *An International Journal of Cybernetics and Systems*, vol. 29, pp. 409-419.

ELSEVIER
IFAC
PUBLICATIONS
www.elsevier.com/locate/ifac

THE BEST LINE CHOICE FOR TRANSMISSION SYSTEM EXPANSION PLANNING ON THE SIDE OF THE HIGHEST RELIABILITY LEVEL

Sungrok Kang[1], Trungtinh Tran[1], Jaeseok Choi[1] Junmin Cha[2] Daeseok Rho[3] Roy Billinton[4]

[1]*Power System Lab., Department of Electrical Engineering*
Gyeongsang National University, Chinju, GN, KOREA
Tel: +82-55-751-5347, Fax: +82-55-759-2723, E-mail: jschoi@nongae.gsnu.ac.kr

[2]*Power System Lab., Department of Electrical Engineering*
Daejin University, Pocheon, KG, KOREA
Tel: +82-31-539-1915, Fax: +82-31-535-3579, E-mail: chamin@road.daejin.ac.kr

[3]*Power System Lab., Department of Electrical Engineering*
University of Technology and Education Chonan, CN, KOREA
Tel: +82-41-560-1167, Fax: +82-41-564-3261, E-mail: dsrho@kut.ac.kr

[4]*Department of Electrical Engineering*
University of Saskatchewan, Saskatoon, SK, CANADA
E-mail: roy_billinton@engr.usask.ca

Abstract : This paper presents a new method for making a choice of the best line for transmission system expansion planning on the side of the highest reliability level of transmission system. Conventional methodologies for transmission system expansion planning have been mainly focused on economics, which is minimization of construction cost. However, quantitative evaluation of transmission system reliability is important because successful operation and planning of an electric power system under the deregulated electricity market depends on transmission system reliability management. Therefore, it is expected that the development of methodology in order to choice best lines on side of the highest transmission system reliability level considering uncertainties of equipments of transmission system is useful in future. The characteristics and effectiveness of the proposed methodology are illustrated by the case study using a MRBTS. *Copyright © 2003 IFAC*

Keywords: Transmission system expansion planning, Transmission system reliability, Best transmission line choice.

1. INTRODUCTION

Deregulation with open access to the transmission system has become a hot issue in electricity energy industries. Deregulation with open access has moved the industry from conventional monopolistic electricity markets to competitive markets. In a competitive market, the price of the delivered energy and the quality of electrical energy including voltage quality and reliability of service are the main factors for business success. Transmission system operators and decision maker for planning have long dealt with system reliability issues and the question of quantifying system reliability. However, a key factor in today's competitive environment is an orientation toward customer's needs and willingness to pay for quality. Methodologies for transmission system expansion planning based on appropriate concepts of transmission system reliability evaluation, therefore, still have to be developed for independent transmission system planners under competitive electricity environment.

This paper presents a new method for making a choice of the best line/plan for transmission system expansion planning under given some candidate lines/plans on the side of the highest reliability level of transmission system. A methodology for assessing and evaluating nodal probabilistic reliability indices of a transmission system is proposed in order to choice best line/plan on the side of the highest reliability level of transmission system. The proposed method for evaluating the transmission system reliability is based on the concept that the reliability level of a transmission system is equal to the difference in the reliability levels of the composite power systems (HLII) and generation system (HLI). The risk indices of the composite power system are larger than those of the generation system because the transmission system includes the uncertainties (forced outage rates).

The effective load duration curve of the composite power system (CMELDC) developed earlier by the authors is used in order to obtain the reliability indices of composite power systems at HLII. The reliability indices of the generation system at HLI can be evaluated using the conventional effective load duration curve (ELDC).

The proposed algorithm can reveal information such as the sensitivity of the reliability level of transmission lines to the transmission system planning and provide useful data for reasonable transmission system expansion planning and operation to the electricity market, to an ISO and to the customers at the load points. It is expected that the proposed method can be used valuably not only best lines/plan choice on the side of highest reliability level but also determination of an optimal reliability level of transmission system under deregulated electricity market in future. The characteristics and effectiveness of this proposed methodology are illustrated by a case study of the MRBTS (Modified Roy Billinton Test System).

2. RELIABILITY EVALUATION OF POWER SYSTEMS HIERARCHICAL LEVELS

2.1. Hierarchical Level of Power Systems

The basic techniques for adequacy assessment can be categorized in terms of their application to segments of a complete power system. These segments are shown in Fig. I and can be defined as the functional zones of generation, transmission and distribution systems. The target of this study is not reliability evaluation of HLII but only transmission system. We can take an idea from the hierarchical level diagram that reliability indices of a transmission system is equal to difference of reliability level between HLII and HLI only if the reliability indices of HLII and HLI are compatible.

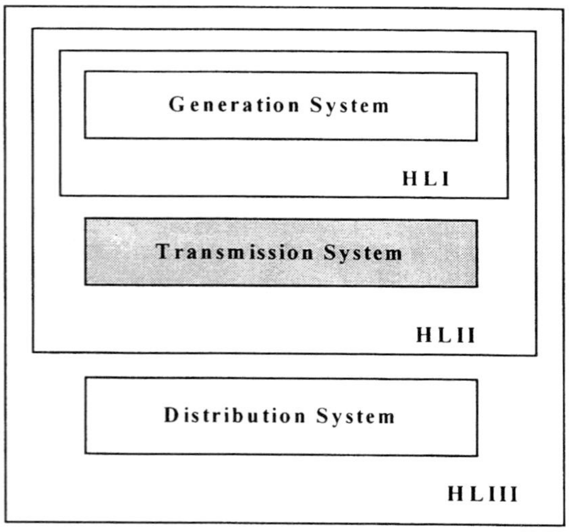

Fig. I. Power system hierarchical level diagram.

2.2. Reliability Evaluation of HLI

Reliability indices of $LOLE_{HLI}$ (Loss of load expection) and $EENS_{HLI}$ (Expected energy not served) of only generation system using ELDC(Effective load duration curve) ${}_{HLI}\Phi(x)$ of HLI are calculated by Eq. (1) and Eq. (2) respectively.

$$LOLE_{HLI} = {}_{HLI}\Phi(x)\big|_{x=IC} \quad \text{[days]} \tag{1}$$

$$EENS_{HLI} = \int_{IC}^{IC+Lp} {}_{HLI}\Phi(x)dx \quad \text{[MWh]} \tag{2}$$

Where, IC : total installed capacity of generators [MW]

$$ {}_{HLI}\Phi_i(x_e) = {}_{HLI}\Phi_{i-1}(x_e) \otimes {}_{HLI}f_{oi}(x_{oi}) = \int {}_{HLI}\Phi_{i-1}(x_e - x_{oi})\,{}_{HLI}f_{oi}(x_{oi})dx \tag{3}$$

Where, $\otimes$: operator meaning convolution integral

${}_{HLI}\Phi_o(x_e - x_{oi}) = {}_{HLI}\Phi(x_L)$

${}_{HLI}f_{oi}(x_{oi})$: the probability distribution funtion of outage capacity of generator #i

Lp : peak load of system [MW]

2.3. Reliability Evaluation of HLII

Many of reliability indices for HLII have been developed until now. The indices of HLII can be classified mainly for two kinds, as it is, for load point indices and bulk system indices according to object of evaluation. Compatible indices for HLII and HLI of them have to be used in order to evaluate reliability indices of a transmission system is equal to difference of reliability level between HLII and HLI. Therefore, *LOLE*, *EENS*, *EDNS*(Expected demand not supplied) and *ELC*(Expected load curtailments) based on *LOLE* and *LOEE*(Loss of energy expectation) indices concepts can be used in this study.

1) Reliability indices at load points

Reliability indices and CMELDC at load point #k is shown as in Fig. II. Where, L_{pk} and AP_k of the horizontal axis expresses peak load and arrival power at load point #k respectively. In this figure, the load point reliability indices, $LOLE_k$ and $EENS_k$ can be calculated using Eq. (4) and Eq. (5) with ${}_k\Phi_{NG}(x)$. T is in Fig. II is study period.

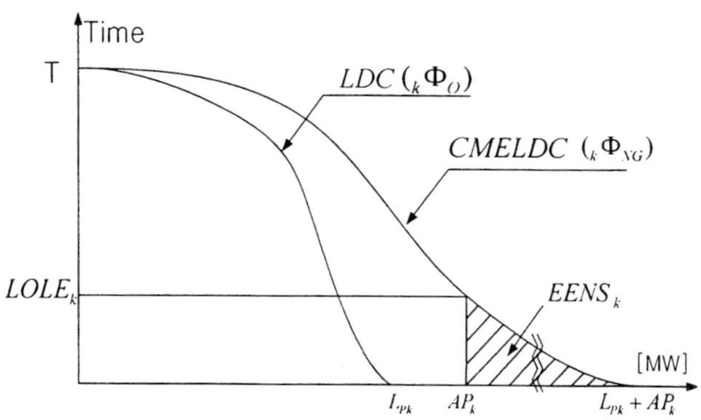

Fig. II. Reliability indices and ELDC at load point #k.

$$LOLE_k = {}_k\Phi_{NG}(x)\big|_{x=AP_k} \quad \text{[day]} \tag{4}$$

$$EENS_k = \int_{AP_k}^{AP_k+Lp_k} {}_k\Phi_{NG}(x)dx \quad \text{[MWh]} \tag{5}$$

Where, AP_k : maximum arrival power at load point # k [MW]

$$
{}_k\Phi_i(x_e) = {}_k\Phi_o(x_e) \otimes_k f_{osi}(x_{oi}) = \int {}_k\Phi_o(x_e - x_{oi})_k f_{osi}(x_{oi}) dx_{oi} \quad (6)
$$

where, $\otimes$: the operator meaning convolution integral

${}_k\Phi_o$ = LDC at load point # k

${}_k f_{osi}$: outage capacity pdf of synthesis fictitious generator operated by generators from #1 to # i at load point # k

State Probability

Total contingency enumeration could require 2^{100} states to be considered for a system composed of 100 generators and transmission lines. This is obviously impractical. Fortunately, the probability of a relatively large number of generators and transmission lines being failed at the same time is virtually zero. It is not practical to consider all states for an actual system. Eq.(7) is more useful for a practical system and can be used to calculate the state probability.

$$
{}_k q_j = \begin{cases} \sum_{\forall j \in \bar{e}_j} [P(B_j) P_{lj}] & n(\bar{e}_j) \le 8 \\ 0 & n(\bar{e}_j) > 8 \end{cases} \quad (7)
$$

where, $\bar{e}_j$: set of elements on outage of system state # j

$n(\bar{e}_j)$: number of elements on outage of set, $\bar{e}_j$

$P(B_j)$: probability of the outage capacity B_j

$P(B_{lj})$: loss of load time probability at state # j

Arrival Power Evaluation

Reliability indices of HLII can be obtained differently according to objective function of optimal power flow in order to evaluate probability distribution function of outage capacity of fictitious generator at the load point. In this study, the objective function was established to minimize the maximum outage power rate at load points under assuming that transmission line losses are ignored and effective power only is considered as following Eqs.

(1) Objective function

$$
Minimize \ \{max(L_{pk} - x_k)/L_{pk}\} \quad k \in B_L \quad (8)
$$

where, L_{pk} : peak load power at load point # k

B_L : set of loads buses

max : abbreviation of *maximum*

(2) Constraints

(a) constraint of incident circuit

$$
\sum_{l=1}^{NB} a_{il} x_l \le CG_i \quad i \in B_B \quad (9)
$$

where, a_{il} : node - incidence matrix

B_B : set of generator buses

NB : total number of branches

CG_i : generation capacity at bus # i [MW] (= 0 for load buses)

(b) constraint of transmission line capacity

$$
-CT_{l\max} \le x_l \le CT_{l\max} \quad (10)
$$

where, $CT_{l\max}$: capacity of transmission line # l [MW]

x_l : control variable meaning effective power flow of branch # l

B_T : set of transmission lines

The CMELDC can be calculated by convoluting the original load duration curve with the probability distribution function of the not served powers at the load points using Eq. (11). Lambda in Eq. (11) is parameter.

$$
\begin{aligned}
&Minimize \quad \lambda \\
&Subject \ to \\
&\left.\begin{aligned}
&\sum_{l=1}^{NB} a_{il} x_l \le CG_i \quad && i \in B_B \\
&-CT_{l\max} \le x_l \le CT_{l\max} \quad && l \in B_T \\
&(L_{pk} - x_k)/L_{pk} \le \lambda \quad && k \in B_L
\end{aligned}\right\} \quad (11)
\end{aligned}
$$

where, B_T : set of transmission lines

We can obtain the AP_{kj}, arrival power at load point # k as like as Eq.(12) for system state # j with optimal solution x_k^* of Eq.(11).

$$
AP_k = x_k^* \quad k \in B_L \quad (12)
$$

Therefore, probability distribution function of outage capacity of fictitious generator at the load point can be constructed and reliability indices of HLII can be obtained as Eq.(4) and Eq.(5) using Eq.(6).

2) Reliability indices of bulk system

While the $EENS_{HLII}$ of bulk system is equal to summation of $EENS_k$ at load points as Eq.(13), *LOLE* of bulk system is entirely different from summation of $LOLE_k$ at load points. But, as the ELC_{HLII} of bulk system is equal to summation of ELC_k at load points, $LOLE_{HLII}$ of bulk system can be calculated the divided by ELC_{HLII} as Eq.(15).

$$
EENS_{HLII} = \sum_{k=1}^{NL} EENS_k \quad [\text{MWh}] \quad (13)
$$

$$ELC_{HLII} = \sum_{k=1}^{NL} ELC_k \quad (14)$$

$$LOLE_{HLII} = EENS_{HLII} / ELC_{HLII} \quad [pu] \quad (15)$$

Where, NL : number of load point

R : set of states of not suppiled powers

$ELC_k = EENS_k / LOLE_k$ [MW/cur.yr]

3. RELIABILITY EVALUATION OF TRANSMISSION SYSTEM

LOLE and *EENS* of only transmission system using reliability indices of HLI and HLII are calculated as Eq. (16) and Eq.(17) respectively.

$$LOLE_{TS} = LOLE_{HLII} - LOLE_{HLI} \quad [days] \quad (16)$$

$$EENS_{TS} = EENS_{HLII} - EENS_{HLI} \quad [MWh] \quad (17)$$

Also, EDNS(expected demand not supplied) and ELC(Expected load curtailment) of transmission system can be calculated using Eq.(18) and Eq. (19) respectively.

$$EDNS_{TS} = EENS_{TS} / T \quad [MW/yr] \quad (18)$$

$$ELC_{TS} = EENS_{TS} / LOLE_{TS} \quad [MW/cur.yr] \quad (19)$$

Where, T is total period for study [year]

4. CASE STUDY

The program has been applied to the MRBTS in order to demonstrate the effectiveness of the proposed method. Capacities and forced outage rates of generators, transmission lines are represented in Fig. III.

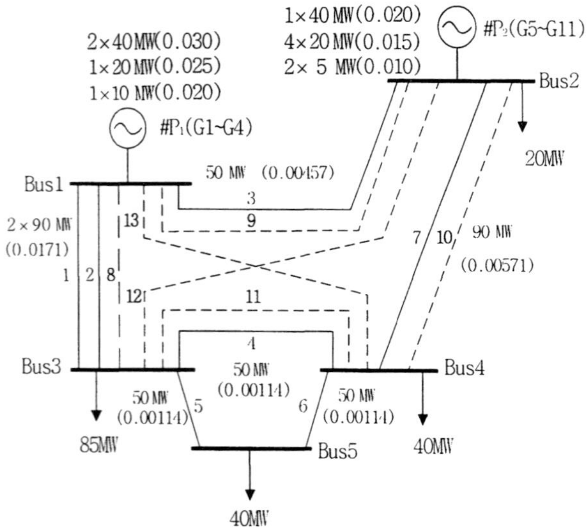

Fig. III. MRBTS for case study.

The difference and reliability indices evaluated in HLI and HLII are shown in Table 1. The difference means reliability level of only transmission system. In this study, load of one day has used for convenience and evaluated for study time. In order to illustrate the effect of adding transmission facilities selected additions were analysed.

Table 1. Reliability Indices - Case 1(base case)

	HLII	HLI	Transmission system	RGTAI
LOLE [hrs/day]	0.0459	0.0277	0.0181	1.5302
EENS [MWh/day]	1.1599	0.06659	0.4940	1.3480
ELC [MW/cur.day]	25.2828	24.0009	27.2444	0.8809

Table 2. Reliability Indices - Case 2 (case of a cct added at T/L8 between buses #1 and #3)

	HLII	HLI	Transmission system	RGTAI
LOLE [hrs/day]	0.0454	0.0277	0.0176	1.5752
EENS [MWh/day]	1.1305	0.6659	0.4646	1.4331
ELC [MW/cur.day]	24.9248	24.0009	26.3800	0.9097

Table 3. Reliability Indices - Case 3 (case of a cct added at T/L9 between buses #1 and #2)

	HLII	HLI	Transmission system	RGTAI
LOLE [hrs/day]	0.0287	0.0277	0.0010	29.1765
EENS [MWh/day]	0.7120	0.6659	0.0461	14.4376
ELC [MW/cur.day]	24.8128	24.0009	48.5026	0.4948

Table 4. Reliability Indices - Case 4 (case of a cct added at T/L10 between buses #2 and #4)

	HLII	HLI	Transmission system	RGTAI
LOLE [hrs/day]	0.0279	0.0277	0.0001	187.2202
EENS [MWh/day]	0.06724	0.6659	0.0066	101.4113
ELC [MW/cur.day]	24.1088	24.0009	44.3091	0.5417

Table 5. Reliability Indices - Case 5 (case of a cct added at T/L11 between buses #3 and #4)

	HLII	HLI	Transmission system	RGTAI
LOLE [hrs/day]	0.0457	0.0277	0.0180	1.5408
EENS [MWh/day]	1.1538	0.6659	0.4879	1.3647
ELC [MW/cur.day]	25.2200	24.0009	27.0984	0.8857

Table 6. Reliability Indices - Case 6
(case of a cct added at T/L12 between buses #2 and #3)

	HLII	HLI	Transmission system	RGTAI
LOLE [hrs/day]	0.0280	0.0277	0.0003	101.6803
EENS [MWh/day]	0.6784	0.6659	0.0125	53.2740
ELC [MW/cur.day]	24.2133	24.0009	45.8088	0.5239

Table 7. Reliability Indices - Case 7
(case of a cct added at T/L13 between buses #1 and #4)

	HLII	HLI	Transmission system	RGTAI
LOLE [hrs/day]	0.0453	0.0277	0.0175	1.5832
EENS [MWh/day]	1.1255	0.6659	0.4596	1.4488
ELC [MW/cur.day]	24.8630	24.0009	26.2279	0.9151

Table 1 is a standard base case(Case 1). These results presented from Table 2 to Table 7 summarize the effect of adding transmission lines between buses for reliability indices of transmission system. From the results, transmission line between buses #2 and #4 can be made a choice for the best line on the side of reliability level because the RGTAI is the highest value. The RGTAI in Tables means Ratios of Generation-Transmission Adequacy Indices. Figure IV presents the new grid on which the best line is added.

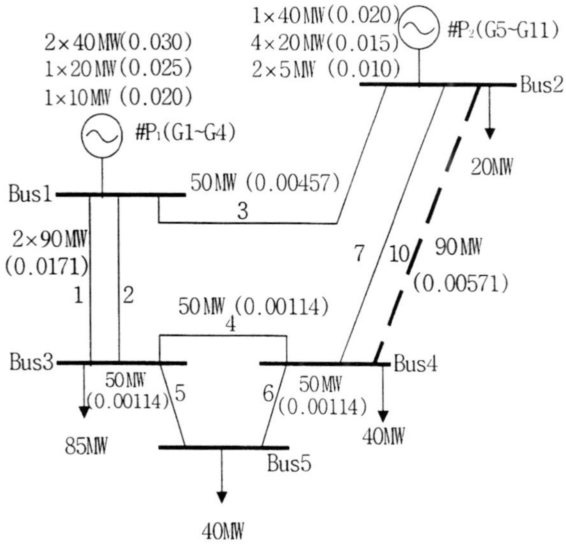

Fig. IV. The best line on the side of reliability level of MRBTS.

5. CONCLUSIONS

This paper presents a new methodology for making a choice of the best line for transmission system expansion planning on the side of the highest reliability level. A methodology for assessing and evaluating nodal probabilistic reliability indices of a transmission system is proposed in order to choice best line/plan on the side of the highest reliability level of transmission system. The proposed method for evaluating the transmission system reliability is based on the concept that the reliability level of a transmission system is equal to the difference in the reliability levels of the composite power systems (HLII) and generation system (HLI). The risk indices of the composite power system are larger than those of the generation system because the transmission system includes the uncertainties (forced outage rates). In order to obtain the reliability indices of a composite power system at HLII, the CMELDC previously developed by the authors is used in this study. Development of methodology on side of reliability of transmission system as well as reasonable nodal reliability criteria and furthermore economics under the deregulated electricity market will be remained for future works.

6. ACKNOWLEDGEMENTS

This work was supported by grant No. 2000-1-30200-006-3 from the Basic Research Program of the Korea Science & Engineering Foundation.

7. REFERENCES

CEA (March 2000). Industry Restructuring - News & Info. CEA connections online news.

IMO (November 1998). Participating in Ontario's New Competitive Electricity Market. Overview Report by IMO.

J. Toyoda, L. Chen, T.Minakawa and T. Yamada (1991). New Indices and Computational Techniques for Composite System Reliability Evaluation. *Cigre Symposium*, Montreal, A1-01.

Jaeseok Choi, Daeho Do, Seungpil Moon and Roy Billinton (June 1999). Development of a Method for ELDC Construction in a Composite Power System. *LESCOPE '99 proceedings*, Halifax.

L. Chen, H. Suwa and J. Toyoda (Feb. 1991). Power Arrival Evaluation of Bulk System Including Network Constraints Based on Linear Programming Approach. *IEEE Trans.*, Vol. 6, No. 1, pp. 37-42.

M. Ilic, et al (1998). *Power Systems Restructuring; Engineering and Economics*, Kluwer Academic

Roy Billinton and Wenyuan Li (1994). *Reliability Assessment of Electric Power Systems Using Monte Carlo Methods*, Plenum Press.

Roy Billinton and A. Sankarakrishnan (1995). A Comparison of Monte Carlo Simulation Techniques for Composite Power System Reliability Assessment. *IEEE WESCANEX '95 proceedings*, pp.145-150.

Roy Billinton and Ronald N. Allan (1996). *Reliability Evaluation of Power Systems*, Second Edition, Plenum Press.

Roy Billinton and W. Zhang (1996). Enhanced Adequacy Equivalent for Composite Power System Reliability Evaluation. *IEE, Proc.-GTD 143(5)*, pp. 420-426 .

S. Niioka, R. Yokoyama K. Okada & H. Asano (January 2000). Impact Evaluation of Reliability Management and Operation under the Deregulated Electric Power Market. *IEEE PES WM2000*, Singapore.

W.S Read, W.K. Newman, I.J. Perez-Arriaga, H.Rudnick, M.R. Gent & A.J. Roman (December 1999). Reliability in the New Market Structure(Part1). *IEEE Power Engineering Review*, p.4-14.

W.S Read, W.K. Newman, I.J. Perez-Arriaga, H.Rudnick, M.R. Gent & A.J. Roman (January 2000). Reliability in the New Market Structure(Part2). *IEEE Power Engineering Review*, p.10~16.

X. Wang. and J.R. McDonald (1994). *Modern Power System Planning*, McGraw-Hill.

8. BIOGRAPHIES

Sungrok Kang was born in Jinju, Korea on 1975. He obtained B.Sc. degrees from Gyeongsang National University in 2002. His research interests include Expansion Planning, Probabilistic Production Cost Simulation, Reliability Evaluation, Outage Cost Assessment and Fuzzy Applications of Power Systems. He is now working toward a M.Sc. degree at Gyeongsang National University.

Trungtinh Tran was born in LongAn, Vietnam in 1973. He obtained B.Sc. degrees from Cantho University in 1997. His research interests include Expansion Planning and Reliability Evaluation. He is now working toward a M.Sc. degree at Gyeongsang National University.

Jaeseok Choi (M88) was born at Kyeongju, Korea on 1958. Obtained B.Sc., M.Sc. and Ph.D. degrees from Korea University in 1981, 1984 and 1990 respectively. His research interest includes Fuzzy Applications, Probabilistic Production Cost Simulation, Reliability Evaluation and Outage Cost Assessment of Power Systems. He has been a Post-Doctor at University of Saskatchewan in Canada on 1996. Since 1991, he has been a faculty of Gyeongsang National University where is now a professor.

Junmin Cha (M96) was born in Inchon, Korea on 1964. Obtained B.Sc., M.Sc. and Ph.D. degrees from Korea University in 1989, 1991 and 1996 respectively. His research interest includes Fuzzy Applications, Probabilistic Production Cost Simulation, Reliability Evaluation, Outage Cost Assessment of Power Systems and Operation and Planning of Deregulated Power Market. Since 1996, he has been a faculty of Daejin University where is now an associate professor.

Daeseok Rho was born in Korea in 1964. He obtained his B.Sc. and M.Sc. degrees from Korea University in 1985 and 1987, respectively, and his Ph.D. degree from Hokkaido University of Japan in 1997. His research interest includes Reliability Evaluation of Distribution System and Operation and Planning of Deregulated Power Market. Since 1993, he has been on the faculty of the University of Technology and Education where he is now an associate professor.

Roy Billinton (F78) came to Canada from England in 1952. Obtained B.Sc. and M.Sc. Degrees from the University of Manitoba and Ph.D. and D. Sc. Degrees from the University of Saskatchewan. Worked for Manitoba Hydro in the System Planning and Production Divisions. Joined the University of Saskatchewan in 1964. Formerly Head of the Electrical Engineering Department. Presently C.J. Mackenzie Professor of Engineering and Acting Dean, Graduate Studies, Research and Extension of the College of Engineering. Author of paper on Power System Analysis, Stability, Economic System Operation and Reliability. Author or co-author of eight books on reliability. Fellow of the IEEE, the EIC and the Royal Society of Canada and a Professional Engineer in the Province of Saskatchewan.

ELSEVIER
IFAC
PUBLICATIONS
www.elsevier.com/locate/ifac

BIFURCATION CONTROL APPLICATION TO VOLTAGE COLLAPSE IN A MODEL POWER SYSTEM

Taihyun Kim* Tae-Kyoo Oh* Kyung-Soo Kook*
Young-Hwan Moon*

* *Korea Electrotechnology Research Institute*
665, Naeson-dong, Euiwaang-city, Kyonggi-do, Korea

Abstract: In this paper, we have proposed bifurcation control of voltage collapse in the model power system. Bifurcation theory is applied to analyze the voltage collapse in a model power system. Controls using existing power system facilities have suggested to mitigate voltage collapse phenomena. Nonlinear control law is designed to stabilize periodic solutions which emerges after voltage collapse phenomena. Moreover, a control law, which achieves reducing amplitude of the bifurcated periodic solution, is designed. *Copyright © 2003 IFAC*

Keywords: voltage collapse, bifurcation, control

1. INTRODUCTION

Voltage collapse phenomena in power systems have been one of the well known dynamic behavior of power systems and studied by numerous researchers. One of approach to study voltage collapse is using bifurcation theory. Dobson and Chiang (Dobson and Chiang, 1989) suggested that voltage collapse phenomena could be understood as a characteristic changes of system dynamics. Moreover, they suggested that these changes are caused by varying reactive power demand. This approach agrees with a view in electric power system community that voltage collapses are instability caused by lack of reactive power support(Taylor, 1994) and explains voltage collapse in dynamical terms. In that paper, they showed that the model power system undergoes voltage collapse at saddle-node bifurcation point of a equilibrium points. Ajjarapu and Lee (Ajjarapu and Lee, 1991) showed that Hopf bifurcation precedes saddle-node bifurcation in a parametric space for the same model power system. Later, Wang et. al. (Wang *et al.*, 1994) took the detail analysis for the same model power system with slight variations in some parameters. In that paper, they found out rich dynamic behaviors leading to voltage collapse such as period doubling, boundary crisis, etc.

To control voltage collapse, Wang and Abed (Wang and Abed, 1993) have showed that linear and nonlinear control could achieve mitigation of voltage collapse phenomena based on bifurcation control theory.

In this paper, we suggest new way to mitigate voltage collapse phenomena of the model power system. Here, new way means using actuators different from the Wang and Abed (Wang and Abed, 1993). In Wang and Abed (Wang and Abed, 1993), they considered the control input (i.e. actuator) which has direct effect on voltage dynamics of the load. This load side control generally requires additional facilities such as STACOM, TCSC, etc. However, new investment to power system might not be easily implemented due to financial or other limiting reasons. Thus, we suggest new control method which can be readily implemented to power systems by adding simple

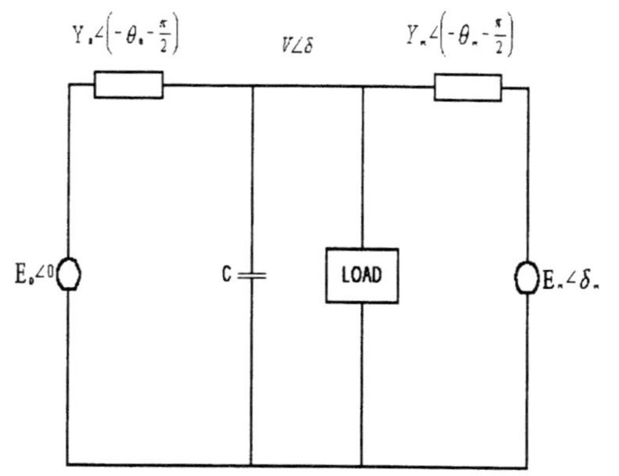

Fig. 1. Model Power System

control law to existing facilities. Moreover, we seek nonlinear control law which not only stabilize the bifurcated periodic orbit but also reduce the amplitude of the bifurcated periodic orbit. By doing so, it will reduce the possible damage to power system caused by oscillations.

This paper proceeds as follows. In section 2, model power system and its analysis will be presented. In section 3, nonlinear control using the existing facilities is proposed to mitigate voltage collapse and this control law will be applied to a model power system. In section 4, new nonlinear control law will be proposed to reduce the amplitude of the bifurcated periodic solution and will be applied to a model power system. Conclusion will be given in section 5.

2. POWER SYSTEM MODEL AND ITS BIFURCATION ANALYSIS

In this section, we briefly review the model power system which is going to be used throughout in this paper. In addition, bifurcation analysis will be taken for the model power system.

Figure 1 shows layout of the model power system. It consists of a infinite bus, one generator, nonlinear load including a dynamics of a induction motor, and a capacitor for the purpose of voltage support.

Governing dynamics of the given model power systems are as follows. Detail derivations are in (Dobson and Chiang, 1989).

$$\dot{\delta}_m = \omega \tag{1}$$

$$M\dot{\omega} = -d_m\omega + P_m + E_mY_mV \sin(\delta - \delta_m - \theta_m) + E_m^2Y_m\sin(\theta_m) \tag{2}$$

$$K_{qw}\dot{\delta} = -K_{qv2}V^2 - K_{qv}V + Q(\delta, V) - Q_0 - Q_1 \tag{3}$$

$$\begin{aligned} TK_{qw}K_{pv}\dot{\delta} &= K_{pw}K_{qv2}V^2 \\ &+ (K_{pw}K_{qv} - K_{qw}K_{pv})V \\ &+ K_{qw}(P(\delta, V) - P_0 - P_1) \\ &- K_{pw}(Q(\delta, V) - Q_0 - Q_1) \end{aligned} \tag{4}$$

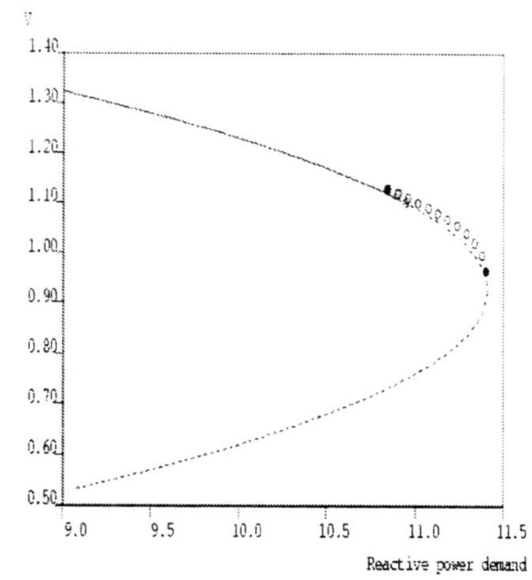

Fig. 2. Bifurcation Analysis of the Model Power System

where $P(\delta, V)$ and $Q(\delta, V)$ are defined as

$$\begin{aligned} P(\delta, V) &= E_0'Y_0'V\sin(\delta + \theta_0') - E_mY_mV \\ &\sin(\delta - \delta_m + \theta_m) + (Y_0'\sin(\theta_0') \\ &+ Y_m\sin(\theta_m))V^2 \\ Q(\delta, V) &= E_0'Y_0'V\cos(\delta + \theta_0') + E_mY_mV \\ &\cos(\delta - \delta_m + \theta_m) - (Y_0'\cos(\theta_0') \\ &+ Y_m\cos(\theta_m))V^2 \end{aligned}$$

Here, δ_m denotes generator angle relative to infinite bus and ω denotes its angular velocity of the generator. V and δ denote load side voltage and its angle relative to infinite bus, respectively. Moreover, Q_1 denotes a load side reactive power demand. For the parameter values used in this paper readers are referred to (Dobson and Chiang, 1989). Figure 2 shows a result of bifurcation analysis of the model power system using load side reactive power demand(Q_1) as the bifurcation parameter. Note that we use AUTO97 very first version of which was introduced in (Doedel, 1981) to get this figure. In this figure, solid line denotes stable equilibrium point and dotted line denotes unstable equilibrium. Moreover, empty circle represents unstable periodic orbit and solid circle represents stable periodic orbit. As a load side reactive power demand passes 10.946, new unstable periodic orbit emerges and equilibrium point loses its stability. This implies that subcritical Hopf bifurcation occurs and the model power system does not have any local stable equilibria after this critical parameter value. Thus, we can conclude that the model power system with a load side reactive power demand greater than 10.946 will experience voltage collapse.

3. BIFURCATION CONTROL OF VOLTAGE COLLAPSE PHENOMENA IN THE MODEL POWER SYSTEM

In this section, we consider bifurcation control which can mitigate voltage collapse in the model power system.

As we have noted in the previous section, voltage collapse in the model power system can be related to the subcritical Hopf bifurcation. *Bifurcation control* can be classified in two parts. First one is achieving delay of voltage collapse in parametric space (i.e. load side reactive power demand Q_1). Second one is stabilizing emerged periodic orbits after Hopf bifurcation. By doing so, even if a normal operating equilibrium loses its stability the controlled model power system maintains its stability due to existence of local stable periodic orbits. Thus, voltage collapse can be prevented. In terms of a bifurcation control, this can be explained as rendering subcritical Hopf bifurcation to supercritical Hopf bifurcation (Abed and Fu, 1986). Here, we limit ourselves to the second approach. However, it should be noted that the first approach can be readily implemented.

In (Abed and Fu, 1986), it is proved that if linearized system is controllable then cubic order state feedback is enough to render subcritical Hopf to supercritical Hopf. Moreover in (Wang and Abed, 1993), it has proposed to use ω as the state feedback. Using ω as a state feedback has two advantages. One is that ω has value 0 for all the interested equilibrium points of the model power system. This guarantees that additive control law using ω as a state feedback does not change values of the uncontrolled model power system's equilibrium points. The other is that generator angle velocity ω can be readily measured and used in power system control such is the case for power system stabilizer (PSS). Based on these reasons, we suggest nonlinear control as form of the following equations.

$$u_n = k_n \omega^3 \qquad (5)$$

where k_n are constant scalar feedback gains.

Unlike Wang and Abed (Wang and Abed, 1993), in this paper two different control input will be used to implement the suggested control law (5). First, we consider mechanical power input to generator (i.e. P_m in (2)) as a control input. From now on we name this as the *Case 1*. It should be noted that P_m can be controlled by a governor in generator system. Although it is desirable including governor dynamics, we omit its dynamics here to simplify problem and make clear the purpose of this paper. However, including governor dynamics is just simple extension of the results presented here. Thus, P_m in (2) including control becomes

$$P_m = \bar{P}_m + u \qquad (6)$$

where $\bar{P}_m$ is a normal mechanical power input and u is a control input which has a form of the (5).

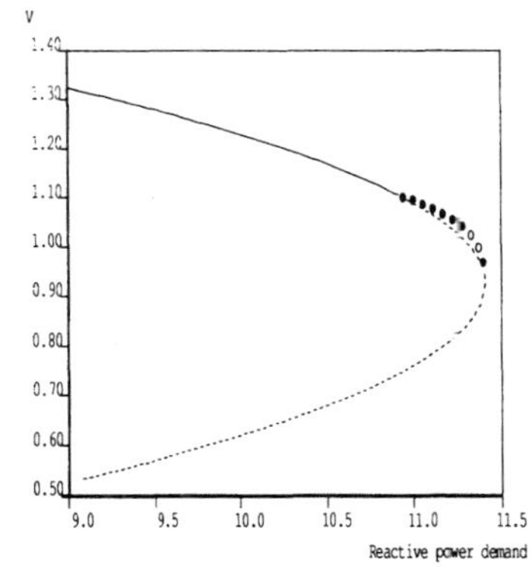

Fig. 3. *Case 1* nonlinear control with $k_n = -0.5$

Note that we will employ same notation for bifurcation diagram as in the previous section throughout this paper. Figure 3 shows that the proposed nonlinear control stabilizes periodic orbit emerged after Hopf bifurcation point. Note that the critical parameter values are same for both the controlled system and the uncontrolled system. It should be noted that period doubling(PD) bifurcation occurs at $Q_1 = 11.31$ and periodic orbit after the PD is no longer stable. Thus, voltage collapse can be avoided between $Q_1 = 10.946$ and $Q_1 = 11.31$ for the proposed nonlinear control.

Now, we consider generator terminal voltage (i.e. E_m in (2)~(4)) as a control input. From now on we name this as *Case 2*. It should be noted that E_m can be easily controlled by automatic voltage regulator (AVR) and exciter in generator systems. As in the mechanical input case, it is desirable including exciter and AVR dynamics. However we omit their dynamics here and including those dynamics is just simple extension of the result presented here. Thus, E_m including control becomes

$$E_m = \bar{E}_m + u \qquad (7)$$

where $\bar{E}_m$ is a normal generator terminal voltage and u is a control input which has a form of the (5). It should be noted that E_m has nonlinear relation with other states within equations (2)~(4). However, detail relations are omitted here due to space consideration.

Figure 4 show that the proposed nonlinear control accomplish their designed purposes. Note that *Case 2* has smaller stabilized periodic orbit in amplitudewise than *Case 1*. From these observations we can deduce that using generator terminal voltage as a control is more desirable than using mechanical power input to generator as a control.

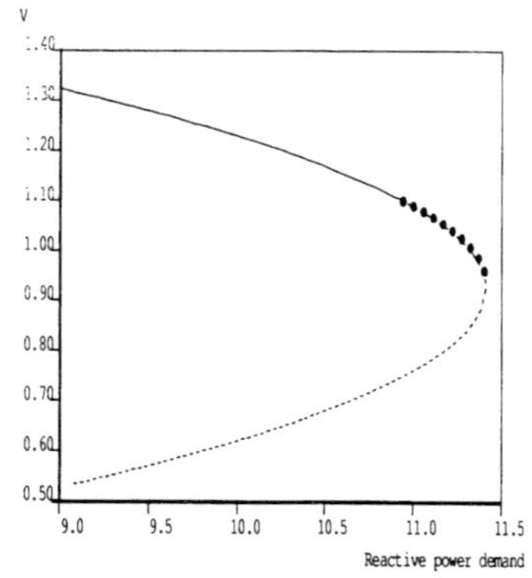

Fig. 4. *Case 2* nonlinear control with $k_n = 0.5$

4. DESIGN OF AMPLITUDE REDUCING CONTROL LAW

In this section, control law will be designed to reduce the amplitude of the bifurcated periodic solution and time domain simulations of the model power system will be taken to illustrate the proposed control law.

4.1 Preliminaries

Next discussion follows closely (Abed and Fu, 1986). Let's consider following nonlinear dynamical system with its linearized system.

$$\dot{x} = f(x, \mu, u) \quad (8)$$

$$\dot{z} = A_\mu z + b_\mu u \quad (9)$$

where μ is bifurcation parameter, u is control input. Let's assume that (8) undergoes Hopf bifurcation at the critical parameter value μ_c with $u \equiv 0$. Then, criticality of that Hopf bifurcation is determined by following equation.

$$\beta_2 = 2Re(2lQ(r, a) + lQ(\bar{r}, b) + \frac{3}{4}C(r, r, \bar{r}) \quad (10)$$

where r and l denote right and left eigenvector of A_{μ_c} with eigenvalue $i\omega_c$, respectively. In (10), $Q(x, x)$ and $C(x, x, x)$ denote symmetric bilinear and trilinear form generated from (8). Moreover, $Re(y)$ and $\bar{y}$ denotes real part of the y and complex conjugate of y, respectively. If β_2 is less than 0 then (greater than 0)bifurcated periodic solution is stable (unstable). Moreover, it should be noted that amplitude of the bifurcated periodic solution near by bifurcation point depends on β_2. Less value of the β_2 results in smaller magnitude of the bifurcated periodic solution. Considering nonlinear input $u = x^T Q_u x + C_u(x, x, x)$ which is only composed of quadratic and cubic term. Then, β_2 becomes

$$\beta_2^* = \beta_2 + 2Re\Delta \quad (11)$$

where Δ is the contribution of the input to criticality and its definition can be found in (Abed and Fu, 1986). Thus, under a certain condition we can set β_2 to any value we want by changing input u . For a linearly controllable system, Δ can be simplified to following equation by setting $Q_u \equiv 0$

$$\Delta = \frac{3}{4}(C_u(r, r, \bar{r}))lb_{\mu_c} \quad (12)$$

This shows that for a linearly controllable system using only cubic term as a input can stabilize the system.

In summary, amplitude of the bifurcated solution can be controlled by modifying β_2. In addition, for linearly controllable system cubic term is enough to control β_2. Based on these observation, control law can be designed, which reduce amplitude of the bifurcated solution, by carefully selecting C_u.

4.2 Application to a model power system

To apply discussions of the previous section to the model power system, we made following simplifications. Here, we limit our discussions on mechanical input case (*Case 1*). However, exciter case (*Case 2*) is just a simple extension of following discussions. Due to the reasons we described in section 3 we only consider cubic input with at least square of ω. Moreover, including ω with less than square degree would result in change of the Jacobian and it will complicate control design problem. Thus, state feedback input becomes

$$u \equiv C_u = k_n \omega^2 x_i \quad (13)$$

where x_i is one of the state variable in dynamic equations (1) (4) and k_n denote constant control gain. In the *Case 1*, $b_\mu$of the equation(9) is $[0, 1, 0, 0]^T$ for all parameter values of μ. Since we limit inputs as above descriptions, linearized system will not be affected and normal equilibrium points will be maintained. Thus, r and l will be same for the controlled system. We have following values for r for a model power system for the given parameters and at the bifurcation point $Q_1 = 10.946$.

$$r = [\frac{-1 - 2577i}{9638}, 1, \frac{93 - 565i}{9638}, \frac{74 + 370i}{9638}]^T \quad (14)$$

Note that linear system is controllable so that $lb_{\mu_c} \neq 0$.

From (13) and (14), for the fixed magnitude of control gain (k_n) with allowing sign variations ω^3 results in the least value from the possible 4 choices. Following table shows numerical values for each possible C_u.

C_u	Real part of $C_u(r, r, \bar{r})$
$\omega^2\delta_m$	$1\times1\times \frac{-1}{9638}$
ω^3	$1\times1\times 1$
$\omega^2\delta$	$1\times1\times \frac{93}{9638}$
$\omega^2 V$	$1\times1\times \frac{74}{9638}$

To illustrate above results, we take time simulations with load side reactive power demand (Q_1) is set to 11.2. Note that at this parameter value voltage collapse occurs for uncontrolled system. Figure 5 shows that voltage collapse occurs at $t = 64.25$ for the uncontrolled model power system. Now, consider $5\omega^3$ and $5\omega^2\delta_m$ as a control

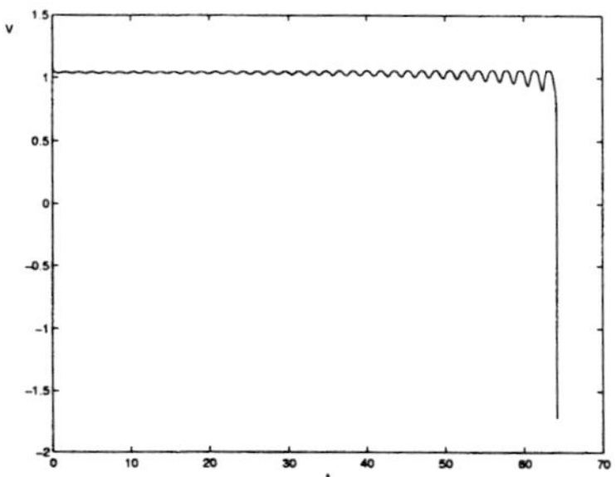

Fig. 5. No control

input. From above discussion $5\omega^3$ must have lesser amplitude periodic solution. Figure 6 and 7 shows that amplitude reducing can be achieved through the discussed control law. Note that we use same magnitude of control gain to get following figures.

5. CONCLUSION

In this paper, we have proposed bifurcation control of a voltage collapse in the model power

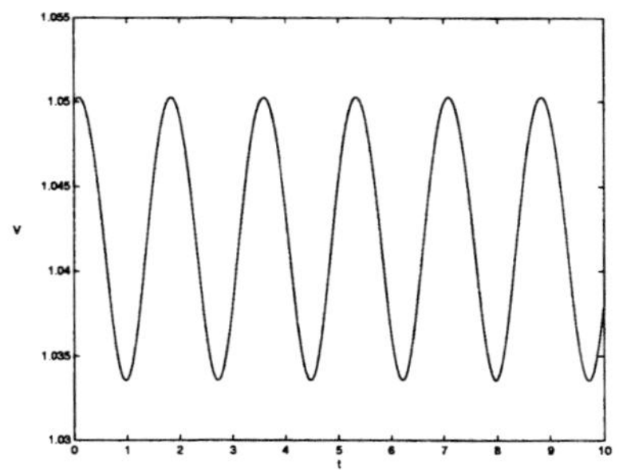

Fig. 6. $5\omega^3$

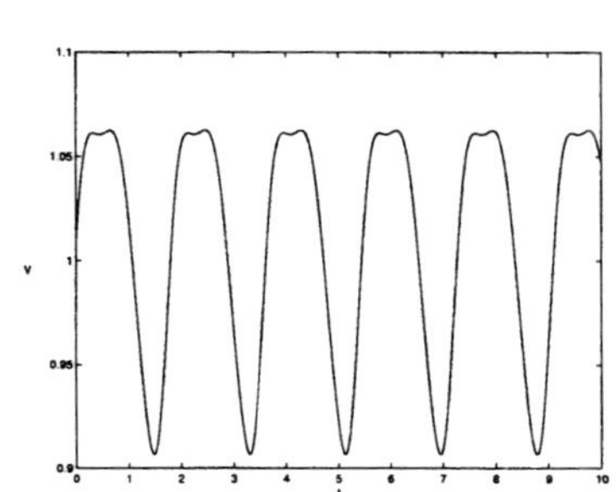

Fig. 7. $5\omega^2\delta_m$

system. Two newly proposed control using existing power system facilities have successfully mitigates voltage collapse. In addition, amplitude reducing control law was designed from theoretical backgrounds. Time domain simulations have been taken to illustrate this approach.

REFERENCES

Abed, E. H. and J.-H. Fu (1986). Local feedback stabilization and bifurcation control, i. hopf bifurcation. *Systems and Control Letters* **7**, 11–17.

Ajjarapu, V. and B. Lee (1991). Bifurcation theory and its application to nonlinear dynamical phenomena in a electrical power system. *IEEE Trans. on Power Systems* **7**, 424–431.

Dobson, I. and H.-D. Chiang (1989). Toward a theory of voltage collapse in electric power systems. *Systems and Control Letters* **13**, 253–262.

Doedel, E. J. (1981). Auto: A program for automatic bifurcation analysis of autonomous systems. *Cong. Num.* **30**, 265–284.

Taylor, C. (1994). *Power System Voltage Stability.* McGraw-Hill Inc. New York.

Wang, H. O. and E. H. Abed (1993). Control of nonlinear phenomena at the inception of voltage collapse. *Proceedings of the American Control Conference* pp. 2071–2075.

Wang, H. O., E. H. Abed and A. M. A. Hamdan (1994). Bifurcations, chaos and crises in voltage collapse of a model power system. *IEEE Trans. on Circuits and Systems I* **41**, 294–302.

ELSEVIER
IFAC
PUBLICATIONS
www.elsevier.com/locate/ifac

ANALYSIS OF SYSTEM IMPACT OF THE DISTRIBUTED GENERATION USING EMTP WITH PARTICULAR REFERENCE TO VOLTAGE SAG

S. M. Yeo, C. H. Kim, Y.K.Kim, M. C. Shin

Information and Computer Engineering, Sungkyunkwan University
300, Cheoncheon-dong, Jangan-gu, Suwon-city, Gyeonggi-do, Korea

S.P.Ahn, D.J.Lee

KERI
665, Naeson 2-dong, Uiwang-si, Gyeonggi-do, Korea

Abstract: With the advent of distributed generation, the overall power system are impacted on stability and power quality. The distributed generation has a positive impact on system restoration following a fault, higher reliability, and mitigation of effect due to voltage sag. However, the distributed generation also has a negative impact on decrease of reliability such as changes of protective device setting and mal-operation.
Because bulk power systems consist of various sources and loads, it is complicated to analyze power system with distributed generation. The types of distributed generation usually are classified both rotating machinery and inverter-based system.
In this paper, the distributed generation is designed by rotating machinery, and the distributed system with model of the distributed generation is simulated using EMTP. In addition, this paper shows the simulation results according to the types of distributed generation.

Keywords: Distribution system, Electric power systems, Fault, Generator, Power-system voltages, Voltage

1. INTRODUCTION

Voltage sag is defined as a momentary decrease to between 0.1 and 0.9 pu in the RMS voltage magnitude for durations from 0.5 cycle to 1 minute, usually caused by a remote fault somewhere on the power system. Voltage sags are the most important Power Quality (PQ) problems facing many industrial customers. Equipment used in modern industrial plants (process controllers, programmable logic controllers, adjustable speed drives, robotics) is actually becoming more sensitive to voltage sags as the complexity of the equipment increases. Even relays and contactors in motor starters can be sensitive to voltage sags, resulting in shutdown of a process when they drop out. Because of the increased use of sensitive electronic equipment, these disturbances have a greater impact on customers than ever before. As a result, monitoring and assessment of the system performance at both the transmission and distribution voltage levels are becoming increasingly important (Kim, *et al.*, 2000; Dugan, *et al.*, 1996; Bollen, 1999). Recently there has been a considerable issue connecting new generation to the distribution network and this is known as embedded, distributed or dispersed generation (Nick, *et al.*, 2000; Schweer, 1997).

Such distributed generation has some advantages and disadvantages. When voltage sags occur in power system, distributed generation react on voltage sag, voltage sag may be mitigated or distributed generation may be experienced out-of-step. Distributed generation in distribution system has various effects such as network voltage changes, increase fault levels.

Many researchers have studied distribution network including distributed generation (Milanovic and Gnativ, 1997; Girgis and Brahma, 2001; Miller, *et al.*, 2001). Particularly, J. V. Milanovic and R. Gnativ (1997) studied characteristics of voltage sags in radial network with dynamic loads and embedded generator, and A. Girgis and S. Brahma (2001) studied effect of distributed generation of protective device coordination.

In this paper, the distribution system with distributed generator is simulated for various system conditions in order to analyze and is analyzed system impact of the distributed generation with particular reference to voltage sags. So, the distributed generation is designed by rotating machinery, and the distributed system with model of the distributed generation is simulated using EMTP. And, this paper shows the simulation results according to the types of distributed generation.

2. DISTRIBUTED GENERATION

Distributed generation was investigated be working groups, CIGRE (The International Conference on Large High Voltage Electric Systems) and CIRED (The International Conference on electricity Distribution Networks). Distributed generation consists of synchronous or induction generator, or power electronic device. The CIRED survey and the CIGRE report represented reasons of encouraging distributed generation. Some reasons are followed (Schweer, 1997).

(1) Reduction in gaseous emissions
(2) Energy efficiency or rational use of energy
(3) Deregulation or competition policy
(4) Diversification of energy sources
(5) Ease of finding sites for smaller generators

Because environmental impact is a major factor in the consideration of electrical power scheme, distributed generation is classified into combined heat and power generation and renewable energy resources such as wind power, micro-hydro, solar photovoltaic, landfill gas, etc. The different classification of distributed generation is for types of generation, such as synchronous, induction generator or inverter-based system.

A distributed generator is connected to an electrical distribution network. Distributed generator is used to supply electric power for consumer typically. Sometimes, electric power generated by distributed generator flows reversely in distributed network. So, it is important to check that they will not degrade the quality of supply for the consumer. Fig. 1 shows power flow at distribution network with distributed generation.

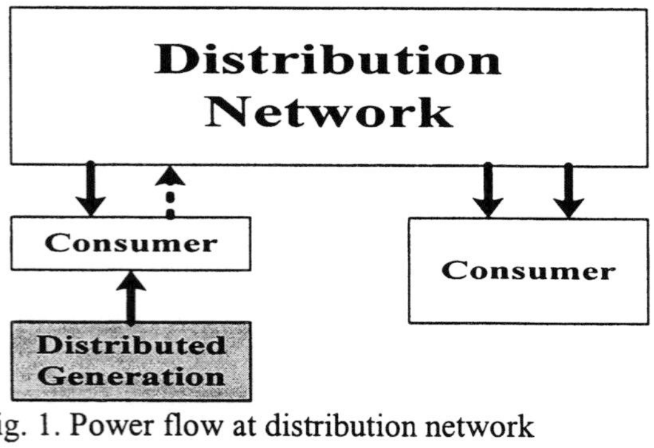

Fig. 1. Power flow at distribution network

In this paper, we design for a model of synchronous generator using EMTP for simulation.

3. VOLTAGE SAG

Voltage sags are typically caused by fault conditions. Fault resulting in voltage sags can occur within the plant or on the utility system. The voltage sag condition lasts until the fault is cleared by a protective device. In the plant, this will typically be a fuse or a plant feeder breaker. On the utility system, the fault could be cleared by a branch fuse or a substation breaker. If reclosing is used by the utility, the voltage sag condition can occur multiple times.

Utility system faults can occur on a distribution system or on a transmission system. Fig. 2 illustrates a typical distribution system configuration with a number of feeders supplied from a common bus. A fault on Feeder F1 will cause an interruption, which will affect the customers on that particular feeder. However, most of the customers on the others three parallel feeders will also experience voltage sag while the fault is actually on the system.

Faults on a transmission system can affect even more customers. Customer miles away from the fault location can still experience voltage sags resulting in equipment mal-operation when the fault is on the transmission system.

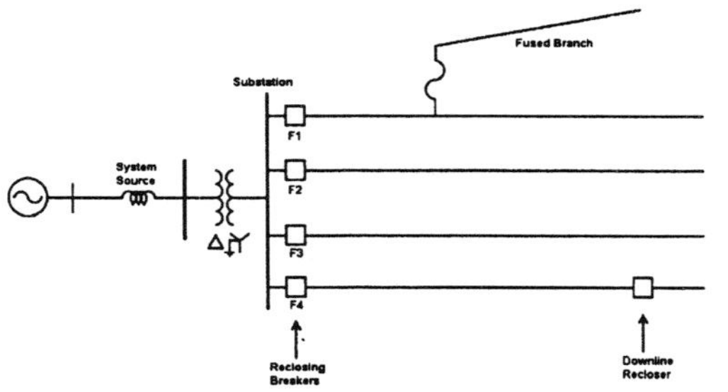

Fig. 2. Typical distribution system configuration

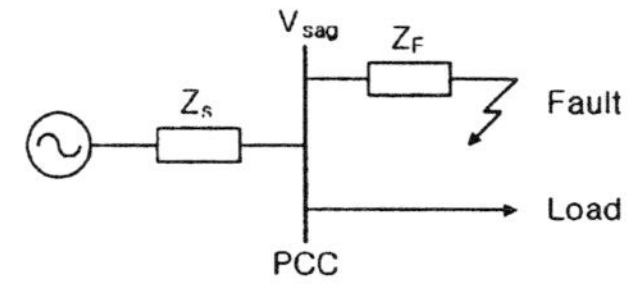

Fig. 3. Voltage divider model for voltage sag

To quantify sag magnitude in radial systems, the voltage divider model, shown in Fig. 3 can be used. This might appear a rather simplified model, especially for transmission systems. In Fig. 3, two impedances: Z_S is the source impedance at the point-of-common coupling (PCC); and Z_F is the impedance between the PCC and the fault. The point-of-common coupling is the point from which both the fault and the load are fed. In other words, it is the place where the load branches are off from the fault current. In the voltage divider model, the load current before as well as during the fault is neglected. There is thus no voltage drop between the load and the PCC, and thus the voltage at the equipment terminals, can be found from equation (1).

$$V_{sag} = \frac{Z_F}{Z_S + Z_F} E \qquad (1)$$

If the pre-event voltage is exactly 1 pu, then E = 1. This results in the equation (2) for the sag magnitude.

$$V_{sag} = \frac{Z_F}{Z_S + Z_F} \qquad (2)$$

Any fault impedance should be included in the feeder impedance Z_F. We see from equation (2) that the sag becomes deeper for faults electrically closer to customer (when Z_F becomes smaller), and for systems with a smaller fault level (when Z_S becomes larger). Note that a single-phase model has been used here, whereas in reality the system is three-phase. That means that this equation strictly speaking only holds for three-phase faults.

4. SIMULATION AND RESULTS

4.1 System model studied

Model system of distribution network for simulation is shown Fig. 4. Model system has 5 buses, 3 transformers, and 3 loads. Total simulation time is 1.5[sec], and sampling frequency is 3840[Hz].

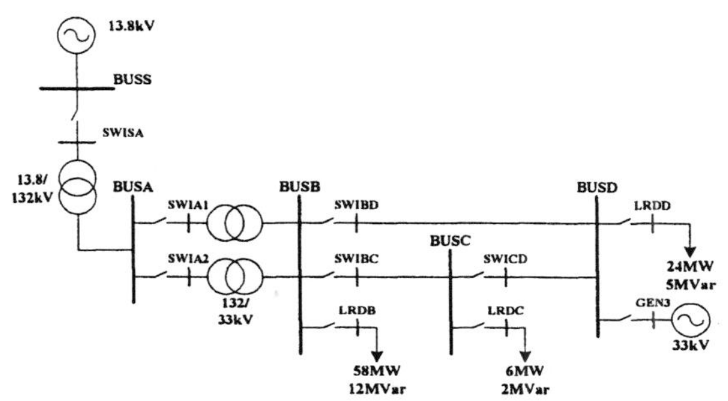

Fig. 4. System diagram for maximum load

Model system is introduced by reference (Nick, *et al.*, 2000) BUSS has been chosen as the slack bus, this generator will supply the active power required to balance the system. So, BUSS is taken as ideal source. Table 1 is system parameters.

Table 1 System parameters for networks

From bus	To bus	Type	Resistance	Reactance
BUSS	BUSA	transformer	0.0	0.06670
BUSA	BUSB	transformer	0.00994	0.20880
BUSA	BUSB	transformer	0.00921	0.21700
BUSB	BUSC	line	0.04460	0.19170
BUSB	BUSD	line	0.21460	0.34290
BUSC	BUSD	line	0.23900	0.41630
BUSD	BUSE	transformer	0.00994	0.20880

Simulation was performed in 3 types of system condition.

(1) Maximum load, no distributed generation (DG)
(2) Maximum load, synchronous generator at BUSD
(3) Maximum load, synchronous generator at BUSE

And, fault condition is simulated for analyzing system impact of the distributed generation under fault condition. Table 2 shows fault types and fault locations. Fault starting time is 0.25[sec].

Table 2 Fault types and locations

Fault type	Single line-to-ground (SLGF) 3-phase fault
Fault location	BUSA, BUSC, LineBD (line between BUSB and BUSD)

Load as shown Fig. 4 is maximum load. Fig. 5 shows different type of system. The distributed generation has been moved to BUSE.

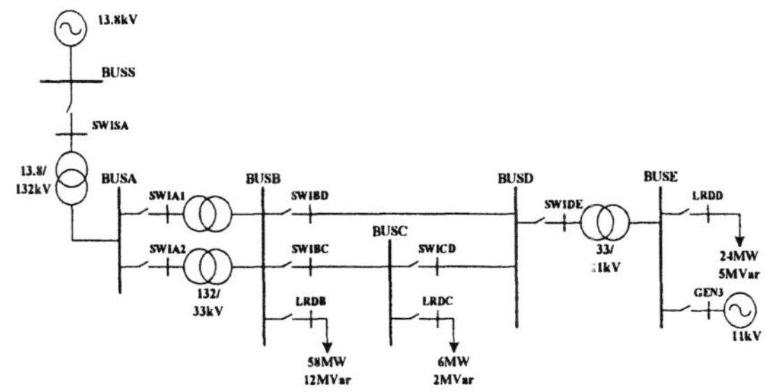

Fig. 5. System diagram for minimum load

4.2 No distributed generation system

At each bus and line, voltages and currents can be obtained, so RMS values are calculated. Per unit quantities are introduced, and baseMVA is 50MVA.

Figure 6 shows waveforms of voltage at each bus under normal condition. As shown in figure 6, because voltage was dropped by loads and line impedances, a bus which was far from source has lower voltage than 1[pu]. For example, voltage at bus BUSD is about 0.98[pu].

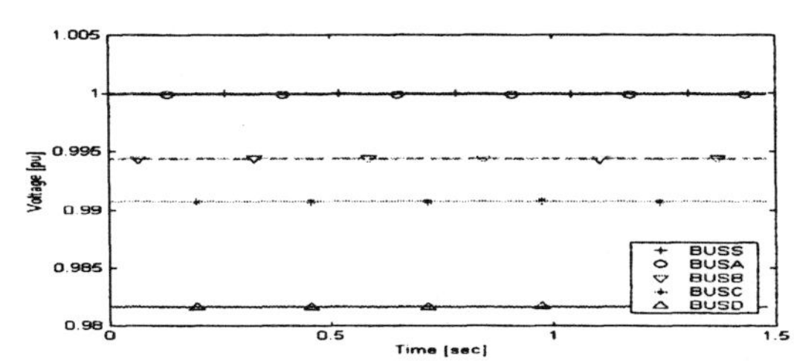

Fig. 6. Voltages for each bus under normal condition, no DG

Figure 7 presents waveforms of current through each switch. Currents flow from bus BUSS to each load at buses, and each current has correlation as shown in the following equation (3).

$$\begin{aligned} i(SWISA) &= i(SWIA1) + i(SWIA2) \\ &= i(SWIBC) + i(SWIBD) + i(SWICD) \end{aligned} \qquad (3)$$

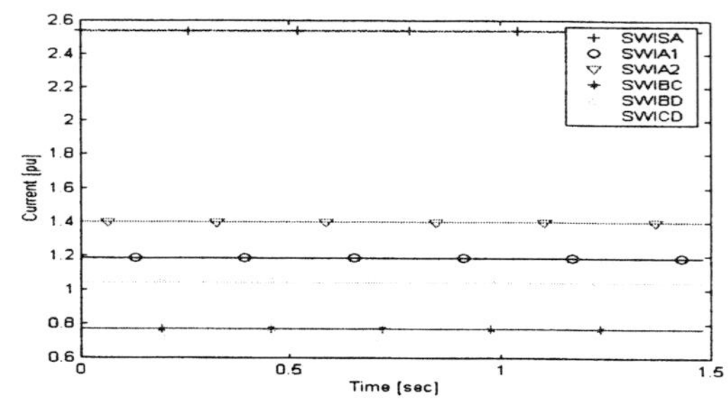

Fig. 7. Currents through each switch under normal condition

Figure 8 presents waveforms of voltage of phase A at each bus under single line-to-ground fault at bus BUSC.

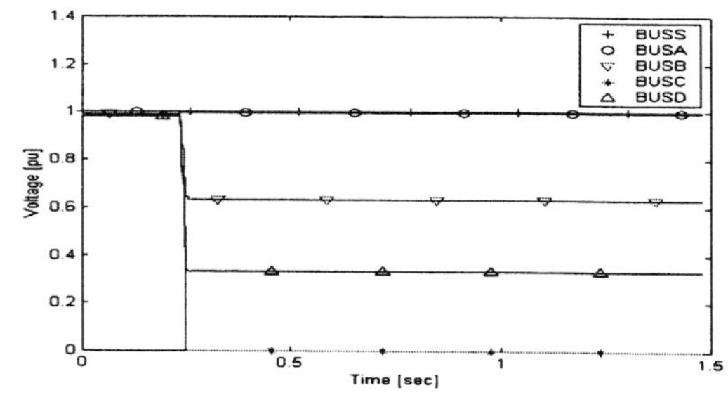

Fig. 8. Voltages at each bus under SLGF (BUSC)

After fault, voltage at faulted bus BUSC, is dropped to 0.0 [pu], and voltage sag occurs in other buses, i.e. BUSB, BUSD. As shown in figure 9, we can know similar results under 3-phase fault.

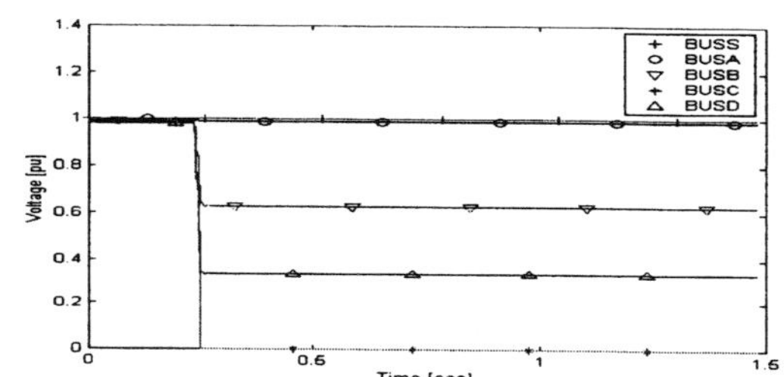

Fig. 9. Voltages at each bus under 3-phase fault (BUSC)

4.3 Maximum load, synchronous generator at BUSD

Figures 10~11 show waveforms of voltage and current at each bus in system connected with synchronous generator (capacity: 10MVA) as distributed generation at bus BUSD.

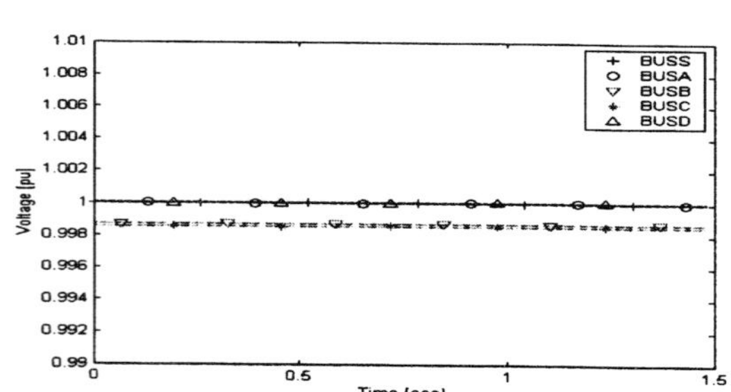

Fig. 10. Voltages at each bus with DG (BUSD)

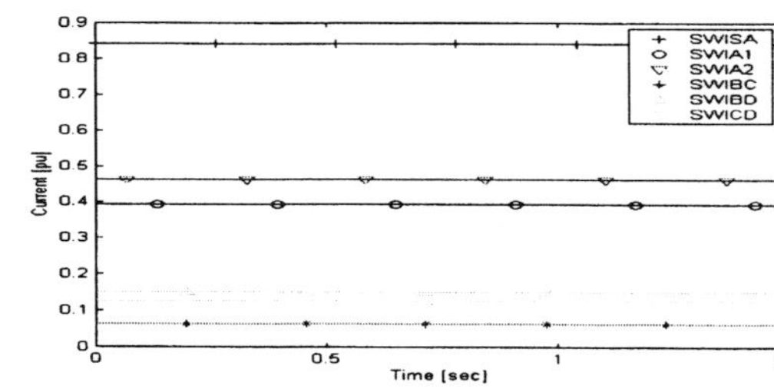

Fig. 11. Currents through each switch with DG (BUSD)

As shown in figures 10~11, voltage connected with synchronous generator at bus BUSD is 1.0[pu], and currents through each switch are small value against the case of no distributed generation.

Figures 12, 13 show the variation of bus voltage in system connected with distributed generation under single line-to-ground and 3-phase fault, respectively. As shown in figure 12, we can observe small variation of voltage at bus BUSD by synchronous generator in comparison with figure 8. This variation is made by speed regulator in synchronous generator, and it is large at the bus connected with distributed generation.

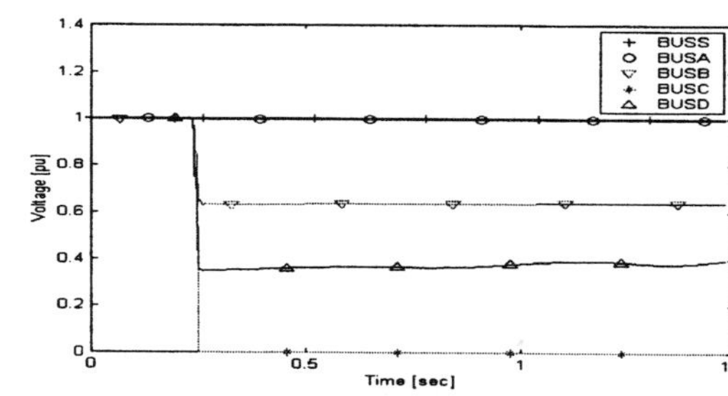

Fig. 12. Voltages at each bus under SLGF (BUSC) with DG

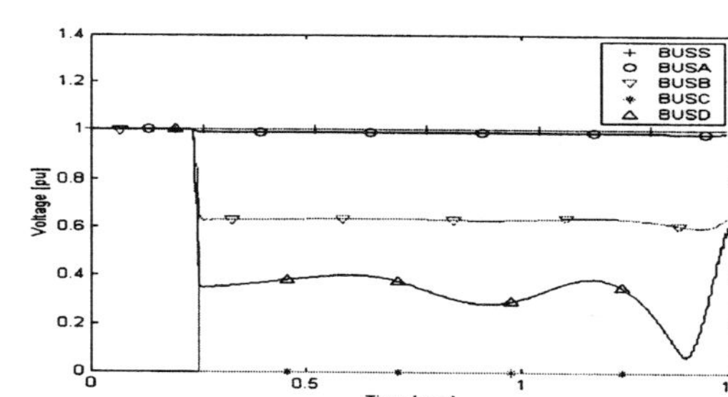

Fig. 13. Voltages at each bus under 3-phase fault with DG

As shown Fig. 13, voltage at BUSD is increasing fluctuation. We can show phase angle of generator from Fig. 14. Magnitude of phase angle is increasing. This means that generator losses synchronism.

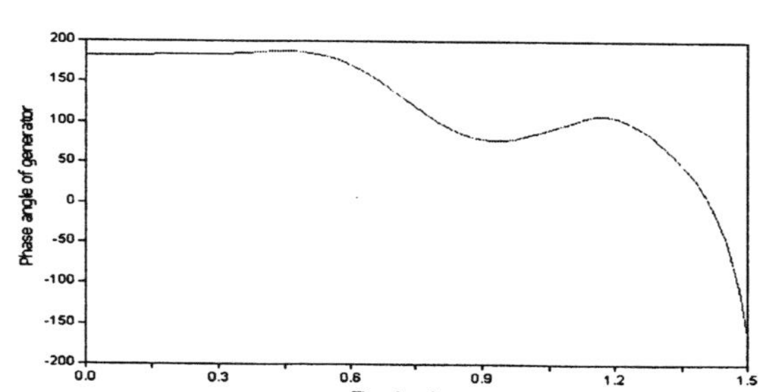

Fig. 14. Phase angle of synchronous generator

4.4 Maximum load, synchronous generator at BUSE

For analysis of impacts on distributed generation connected with low level bus, the system in figure 5 is simulated. At this, Voltage at bus BUSE connected with distributed generation maintains 1.0[pu] under normal condition, but also observe that voltage at bus BUSD is 0.92[pu].

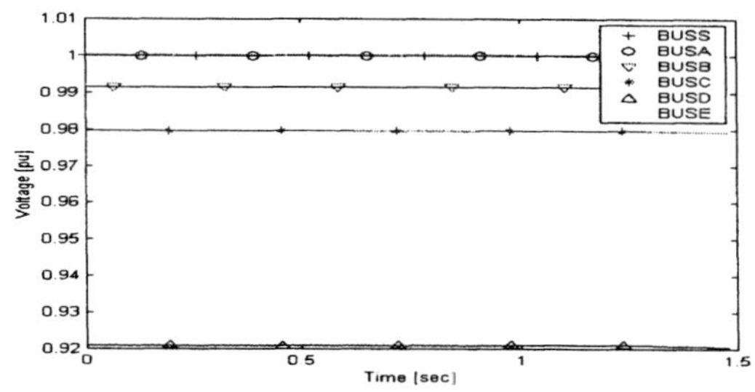

Fig. 15. Voltages at each bus with DG (BUSE)

Figure 16 shows waveforms of voltage under single line-to-ground fault at bus BUSC.

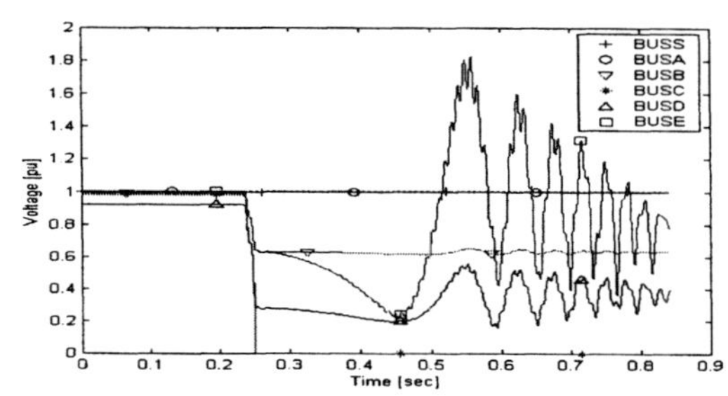

Fig. 16. Voltages at each bus under SLGF with DG (BUSE)

After fault, at 0.45[sec] synchronous generator losses synchronism. Voltage at bus BUSC connected with synchronous generator, fluctuate between 0.6[pu] and 1.8[pu], affects voltages at other buses, i.e. BUSB, BUSD.

Figure 17 shows waveforms of voltage of phase A at bus BUSE under various capacity of distributed generation. The different capacity of generator has been simulated and the result is summarized in figure 17.

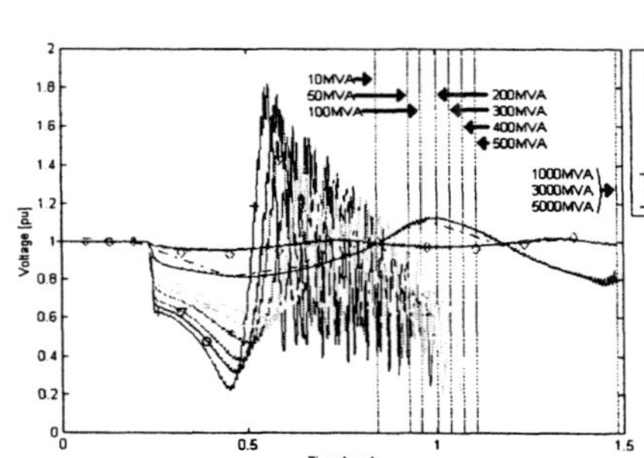

Fig. 17. Voltages of phase A at BUSE under various capacity of distributed generation

As shown in figure 17, if capacity of generator is larger, distribution system well endures to voltage sag or fault. When the speed of generator is very low, EMTP ends simulation. The horizontal lines in figure 17 indicate ending time of simulation. There exist several facts. When the capacity of generator is large, first, fluctuation of voltage is small, second, simulation time increase. Figure 18 shows waveforms of voltage at each bus under single line-to-ground fault in system connected with synchronous generator at bus BUSE.

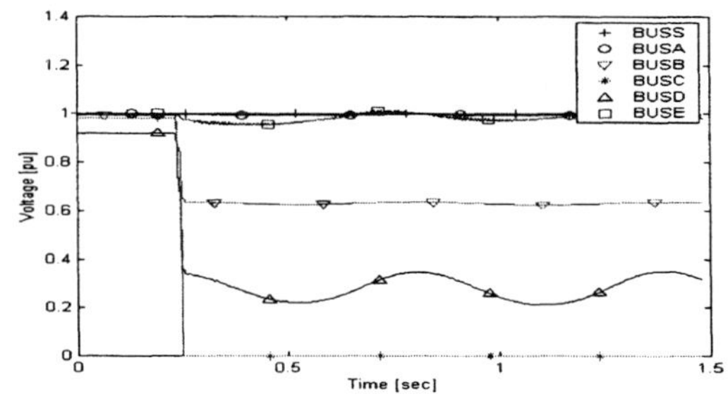

Fig. 18. Voltages at each bus under SLGF (BUSC) with DG (BUSE, 5000MVA)

As shown in figure 18, if the capacity of generator is enough to supply fault current, and then synchronous generator don't loss synchronism. But, fault still remains, so, voltages of other buses are small value.

4.5 Summary

Voltage sags by SLGF on the BUSB and BUSD are shown in Figure 19. When the DG is connected with the distribution system, voltages will have a little increment at each bus during a fault. The magnitude of voltages on the BUSB and BUSD, has increased under various capacity of DG, and shown in Figures 20, and 21.

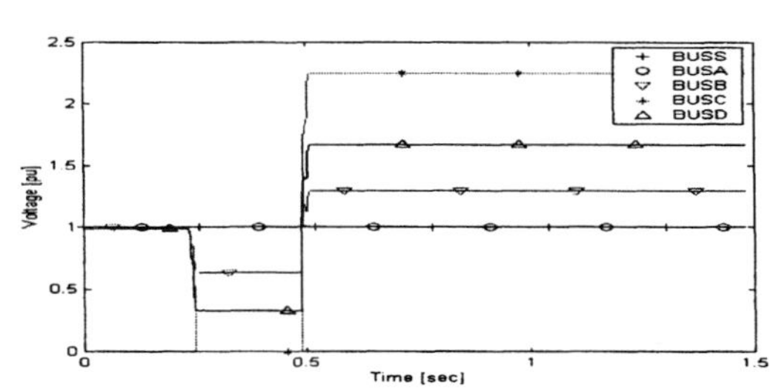

Fig. 19. Voltages at each bus under transient fault without DG

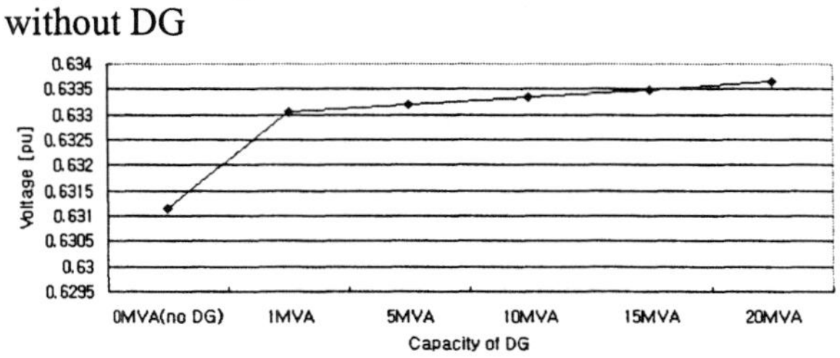

Fig. 20. Voltage at BUSB by various capacity of DG

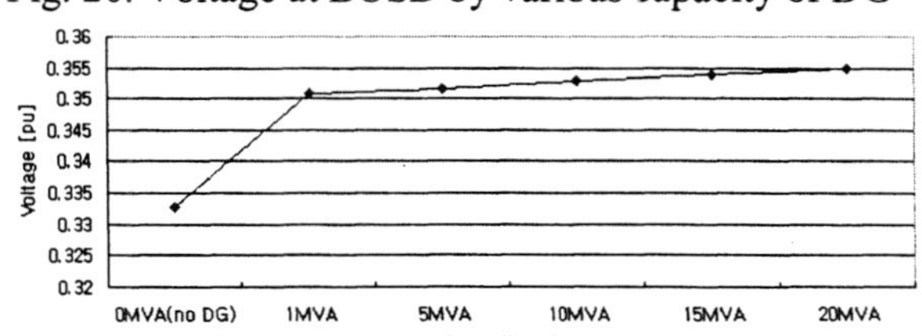

Fig. 21. Voltage at BUSD by various capacity of DG

Because the capacity of DG is smaller than central power plant, a voltage increment is small. But each bus still suffers from voltage sag.

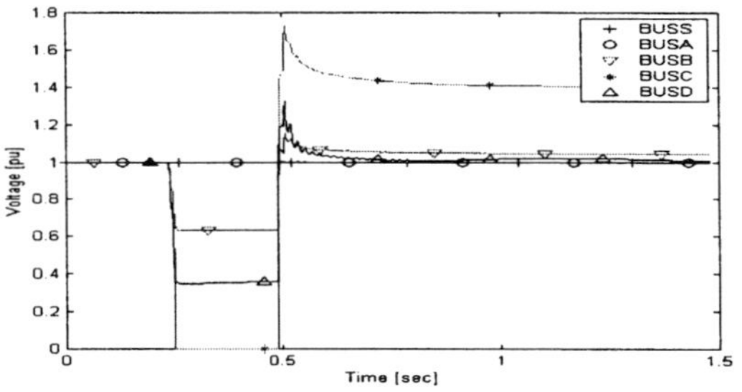

Fig. 22. Voltages at each bus with DG (10MVA)

As shown in figure 22, in the system connected with distributed generation, post-fault voltage approaches to pre-fault voltage at each bus.

The distributed generation is affected with voltage sag, if voltage sag has very large magnitude, the system is unstable and synchronous generator losses synchronism. As shown in figure 17, when synchronous generator has large capacity, system well endures voltage sag. Also, if voltage sag has short duration, generator takes small effects. Therefore, if not preparing for voltage sag, synchronous generator losses synchronism and take damage until voltage sag. Also, synchronous generator induces over-voltage when generator losses synchronism, so it can lead to serious damage to power system equipment. Figure 23 shows waveforms of voltage at bus BUSD for various fault location.

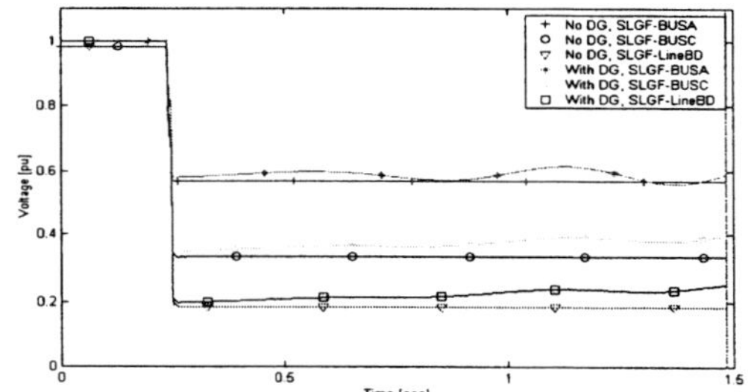

Fig. 23. Voltages for various conditions

As shown in figure 23, magnitude of voltage sag at bus connected with distributed generation is in inverse proportion to distance from fault position. But, power swing exists by synchronous generator and when voltage sag is severe, synchronous generator losses synchronism.

5. CONCLUSION

Voltage sag impacts on various system components, such as generator, motor, power electronic device, computer, etc. When system is normal condition, distributed generation have good effect as maintaining system voltage.

Distribution network without/with distributed generation is simulated by using EMTP in order to analysis of system impact. When system is stable, synchronous generator has good effect for system voltage, but when fault or voltage sag occur, synchronous generator is unstable and has bad effect as system unstable. This simulation is shown serious impact of distributed generation under fault, such as loss synchronism.

6. ACKNOWLEDGEMENT

The authors would like to thank Korea Ministry of Commerce, Industry and Energy and Korea Electric Power Research Institute for their support.

REFERENCES

Bollen M.H.J. (1999), *Understanding Power Quality Problems*, IEEE Press.

Dugan R.C., McGranaghan M.F. and Beaty H.W. (1996), *Electrical Power Systems Quality*, McGraw-Hill.

Girgis A. and Brahma S. (2001), Effect of Distributed Generation on Protective Device Coordination in Distribution System, *Power Engineering, LESCOPE '01*, pp. 115-119.

Kim C.H., Lee M.H., Aggarwal R.K. and Johns A.T. (2000), Educational Use of EMTP MODELS for the Study of a Distance Relaying Algorithm for Protecting Transmission Lines, *IEEE Trans Power Systems*, **Vol. 15, No. 1**, pp. 9–15.

Milanovic J. V. and Gnativ R. (2001), Characteristics of Voltage Sags in Radial Networks with Dynamic Loads and Embedded Generators, *2001 IEEE Porto Power Tech Conference, Porto, Portugal*, **Vol. 1**.

Miller N.W., Walling R.A. and Achilles A.S. (2001), Impact of Distributed Resources on System Dynamic Performance, *Transmission and Distribution Conference and Exposition, 2001 IEEE/PES*, pp. 951-952.

Nick Jenkins, Ron Allan, Peter Crossley, David Kirschen and Goran Strbac (2000), *Embedded Generation*, IEE Power and Energy Series, London, United Kingdom.

Schweer, A. (1997), Special report for Session 3, *CIGRE Symposium on Impact of demand side management, integrated resource planning and distributed generation, Neptun, Romania*, Paper no. 300-00.

ELSEVIER
IFAC
PUBLICATIONS
www.elsevier.com/locate/ifac

APPLICATION OF 3D VISUALIZATION TO POWER SYSTEM TROUBLESHOOTING GUIDE

Chang-Hyun Park, Young-Beom Jung, Gilsoo Jang, Byoungjun Lee, Hongrae Kim*

NPTC/ School of Electrical Engineering, Korea University, Seoul, South Korea
*/School of Electrical & Electronic Engineering, Soonchunhyang University, South Korea**

Abstract: When power systems have troubles, inexperienced personnel have difficulty finding countermeasures and understanding the situation quickly. This paper presents an effective method to visualize power system Equipment in normal & abnormal condition using 3D graphic animations & virtual reality. It allows operators to understand what is going on with the equipment and the state variation of power system equipment can be represented using various methods.
Additionally, this guide suggests suitable countermeasures to fix what happened in order to minimize effects of the incident by presenting a effective troubleshooting process. *Copyright ©2003 IFAC*

Keywords: Troubleshooting guide, WorldToolKit, 3D Visualization, Preventive Maintenance, 3D Max

INTRODUCTION

Present day power systems are becoming more complex and large, and major system failures are rarely the result of a single disturbance causing the collapse of a secure system. Therefore, it is crucial to prevent possible problems in advance and, if they happen, to minimize the effect of a problem by the prompt action of operators. Since, stable operation of power systems depends on rapid and right decisions of operators, efficient notification of the present conditions in power system is needed. Due to worldwide power system deregulation, power systems are being operated by fewer and less experienced personnel than in the past. On the other hand, the consequences of maintenance errors and inefficiencies are as significant as ever. In this situation, there is strong need for more effective tools for enhancing and assuring the proficiency of maintenance personnel. Better tools are needed to bring maintenance and troubleshooting personnel more rapidly to adequate levels of proficiency, to assure that personnel assigned to critical tasks are capable of performing those tasks, and to provide more effective methods of rehearsing critical but infrequently performed tasks. In this paper, a maintenance guide is developed as an electronically-created application in order to provide more efficient data quickly.

BACKGROUND

A. 3D Visualization

With the fast advancement of computer technology and various demand of multimedia simple 2D information is represented by 3D technology. Such technology is used to create virtual model houses, 3D virtual movies or flight simulators. Interactive three-dimensional graphics and animation will be an important part of future user interfaces directed toward business, consumers, and entertainment applications. Rapid improvements in computing power today have opened the way for desktop realistic 3D visualization. Almost all PC systems now include 3D graphics accelerators for games, CAD, or visualization applications. Also, as the technologies of 3D visualization and virtual reality evolve, the applications of VR become literally unlimited. It is assumed that VR will reshape the interface between people and information technology by offering new ways to communicate information, visualization of processes, and the creatively express ideas.

Fig 1. 3D graphic Example

In this paper, 3D graphics are used to enhance operator's perception.

B. Device Selection & 3D Modeling

Currently substation operator terminals don't give notice of the situation rapidly or efficiently enough. When a fault happens in a substation, the successive warning signal on the monitor and alarm sounds may confuse the operators. In this situation, it is hard for inexperienced operator to find a means of settling the problem. The 3D guide which is proposed here is to help the operators effectively understand what is going on and to suggest solutions one by one. Since every substation has transformers and it can be considered as main equipment in the substation, transformer is selected as the troubleshooting guide's target.

In this paper, the transformer is modelled as a 3D object in order to give a more realistic expression to the operator. 3D studio MAX has been used to develop the transformer model. Figure 2 shows an example of 3D modelling.

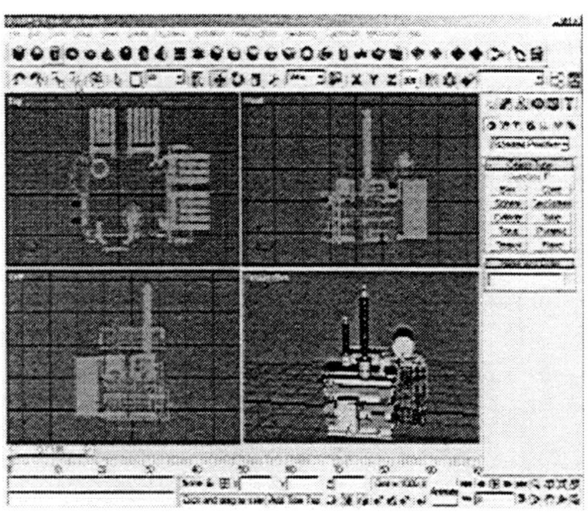

Fig 2. Example of 3D Modeling

C. Simulation & Visualization Tool

To make a more interactive simulation using 3D modeling, often a real-time 3D graphic library is used as Open GL from Silicon Graphics & Direct X from Microsoft. In this paper, the 3D simulation is made by WorldToolKit (WTK) a development tool for virtual reality. WorldToolKit is an Open GL based tool. Also, troubleshooting guide's mainframe is made with MFC (Visual C++).

WorldToolKit (or WTK) is a portable, cross-platform development system for building high performance, real-time, integrated 3D applications for scientific and commercial use and it has a library of over 1000 functions written in C that enable rapidly develop of new virtual reality applications. One function call can do the work of hundreds of lines of C code.

Also, it is structured in an object-oriented way, even though it does not use inheritance or dynamic binding. WTK functions are object-oriented in their naming convention, and are organized into over 20 classes. These classes include Universe (which manages the simulation and contains all other objects), Geometries, Nodes, Viewpoints, Windows, Lights, Sensors, Paths, Motion Links, and others. Functions are included for device instancing, display setup, collision detection, loading object geometry from a file, dynamic geometry creation, specifying object behavior, and controlling rendering.

The architecture of WorldToolKit (WTK) has been designed to incorporate the power of scene hierarchies. WTK allows the building of a simulation by assembling geometry nodes into a hierarchical *scene graph, which dictates how the simulation is rendered and allows all of the efficiencies of a state-preserving, stack-oriented rendering architecture.* WTK applications differ from conventional window system in that they are intended for use in situation where the user' s viewpoint or objects in the universe are continuously changing. At the core of an application written using WTK is a simulation loop that reads input sensors, updates objects, and renders a new view onto the display. WTK is designed to be used in real-time applications such as simulations, where frame rates on the order of 5 to 30 frames per second are maintained.

In WTK, a scene is made in each simulation Loop. When a scene is made, WTK use a scene graph. Figure 3 shows an example of a WTK scene graph.

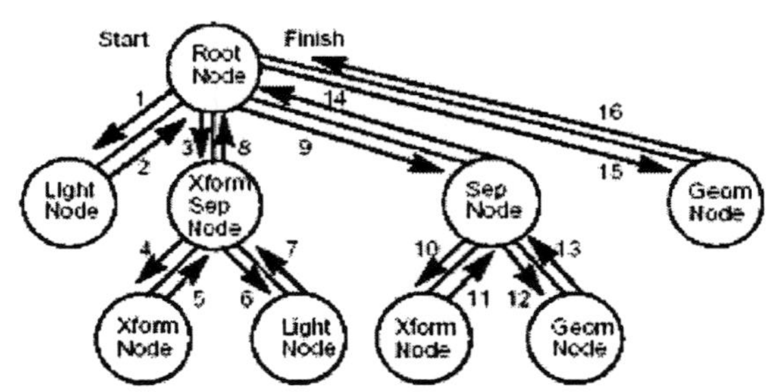

Fig 3. Traversing the Scene Graph

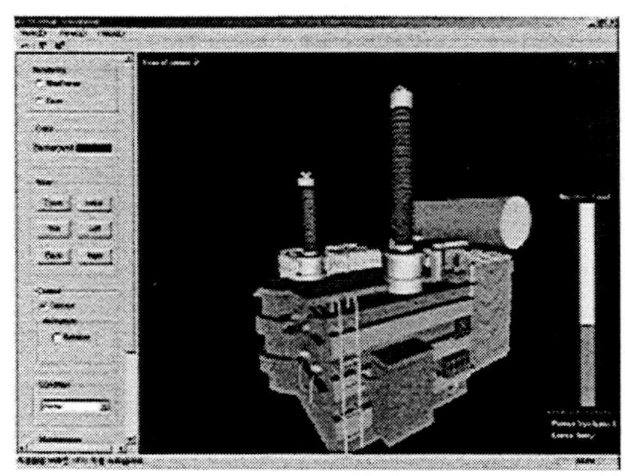

Front

Top

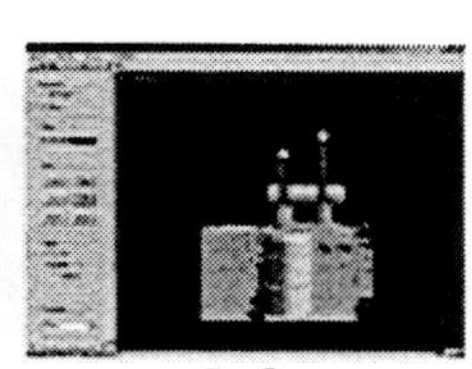

Left

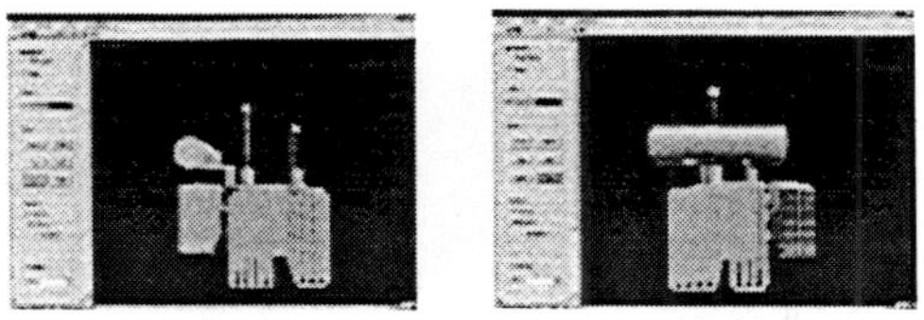

Back Right

Fig 4. Transformer Model in Virtual Space

IMPLEMENTATION

A. Troubleshooting Guide using 3D

In this work, the 3D guide for maintenance and troubleshooting has a mainframe structure as MFC SDI (Single Document Interface).

Figure 5 shows a one-line diagram of a substation in the developed 3D guide. When a problem occurs at a transformer, it causes the transformer on a one-line diagram to twinkle.

With a click on the transformer, Figure 6 appears on the operator's screen. Then, the operator can easily find where the problem is. In order to find what caused the problem, the operator needs to click the red part of the virtual transformer.

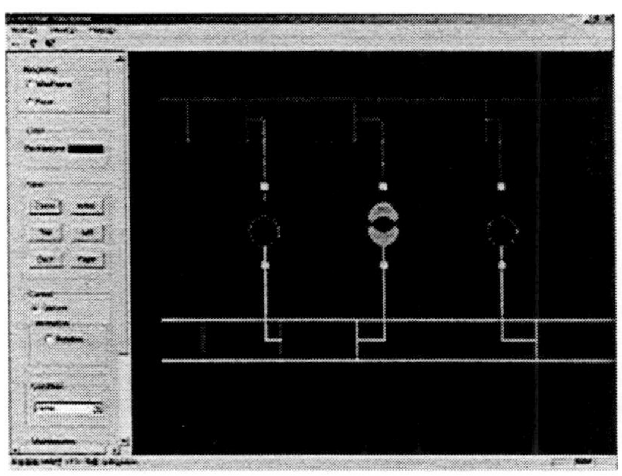

Fig 5. 2D One-line diagram

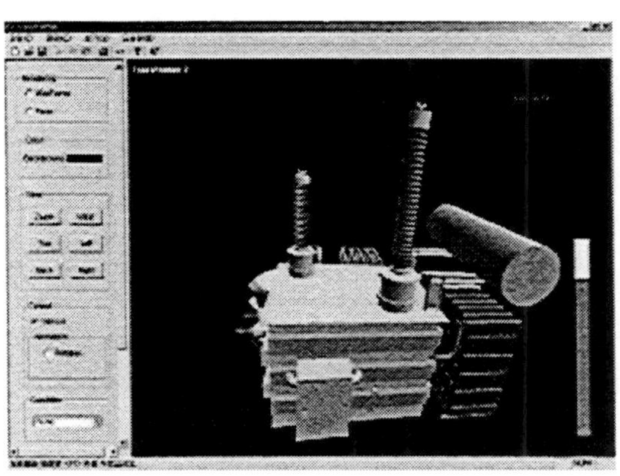

Fig 6. Scene of Virtual Transformer Visualization

B. Transformer State Visualization at Overload & Breakdown

The 3D guide represents the requirements and limits of an overload operating condition in the state of overload. And guide notifies complementary measures. Also, when a transformer has broken down, the guide represents the transformer's 3D model with the broken-down part showing a colour change and quickly giving an alarm sound. Then, the guide informs the operator of the possible causes of the problem and suitable countermeasures for it. This information is helpful in deciding the correct course of action quickly. Figure 7 and 8 show the possible causes of the increasing temperature of insulating oil at overload.

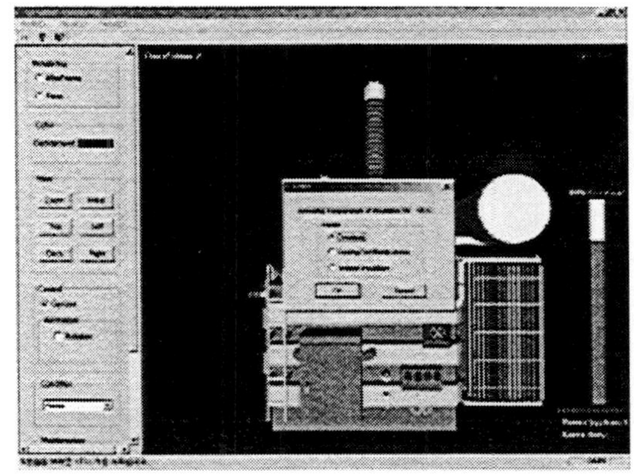

Fig 7. Screen showing the increasing temperature of the insulating oil at overload.

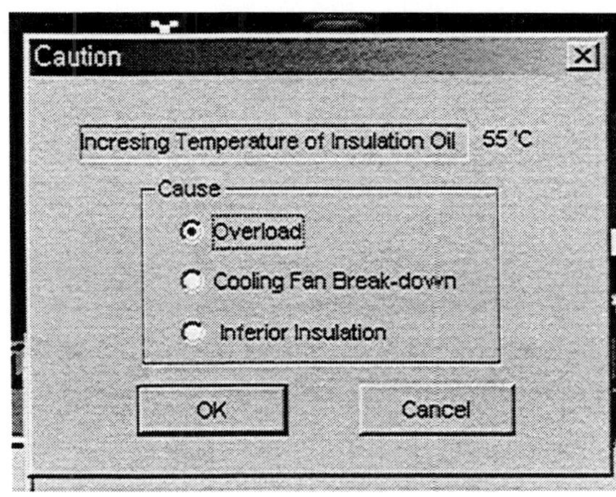

Fig 8. Possible causes of increase of the oil's temperature

Figure 9 and 10 show the suitable countermeasures for a operating condition at 80% overload..

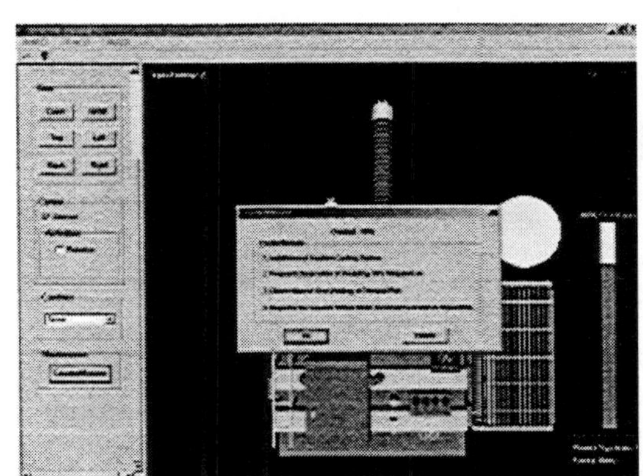

Fig 9. Screen informing the operator of the recommended countermeasure at 80% overload

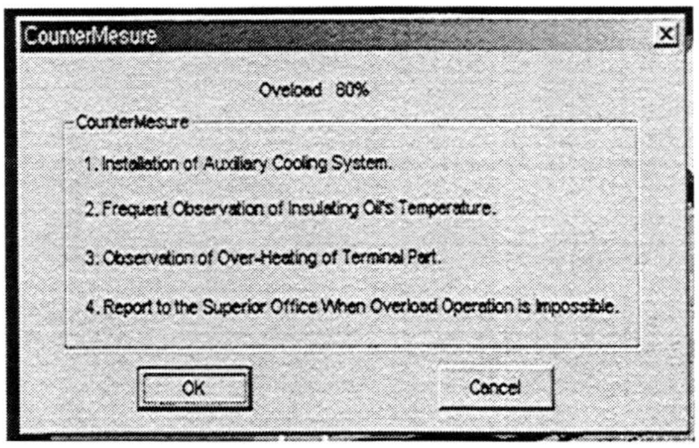

Fig 10. Countermeasures for a 80% overload

Suitable countermeasures for a 80% overload operating condition are

1. Installation of auxiliary cooling system.
2. Frequent observation of insulating oil's temperature.
3. Observation of over-heating of terminal part.
4. Report to the superior office when overload operation is impossible.

The guide informs the operator of the current condition and suitable countermeasure when he/she clicks on the breakdown part in the 3D model. Figure 11 shows that there is a problem in the right cooling system.

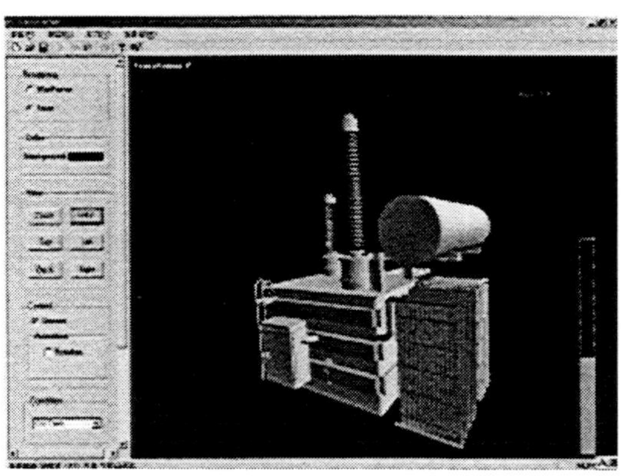

Fig 11. Screen showing the breakdown warning of a transformer with an abnormal part (shown in red)

C. Transformer Preventive Maintenance

A transformer's history of breakdown and various test results are managed, by this database system which they informs the right time and place for transformer maintenance.
This preventive maintenance contributes to reliable substation operation. With this system, operators can recognizes the appropriate inspection times of a transformer by clicking the mouse on a part of the 3D transformer. Operators can see former maintenance dates, inspection results and following inspection times by progress a bar, which will help secure and stabilize the operation of the transformer. Figure 12 and Figure 13 shows a bushing maintenance data dialog box.

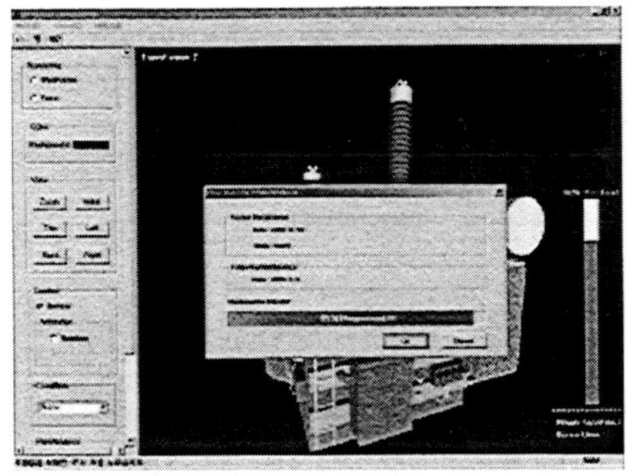

Fig 12. Screen informing the bushing maintenance data

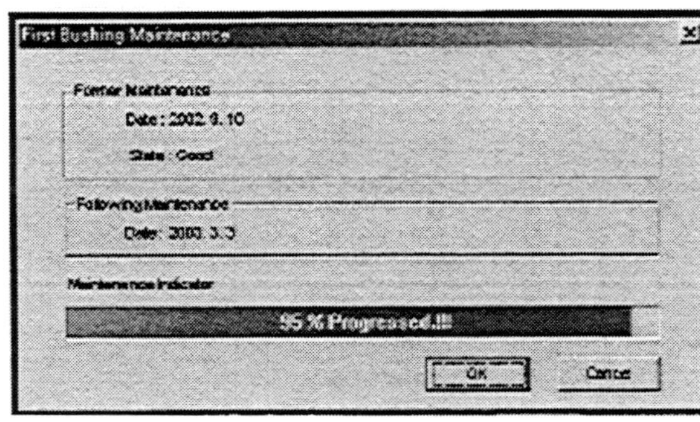

Fig 13. The bushing maintenance data..

CONCLUSION

In this work, a well-visualized 3D guide is developed as an interactive window application for a system operator to recognize the state of systems and equipments more quickly and efficiently using a 3D graphic engine. It is helpful in reducing the mistakes of unskilled operators.
Additionally, this 3D guide is expected to offer a guideline of development of training system to operators.

ACKNOWLEDGMENTS

Authors would like to thank the Korea Ministry of Science and Technology and the Korea Science and Engineering Foundation for their support through ERC program.

REFERENCES

P.M. Anderson and A.A. Fouad, "Power System Control and Stability", IEEE Press, (1994)

Grigore Burdea, Philippe Coiffet, "Virtual Reality Technology", JOHN WILEY & SONS, INC. (1994)

David J.Keuglinski "Inside Visual C++" Microsoft Press (1995)

Korea Electric Power Corporation, "FIELD GUIDE BOOK OF SUBSTATION-FACILITIES", (1999)

ENGINEERING ANIMATION, INC. SENSE8 PRODUCT LINE, "WorldToolKit Reference Manual" version April (1999)

AUTODESK, INC. "3ds max reference", 2001

ELSEVIER
IFAC
PUBLICATIONS
www.elsevier.com/locate/ifac

NEURAL NETWORK APPROACH TO TRANSFORMER FAULTS DIAGNOSIS USING DISSOLVED GAS COMPOSITION

J. H. Sun[1], K. H. Kim[2]

Korea Electrotechnology Research Institute

Abstract: This paper deals with neural network approach to transformer faults diagnosis using dissolved gas composition. The percentage of each gas to the summation of 5 gases(H_2, C_2H_2, C_2H_4, CH_4, C_2H_6) were calculated and gas composition patterns for the 3 types of faults(arcing, overheating, partial discharge) were curved. Two types of composition patterns for the arcing, three types for overheating, one type for partial discharge were obtained. Five gases were input and six types of gas composition patterns were used as output vector in train. 4 DGA test data were gathered from faults of the serviced transformers. Diagnosis results using this Neural Network Dissolved Gas Analysis(NNDGA) were well consistence with inspection results.

Keywords: Dissolved Gas Analysis(DGA), Transformer fault diagnosis, , Neural network

1. INTRODUCTION

Incipient abnormalities in power transformers lead to electrical and thermal faults on insulating materials. Power transformer reliability improves when their incipient faults are detected and eliminated before they are developed to severe problem. When sufficient energy dissipated by faults is applied to insulation oil and solid insulation as like cellulosic paper or press board, the bonds between its molecules may be broken. Depending on the level of energy according to fault types(arcing, overheating, partial discharge), different bonds are broken forming different kinds of combustible and non-combustible gases. These generated gases are dissolved into oil and have notable characteristics due to fault types. This has been understood for a long time and several efforts have been made in the development of Dissolved Gas-in-oil Analysis(DGA) techniques for detection of incipient faults in power transformers.

Presently, DGA is a common practice in transformer fault diagnosis. Many electrical utilities and researchers have reported on their developed DGA methods which include key gas method and ratio method. Key gas method relates key gases to fault types and if the amount of specified key gasesbecome excessive, corresponding fault condition can be identified. Ratio methods assign certain combination of gas ratios to a specific fault type. The related gases commonly used in those methods are hydrogen(H2), carbon monoxide(CO), carbon dioxide(CO2), methane(CH4), acetylene(C2H2), ethane(C2H6) and ethylene(C2H4) and ppm concentrations of gases are used widely. If the DGA results are not correctly interpreted, some incipient deterioration may be developed to the fatal problem later and the maintenance costs may be increased.

The key gas methods can simultaneously suggest various fault types which correspond to key gases but the diagnosis accuracy can be decreased because various gases which exceed specified gas ranges can simultaneously be generated in one fault type. The Ratio methods have higher decision ability than key gas method but some combinations of gas ratios may have no diagnostic decision when the ratio does not fall within a certain range due to the error of measuring system or more faults than one. Therefore, the search for a reliable method for fault detection in power transformers using DGA is still a topic of interest in many utilities.

Neural network have been used extensively in applications where pattern recognition is needed. This study describes neural network approach to transformer faults diagnosis using dissolved gas composition.. 78 DGA data which contain ppm concentration of H_2, CH_4, C_2H_2, C_2H_4 and C_2H_6 were gathered from literatures related to DGA studies and the percentage of each gas to the summation of 5 gases were calculated. The gas compositions for DGA data were classified into arc1, arc 2, arc 3, overheat1, overheat 2, partial discharge. Graphic User Interface(GUI) was used for training and simulation. Five Gas concentrations are used as input vectors and output vectors were obtained from six types of gas composition patterns in train. 4 DGA test data were gathered from the serviced transformers whose fault causes were illuminated by

1. Senior researcher at KERI
2. Principal researcher at KERI

visual inspection and were diagnosed by trained NNDGA.

2. EVOLUTION OF GAS IN OIL

Most Power transformers use mineral oil as dielectrics and coolant. Abnormal conditions can occur within a transformer because of electrical, thermal, mechanical, environmental stresses. These conditions result in the occurrence of various types of faults and the principal fault types are arcing, overheating, partial discharge. These faults lead to decomposition of oil and thus to the formation of gases

The various kinds of hydrocarbon gases are generated due to levels of energy with fault types. The energies are mainly dependent on the temperatures on the vicinity of fault. In both low temperature, example for <500 K, and high temperature, example for >2000 K, the end products are the same simple molecules for the process like equation (1) but variations in temperature directly effect on the concentrations of the various gases.

Oil molecules $\rightarrow C_2H_2$, C_2H_4, CH_4, C_2H_6, H_2, C_3H_6, C_3H_8 (gas)+C(solid) (1)

Fig. 1 shows the decomposition gases of oil with fault types. As you can see in Fig. 1, mineral insulating oil are made of a blend of a different hydrocarbon molecules and scission of some of the C-H and C-C bonds may occur as a result of faults like arcing, overheating and partial discharge with the formation of unstable fragments in radical and ionic form, which recombine rapidly into gas molecules such as hydrogen(H-H), methane(CH_3-H), ethane(CH_3-CH_3), ethylene(CH_2=CH_2), acetylene(CH ≡CH). Low energy faults, such as partial discharge favor the scission of the weakest C-H bonds through ionization reactions and hydrogen is the main recombination gas. More and more energy(higher temperature) are needed for the scission C-C bonds and they recombine into gases with a C-C single bond like ethane, a C=C double bond like and a C≡C triple bond like acetylene. In the relation between faults types and their energy levels, the notable hydrocarbon gas for the each faults will be as follows.

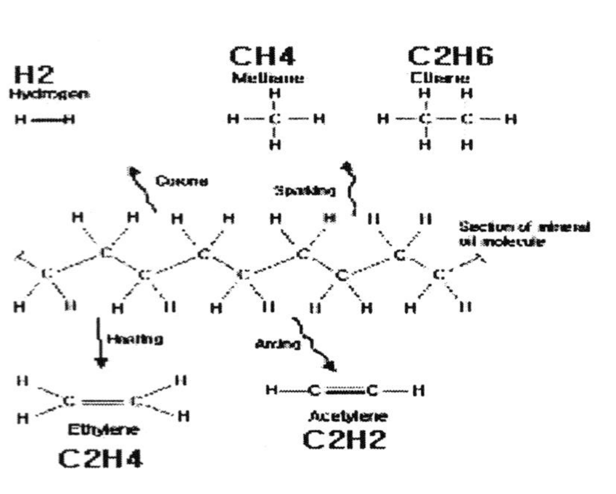

Fig. 1. Decomposition gas of oil

- arcing : hydrogen, acetylene
- overheating : ethylene, ethane, methane,
- partial discharge : hydrogen

3. NEURAL NETWORK MODELS

Conventional DGA methods sometimes have no diagnostic decisions due to ambiguous gas pattern, The NN has been used since the obvious and not so obvious relationships are detected. An NN design includes selection of input, output, network topology and weighted connections. The used NNDGA architecture is shown in Fig. 2. Six gas concentrations in ppm are presented the input nodes and Total gas means the summation of five gases. The input nodes are connected to a layer of 210 hidden nodes. The output layer consists of nodes which represent six types of faults.

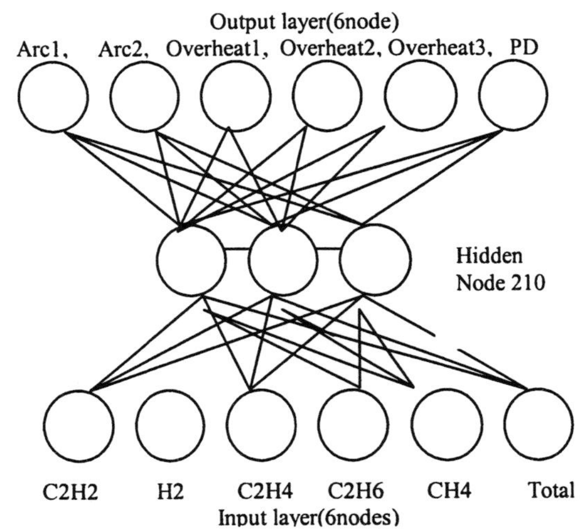

Fig. 2. Architecture of used multi layer perceptron neural network

NNDGA was developed using MATLAB. GUI which is shown in Fig. 3 was used for training and simulation. The GUI panel consists of input windows to type six gases concentrations in ppm and output window to display accuracy as numbers from 0 to 1. Also, accuracy, gas composition patterns and training process are seen in types of bar graph or curves in main window. Scaled gradient descent algorithm was used to train the NNDGA and the training was iterated until reaching the error goal of 0.001. Log sigmoid activation function is used in each node so that outputs range from 0 to 1

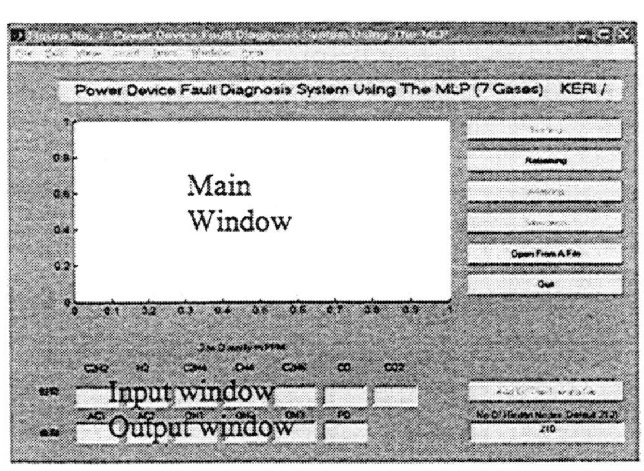

Fig. 3 GUI panel of NNDGA

4. CLASSIFICATION OF DGA DATA

Principally, the oil fault related gases commonly used are hydrogen(H_2), methane(CH_4), ethane(C_2H_4), acetylene(C_2H_2), ethylene(C_2H_6) as like Fig. 1 and also, they have been used in the conventional DGA methods such as key gases, IEC code, Rogers, Dornenburg etc.,. 78 DGA samples which contain ppm concentration of each gas were gathered from literatures related to DGA studies and DGA data measured in Korea. The data are classified in six types of fault of arc1, arc2, overheat1, overheat2, overheat3, partial discharge and table 1 shows the classification result of faults.

Table 1. Classification of faults

Fault types	Fault cases
Arcing 1	11
Arcing 2	13
Overheating 1	27
Overheating 2	14
Overheating 3	7
Partial discharge	6

Fault causes of all samples in table 1 were searched by visual inspection of transformer. The percentage of each gas to the summation of 5 gases in each sample were calculated and gas distribution for the six types of faults were curved. Fig. 4, 5 and 6 show distribution curves of gas composition percentages for each fault type. Fig.4 (a) and (b) are gas composition distribution for arcing. As you can see in these figures, arcing has two types of curves patterns. The main gas formed is acetylene followed by hydrogen, ethylene, methane, ethane. Also, the percentages of hydrogen and ethylene show some difference but are alike. In arcing type II, the main gas formed is hydrogen followed by acetylene, ethylene, methane, ethane. Fig.5 shows gas composition distribution for overheating. As you can see in these figures, overheating has three types of curve patterns like (a), (b) and (c). In overheating type I, the main gas formed is ethylene followed by methane. In type II , the main gas formed is methane followed by ethylene and ethane. In type III , the main gases formed are hydrogen and methane followed by ethane. In gas composition due to overheating, main gases are various such as ethylene, methane, sometimes hydrogen with gas composition types and gas composition have unique pattern for composition type that each of main gas belongs to.

Fig. 6 shows gas composition distribution for partial discharge. As you can see in this figure, partial discharge has only one type of composition curve. The main gas formed is hydrogen followed by methane, ethane, ethylene and acetylene. Main gases and gas compositions with faults observed from Fig. 4, 5 and 6 and composition patterns of gases with faults can be recognized by using any tools, the DGA by gas composition will more increase its accuracy and reliability.

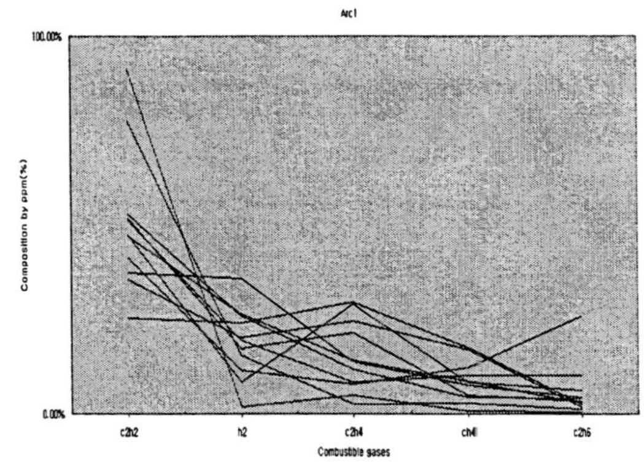

(a) arcing type I

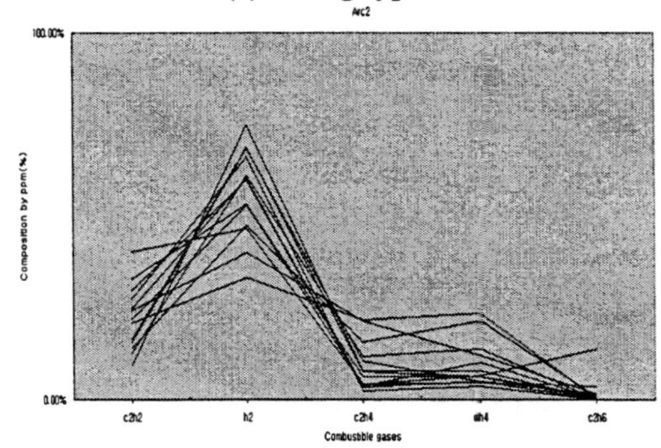

(b)arcing type II

Fig. 4. Gas composition distribution for arcing

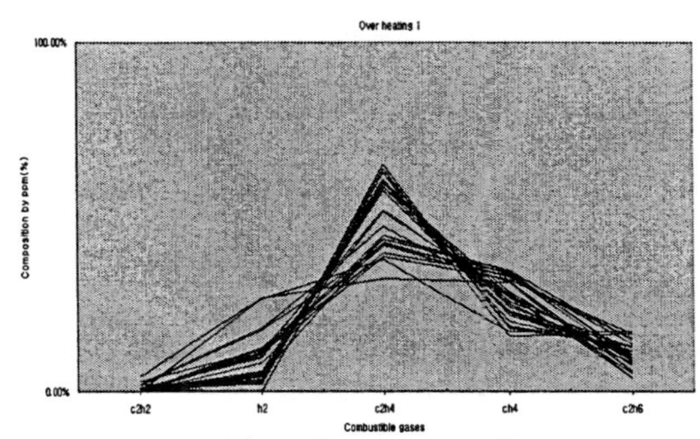

(a)overheating type I

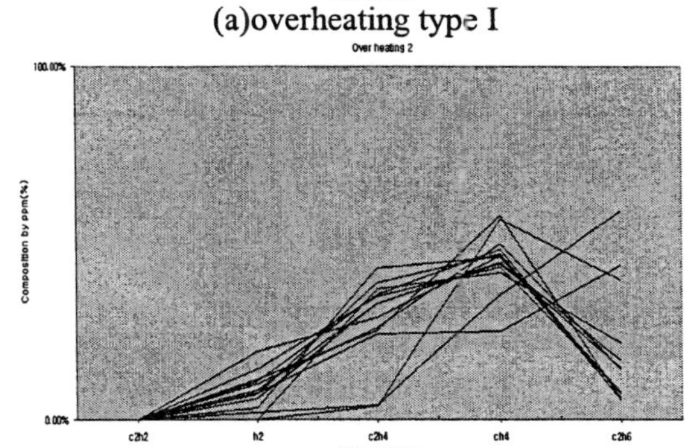

(b)overheating type II

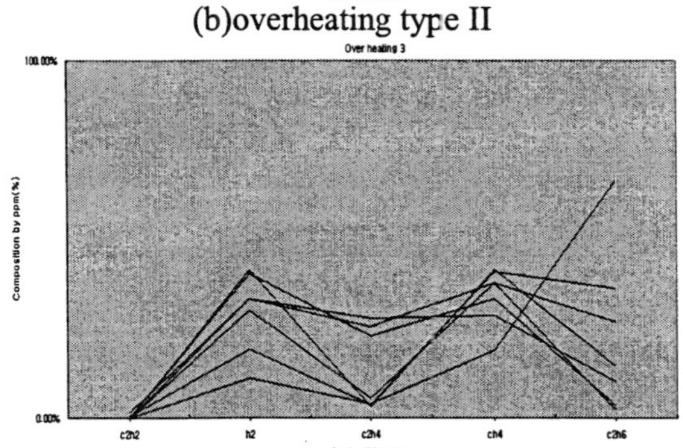

© overheating type III

Fig. 5 Gas composition distribution for overheating

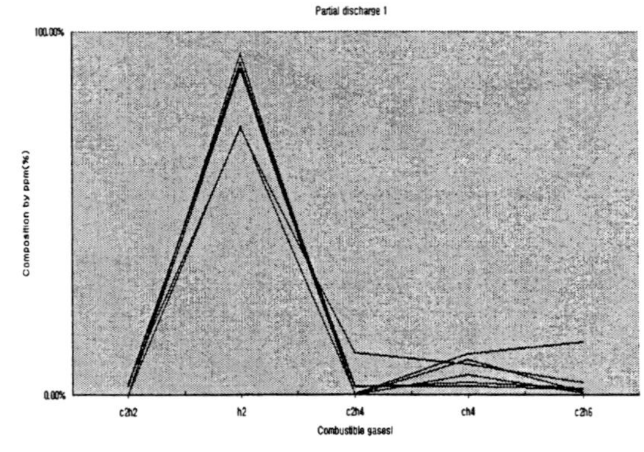

Fig. 6 Gas composition distribution for partial discharge

5. FAULT DIAGNOSIS USING NNDGA

The three layer scaled gradient descent algorithm was trained with the above 78 DGA samples. Fig 7 shows error trend during NNDGA training. After 770 iterations the error reach a value of 0.000995 based on 210 nodes, which is the optimal one for the accuracy, in the hidden layer.

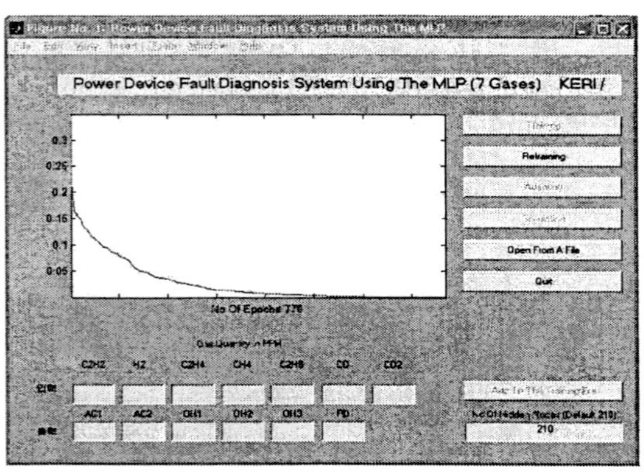

Fig. 7 error trend of NNDGA

4 DGA test data which are not used in training were gathered from the serviced transformers where the fault causes were illuminated by visual inspection and are shown in table 2. Those data were diagnosed by trained NNDGA. Fig. 8 shows gas composition percentage pattern for sample 1 and the pattern is considered to be similar to overheating type II. Fig 9 shows diagnosis results for sample 1 and NNDGA output indicates that it is overheat 2 of accuracy 0.998. NNDGA diagnosis results are listed in table 3.

Table 2 Test sample data

No	1	2	3	4
C2H2	16	65	270	2321
H2	616	0	76	15994
C2H4	878	38	51	0
CH4	4360	28	81	0
C2H6	1119	2	21	99

As you can see in table 3, the NNDGA diagnosis results are well agree with visual inspection results. In case of sample 4, the NNDGA diagnosis and the actual inspection disagree. This may be explained by arc energy level dispersed by fault.

Table 3 NNDGA diagnosis results

Sample No	1	2	3	4
Arc1	0.002	0.776	0.992	0.001
Arc2	0.000	0.084	0.025	0.837
Overheat1	0.000	0.004	0.000	0.001
Overheat2	0.998	0.123	0.009	0.000
Overheat3	0.089	0.013	0.003	0.002
PD	0.000	0.001	0.000	0.521
Diagnosis results	Over-heat2	Arc1	Arc1	Arc2+ PD
Inspection results	Over-heat	Arc	Arc	Arc

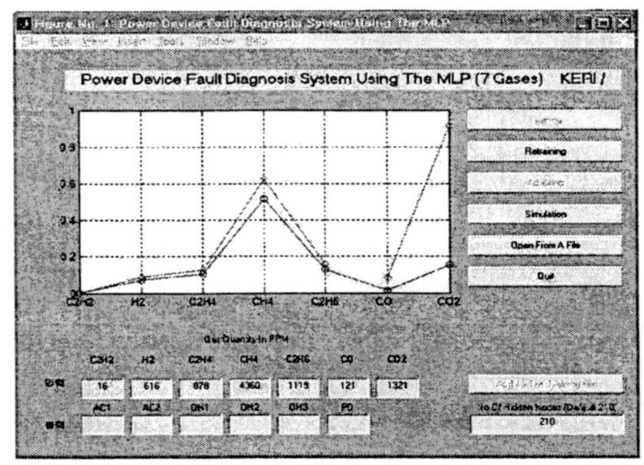

Fig. 8 Gas composition percentage pattern for sample 1

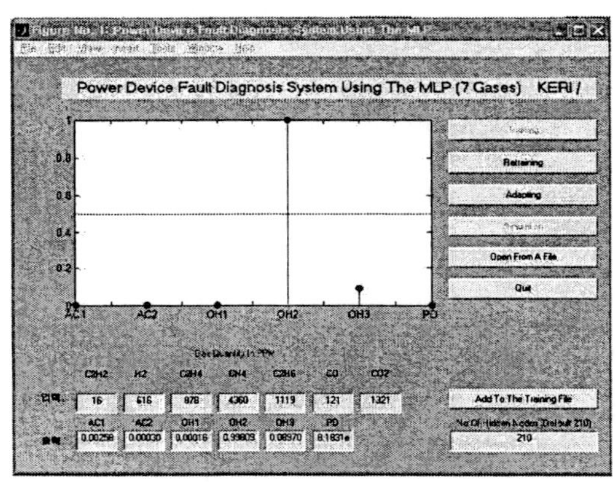

Fig. 9 Test results for sample 1

6. CONCLUSIONS

In this paper, we presented neural network approach to transformer faults diagnosis using dissolved gas composition. Gas composition patterns for 78 samples data were classified in six fault types of arc1, arc2, overheat1, overheat2, overheat3, partial discharge and then six gas composition patterns were obtained. Arcing had two types of curves patterns.In arcing type I, the main gas formed was acetylene and in arcing type II, the main gas formed was hydrogen. Overheating had three types of curve patterns. In overheating type I, the main gas formed was ethylene and in type II, the main gas formed was methane and in type III, the main gases formed were hydrogen and methane. Partial discharge had only one type of composition curve pattern and the main gas formed was hydrogen. NNDGA was designed and GUI was used for training testing. 78 sample data which were categorized in six types faults were trained and 4 sample DGA data which are obtained from faults of the serviced transformer were tested. The diagnosis results by NNDGA were well agree with visual inspection results.

REFERENCES

C.E.Lin(1993). An Expert System for Transformer Fault Diagnosis using Dissolved Gas Analysis, IEEE Transactions on Power Delivery, **Vol. 8. No. 1**. pp. 231-238.

R.R.Rogers(1978). IEEE and IEC Codes to Incipient Faults in Transformers using Gas in Oil Analysis, IEEE Trans. On EI-**13, No. 5**, pp. 349-354.

J.J.Kelly(1980). Transformer Fault Diagnosis by Dissolved-Gas Analysis, IEEE Transactions on Industry Application. **Vol. 16. No. 6**. pp. 777-782, November 1980.

E. Dornenburg(1974). Monitoring oil-cooled transformers by gas analysis, Brown-Boveri Rev. 61, p.238-247.

Guide for the detection and determination of generated gases in oil-immersed transformer and their relation to the serviceability of the equipment, ANSI/IEEE(1978). **C57.104**

Interpretation of the Analysis of Gases in Transformers and Other Oil_Filled Electrical Equipment in service, First Edition, IEC Publication **60599**(1999)

ELSEVIER
IFAC
PUBLICATIONS
www.elsevier.com/locate/ifac

AN OPTIMAL MEASUREMENT PLACEMENT FOR HARMONIC STATE ESTIMATION USING GENETIC ALGORITHMS

Yong-P. Wang, Hee-C. Park, Hyeng-H. Chung

School of Electrical, Electronic & Computer Engineering
University of Dong-A, Busan, Korea

Abstract: The design of a measurement system to perform Harmonic State Estimation (HSE) is a very complex problem. Among the reasons for its complexity are the system size, conflicting requirements of estimator accuracy, reliability in the presence of transducer noise and data communication failures, adaptability to changes in the network topology and cost minimization. In particular, the number of harmonic instruments available is always limited. Therefore, a systematic procedure is needed to design the optimal placement of measurement points.
This paper presents a development of new HSE algorithm which is based on an optimal placement of measurement points using Genetic Algorithms (GAs) which is widely used in areas such as; optimization of the objective function, learning of neural networks, tuning of fuzzy membership functions, machine learning, system identification and control.
This HSE has been applied to the New Zealand lower South Island AC Power System for the validation of the new HSE algorithm. The study results have indicated an economical and effective method for optimal placement of measurement points using Genetic Algorithms (GAs) in the Harmonic State Estimation (HSE). *Copyright © 2003 IFAC*

Keywords: Harmonic State Estimation (HSE), Optimal Placement of Measurement, Genetic Algorithms (GAs)

1. INTRODUCTION

System wide Harmonic State Estimation (HSE) is an efficient and economic tool to obtain harmonic information throughout a real-size power system, from partial and synchronized harmonic measurements at selected buses and lines. Observability analysis is essential for system wide HSE to identify its solvability. A power system is said to be observable if the set of available measurements is sufficient to enable the calculation of the system state variables uniquely. Observability is dependent on the number, locations and types of available measurement, as well as the topology of the network. For a different network topology, or same network topology but different measurement placements, an OA is to be performed in each case.
The emphasis so far has been on system-wide observability, requiring an over-determined or completely determined set of measurements, a condition difficult to be met in HSE given the limited number of harmonic monitoring (Continuous Harmonic Analysis in Real Time: CHART) locations available. However, it is important for OA not only to decide whether the system is observable and hence system-wide HSE can be performed, but also to provide information of the observable/unobservable islands as well as redundant measurement. This allows the re-positioning of measurement points to maximize their usefulness.
This paper presents a new HSE algorithm which is based on an optimal placement of measurement points using Genetic Algorithms (GAs) which is widely used in areas such as; optimization of the objective function, learning of neural networks, tuning of fuzzy membership functions, machine learning, system identification and control. This HSE has been applied to the New Zealand lower South Island AC Power System for the validation of the new HSE algorithm. The study results have indicated an economical and effective method for optimal placement of measurement points using Genetic Algorithms (GAs) in the Harmonic State Estimation (HSE).

2. HARMONIC STATE ESTIMATION

System wide Harmonic State Estimation (HSE) is an efficient and economic tool to obtain harmonic

Three different methods of OA have been proposed, ie Numerical, Topological and Symbolic for HSE.

Numerical observability determination is based on assessing the rank of the measurement matrix by triangular factorization. There are several algebraically equivalent ways of expressing the state estimation equations that have good sparsity and numerical stability for large systems. However, due to ill-conditioning and finite precision arithmetic, numerical problems can occur.

As the name indicates, the topologically approach tells whether a system is topologically observable. Although it is possible for a topologically observable system to be algebraically unobservable, this is unlikely to occur in a practical system as it only happens with a theoretical choice of network admittances. This condition is called parametrically unobservable. The system is topologically observable if there exist a spanning tree of full rank. In this respect, a tree is any interconnected, loop-free collection of branches of the network and a spanning tree is a tree that is incident to every busbar. The number of possible trees for n busbars is n^{n-2}, a very large figure even for a small system. However, the topological method requires procedures that are not needed to compute the state estimation and is combinatorial, thus requiring considerable computational effort.

The symbolic method seeks to overcome the numerical problems associated with floating point operations by replacing them with symbolic calculations, where each entry in the measurement matrix used for OA is either a 1 or a 0. While being extremely fast and simple, the basic method is not capable of finding all the observable islands and redundant measurements.

In harmonic state estimation, due to the cost of suitable harmonic measuring equipment it is impractical to perform complete system-wide estimation. To solve $[A]\underline{x} = \underline{b}$ for the under-determined case, where only observable islands exist, Singular Value Decomposition (SVD) needs to be applied, as standard techniques for solving such equations will fail. For the under-determined case there is of course no unique solution but an infinite number of solutions. SVD, however, will provide a particular solution and a null space vector for each singularity.

The null space vectors can be multiplied by any constant and added to the particular solution to give another valid solution to the set of equations, thereby specifying the infinite number of solutions. Variables corresponding to zeros in all the null space vectors will not be changed by this process and hence are completely specified by the particular solution. These variables corresponding to non-zero elements in the null space vectors are in the unobservable regions as they can not be uniquely determined. Therefore analysis of the null space vectors also provides OA as a by product.

Regardless of whether the HSE is achieved via solution of the Normal equations ($|h^T R^{-1} h| \underline{x} = |h^T| \underline{Z}$) or using SVD to solve the measurement equation, the complex-valued problem is split into real and imaginary components. This is done to take advantage of standard real valued numerical procedures. The converted terms of the measurement equation

$\underline{Z} = [H]\underline{X} + \underline{E}$ are:

$$\begin{cases} x = x_R + jx_I \\ z = z_R + jz_I \\ h = h_R + jh_I \\ \varepsilon = \varepsilon_R + j\varepsilon_I \end{cases}$$

where

$$\underline{Z} = \begin{bmatrix} z_R \\ z_I \end{bmatrix} \qquad [H] = \begin{bmatrix} h_R & -h_I \\ h_I & h_R \end{bmatrix}$$

$$\underline{X} = \begin{bmatrix} x_R \\ x_I \end{bmatrix} \qquad \underline{E} = \begin{bmatrix} \varepsilon_R \\ \varepsilon_I \end{bmatrix}$$

Figure 1 depicts the solution process for an under-determined system. An alternative approach, still using SVD, suggests eliminating the state variables in the unobservable system, ie partitioning into unknown (denoted by subscript u), known measured values (subscript k) and "not desired" current injection estimates (subscript n).

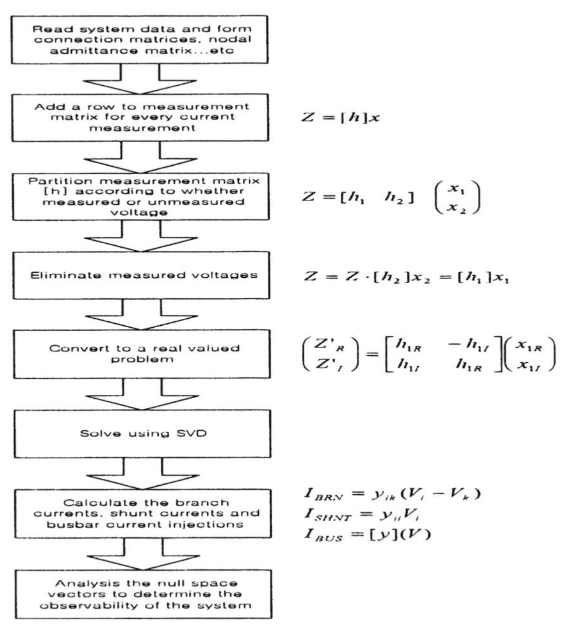

Figure 1. Flow-chart of HSE.

$$\begin{bmatrix} Y_{1u} & Y_{1k} \\ Y_{2u} & Y_{2k} \\ Y_{3u} & Y_{3k} \end{bmatrix} \begin{pmatrix} \underline{V_u} \\ \underline{V_k} \end{pmatrix} = \begin{pmatrix} \underline{I_{bu}} \\ \underline{I_{bk}} \\ \underline{I_{bn}} \end{pmatrix}$$

Sub-vector $\underline{I}_{bu}$ is then solved in the least squares sense, ie

$$\underline{I}_{bu} = Y_{1u}Y_{2u}^{-1}\underline{I}_{bk} + (Y_{1k} - Y_{1u}Y_{2u}^{-1}Y_{2k})\underline{V}_k$$

where Y_{2u}^{-1} is a pseudo-inverse in the case of an under-determined system.

When solving the linear equation $[A]\underline{x} = \underline{b}$ the relative error in A or b can be magnified by as much as the condition number of A. The condition number of the gain matrix $[h^T R^{-1} h]$ is the square of the measurement matrix $[h]$, therefore the normal equation formulation is not well conditioned. Due to finite precision the build-up of round-off error can be a problem. Two approaches to alleviate the numerical ill-conditioning are behind the various methods that have been proposed. The first is to avoid the formulation of the gain matrix. The second is to treat virtual measurements as equality constraints. For an over-determined system the use of SVD can be significantly slower than solving the normal equations and requires more storage, but is less susceptible to round-off error. Moreover, its theoretically fool-proof reliability more than makes up for the speed disadvantage.

3. GENETIC ALGORITHMS

Genetic Algorithms (GAs) are effective search methods based on the principles of natural selection and genetics. They were developed by John Holland to simulate some of the processes observed in natural evolution. Generally, a GA works on a set of potential solutions to a specific problem encoded into chromosome-like data structures. Some of these solutions, chosen on the basis of their performance in solving the problem, are used to create a new set of potential solutions through the use if operator. A GA uses this process repeatedly until a particular criterion is met. GAs are often described in biological terms. Potential solutions are called chromosomes and are represented by binary strings or floating point numbers. A set of chromosomes is called a population and a problem to be solved is represented by a fitness function. The choice of individuals to reproduce is performed in a process called selection which is based on the fitness values assigned to chromosomes. Genetic operators such as crossover and mutation are operators used to create a new population. Crossover permits the exchange of information among individuals in the population and provides the innovative capability of a GA. Mutation ensures need diversity. The flowchart of a Genetic Algorithm is presented in Figure 2.

To minimize the estimation error and measurement system (CHART), the measurement placement is optimized using GA for power system harmonic state estimation. Therefore the object function is consisted as the CHART cost and state estimation error to optimize measurement placement. It is assumed that a standard CHART cost is \$20,000 and the cost of two extra channels CHART is \$2,000.

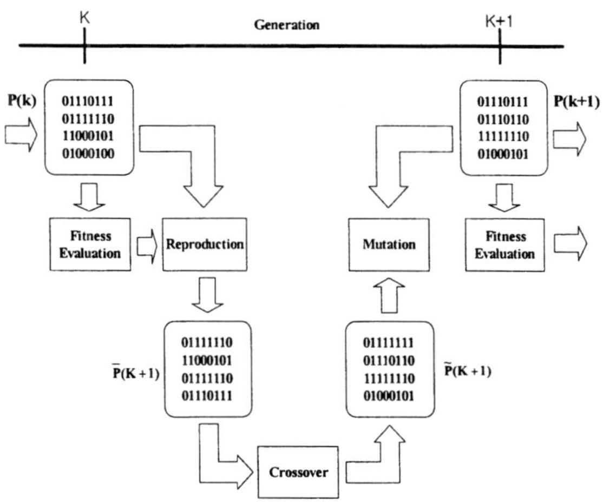

Figure 2. Flow-chart of GA

4. OPTIMIZED MEASUREMENT PLACEMENT USING GA

To minimize the estimation error and measurement system (CHART), the measurement placement is optimized using GA for power system harmonic state estimation. Therefore the object function is consisted as the CHART cost and state estimation error to optimize measurement placement. It is assumed that a standard CHART cost is \$20,000 and the cost of two extra channels CHART is \$2,000.

$$Min\left\{ J = \sum_{i=1}^{n} I_i + E_{busvolt} + E_{buscurrnt} + E_{linecurrent} \right\}$$

i=1…, n, n: number of buses

where,

$E_{busvoltage}$: Bus voltage error

$E_{buscurrent}$: Bus current error

$E_{branchcurrent}$: Line current error

$$I_i = \alpha + \kappa\beta$$

α: Standard CHART cost

β: Two extra channels CHART cost

κ: Number of CHART channels

To optimize the measurement placement, the fitness function is used as fellows:

$$Fitness = \frac{1.0}{1.0 + Min[J]}$$

The parameters of GA are presented in Table 1.

Table 1. Parameters of GA

Parameter	Value
Number of Generations	100
Population Size (N)	70
Crossover Probability (P_c)	0.9
Mutation Probability (P_m)	0.3679

5. ILLUSTRATIVE EXAMPLES

The backbone of the test system illustrated in Figure 3 is taken from the 220 kV network below Roxburgh in the south island of New Zealand. The system contains nine buses, five star-delta connected two-winding transformers, three generators at Manipouri and Roxburgh, two single transmission lines between Roxburgh and Invercargill, and one double transmission lines between Manapouri, Invercargill and Tiwai. It is assumed that the harmonic current sources consist of one linear load at Tiwai and one linear load at Invercargill. To illustrate the use of GA-HSE, let us consider a possible measurement placement such as shown in Figure 3 , and subject it to symbolic observability. The corresponding symbolized measurement matrix[h] involves 9 state variables.

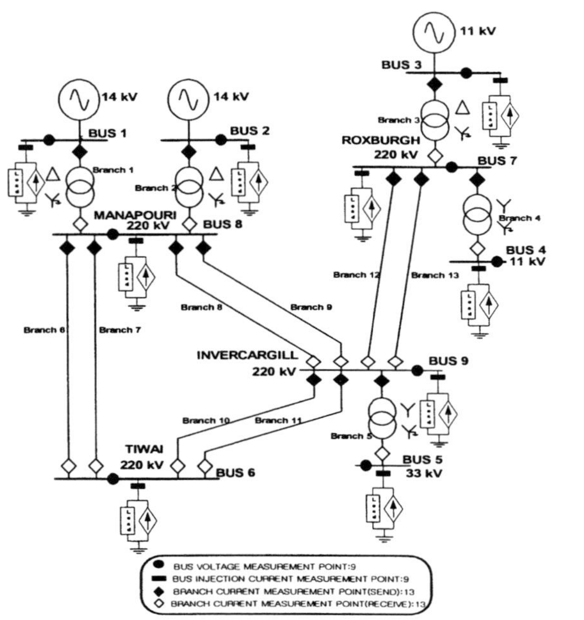

Figure 3. South island system of New Zealand test system

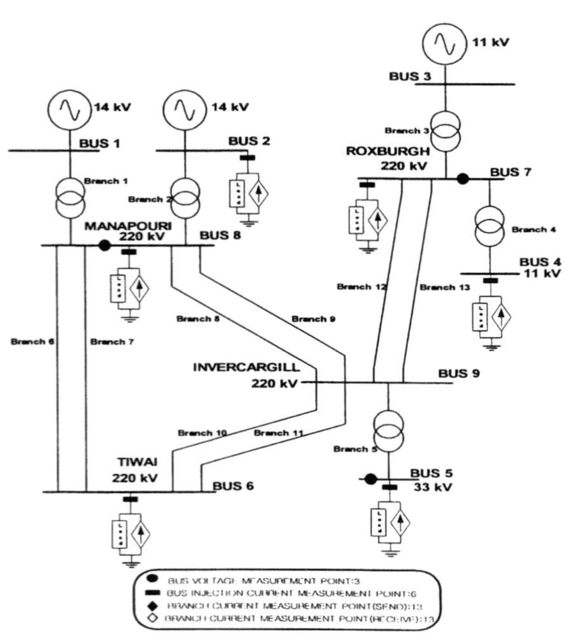

Figure 4. Reduced measurement placement.

Figure 4 depicts the HSE with reduced measurement placement. There are voltage measurements at 5, 7 and 8 buses. There are injection current measurements at 2, 4, 5, 6, 7 and 8 buses. The results of state estimation error are 0%. The line current, bus voltage and bus current have shown in Figure 5, respectively. The standard CHART cost is \$20,000 at 2, 4 and 6 buses, respectively. Also the cost of two extra channels is \$2,000 at 5, 7, and 8 buses, respectively. Therefore, the total cost of CHART is \$126,000.

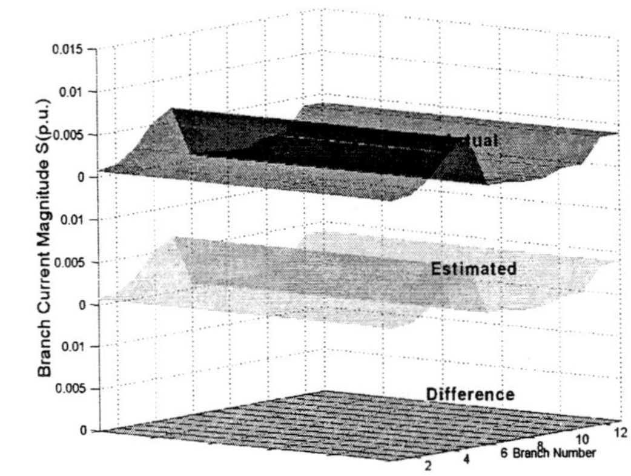

(a) Send Branch Harmonic Current Magnitude

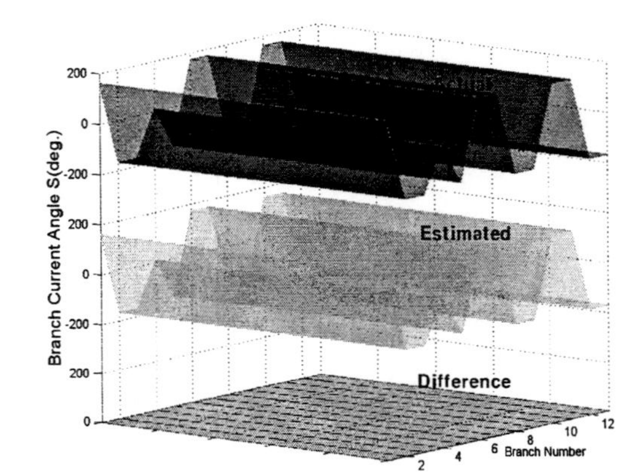

(b) Send Branch Harmonic Current Angle

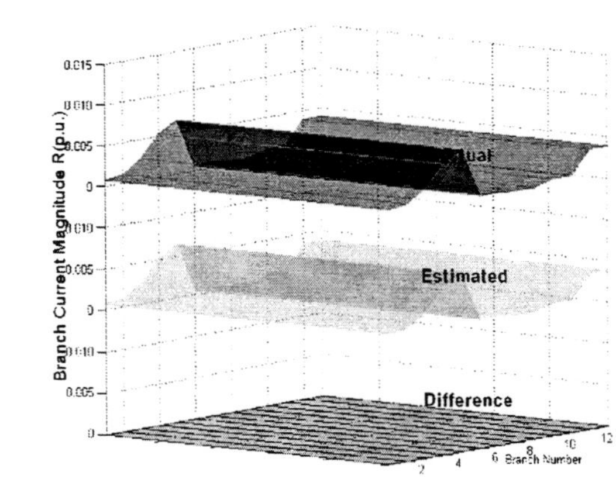

(c) Receive Branch Harmonic Current Magnitude

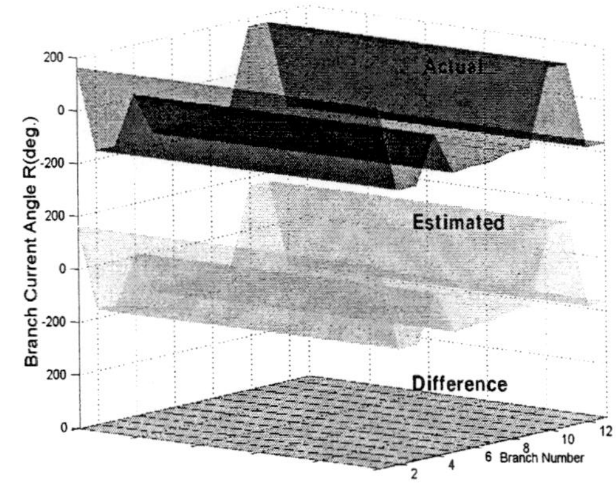

(d) Receive Branch Harmonic Current Angle

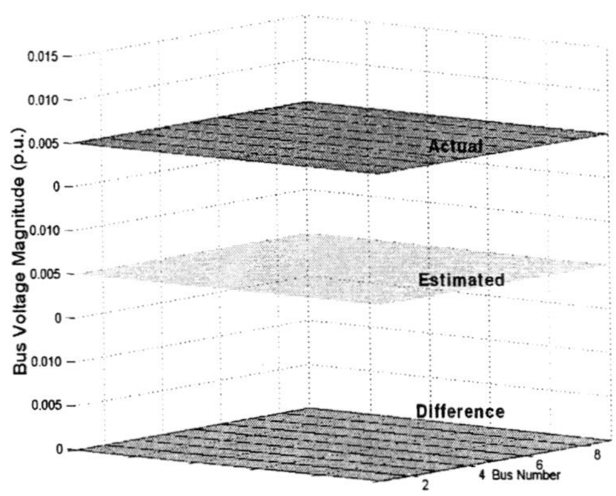

(e) Harmonic Bus Voltage Magnitude

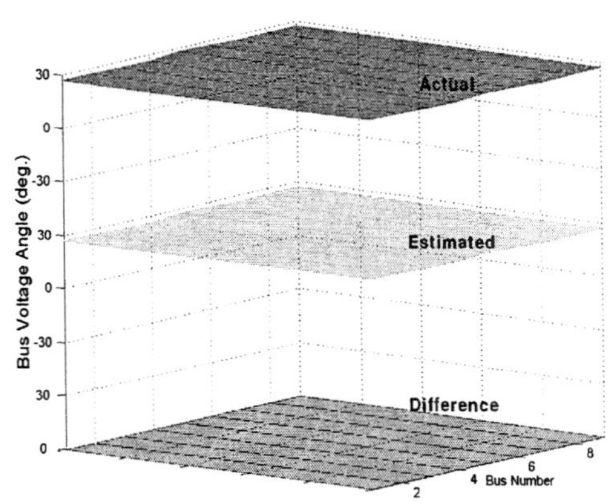

(f) Harmonic Bus Voltage Angle

Figure 6 depicts the HSE with optimized measurement placement using GA. There are voltage measurements at 5, 7, 8 and 9 buses. There are injection current measurements at 4, 5, 6, 7 and 8 buses. The results of state estimation error are 0%. The line current, bus voltage and bus current have shown in Figure 7, respectively. The standard CHART cost is $20,000 at 4 and 9 buses, respectively. Also the cost of two extra channels is $2,000 at 5, 7, 8 and 9 buses, respectively. Therefore, the total cost of CHART is $108,000.

The GA-HSE model is an economical and effective method for optimal placement of measurement points in HSE.

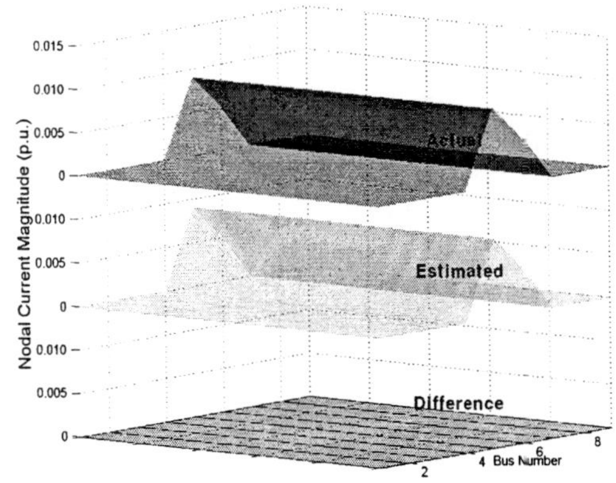

(g) Harmonic Current Injection Magnitude

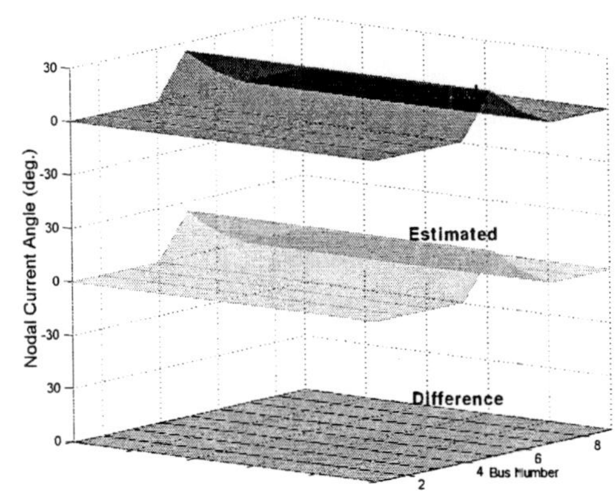

(h) Harmonic Current Injection Angle

Figure 5. Comparison of estimated and simulated level of reduced measurement placement

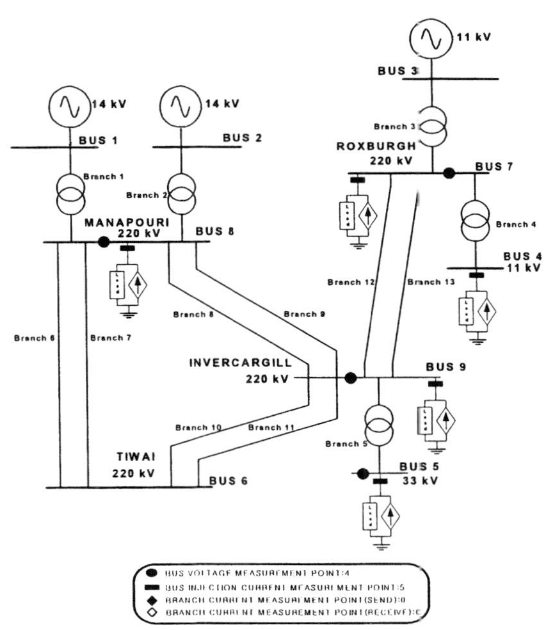

Figure 6. Optimized measurement placement using GA

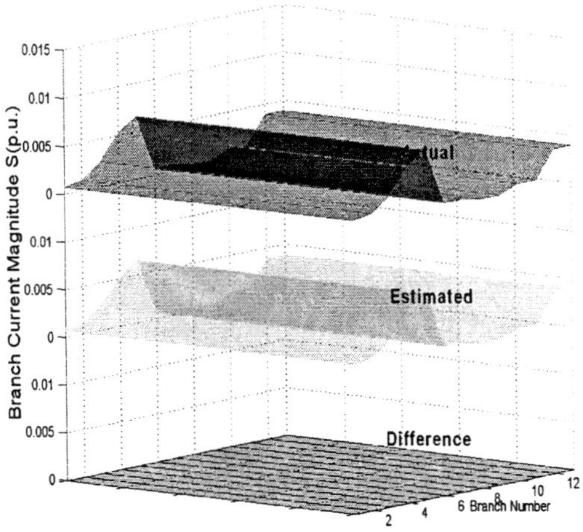

(a) Send Branch Harmonic Current Magnitude

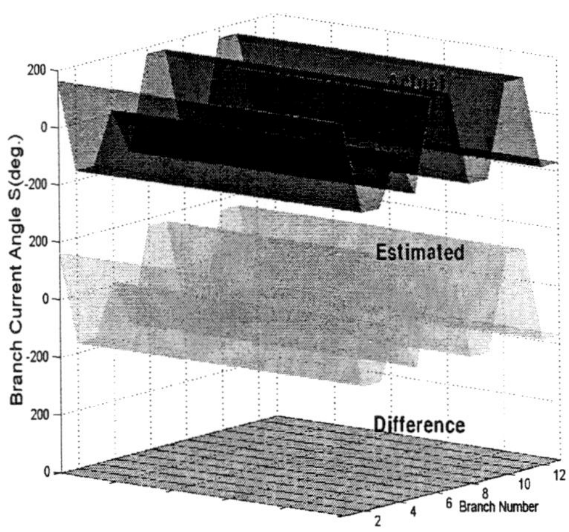

(b) Send Branch Harmonic Current Angle

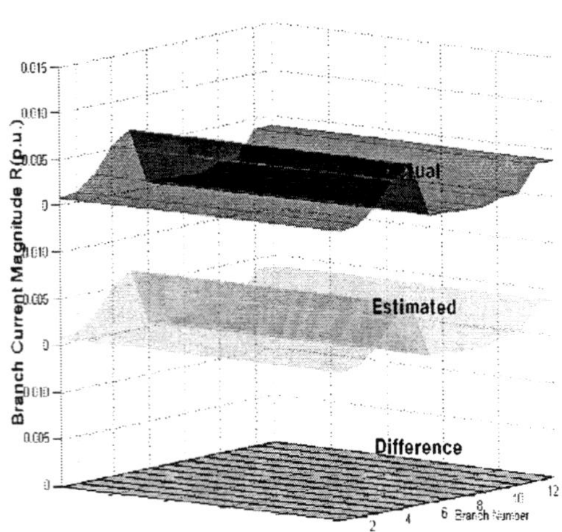

(c) Receive Branch Harmonic Current Magnitude

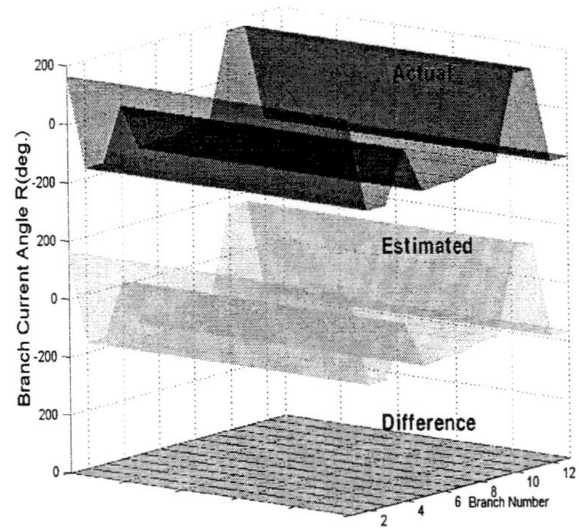

(d) Receive Branch Harmonic Current Angle

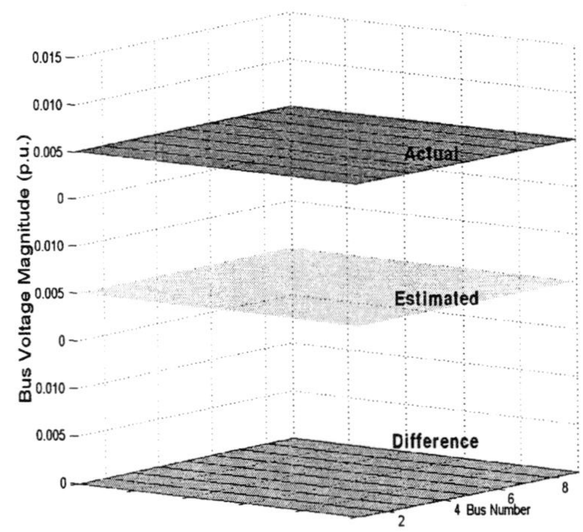

(e) Harmonic Bus Voltage Magnitude

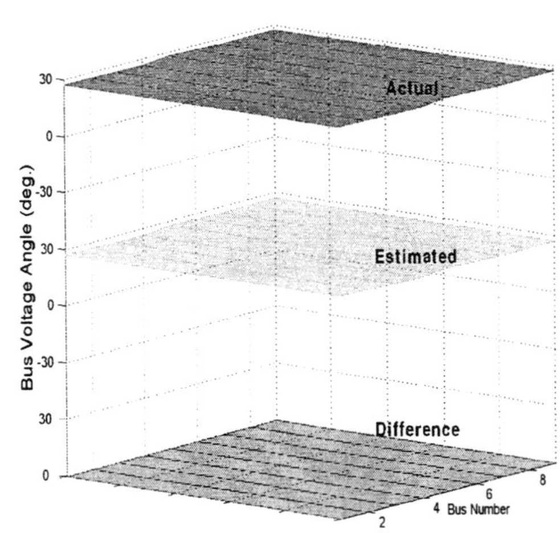

(f) Harmonic Bus Voltage Angle

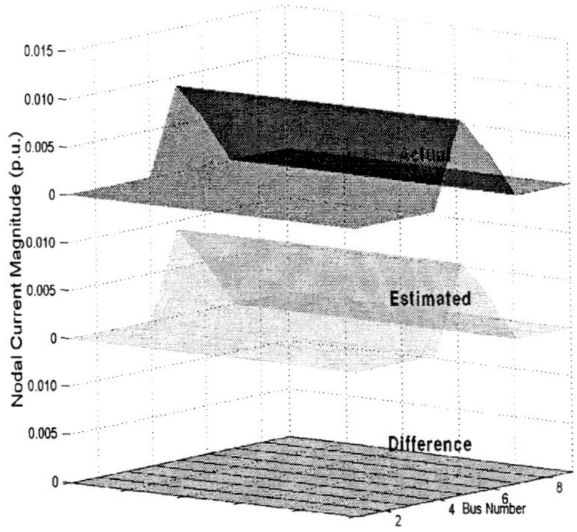

(g) Harmonic Current Injection Magnitude

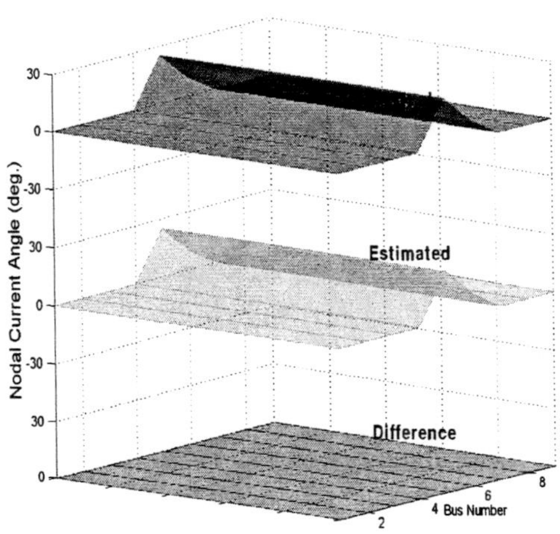

(h) Harmonic Current Injection Angle

Figure 7. Comparison of estimated and simulated level optimized measurement placement using GA

6. CONCLUSIONS

This paper presents a development of new HSE algorithm which is based on an optimal placement of measurement points using Genetic Algorithms (GAs). This HSE has been applied to the New Zealand lower South Island AC Power System for the validation of the new HSE algorithm. The results have shown that:

1) The GA-HSE model can be minimized the number of measurement points for network observability.
2) The GA-HSE model can be minimized or limited measurement system cost.
3) The GA-HSE model is an economical and effective method for optimal placement of measurement points in HSE.

ACKNOWLEDGEMENTS

This work has been supported in part by EESRI (02-JiYeog-01), which is funded by MOCIE (Ministry of commerce, industry and energy).

REFERENCES

Smith, S.E. (1991). Name of book in italics or underlined, page or chapter numbers if relevant. Publisher, Place of publication.

M. Najjar & G.T. Heydt, 'A hybrid nonlinear-least squares estimation of harmonic signal levels in power systems', IEEE Trans. On Power Delivery, 6, no. 1, 1991, pp. 282-88.

H. M. Beides & G. T. Heydt, 'Dynamic state estimation of power system harmonics using Kalman filter methodology', IEEE Trans. On Power Delivery, 6, no. 4, 1991, pp. 1663-70.

A. P. S. Meliopoulos, F. Zang & S. Zellingher, 'Power system harmonic state estimation', IEEE Trans. On Power Delivery, 9, no. 3, 1994, pp. 1701-09.

Z. P. Du, J. Arrillaga & N. R. Watson, 'Continuous harmonic state estimation of power systems', Proc. IEE, 143 Pt.C, no.4, 1996, pp. 329-36.

G. R. Krumphotz, K.A. Clements & P.W. Davis, 'Power system observability: A practical algorithm using network topology', IEEE Trans. On Power Apparatus & Systems, PAS-99, no.4, July/August 1980, pp.1534-42.

D. E. Goldberg, 'Genetic algorithm', Addison-Wesley Publishing Co. Inc., 1989.

ELSEVIER
IFAC
PUBLICATIONS
www.elsevier.com/locate/ifac

A REAL TIME MONITORING SYSTEM OF POWER QUALITY DISTURBANCES USING DIGITAL FILTERING TECHNIQUES

Kyo-Sung Lee, Yong-Jea Lee, Do-Hun Kim, Yang-Mo Kim

Department of Electrical Engineering, Chungnam National University, Korea

Abstract: In this paper, digital filtering techniques using the Wavelet Transform are proposed to monitor various power quality disturbances. Different from those Fourier-based Transforms, the Wavelet Transform is more efficient in detecting power quality disturbances, because the Wavelet Transform provides time information in addition to frequency information. The detection method is tested through the simulation of various disturbances, including voltage sags, voltage swells and interruptions. For real time detection of power quality disturbances, the monitoring system is implemented. And the obtained results are given. *Copyright © 2003 IFAC*

Keywords: Power quality disturbances, Monitoring System, Digital filter, Wavelet transform

1. INTRODUCTION

In recent years, concerns over the power quality (PQ) have been rapidly increasing because of a broader application of nonlinear electronic devices in power apparatus and systems. PQ problems deal with events lasting for short duration, typically from less than one cycle to few seconds.

Poor PQ causes many problems to the loads such as malfunctions, system instabilities, life time reduction of equipment etc. The sources of poor PQ are power line disturbances such as voltage sags, swells, interruptions, impulses, outages and harmonic distortion.

To improve the PQ, various types of equipment are used. Dynamic Voltage Restorer (DVR), Static Compensator (STATCOM) and Uninterruptible Power Supply (UPS) are among those equipments produced for voltage support. DVR is efficient in compensating the negative effect of voltage sags, swells, STATCOM is applied for reactive power compensation and the UPS is expected to maintain the energy during a power supply interruption. To maximize cost and benefit, Unified Power Flow Controller (UPFC), combined system of shunt and series compensator is used.

However, without determining the existing levels of PQ, electric utilities and customers cannot adopt suitable strategies and equipments to improve the PQ. Therefore, the sources and causes of such disturbances must be known before appropriate action can be taken. This can be done by detecting, localizing and classifying different disturbances. To achieve this, real time monitoring is necessary.

To achieve this goal, we used a digital filtering technique based on the Wavelet Transform (WT) that has the ability to analyze different PQ disturbances simultaneously in both time and frequency domains.

The Fourier Transform (FT) has been constructed as the basis of modern spectral analysis. The FT was successfully applied to the stationary signals where the properties of signals did not change in time. However, for those nonstationary signals, the FT is less efficient in tracking the signal dynamics. While the FT relies on a uniform window for spread frequencies, the WT can apply various lengths of windows according to the amount of signal frequencies. So characteristics of nonstationary disturbances were found to be more closely monitored by the WT. Different from the FT, both time and frequency information can be obtained by the WT. It's more attractive point than others'.

In this paper, we apply an approach which uses digital filtering techniques to monitor various PQ disturbances. Two kinds of digital filters are used for noise filtering and the WT. Typical PQ disturbance signals of voltage sags, swells and interruptions have been constructed to simulate the proposed method, and simulation results are presented. And real time monitoring system of the PQ is implemented by using Digital Signal Processor (DSP)-based structure.

This paper is organized as follows: Section 2 presents paradigms of the WT, Section 3 presents the PQ disturbances, Section 4 shows simulation results, Section 5 shows experiment results and Section 6 shows the conclusion.

2. THE WAVELET TRANSFORM

2.1 The Fourier-based Transform

The FT has a serious drawback. Most interesting signals contain numerous nonstationary characteristics. These characteristics are often the most important part of the signal. However, in transforming to the frequency domain, time information is lost. So, the FT is not suitable for monitoring them.

To correct the FT's drawback, the Short-Time FT (STFT) adopts the adapted FT to analyze only a small section of the signal at a time, which is called windowing the signal.

The STFT is similar to the FT except that the input signal x(t) is multiplied by a window function w(t) whose position is translated in time by τ .

$$STFT(f,\tau) = \int_{-\infty}^{\infty} x(t)w(t-\tau)e^{-j2\pi ft}\,dt \qquad (1)$$

For digital implementation of the STFT, the Windowed DFT (WDFT) is used.

$$WDFT[k,m] = \sum_{n} x[n]w[n-m]e^{-j\frac{2\pi kn}{N}} \qquad (2)$$

The WDFT of a signal maps a signal into a two-dimensional function of time and frequency k. The WDFT provides some information about range of time in the signal when particular frequencies have a relatively large magnitude and what frequencies PQ disturbances occur. However this information has a limited precision and that precision is determined by the size of the window. That window is the same for all frequencies. Many disturbances require a more flexible approach. That is, fine time resolution for short duration and high frequency disturbances, and fine frequency resolution for long duration and lower frequency disturbances are needed.

2.2 The Wavelet Transform

Like the FT, the WT deals with expansion of function in terms of a set of basis functions. However, unlike the FT, the WT expands functions not in terms of trigonometric polynomials but in terms of wavelets, which are generated in the form of translations and dilations of a fixed function called the mother wavelet. The wavelet is described as a little wave, little in the sense of being of short duration with finite energy which integrates to zero.

The Continuous WT (CWT) of a signal x(t) is generally defined as

$$W_g f(a,b) = \frac{1}{\sqrt{a}} \int_{-\infty}^{\infty} x(t)\overline{g\left(\frac{t-b}{a}\right)}dt \qquad (3)$$

where g(t) is the wavelet basis function, a is the scaling or dilation factor and b is the time shifting or translation factor.

Like the STFT, the WT has a digital implementation, the Discrete WT (DWT). The DWT is defined as

$$DWT[m,k] = \frac{1}{\sqrt{a_0^m}} \sum_{n} x[n] g\left[\frac{k - na_0^m}{a_0^m}\right] \qquad (4)$$

where g[n] is the mother wavelet, and the scaling and shifting factors are $a = a_0^m$ and $b = na_0^m$, functions of an integer m. The scaling factor gives rise to geometric scaling, i.e. 1, $1/a_0$, $1/a_0^2$, ⋯, and gives the DWT a logarithmic frequency coverage in contrast to the uniform frequency coverage of the STFT.

The DWT of a signal is represented in a two dimension like the WDFT but with very different divisions in time and frequency.

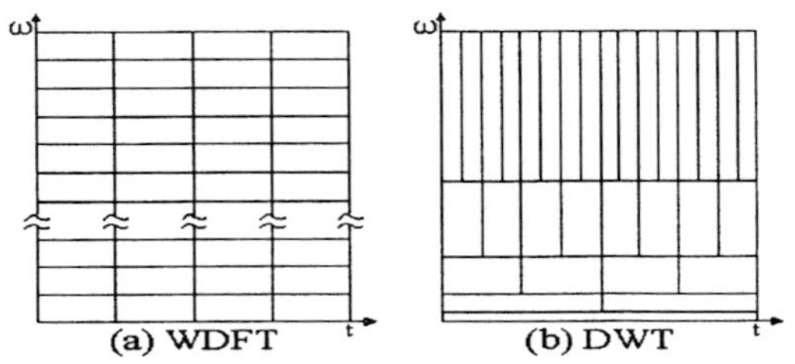

Fig. 1. Time-frequency windows.

Fig. 2 shows the difference between the WDFT and the DWT. Time-frequency windows of the DWT narrow at the lower scales (higher frequencies) and widen at the higher scales (lower frequencies). This illustrates how the WT is suitable for detecting disturbances.

By simple interchange of the variable n and k, (4) can be rewritten.

$$DWT\,[m,n] = \frac{1}{\sqrt{a_0^m}} \sum_{k} x[k] g[a_0^{-m}n - k] \qquad (5)$$

We can see that (5) has the same formation of the convolution equation for the Finite Impulse Response (FIR) digital filters, that is;

$$y[n] = \frac{1}{c} \sum_{k} x[k]h[n-k] \qquad (6)$$

So, $g[a_0^{-m}n-k]$ is the impulse response of the FIR filter. Then by selecting $a_0=2$, the DWT can be implemented by using a multi-stage filter with the mother wavelet as the lowpass filter g[n] and its dual, the highpass filter h[n].

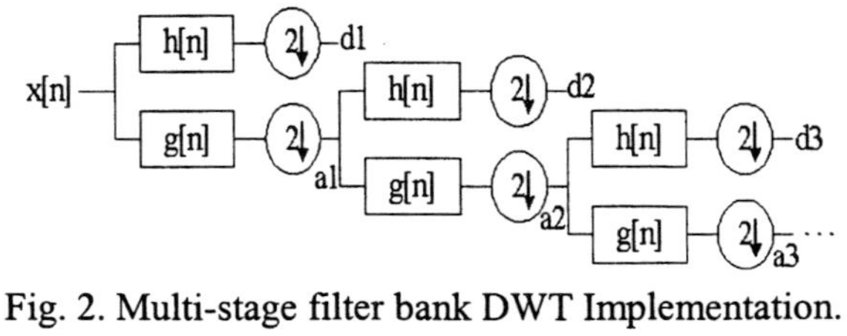

Fig. 2. Multi-stage filter bank DWT Implementation.

By using a multi-stage filter, the signals are decomposed into two other signals; one is the smoothed version of the signals and the other is the detailed version of the signals that contain the disturbance elements.

The WT does a good job in detecting the time of the signal but they give the frequency information in terms of frequency band regions. If the sampling frequency is f_s, then the maximum frequency of the sampled signal is $f_s/2$. This would be seen as the output of the highpass filter h[n], which is the d1 in Fig. 2. The d1 represents the band of frequencies between $f_s/2$ and $f_s/4$. So, the d1 is the most effective to detect the PQ disturbances. Likewise, the d2 shows the band of frequencies between $f_s/4$ and $f_s/8$.

3. THE PQ DISTURBANCES

3.1 Voltage Sags

There is a difference between the description of the voltage sag as given by the IEEE and IEC standards, Fig. 3. The IEEE standard describes voltage sag as a decrease to between 0.1 to 0.9pu in rms voltage (or current) at the power frequency, for the duration of 0.5 cycle to 1 minute, and a sag magnitude refers to the remaining voltage in percent of nominal voltage. The IEC definition for this phenomenon is dip. The IEC standard describes voltage dip as a sudden reduction of the voltage at a point in the electrical system, from 0.5 cycle to a few seconds, and a dip magnitude is the value of the voltage variation in percent of nominal voltage.

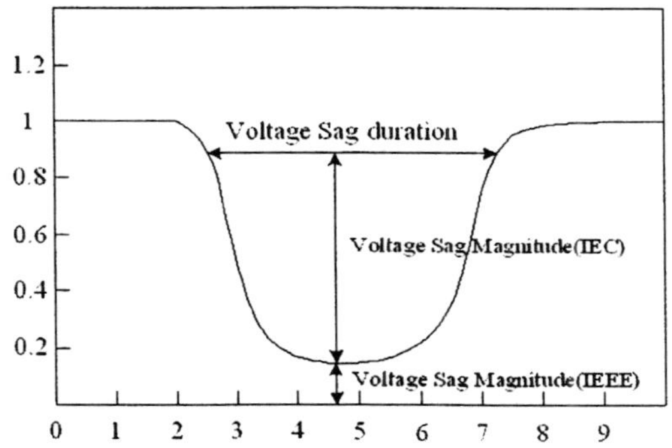

Fig. 3. Voltage sag description.

Voltage sags are usually associated with system faults but can also be caused by switching of heavy loads or starting of large motors. Sag durations are subdivided into three categories, is shown in Table 1.

Table 1. Categories and typical characteristics of voltage sags

Categories	Typical duration	Magnitude
Instantaneous	0.5 ~ 30 cycles	0.1 ~ 0.9 pu
Momentary	30 cycles ~ 3s	0.1 ~ 0.9 pu
Temporary	3s ~ 1 min	0.1 ~ 0.9 pu

Voltage sags occur the most frequently than any other disturbances. So, voltage sags are the most important problem facing many industrial customers.

3.2 Voltage Swells

A swell is defined as an increase in rms voltage or current at the power frequency for duration from 0.5 cycles to 1 min. Typical magnitudes are between 1.1 and 1.8 pu. Swell magnitude is also described by its remaining voltage, always greater than 1.0.

Voltage swells are also associated with system fault condition, but they are much less common than voltage sags. A swell can occur due to a single line-to-ground fault on the system resulting in a temporary voltage rise on the unfaulted phases. Swells are also characterized by their magnitude and duration.

Table 2. Categories and typical characteristics of voltage swells

Categories	Typical duration	Magnitude
Instantaneous	0.5 ~ 30 cycles	1.1 ~ 1.8 pu
Momentary	30 cycles ~ 3s	1.1 ~ 1.4 pu
Temporary	3s ~ 1 min	1.1 ~ 1.2 pu

3.3 Interruptions

The interruption occurs when the supply voltage or load current decreases to less than 0.1 pu for a period of time not exceeding 1 minute. The interruption can be result of power system faults, equipment failures and control malfunctions. Some interruptions may be preceded by voltage sag when these interruptions are due to faults on the source system. The voltage sag occurs between the time a fault initiates and the protective device operates. So, the faulted feeder and loads will experience voltage sag followed immediately by an interruption. The duration of the interruption will depend on the reclosable capability of the protective device.

Table 3. Categories and typical characteristics of interruptions

Categories	Typical duration	Magnitude
Instantaneous	0.5 ~ 30 cycles	Less than 0.1 pu
Momentary	30 cycles ~ 3s	Less than 0.1 pu
Temporary	3s ~ 1 min	Less than 0.1 pu

4. SIMULATION RESULTS

The disturbance detection is the procedure for estimating start and end times of disturbances.

The detail coefficients in Fig. 2 work as weights of the wavelets to represent the signal at these locations and scales. Those detail coefficients with high values indicate the sudden change of the signal. So, in case that disturbance happens, detail coefficient has a high value.

In this paper, digital filtering technique using the WT is used for detecting the PQ disturbances. Because high frequency band filter h[n] can extract disturbances in power system, the d1 represents the

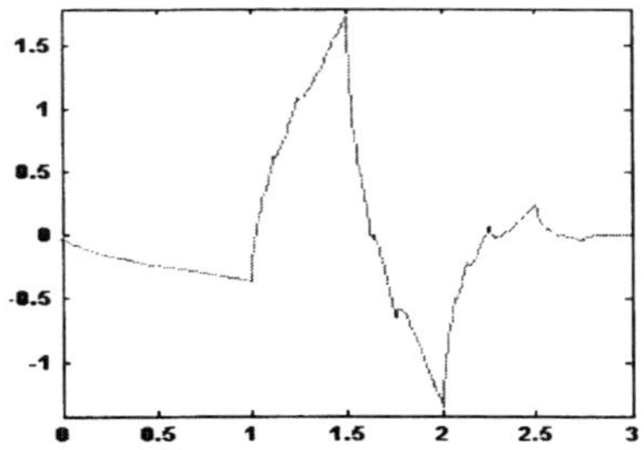

Fig. 4. Daubechies 2 wavelet.

highest frequency band. The Daubechies filter with 4 coefficients was adopted as the discrete mother wavelet for the DWT because of its known suitability in PQ disturbance analysis.

Typical PQ disturbances including voltage sags, voltage swells and interruptions were constructed to assess the efficiency.

4.1 Voltage Sag Detection

A simulated waveform of voltage sag to 0.5pu is shown in Fig. 5(a). Duration time is 0.2 seconds.

In case that the disturbance signal contains only PQ disturbances, the detection result is shown in Fig. 5(b). The high values are found at the start and end of the event. But the signal contains not only PQ disturbances but also electrical noise. Because the WT is very sensitive to the change of the signal, it is difficult to detect disturbances in case that the signal has a noise. So, digital lowpass filter is used for removing a noise. The result using digital lowpass filter is shown in Fig. 5(c).

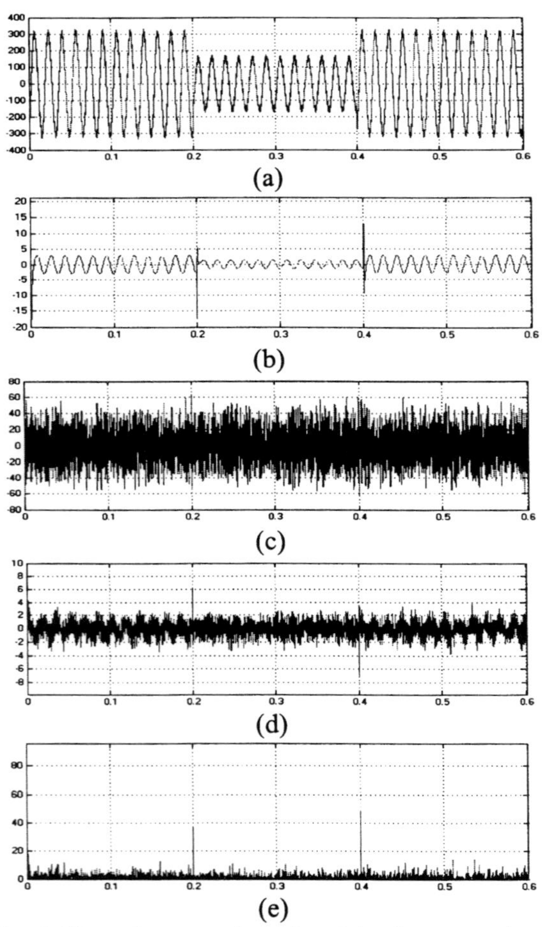

Fig. 5. The voltage sag detection; (a) voltage sag signal, (b) simulation result from d1, (c) simulation result from d1 considering a noise, (d) simulation result from d1 after filtering a noise (e) the SDC from d1 in case that the signal contains a noise.

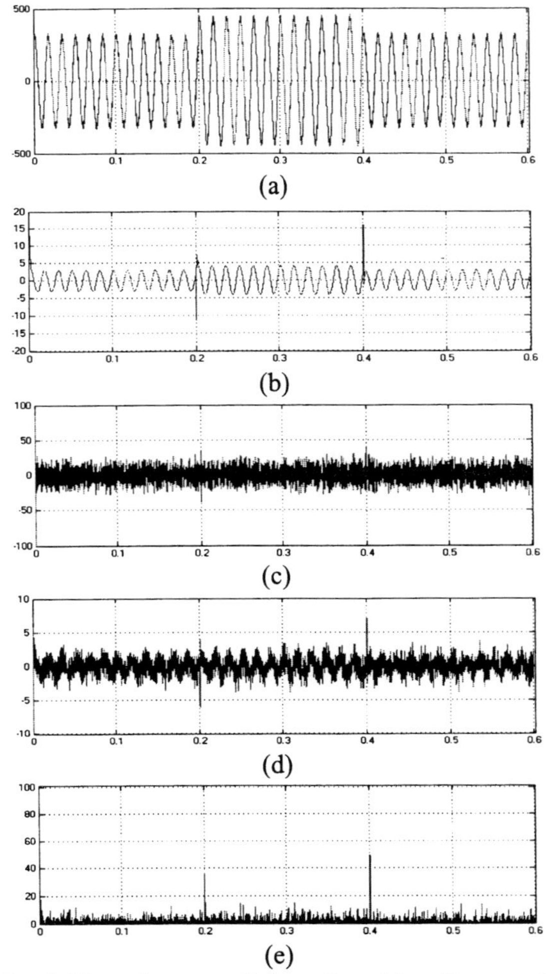

Fig. 6. The voltage swell detection; (a) voltage swell signal, (b) simulation result from d1, (c) simulation result from d1 considering a noise, (d) simulation result from d1 after filtering a noise, (e) the SDC from d1 in case that the signal contains a noise.

However, the detection result in Fig. 5(c) is not easy to observe. Therefore, in order to enhance the detection outcome, we utilized the Squared Detail Coefficients (SDC). Fig. 5(d) shows the SDC at d1. These spikes make disturbances easier to recognize.

4.2 Voltage Swell Detection

The waveform of voltage swell is shown in Fig. 6(a). A swell to 1.5pu was simulated to last for 0.2 seconds. The start and end of event is exactly detected in Fig. 6(b). Like the voltage sag, simulation is also performed in case that the signal contains a noise. After digital lowpass filtering for removing a noise, highpass filtering technique using the WT is performed. Fig. 6(d) shows the result from d1 and the SDC from d1 is shown in Fig. 6(e) respectively.

4.3 Interruption Detection

The waveform of interruption with a duration time of 0.2 seconds is simulated, as shown in Fig. 7. Like the voltage sag and swell, Fig. 7(b), (c), (d) and (e) are represented respectively. Start and end times of disturbances are detected accurately.

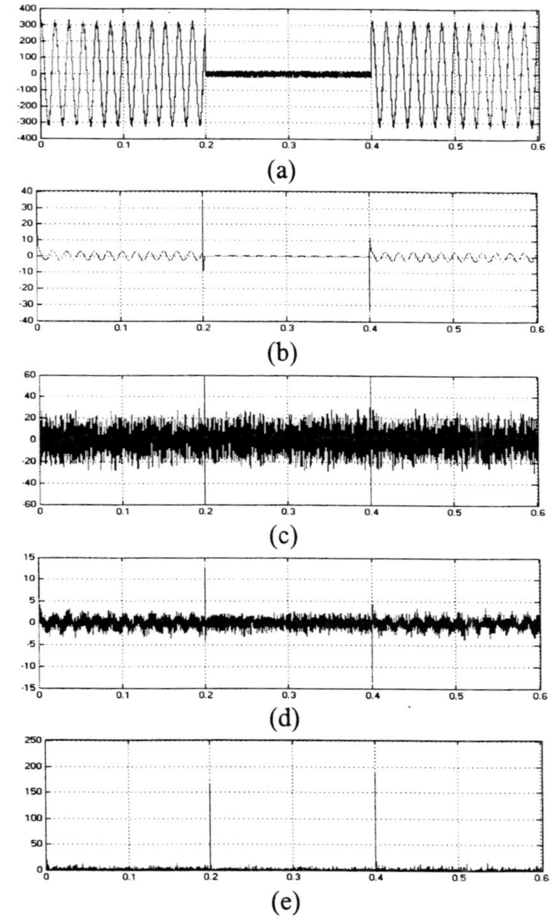

Fig. 7. The interruption detection; (a) interruption signal, (b) simulation result from d1, (c) simulation result from d1 considering a noise, (d) simulation result from d1 after filtering a noise, (e) the SDC from d1 in case that signal contains a noise.

5. EXPERIMENT RESULTS

5.1 Hardware structure

The hardware for the real time PQ monitoring system based on the Digital Signal Processor (DSP) has been implemented. The acquired voltage from the source generator passes through the DSP board for the lowpass filtering to remove a noise and the highpass filtering to perform the WT. The additional process for detecting and classifying disturbances is performed in the DSP board, and the processed data is transmitted to the PC monitoring part for display and stored in file simultaneously. The PC monitoring part displays a kind of disturbances, Root Mean Square (RMS) value, duration time of disturbance and time that data is stored. For the data transmission, serial communication (RS232C) is used. The hardware structure is shown in Fig. 8.

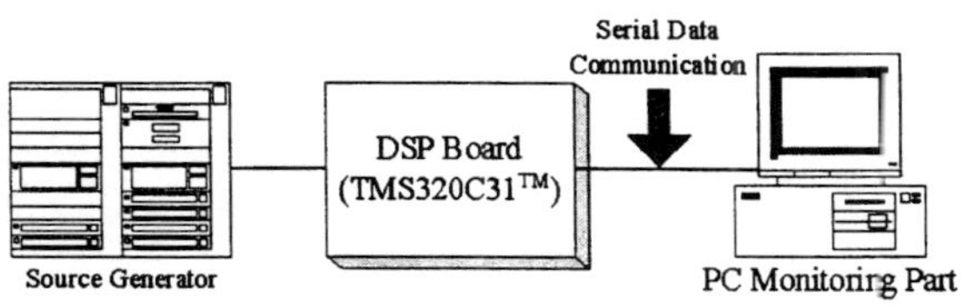

Fig. 8. Hardware structure for the real time PQ monitoring system

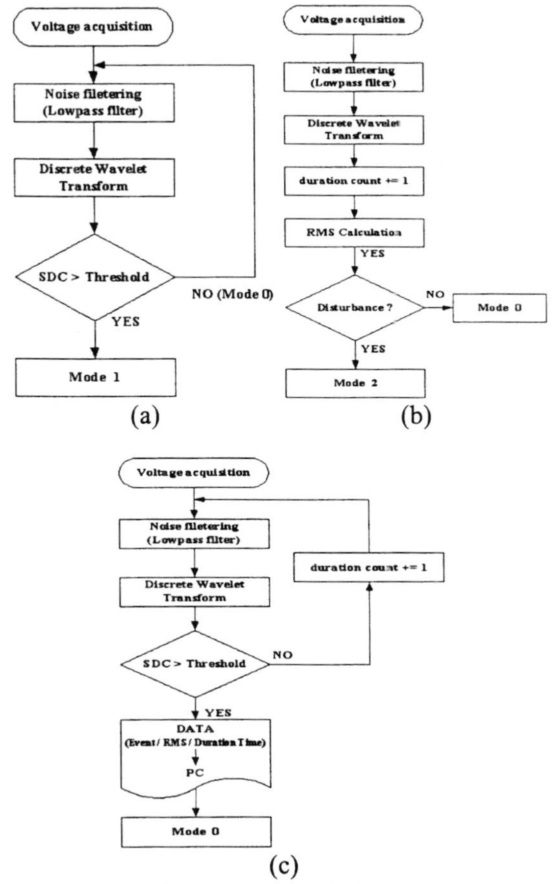

Fig. 9. Overall algorithm of the PC monitoring system for the detection of PQ disturbances; (a) Mode 0 (No disturbance), (b) Mode 1 (Disturbance starting mode), (c) Mode 2 (Disturbance ending mode)

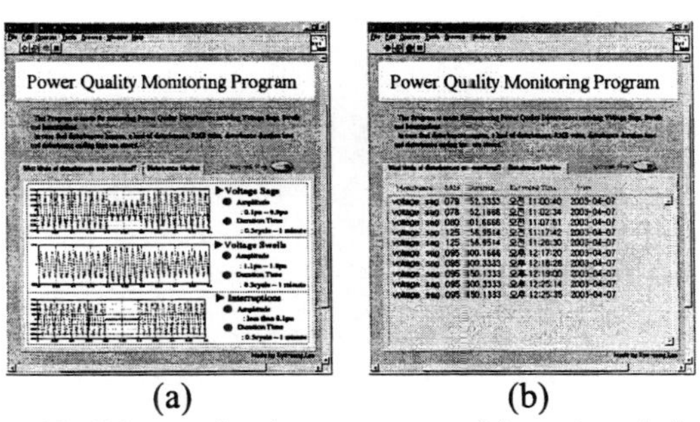

Fig. 10. PQ monitoring program; (a) main window, (b) Data display window

5.2 Detection Algorithm

The overall algorithm for the detection of PQ disturbances is shown in Fig. 9.

In case that the PQ disturbance doesn't happen, Mode 0 is performed. If SDC is larger than threshold, Mode 1 is performed. If Calculated RMS value is not under disturbances, detection procedure goes to the Mode 0. In case that RMS value comes under disturbances, Mode 2 is performed.

5.3 PQ monitoring system performance verification

Several experiments were conducted for different values of disturbance amplitude and duration time. The measured RMS value and duration time were compared to the nominal values set in the source generator. The sampling frequency is 12kHz and the DWT is performed on every other sample.

Table 4. The results of voltage sags detection
(Reference = 150V(RMS))

Source Generator RMS (V)	Source Generator Duration (ms)	Measured value RMS (V)	Measured value Duration (ms)	RMS Error (%)	duration error (cycle)
100	45	103	45.6666	3	0.04
100	45	101	45.1666	1	0.01
100	20	101	20.3333	1	0.02
80	900	80	900.1666	0	0.01
80	900	81	900.6666	1.25	0.04
80	500	81	501.5000	1.25	0.09
80	500	80	501.3333	0	0.08
80	500	81	500.1666	1.25	0.01
80	300	80	301.0000	0	0.06
80	300	82	300.6666	2.50	0.04
80	50	80	50.1666	0	0.01
80	50	81	50.8333	1.25	0.05
60	800	62	801.8333	3.33	0.11
50	200	52	200.8333	4	0.05
50	200	50	200.6666	0	0.04
			Average :	1.32	0.044

Table 5. The results of voltage swells detection
(Reference = 150V(RMS))

Source Generator RMS (V)	Source Generator Duration (ms)	Measured value RMS (V)	Measured value Duration (ms)	RMS Error (%)	duration error (cycle)
220	900	218	902.0000	0.91	0.12
220	586	219	587.6666	0.45	0.10
216	642	215	642.6666	0.46	0.04
216	642	214	643.3333	0.93	0.08
210	495	211	496.0000	0.48	0.06
210	657	208	658.6666	0.95	0.10
190	642	191	643.6666	0.52	0.10
190	600	192	600.1666	1.05	0.01
190	300	191	300.1666	0.52	0.01
185	200	187	200.8333	1.07	0.05
185	200	185	201.0000	0	0.06
			Average :	0.67	0.066

Table 6. The results of interruptions detection
(Reference = 150V(RMS))

Source Generator RMS (V)	Source Generator Duration (ms)	Measured value RMS (V)	Measured value Duration (ms)	RMS Error (%)	duration error (cycle)
10	687	14	687.8333	40	0.05
10	687	13	688.5000	30	0.09
10	687	13	687.8333	30	0.05
5	900	13	900.3333	160	0.02
5	900	12	901.1666	140	0.07
5	900	12	900.8333	140	0.05
			Average :	90	0.055

Table 4-6 summarise the results for the mentioned disturbances. The average error of the duration time is less than 0.06cycle (1/1000 second). Because a decimal fraction is not considered, a little error is shown in the measured RMS value. However, from the result of interruption detection, the error of the RMS value is shown on a large scale because of sampling on a transient region of interruptions during RMS calculation.

6. CONCLUSION

In this paper, the real time PQ monitoring system, based on digital filtering techniques is proposed. It has a good performance in detection, classification and analysis of PQ disturbances. The DSP-based structure and the WT for digital filtering procedure are used. The processed data is transmitted to the PC monitoring program and stored in file. Therefore it is very useful for analyzing PQ disturbances and adopting suitable strategies to improve the PQ.

Continuing work is oriented to enhancing the implemented procedure. Especially, suitable solutions are being studied for interruption detection and more optimal detection algorithm.

ACKNOWLEDGEMENT

The developments reported in this paper are funded by Korea Science and Engineering Foundation (KOSEF) of Korea.
Project number: R01-2000-000267-0

REFERENCES

C.H. Kim and R. Agfarwal (2000). Wavelet transforms in power systems Part 1. *Power engineering journal*, Vol. 15, pp.81-87.

IEEE Std 1159-1995 (1995). *Recommended practice for monitoring electric power quality.*

T.K. Abdel-Galil, E.F. Saadany and M.M.A Salama (2001). Effect of new deregulation policy on power quality monitoring and mitigation techniques. *Proceedings of the IEEE/PES transmission and distribution conference and exposition*, Vol. 1, pp.554-560.

L. Angrisani, P. Daponte, M. D'Apuzzo and A. Pietrosanto (1999). A VXI power quality analyser implementing a wavelet transform-based measurement procedure. *Measurement*, Vol. 26, pp.91-102, Elsevier.

C. Parameswariah and M. Cox (2002). Frequency characteristics of wavelets. *IEEE Transactions on power delivery*, Vol. 17, No. 3, pp.800-804.

M.F. Alves and T.N. Ribeiro (1999). Voltage sag : An overview of IEC and IEEE standards and application criteria. *Proceedings of the IEEE transmission and distribution conference*, Vol. 2, pp.585-589.

S.J. Huang, C.T. Hsieh and C.L. Huang (1998). Application of wavelets to classify power system disturbances. Electric power systems research, Vol. 47, pp.87-93, Elsevier.

S. Santoso, E.J. Powers, W.M. Grady and G.J. Hofmann (1996). *Power quality assessment via wavelet transform analysis. IEEE Transaction on power delivery*, Vol. 11, No. 2, pp.924-930.

D.C. Robertson, O.I. Camps, J.S. Mayer and W.B. Gish (1996). *IEEE Transaction on power delivery*, Vol. 11, No. 2, pp.1050-1058.

A.M. Gaouda, M.M.A. Salama and M.R. Sultan (1998). Automated recognition system for classifying and quantifying the electric power quality. *Proceedings of 8th international conference on harmonics and quality of power*, Vol. 1, pp.244-248.

ELSEVIER
IFAC
PUBLICATIONS
www.elsevier.com/locate/ifac

SENSITIVITY-BASED RESTORATIVE CONTROL FOR POWER SYSTEMS

Seung-Ryle Oh* Young-Sik Baek** Chi-Yun Ok** Jung-Nyun Kim***

*LG Industrial Systems**
*Kyungpook National University***
*LG Cable****

Abstract - Recently, electric power demands tend to be increased continuously. In this situation, current states of power systems must be definitely recognized to operate power systems stably and economically. Also, corrective actions must be performed for the violation of bus voltage and line overload. It is desirable that the situations of violation are promptly removed to operate power systems effectively. This paper deals with the development of guidance system for voltage and overload correction on base case using Generator-Voltage Sensitivity List(GVSL) and Generator Shifter Factor List(GSFL). Also, to show the superiority and economical efficiency of the proposed application, we applied the proposed guidance system to the Kwangyang Steel Works' Power Systems. *Copyright © 2003 IFAC*

1. INTRODUCTION

Efficient operation of power systems on large scale consumer is important problem that affects quality and price of products. But, research in these field wasn't performed up to now. Then real-time control of power systems could be applied due to development of IT technology. The purpose of operation of power systems is to operate economically and to prevent accident beforehand. To acquire these goal, it must be able to recognize system state under the various situations, and proper control was done in case of violations. Up to now, many solutions to control for these violations have been studied, but these results are inefficient to apply to large system for computing time.
There are two constraints to apply Decision Support System that guides operator to judgment of condition. one is that it must be real-time systems, the other is that recognize the system topology.
In this paper, each components that compose of systems are described as object and total system is represented as connections of each objects. and time Also, to the time efficiency, sensitivity between device and control variables was calculated and implemented.
presented methods are applied to Gwangyang Steel Works' Power System to show the feasibility and effectiveness.

2. FORMULATION

The most important part of control system is that it can realize the real time program, make system operator monitor system state, prepare emergency state that can happen in the future, and operate system more safely. In the proposed system, restorative controls for bus voltage and line flow are achieved with relevant devices on sensitivity.

2.1 Power Flow

The power flow is the basic field of the power system analysis, and the proposed paper used Fast Decoupled Power Flow to acquire initial state fast and exactly. Sparsity technique and ordering of bus order is applied to save computing time. Equation (1) and (2) are power mismatch equations used in Fast Decoupled Load Flow Method.

$$\Delta P/V = -[B_P]\Delta\theta \tag{1}$$

$$\Delta Q/V = -[B_Q]\Delta V \tag{2}$$

where,

$\Delta P, \Delta Q$: mismatch of active and reactive power mismatch

$\Delta\theta, \Delta V$: correction vectors of voltage angle and magnitude

B_P : imaginary part of Y_{Bus} (NB-1) × (NB-1)

B_Q : imaginary part of Y_{Bus} (NB-NPV-1) × (NB-NPV-1)

NB : numbers of total bus

NPV : numbers of voltage controlled bus

At each iteration step, voltage magnitude and angle of each bus is modified according to equation (3) and (4)

$$\theta_i^{k+1} = \theta_i^k + \Delta\theta_i \tag{3}$$

$$V_i^{k+1} = V_i^k + \Delta V_i \tag{4}$$

Finally, Power flow calculations converges on desired results and is stopped when $\sum\Delta V_i$ and $\sum\Delta\theta_i$ became small enough.

2.2 Control Variables for Voltage Violations

Generator Reactive Power

Bus voltage varies according to reactive generation of generator mainly. To obtain optimal bus voltage, demand and supply condition of reactive power must be satisfied. At this viewpoint, reactive power variation of real generator brings variation of load bus voltage, then can control load bus voltage by changing reactive power generation according to sensitivity between bus voltage and reactive power generation. There are two modes in real generator operating mode, one is Automatic Power Factor Regulation(APFR), the other Automatic Voltage Regulation(AVR) mode. In APFR mode, voltage control of load bus is achieved with generator reactive power amount, namely generator power factor is adjustment. In equation (2), equation (5) is formed between reactive generation ΔQ_j of generator bus and voltage deviation ΔV_i of load bus.

$$\Delta V_i = S \cdot \Delta Q_j = -[B^{ij}_{exp}]^{-1} \cdot \Delta Q_j / V_j \qquad (5)$$

B_{exp}: imaginary part of Y_{bus} (NB-NAVR-1) × (NB-NAVR-1)

NAVR : Number of voltage controled generator bus

Generator Voltage

Operating generator can control voltage of load bus with changing generator terminal voltage within reactive power limit. Sensitivity about voltage change of load bus and voltage change of voltage controled bus can be acquired by indirect way. Injected reactive power of linked load bus is changed owing to change of generator terminal voltage. Injected reactive power can be express from sensitivity in equation (4). Therefore, finally there is voltage deviation ΔV_k of load bus to be controlled in equation(7).

$$\Delta Q_i = -b_{ij} \cdot V_j \cdot \Delta V_j \qquad (6)$$

$$\Delta V_k = \sum_{i \in \Lambda} S \cdot V_i = -\sum_{i \in \Lambda} [B^{ik}_{exp}]^{-1} \cdot \Delta Q_i \cdot V_i \qquad (7)$$

b_{ij} : line susceptance between i and j bus

Λ : Load bus linked with voltage control bus

Figure 1 can help you understand definition of Λ bus. In following figure, bus 3, 5, 6 is Λ bus about voltage controlled bus, 2.

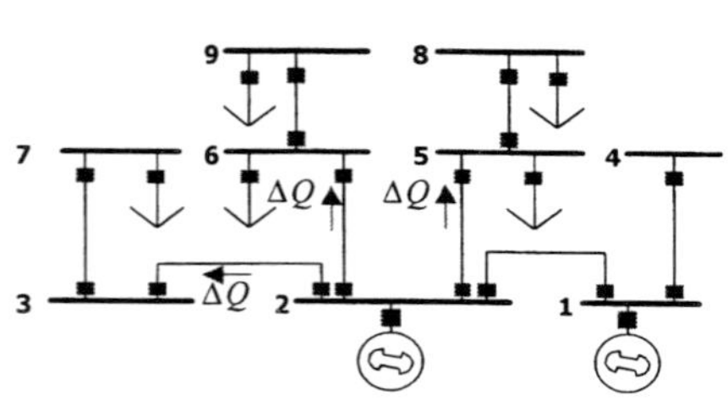

Fig. 1 Definition of Λ bus

In equation (6) and (7), deviation of generator terminal voltage brings deviation of Λ bus voltage. It means that deviation of Λ bus voltage brings reactive power deviation of generator operated by AVR. That is, voltage deviation of Λ bus occurs incidentally in case deviation of load bus voltage. Therefore, voltage deviation of Λ bus that is happened by their process must consist within extent that do not cause violation of reactive power up and low limit of AVR driving generator.

Use sensitivity between voltage change of Λ bus and AVR driving generator reactive power for this and need to control the change amount beforehand lest violation should happen. reactive power change amount ΔQ_k free AVR driving generator to the change amount ΔQ_i of reactive power that is flowed in Λ bus can appear with equation [9].

$$\Delta V_i = -[B^{ij}_{exp}]^{-1} \cdot V_i \cdot \Delta Q_i \qquad (8)$$

$$\Delta Q_k = \sum_{j \in \Lambda} S \cdot \Delta V_i = \sum_{j \in \Lambda} (-b_{jk} \cdot V_i) \cdot \Delta V_i \qquad (9)$$

Shunt VAR Compensator

As SVC (Shunt VAR Compensator) supplies or assimilate reactive power in system, achieves role by value that keep bus voltage changelessly or want. That is, also the reactive power development amount also of generator changes as that change of SVC value changes voltage of Λ bus at the same time that bring voltage change of relevant bus as that change reactive power inflow of load bus. Use above way (2) and sensitivity between voltage change amount ΔV_j of load bus about SVC's ΔQ_k change can appear with way (10).

$$\Delta V_j = S \cdot \Delta Q_k = -[B^{jk}_{exp}]^{-1} \cdot \Delta Q_k / V_j \qquad (10)$$

Transformer Taps

OLTC (On Load Tap Changing) transformer between $i-j$ bus achieves function that keep voltage of bus by optimum level using variableness tap. Tap change of transformer must consist within extent that voltage regulation that use this because bring change of the reactive power generation amount of generator does not escape reactive power upper limit of generator and restriction of low limit.

Change of each $i-j$ bus $-b_{ij} \cdot \Delta t / t$ reactive power of transformer tap change between $i-j$ bus happens. Use this and voltage change amount ΔV_k in free load bus k by transformer tap change of $i-j$ line can appear with way (11).

$$\Delta V_k = S \cdot \Delta t = \left([B^{ik}_{exp}]^{-1} \cdot b_{ij}/t - [B^{jk}_{exp}]^{-1} \cdot b_{ij}/t\right) \cdot \Delta t \qquad (11)$$

2.3 Control Variables for Line Overload

Generator active power

Because ripple effect on whole system of falling off

accident of connection line (Tie Line) between 154 [kV] substations is big in steelworks, generation of active power must consist within range that overload does not happen to these line. Change line flow between change $i-j$ bus that change ΔP_k of the active power of PV bus. This brings each i, j bus phase change by ΔP_k, and i−j bus line flow change happens thereby. This time, if the line flow change amount is ΔP_{ij} sensitivity S

$$S = \frac{\Delta P_{ij}}{\Delta P_k} = \frac{\partial P_{ij}}{\partial P_k} = \frac{\partial (V_i V_j \sin(\theta_i - \theta_j)/x_{ij})}{\partial P_k} \tag{12}$$

here, If suppose that is $|V_i| \approx 1,\ |V_j| \approx 1,\ \sin(\theta_i - \theta_j) \approx (\theta_i - \theta_j)$, way (12) can express with way (13).

$$S = \frac{1}{x_{ij}} \cdot \frac{\partial(\theta_i - \theta_j)}{\partial P_k} = \frac{1}{x_{ij}} \cdot \frac{\partial \theta_i}{\partial P_k} - \frac{1}{x_{ij}} \cdot \frac{\partial \theta_i}{\partial P_k} = \frac{x_{ki}}{x_{ij}} - \frac{x_{ki}}{x_{ij}} = \frac{x_{ki} - x_{ki}}{x_{ij}} \tag{13}$$

After substitute way (13) to way (12), arrange about ΔP_{ij}, and can get way such as way (14) if express as B_P matrix's element.

$$\Delta P_{ij} = S \cdot \Delta P_k = ([B_P^{ik}]^{-1} - [B_P^{jk}]^{-1}) \cdot b_{ij} \cdot \Delta P_k \tag{14}$$

Phase Transformer(Phase Shifter)

The size of flowing active power is decided by phase than dimension of bus voltage of line both end to line. When see in this aspect, phase change of line both end bus brings change of flowing active power to line. Phase change of phase transformer between i−j bus is connected by phase change of terminal bus and this brings change of flowing active power to line. Free k bus phase change $\Delta \theta_k$ about unit phase change of phase transformer can express as following.

$$\Delta \theta_k = ([B_P^{ik}]^{-1} - [B_P^{ik}]^{-1}) \cdot b_{ij} \tag{15}$$

Therefore, line flow change amount ΔP_{ij} about $\Delta \phi$ phase change of phase transformer has following sensitivity

$$\Delta P_{ij} = S \cdot \Delta\phi = \{(\Delta\theta_i - \Delta\theta_j) - N\} \cdot b_{ij} \cdot \Delta\phi$$
$$= \{(([B_P^{ii}]^{-1} - [B_P^{ji}]^{-1}) - ([B_P^{ij}]^{-1} - [B_P^{jj}]^{-1}) - N\} \cdot b_{ij} \cdot \Delta\phi \tag{16}$$

N are 1 and amount to 0 in opposite case here if phase transformer exists to relevant line.

3. REPRESENTATION OF SYSTEMS

Need that recognize present system state compulsorily to analyze electric power system. This treatise displayed state of system as connection between each equipments. There is bus, line, transformer, generator, load, SVC, breaker and disconnect switch etc. as for element that compose system. Expressed each system component as object and mix object that is composed and express whole system to do so that can recognize connection state between each equipment in program in this treatise. Disconnect switch, breaker and so on and connection states of bus interior each Bus Bar, Bus Tie, Bus Section are with more complicated structure bus-bar in actuality system. Disconnect switch, breaker and so on and connection states of bus interior each Bus Bar, Bus Tie, Bus Section are with more complicated structure bus-bar in actuality system. For example, line and connection of bus are consisting through breaker and disconnect switch, and connection consists by bus bar(busbar) and bus tie(breaker), bus section (breaker) etc as well as bus interior. Connection between objects connected by pointer (Pointer) with picture below and each objects can exchange message with other individuals through these path. State of actuality mother ship interior building blocks alters composition of system according to situation with mother ship singleness/separation driving or transfer of load in program. This time, done phenomenon produces preponderance of flow on what part in any bus. Make by breaker (BS, BT), bus interior constituent through disconnect switch (DS) and structure that can connect with picture below as suitable structure to examine closely such a phenomena, and have connectable structure breaker and line to bus outside.

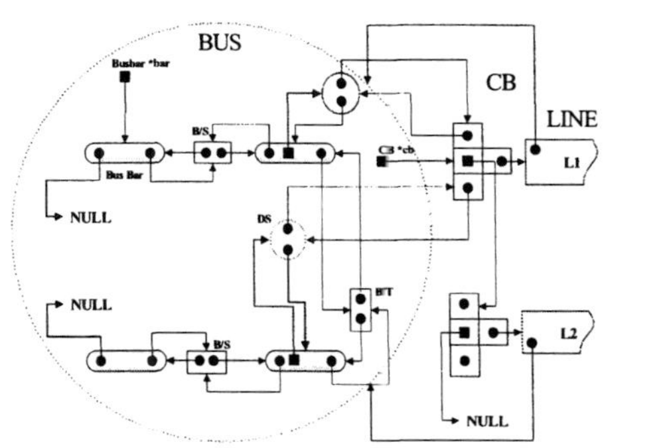

Fig 2 Connection state between bus objects and line object .

Figure 3 and 4 display substation using expression method of system to state above over shortly. If examine connection situation between transformer line that is linked with bus to display connection state between object each other concretely, each transformer line is linked in the bus bar which is fallen into bus tie and bus section through breaker and disconnect switch.

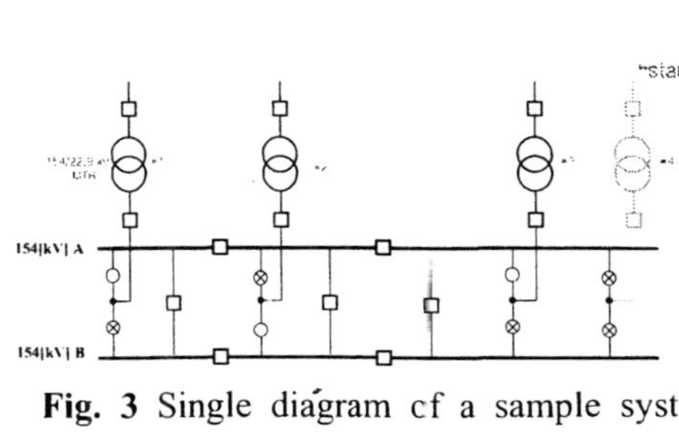

Fig. 3 Single diagram of a sample system

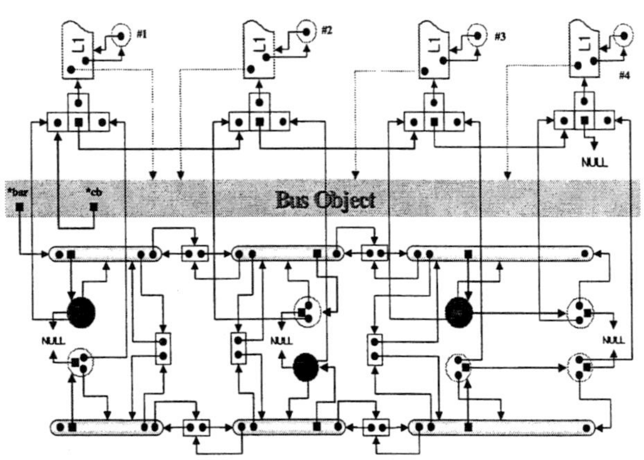

Fig. 4 Object representation of a sample system

4. SYSTEM CONFIGURATION

4.1 System Configuration

Whole structure of composition guidance system is same with figure 5. Is stored to SQL server by real time by data acquisition device from been scattered system equipment. To SQL server periodically in computer that achieve guidance access data use OLE DB and data that need in calculation achieves electric power flow about ground state (base case) after input Calculate voltage of bus and line flow by electric power flow calculation and result is stored to DB. Also, store action item for violation item to DB gradually notifying to driver at violation appearance of bus voltage and line flow.

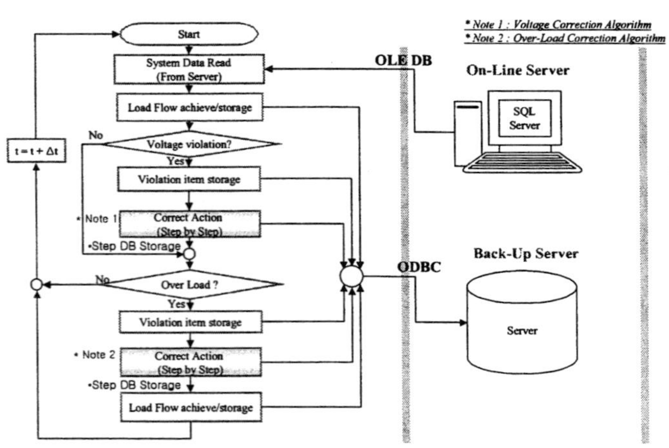

Fig. 5 Structure of the Guidance System

4.2 Algorithm to Correct Voltage Violations

When present power system state escapes voltage extent of optimum level, action about voltage violation by following algorithm is achieved. The voltage change amount by power factor and tab setting of generator is depended on sensitivity at this step, performed all procedures are stored up database step by step, and this action acts as guidance that can solve present voltage violation state. Also, it generates load interception message in case that voltage control is impossible by available all generators. This time, load that must be intercepted is depended on precedence. There is algorithm to solve voltage violation in figure 6 below.

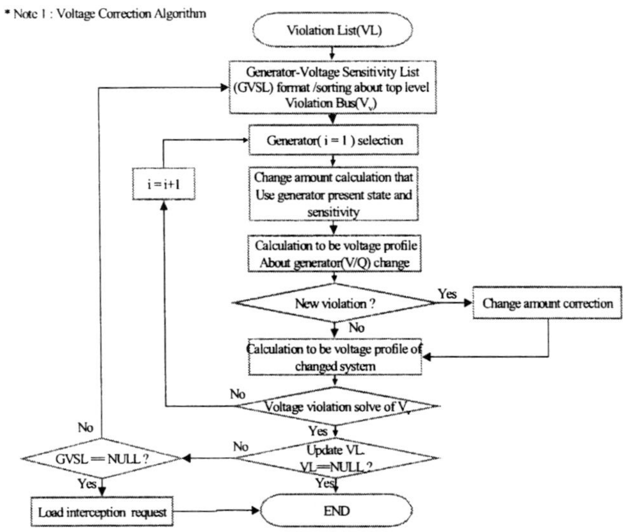

Fig. 6 Flow chart of voltage correction

Step 1 : It composes VL (Violation List) and sort according to size of the violation amount when voltage violation happens.

Step 2 : It composes GVSL (Generator-Voltage Sensitivity List) about maximum violation bus V_v in VL and sort together by size of sensitivity.

Step 3 : Calculate generator reactive power amount and terminal voltage change amount ΔC for voltage violation solution of maximum violation bus V_v.

Step 4 : Correct the change amount ΔC that consider generator reactive power upper/low limit, terminal voltage upper/low limit.

Step 5 : Correct the change amount ΔC by sensitivity in case correcting action aggravates more present violation bus state or new voltage violation bus is added at change.

Step 6 : Calculate reactive power change amount ΔC at change, if generation amount is over the upper/low limit restriction, correct amount ΔC by sensitivity.

Step 7 : Calculate system profile about modified change ΔC lastly.

Step 8 : Go to step 1 when available generator remains in GVSL, namely, GVSL $\neq$ NULL, on the other hand it creates load interception request message when do not remain, namely, GVSL = NULL.

4.3 Algorithm to Correct Line Overload

When system driving state escapes from optimum level of line or transformer capacity, action about overload violation by following algorithm is achieved. Overload solution is depended on real power generation of generator and phasor sensitivity of phasor shifter and performed procedures are stored to database at all voltage violations step by step. Algorithm to solve overload violation is following with figure 7.

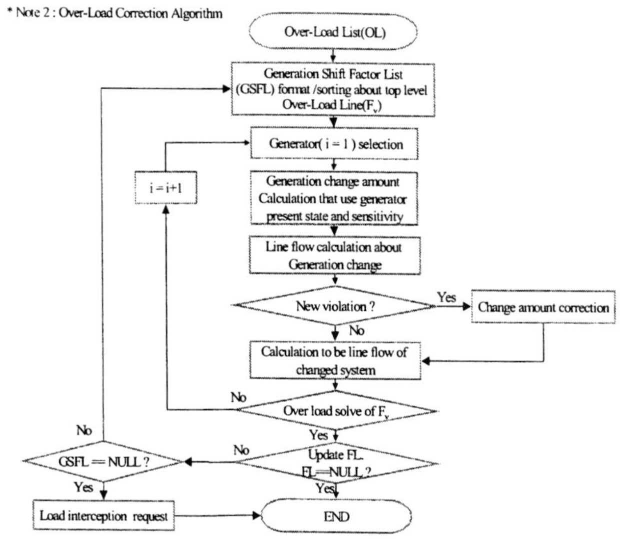

Fig. 7 Flow chart of overload correction

Step 1 : Composite OL(Overload List) and sort OL in overload degree when line overload happens.

Step 2 : Compose GSFL (Generation Shift Factor List) about maximum violation line F_v in OL and sort according to size of sensitivity.

Step 3 : Calculate the real power change amount ΔC for solution of bus voltage violation F_v.

Step4 : Correct the change amount ΔC that consider generator active power upper/low limit, tab setting upper/low limit.

Step 5 : When another line overload happens or it aggravates more present line overload state, corrects the change amount ΔC with using the sensitivity.

Step 6 : Calculate system profile about modified change amount ΔC lastly.

Step 7 : Go to step 1 when available generator remains in GSFL, namely, GSFL $\neq$ NULL, on the other hand it creates load interception request message when do not remain, namely, GSFL = NULL.

4.4 Database

In client/server computing system environment, it is important that user efficiently operates server that data is focused on. At this viewpoint we can say that OLE DB used for access in guidance system is one way to relieve burden of the server.

OLE DB is based on COM object that make use of already compiled component. So it can take advantage of all merits that COM provides and it offer more efficiency than used Database as it divides consumer roll into provider roll.

Also OLE DB is designed to utilize in application that is based on the web, so is designed to load less to approach on Data Source and connect to Internet .

As referred before, these structures of OLE DB can be effective method in side for stable server operation or minimum of client resource.

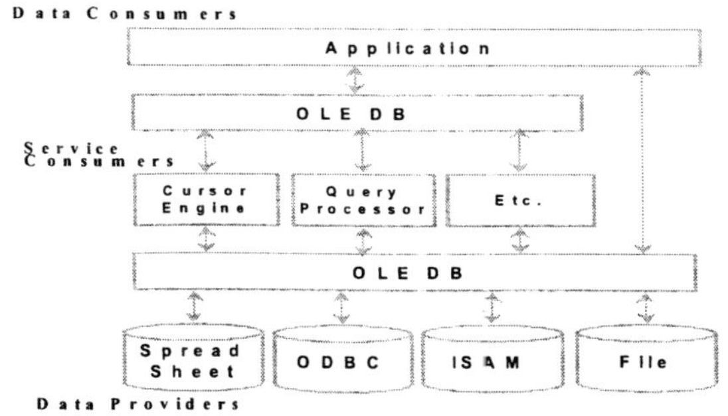

Fig.8 OLE DB Architecture

5. EXAMPLE

5.1 Test System

We applied developed system to GwangYang steelworks. Electric power systems of GwangYang steelworks are consisted of Su-joen substation of 154[kV] or 345[kV] and 4 substations of 154[kV] and working voltage distribution line of 154 [kV], 22.9 [kV], 6.6 [kV]. When we see present GwangYang steelworks load characteristic, it is classed by variable load and fixing load greatly, and fixing load is composed of seacoast area load of a shaft furnace and coke equipment and variable load of rolling area is composed of hot rolling and mini-mill etc. Maximum moment electric power of load is 1200 [MVA] considering characteristic of variable load that influence in system driving stability, and minimum moment load electric power is 580 [MVA]. Also it is driven by parallel AVR mode and APFR mode by plant to keep optimum voltage level of 154 [kV] bus. Figure 9 displays GwangYang steelworks single line picture.

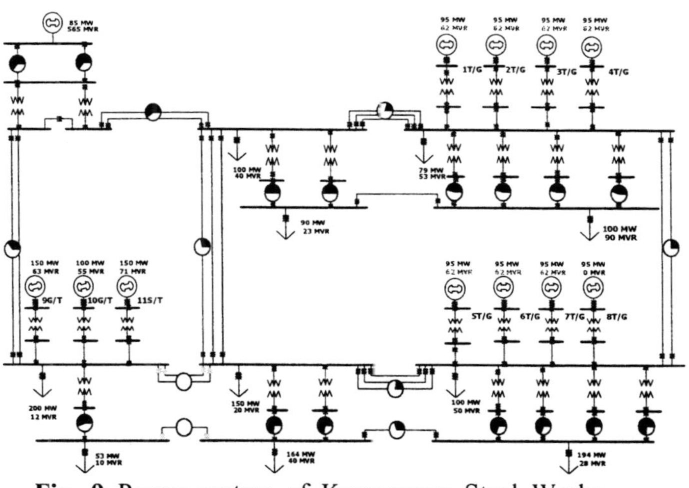

Fig. 9 Power system of Kwangyang Steel Works

We create guidance system input data that was changed arbitrary to accomplish violation state in present GwangYang steelworks system server's.

Also, execution condition of guidance system is that an voltage change extent of 154[kV] bus is over +/-2[%] or flow of connection line is over 65 [%] of capacity. There is a table displays simulation result about GwangYang steelworks 134 bus-bars below.

At each step, after select generator that has maximum sensitivity to maximum violation bus or line, decide control value according to driving way of selected generator.

5.2 Results

As result of system simulation, 6 voltage violation buses and 4 overload lines happened in ground state. All driving conditions are satisfied with correcting procedure seeing at table 1.

Table 1 Comparison of results

	UV No.	OV No.	Linked OL	Time [sec]
Before Correction	6	0	4	1.3
After Correction	0	0	0	2.7

In case violation of driving condition happened, expressed correction procedure at each step in table 2 like table 1. Bus voltage violation and line overload that happen in ground state are solved flowing 3 steps correction action about each, and can see that satisfy driving condition as solution process is proceeded step by step. At series process, time that take generator moment or terminal voltage change amount calculation and accordingly calculation about system Profile could shorten using sensitivity between each object.

GNO	IS READ	EX TIME	G TYPE	OV	UV	OL	MAX VIO NAME	VALUE	CTRL POS	CTRL TYPE	PRE STATE	PRE PGEN	PRE CTRL VALUE	POST STATE	POST PGEN	POST CTRL VALUE	REQ LOAD SHED
1	□		1	0	6	0	제선 변전소 154[kV]	149.25	3 T/G 11[kV]	발전기 역률[%]	0	0	0	1	70	90	□
2	□		1	0	6	0	제선 변전소 154[kV]	150.02	1 T/G 11[kV]	발전기 단자 전압[kV]	0	0	0	1	50	11	□
3	□		1	0	4	0	제선 변전소 154[kV]	150.54	2 T/G 11[kV]	발전기 단자 전압[kV]	0	0	0	1	50	11	□
4	□		1	0	0	0	없음	0			0	0	0	0	0	0	□
5	□		2	0	0	4	신수천 <-> 서부변전소 1L	72.06	NO.9 G/T 11[kV]	유효 발전량 [MW]	1	70	70	1	112.5	112.5	□
6	□		2	0	0	2	임연<-> 서부 변전소 1L	66.77	5 T/G 11[kV]	유효 발전량 [MW]	1	90	90	1	100	100	□
7	□		2	0	0	2	임연<-> 서부 변전소 1L	65.09	6 T/G 11[kV]	유효 발전량 [MW]	1	90	90	1	100	100	□
8	□		2	0	0	0	없음	0			0	0	0	0	0	0	□

Table 2 The Procedure of Correction

Figure 10, 11 shows progress that satisfy driving condition of system passing through correction procedure of preceding descriptions. When violate driving condition by 149.25 [kV] in system that is driving by smallest 150.02 [kV], maximum 157.08 [kV] in ground state, can see that satisfy driving condition passing through 3 steps in figure 10.

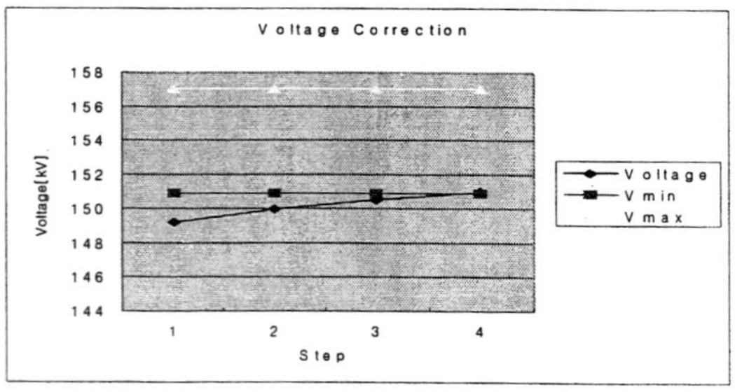

Fig. 10 Results of voltage correction

At figure.11 we can see process that satisfy driving condition passing through 3 steps correction procedure when violate driving condition, 72.06% in system that is driving line capacity 65% in ground state.

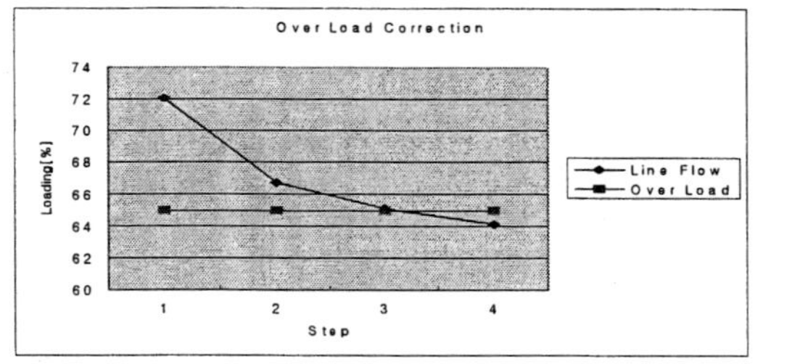

Fig. 11 Results of overload correction

6. CONCLUSION

This paper explained restorative control system that can help operator judge to lead safe and efficient system operation.

Proposed algorithm showed the efficiency and validity by applying real system and following results could be obtained.

First, proposed system was enable to apply real power system by sensitivity based control action that can shorten computing time and could obtain efficiency on operating system

Second, fast decoupled algorithm was applied to proposed system to shorten computing time and to use memory efficiently.

Third, components of power system was represented by objects and total system is consisted of connections between each objects. therefore this structure can be changed dynamically to cope with system change

[Reference]

[1] Show-Kang Chang, " Adjusted Solution In Fast Decoupled Load Flow", IEEE Trans. on Power System, Vol. 3, No.2 May.1998.

[2] Bansilal, D. Thukaram, K. Parthasarathy, " An Expert System For Voltage Control In A Power System Network", IEEE Catalogue No. 95TH8130 1995.

[3] B. Hakavik and A.T. Holen, " Power System Modelling and Sparse Matrix Operations Using Object-Oriented Programming",IEEE Trans. on Power System, vol. 9, No. 2, May 1994.

[4] James A. Momoh, Arunsi U. Chuku, "Application of expert system for improved contingency analysis and optimal correction" Electrical Engineering Department Howard University.

[5] Carlos A.Castro, Anjan Bose, " Correctability In On-Line Contingency Analysis", Department of Electrical Engeneering Arizona State University.

[6]M Moghavvemi, O. Faruque, " Real-Time Contingency Evaluation And Ranking Technique", IEEE Proc-Gener, Transm. Distrib, Vol. 145, No.5, September 1998.

ELSEVIER
IFAC
PUBLICATIONS
www.elsevier.com/locate/ifac

PETRI-NET APPROACH TO FAULT DIAGNOSIS IN POWER SYSTEMS USING THE TIME SEQUENCE INFORMATION OF PROTECTION SYSTEM

Myong-Gyun Roh Sang-Eun Hong

Department of Information Technology Engineering,
College of Engineering, Soonchunghyang University,
Asan 336-745, Korea

Abstract: In this paper we proposed backward Petri-nets to diagnoses fault on protection system of power system. The Petri-nets models of protection system are composed of the operating process of protective devices and the fault diagnosis process. Fault diagnosis model, which makes use of the nature of Petri-nets, is developed to overcome the drawbacks of methods that depend on operator knowledge. The proposed method can reduce processing time and increase accuracy when compared with the traditional methods. And also this method covers online processing of real-time data from SCADA (Supervisory Control and Data Acquisition). *Copyright © 2003 IFAC*

Keywords: Petri-nets, fault diagnosis, protection, modelling, power system

1. INTRODUCTION

To reduce outage time and to enhance service reliability, it is essential for operators to find the fault location in power system. However, an increasing number of system variables and alarms can now be monitored and processed. Few operators now handle system control. After the fault occurrence, operators would have to respond to an avalanche of alarm messages and the fault diagnosis can be very difficult.

We consider fault protection system as discrete event system. Even though power systems are continuous-time systems from a macro point of view, some operating procedures, such as system change from one state to another state, relay or circuit breaker operation, can be viewed as a set of discrete event. Therefore, the systems consisting of those procedures can consider as discrete event systems, such as protection system and switching sequence operation, from micro operating point of view.[1]

When fault occurs, the automatic fault diagnosis system suggests the possible way to remove fault and assists an operator to protect the systems with the best way in that situation. Fault diagnosis is also to identify fault location in power system by using information on operation of relay and circuit breakers, which is available from SCADA systems.[2]

Many methods have been adopted to solve this problem. Among these, the logic based method, the expert system approach and the artificial neural network were researched earlier.[3][4][5]

This paper describes a new method of the modeling of protection system and fault diagnosis in power systems using Petri nets. The Petri net models of protection system are composed of the operating process of protective devices and the fault diagnosis process. Fault diagnosis model, which makes use of the nature of Petri net, is developed to overcome the drawbacks of methods that depend on operator knowledge. The proposed method can reduce processing time and increase accuracy when compared with the traditional methods. And also this method covers online processing of real-time data from SCADA (Supervisory Control and Data Acquisition).

2. PETRI NETS THEORY

2.1 Basic models

A Petri net is a directed graph consisting of two types of nodes, places and transitions. The places are always followed by transitions, and vice versa. The places that signify activities represent by circles. The transitions which imply events are represented by bar. The state of Petri nets (its current marking) is denoted by the distribution of tokens in the places. Each token represents an instance of entity and the place where it is located is its current state. A place containing n tokens describes the fact that n instances of the same class of entity are in the same state.[6]

A Petri net is a 5-tuple, PN=(P, T, F, W, M0) where:

$P = \{p_1, p_2, \cdots, p_m\}$ is a finite set of places,
$T = \{t_1, t_2, \cdots, t_n\}$ is a finite set of transitions,
$F \subseteq (P\times T) \cup (T\times P)$ is a set of arcs (flow relation),
$W : F \rightarrow \{1, 2, 3, \cdots\}$ is a weight function,
$M_0 : P \rightarrow \{0, 1, 2, 3, \cdots\}$ is the initial marking,
$P\cap T = \varnothing$ and $P\cup T \neq \varnothing$.

A Petri net structure N=(P, T, F, W) without any specific initial marking is denoted by N. A Petri net with the given initial marking is denoted by (N, M_0).

2.2 Firing rule

A firing results in removing one token from each of its input places and adding one token to each of its output places. A transition may fire if it is enabled. A transition is enabled if each of its input places has many token in it as arcs from the place to the transition: P→T. A transition fires by removing all of its output places one token for each arc from the transition to the place: T→P.

2.3 Linear algebra

In modeling of discrete event systems, transition firing is represented system operation. Place is the condition of system. Change of marking can be obtained from the state transition matrix.[7][8]
Marking vector M(k) is defined in time k as below,

$$M(k) = [p_1(k), p_2(k), \cdots, p_n(k)]^T \qquad (1)$$

$p_j(k)$ is marking state of p_j in time k (j=1,2,...,n).
Control vector U(k) is defined in time k as below.

$$U(k) = [t_1(k), t_2(k), \cdots, t_m(k)]^T \qquad (2)$$

$t_i(k)$ is fire able (1) or not (0) in time k (i=1,2,...,m)
In any marking state, after firing of some transition, is represented for the next equation.

$$M(k+1) = M(k) + [B^+ - B^-] \times U(k) \qquad (3)$$

M(K+1) : Marking vector in time k+1, after U(k) fired in time k.
B^+ : incidence matrix, place to transition
B^- : incidence matrix, transition to place
$B = B^+ - B^-$: incidence matrix

3. MODELING

In this section, we describe a modeling strategy and fault diagnosis method for protection scheme of simple power system.[9][10]

3.1 Operating process of protective device

Since the fault occurrence and clearance process in power systems can be regarded like as discrete event system, Petri nets can model series of that operation. In the Petri nets model, the places denote the state of relay, circuit breaker, and bus and line faults, the transitions are formed by the operation of relay and circuit breaker.

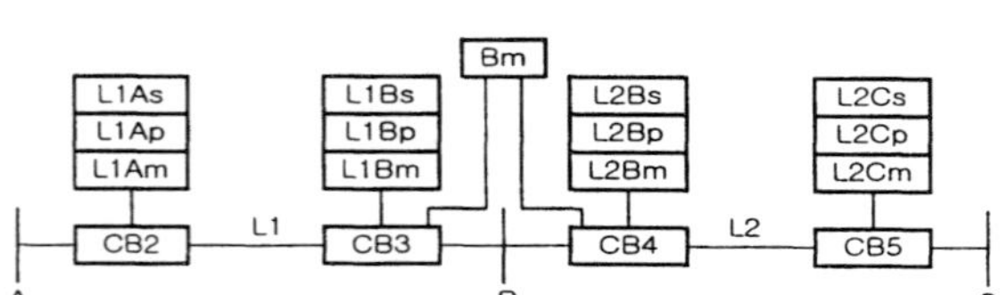

Fig 1. Simple example

Location of the token in each place is represented available state of protection relay, operated state of circuit breaker, and fault state on transmission line and busbar.
In general, protection systems are composed of main protection relay and some backup protection relays. Backup protection relays consist of primary and secondary protection relay. The operating process of protective devices(relay and circuit breaker) has some operating logic. For instance, when a fault occurs, main protection relays are operated, and if main protection relays are failed operation, then backup protection relays operate to clear the fault.

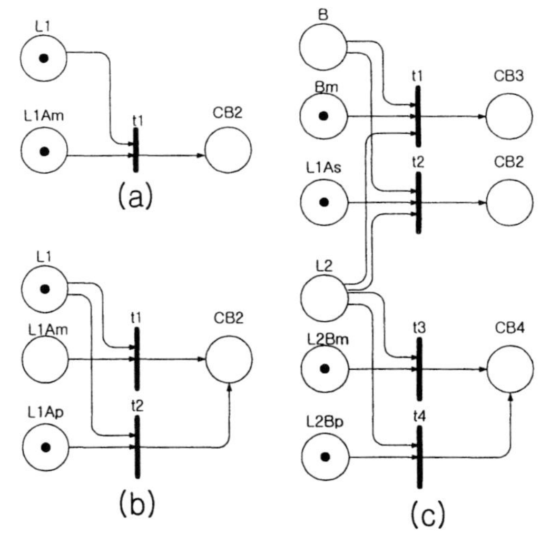

(a) Main protection (Fault on L1),
(b) Primary backup protection (Fault on L1),
(c) Secondary backup protection (Fault on L1 or B or L2)

Fig 2. Modeling of Fig 1.

Table 1. Operating logic of protection relays

Protection relay	Operating logic
L1Am	Fault occurred on L1 ; Normal operation : L1Am Trip operation : CB2.
L1Ap	Fault occurred on L1 ; Normal operation : Failed Non-operation : L1Am Backup operation : L1Ap Trip operation : CB2.
L1As	Fault occurred on B ; Normal operation : Failed Non-operation : CB3. Fault occurred on L2 ; Normal operation : Failed Non-operation : CB3, CB4 Backup operation : L1As Trip operation : CB2.

When a fault has occurred at L1 as shown in Fig 1. L1Am and L1Bm should operate to trip CB2 and CB3. If both CB2 and CB3 do not operate, then backup protection relays(L1Ap and L1Bp) should operate to trip circuit breakers. If a fault occurred on A and CB2 did not trip or if a fault occurred on L2 and both CB3 and CB4 did not trip, then L1AS should operate to trip CB2. And also if a fault occurred on A, then Am should operate to trip CB1 and CB 2.

When a fault occurred on L2, Petri net models are showed in Fig2. Table 1. is shown operating logic of protection relays.

3.2 Fault diagnosis process by backward Petri nets

We propose backward Petri nets to fault diagnosis. The fault diagnosis process works essentially in the reverse direction to the operating process. It makes use of the same Petri net model with the directions of all the arrows reversed. The original Petri net is called a forward Petri net and the latter type is called backward or reverse Petri net. Fig. 3 shows the backward Petri net of Fig 2. In this Petri net models the initial marking is received the information from SCADA.

For Fault diagnosis process, we will find some rule. First, since the final marking for operating results of protection relay has the value of '0'.

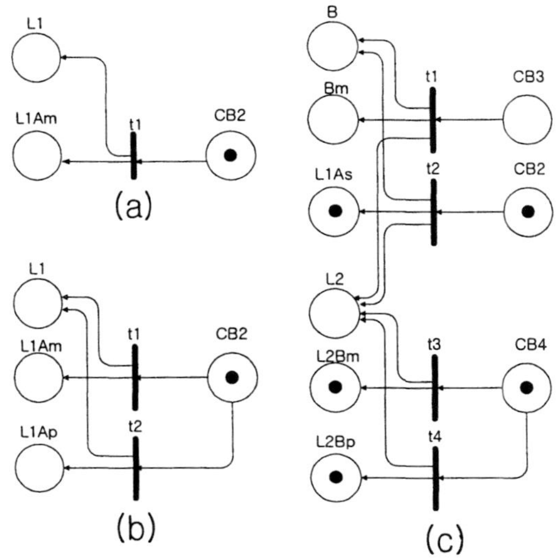

Fig. 3. Backward Petri nets model of Fig. 2.

Therefore, if initial and final marking of protection relay operation have [0 1], this denotes the normal operation of protection relay or the occurrence of fault. In this way, we can find the fault section, which is at the busbar and the transmission line. Second, the fault location is detected from the operating information of secondary backup protection relay. In this case, the strategy is the same as first rule. However, because the secondary backup protection has the delay time characteristics, we can easily find the fault location using these characteristics. Third, if all elements of control vector U(k) set 1, then all transition can make the fire which probability is in 100%. From this change of marking, we can detect the operating relay.

For example, we assume the received information (L1Ap, L1Am, Bm, CB2, and CB4) from SCAD available. Initial marking is the same as Fig 3. and we get fault location (L1) by marking vector of backward Petri nets. At this point, CB3 is not operated. Here, we show the initial and the final marking vector of Fig. 3 and for next circuit breaker CB5 and CB3, influence of fault L1 as follow.

- CB2, L1Am, L1Ap

$$\begin{array}{lcccc} & L1 & L1Am & L1Ap & CB2 \\ M_0 = [& 0 & 0 & 0 & 1]^T \\ M_1 = [& 2 & 1 & 1 & -1]^T \end{array}$$

- CB2, L1As

$$\begin{array}{lccccccccc} & B & Bm & L1As & L2 & L2Bm & L2Bp & CB3 & CB2 & CB4 \\ M_0 = [& 0 & 0 & 1 & 0 & 1 & 1 & 0 & 1 & 1]^T \\ M_1 = [& 2 & 1 & 2 & 4 & 2 & 2 & -1 & 0 & -1]^T \end{array}$$

- CB3

$$\begin{array}{lcccc} & L1 & L1Bm & L1Bp & CB3 \\ M_0 = [& 0 & 0 & 0 & 0]^T \\ M_1 = [& 2 & 1 & 1 & -2]^T \end{array}$$

- CB5, L2Cs

$$\begin{array}{lccccccccc} & B & Bm & L2Cs & L1 & L1Bm & L1Bp & CB4 & CB5 & CB3 \\ M_0 = [& 0 & 0 & 1 & 0 & 0 & 0 & 1 & 0 & 0]^T \\ M_1 = [& 2 & 1 & 2 & 4 & 1 & 1 & 0 & -1 & -2]^T \end{array}$$

4. CASE STUDY

In this section, we choose the system as in Fig 4., which represent 16 areas, 56 protection relays and 28 circuit breakers. Table 2. is shown simulation results of backward Petri nets for case study.

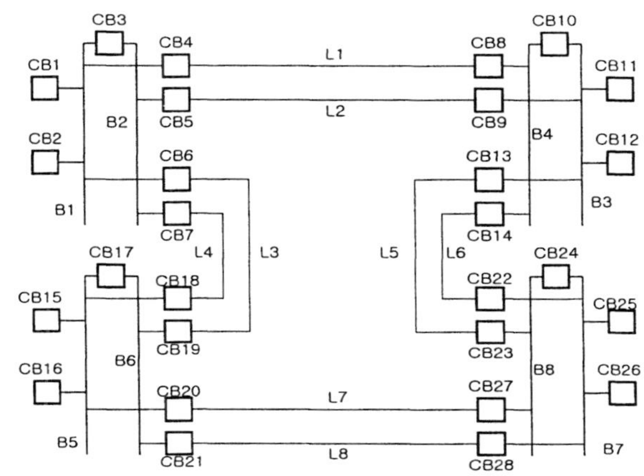

Fig 4. Application system

Table 2. Simulation results

Case	Information of SCADA	Simulation result (fault location)
Case1	B1m,L1B4p,L1B1m;CB1,CB2 ,CB3, CB4, CB6, CB8	B1, L1
Case2	B1m,B2m,L1B4m,L1B1p, L2B2p,2B3m;CB1,CB2,CB3, CB4,CB5,CB6,CB7,CB8,CB9	B1, B2, L1, L2
Case3	L1B4m,L1B1p,L2B2p,L2B3p, L7B8p,L7B5m,L8B6m,L8B7 m;CB4,CB5,CB8,CB9,CB20, CB21,CB27,CB28	L1, L2, L7, L8
Case4	L2B3m;CB5,CB9	L2

CASE 1 : A fault occurred on B1 and L1.

Fault occurred on B1

Normal operation : B1m

Trip operation : CB1, CB2, CB3, CB4, CB6

Fault occurred on L1
Normal operation : Failed(B4 side)
Non-operation : L1B4m
Backup operation : L1B4p
Trip operation : CB8
Normal operation : L1B1m(B1 side)
Trip operation : CB4.

Case 2 : A fault occurred on B1, B2, L1 and L2.
Fault occurred on B1
Normal operation : B1m
Trip operation : CB1, CB2, CB3, CB4, CB6

Fault occurred on B2
Normal operation : B2m
Trip operation : CB3, CB5, CB7

Fault occurred on L1
Normal operation : Failed(B1 side)
Non-operation : L1B1m
Backup operation : L1B1p
Trip operation : CB4
Normal operation : L1B4m(B4 side)
Trip operation : CB8.

Fault occurred on L2
Normal operation : Failed(B2 side)
Non-operation : L2B2m
Backup operation : L2B2p
Trip operation : CB5
Normal operation : L2B3m(B3 side)
Trip operation : CB9.

Case 3 : A fault occurred on L1, L2, L7 and L8.
Fault occurred on L1
Normal operation : L1B1m(B1 side)
Non-operation : L1B4m
Backup operation : L1B1p
Trip operation : CB4
Normal operation : L1B4m(B4 side)
Trip operation : CB8.

Fault occurred on L2
Normal operation : Failed(B2 side)
Non-operation : L2B2m
Backup operation : L2B2p
Trip operation : CB5
Normal operation : Failed(B3 side)
Non-operation : L2B3m
Backup operation : L2B3p
Trip operation : CB9.

Fault occurred on L7
Normal operation : Failed(B8 side)
Non-operation : L7B8m
Backup operation : L7B8p
Trip operation : CB27
Normal operation : L7B5m(B5 side)
Trip operation : CB20.

Fault occurred on L8
Normal operation : L8B6m(B6 side)
Trip operation : CB21
Normal operation : L8B7m(B7 side)
Trip operation : CB28.

CASE 4 : A fault occurred on L2.
Fault occurred on L2
Normal operation : Failed(B2 side)
Miss-operation : L2B2m
Trip operation : CB5
Normal operation : L2B3m(B3 side)
Trip operation : CB9.

5. CONCLUSION

In this paper, we proposed strategy of diagnosis for detecting the fault location in power system. At First, we focused on the fault procedure models of power system using Petri nets. And then, we constructed the backward Petri nets for the fault diagnosis.

REFERENCES

[1] L. Jenkins and H. P. Khincha, "Deterministic and Stochastic Petri Net Models of Protection Schemes", IEEE Trans. on Power Delivery, Vol.7, No.1, pp.84-90, January 1992.
[2] J.Tang and F.Wang, "Modeling of a transmission network protection system using Petri nets", Electric Power Systems Research Vol. 44, pp.175-181, 1998.
[3] K. L. Lo and H. S. Ng and J. Trecat, "Power systems fault diagnosis using Petri nets", IEE Proc.-Gener. Trans. Distrib., Vol. 144, No.3, pp.231-236, May 1997.
[4] C. Rodriguez, S. Rementeria, J. I. Martin, A. Lafuente, J. Mugureza, and J. Perez, "A Modular Neural network approach to fault diagnosis", Trans. on Neural Networks, Vol. 7, No. 2, pp.326-340, March 1996.
[5] Hyun-Joon Cho and Jong-Keun Park, "An Expert System for Fault Section Diagnosis of Power Systems using Fuzzy Relations", IEEE Trans, on Power systems, Vol. 12. No. 1. Feb. 1997
[6] James L. Peterson, "Petri net theory and the modeling of system", Prentice-Hall
[7] C. S. Hwang and J. M. Lee, "Analysis of Matrix Equation Based on Petri Net for Discrete System Control", Proceedings of the 29th SICE Annual Conference International Session, p639-696, July 1990.
[8] K. P. Brand and J. Lopainsky, "Principles and engineering of process control with Petri Nets", IEEE Trans. on Automatic Control, Vol.33, No.2, pp.138-149, Feb. 1988.
[9] C.S.Chang and L.Tian, F.S.Wen, "A new approach to fault section estimation in power systems using Ant system", Electric Power Systems Research 49 (1999) 63-70
[10] F.S.Wen and C.S.Chang, "Probabilistic approach for fault section estimation in power systems based upon a refined genetic algorithm, IEE Proc, Generation, Transmission Distribution, UK, 144(1) (1997) 160-168

ELSEVIER
IFAC
PUBLICATIONS
www.elsevier.com/locate/ifac

A RESTORATION AID EXPERT SYSTEM FOR DISTRIBUTION NETWORKS

Heung-Jae Lee, Kyeong-Seob Lee, Sung-Min Park

Department of Electrical Engineering, Kwangwoon University
447-1, Wolgae-Dong, Nowon-Gu, Seoul, 139-701, Korea
Next Generation Power Technology Center

Abstract: When faults occur in a distribution network, blackout loads must be restored as fast as possible to minimize interruption of electric service. This paper presents an expert system to restore distribution networks.
The system is designed to minimize switching operations in a heuristic sense by using heuristic rules. The restoration process is converted to the feasible path finding problem in the state space of expert system by application of general topological knowledge base. Multiple load transfer algorithm is proposed to deal with complex situations. *Copyright © 2003 IFAC*

Keywords: Distribution network, Expert system, Heuristics, Inference processes

1. INTRODUCTION

A major aim of restoration is fast recovery of blackout area as much as possible prior to the network loss minimization with minimum switching operations. The restoration of distribution network is a non-linear combinatorial optimization problem to find a set of switching operations. Of course, it is desirable to minimize the number of switching operation.

Many papers have been presented and that can be classified into two categories. One is the analytical method (Aoki *et al.*, 1989; Dialynas and Michos, 1989; Shirmohammadi, 1992; Sarma and Prasad, 1994; Aoki *et al.*, 1987), which utilizes a mathematical model, and the other is the heuristic method (Liu *et al.*, 1988; Hsu *et al.*, 1992; Susheela Devi and Anandalingam, 1995; Morelato and Monticelli, 1989), which utilizes experience and knowledge of human operator. In the analytical method, if the exact mathematical model of problem is given, optimal solution can be obtained by specific optimal algorithm. However, the derivation of exact mathematical model is difficult. Even if the model is derived, usually it takes considerable calculation time.

Meanwhile the heuristic method has a benefit in calculation time than analytical methods. But it may not get optimal feasible solutions. Unlike reconfiguration problem, which intends loss minimization at normal state of network, the aim of restoration is quick recovery. Thus, the heuristic method has more advantages than the analytical method.

In the heuristic based researches, Liu, *et al.* (1988) proposed the restoration expert system, which divided whole blackout area into sub-groups by grouping rules and restored each groups, Hsu *et al.* (1992) proposed branch restoration method for the Taiwan distribution system, and Susheela Devi and Anandalingam (1995) utilized breadth-first searching technique to reduce the number of switching.

In this paper, we propose new expert systems to generate an effective restoration solution. Inference process and knowledgebase are designed to reduce switching operations. In addition, multiple load transfer strategy is developed to deal with rare situation.

2. REPRESENTING DISTRIBUTION NETWORK

Distribution network consists of many types of switches and zones (Liu, *et al.* 1988) that can be divided by switches.

Actually, a restoration process is non-commutative proper sequence of switching operations depending on the network topology and constraints such as avoiding feeder overload and maintaining radial structure.

In this paper, new representations of each part in distribution network are defined to manipulate network topology efficiently.

As shown in Fig. 1, distribution network has a radial structure composed with feeders and zones, all components are combined with switches and circuit breakers forming hierarchy in order of circuit breaker, feeder, and zones.

Herein each part of the distribution network is represented in Prolog language format as follows.

- FEEDER(*name, capacity, cb_name, flow*)
- CB(*name, state*)
- SW(*name, state*)
- ZONE(*name, load, [sw_list], type*)
- GROUP(*name, load, [zone_list], type*)

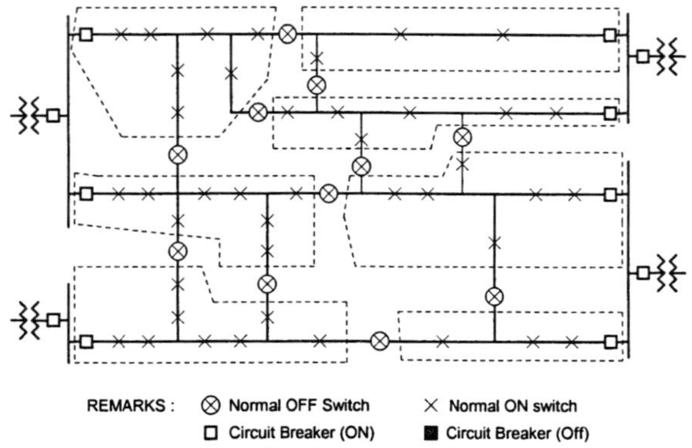

Fig. 1. Sample distribution network

In these representations, '***name***' in every predicates means a name of that element, '***capacity***' and '***flow***' mean maximum capacity and current flow of feeder, '***cb_name***' means circuit breaker name, which connects bus and feeder, '***state***' means on-off states of each circuit breaker or switch. In the 'zone' predicate, '***load***' means estimated load in zone and '***[sw_list]***' means a set of switches connected zone directly and includes circuit breaker. '***type***' means a kind of zones, which is node, branch, or end, which is determined by the number of '***[sw_list]***'. '***[zone_list]***' is a list of zones.

3. RESTORATION PROCESS

To reduce the switching operation during restoration, whole blackout area must be restored in one group if it is possible, because apparently it requires only one switching operation. However, if it is not possible, whole blackout area must be divided into a few feasible subgroups. Of course, the number of groups should be as small as possible. It should be noted here that actually the 'division into feasible subgroups' process is the actual solution finding process exactly in the state space of the expert system. In cases of load transferring, though these will be practically very rare situation, the switching operation is proportional to the number of load transferring.

In the proposed system, overall restoration process is composed with restoration planning stage and load transferring stage. Load transferring should be considered to increase the capacity of feeders when the restoration is impossible with available feeder capacity. Both of restoration planning and of load transferring inference procedures are designed to reduce switching operations for restoration by using decision-making heuristics. Some major rules are as follows.

1) Total available feeder capacity must be larger than total blackout load. If total blackout load is more than total available feeder capacity, complete restoration is not possible.

2) If any one feeder has sufficient capacity to restore whole blackout area, restore the blackout area using one feeder after checking overload.

3) If one feeder has not sufficient capacity to restore whole blackout area, the blackout area is divided into two parts and restored by using two feeders. But if no two feeders have sufficient capacity, divide the blackout area one by one.

4) If load transfer is indispensable, it should be small as possible. For this action, actually the expert system searches the state space from the first level, second level etc. forming the breadth first search strategy.

3.1 Restoration Planning

To describe the restoration planning, some definitions are proposed for convenience as follows.

L_T : Total blackout load

n : The number of neighbour feeders of blackout area

F_i : The *i*-th neighbour feeder of blackout area

I : Set of F_i, $I = \{i \mid F_i, 1 < i < n\}$

$C(F_i)$: Capacity of F_i

C_T : Total capacity of neighbour feeders,

$$C_T = \sum_{i=1}^{n} C(F_i)$$

C_{LT} : Capacity of load transfer

$C_{T(new)}$: Total capacity with load transfer,

$$C_T + C_{LT}$$

Whole restoration planning consists of four processes, and each process reflects heuristic rules stated in previous section. Functions and procedures of restoration planning are explained in detail as following.

Process 1:

- If $L_T > C_T$ then goto ***Load Transfer***
- If $L_T \le C_T$ then goto ***Process 2***

For successful restoration, total capacity of all neighbour feeders must be more than total blackout load. If not, load transfer must be performed for increasing the capacity of feeders. So, at first the comparison of L_T and C_T is performed in process 1.

Process 2:

- If $\exists i, L_T \le C(F_i)$ then goto ***Result out***
- If $\forall i, L_T > C(F_i)$ then goto ***Process 3***

By comparing blackout load and capacity of each feeder, process 2 checks whether overall blackout load is restored by only one neighbour feeder or not. If $C(F_i)$ is more than L_T, restoration is completed by using neighbour feeder F_i and it requires only one switching operation. If not, process 3 is performed.

Process 3:

- If $\exists M, L_T \leq \sum_{m \in M} C(F_m), (M \subset I)$ then
  - If it is radial structure, goto ***Result out***
  - If it is not radial structure, goto ***Divide(N)***
- If no more subset M exists then goto ***Load Transfer***

Divide(N) is a recursive process, In this process, whole blackout area is divided into more than two parts considering each feeder capacity, and blackout load is compared to the capacity sum of neighbour feeders that related to divided parts. ***N*** is the number of groups, and it is increased one by one until finding solution.

From process 3, though $C(F_i) + C(F_j)$ is more than L_T, it is impossible to restore that blackout area because of violating radial structure. Fig. 2 shows this case.

In Fig. 2, blackout area has 6 zones, if load of each zone is identically 10 and each capacities of F_i and F_j are 40 and 30, L_T equals 60 and $C(F_i) + C(F_j)$ is more than L_T. However, because of network structure, this case cannot maintain radial structure. Therefore, the area should be divided into two groups at first. In this division process, the feeder-balancing algorithm is applied.

3.2 Load Transferring

In process 1 and process 3, load transfer may be performed to restore blackout area. By load transferring, capacity of the neighbour feeder can be increased and restoration with insufficient capacity of neighbour feeders can be possible. To achieve a solution, multiple load transfer strategy is adopted here and the searching order of load transfer is the number of load transferring. The process is as follows.

Load Transfer:

- Perform load transfer in order
- If $L_T \leq C_{T(new)}$ then goto ***Process 2***
- If $L_T > C_{T(new)}$ then goto ***Load Transfer***

The order of load transfer is important to reduce switching operation and determined by the number of load in order of increasing the number of load transfer such as one time, two times, and so on. This process is the breadth first search in the searching space. If one load transferring cannot obtain sufficient capacity for restoration, one more load transfer can be performed. If capacity is not sufficient, another load transfer is performed continuously until sufficient capacity is obtained.

Overall restoration process is shown in Fig. 3 in the flowchart form.

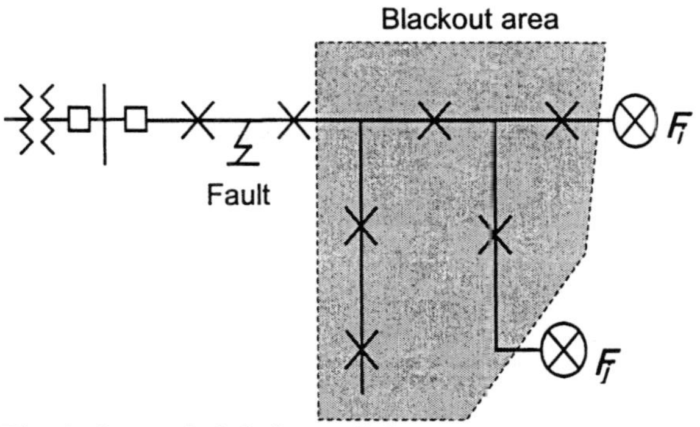

Fig. 2. Case of violating radial structure

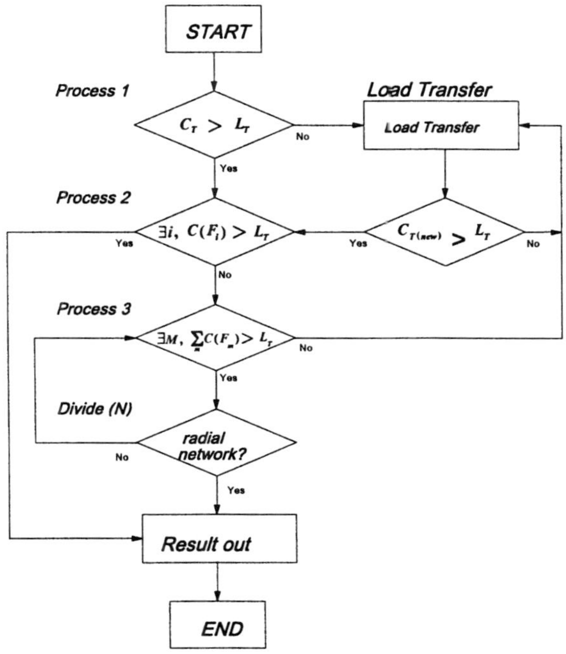

Fig. 3. Inference flow of the restoration process

4. EXAMPLES

Two case studies are introduced here. Fig. 4 shows the state after a fault occurs at zone"z23" on a distribution network.

The sample distribution network consists of 5 feeders, 80 zones, and 85 switches. It is assumed that each zone load is identically 10 without loss of generality. By a fault, the total number of blackout zone is 14.

4.1 Case 1

The whole blackout load is 140 and capacity sum of neighbour feeders is 165. Though blackout load is less than capacity sum, load transfer must be performed because of violating radial structure.

Table 1 and Fig. 5 are the results of example based on proposed expert system.

In the result, the number of switching operations is five, blackout area is divided into two parts, and load transfer is performed one time. In this example, proposed system restores blackout area with minimum switching operations.

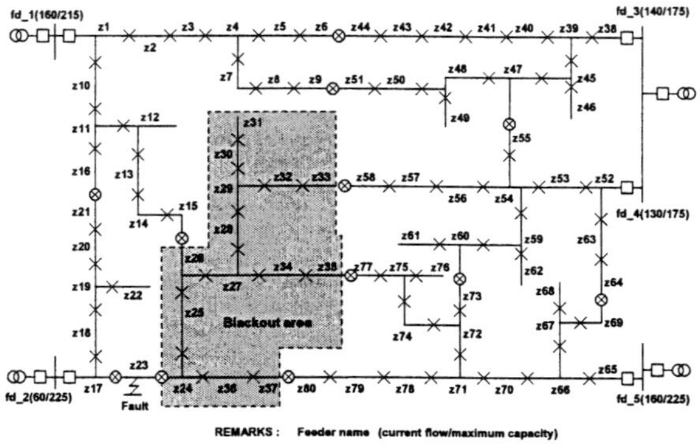

Fig. 4. Example network after a fault for case 1

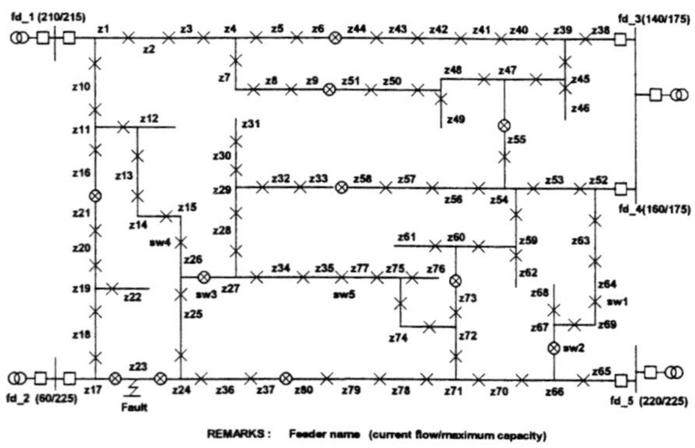

Fig. 5. Restoration result of case 1

Table 1 Switching operation of case 1

Switch Name	State	Usage of Switching
sw1	ON	Load transfer {z67, z68, z69} from fd_5 to fd_4
sw2	OFF	Load transfer {z67, z68, z69} from fd_5 to fd_4
sw3	OFF	Divide blackout area into two parts. {z24, z25, z26, z36, z37}, {z27, 28, z29, z39, z31, z32, z33, z34, z35}
sw4	ON	Restore {z24, z25, z26, z36, z37}
sw5	ON	Restore {z27, z28, z29, z39, z31, z32, z33, z34, z35}

4.2 Case 2

Fig. 6 is same as Fig. 4, but capacity of neighbour feeders is 125 because of increasing the customer in fd_1 and fd_5 by 20 respectively. Since capacity sum is less than blackout load, load transfer must be performed first to restore the network. Table 2 and Fig. 7 are the results of example based on proposed expert system.

In the result, the number of switching operations is 9, blackout area is divided into three parts, and load transfer is performed two times. In this case, proposed system restores blackout area with minimum switching operations too.

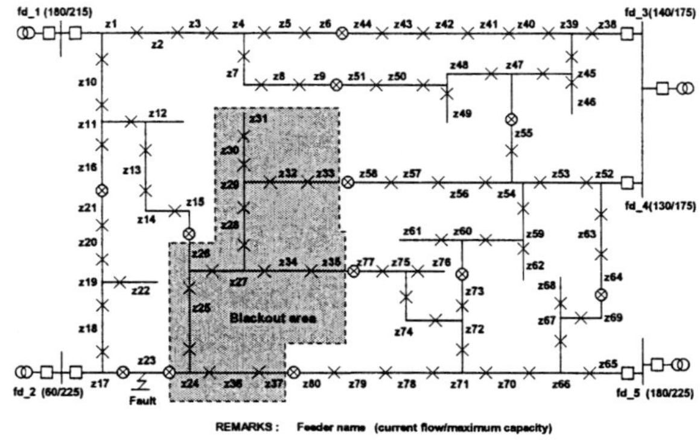

Fig. 6. Example network after a fault for case 2

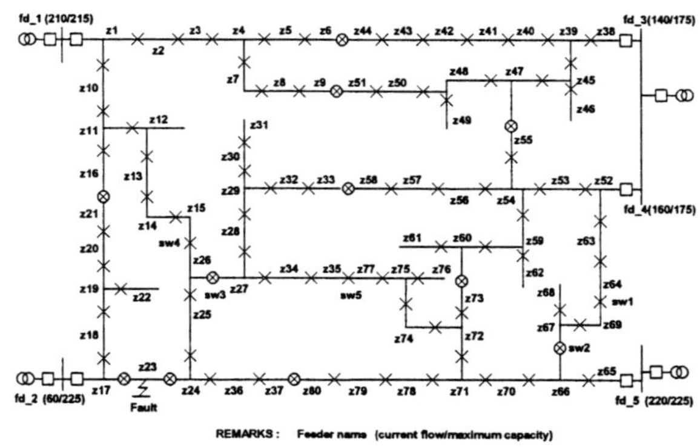

Fig. 7. Restoration result of case 2

Table 2 Switching operation of case 2

Switch Name	State	Usage of Switching
sw1	ON	Load transfer {z6, z5} from fd_1 to fd_3
sw2	OFF	Load transfer {z6, z5} from fd_1 to fd_3
sw3	ON	Load transfer {z55} from fd_4 to fd_3
sw4	OFF	Load transfer {z55} from fd_4 to fd_3
sw5	OFF	Divide blackout area into three parts. {z26, z25, z24, z36, z37}, {z35, z34, z27, z28}, {z33, z32, z29, z30, z31}
sw6	OFF	Divide blackout area into three parts. {z26, z25, z24, z36, z37}, {z35, z34, z27, z28}, {z33, z32, z29, z30, z31}
sw7	ON	Restore {z26, z25, z24, z36, z37}
sw8	ON	Restore {z35, z34, z27, z28}
sw9	ON	Restore {z33, z32, z29, z30, z31}

5. CONCLUSION

In this paper, a restoration aid expert system is presented that generates effective and reliable restorative process. The expert system utilized heuristic rules and adopted multiple load transfer strategy.

As shown in the examples, proposed system could restore blackout area with relatively small switching operations though available capacities of neighbour feeders were insufficient. As the processing time is just a few seconds, it shows promising possibility for practical application.

ACKNOWLEDGMENT

Authors would like to thank Korea Ministry of Science and Technology and Korea Science and Engineering Foundation for their support through ERC program.

REFERENCES

K. Aoki, H. Kuwabara, T. Satoh, M. Kanezashi (1987). Outage State Optimal Load Allocation by Automatic Sectionalizing Switches Operation in Distribution Systems. *IEEE Trans. on Power Delivery.* Vol. PWRD-2, No. 4, pp. 1177-1185.

K. Aoki, K. Nara, M. Itoh, T. Satoh (1989). A New Algorithm for Service Restoration in Distribution Systems. *IEEE Trans. on Power Delivery.* Vol. 4, No. 3, pp. 1832-1839.

E. N. Dialynas, D. G. Michos (1989). Interactive Modeling of Supply Restoration Procedures in Distribution System Operation. *IEEE Trans. on Power Delivery.* Vol. 4, No. 3, pp. 1847-1854.

Y. Y. Hsu, H. M. Huang, H. C. Kuo (1992). Distribution System Service Restoration Using A Heuristic Search Approach. *IEEE Trans. on Power Delivery.* Vol. 7, No. 2, pp. 734-740.

C. C. Liu, S. J. Lee, S. S. Venkata (1998). An Expert System Operational Aid for Restoration and Loss Reduction of Distribution Systems. *IEEE Trans. on Power Delivery.* Vol. 3, No. 2, pp. 619-626.

A. L. Morelato, A. Monticelli (1989). Heuristic Search Approach to Distribution System Restoration. *IEEE Trans. on Power Delivery.* Vol. 4, No. 4, pp. 2235-2241.

N. D. R. Sarma, V. C. Prasad (1994). A New Network Reconfiguration Technique for Service Restoration in Distribution Networks. *IEEE Trans. on Power Delivery.* Vol. 9, No. 4, pp. 1936-1942.

D. Shirmohammadi (1992). Service Restoration in Distribution Networks via Network Reconfiguration. *IEEE Trans. on Power Delivery.* Vol. 7, No. 2, pp. 952-958.

V. Susheela Devi, G. Anandalingam (1995). Optimal Restoration of Power Supply in Large Distribution Systems in Developing Countries. *IEEE Trans. on Power Delivery.* Vol. 10, No. 1, pp. 430-438.

ELSEVIER
IFAC
PUBLICATIONS
www.elsevier.com/locate/ifac

Transient Analysis and Reduction Method of Sheath Circulating Current in Underground Power Cable Systems

Chae-Kyun Jung*, Jong-Beom Lee*, Ji-Won Kang**

Wonkwang University, KEPRI***
Korea

Abstract: Sheath circulating current is increased by the electrical unbalance in underground power cable systems. Such unbalance is caused by the mixed burying formation and length unbalance between cross bonding minor sections. If the sheath circulating current rises in power cable systems, the loss caused by the current will be increased and the permissible current of cable will be reduced. The high sheath current may also cause the danger on person and cause the fault by the breakdown of insulation. Therefore, in this paper, the authors analyse the sheath circulating current in transient state and propose the current reduction device using reactor. Reduction effect is proved by the measurement of current and simulation using EMTP. In order to protect the reduction device in case of transient state, the authors propose the protection methods for the reduction device. *Copyright © 2003 IFAC*

Keywords: Sheath circulating current, Cables, Transient states, Protection, Reduction device

1. Introduction

In high voltage power transmission cables, high sheath circulating current usually causes the increase of the power loss in the sheath. If the sheath loss rises, the total thermal resistance of the power cable will be increased. Then the permissible current will be reduced. At the same time, the high sheath current may cause the danger of cable maintenance personnel and cause the faults by the breakdown of insulation or cable jacket. Therefore, the sheath circulating current must be reduced in underground power cable systems.

The increase of the sheath circulating current is mainly caused by the mixed burying formation and length unbalance between cross bonding minor sections. The mixed burying formation and length unbalance cause the unbalance of mutual impedances between sheath and conductor as well as sheaths.

In this paper, we analysed the characteristic of sheath circulating current in actual underground power cable systems which have the problem of the high sheath current. The sheath circulating currents are measured directly on field. Based on these results, we proposed a current reduction device which is actually a reactor of 2.65mH. The reduction effect is proved by current measurement and simulation using EMTP. At the same time, the authors extensively studied the transient phenomena on the sheath and current reduction device and a protection method is proposed to protect the device when the line-to-ground fault is occurred and the lightning strikes on the power cables.

2. Characteristic of the sheath circulating current

2.1 Calculation of sheath circulating current

The equation for sheath circulating current is expressed as equation (1) on a major section of cross bonding.

$$l\cdot[z1][Isi]+m\cdot[z2][Isi]+n\cdot[z3][Isi]+[Vs]+2\mathrm{Re}\sum_{i=1}^{n}Isi=0 \quad (1)$$

Where,
l,m,n : length of each cross-bonded minor section [km]
[z1],[z2],[z3] : sheath impedance matrix of earth return path in each minor section
[Isi]: sheath circulating current matrix[A]
[Vs]: induced sheath voltage matrix[V]
Re : ground resistance[Ω]

[Vs] of (1) is expressed as equation (2).

$$[Vs]=l\cdot[Z1][Ii]+m\cdot[Z2][Ii]+n\cdot[Z3][Ii]=0 \quad (2)$$

Where,
[Z1],[Z2],[Z3] : Mutual impedance matrix between conductor and sheath in each minor section
[Ii]: Load current matrix[A]

Specific impedance calculations can be also written as eqs. (3) – (6).

○ mutual impedance between conductor and sheath

$$Zii = 0.000989\,f + j0.0029\,f\cdot\log\frac{De}{r} \quad [\Omega/\mathrm{km}] \quad (3)$$

○ mutual impedance between conductor and another phase sheath

$$Zij = 0.000989\,f + j0.0029\,f \cdot \log\frac{De}{S} \quad [\Omega\ /km] \tag{4}$$

○ sheath self impedance

$$zii = Rs + 0.000989\,f + j0.0029\,f \cdot \log\frac{De}{r} \quad [\Omega\ /km] \tag{5}$$

○ sheath mutual impedance

$$zij = 0.000989\,f + j0.0029\,f \cdot \log\frac{De}{S} \quad [\Omega\ /km] \tag{6}$$

Where,

De : equivalent depth of earth return path
f : frequency [60Hz]
r: radius of sheath [mm]
S: spacing between conductor center points [mm]
Rs : AC resistance of sheath [Ω /km]

2.2 Correlation of sheath circulating current and burying formation`

The authors change the spacing between conductor center points at the cross bonding minor sections between normal joints(NJs), then analyse the characteristic of sheath circulating current with varying impedance. Fig. 1 shows a typical burying formation of power cable and the section length between NJ-a and NJ-d. Load current is 300A.

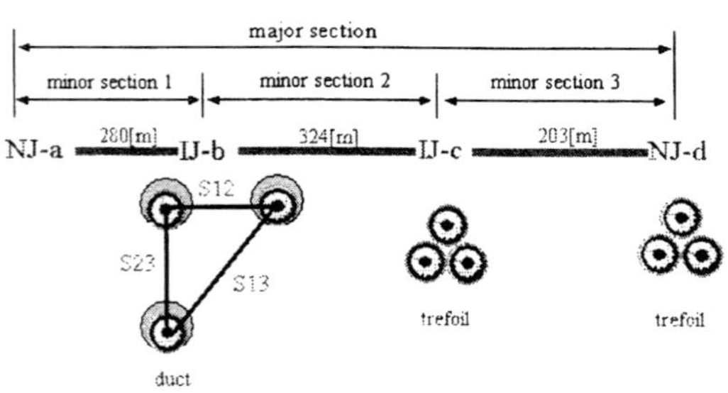

Fig. 1. Burying formation within a major section

Fig. 2 shows the amplitude and phase angle of sheath circulating current in each phase, when S12 and S13 in Fig.1 are changing from 100mm to 500mm as indicated by arrow. The left graph shown in Fig. 2 means the sheath current with variation of S12. The amplitude of sheath current(is1) in phase A decreases slowly to 120A, and then increases up to maximum of 140A. The angle is steeply increasing with the spacing of S12. However, the amplitude of is2, in phase B, increases more steeply than the variation of phase angle. The right graph shown in Fig. 2 shows the sheath current with variation of S13. The amplitude of isl changes according to the phase angle. However, the amplitude of is3 changes more slowly than the increment of angle.

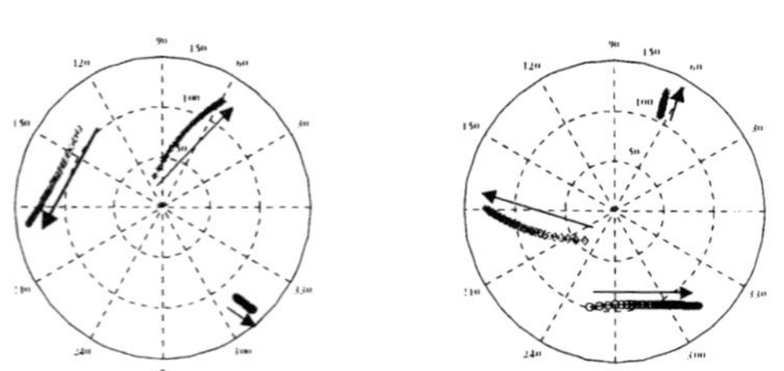

Fig. 2. Sheath circulating current in varying S12 and S13

From these results, the authors concluded that the sheath circulating current can be reduced by the equal burying formation.

2.3 Correlation of the sheath circulating current and length between minor sections

2.3.1 Unified burying formation

The sheath circulating current is also analyzed in case of length unbalance between minor sections, in unified formation of trefoil or duct. The sheath current is theoretically zero at the equal burying formation and section length.

However, the high sheath current will be produced on sheaths because of the length unbalance between minor sections within each major section as well as the mixed burying formation.

Fig. 3 shows the change of sheath circulating current with length unbalance of minor sections at trefoil and duct formation. As shown in Fig. 3, the sheath circulating current in duct formation is higher than in trefoil formation. In addition, the sheath currents have the minimum value at unbalance rate of zero. Then the current rises with the increase of the unbalance rate. Such analysis results can be used to estimate the sheath current of the actual systems and establish the current reduction methods.

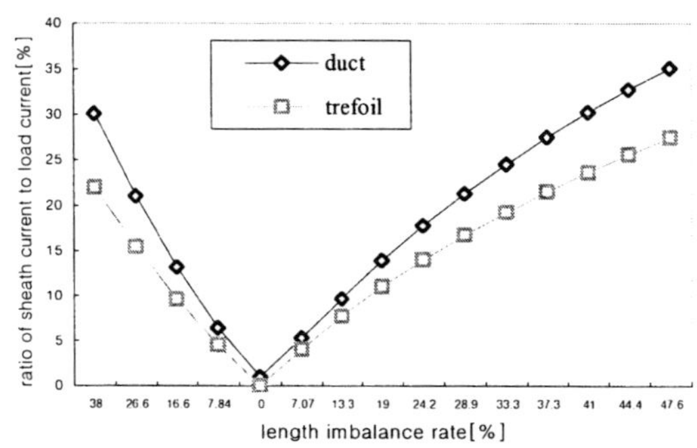

Fig. 3. Sheath circulating current with length imbalance rate

2.3.2 Mixed burying formation

The mixed burying formation as well as length unbalance in a cross bonding major section influence the sheath impedance. Sheath current rises because of the change of the impedance. Sheath circulating currents are simulated in case of length unbalance and mixed burying formation as in Table 1.

Table 1 Examples of mixed burying formation

Case	Burying formation of major section		
	minor 1	minor 2	minor 3
Case 1	Direct burying	Duct	Trefoil
Case 2	Trefoil	Duct	Trefoil
Case 3	Trefoil	Duct	Duct

Fig. 4 shows the ratio of sheath current to load current with the length unbalance rate in each minor section in terms of Case 1. As shown in Fig. 4, the ratio of sheath current to load current is 35.1% at unbalance rate of zero. For the unbalance rate below zero, the sheath currents in all three minor sections rises with the increase of length unbalance. For the unbalance rate higher than zero, the sheath currents in minor section 1 and minor section 2 rise with the increase of the unbalance rate. However, in minor section 3, the sheath current decreases with the increase of the unbalance rate.

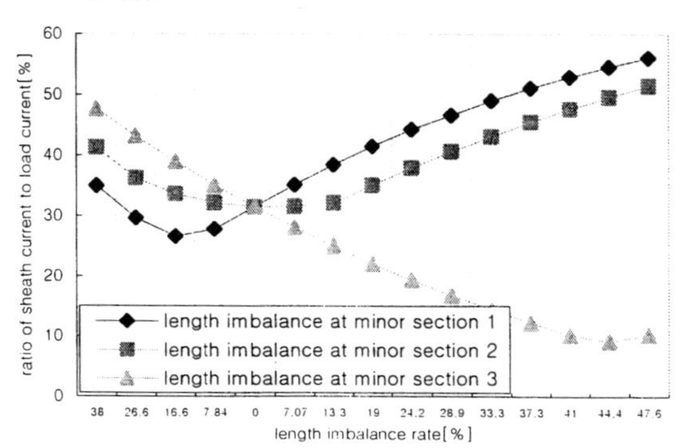

Fig. 4. Sheath current with length imbalance at Case 1

Fig. 5 shows the sheath currents in Case 2. As shown in Fig. 5, for the unbalance rate below zero, the sheath currents in minor section 1 and section 3 increase with the increase of the unbalance rate. However, the unbalance rate has little effect on the sheath currents in these two sections when the unbalance rate is higher than zero. But for minor section 2, the sheath current has different situation. It decreases with the unbalance rate when it's below zero, but increases with the unbalance rate when it's higher than zero. The maximum ratio of sheath current to load current is 60.4 % at the unbalance rate of 47.6%.

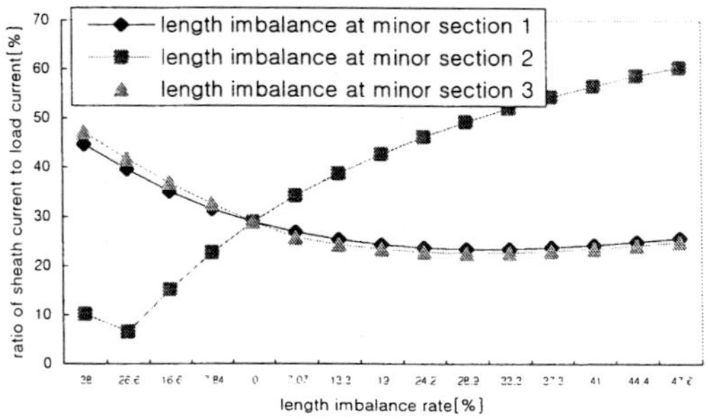

Fig. 5. Sheath current with length imbalance at Case 2

Fig. 6 shows the sheath currents in Case 3. The sheath currents in minor sections 2 and 3 have the same characteristics as the sheath current in minor section 2 in Case 1. The sheath current in minor section 1 is similar to the sheath current in minor section 3 in Case 1.

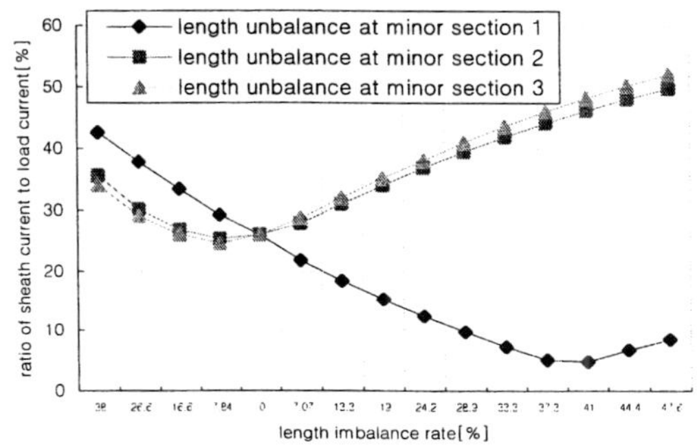

Fig. 6. Sheath current with length imbalance at Case 3

Sheath circulating current steeply rises because of the mixed burying formation and length unbalance between minor sections. At the same time, the sheath current can rise by the bad connection of cross bonding lead and be affected by the distribution cables connected to the transmission cables.

2.4 Correlation of the sheath circulating current and permissible current

According to IEC Std. 287, the permissible current of an AC cable is written as equation (7) in buried cables where drying out of the soil does not occur or cables installed in air.

$$I = \sqrt{\frac{\Delta\theta - W_d[0.5T1 + n(T2+T3+T4)]}{RT1 + nR(1+\lambda_1)T2 + nR(1+\lambda_1+\lambda_2)(T3+T4)}} \qquad (7)$$

Where, λ_1 is the ratio of losses in the sheath to total losses in all conductors in that cable, λ_2 is the ratio of losses in the armouring to total losses in all conductors in that cable. Power loss in the sheath λ_1, is expressed as equation (8).

$$\lambda_1 = \lambda'_1 + \lambda''_1 \qquad (8)$$

Where,

λ'_1 : loss caused by sheath circulating current

λ''_1 : loss caused by eddy current

If the sheath circulating current rises, the loss caused by sheath circulating current will increase, and then the ratio of loss dissipated in sheath per unit length to loss in conductor per unit length will increase, too By such effect, the total thermal resistance of the cable is increasing, and the permissible current is reduced. Therefore, the sheath circulating current must be reduced for the increment of transmission capacity. In addition, the high sheath current has an effect on person and may cause the fault by the breakdown of insulation.

3. Reduction device of the sheath circulating current

Reduction method of sheath circulating current is established through various methods such as direct measurement, EMTP simulation and calculation. From these results, reduction device of sheath circulating current is developed.

3.1 Analysis of the sheath circulating current under the normal operating condition

The double circuits of 154kV underground power cable system used in this paper are shown in Fig. 7. It is composed of three phases and a single core oil-filled cable. The whole length of the cable is 6.245km. Load current is 300A. The burying formation of this system is mixed duct with trefoil. The sheath voltage limiters(SVLs) are used to protect the sheath against the transient overvoltage on each joint(IJ). This system is currently operating in Korea.

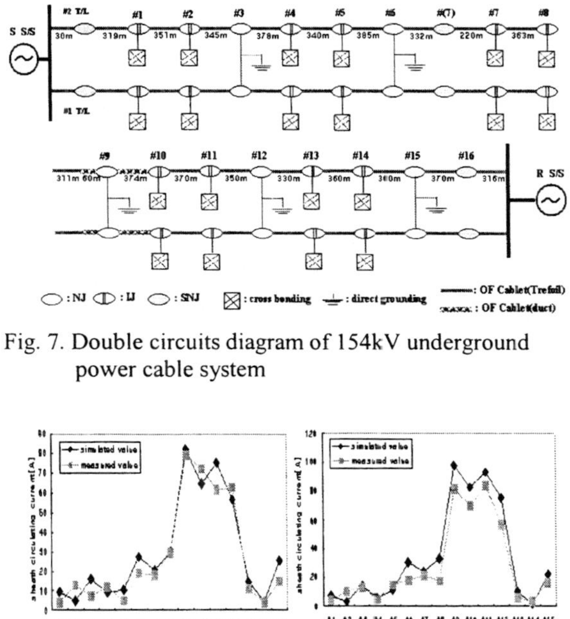

Fig. 7. Double circuits diagram of 154kV underground power cable system

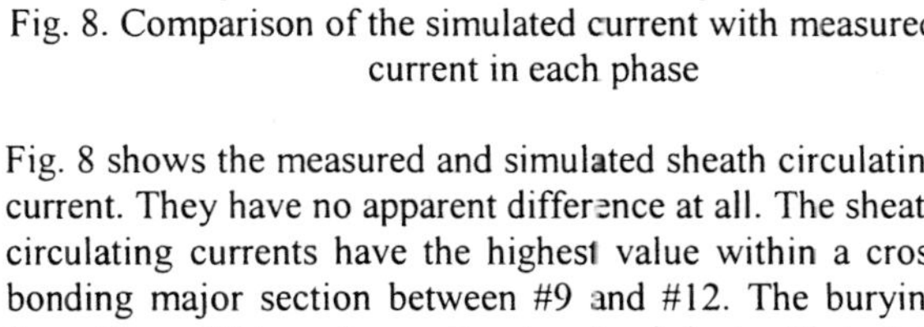

(a) In phase B (b) In phase C

Fig. 8. Comparison of the simulated current with measured current in each phase

Fig. 8 shows the measured and simulated sheath circulating current. They have no apparent difference at all. The sheath circulating currents have the highest value within a cross bonding major section between #9 and #12. The burying formation of this major section is mixed duct with trefoil. The mixed formation within each major section can generate the spacing change between conductor center points. This causes the unbalance of the mutual impedance between conductor and another phase sheath as well as sheath mutual impedance.

3.2 Reduction method of sheath circulating current

The high sheath circulating current could be produced in underground power cable systems because of the length unbalance, mixed burying formation between minor sections, bad connection of cross bonding lead and the distribution cables connected to the transmission cable. Such sheath current has bad effects on the permissible current as well as the person, and causes the fault by breakdown of insulation. Therefore, the authors want to reduce the sheath circulating current. From the results of many studies, the best current reduction method using the reactor was proposed. The reactor of 2.65mH is serially connected to the cross bonding lead as shown in Fig. 9.

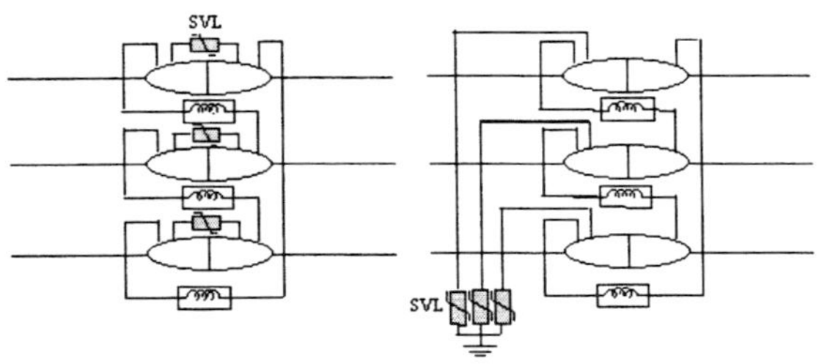

[In non-grounding type with cross-connected to SVL] [In direct grounding type connected to SVL]

Fig. 9. Connection diagram of reduction device

The reactors are installed at # 11 joint of # 2 T/L. Reduction effect was analysed by using EMTP simulation and measurement. Table 2 shows the reduction effect of sheath circulating current. As shown in Table 2, the sheath current exceedingly decreases at each phase within a major section. The average reduction rate is 84.98%.

Table 2 Sheath circulating current when reactors install to J/B #11

T/L	#2 T/L			
joint box	Sheath circulating current[A]			Average [%]
	A-C reduction rate[%]	B-A reduction rate[%]	C-B reduction rate[%]	
#9(NJ)	90.9	79.4	73.5	81.26
#10(IJ)	94.99	94.38	91.3	93.55
#11(IJ)	89.04	86.56	89.63	88.21
#12(NJ)	95.43	62.64	72.7	76.9
Average[%]				84.98

The voltage between sheath and earth and the voltage between both ends of reactors are shown in Table 3. The maximum sheath voltage occurred in this system is 24.6V when the reactors of 2.65mH are installed. Therefore, the sheath voltage is below the limited sheath voltage of 50V. That confirms that after the installation of the reactors on the underground power cable system, there have no problems under the normal operation condition.

Table 3 Voltage induced between sheath and earth, both ends of each reactor

# 2 T/L				
voltage between sheath and earth[V]			voltage between both ends of each reactor[V]	
reactor	without	# 11	# 11	
# 9(NJ)	0.85	0.93	A-C	9.2
# 10(IJ)	15.1	24.6		
# 11(IJ)	8.6	19.4	B-A	13.4
# 12(NJ)	1.23	1.23		
			C-B	9.5

3.3 *Development of reduction device and analysis of reduction effect through the measurement*

In this paper, the reactor of 2.65mH is proposed to decrease the sheath current. The specification of the reactor is listed in Table 4. On the actual underground power cable system, the reduction device of the reactors are installed as shown in Fig. 10, where the sheath circulating currents are measured directly.

Table 4 Specification of the reactor

Items	Specification
Characteristic	waterproof, indoor
Rating impedance	1Ω (60Hz, 2.65mH)±3% at 20℃
Rating current	100A, 60Hz
Capacity	10kVA
Insulation resistance	2000 MΩ

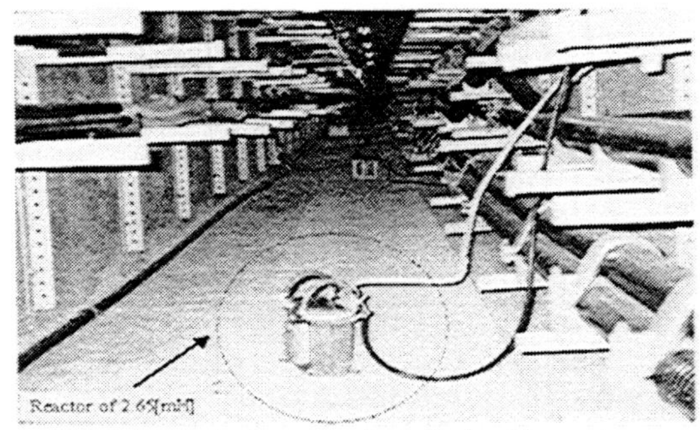

Fig. 10. Actual system installed with reactors

The sheath currents of the system with and without reduction device of reactors are obtained using the actual measurement and EMTP simulation, and shown in Table 5. It is proved that sheath current can be steeply reduced by installation of reactors in both cases of measurement and the simulation.

Table 5 Analysis of the reduction effect through measurement and simulation

J/B	phase	Sheath circulating current[A]			
		measurement		simulation	
		without	with	without	with
# 10 (IJ)	A-C	56.1	7.2	53.9	2.7
	B-A	72	5.7	64.3	3.61
	C-B	69.9	7.2	82.3	7.17
# 11 (IJ)	A-C	61.6	1.7	75.2	8.24
	B-A	83.8	1.6	92.7	12.46
	C-B	78.35	1.4	64.9	7.12
# 12 (NJ)	A-C	56.8	9.5	75	3.43
	B-A	72.4	8.6	44.7	16.7
	C-B	62.55	8.8	56.4	15.4

4. Analysis of the sheath current and reduction device in transient state

4.1 Protection methods of reduction device

Reduction devices are serially connected to the cross bonding leads as shown in Fig. 9. Accordingly, if the fault occurs in underground power cable system with the reduction device of reactor, high overvoltage is induced between the both ends of the reactor. Therefore, the reduction device may be broken. If the reduction device is broken by the transient surge such as line-to-ground fault or lightning surge, the cross bonding lead cable will be disconnected, its function will be lost. Because of the disconnection, the fault current cannot return through the sheath. That will cause the overvoltage produced in the sheath and the joint. The whole cable system may be dangerous.

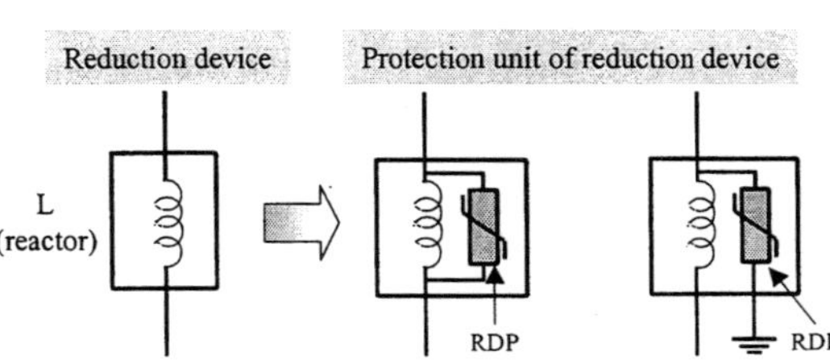

Fig. 11. Protection methods of reduction device

Therefore, the authors propose a device, called Reduction Device Protector(RDP), to protect the reactor against the transient overvoltage as illustrated in Fig. 11. The function of RDP is very similar to that of SVL. The RDP operates when the ovevoltage exists between both ends of the reactor. In this paper, two methods to protect the reactor are proposed. The first method is that the RDP is connected with the reactor in parallel. The second is that the single end of RDP is grounded. The protection effect of RDP is extensively tested by EMTP simulation. Table 6 presents the eight methods to protect the reduction device with SVL connection types. SVL is usually connected as shown in Fig. 9.

Table 6 Protection methods of reduction device

reduction device	SVL connection type	name	protection methods
rector	direct grounding type connected to SVL(D)	DK	D+K
		DKC	D+K+C
		DC	D+C
		DCG	D+CG
	non-grounding type with cross-connected to SVL (K)	KD	K+D
		KDC	K+D+C
		KC	K+C
		KCG	K+CG

* C : RDP is connected with reduction device in parallel
* CG : The single end of RDP is grounded

4.2 *Single line to ground fault*

The single line-to-ground fault usually occurs by the dielectric breakdown between the conductor and sheath in underground power cable system. In this paper, authors analyse the transient phenomena as the single line-to-ground fault occurs at 1.5km, 3km, 4km and 5km from S S/S, # 2 T/L, in Fig. 7. The reactors are installed at # 11.

Table 7 Sheath voltage according to reduction device installation

single line to ground fault position	reactor installation in J/B #11(IJ)							
	D				K			
	voltage between sheath and earth [kV]		voltage between joint boxes[kV]		voltage between sheath and earth [kV]		voltage between joint boxes[kV]	
	#10	#11	#10	#11	#10	#11	#10	#11
1.5km	8.46	8.6	2.51	5.37	16.1	15.4	2.56	5.57
3km	8.12	8.4	3.17	5.3	15.9	15.3	4.15	6.1
4km	9	11	3.9	7.5	16.5	16.4	4.2	8.01
5km	8.71	9.8	2.31	8.3	14.4	14.2	1.16	5.8

* D : Direct grounding type connected to SVL
* K : Non-grounding type with cross-connected to SVL

Table 7 shows the voltage between sheath and earth, the voltage between joints of # 10 and # 11 with SVL when the single line-to-ground fault occur at each point. As shown in Table 7, in terms of ground fault at 4km point, overvoltage is the highest at all points. The reduction devices of reactor are installed in # 11 which is close to 4km point.

Fig. 12 shows the sheath currents with different SVL connection types at each point. The sheath current is 0.82kA higher in K type than in D type. In K type, in terms of the ground fault at 4km point, the sheath current of 11kA is the highest at all points.

Table 8 shows the voltages between sheath and earth, and the voltages between joint boxes with different protection methods after the application of the reduction device when the single line to ground fault occurs at 4km point.

In D type, the voltages between the joint boxes are somehow reduced in all protection methods, and the voltage between sheath and earth is reduced by 3.1kV in DCG, and 2.12kV in DK.

In K type, when there is no protection, the voltage between sheath and earth is 16.4kV as shown in Table 7.

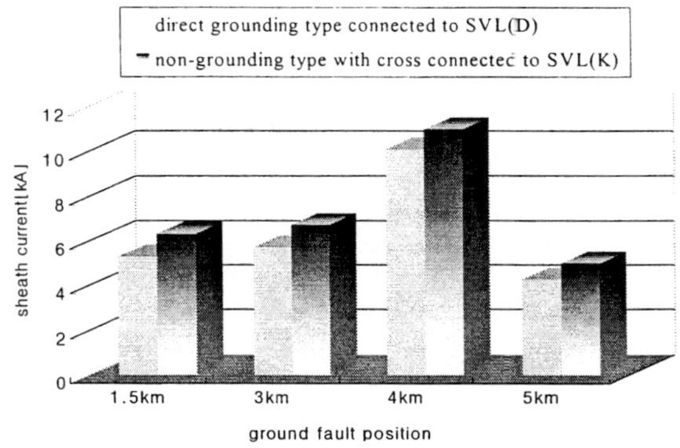

Fig. 12. Current within the reduction devices of reactor

Table 8 Sheath voltage by protection methods

SVL connection type	Protection methods	single line to ground fault(4km)	
		reactor installation to J/B #11(IJ)	
		voltage between sheath and earth [kV]	voltage between joint boxes[kV]
D	DK	8.88	6.5
	DKC	8.93	6.44
	DC	8.94	6.59
	DCG	7.9	6.4
K	KD	9.47	7.45
	KDC	9.46	6.18
	KC	16.1	7.2
	KCG	6.6	7.03

However, when the reduction methods are applied, the reduction effect of KCG is the best. It reduced the voltage of 9.8kV, while in KD and KDC show the reduction effect of 6.93kV and 6.94kV,respectively.

Fig. 13 shows reactor current with the application of various protection methods. As shown in Fig. 13, in D type, the current in DCG is 9.35kA, and DK, DKC, DC are nearly at the same value, around 9.5kA. Therefore, the big difference does not appear among all methods. In case of K type, the current in KC is 10.95kA, and KCG and KD are 9.6kA and 9.51kA, respectively.

The transient characteristic is analysed to develop the protection method of the reactor when the single line-to-ground fault occurs in underground power cable system. From the simulation results, in case of D type, the reduction effect is sequenced by DCG, DKC, DK and , DC order. In case of K type, there was no reduction effect of KC because the voltage between sheath and earth, the reactor current had the highest value. The protection effects of the other methods are sequenced by KDC, KD and KCG order.

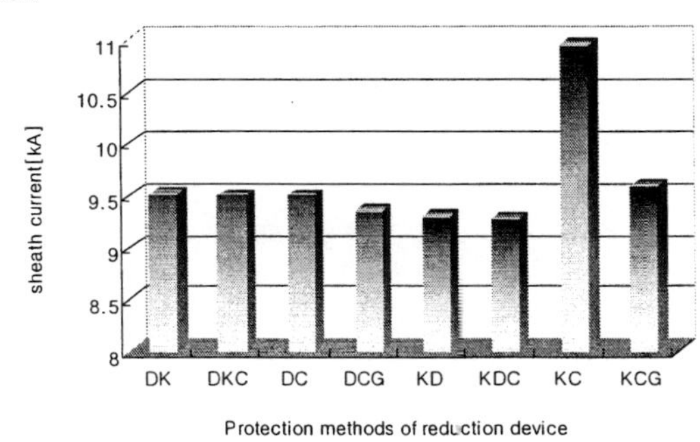

Fig. 13. Reactor current with the protection method application

4.3 *Lightning surge*

The waveform of lightning surge used in this paper is 2/70 μs. The crest value of lightning current is 60kA. Lightning surge is assumed linearly rising and damping. Combined transmission system with underground power cable system is used to analyse the influence of lightning surge. The total length of overhead transmission line is 20km. It is assumed that the lightning surge strikes on phase A, 1km away from cable head in # 2 T/L.

4.3.1 Transient characteristic of the reactor

When the lightning surge strikes on combined transmission system, the voltages between sheath and earth, and the voltages between joints are simulated with and without the reduction device of reactor. As shown in Table 9, in D type, the voltage between joint boxes rises up to 175kV because the impedance of reactor is increased by the effect of high frequency. Therefore, the protection methods of reduction device must be applied on the underground power cable system. In K type, the sheath voltages rise a little because of the operation of SVL.

Table 9 Sheath voltage according to reduction device installation

reduction device	D #11(IJ) voltage between sheath and earth [kV]	D #11(IJ) voltage between joint boxes [kV]	K #11(IJ) voltage between sheath and earth [kV]	K #11(IJ) voltage between joint boxes [kV]
Without	8.43	16.9	8.87	8.68
Reactor	10.53	175	13	9.77

4.3.2 Application of the protection methods

When the lightning surge strikes on the combined transmission system with the installed reactors, the voltages between sheath and earth, and the voltages between joint boxes with the application of protection methods are shown in Table 10

Table 10 Sheath voltage by the application of protection methods

SVL type	protec-tion methods	#10(IJ) voltage between sheath and earth [kV]	#10(IJ) voltage between joint boxes [kV]	#11(IJ) voltage between sheath and earth [kV]	#11(IJ) voltage between joint boxes [kV]
D	DK	9.13	10.24	9.75	9.58
	DKC	8.13	9.91	10.39	10.16
	DC	9.5	17.5	9.3	18.5
	DCG	9.5	17.7	8.38	170
K	KD	8.22	9.02	10.3	9.43
	KDC	8.75	9.09	9.8	9.1
	KC	10.89	9.18	11.9	9.24
	KCG	10.1	9.47	9.45	12.8

In D Type, voltage between joint boxes temporarily rises to 170kV in case there are no protection methods. However, after the application of protection methods, these voltages in DK, DKC are 9.58kV and 10.16kV, respectively. From these results, authors find that the protection effect by RDP and SVL is superior. But, there is no protection effect in DCG. In K type, the protection effect of KDC is the best, but KD is also good.

Fig. 14 shows the flowing current within the reactor according to SVL connection types. As shown in Fig. 14, in each SVL connection type, DKC and KDC show the best protection effect against the lightning current. However, the reduction effects of DCG, KD and KCG aren't big.

In this section, we analyse the protection effect of reduction device against the lightning current when the lightning surge strikes on the combined transmission system installed with the reduction device of reactors.

In D type, there is no protection effect in DCG, but other methods are shown by DKC, DK and DC order, In K type, the voltage and current of sheath are lower than D type. The reason is that SVL in K type is connected with joint box and reduction device in parallel.

Therefore, SVL and RDP can protect the reduction device against the high overvoltage. The protection effects in K type are shown by KDC, KC, KD and KCG order. From these results, authors find that the protection effects of DKC and KDC are the best.

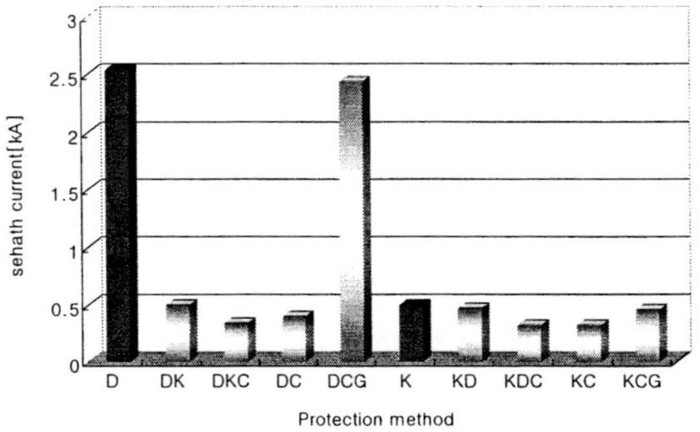

Fig. 14. Reactor current by the application of protection methods

5. Conclusions

In this paper, the authors directly developed the reactor of 2.65mH for the reduction of sheath circulating current, and analysed the effect of the reduction device through the EMTP simulation and measurement. We also analysed the transient characteristic of sheath system and reduction device when the single line to ground fault and lightning surge occur in underground power cable system installed with the reactors.

The reduction devices are serially connected to the cross bonding lead. Therefore, if the fault occurs on underground power cable system installed with the reduction device of reactor the high overvoltage is induced between the both ends of the reactor. So, the reduction device may be broken. If the reduction device is broken by the transient, the cross bonding lead will be disconnected, and the cross bonding function will be lost. Because of the disconnection of the bonding lead, the fault current does not return through the sheath, and the overvoltage will be produced in the sheath and joint box. The whole cable system may be dangerous. Therefore, in this paper, the authors propose the protection method for the reduction device using RDP. The proposed eight methods are analysed by EMTP simulation. From these results, the authors find that the protection effect of DKC and KDC is best among the all protection methods.

REFERENCES

Chae-Kyun Jung, Jong-Beom Lee, Ji-Won Kang, Tae-In Jang(2001). Analysis and Reduction Methods of Sheath Circulating Current in Underground Transmission Systems, KIEE, Vol. 50A, No. 11.

IEC 287-1-1(1995). Electric Cables – Calculation of the Current Rating, Part 1 : Current rating Equations (100% Load Factor) and Calculations of Losses, Section 1 : General.

IEEE(1998). IEEE Guide for the Application of Sheath Bonding Methods for Single-conductor Cables and Calculation of induced Voltage and Current in Cable Sheaths, ANSI/IEEE Std. 575.

Ji-Won Kang, Hai-Won Yang(2002). A Study on the Reduction Methods of Sheath Circulating Current Using the Reduction Equipment in Underground Transmission Systems. KIEE, Vol. 51A, No. 9.

J.S.Barrett, G.J.Anders. Circulating Current and Hysterisis Losses in Screens, Sheaths, and Armour of Electric Power Cables Mathematical Models and Comparison with IEC Std.

ELSEVIER
IFAC
PUBLICATIONS
www.elsevier.com/locate/ifac

PROPERTIES OF PORCELAIN SUSPENSION INSULATOR FOR ULTRA HIGH VOLTAGE TRANSMISSION LINE

In-Hyuk Choi[1], Jang-Hyun Choi[1], Dong-Il Lee[1], Gil-Jo Jung[1], Yeon-Gye Choi[2].

[1] *Power System Research Lab. Korea Electric Power Research Institute. Taejon, Korea*
[2] *Korea Insulator Industrial Co. Ltd. Masan, Korea*

Abstract: This paper was the research of 400 kN suspension insulator for ultra high voltage that included alumina of 17 wt% . The slurry was fabricated after ball milling were mixed raw materials. Test results show that the sintering density was reached 97% of theoretical density, and the values of bending strength, hardness, and fracture toughness were 1658 kgf/cm^2, 27.5 Gpa, and 2.3 MPa·m$^{1/2}$, respectively. In the case of pin diameter 51 mm and pin bottom formed multistep and straight in the suspension insulator, the overvoltage break load of 400 kN insulator appeared 52 ton of the highest value.

Keywords: Transmission line, Porcelain insulator, Suspension insulator, Alumina, XRD.

1. INTRODUCTION

Today, the demand of electrical energy has been suddenly increased because of the development of industry and cultural life. In order to cope with the increase of these power demand, a 765 kV transmission is in progress and partly being operated at present. The number of conductors was increased and the weight of lines was also increased because of the rise of transmission voltage, therefore the development of insulator for ultra high voltage is demanded.

Consequently, the insulators with powerful mechanical strength are required. The porcelain insulator for ultra high voltage can be classified into feldspar, cristobalite and alumina insulator according to raw materials. The bending strengthes of these insulators are in the ranges of 600 to 1000 kg/cm^2 for feldspar, 1000 to 1400 kg/cm^2 for cristobalite, and 1200 to 1700 kg/cm^2 for alumina, respectively (Carty, *et al.*, 1990). So the alumina insulator is presented the most mechanical strength among these insulators. The alumina insulator is manufactured by adding alumina at the range of 10 to 40 wt% in basic sample to improve mechanical strength of feldspar insulator.

To improve mechanical strength of the existing Al_2O_3 insulator included alumina of 30-50 wt%, Zivanovic (Khandelwal, *et al.*, 1961) investigated the result after adding annexes of tufa, talc, TiO_2, MnO_2, Fe_2O_3, ZrO_2, etc. As the result of that, the suspension insulator included alumina of 17 wt% had the most excellent property.

In this study, we developed 400 kN insulator for ultra high voltage more excellent than that of existing feldspar or cristobalite insulator (Kim, 1995) that is, the insulator added 17 wt% alumina to feldspar insulator specimen was improved the physical and electrical properties. This paper also was investigated of the influence of pin forms in various performance tests of suspension insulator.

2. EXPERIMENT

2.1 Starting raw materials and manufacture of insulator

In this study, the porcelain sample is fabricated with raw materials that are composed of pottery stone, feldspar, clay(Kyoritsu Ceramic, Japan) and alumina(Nippon Light metal, Japan) and Zirconia(MZY-3, Future Material, Korea) using a domestic insulator manufacturing company. A specimen has done ball milling for 25 hours after being weighted raw material with mixing rate. The specimen is fabricated with extrusion after the ball milling slurry was eliminated iron and moisture. This formed frame is sintered at 1300°C for 50 minutes in tunnel kiln after dried at 60°C for 48 hours in oven.

2.2 Analysis and measurement

Particle size analysis; A particle size of grinded powders is measured with mastersizer(Malvern, England) using a laser. Slurry is done stability as electrostatic to measure the particle size and is used each 2~3 wt%.

Particle size analysis X-ray diffraction analysis; To observe change of crystal phase according to each specimen, the measurement is performed under condition of a CuKa, 40 kV-30 mA, and a scan speed 3°/min and the angle diffraction(2θ) at the range of 10 to 70° as a X-ray diffraction analysis machine (X'Pert, Philips, Netherlands) after the mixed and grinded specimen is treated with heat at 1300°C

Measurement of electrical property; The porcelain is cut to a thickness of 2 mm and a diameter of 3.5 cm to measure dielectric break test. Dielectric break voltage is measured using a ATM-106 (SOUKOU, Japan) which has a maximum measuring voltage of 60kV(60Hz) and a rise voltage of 3 kV/sec. The characteristics of dielectric is measured using DEA 2970(TA Instruments, U.S.A). At this time, a measuring frequency is 30 Hz ~ 3 kHz and a measuring temperature is 25~150°C.

Measurement of mechanical property; Sintered density is measured with ASTM C20 using the principle of Archimedes. MOR-bar specimen, a diameter of 15mm and a length of 150 mm, is fabricated to measure a bending strength and a 3-point bending strength is measured by a bending strength machine of AFG-250 N (Mecmesin, England). At this time, the velocity of cross-head is 0.1 mm/min and the used specimens are each fifteen. To measure hardness and fracture toughness, these were done a polishing in the order of 15, 9, 6, 3, 1 μm at a diamond paste. Hardness(H_v) is measured using a vickers hardness meter (Mitutoyo, Japan) from indentation which is formed at the load of 10 kgf and the duration time of 10 sec. The fracture toughness(K_{Ic}) is measured with ICL (136° Vickers indenter, Mitutoyo, Japan). At this time, the settings are the load of 10 kgf and the duration time of 10 sec. Hardness and fracture toughness is calculated using the following formulas (1)~(2).

$$H = \frac{F}{2a^2} \qquad (1)$$

$$K_{IC} = 0.032H\sqrt{a}\left(\frac{E}{H}\right)^{(1/2)}\left(\frac{c}{a}\right)^{(-2/3)} \qquad (2)$$

F : load a : length in diagonal of Vickers indent
c : crack length E : Young's Modulus
H : Hardness K_{IC} : fracture toughness

3. RESULTS AND DISCUSSION

3.1 Physical Characteristics

In this study, the jiggering method, one of the forming processes of insulator, has to a granularity distribution at the range of sub μm ~ several μm to mold. So the ball milling of specimens was performed for 6 hours to be a possible forming granularity distribution. The granularity distribution of specimens after ball milling for 25 hours was shown in Fig.1. This granularity distribution was possible to be molded by the jiggering method which is a wet forming method because the grain below 10 μm is about 82 %.

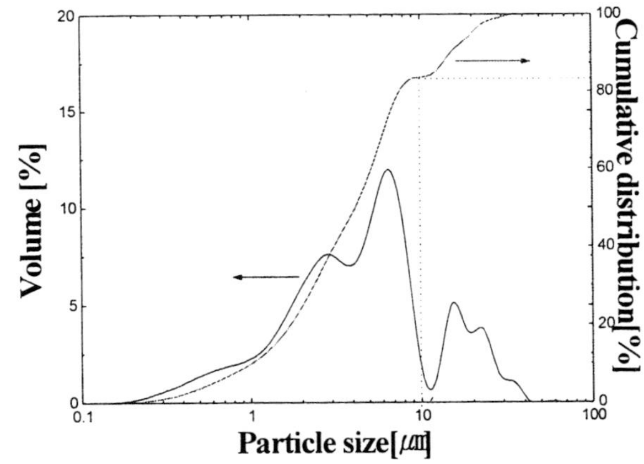

Fig.1. Particle size distribution of specimens after ball milling for 25 h.

The porcelain was fabricated by the jiggering method. To observed change of density as the temperature, the specimen of porcelain was sintered at 1250~1325°C for 60 minutes. A relative density of sintered specimen was calculated by the theoretical density which is measured with gas pycnometer. The porcelain of relative density as sintering temperature was shown Fig. 2. According as a sintered density was increased with a temperature of plasticity, a sintered density was reached about 96 % of theoretical density.

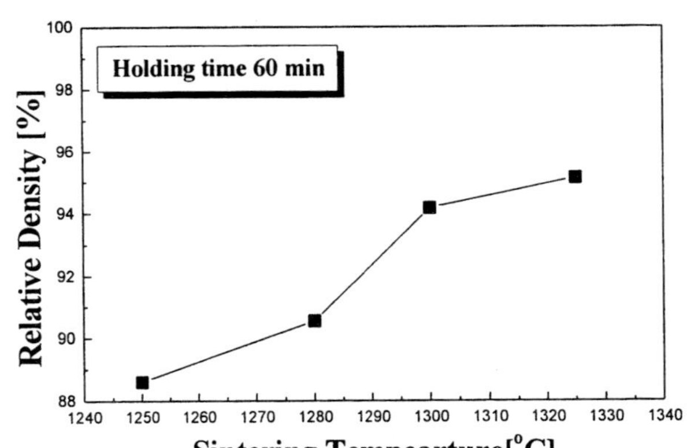

Fig. 2. Change of density according to sintering temperature.

The x-ray diffraction pattern of specimens as heated temperature for one hour was shown Fig. 3. At room temperature, the orientation of crystallization was observed generally quartz, kaolinite albtie, and alumina. On the other hand, the more annealing temperature was increased, the more orientation of crystallization was transformed quartz, albite, and alumina. Duncan (1969) and Chaktaborty (1978) studied phase in thermal state of the kaolinite.

$Al_2O_3{\cdot}2SiO_2{\cdot}2H_2O \rightarrow Al_2O_3{\cdot}2SiO_2+2H_2O$:500 ~ 600°C (3)
(kaolinite) (metakaolinite)

$2(Al_2O_3{\cdot}2SiO_2) \rightarrow Si_3Al_4O_{12} + SiO_2$: 980°C (4)
(metakaolinite) (Si-Al spinel)

$3(Si_3Al_4O_{12}) \rightarrow 2(3Al_2O_3 \cdot 2SiO_2) + 5SiO_2$: 1150°C (5)
(Al-Si spinel) (mullite) (cristoablite)

Metakaolinite was formed for dehydration process from 500°C to 600°C that showed from equation(Eq.3). After exothermic reaction at the 890°C temperature, the spinel of major quantity and the mullite of minor quantity were formed silica of amorphous that included Al-Si(Eq.4). And, mullite and cristobalite were formed for exothermic reaction at 1150°C(Eq.5). The porcelain of sintered at 1300°C for 50 minutes is composed of amorphous of noncrystalloidal, quartz, mullite, and corundum which is a basic crystal phase. These crystal phases improved the electrical and mechanical characteristics of insulator. Especially, a corundum crystal was improved the withstand arc and the withstand impulse of insulator[6].

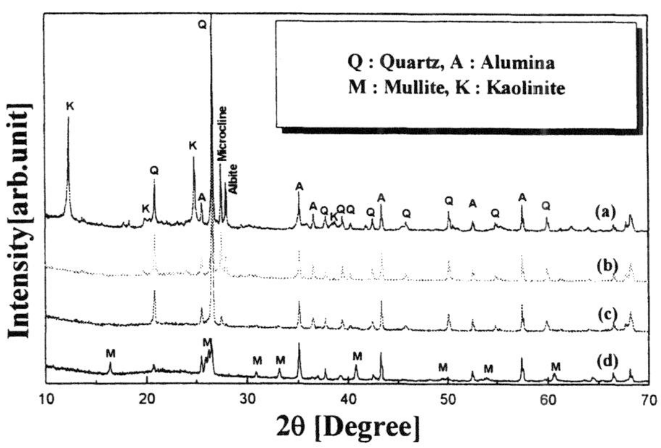

Fig. 3. X-ray diffraction patterns of specimens sintered at various temperature for 50 minutes, (a) room temperature, (b) 750°C, (c) 1000°C, (d) 1300°C.

Fig. 4 shows the distribution of vesicles which were observed by the scanning electron microscope after the surface of specimen sintered at 1300°C for 50 min in the tunnel kiln was grinded in detail. The air porosity of insulator surface was widely distributed(a), moreover, mullite, quartz, and alumina crystalline observed as same the XRD diffraction(b).

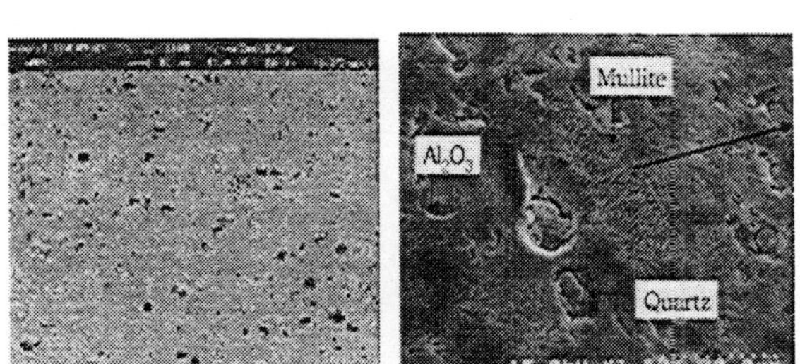

(a) 500 X (b) 50000 X

Fig. 4. SEM photographs of porcelain sintered at 1300°C for 50 min in the tunnel kiln.

Table 1 shows mechanical characteristics of 400 kN suspension insulator. The 3-point bending stress of 400 kN porcelain including alumina of 17 wt% and hardness and fracture toughness were 1658 kgf/cm^2 and 27.5 Gpa and 2.3 MPa·m$^{1/2}$, respectively. This result showed the same excellent value as that of included Al_2O_3 – 40 wt%[6].

Table 1. Electrical characteristics of insulator sintered at 1300°C for 50 minutes.

Mechanical Function of insulator	Value
Absolute Density [g/cm^3]	2.58
Relative Density [%]	97.7
Flexural Strength [kgf/cm^2]	1658
Young's Modulus [GPa]	102
Vickers Hardness [GPa]	27.5
Fracture Toughness [MPa·m$^{1/2}$]	2.3

3.2 Electrical Characteristics

Fig. 5 shows the dependence upon temperature of tanδ as the frequency of specimen, where specimen was made by slurry titrated pH 7.8. The more temperature is increased, the more dielectric constant and loss factor are increased. Moreover dielectric loss is increased by increasing a conductivity loss and ion jump.
DIN and ASTM presented that the dielectric loss was taken stable characteristics up to 100°C in the range of 120×10^{-3}. In the result of this experiment, dielectric loss is excellent in the range of about 17×10^{-3} to 61×10^{-3} at 100°C.

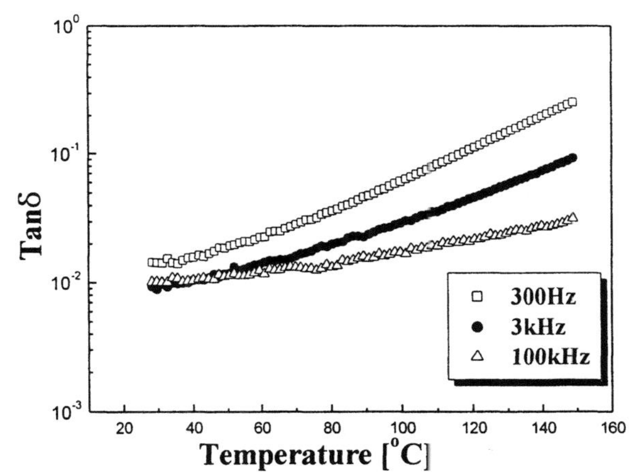

Fig. 5. Dependence upon temperature of Tanδ.

3.3 Electrical performance test of 400kN insulator

In the study, we observed that the form of pin influenced overvoltage brake load in various performance tests of suspension insulator. The overvoltage break load is improved the disperse stress concentrated into the head by changing form of multistep pin. Therefore, in this study, the overvoltage break load test is performed after pin is constructed by changing a diameter of pin up to 40~53 mm and bottom form of multistep pin into straight and round as Fig. 6.
Fig. 7 shows the result of overvoltage break load test performed after we each constructed pin illustrated in Fig. 6 for 400 kN insulator.

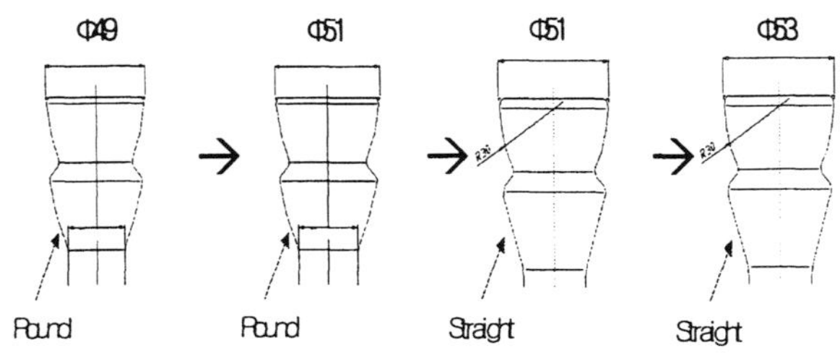

Fig. 6. Form of pin used in a 400kN insulator.

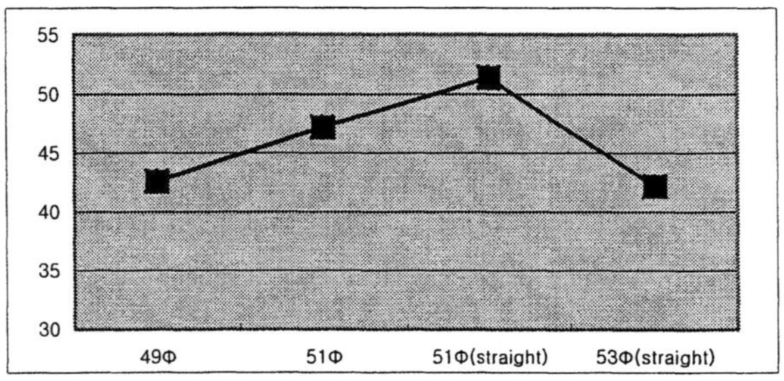

Fig. 7. Overvoltage break load according to form of pin.

When a diameter of pin is 51 mm and bottom form of multistep pin is constructed with straight pin, the overvoltage break load of 400 kN insulator appeared 52 ton of the highest value. So, overvoltage break load was improved about 15% than that of insulator with 51 mm of head diameter and multistep rounded bottom form.

4. CONCLUSION

In this paper, we improved physical characteristics of insulator by adding Al_2O_3 of 17 wt% in feldspar insulator specimen. The sintered density was reached 97 % of theoretical density. The bending strength, the hardness, and the fracture toughness were 1658 kgf/cm^2, 27.5 Gpa, and 2.3 MPa·m$^{1/2}$, respectively. The more temperature was increased, the more dielectric constant and loss factor were increased in the specimen of 400 kN porcelain. The idea suspension insulator showed that the diameter of pin is 51 mm and the bottom form of multistep pin is constructed with straight. The overvoltage break load of 400 kN insulator appeared 52 ton of the highest value and 51 mm pin also was the same excellent value as that of included Al_2O_3 – 40 wt%.

REFERENCE

Carty, W. M. and U. Senapati (1990). Porcelain-raw Materials, Processing, Phase Evolution and Mechanical Behavior, *J. Am. Ceram. Soc.*, **81(1)**, pp.3-20.

Khandelwal, S. K. and R. L. Cook (1970). Effect of Alumina Additions on Crystalline Constituents and Fired Properties of Electrical Porcelains, *Am, Ceram. Soc. Bull.*, **49(5)**, pp. 522-526.

Kim, H. S (1995). *Development of Procelain Suspension Insulator with High Reliability for the Distribution Line*, Kepco.

Duncan, J. F. and K. J. D. Mackenzie (1969). Kinetics and Mechanism of High-temperature Reaction of Kaolinite Minerals, *J. Am. Ceram. Soc.*, **52(2)**, pp, 74-77.

Chakraborty, A. K. and D. K. Ghosh (1978). Reexamination of the Kaolinite-to-mullite Reaction Series, *J. Am. Ceram. Soc.*, **61(3-4)**, pp. 170-173.

ELSEVIER
IFAC
PUBLICATIONS
www.elsevier.com/locate/ifac

ELECTROMAGNETIC RADIATION OF PARTIAL DISCHARGES IN INSULATION OF THE HV DEVICES

N.V.Kinsht, M.A.Katz, N.N.Petrun'ko

Electrophysics And Power Laboratory,
The Institute for Automation and Control Processes,
Far-Eastern Branch, Russian Academy of Sciences,
5 Radio Street, Vladivostok, 690041, Russia, kin@iacp.dvo.ru

Abstract: Properties of own high-frequency electromagnetic radiations (EMR) of the HV devices are analyzed. EMR of the HV device elements are dependent on the size, form and site of inserts in insulation as well as the level of the operating voltage, disruptive voltage of insulation, a tangent of losses, insulation heterogeneous . Also it is theoretically and experimentally shown that spectra EMR have polymodal character and wavy depend on time. *Copyright © 2003 IFAC*

Keywords: electromagnetic fields, insulation, power devices, spectra

1. THE INSULATION PARAMETERS AND SOURCES OF THE ELECTROMAGNETIC RADIATION

The electromagnetic fields of the high-voltage electrical power equipment (HV devices) have the following basic features. The pseudo-steady (primary) component of field of industrial frequency is generated by power stations and is intended for transfer and distribution of the electrical power. Its major property is in exist of the nearest zone around current-carrying equipment components. Its electric and magnetic components fall fast, with rate, at least, as $1/R^3$. The component is not beamed in space.

The second component is generated by special equipment, and provides activity of control systems and communication.

At last, the third component — own high frequency electromagnetic radiations – are generated by the very devices in natural modes of its activity. Usually it is interpreted as integral parasitic component of process with the use of electrical power, which hinders to receive the information radio signals. The second and third components are beamed in the space.

The electric discharges, just as they arise by normal activity HV devices, so at a degradation of insulation and other structural elements are the emitters of the own EMR. The link between own EMR and technical state of HV devices until recent times remained outside of area of the explorers attention. It is obviously important to analyze processes in insulation from the view point of explicitly and to direct efforts on theoretical and experimental researches of the HV devices EMR.

To the greatest degree technical state of HV devices insulation is connected with availability, intensity and distribution of partial discharges (PD). PD are generated in small inserts, which inductivity and electric strength are distinct from the characteristics of the main insulation. In each of elementary inserts there is a PD series , basically determined bound to value of electric intensity of a primary electric field in its neighborhood E_1(x,y,z) and insulation parameters. The elementary discharge current that is connected with PD generates practically an ideal infinitesimal Hertz dipole.

The time of flow of PD current estimates by the order $10^{-8} \dots 10^{-9}$ seconds, and, therefore, EMR frequency band extends down to a centimeter waves (about 10^{10} Hz). The charge current of a Hertz dipole is sluggish piece, its influencing on EMR is left out of account, it does not create radiation, and, it can be neglected in primary current piece in view of a smallness.

Usually designers of the HV equipment try to minimize volume of insulation which influences as much as possible allowable intensity of electric field. Any deformation of HV structural pieces may result in local disturbances of a pseudo-steady field causing, as a rule, to ascending electric intensities. There are the new gas inserts in an ageing process of insulation, and also its electrophysical characteristics are changed. Arising inserts PD accelerate the ageing process of the insulation, and this process can develop as chain reaction. These phenomena, being superimposed against each other, are accompanied by local temperature overheating causing only to increase of intensity PD. Thus the total power of EMR is augmented. The electrophysical processes in power contacts, basically, are similar.

The quality of insulation is as a whole depended on the following parameters: disruptive voltage of insulation, tangent of losses, heterogeneous of dielectric, availability of gas or other inserts. These inserts, in turn, are characterized by quantity, sizes, forms and sites. At least theoretically, if all these parameters in each element of insulation of a device are known, it is possible to calculate quantity, intensity, power characteristics PD, and also parameters of a degradation of a defect. Actually such information misses; also there are no hopes that the full information can be determined. At the same time each of the listed factors finds the reflection in own EMR of equipment components, and the own EMR bear in themselves the valuable information on technical state of equipment components, and this information is transmitted in the space spontaneously and continuously.

Own EMR of the HV devices are reflection of electromagnetic processes in the electrical power equipment. These processes in the basis have some components. Maiden component, leading from them, has frequency of transmitted power of 50 (60) Hz. The second one is relaxing, stating some periodicity during inserts charge and discharge as capacities charge and discharge. The third component is stochastic, bound that process of formation and life of a PD spark can't be precisely described by the determined mode. It is necessary to add also the structurally — mechanical factors. Here it means, that, for example, chattering of transformer windings. The chattering, being transmitted to insulation, also makes a certain modulation of PD currents. Allowing that the most fast–flowing process is the process of spark producing PD, it is possible conditionally to consider it as a carrier process , and remainder processes - as a modulation

2. SPECTRA OF THE PD CURRENTS

The element current of the PD i_0(t) is a function of time, coordinates (x, y, z), size (diameter) of insert, steady-state electric intensity component E_1 and electrophysical characteristics of insert (inductivity ε and disruptive electric intensity E_d). The Fourier transformation of a function i_0(t) is $I_0(j\omega)$. For any defect insert with coordinates (x, y, z) the general spectrum of the PD current will be generated by a sequence of single PD's with amount N(E_1(x,y,z)), on set of instants $\{\tau_k\}$:

$$F(j\omega) = I_0(x, y, z, E_1, j\omega) \sum_{k=1}^{N} \exp(-j\omega\tau_k) \quad (1)$$

If in volume of insulation presents M inserts expression (1) is fair for everyone of inserts (m = 1 ... M this index will appear at the appropriate variables).

There are the following links between PD current (and its spectrum), on the one hand, and electric intensity E_1, with another hand: function M(E_1) monotonously grows, increasing an average current and capacity of EMR; the spectrum of an element current is multiplied for sum of N oscillators, creating the polymodal characteristic. The kind of a spectrum (1) can be attributed to all oscillators, taking place in layer E_1 = const of insulation volume.

An electric component of spectrum can be received in a point of supervision (x_0, y_0, z_0) by summarize of all inserts EMR in view of weight function of complex spatial transfer factor h$(x_0, y_0, z_0, x_m, y_m, z_m, j\omega)$ which reflect boundary conditions.

$$\bar{E}_{EMR}(x_0, y_0, z_0, j\omega) = \\ = \sum_{m=1}^{M} h(x_0, y_0, z_0, x_m, y_m, z_m, j\omega) F_m(j\omega) \quad (2)$$

3. PROPERTIES OF THE OWN PD EMR

EMR, generally speaking, it is possible to log with the help of a radio receiver, which one by a natural image executes some preliminary mathematical processing of a signal. Structurally this radio receiver is represented, for example, by the way of spectrum analyzer.

Let's find out the basic properties of EMR signals as functions of time. In the elementary case it is possible to consider that the PD current as function of time looks like:

$$i(t) = I_0 \exp(-t/\tau_s),$$

Where I_0 – amplitude of PD current and τ_s – a constant of time of the spark burning. The spectrum of a PD current has a complex image

$$I_0(j\omega) = \frac{I_0}{1 + j\omega\tau_s} \quad (3)$$

and its module (fig. 1, the upper line)–

$$I_0(\omega) = \frac{I_0}{\sqrt{1 + (\omega\tau_s)^2}}.$$

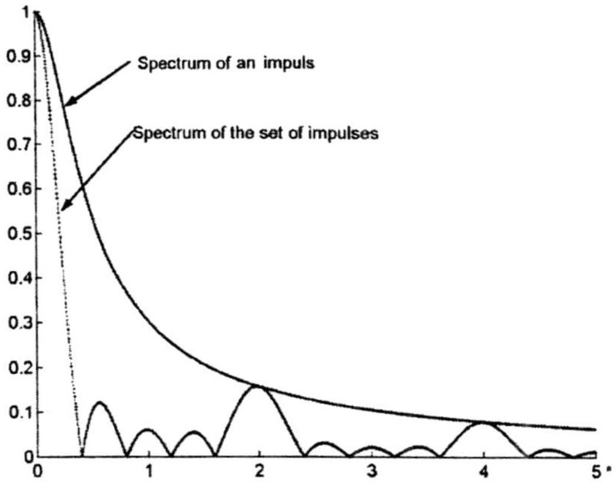

Fig.1. Spectra of single impuls and of the set of impulses

We shall remark that the spectrum of the single PD monotonically damps also width it, captured on some level, depends on process PD time constant. The zone of radiation of a spectrum is in a high-frequency band.

At addition of two pulses PD spectra shifted on an interval τ_d it is resulted a complex spectrum:

$$I_2(j\omega) = I_0 \frac{1 + \exp(-j\omega\tau_d)}{1 + j\omega\tau_s}.$$

This spectrum has already polymodal nature and the arrangement of zero points is determined from a relationship τ_s and τ_d. At repetition several (N) of PD impulses , shifted in time for an interval τ_d, the general spectrum gains a view:

$$I_2(j\omega) = \frac{I_0}{1 + j\omega\tau_s}(1 + \exp(-j\omega\tau_d) + \dots$$

$$\dots + \exp(-j(N-1)\omega\tau_d)) =$$

$$= I_0 \frac{1 - \exp(-jN\omega\tau_d)}{(1 - \exp(-j\omega\tau_d))(1 + j\omega\tau_s)} \quad (4)$$

In a fig. 1 the dotted line shown a spectrum at $N = 5$.

The arrangement and quantity of zero points on some finite interval of frequencies, and also quantity and arrangement of maximums is determined by properties of function:

$$\frac{1 - \exp(-jN\omega\tau_d)}{1 - \exp(-j\omega\tau_d)}, \quad (5)$$

which one conventionally here we shall call "a polymodality factor".

The relation (2) is fair for a series of identical equidistant impulses; if the impulses are distributed in time unevenly, the symmetry of a spectrum, as a rule, is losted. The polymodality remains, but quantity of maximums will not be increased; the spectrum can accept a view, for example, figured on a fig. 2.

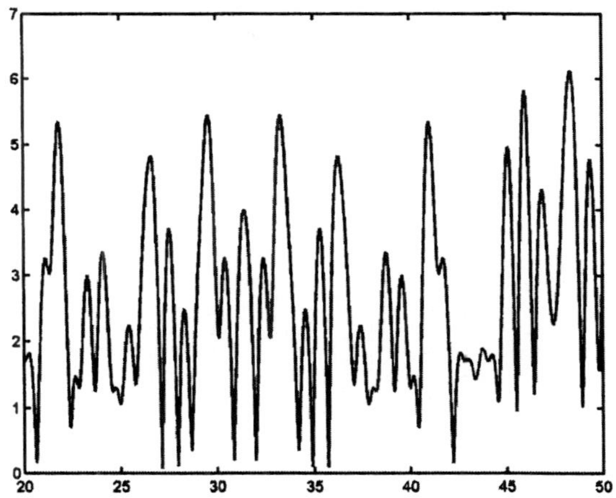

Fig.2. Factor of poly-modality for 10 unequally spaced impulses (on an abscissa axis of number of design points, 1000 points corresponds 2π radians).

It is clear, if the impulses have different amplitudes and time constants, basically conclusion remains former: the spectrum EMR, induced by impulse train PD, has polymodal nature.

At research of actual spectra of signals it is necessary to be considered that the real time of signals integrating is not indefinite large, it is limited. At the same time the period of relaxing process PD can make value about a few or tens seconds. It is clear, the spectrum EMR will have variability (periodicity) of the same order.

At last, it is necessary to add that the amplitude of PD current depends on a level of the operation voltage; the charging interval (and, therefore, relaxing process period) depends also on this level.

The technical state of HV devices (and, therefore, resource of insulation) depend on the random personal factors. There are total number of inserts and for each of insert - his size, geometrical site, electric intensity of a non-perturbed electrical field, ambient it, and inductivity of dielectric, ambient insert.

So, anticipated properties PD EMR spectra are: polymodality, capability of wave-form time-dependence with the period on the order of a few or tens seconds, damping of the spectra spotted by time constants of PD combustion and relation to an operation voltage. In remaining its property are determined by a personal picture of its availability index of insulation.

4. EXPERIMENTAL RESULTS

Experimental researches of the EMR of the HV devices in real industrial conditions under a working voltage were made by laboratory researchers on the Far-Eastern substations 500 kV of the UES of Russia. The basic properties of EMR signals as functions of time were found out.

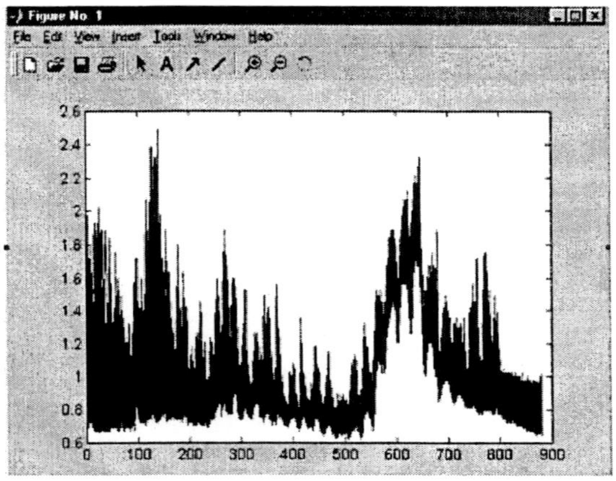

Fig.3. Imposing of diagrams of a spectrum of the high-voltage transformer (experiment within 1 minutes).

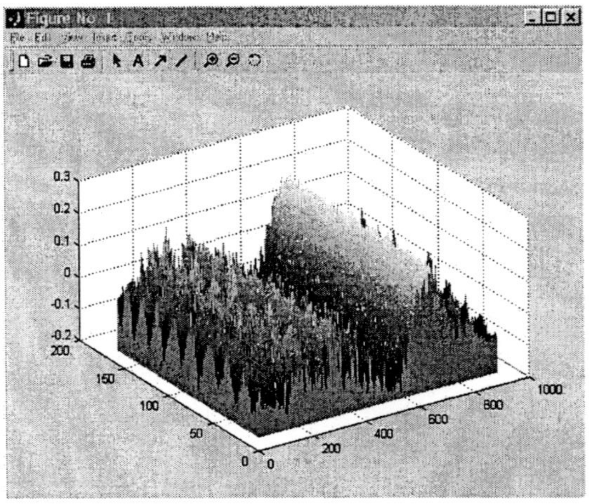

Fig.4. The 3-d spectrum of the transformer 500 kV).

The first experiments have allowed to reveal some obviously expressed characteristic properties of spectra EMR of the HV equipment. The first visually found out property is a certain periodicity of a spectrum with the period of several seconds. This feature was observed practically in all experiments, irrespective of type of an element of the equipment. The second clearly expressed feature of a spectrum is its polymodality. So if to impose all frames of spectra received on experiment during approximately 1 min., it is possible to receive the image of the kind represented on fig. 3.

Fig. 4 demonstrates the three-dimensional image of dependence of a spectrum of the power(force) transformer 500 kV from time. The numbers of points corresponding to frequencies from 40 up to 170 MHz are plotted on a horizontal axis to the right. The numbers used at construction of figure of the frame are plotted on an axis to the left. 150 frames correspond to experiment within 1 minute. Intensity of a spectrum is represented in relative logarithmic units.

For creation of database of the HV devices EMR which focused on the comparative analysis EMR and the technical state of the equipment elements it is necessary to develop classification of the attributes of EMR spectra. Basically it is an uneasy problem because of practically observed spectra EMR have complex character. Generally speaking the EMR are casual processes with the unexplored characteristics. There are preconditions to study the period infra-low fluctuations and property of a polymodality of the EMR spectrum as potentially diagnostic parameters.

REFERENCES

Киншт Н.В., Кац М.А. (1999). Диагностика точечных источников электромагнитных шумов. *Электричество*, **№ 4** с.40-42

Kinsht, N.V., M.A. Katz , N.V. Silin , O.V. Preobragenskaya and Y.B. Petropavlovsky (2000) Application of Electromagnetic Radiation of High-Voltage Equipment for Diagnostics of its Technical State.*Int.Symp. on EMC, Wroclaw, June 2000, pp.37-38*

Copyright © IFAC Power Plants and Power Systems Control,
Seoul, Korea, 2003

ELSEVIER
IFAC
PUBLICATIONS
www.elsevier.com/locate/ifac

Comparison of Electrical-Environment Characteristics between Two Different Types of EHV Transmission Lines

Tae-In Jang, Seong-Doo Lee, Koo-Yong Shin, Dong-Il Lee

Korea Electric Power Research Institute, Daejeon, 305-380, Korea
E-mail : angeljti@kepri.re.kr, joyful@kepri.re.kr, kyshin@kepri.re.kr, dilee@kepri.re.kr

Abstract This paper represents the analysis results by comparison of characteristics on electrical environment between different EHV transmission lines, such as AN (Audible Noise), RI (Radio Interference), TVI (Television Interference), EMF (Electric and Magnetic Field). In Korea, the transmission lines with nominal voltage of 765-kV have been operating since 2002, and Hungary energized a 750-kV line in 1978. However both EHV transmission lines have different line configuration, Korea is a prime example of the first double circuit 765-kV line in the world, and Hungary has been operating one of the single circuit 750-kV.

Keywords : Audible Noise, Radio Interference, Television Interference, EMF, Transmission Line, and Extra High Voltage

I. Introduction

Electricity utilities involved in development, maintenance and operation of transmission lines face increasingly severe challenges as results of heightened environmental awareness, more comprehensive legislation and standards, and growing public expectations and concerns. These trends will have a major influence on the approach of companies to the management of both new projects and existing transmission assets. Both countries, Korea and Hungary studied the electrical environmental effects of both EHV transmission lines using the design parameters, real measured data through exchanging visits sponsored by government.

Due to increase of peak demand by more than 10 % per annum, Korea have constructed the first project 765-kV transmission lines of total 340 km, and is preparing the second 765-kV project. In order to meet the need for economical bulk power transmission over the limited right-of-way with the least environmental impact to the people who live in the vicinity of these lines, the 765-kV transmission lines of Korea adopted double-circuit vertical-line type with low reactance arrangement. KEPRI designed optimally 765-kV double circuit in the viewpoint of electrical environment covering audible noise, radio interference, television interference, magnetic field and electric field using a corona cage and a full scale 765-kV test line.

However in case of Hungary, a single horizontal circuit of 750-kV transmission lines was adopted in 1978 due to the electric demand and different social circumstances. Several countries have been using this type of single circuit transmission lines in worldwide. In this paper, as the beginning stage of construction of the first double circuit 765-kV transmission line, two different types of EHV transmission lines was analyzed by comparison of characteristics of electrical-environment items such as AN, RI, TVI and EMF

II. 765-kV DOUBLE CIRCUIT VERTICAL TRANSMISSION LINE OF KOREA

A. Design of the transmission line

From insulation coordination analysis, mechanical load study, environmental study etc., KEPCO (Korea Electric Power Corporation) made a design guide of 765-kV double circuit transmission line as Table 1.

Table 1. Design Criteria of 765-kV transmission line

Items		Criteria	Condition
Voltage		765/800 kV	Nominal/Max.
Tower Height		95 m	Average
Insulation Clearance		19 m	Phase to Phase
Insulator		30 x 2 strings/ 28 x 3strings	Suspension(300 kN)/Tension (400 kN)
Conductor		Cardinal x 6B	Dia. 3.04 cm
ENVIRON	AN	Below 50 dBA	Residential area
	RI	Above SNR 24 dB	All area
	TVI	Above SNR 40 dB	All area
	EF	Below 3.5/7.0 kV/m	Residential/ Other area
	MF	1,000 mG	
	R.O.W	3 m	From the direct under conductor

In order to make the guideline of 765-kV double circuit vertical transmission line as Table 1., KEPRI had been studying from 1984 to 1996 using a corona cage and a full scale 800-kV class test line as shown in Fig. 1.

B. Facilities and instrument system of 765-kV test line

Maximum test voltage of test line which has total 700 m length with 3 spans of 200 m, 300 m and 200 m, is 850-kV line-to-line. The voltage to the test line is supplied by 3 single-phase auto-transformers. Each auto-transformer is rated 3 MVA, with a BIL of 1425 kV.

(a) 800-kV class test line

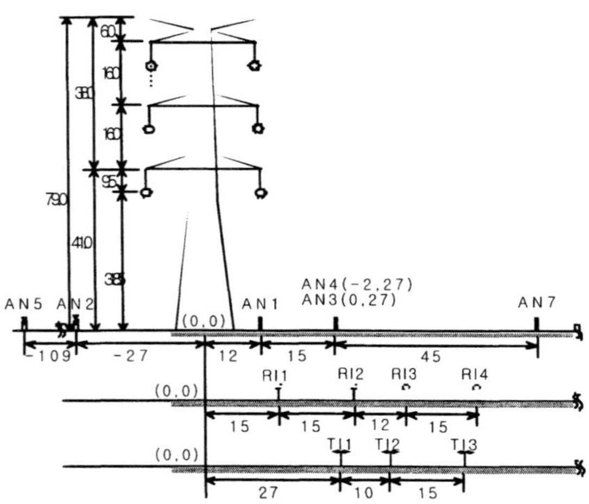

(b) Sensor arrangement around 800-kV test line

Fig. 1. Kochang test line of KEPCO

The instrument system consists of Data Acquisition Equipments and AN, RI, and TVI sensors which are installed along a lateral profile at mid-span and were connected to their related instruments of the measuring house by long coaxial cables. To measure AN, Bruel & Kjaer type 4184, 1/2" weatherproof microphone units were used [1]. RI data have been collected with the test receiver's Rohde & Schwarz type ESHS 30 using Quasi Peak detectors. Active loop antennas, R/S type HFH2-Z2 are installed at 2 m above ground. The frequency set for long-term measurement of RI is 0.475 MHz.

To measure television interference, biconical antennas, R/S type HK116 are installed at 3 m above ground. The test receivers using QP detectors, R/S type ESVS30 are tuned to 75 MHz according to CISPR standard [2].

C. Results of Electrical-Environment Study using the 765-kV Full Scale

In this paper, we evaluated the environmental database obtained from test line from Oct.' 95 to Oct.' 99 (49months) with KEPCO's 765-kV transmission line design criteria as Table 1. The summarized test results of Table 2. satisfy the design criteria of Table 1. Fig. 2. shows the cumulative distribution of audible noise in rainy weather at 15 m from outer conductor. The curves have similar gradient for nominal voltage 765-kV and maximum voltage 800-kV. The cumulative curve of RI of Fig. 3 shows the characteristics of lateral profile.

Table 2. Summarized results of 765-kV full scale test

Items	Sensor Name	All Weather		
		Rain		Fair
		L5%	L50%	L50%
Audible Noise	AN3 [dBA]	52.6	47.3	42.8
Radio Interference	RI2 SNR [dB]	SNR 0.4	SNR 8.1	SNR 24.3 (1)
TV Interference	TVI SNR [dB]	SNR 37.5	SNR 40.0 (2)	SNR 42.8
Electric Field	Portable HI-3604	5.8 kV/m (3)		
Magnetic Field	Portable HI-3604	42.6 mG (4)		

(1) Evaluation Datum : SNR=71(Signal)-46.7(Noise) =24.3
(2) Evaluation Datum : SNR=54(Signal)-14.0(Noise) =40.0
(3) Maximum Voltage : 800kV case
(4) Current : 1000 A case

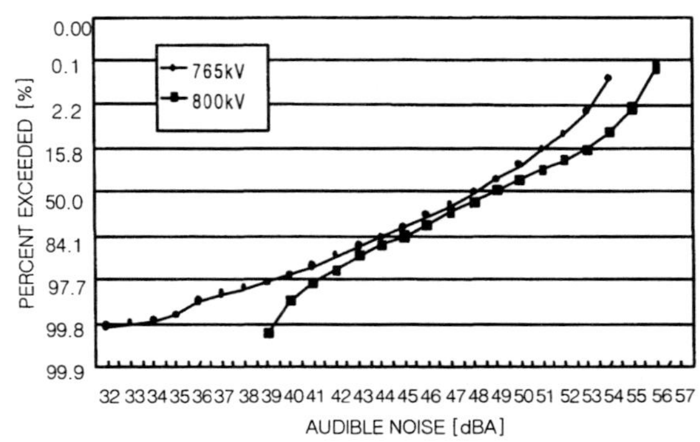

Fig. 2. Cumulative distribution of audible nois e

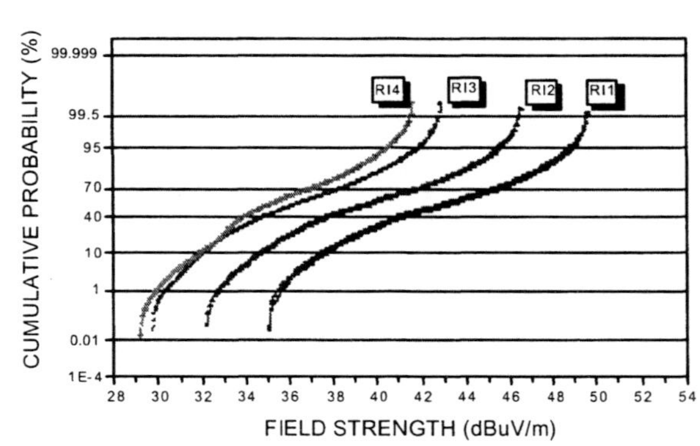

Fig. 3. Cumulative probability of radio interference at 475 kHz

TVI cumulative frequency distribution at 75 MHz in foul weather of T11 sensor is as shown in Fig. 4. The electric field along the lateral line was measured and compared with the calculated as shown in Fig.5. The magnetic field was calculated with 1,000 A current condition. The peak value of magnetic field shows 42.6 mG at 1m height above the ground on the center of test line as shown in Fig.6 [3].

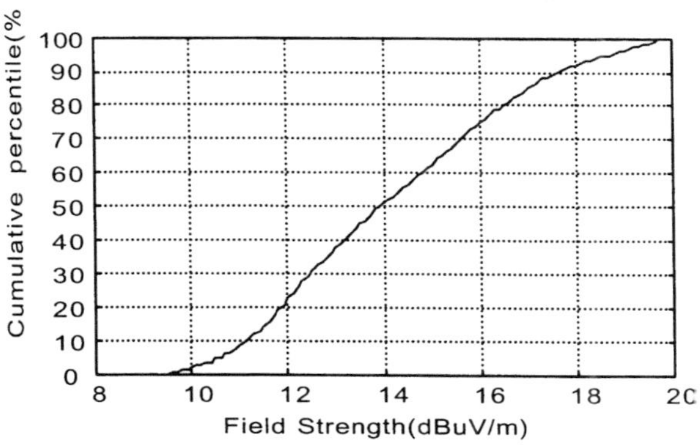

Fig. 4. Cumulative distribution of TV interference at 75 MHz in foul weather

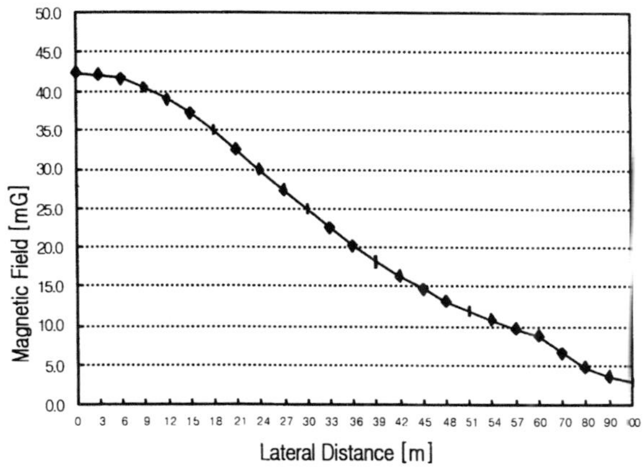

Fig. 5. Lateral profile of electric field strength

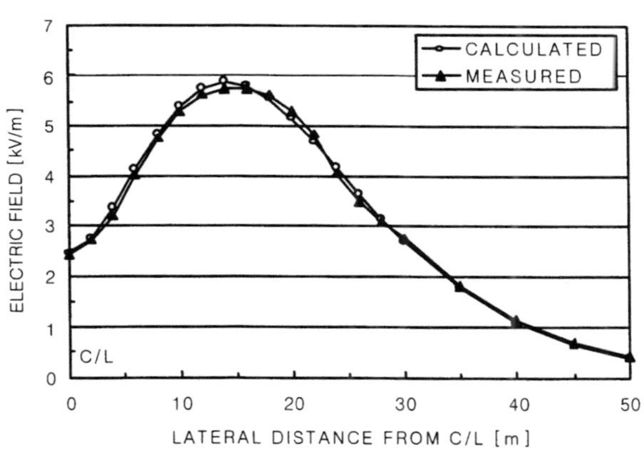

Fig. 6. Lateral profile of magnetic field strength

D. Survey on TV Ghost Interference Caused by 765kV Transmission line

The survey on TV ghost interference before and after constructing 765kV T/L has been performed at 123 points from December 1997 to February 2000 [4]. Main equipments for surveying TV ghost interference are a TV ghost analyzer to measure PDUR and an electric field strength meter to measure the received electric field strength of TV waves. There also are TV monitor to evaluate picture quality included in TV ghost and a portable PC which a DAS (data acquisition system) for automatic measurement is mounted on. Fig. 7 illustrates the configuration diagram of this instrumentation system.

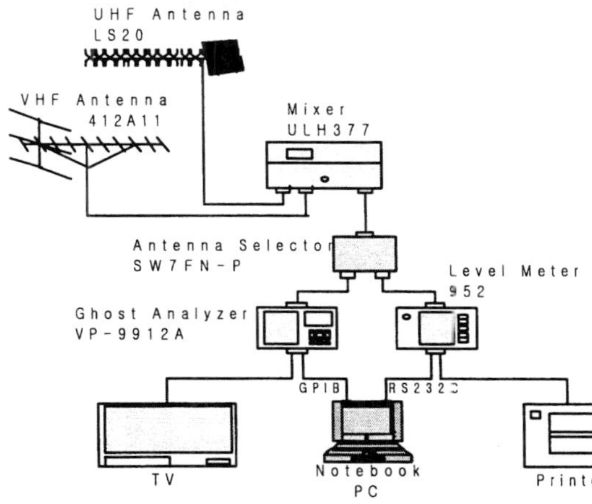

Fig. 7. Configuration Diagram of Instrumentation System for Surveying TV Ghost Interference

The above equipments are installed in a vehicle as shown in Fig. 8. the DAS program developed appropriately for Korean broadcasting is used for surveying TV ghost. Consequently, measured database can be constructed.

Fig. 8. Vehicle for TV Ghost Interference Measurement

Table 3. Survey Results on TV Ghost Interference According to TV Frequencies

Items	Measuring Points	Lateral Profile, m 100	150	200	299	300
CH.7	PDUR, dB	17.6	17.5	24.4	21.8	23.0
	Electric Field Strength, dBμV/m	50.3	50 6	53.0	53.7	54.5
CH.11	PDUR, dB	25.8	13.5	17.6	20.0	21.1
	Electric Field Strength, dBμV/m	53.2	44.4	45.9	52.4	52.8
CH.25	PDUR, dB	19.9	20.8	25.7	-	-
	Electric Field Strength, dBμV/m	51.9	50.5	52.7	-	-
CH.27	PDUR, dB	25.5	20.7	25.0	-	-
	Electric Field Strength, dBμV/m	52.7	49.5	50.4	-	-

Table 3 shows the results which TV reception qualities have been measured around 765kV T/L. According to lateral profiles, TV ghost interference ranges approximately 200 m to 300 m from the outermost phase of the line in VHF channel 7 and 11. However it does not appear in UHF channels even though approaching 100 m near the line. Therefore it seems evident that TV ghost interference occurs little in UHF channels since its range by T/L is very narrow in TV waves transmitted from the same direction. However the ghost interference can be easily generated in VHF channels.

The result shown in Table 4 is obtained around 765kV T/L that both the electric wave directed vertical incidence to the line and the wave of an oblique incidence can be measured to survey TV ghost interference according to the incidence angle to T/L.

Table 4. Survey Results on TV Ghost Interference According to Incidence Angles to the Line

Channel		CH. 8	CH. 22	CH. 10	CH. 24	CH. 28
		MBC	SBS	KBS1	KBS2	SBS
PDUR, dB	1	21.0	32.2	18.9	15.2	15.6
	2	20.4	26.8	16.4	13.9	16.6
	3	19.1	27.8	20.3	12.6	16.7
Electric Field Strength, dBμV/m		48.8	43.5	51.3	41.1	42.1
Incidence Angle		0°	0°	40°	40°	40°
Transmitter		A	A	B	B	B

TV field strength in UHF channels is measured almost similarly at the measuring points. However inferior TV pictures are observed in the broadcasting channels from transmitter B. From this fact, it is shown that the incidence angle of wave reached to the line from transmitter B is approximately 40° and TV ghost interference caused by the reflected wave is lager in channels of transmitter B than that of transmitter A, as the waves from transmitter A are received almost in a vertical incidence. Therefore the levels of PDUR by transmitter B are lower , and also TV pictures are not good. By these facts, it may be verified that the range of TV ghost interference becomes wider as the incidence angle to the line gets bigger. And the ghost interference is generated little for TV wave of a vertical incidence.

III. 750-kV SINGLE CIRCUIT HORIZONTAL TRANSMISSION LINE OF HUNGARY

A. Design of the transmission line

In the 1970s, transmission lines and substations of large capacity were needed to be built to extend the cooperation of the interconnected electrical power systems of the East European countries. Therefore according to the requirement and to the connecting line parameters of that time, the single circuit horizontal line was built.

Table 5. Design Criteria of 750-kV transmission line

Items		Criteria	Condition
Voltage		750/787 kV	Nominal/Max.
Tower Height		42 m	Average
Insulation Clearance		17.5 m	Phase to Phase
Insulator		44 x 2 strings 35 x 3 strings	Suspension(120 kN) / Tension (220 kN)
Conductor		ACSR 500/65x4B	Dia. 3.105 cm
ENVIRON	RI	50 dB	All area
	EF	15 kV/m/ 10.5 kV/m	Residential/ Road crossing
	R.O.W	40 m	From the direct under conductor

At that time the radio interference and the electric field were taken into account as environmental parameters. In the course of the operation, all the environmental characteristics above were checked under the existing line.

B. Measuring apparatuses of the 750 - kV transmission line

The radiated electromagnetic waves were detected in the frequency range between 0.2 - 30 MHz by a ferrite aerial and by a selective micro-voltmeter. For TVI measurements the meter used met the prescription of CISPR and had a bandwidth of 120 kHz. A real-time analyzer and an MV 102 type microphone were used to the determination of the acoustic noise.

C. Results obtained on actual 750 - kV transmission line

In order to make a statistical evaluation of the radio interference caused by actual line, a number of measurements were performed. Fig. 9. demonstrates general results of the 750-kV line 20 m from the outer phase in different spans. In every case the maximal gradient is given, which was calculated taking into account the line parameters and the actual voltage.

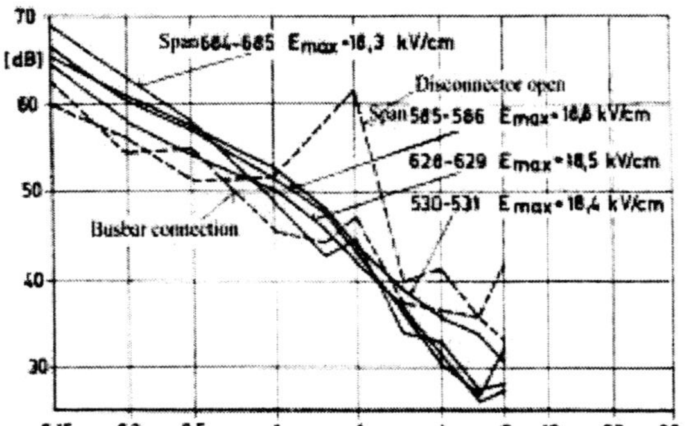

Fig. 9. Radio interference of 750-kV transmission line in function of the frequency

The interference on the line is the most intensive below 1 MHz. The disturbances observed in substation vary randomly in a relative wide range.

Frequency spectra of TVI at four distances from 2C to 100 m from the transmission line were obtained at one mid-span of the 750-kV line caused by a micro-gap, see Fig. 10.

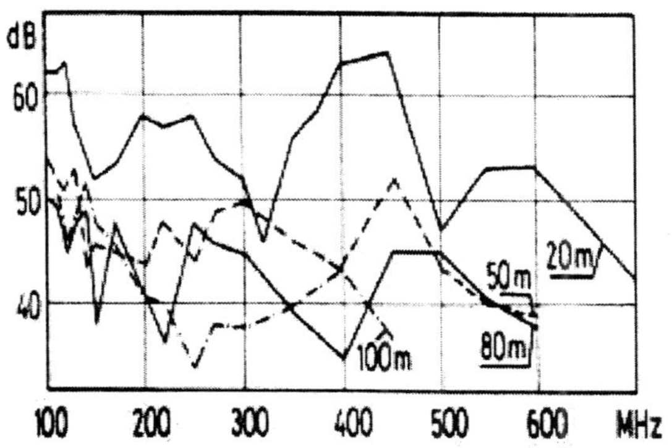

Fig. 10. Frequency spectra of TVI generated by gap discharges in the midspan of the 750-kV transmission line

Actual line measurement can practically be performed for audible noise at EHV level in fair weather. The corona generating audible noise as shown in Fig. 11. was recorded by micropho:e placed in a height of 1.8 m above ground.

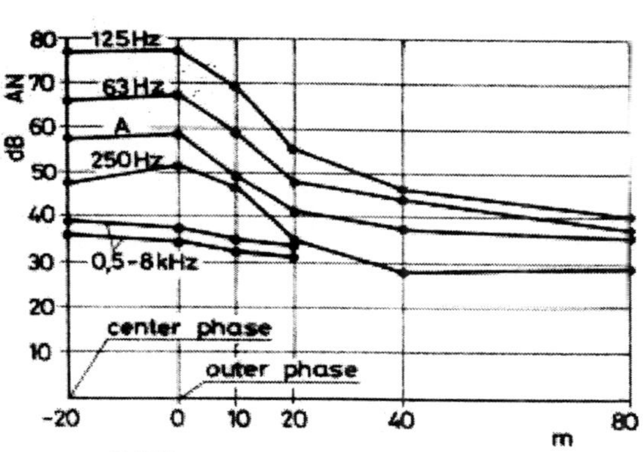

Fig. 11. Audible noise octave-band and A-weighted levels in :he vicinity of a 750-kV transmission line

From Fig. 11. it can be seen the most noticeable AN tone is the hum. Neither the other harmonic components nor the A-weighted noise play an important role. From the viewpoint of audible noise, the line can be characterized by the A-weighted noise level measured at the edge of R.O.W. In this case the AN level of the 750-kV line is equal to 37.5 dB(A) in fair weather.

Electric and magnetic field problems raised from the transmission lines and substations have been studied in Hungary practically from the erection of the 750-kV line in the last twenty years. At the beginning the activity was focused to development of measuring device for electric field then later for magnetic field. Fig. 12. gives the geometric configuration of the 750-kV lines at the place where measurements were performed.

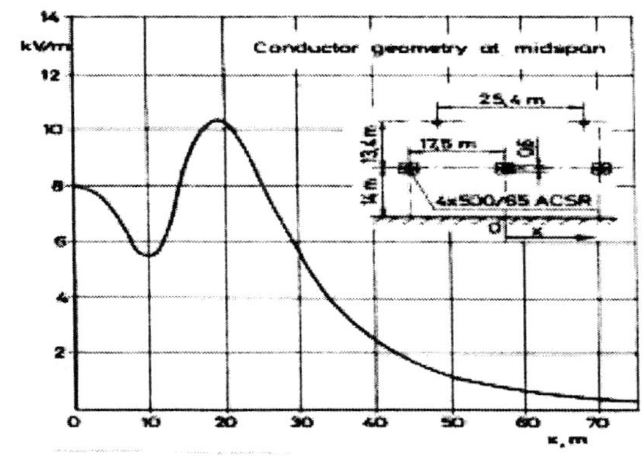

Fig. 12. Electric field strength at the level cf 1.8 m above the ground for the 750-kV transmission line

The magnetic induction as a function of distance measured from the axis of the central conductor projected to ground has been measured at points normal to the axis of the line at a height of 1.5 m above ground at the maximal sag. Change of the magnetic induction in the function of distance is shown in Fig. 13. Then the line operated with about a half capacity.

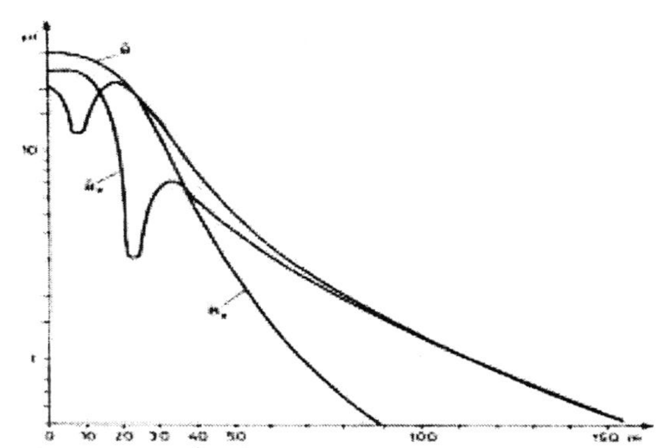

Fig. 13. 50 Hz magnetic induction in the function of the distance in the vicinity of the 750-kV line.

Almost the same value could be founded in dwelling houses in the vicinity of transmission line, i.e. there was practically no difference between the values of magnetic induction inside and outside of the building.

With respect to the personal carrying out live line work, the following effect developed in the body on 750-kV line in case of a current of 1,000 A: 25.1 μT working on vertical structure as well as 45 μT on horizontal structure.

Considering the mitigation of electric field measures is taken in the 750-kV substation where the corridors on the field are screened by grounded wires. The live line work is performed in electrostatic clothes. The agricultural machines have a grounding chain against charging the large metal body.

IV. Conclusions

In this paper, comparison on electrical-environment characteristics between two EHV transmission lines with significantly different design parameters, design philosophy, and technology was made and surveyed. Each experimental result obtained here from the EHV test lines could attribute to real commercial line constructions.

In this comparison results, the difference of conductor arrangements will make it somewhat possible to estimate electromagnetic effects in other future projects. Of course, the long-term monitoring is more excellent method and the cumulative results of each parameter would give more details on electrical-environment characteristics.

As a guideline of permissible values for general regulation about the environmental aspect of EHV lines, IRPA prescription may be used in practice. However, further comprehensive research and analysis should be conducted to establish appropriate permissible values.

V. References

[1] Jeong-Boo Kim , Dong-Il Lee, Koo-Yong Shin, "Electromagnetic Interference from a Three Phase Double Circuit 765-kV Test Line", PE-236-PWRD-0-04-1998, *IEEE Transactions on Power Delivery, Vol. 14, No. 1, January* etc., *1999, pp 266-271*

[2] Dong-Il Lee, Jeong-Boo Kim, Koo-Yong Shin, etc., "Audible noise performance of 6-Rail conductors on a 765-kV double circuit test line", 96 SM 461-4 PWRD, *IEEE Transactions on Power Delivery, Vol. 12, No. 3, July 1997, pp 1343-1351*

[3] Jeong-Boo Kim, Dong-Il Lee, Koo-Yong Shin, etc., "Environmental Effect Analysis of a 765-kV Double Circuit Transmission Line". 10th ISH, Montreal, Canada, Aug. 1997. pp 157-160

[4] Mun-No Ju, Kwang-Ho Yang, Dong-Il Lee, Koo-Yong Shin, Sung-Du Lee, "Survey on TV Ghost Interference Caused by 765kV Transmission Line", ICEE2K Proceedings, July 24-28, 2000, pp 858-861

[5] Jermendy, L., Propagation of Electromagnetic Interferences Caused by HV Lines, *Euro Electromagnetics*, Bordeaux, May 30 - June 4, 1994. Paper MOp-06-05, pp. 1-8.

[6] Jermendy, L., Fogarasi, I., Model Investigation of the Effect of Corona, *ISEMC 94.*, Sao Paulo, Dec. 4 - 9, 1994. Pp. 1-6.

ELSEVIER IFAC PUBLICATIONS
www.elsevier.com/locate/ifac

IMPLEMENTATION OF FUZZY LOGIC CONTROLLER FOR HVDC USING GENETIC ALGORITHM

Jong-Bo Ahn* Gi-Hyun Hwang J. H. Park****

* Korea Electrotechnology Research Institute, Changwon 641-120, Korea
** Dept. of Electrical Engineering, Pusan National University, Pusan 609-735, Korea

Abstract: This paper presents an optimal tuning method for Fuzzy Logic Controller (FLC) of current controller for HVDC using Genetic Algorithm (GA). GA is probabilistic search method based on genetics and evolution theory. The scaling factors of FLC are tuned by using real-time GA. The proposed tuning method is applied to the scaled-down HVDC simulator at Korea Electrotechnology Research Institute(KERI). Experimental result shows that disturbances are well-damped and the dynamic performances of FLC have the better responses than those of PI controller for small and large disturbances such as ULTC tap change, reference DC current change and DC ground fault. *Copyright © 2003 IFAC*

Keywords: Fuzzy Logic Controller, HVDC, Genetic Algorithm

1. INTRODUCTION

High Voltage Direct Current (HVDC) converts AC into direct current (DC) in a rectifier and transmits DC power through the transmission line, and then reconverts DC into AC power in a inverter and supplies the power. Since voltage, current, and transmission power in the DC transmission can be controlled rapidly relative to the AC transmission, it is robust against a disturbance and increases a dynamic characteristic of AC power system, while decreases a short-circuit capacity.

Power system controllers based on the modern control theory such as optimal and modal controller [1-2] show better performances to the dynamic disturbances than conventional PID controllers. However, the practical application of the modern control theory in real HVDC system is difficult because it is a highly nonlinear system dependent on the ignition angle of converter and AC system voltage.

The conventional controllers [1-4] used to improve a dynamic characteristic are a lead-lag controller and a Proportional-Integral (PI) controller. They are designed by the root locus and Bode plot method, using the linearized model at a specific operating point. As they are designed using the linearized model at a specific point and linear control theory, they can show a good performance at a specific point, however, they are not to be expected to obtain a good performance, in cases where operating conditions of power system, such as load variation and system default, change. Furthermore, methods [5-7] are proposed. They have an adaptive capability against time-variation and nonlinear of power system using an adaptive control theory. They increase the dynamic characteristic of power system, but cannot control in real-time. Recently, intelligent controls [8], such as fuzzy logic controller (FLC) and neural network controller as HVDC system controller, are proposed, such that it improves the dynamic characteristic of HVDC system[9,10].

The HVDC systems have the advantages of fast power controls and power modulations[11,12]. Therefore, the control characteristics of the HVDC system are very important for the stabilization of the HVDC system. In this paper, we propose an FLC current controller for HVDC using genetic algorithms(GA). To evaluate the effectiveness of the proposed FLC, HVDC power system simulator is utilized in Korea Electrotechnology Research Institute(KERI). As a result of the experiment, the FLC shows the better performance than PI controller in regard to the disturbance used in designing the FLC. Also, it shows better robustness than PI controller in regard to disturbances, which is not used in designing the FLC for evaluating the robustness of the proposed FLC.
Text of paper, 76 mm (3in) column width, with 8 mm (.3in) space between. Use full 253 mm (10 in) column length. Paragraphs should be justified, using single spacing, with no paragraph indentation. Use Times Roman font, 10 point. Leave one clear line between paragraphs within a section; two clear lines before a main or secondary heading.

2. SYSTEM MODEL

2.1 HVDC system

Fig. 1 shows the configuration of HVDC simulator at KERI which is used to evaluate the validity and to design FLC using GA. This HVDC simulator is a scaled-down model of 300MW of actual HVDC system that links the mainland Haenam and island Jeju. The rated capacity is 5[kW] on the output side of DC reactor, where its voltage is ±250[V] and current is 10[A]. The HVDC system is configured in bi-polar mode by two conductors, where the conductors have opposite poles to each other in order to control the two poles independently.

The advantage of bi-polar configuration is that it has a capability of operating in mono-polar mode when one of the two poles is damaged. Thyristor valve has 2 bridges per pole and 6 valves per bridge, which outcomes total 48 valves and each will be replaced when it becomes damaged. The model of the DC line can simulate the actual line of up to 500[km] and is composed of two equivalent variable resistors. Total four of step-down transformers has the capability of two-way DC transmission at 0～150[%] load condition while maintaining the variation range of AC voltage within ±10[%]. Furthermore, for having easier control of two-way power flow, the transformers secondary voltage is calculated according to the various operating condition and then ULTC is utilized, in order that it maintain the maximum voltage in transformer and minimum voltage in inverter when the transformer is 150[%] overloaded. The voltage deviation between the taps in the transformer is 1.5[%] each, and total 24 taps are installed. The range of tap voltage falls from -18.0[%] to +16.5[%]. The compensation of reactive power is shared by the high pass/band pass filters, capacitor banks and the ac power system. For the purpose of having the compensated capacity of each 11/13th harmonic filter and each high-pass filter to meet the required capacity at the minimum load, the transformer is designed, where unit capacity of the filters is set to 0.6[kVar] and the overall capacity of reactive power compensation becomes 4.8[kVar] per each terminal. Consequently, it is set so the reactive power can be supplied from the AC power only during the excessive occurrence of AC under-voltage and of firing angle. Two capacitors, on the other hand, are installed in 380[V] ac bus, where the AC voltage variation is considered as 10[%].

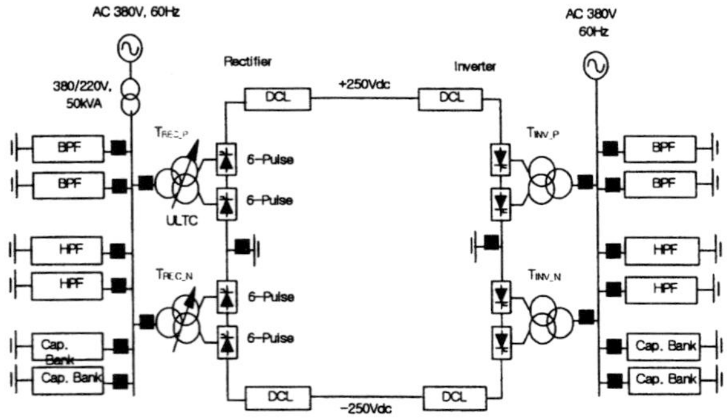

Fig. 1 Configuration of HVDC simulator in KERI

3. THE DESIGN OF FLC USING GA

Fig. 2 shows the overall configuration of the on-line GA-based FLC tuning system with HVDC simulator. As shown in the figure, the system consists of three parts; master controller, current controller and HVDC simulator. Master controller (personal computer) performs generating reference signal, gathering data from controller and computing GA. The DSP-based current controller receives reference signal from master controller, performs FLC current control and finally generates the firing signal of HVDC simulator. The controller uses analog-to-digital converter(ADC) to feedback the dc current. HVDC simulator that received the firing angle signal actually triggers thyristors and flows DC current. Data communication between master and current controllers is via ethernet/TCP network.

After HVDC starts up and builds the dc current at low current level stably, GA tuning procedure starts. Master controller sends step reference change at every 2[sec] with the evaluated FLC parameters. The current controller calculates the error at every sampling interval and sends back to master controller. Master controller evaluates fitness using fitness function and updates parameters through the tuning procedure that will describe hereafter.

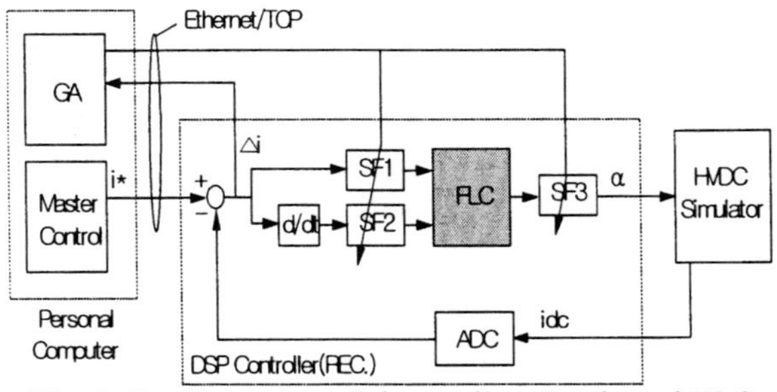

Fig. 2 Configuration of the on-line GA-based FLC tuning system

In this paper, we use GA-based FLC on-line tuning for current controller of HVDC system. The parameters to be tuned are the input and output scaling factors. Fig. 3 shows a triangular membership function used, where we set the center of ZE to 0 and positive and negative membership functions are constructed to be symmetrical of 0. Here, linguistic variable such as NB denotes Negative Big, NM denotes Negative Medium, NS denotes Negative Small and so on. Fig. 4 shows the configuration of population for tuning the scaling factors of FLC.

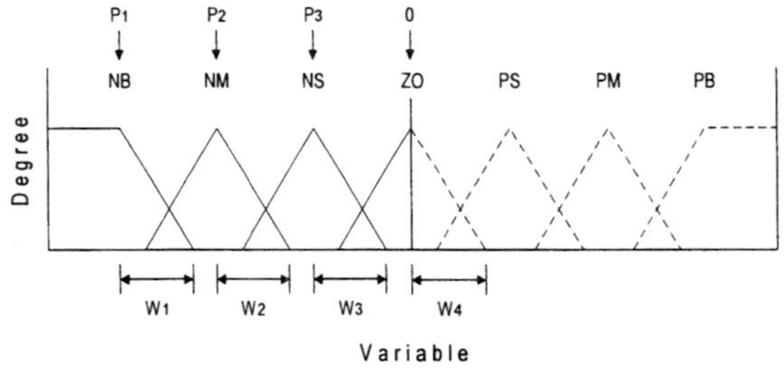

Fig. 3 Symmetrical membership function

S1	SF11	SF12	SF13
S2	SF21	SF22	SF23
		.	
		.	
Sn	SFn1	SFn2	SFn3

where, n : population size

SFij : Scaling factors

Fig. 4 Strings architecture for tuning scaling factors

The fitness function used to evaluate the each string of population is a sum of the absolute error of direct current as shown in (1). The error data is obtained by digital controller for 2[sec] from the time to change the reference by sampling frequency of 1.8[kHz]. 10[%] step reference is applied to controller at the every generation.

$$fitness = \frac{a}{b + \int_{t=1}^{T} | I_{ref}(t) - I_{d}(t) |} \qquad (1)$$

where , $I_{dc}(t)$: measured direct current

$I_{ref}(t)$: reference direct current

T : specified time that data is acquired (2[sec])

a, b : constants

4. EXPERIMENTAL RESULTS

To evaluate the effectiveness of the proposed FLC, HVDC power system simulator is utilized in the KERI. The overview of HVDC simulator and controllers is shown in Fig. 5. Master controller and rectifier current controller is shown at left in the figure. Fig. 6 shows the detail configuration of current controller. In the figure, DIO, TC, SC, ADC, GD represent digital input/output, timer counter, signal conditioner, analog-to-digital converter and gate driver respectively.

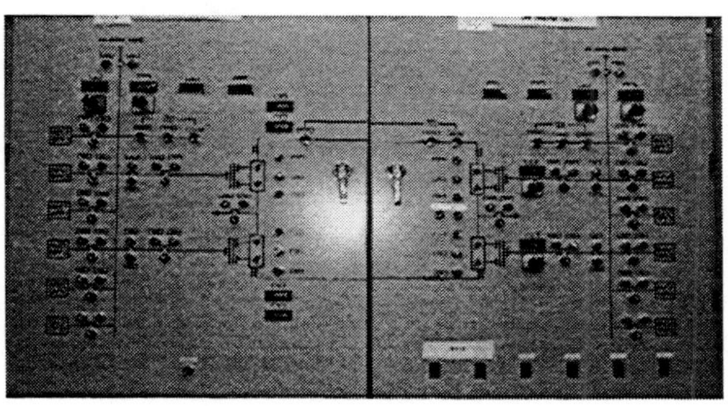

Fig. 5 Overview of HVDC Simulator in KERI

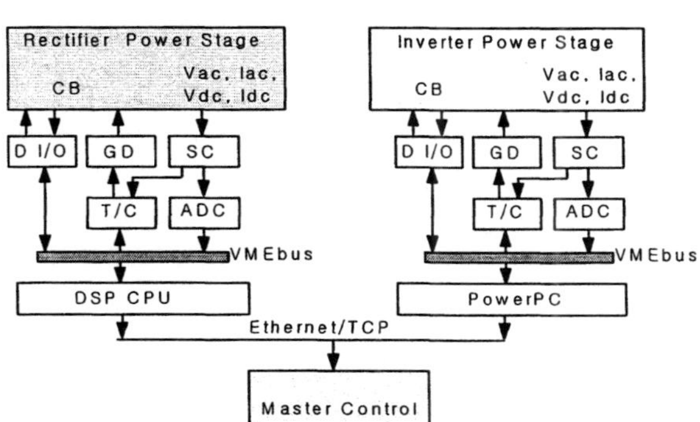

Fig. 6 Hardware configuration of HVDC control system

As described before, HVDC simulator is scaled-down bipolar model and control system is fully digitalized. Therefore, tuning process can be done on-line automatically. First of all, Fig. 7 shows the tuning process scaling factor of FLC using GA. From the top in the figure represents DC current reference, filtered feedback DC current, actual DC current and firing angle of rectifier respectively. Each step of reference signal corresponds to the response with parameters that determined at the each population. Because each population has their own parameters, i.e. the scaling factors, figure shows different step responses for every step.

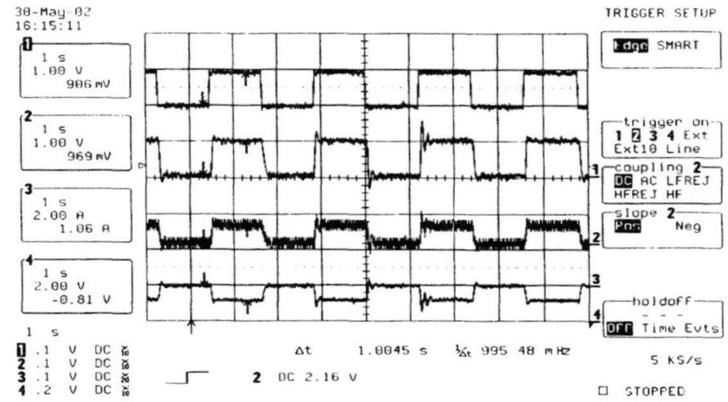

(Ch 1: current reference, Ch 2: filtered DC current, Ch 3: actual DC current, Ch 4: firing angle, time scale : 1[sec/col], y scale : scaled)

Fig. 7 waveform during tuning process

The Fig. 8 shows tuning results after 20 generations. The fitness trend through each generation is displayed in (a) of the figure. Fitness increases gradually and converges as the generation progresses. The Fig. 8(b) represents the gains of input and output of FLC through each generation. As shown in this figure, the gains of input and output of FLC converges to a certain value as the generation increases.

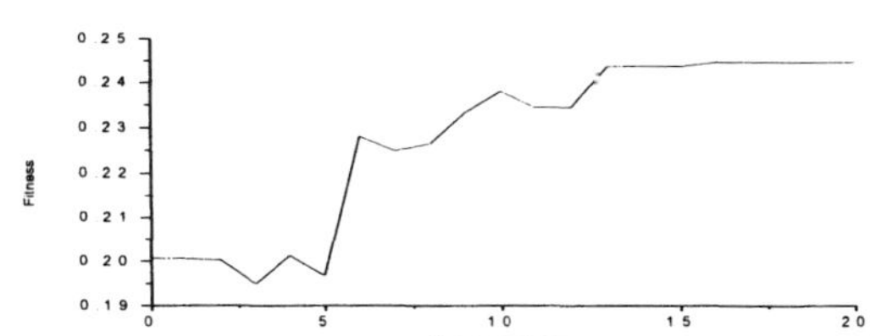

(a) Fitness value

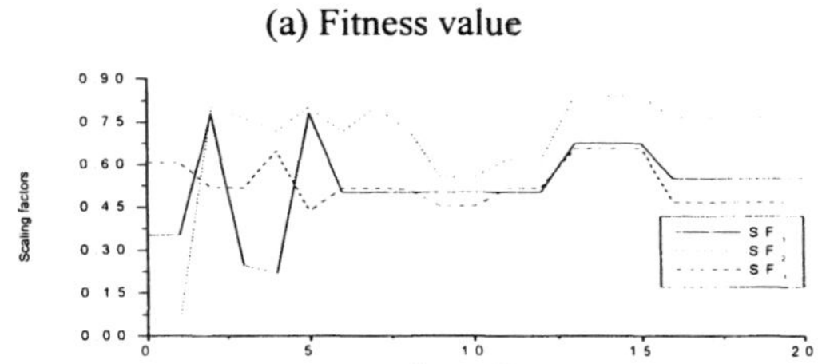

(b) Scaling factors

Fig. 8 Fitness and scaling factors in each generation

Figure 9 shows the response of the proposed FLC and of the PI controller to ULTC tap sudden change at 0.5pu current command. In this case, since ULTC tap changes abruptly in the direction of decreasing ac voltage, current drops temporarily and restores by the current control action. As presented in this figure, proposed FLC also shows better performance than PI controller in the rising time and the settling time, although slight overshoot exists.

Figure 10 shows response to large disturbance. During the HVDC system is operating at 0.2[pu] current reference, ground fault with some resistance at the dc line is happened for about 0.1[sec]. FLC shows better performance in viewpoint of limiting fault current and restoring after fault clearing than PI controller. Figure 11 shows response to large step reference change of 0.6[pu]. Current reference steps from 0.3[pu] to 0.9[pu] and current response shows the same as the point that tuning was performed, i.e. 0.3[pu]. And this proofs that the tuned parameters are valid for wide operating ranges.

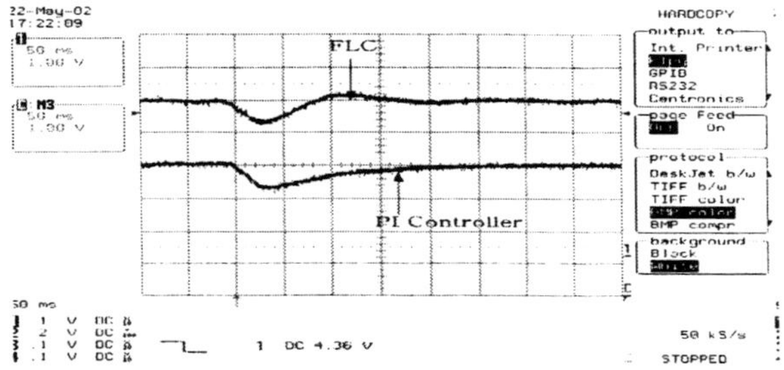

Fig. 9 Response to ULTC tap change (at 0.5[pu], time: 50[ms/col])

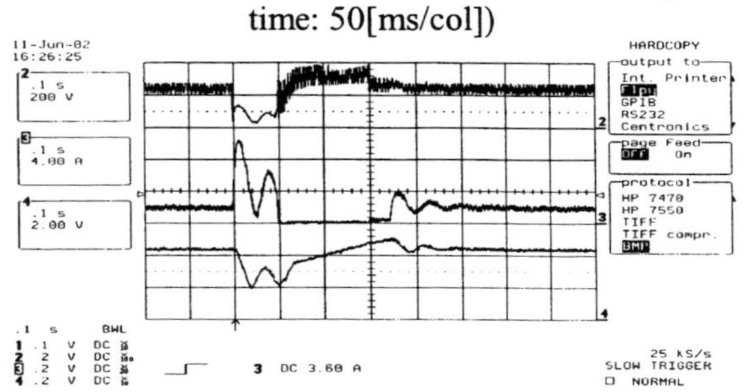

(a) PI controller

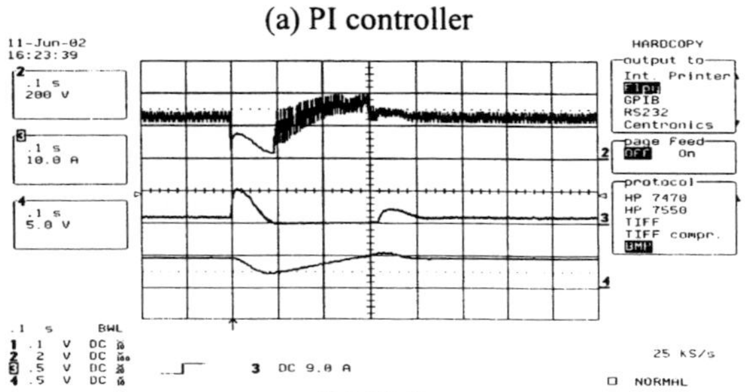

(b) FLC

(top : dc volage, mid : dc current, bot : firing angle)

Fig. 10 Response to DC Line Ground Fault

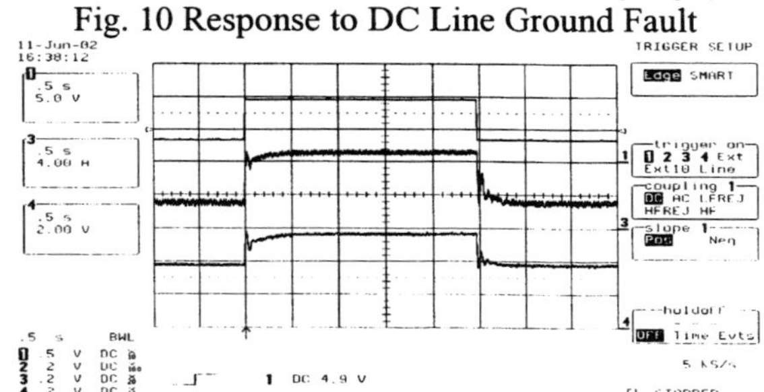

(a) PI controller

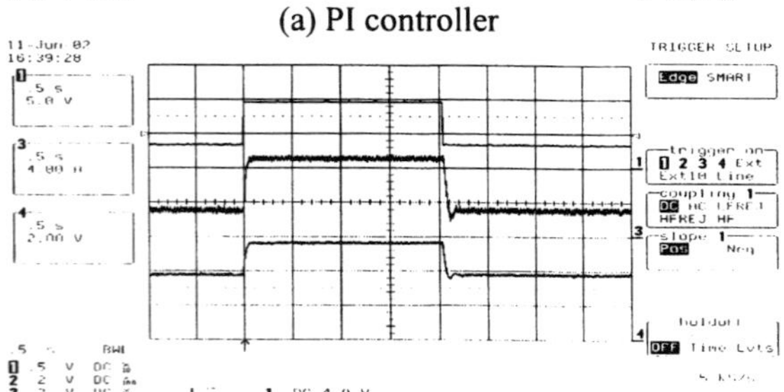

(b) FLC

(top: current reference, mid: dc current, bot:firing angle)

Fig. 11 Response to large step reference change (0.3->0.9[pu])

5. CONCLUSIONS

In the paper, we propose a fuzzy current controller tuned by the on-line GA. It tunes the three scaling factors of FLC, two input scaling factors and a output one using GA optimization technique. As the result of the experiment using the scaled-down HVDC simulator in KERI in order to compare the proposed FLC with PI controller, the proposed FLC shows the better control performance than PI controller under the various disturbance conditions. Therefore, we concludes as follows;

(1) We verifies that the FLC current controller tuned on-line under the limited operating condition, that is to say, at the low but relatively stable current reference region(0.3[pu]) shows the better performance than conventional PI controller under various disturbances. That means the tuned FLC guarantees the robustness under different operating points.

(2) By implementing on-line tuning system by GA technique using the scaled-down HVDC simulator, the proposed tuning methods can be applied to actual HVDC system.

(3) Further works will be necessary in the areas such more complicated fitness function establishment to classify the difference of response and shortening the tuning time guaranteeing the stability and performances. To evaluate the effectiveness

REFERENCES

[1] C. E. Grund and C. M. Pollard, "Power modulation controls for HVDC systems" *CIGRE* 14-03, 1984.

[2] S. Lefebvre, "HVDC controls for system dynamic performance", *IEEE PWRS-6*, pp. 743-752, 1991.

[3] M. A. Chudhury, A. S. Emarah, K. A. Ellithy and G. D. Gallanos, "Stability analysis of a modulated ac/dc system using the eigenvalue sensitivity approach", *IEEE Trans. PWRS-1*, pp. 128-137, 1986.

[4] Y. Y. Hsu and L. Wang, "Damping a parallel ac/dc system using PID power system stabilizers and rectifier current regulators", *IEEE Trans. Energy Converson*, vol. 3, no. 3, pp. 540-548, 1988.

[5] S. Lefebvre, M. Saad, and A. R. Hurteau, "Adaptive control for HVDC power transmission systems", *IEEE Trans. Power Apparatus Syst.*, vol. PAS-104, no 9, pp. 2329-2335, Sept. 1985.

[6] W. J. Rugh, "Analytical framework for gain scheduling", *IEEE Control Syst. Mag.*, vol. II, no. 1, pp. 79-84, Jan. 1991.

[7] J. Reeve and M. Sultan, "Gain scheduling adaptive control strategies for HVDC systems to accommodate large disturbances", *IEEE Trans. Power Syst.*, vol. 9, no. 1, pp. 366-372, Feb. 1994.

[8] A. Daneshpooy, A. M. Gole, D. G. Chapman and J. B. Davies, "Fuzzy Logic Control for HVDC Transmission", *IEEE Trans. Power Delivery*, vol. 12, no. 4, pp. 1690-1697, Oct. 1997.

[9] Gi-Hyun Hwang, J. H. Park, H. T. Kang and Sungshin Kim, "Design of Fuzzy Power System Stabilizer Using Adaptive Evolutionary Algorithm", *Proceedings of IEEE International Symposium on Industrial Electronics*, Vol. 2, pp. 42-47, December, 2000.

[10] Gi-Hyun Hwang, June Ho Park, Jae Young Yoon, "Genetic Algorithm Approach to Design the Optimal Fuzzy Logic Controller for Rectifier Current Control in HVDC System", *Journal of Electrical Engineering and Information Science,* Vol. 3 No. 6, pp. 792-797, 1998,

[11] C. E. Grund and C. M. Pollard, Power modulation controls for HVDC systems, *CIGRE 14-03*, 1984

[12] S. Lefebvre, HVDC controls for system dynamic performance, *IEEE PWRS-6,* 1991, 743-752

[13] Z. Michalewicz, Genetic Algorithms + Data Structures = Evolution Programs, *Springer-Verlag*, 1992.

[14] Mitsuo Gen and R. Cheng, Genetic Algorithms & Engineering Design, *A Wiley-Interscience Publication*, 1997.

[15] P. Kundur, Power System Stability and Control, *McGraw-Hill*, Inc., 1994

ELSEVIER
IFAC
PUBLICATIONS
www.elsevier.com/locate/ifac

ESTIMATING FACTOR FOR IMPROVING ARC EXTINCTION CAPABILITY OF DC MAGNETIC SWITCH

Hyun-Kil Cho, Eun-Woong Lee

**LG Idustrial systems. Ltd. Co ,*
Dept. of Electrical Engineering, Chung-nam Nat'I Univ.

Abstract: In this paper, the electromagnetic force acting on the arc column of 3 different arc extinction units is compared using the FEM(Finite Element Method) and the arc velocity is calculated by the dragforce of the fluid mechanics.
3 type arc extinction units have been made and the breaking test was perfomed at 100 Voltage in order to measure the arcing time. And the result of the breaking test was compared with computing values. So, the theory was demonstrated validly. As the result, this paper proposes the divided U-shaped grid having the external magnetic field to shorten arcing time and improve the electric performance and suggests a methodology for comparing and analyzing the results obtained by simulation and experiment.

Keywords: Switching arc, Finite element field simulation, Electromagnetic field computation

1. INTRODUCTION

Magnetic contactors are used for switching off load currents in low-voltage circuits because of their low cost. Since this situation is expected to continue for many years, it is necessary to improve the performance of low-voltage magnetic contactors. The main subjects of magnetic contactors are the shortening of arcing time for the interruption of currents in low-voltage circuit and reduction of contact wear. Therefore many studies on arc extinction and contacts wear have been performed (Ragnar Holm, 1981; T. Mori and H. Okado, 1989; Teijiro Mori et al., 1991; K. Kolacinski, 1992; J. Shea and Yun-Ko, 1993, Shokichiro Ito and Yoshiaki Takato, 1998)
Horinouchi presented a simple practical method of simulating the behavior of arc driven by magnetic force to study the arc extinguishing ability.(Horinouchi et. al, 1996) In his model, the arc is assumed to be a chain of small rigid cylindrical current elements. Each element receives magnetic force (Lorentz force) for the magnetic field and fluid drag from the surrounding gas. The behavior and the shape of the arc elements can be predicted by combining these force.
And, the power current is intend to change DC from AC in the Anan Japane. So, the studies of DC insulators and switching method are required before normally operating DC power transmission system (Mesut Baran and Nikhil R Mahajan, 2002).

Therefore, this paper makes to reference the improvement of the performance of DC switching device in the air and suggest new interrupting method at the DC circuit. The electromagnetic force acting on the arc column of 3 different arc extinction units is compared using the FEM(Finite Elelement Method) and the arc velocity is calculated by the dragforce of the fluid mechanics.
Three types of arc extinction units have been made and the breaking test was perfomed at 100 Voltage in order to measure the arcing time. And the result of the breaking test was compared with computing values. So, the theory was demonstrated with validity. As a result, this paper proposes the divided U-shaped grid having the external magnetic field to shorthen arcing time and improve the electric performance and suggests a methodology for comparing and analyzing the results obtained by simulation and experiment.

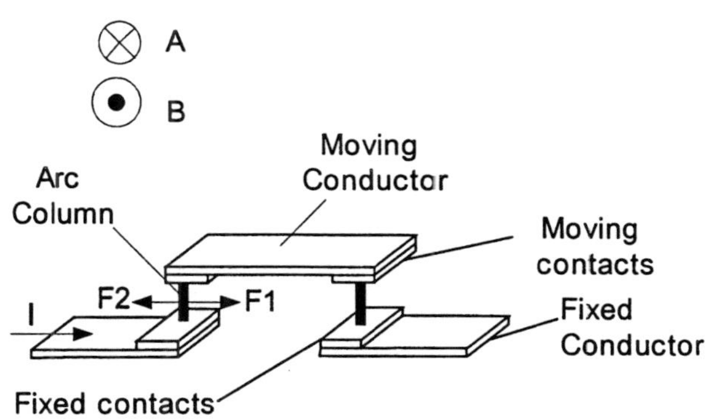

Fig. 1 Mechanism of extinguishing

2. ARC EXTINCTION

2.1 Construction and operation

Fig. 1 shows the switching mechanism in the magnetic contractor. The switching mechanism consists of fixedcontacts, fixed conductors, moving contacts and moving conductor. When the moving contacts are separated from the fixed contacts, arc columns arc formed between contacts.

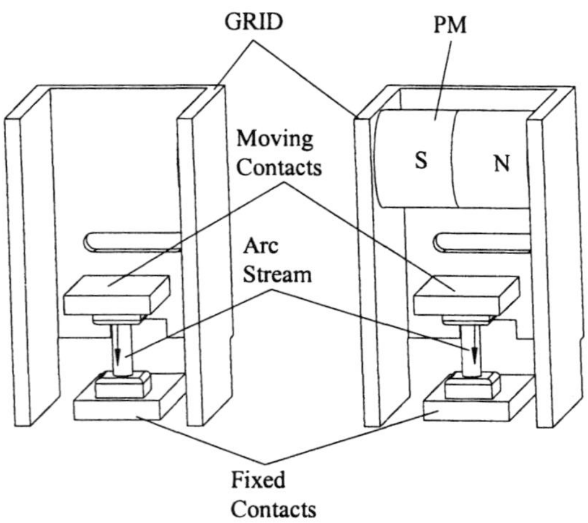

(a) Model 1 (b) Model 2

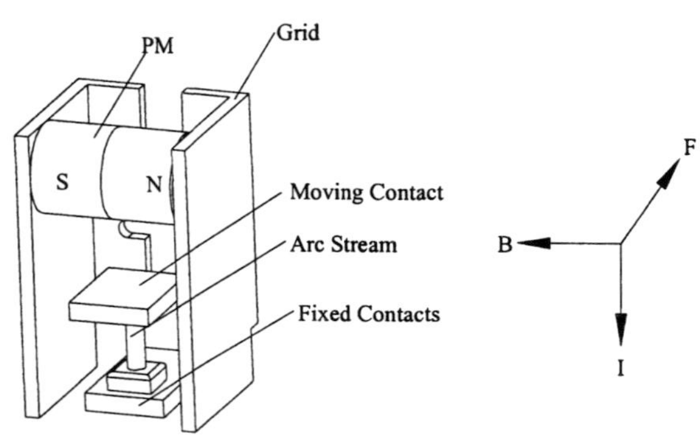

(c) Model 3 (d) Vector directions

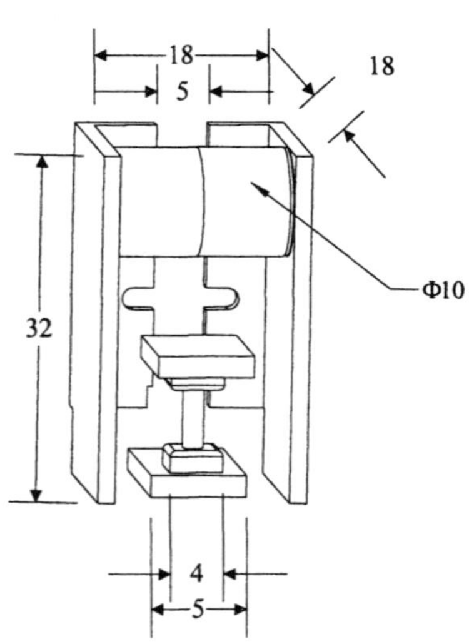

(e) Dimension of arc quenching unit

Fig. 2. Model of Arc quenching unit

Arc column receives forces F1 and F2. Force F1 and F2 is caused by the magnetic field due to the arc current. The magnetic field of F1 side directs to A and the magnetic field of F2 side directs to B.

So that, if the magnetic distortion does not occur, force F1 approximately equals force F2. Therefore, in order to increase the magnetic flux density effecting force F2, U-shaped grid is used in an arc chamber in the magnetic contactor. Also, an aim of the optimizing grid is to quickly extinguish the arc between contacts. Various grids is developed by manufactures.

2.2. 3 types of arc extinction unit

Three kinds of grid models are showed in Fig.2. The arc chamber is symmetrically arranged. So, half models of a real mechanism are expressed in Fig. 2 (a), (b), (c). The arc extinction model 1 of Fig. 2(a) employs a U-shaped grid used in the arc chamber of an AC magnetic contactor. And the model 2 of Fig. 2(b) has an U-shaped grid and permanent magnet to increase the magnetic flux crossing the arc current. The model 3 is the new interruption method. This has divided grid and permanent magnet to increase the magnetic flux crossing the arc current. The vector direction is showed in Fig. 2(d). The volume of the arc extinction unit and the size of the permanent magnet are shown in Fig. 2(e). And all dimensions of 3 magnetic contactors of different arc extinction unit are the same size. The material of grids is a SCP1 steel plate and the permanent magnet is an Alnico5.

2.3. *Electromagnetic blow-out force*

Fig. 3(a) shows the 2-D model of the modell. The magnetic flux by arc current concentrates in the grid (magnetic plate) and passes the open part of the grid. The arc current flows toward the z direction column on the origin of the x-y coordinate and the magnetic field by the arc current confirms Ampere's low. Accordingly, the magnetic flux density B_Y is lower closing area enveloped in the U-shaped grid than opening area.

Fig. 3 (b) shows the 2-D equivalent magnetic circuit of model 2. The magnetic flux by the arc current and most magnetic flux by permanent magnet flow along grid and the open part of the U-shaped grid. The leakage flux of permanent magnet passes the inside of U-shaped grid. On the left side Y-axis, magnetic flux density is low due to magnetic flux by the arc current opposites to leakage flux of permanent magnet. On the other side, magnetic flux density is high due to adding up magnetic flux density by arc current and that by permanent magnet.

Fig. 3(c) shows the divided U-shaped grids as the model three. The amount of the permanent magnet flux flows along grid and goes across the inside of divided grids. If it fades away from the permanent magnet to the distance, the magnetic flux density inside area of divided grids reduces gradually. As the results, the magnetic flux on left side of Y-axis reduces due to opposite face. On the right side, it increases owing to face the same direction. Resultantly, the blowout force of arc column is proportional to the difference of the magnetic flux density.

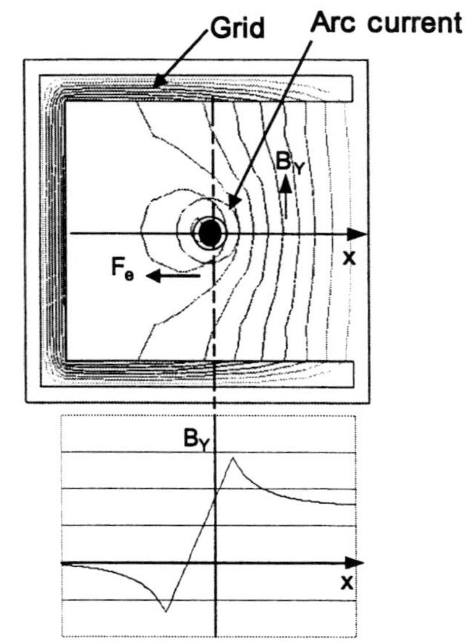

(a) Flux path and flux density Model 1

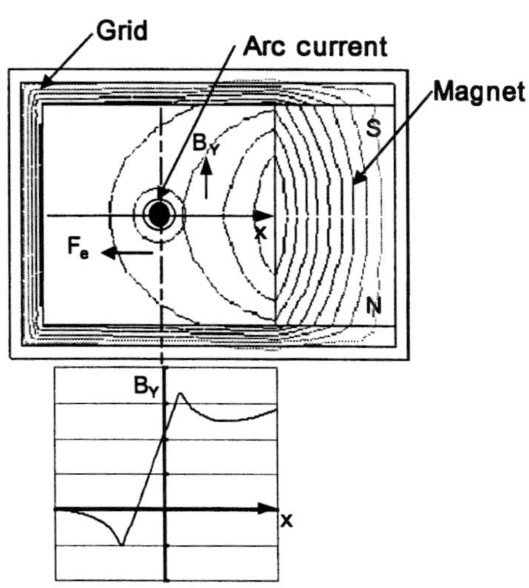

(b) Flux path and flux density of Model 2

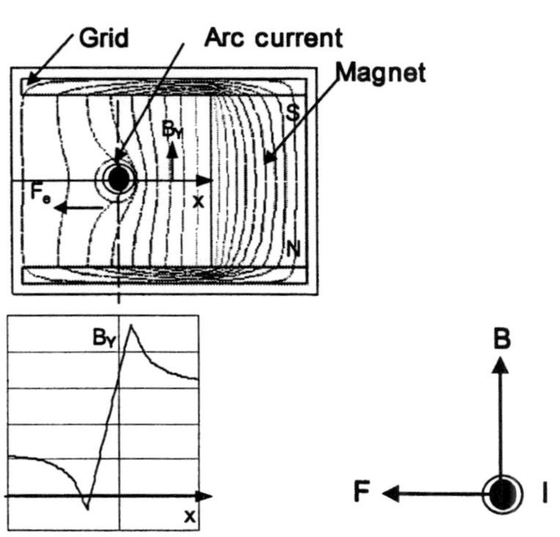

(c) Flux path and flux density Model 3

Fig. 3 Flux path and flux density of the 2D Model.

The electromagnetic force F acting on the arc can be obtained by Eq. (1)

$$F_e = \int_{V_a} \bar{J}_a \times \bar{B}_a dV \qquad (1)$$

where $\bar{J}_a$: current density, V_a : arc area , $\bar{B}_a$: magnetic flux density

2.4. Drag Force and arc motion

Arc flow by the blowout forces is prevented drag force by frictions of the air viscosity on the front edge of the arc column. On the rear edge of the arc column, an in ward suction force to suck the air from the surrounding area is generated. The force acting on the arc column by the surrounding air is named the drag force. (Horinouchi et. al, 1996)

An arc is a current path, which is a stream of ions and electrons. And an arc column is deformed freely by the electromagnetic force and the drag force. Because it is very difficult to obtain the motion of all particles in the arc, arc model is simplified as follows. The center of an arc contains strongly ionized gas at a high temperature and the boundary of arc column is not clear from surrounding gas. But, in order to calculate drag force, we assume the arc region is clear from surrounding gas. Air layer enclosing arc column is not driven by magnetic force because of neutral particles

$$F_d = \frac{1}{2} C_D \rho S u^2 \qquad (2)$$

where S is the frontal area of the column exposed to the flow and u is the moving velocity of the arc column. C_D is the drag coefficient $C_D = 0.62\left(\frac{L}{d}\right)^{0.12}$.

The equation of arc motion is given by

$$F_e - F_d = m\frac{du}{dt} \qquad (3)$$

F_d is the frictional force of the arc, m is the mass of the arc.

On the assumption that the arc consists of electrons and singly ionized silver atoms, the mass of the arc is

$$m = \left(\frac{M}{N_0}\right) \times \left(\frac{\pi}{4} D^2 L\right) \times \left(\frac{P}{2kT}\right) \qquad (4)$$

Where M is the atomic weight of ionized atoms, No is Avogadro's number, k is Boltzmann's constant, T is the arc temperature, and p is the arc pressure.

The contact radius, r, is obtained by Eq. (5) from Holm's equation for the modeling of arc, where P is contact pressure[N], H_d is the Brinell hardness and ζ is a constant determined by the condition of the contact shape. (Ragnar Holm, 1967; Shokichiro Ito and Yoshiaki Takato, 1998)

$$r = \sqrt{\frac{P}{\pi \zeta H_d}} \qquad (5)$$

The table 1 shows the material properties used in the simulation and the experimental.

Table 1. The actual values of contacts model

Material of contacts	H_d	Z	P[kgf]	r[mm]
AgCdO	90	0.5	0.7	0.25

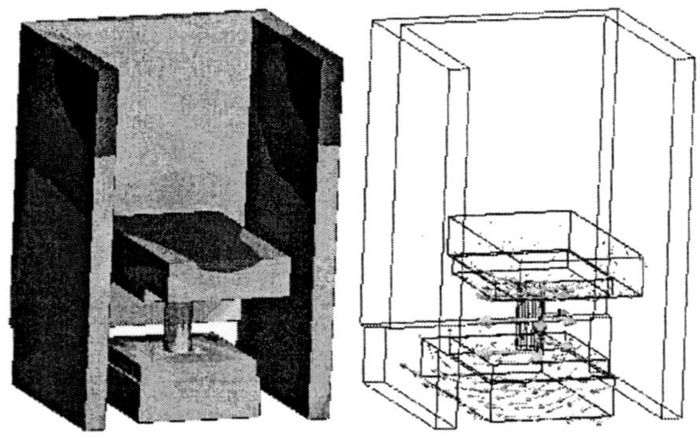

(a) Flux density and field vector of Model 1

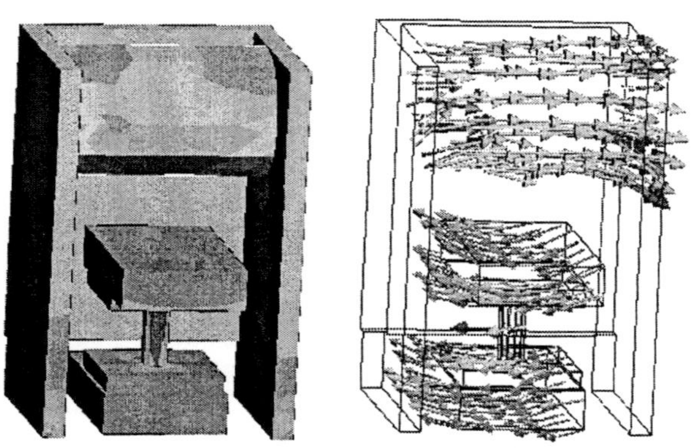

(b) Flux density and field vector of Model 2

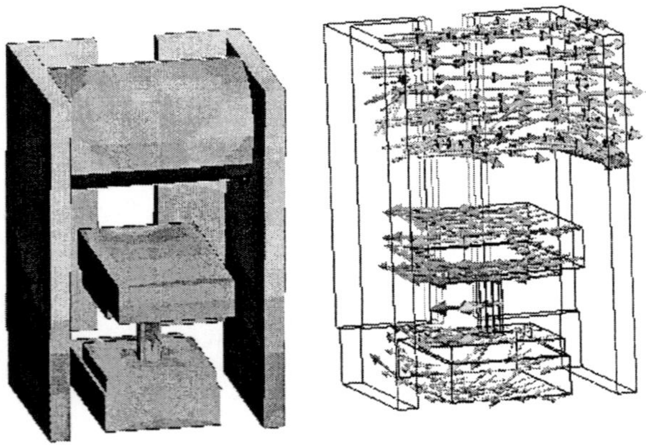

(c) Flux density and field vector of Model 3

Fig. 4. The analysis results of arc quenching unit using FEM

3. COMPARATION OF SIMULATIONS AND TESTS

3.1 Results of simulations

Fig. 4(a) shows the distribution of magnetic flux density generated by an arc current and the direction of the magnetic flux vector.

The magnetic flux vector around the arc column is rotating through the arc column. The blowout force acting on the arc stream is 0.002[N].

As is shown in model 2, the magnetic flux vector is heading toward the reverse direction of the magnetic field inside the permanent magnet and the magnetic flux density of the grid is increased. The resultant arc blowout force is 0.014[N].

Fig. 4(c) shows the distribution of magnetic flux density and the direction of magnetic flux in model 3. The resultant arc blowout force is 0.024[N].

Magnetic flux of (c) which crosses the arc stream and increases the arc blowout force is the greatest among (a), (b) and (c) of Fig. 4.

It is due to the magnetic flux passing only through the grid and forming a closed loop is decreased and that the magnetic flux passing through the arc extinction space is increased compared to model 2. And the magnetic flux density around the arc stream is shown to increase.

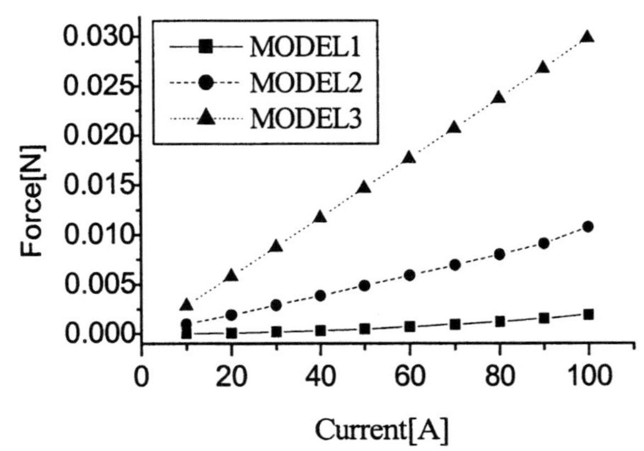

(a) Driving force for arc as a function of current

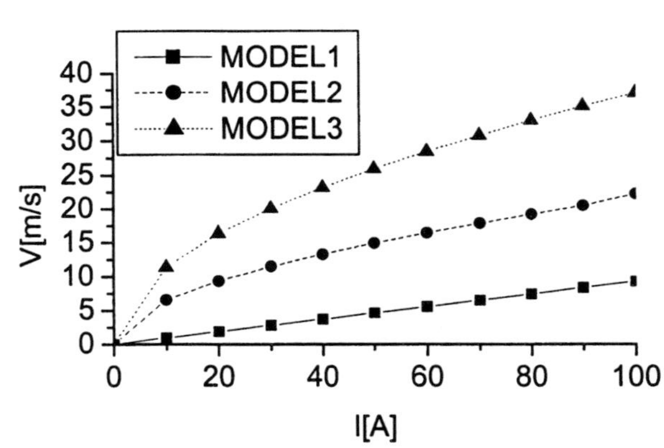

(b) Blow-out Velocity of each Model

Fig.5 Results of Analysis

Fig.5 Results of Analysis; (a) Driving force for arc as a function of current, (b) Blow-out Velocity of each Model

Fig. 5(b) shows driving velocity acting on arc as a function of current on each models of silver vapor arc.

The arc velocity depends on magnetically driving force and drag force acting on arc column for each models. Driving velocities of arc are 9.3 , 22 and 36.9 m/s respectively model 1, 2 and 3 when contacts are opening at I=100A. Teijiro Mori reported that the calculated velocities of a 100A arc with a 5mm length under the magnetic flux densities of 10mT were 25m/s. Our calculation result agrees approximately with value calculated by Teijiro Mori.(1992)

3.2 Experimental Condition

All tested contactors including the new interrupting mechanism were used in this study. Dimensions of the principal parts in the arc chamber of the new interruption method are shown in fig 2. Volumes of the conventional mechanism are the same as those of the new interrupting mechanism, except volume of the included permanent magnet and the gap in divided grids. For all experimental contactors, the contact material is a silver-cadmium alloy and all operating mechanisms are the same. The contact opening speed and contact gap are about 1m/s, 4.5mm, respectively. The test was conducted the range to 110 voltage from 60voltage by the shunt motor with 10.2 [ms] time constant.

3.3 Experimental results

The increase of arc blow-out force results in the improvement of arc extinction performance and the shortening of arc dwelling time in an arc extinction chamber. Hence, waveforms were measured at the precise moment of the load of electromagnetic contactor being turned off to get a relationship between the arc quenching time and the arc blow-out force.

Fig. 6 shows the shape of the current wave measured with an oscilloscope. As showing Fig. 6 (a), the arc current is smoothly reduced and the arc voltage is evenly increased along changing the arc current. Fig.6 (b) show that the arc current is rapidly reduced and the arc voltage has a steep slope. And Fig. 6 (c) show that the arc current rapidly comes down and the arc voltage has many peak voltages with having steep slope due to rapidly reducing arc current.

The arcing time of models 1,2 and 3 are 20.9[ms], 10.1[ms] and 5.2[ms], respectively. The arc voltage rises in the proportion to the variation of arc current and arcing time.

Measured results of arc time are showed by Fig. 7(a) . Interrupting tests were made in dc circuits having line voltages of 60V, 100V and 110V. Interrupting tests were repeated 30 times for each models and each voltage levels. There was significant difference of arcing time in the 3 type models. For the model 1, average arcing time was increased largely with line voltage. On the other hand, for the model 3, average arcing time was increased gradually with line voltage. At he voltage of 100V, average arcing time of model 3 was about 38% of that the model 1. Fig. 7(b) showed estimating method of comparing results of analysis with measured results. Velocities of moving arc for each model were carried out simulation at the current 100A. Arc extinction time and velocity of arc driving force can be compared relatively because of same arc quenching room. The result shows that bigger arc blow-out forces lead to shorter arc extinction time, which coincides with the FEM analysis and drag force

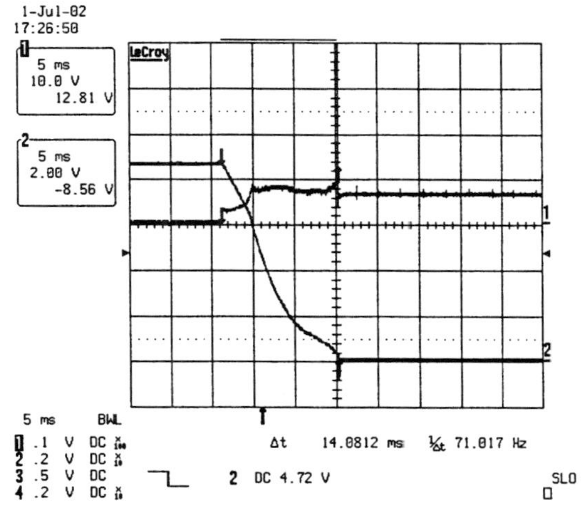

(a) Waveforms of voltage and arc current on the Model 1

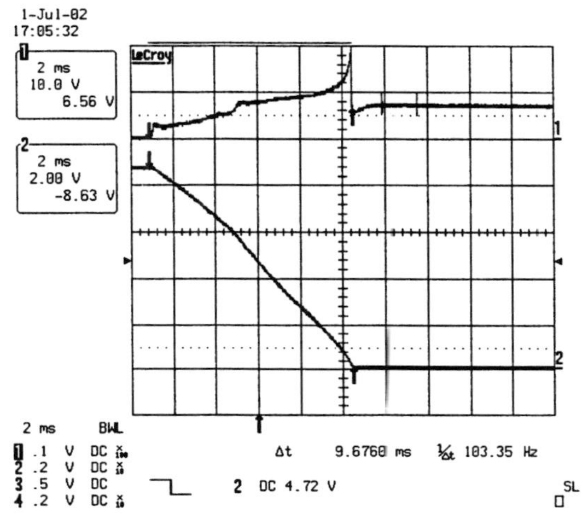

(b) Waveforms of voltage and arc current on the Model 2

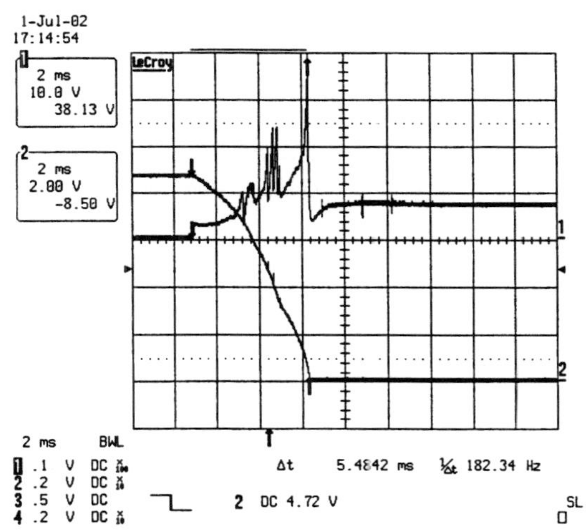

(c) Waveforms of voltage and arc current on the Model 3

Fig. 6 Waves of arc current and voltage

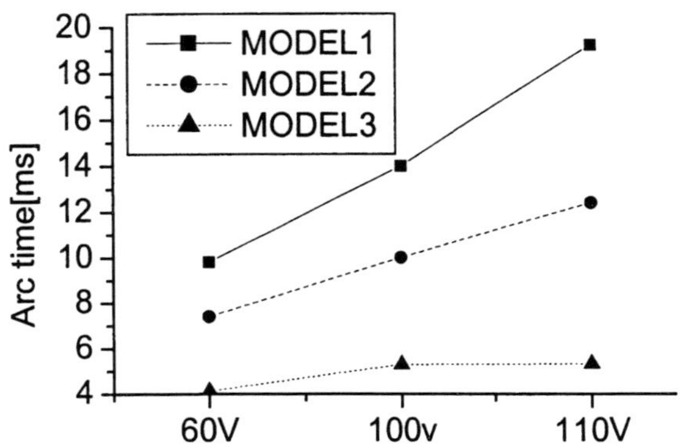

(a) measurement of Arc extinction time

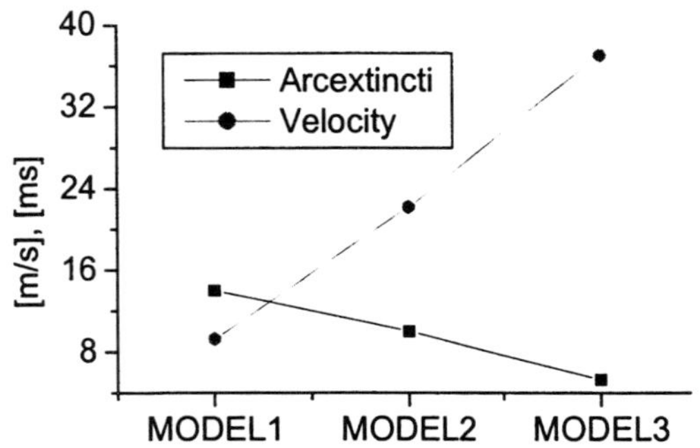

(b) Comparing calculated velocities and measured arcing time(I=100A)

Fig. 7 measurement of Arc extinction time

Measured results of arc time are showed by Fig. 7(a) . Interrupting tests were made in dc circuits having line voltages of 60V, 100V and 110V. Interrupting tests were repeated 30 times for each models and each voltage levels. There was significant difference of arcing time in the 3 type models. For the model 1, average arcing time was increased largely with line voltage. On the other hand, for the model 3, average arcing time was increased gradually with line voltage. At he voltage of 100V, average arcing time of model 3 was about 38% of that the model 1. Fig. 7(b) showed estimating method of comparing results of analysis with measured results. Velocities of moving arc for each models were carried out simulation at the current 100A. Arc extinction time and velocity of arc driving force can be compared relatively because of same arc quenching room. The result shows that bigger arc blowout forces lead to shorter arc extinction time, which coincides with the FEM analysis and drag force.

4. CONCLUSION

In this research, three different models of DC arc extinguishing mechanism have been formed and the appropriate contact area between the moving contacts and fixed contacts were estimated to implement three different types of arc model and to analyze them with 3D FEM. The arc blow-out force acting on the arc stream and the arc velocity were computed accordingly and the arc breaking time was measured employing an arc extinction chamber which was equivalent to the one used in FEM analysis. This paper showed that arc blowout force is a main factor for arc velocity and the arc breaking performance by demonstrating that the arc breaking time is inversely proportional to the arc velocity through the comparison of arc velocity with the arc breaking time obtained by experiment. Also, a model for improving the performance of DC contactor is suggested.

REFERENCES

John. Shea and Yun-ko Chien (1993), Blow-open Forces on Double-Break Contacts, *IEEE Holm Conference*, pp103-109.

K. Horinouch et al. (1997), A Method of Simulating Magnetically Driven Arcs, *IEEE Trans. Power Delivery*, **Vol. 12, No**. 1, pp. 213-218.

Mesut and Nikhil R Mahajan (2002), *DC Distribution for Industrial Systems: Opportunities and Challenges*, IEEE/PES ,**Vol. 1**. pp. 38-74

Shokichiro Ito and Yoshiaki Takato(1998), Numerical Analysis of Electromagnetic Forces in Low Voltage Circuit Breakers using 3-D Finite Element Method into account eddy Currents, *IEEE Transaction on Magnetics*, **Vol.34, No.5**, pp.2597-2600.

Ragnar Holm(1981), Electric Contacts, *4th edition*, pp. 275-366, springer-verlag, New York

T. Mori et al. (1989), New Interruption Method for Low-Voltage, Small-capacity, air-break contactors, *Holm Conference*, pp. 2308-2313.

Teijiro Mori, Kenichi Koyama(1991), A new interruption Method for low-voltage, small-capacity, air-break contactor, *IEEE Industy Application*, **Vol. 27. No. 1**. PP.161-165.

Z. Kolacinsk, et al.(1992), Spiral Arc Quenching, IEEE Trans. Power Delivery, Vol. 7, No. 2, pp. 822-828.

ELSEVIER
IFAC
PUBLICATIONS
www.elsevier.com/locate/ifac

A STUDY ON THE MODELING OF SUPERCONDUCTING FAULT CURRENT LIMITERS USING EMTDC

Jaedeuk Lee
Dept. of Electrical Engineering
Changwon National University
Changwon, Korea 641-773

Minwon Park
Center for Applied
Superconductivity Technology
Changwon, Korea 641-120

In-Keun Yu
Dept. of Electrical Engineering
Changwon National University
Changwon, Korea 641-773

Abstract: An effective modeling and simulation schemes of a resistive type Superconducting Fault Current Limiter (SFCL) and DC reactor type SFCL using PSCAD/EMTDC are proposed in this paper. In case of resistive type SFCL, current limiting is implemented by the ultra-fast transition characteristics from the superconducting (non-resistive) state to the normal (resistive) state by overstepping the critical current density. The states can generally be divided into three sub-states: the superconducting state, the quench state and the recovery state, respectively. In case of DC reactor type SFCL, it is composed of a superconducting coil and a diode bridge through which AC line current is rectified, and the increase in fault current is limited by the inductance of a superconducting coil. An actual experiment based component model of the real resistive type SFCL is developed and applied for the simulation. The proposed modeling and simulation schemes can be implemented to the grid system readily under various system conditions including sort of faults and the system capacity as well. The simulation results demonstrate the effectiveness of the proposed model and simulation schemes. *Copyright © 2003 IFAC*

Keywords: Resistive type SFCL, DC reactor type SFCL, EMTDC, fault current limiter

1. INTRODUCTION

The need for fault current limiters (FCL) is associated with the continuous growth and interconnection of modern power systems and increase in dispersed generation facilities, which result in progressive increase in the short circuit capacity far beyond their original design capacity.
Fault currents are usually momentary and originated from a variety of source such as: equipment failures, lightning, tree branches, or animals shorting out circuits. In all cases the resulting current surges can easily damage transmission line and substation components. Under certain conditions faults can destabilize a regional power grid producing blackouts extending over several regions.
Conventional alternative methods for current limiting include; replacing the existing circuit breakers and other equipment with more expensive components with higher short circuit ratings, splitting and reconfiguring the system and consequently reducing the operational flexibility and system stability; using cost ineffective transformers with higher short circuit impedance; using series reactors which incur losses during the normal system operation; and using single operation devices, such as fuses and pyrotechnic current limiters, which have to be replaced after each operation.
An ideal FCL has to have characteristics of zero resistance or zero impedance at normal operation, no power loss during normal operation as well as current limitation, large impedance under fault condition, a quick appearance of impedance at an occurrence of faults, fast recovery to normal state after a fault removal, reliable current limitation at the defined fault current and low voltage operation besides reliability and cost performance.
The SFCLs with High Temperature Superconductor (HTS) will be introduced into electric power system as an effective countermeasure for the conventional alternative methods. With respect to cost, reliability and compatibility, the SFCLs have to fulfil the same requirements as conventional devices. They must be cost effective, as reliable as conventional devices for limiting short-circuit currents and compatible to existing power systems.
There are typical three types in a SFCL, a resistive type, an inductive type and a hybrid type. The most developed SFCL is the resistive type. Still at the prototype stage, some realizations have already shown satisfying operation at significant power levels. It is expected that the resistive type SFCL of the superconducting power devices will be commercialized quickly near in the future.
Before the commercialization of the SFCL, at first, the simulation study of the SFCL in the hypothetical power system is needed and strongly recommended.
For the purpose of providing an alternative scheme, effective modeling and simulation methods of resistive type and DC reactor type SFCLs using PSCAD/EMTDC is proposed in this paper. An actual experiment based component model of the real resistive type SFCL is developed and applied for the simulation. The proposed modeling and simulation schemes can be implemented to the grid system readily under various system conditions including

sort of faults and the system capacity as well. The simulation results demonstrate the effectiveness of the proposed model and simulation schemes.

2. PRINCIPLE OF RESISTIVE TYPE SFCL

Fig. 1 shows a simplified phase diagram of superconducting state. It is divided into three regions namely, the "superconducting regime" ($\rho = 0$), the "flux-flow regime" ($\rho = \rho(j)$), and the "normal conducting regime" (ρ = constant). The most straightforward concept of an SFCL is the resistive one, in which the superconductor is directly connected in series to the line. The cross-section of the superconductor is so determined that at I_n the superconductor is operated inside the superconducting state, where its interference with the electrical network is negligible. Even though, for ac applications the superconductor has a certain reactance and exhibits ac-losses. Both depend very strongly on the geometry of the superconductor, and can be minimized by optimal conductor-architecture.
For SFCL application the superconducting-materials have to meet three requirements : (1) low ac-losses in normal operation to minimize cooling costs, (2) high mechanical strength to withstand the thermo-mechanical and magnetic forces during the limitation process, and (3) good thermal stability to avoid excessive heating at "hot-spot", which may develop during the limitation process and can lead to a burn-through of the superconductor.
The SFCL based on LTS (Low-Temperature Superconductor) usually utilize "ac = wire", where LTS filaments are embedded in a normal conducting matrix. These materials are of good mechanical quality and exhibit low ac-losses. Because of their low specific heat, high thermal conductivity and high j_c, hot sports spread very fast (several 100m/sec), even at j values below j_c. Thus in case of a "quench", a rather homogeneous heating of the conductor is guaranteed. The LTS based SFCLs are usually of the "fast heating" type.

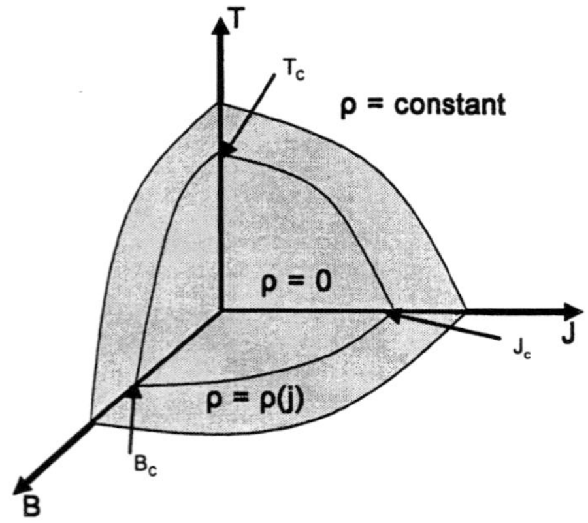

Fig. 1. Simplified phase diagram of superconducting state

Due to the ceramic nature of the HTS, conductors with sufficiently low ac-losses for arbitrary directions of the magnetic field have not been demonstrated. Most HTS are fabricated in tapes or plates or tubes. Decreasing the conductor dimensions transverse to the local magnetic field usually minimizes ac-losses. The rather brittle HTS needs to be mechanically stabilized, e.g. by supporting substrate, which might be metallic or insulating. HTS, in contrast to LTS, are very poor thermal conductors, and at 77K have rather high specific heat, thus, hot spots spread very slowly or even contract if j drops below j_c. In such a case, the whole grid voltage would drop across a decreasing part of the conductor, which eventually will lead to a burn-through in the HTS. A common measure to reduce the problem is the application of a normal conducting "electrical bypass". Thus allowing the current to bypass the hot spot. Most HTS components for SFCL are composites, comprising the HTS, a mechanical substrate or supports, and an electrical bypass. The current limiting performance of the composites largely depends on the parameters critical current density, I-V characteristics, thermal conductivity, thermal mass, and bypass. Among HTS there are three major material systems under intensive research for SFCL application: i.e. Bi2223-wires, YBCO-films, and Bi2212-bulk.

3. PRINCIPLE OF DC REACTOR TYPE SFCL

Fig. 2 shows the specific diagram of single-phase dc reactor type SFCLs. A high temperature-superconducting coil is used and it is connected in the diode bridge. In the case of a line fault, the diode bridge switches it and the fault current flows through the dc reactor coil. Thus, the increase rate of a fault current can be limited to a determined value by the inductance of the coil. The load current is usually flowing through the dc reactor coil and the current change is limited by the coil inductance.
At the occurrence of a line to ground fault, the primary current in the transformer will increase to fault current and the primary voltage will rise to the line voltage. The secondary currents and voltages will also increased to the values corresponding to the transformer ratio.

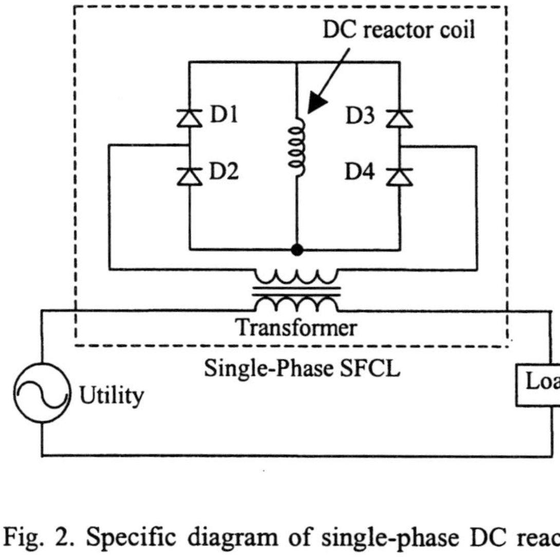

Fig. 2. Specific diagram of single-phase DC reactor type SFCL

4. MODELING AND SIMULATION OF A RESISTIVE TYPE SFCL

In this paper, three cases: single line to ground fault, two lines to ground fault and 3-phase short-circuit, are simulated and the results are given.

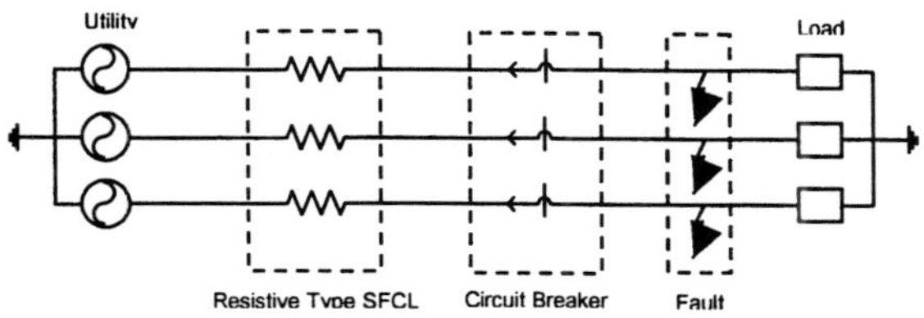

Fig. 3. Simulation diagram of resistive type SFCL using EMTDC

Fig. 3 shows simulation diagram of resistive type SFCL using EMTDC. This diagram consists of three-phase source, resistive type SFCL, circuit breaker, and load.

Table 1 shows the simulation parameters. For this simulation, a utility that the rated voltage is 6600[V] and the rated current is 200[A], are considered. The resistance curve of the SFCL is fitted using experimental R (T. I) curve of YBCO film SFCL.

In addition, the EMTDC component of resistive type SFCL is also studied by using fitted equations. As given in Fig. 4, the EMTDC component of resistive type SFCL consists of four inputs and one output. The inputs are current, fault signal, breaker signal and RMS value of line current. The output is resistance.

Table 1 Simulation conditions

Duration of Simulation	1.0 [sec]
Duration of Fault	0.3 [sec]
Time of First Breaker operation	0.264 [sec]
Time of 2nd Breaker operation	0.664 [sec]
Line-to-Line Voltage	4.67 [kV]
V_{RMS}	3.81 [kV]
Line Current	0.2 [kA]
Fault Resistance	0.5 [Ω]
Critical Current	2 [kA]

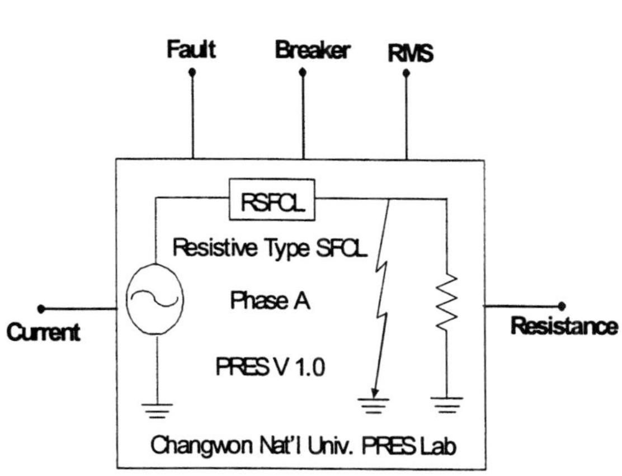

Fig. 4. EMTDC component of resistive type SFCL

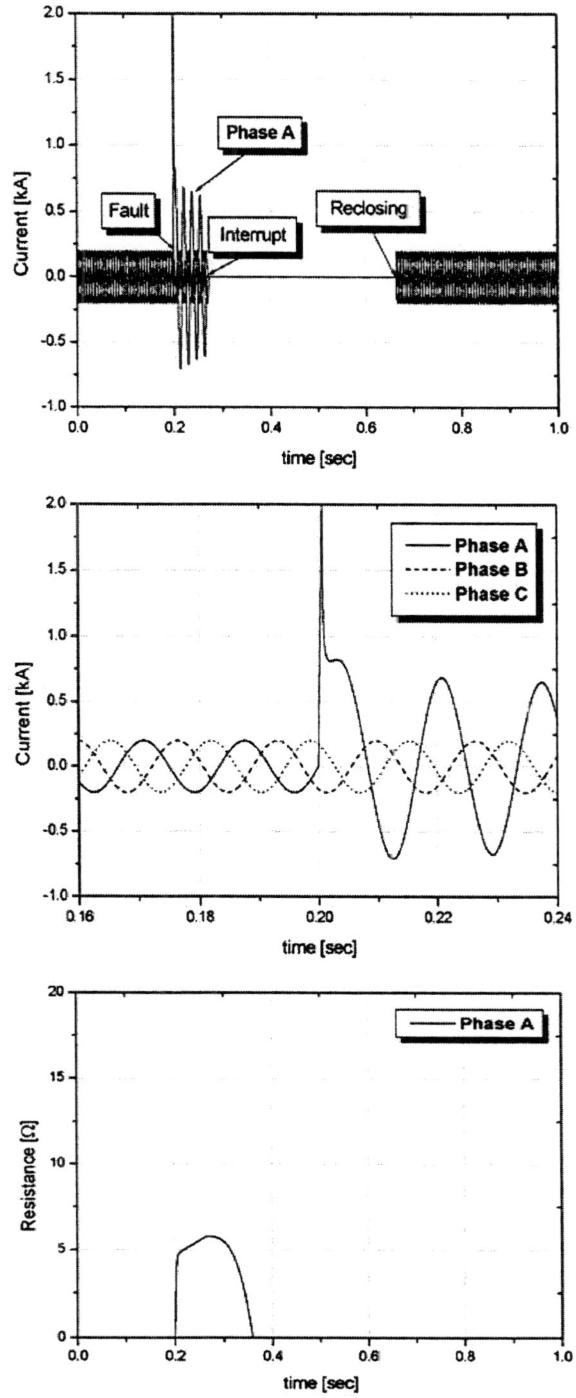

Fig. 5. Waveforms of single line to ground fault with resistive type SFCL (Phase A)

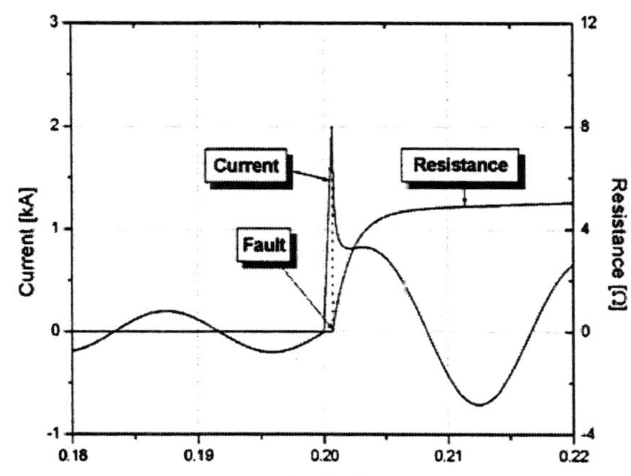

Fig. 6. Waveform of current and resistance at the quench

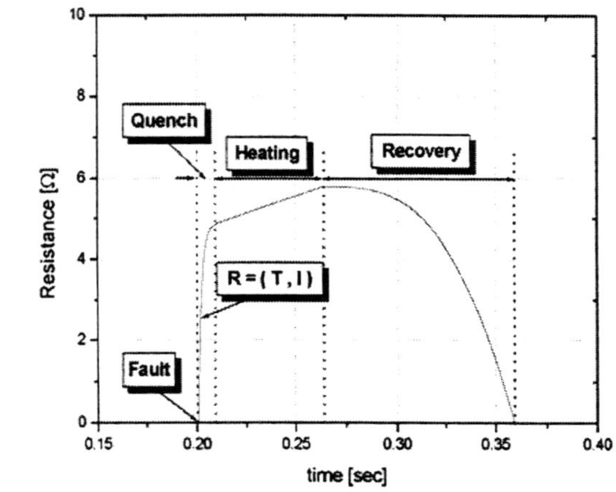

Fig. 7. The S/N transition of resistive type SFCL with recovery

Fig 5 shows the waveforms of single line-to-ground fault with resistive type SFCL at Phase A. The results show that the currents are limited to 2 [kA] after fault has been occurred. Fig. 6 shows the waveforms of the current and resistance at the quench in detail. When the single line-to-ground fault occurs in utility, the rated current increases over critical current, and SFCLs resistance increases due to the normal conductor resistance (quench).

Fig. 7 shows the S/N transition of resistive type SFCL with recovery in this simulation. The recovery behavior, which is determined by heat flux within the substrate and into the bath of LN_2, is represented by the decay of the resistance. The resistance decrease within 100 [ms] does not necessarily mean that the complete superconductor has cooled down to a temperature below its critical value, and that the nominal current can be newly carried after 100 [ms], but it is at least an indication of the cooling process.

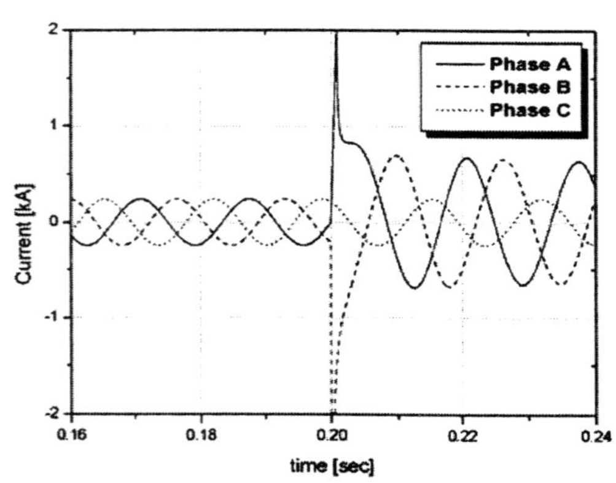

(a) Phase A and Phase B

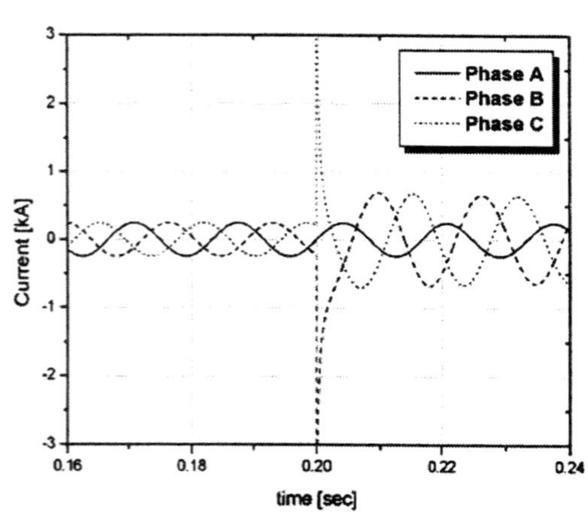

(b) Phase B and Phase C

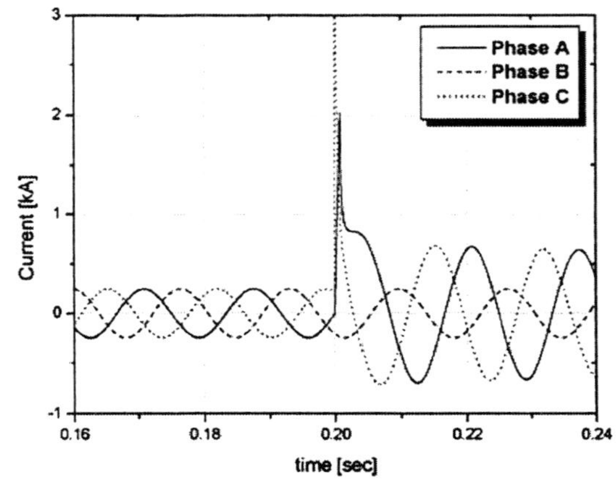

(c) Phase A and Phase C

Fig. 8. Waveforms of two line-to-ground fault current

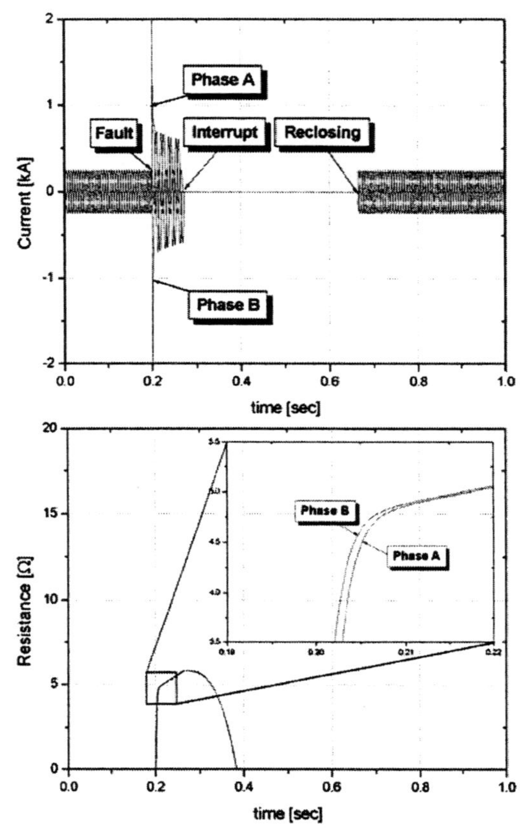

Fig. 9. Two line-to-ground fault current waveforms and the SFCL resistance variations due to circuit breaker operation. (Fault occurred at 0.2 [sec], interruption at 0.264 [sec] and reclosing at 0.664 [sec])

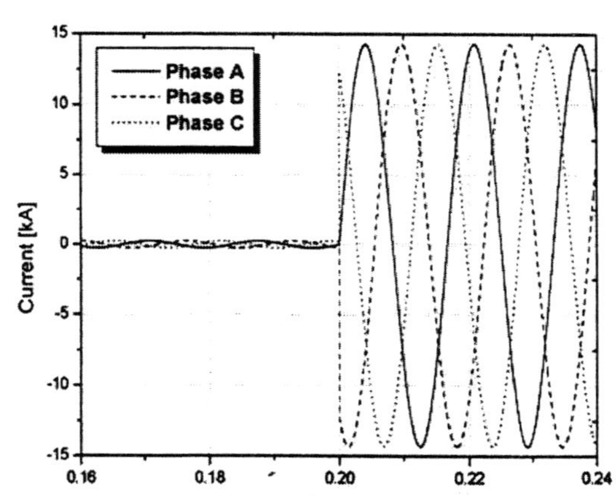

(a)

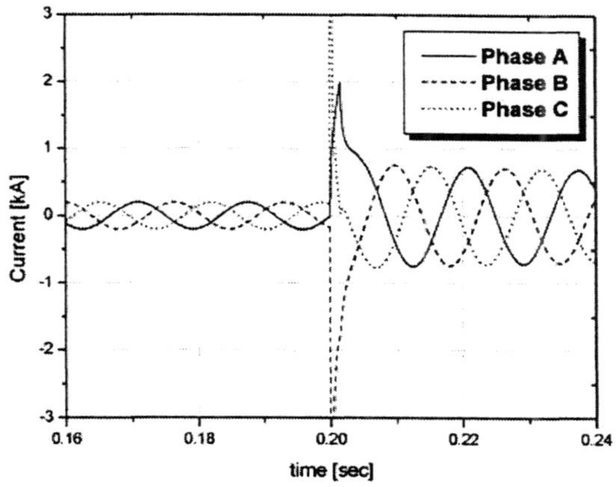

(b)

Fig. 10. (a) Waveforms of three phase short-circuit current without SFCL (b) Waveforms of three phase short-circuit current with SFCL

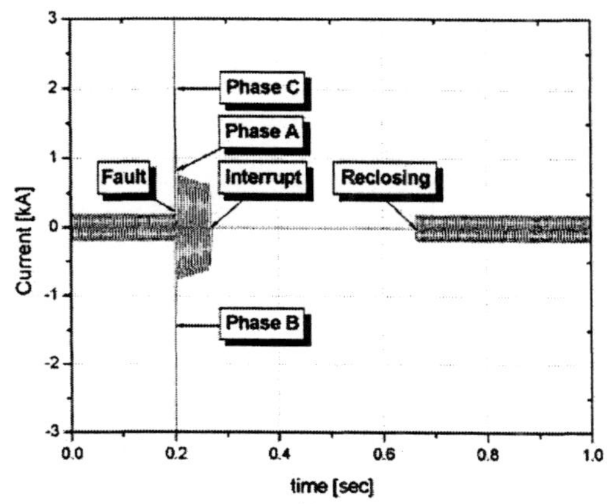

(a)

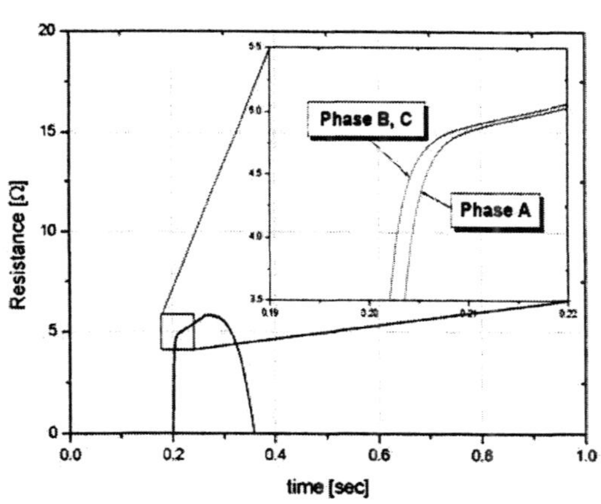

(b)

Fig. 11. Three-phase short-circuit current waveforms and the SFCL resistance variations due to circuit breaker operation. (Fault duration 0.2 [sec], interruption at 0.264 [sec] and reclosing at 0.664 [sec])

Fig. 8 shows waveforms of two line-to-ground fault current with SFCL. The results show that the currents are limited to 2 [kA] after fault has been occurred. Fig. 9 shows the simulation results of the two line-to-ground fault with circuit breaker operation.

Fig. 10 (a) and (b) represent the current waveforms of three phase short-circuit fault in the 3-phase power system without and with resistive type SFCL, respectively. It is confirmed that the fault current is limited to under 2 [kA] from 15 [kA]. Fig. 11 shows the simulation result of the three phase short-circuit fault with circuit breaker operation. Three phase short-circuit fault occurred at 0.2 [sec], in this time, the current is exceeded the critical current of 2[kA], the HTSC starts to heat up. The fault current pushed the superconductor into the resistive state and resistance R appears in the SFCL. That resistance is sufficient to reduce the fault current below the critical current. The circuit breaker operated at 0.264 [sec] after 4 cycles.

5. SIMULATION OF A DC REACTOR TYPE SFCL

Table 2 shows the simulation parameters. For this simulation, the utility voltage of 6600 [V], load resistance of 3.81 [Ω] and the ratio of transformer 1:1 are used, respectively. The DC reactor coil capacity is 52.5 [mH] (m = 4, n = 9).

Fig. 12 depicts the fault current and the DC reactor coil current during the fault. The result shows that the dc reactor coil current increases at certain rate after fault inception. The peak of the fault current coincides with the inductor current and both have the same rising rate.

Table 2 Simulation Parameters of DC reactor type SFCL

DC reactor coil	52.5[mH]
Ratio of transformer	1:1
Fault resistance	0.1[Ω]
Source Voltage	6.6[kV]
Source Current	1000[A]

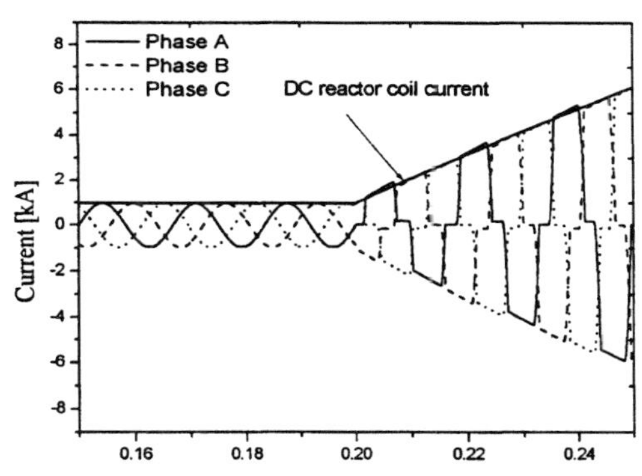

Fig. 12. Current waveforms during the fault of the DC reactor type SFCL

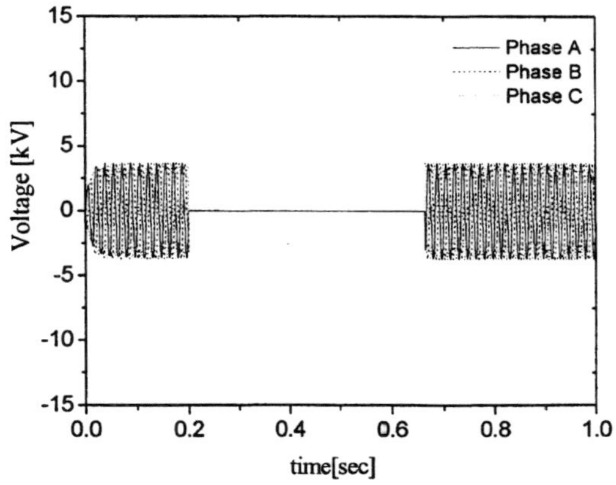

(a) Line-Line Voltage

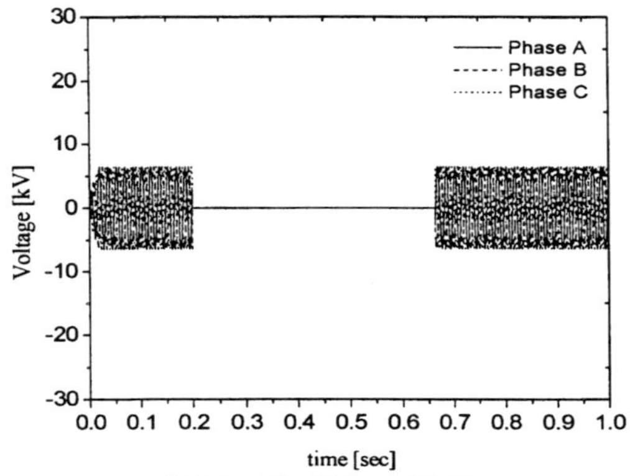

(b) Line-Line Voltage (RMS)

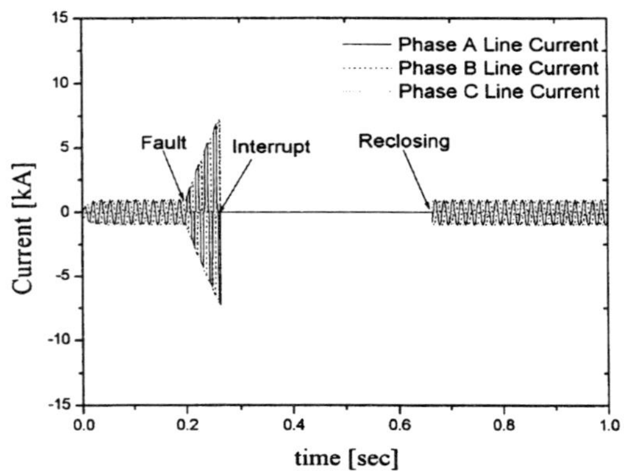

(c) Line Current

Fig. 13. Voltage and current waveforms for the simulation period

Fig. 13 shows the long-term simulation result of the voltage and the current waveforms due to circuit breaker operation of the reclosing. In this case, the fault occurred at 0.2 [sec]. After 4 cycles, the circuit breaker was opened, and reclosed at 0.664 [sec].

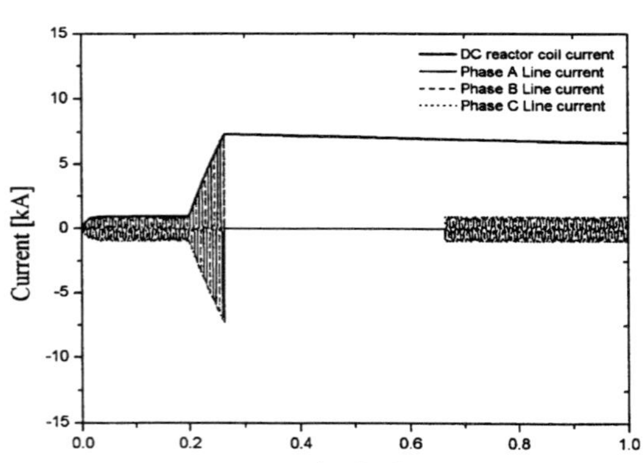

Fig. 14. DC reactor coil current and line current

Fig. 14 shows the DC reactor coil current and line current. The line current increases due to the fault, and the DC reactor coil current increases with the same shape. After the circuit breaker is opened at 0.264 [sec], the inductor current remains at higher value and slowly decreases to the steady state value.

6. CONCLUSIONS

An effective modeling and simulation schemes of both a resistive type SFCL and DC reactor type SFCL using PSCAD/EMTDC are proposed in this paper. Especially, an actual experiment based component model of the real resistive type SFCL is developed and applied for the simulation. The proposed simulation scheme can be implemented to the grid system readily under various system conditions including sort of faults and the system capacity as well. The simulation results demonstrate the effectiveness of the proposed model and simulation scheme. In addition, the simulation method actually contributes to the time and cost saving for the application study of the SFCLs.

ACKNOWLEDGEMENTS

This research was supported by a grant from Center for Applied Superconductivity Technology of the 21st Century Frontier R&D Program funded by the Ministry of Science and Technology, and in part by the Korea Science and Engineering Foundation (KOSEF) through the Machine Tool Research Center at Changwon National University.

REFERENCES

E. M. Leung, A. Rodriquez, G. W. Albert, etc. "High Temperature Superconducting Fault Current Limiter Development", *IEEE Transactions of Applied Superconductivity*, Vol. 7, No.2, June 1997.

K. TEKLETSADIK and M.P. SRAVOLAC, "Development of a 7.5 MVA Superconducting Fault Current Limiter", *IEEE Transactions on Applied Superconductivity*, Vol.9, No.2, June 1999.

M, Lindmayer, H. Mosebach "Quenching of High-Tc-Superconductors and Current Limitation Numerical Simulations and Experiments", *IEEE Transactions on Applied Superconductivity*, Vol.7, No.2, June 1997.

M P Saravolac, K Tekletsadik, A T Rowley, "High Temperature Superconducting Current Limiting Series Reactors", *CIGRE session - 1998*, paper 12-307, Paris, Sep. 1998.

M.Yamaguchi, S. Fukui, T. Satoh, Y. Kaburaki and T. Horikawa, "Performance of DC Reactor Type Fault Current Limiter Using High Temperature Superconducting Coil", *IEEE Transactions on Applied Superconductivity*, Vol. 9. No. 2. June 1999.

P.Tixador, "Superconducting Current Limiters Some Comparisons and Influential Parameters", *IEEE Transactions on Applied Superconductivity*, Vol.4, No.4, pp.190-198, December 1994.

Takahiro Nomura, Mitsugi Yamaguchi, Satoshi Fukui, Kazuya Yokoyama, "Single DC Reactor Type Fault Current Limiter for 6.6kV Power System", *IEEE Transactions on Applied Superconductivity*, Vol. 11. No. 1. March 2001.

ELSEVIER
IFAC
PUBLICATIONS
www.elsevier.com/locate/ifac

EXPERT SYSTEM FOR DIAGNOSING SILICONE OIL-FILLED TRANSFORMERS

Jong-Fil Moon Jae-Chul Kim Tae-Hoon Im

Department of Electrical Engineering in Soongsil University
Seoul 156-743, Korea

Abstract: In this paper, expert system for diagnosing silicone oil-filled transformer was developed using dissolved gas analysis (DGA). DGA was used to the proposed expert system since it has been verified that it is very efficient diagnostic method for oil transformer. In addition, it is reasonable that fuzzy rule, degree of inclusion and fuzzy measure must be considered to handle the uncertainty of gas boundary and rules. The proposed expert system consists of knowledge based module, inference engine module and human-machine interface (HMI) module. The knowledge base consists of the knowledge using the rule. The inference engine module is used to the fuzzy rule. The history of the transformer gas data is managed by the database. The effect of the proposed expert system is verified by case studies. *Copyright © 2003 IFAC*

Keyword: silicone oil-filled transformer, expert system, dissolved gas analysis, fuzzy rule, database

1. INTRODUCTION

Since its introduction in 1974, silicone oil has become widely accepted for power transformers and distribution transformers and used in the transformers of railway vehicles, etc in Korea. As it is used about 30 years, the number of in-service transformers grows and their average age increases, the need for early detection of incipient fault conditions becomes more important (Chatterton and Goudie, 2000; IEEE Std, 1999).

The transformers are faulted because of electrical and mechanical performance decline resulted from aging under operation. It is necessary to detect the symptom of the matter at the first stage, grasp the matter exactly and work out the countermeasures for preventing this fault previously (Chatterton and Goudie, 2000; IEEE Std.,1999; Toskey, *et al.*,1995).

In this paper, thus, expert system for diagnosing silicone oil-filled transformer was developed using DGA. First, the dissolved gas characteristics and threshold level of silicone oil were compared to it of the mineral oil. By applying DGA to silicone oil, we made the fault of transformers detected effectively. The proposed expert system consisted of knowledge base module, inference engine module and human-machine interface (HMI) module. The knowledge based module consisted of the knowledge using the rule. The fuzzy rule was used to the inference engine module for solving the uncertainty of the boundary of danger level and rules, and it was gotten the belief measure and plausibility measure by using Dempster- Shafer theory. The HMI module is programmed to GUI environment by using the BASIC language for the field engineer can use it easy. Also diagnosis can be performed more effectively by the history of each gas data of transformers was managed by the database.

For verifying this expert system, we performed the case studies using the data that IEEE proposed. By analyzing the relationship of the real data and the data inferred from the expert system, the usefulness of the expert system is verified.

2. THE METHOD OF DGA

2.1 The Method of DGA

If the transformers are used during the long time, the partial temperature rise and the aging of insulation oil is occurred in transformers by the increase of moisture, combustive gas and solid impurities. If the matter inside of transformers, that is the partial discharges, corona and partial overheat, is occurred, the heat follows and the insulation materials, insulation oil, insulation paper and pressboard etc, attached to this heat source is reduced and generate hydrogen (H_2), methane (CH_4), ethane (C_2H_6), ethylene (C_2H_4), acetylene (C_2H_2), carbon monoxide (CO) and carbon dioxide (CO_2). These gases are mostly dissolved in insulation oil (Jeon, 1996; Kim, *et al.*, 1998).

Thus, by analyzing the dissolved gas in insulation oil of transformers, we can recognize existence and degree of the matter according to gas type and gas amount. Up to date, it has been the many studies about DGA of mineral oil. But it has not about the silicone oil, it is the merely IEEE guide and study that the Dow corning co. and Doble co. did.

2.2 The Dissolved Gas Characteristics of Each Fault Type

The major gases of the DGA of silicone oil-filled transformers are the following: H_2, CH_4, C_2H_6, C_2H_4, C_2H_2, CO and CO_2. The elements for judging the transformer the matter among these gases are H_2, C_2H_6, C_2H_4 and C_2H_2, and the elements for judging the insulation oil the aging are CO, CO_2 and CH_4. The CO and CO_2 is related to the overheat of insulation oil, additionally, the existence of C_2H_6 and C_2H_4 divides the fault between the overheat of insulation oil and solid insulation material. And H_2 and CH_4 is related to arc fault, additionally, the quantity of C_2H_2 and CO divides the fault between the arc in insulation oil and solid insulation material. These contents are presented at the Table 1 by the comparison of mineral and silicone oil.

Table 1 Dissolved gas characteristics of mineral and silicone oil

Gas	Insulation oil overheat		Solid insulation material overheat		Arc dissolving in insulation oil		Arc dissolving in solid insulation material	
	MO	SO	MO	SO	MO	SO	MO	SO
H_2	○	○	○	-	◎	◎	◎	◎
CH_4	◎	-	◎	-	○	◎	○	◎
C_2H_6	○	○	○	-	-	○	-	-
C_2H_4	◎	◎	◎	-	○	-	○	-
C_2H_2	-	-	-	-	◎	○	◎	○
CO	-	◎	◎	◎	-	○	◎	◎
CO_2	-	◎	◎	◎	-	-	◎	-

NOTES.
MO : Mineral Oil, SO : Silicone Oil
○ : generated gas, ◎ : major generated gas
by IEEE Std. C57.104 & IEEE Trial Use Guide P1258

2.3 Dissolved Gas Threshold Level

Now, IEEE Std. expresses the dissolved gas threshold level by the ppm and classifies into the danger when the specified level is exceeded. The level for the mineral oil is divided into danger, warning and normality at the international standard, but the silicone oil is divided into danger, normality. Thus, for effective comparison, only the danger level for mineral and silicone oil was compared.
Generally, the H_2 and CH_4 level for mineral oil is higher to silicone oil, but the CO and CO_2 level is lower. Also, C_2H_6, C_2H_4 and C_2H_2 level is low in both sides. The dissolved gas threshold level of mineral and silicone oil is presented at the Table 2.

Table 2 Dissolved gas threshold level of mineral and silicone oil

Gas	Threshold level [ppm]	
	Mineral Oil	Silicone Oil
H_2	1,800	200
CH_4	1,000	100
C_2H_6	150	30
C_2H_4	200	30
C_2H_2	80	1
CO	1,400	3,000
CO_2	10,000	30,000

2.4 The Problems and Countermeasures of DGA

Although the exact diagnosis of transformers is performed more exactly using the DGA relative to the others, some problems exist for doing the real application. That is, the DGA is affected easily by establishment time, construction, capacity and oil quantity of transformers. In addition, the measured value has the different value according to the measured methods. Also when it is constructed to the expert system using main gas analysis, the collision of rules is occurred (Jeon, 1996; Kim, *et al.*, 1998).
For example, if C_2H_2 exceeds the 30 [ppm], it is danger condition. At this time, 29 [ppm] is judged to normality and 31 [ppm] is judged to danger. It is classified to normality and danger because of the difference of only 2 [ppm]. But it is conflict on classifying to normality and danger according to this very small difference like it because the error of measure exists always.
Also, because the uncertainty always exists at the knowledge, the uncertainty always exists at the rules of the expert system. For solving these problems, the fuzzy set was used at the uncertainty of the data and the boundary, the fuzzy measure was used at the uncertainty of the rules.

3. EXPERT SYSTEM

3.1 The Definition of Expert System

The specialist is the man that can memorize the expert knowledge, infer the conclusion according to the logical inference ability and explain something persuasively. The expert system is the software that has these abilities. That is, as a field of artificial intelligence, the expert system is the computer system that imitates and embodies the ability of the thought of expert and the required knowledge for it that can settle the complicated problems of the special field, ultimately that the many general man can substitute to the a small number of expert (Giarratano, 1989; Lee, *et al.*, 1998).
Unlike the general software, expert system can apply the knowledge consistently because of the division of the knowledge base and the inference engine. Also the expert system has the infinite life and can copy and move the knowledge easily.
The expert system is mainly divided to knowledge based module, inference engine module and HMI module but the whole expert system includes working memory, knowledge acquisition assistance system and explanation assistance system like Fig. 1.

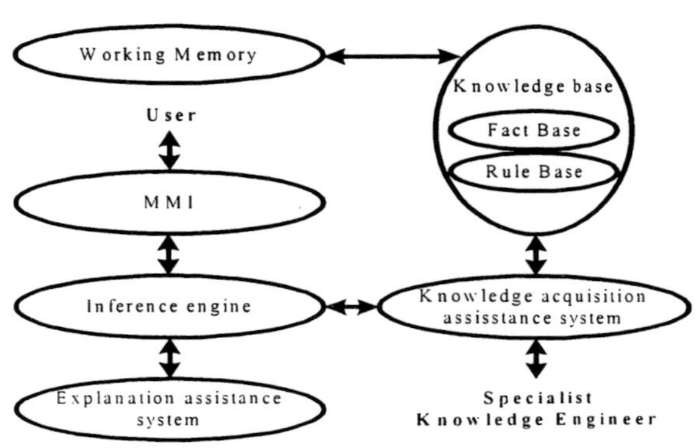

Fig. 1. Configuration of expert system

3.2 Knowledge Based Module

The knowledge is the required information for computer to be able to settle some problem intellectually. The information is the form of fact or rule. The expert system is the whole of this knowledge.
The fact is the information that the specialist acknowledges and many people own jointly. The rule is something for the reasonable decision for the private and little-discussed problems for the decision of the level of the specialist does and it is said to heuristic.
The knowledge based module is the database of the knowledge for the special subject used in the expert system and this database consists of the facts needed in solving the problems and the rules used for the data of the some object in these facts (Giarratano, 1989; Lee, *et al.*, 1998).
The rule, semantic net, frame and logic are the methods for representing the knowledge base. In this paper, The rules was used for it. This method is expressed the condition and the performed action when the condition is satisfied, generally 'If <condition> Then <result>'. The form of the used rules is equal to the following.

Rule rule (R_i)
If condition (A_j)
Then result (B_k)
Required belief (m)

3.3 Inference engine module

The inference engine has the knowledge for the settling overall problem and it is the most important part that supervises overall expert system. And the object of the expert system is that it gets the results like specialist. But because the knowledge of specialist is abstract, thus expert system has the defect of inaccuracy, uncertainty and imperfection (Giarratano, 1989; Lee, *et al.*, 1998). Also there is uncertainty in the dissolved gas threshold level and inference rules. Thus, for transforming the abstract knowledge of specialist to a numerical formula properly, based on the dissolved gas data of the knowledge base, the fuzzy rules were applied to the inference engine module and the belief measure of normality and fault was gotten. The block diagram of the inference engine module is presented at Fig. 2.

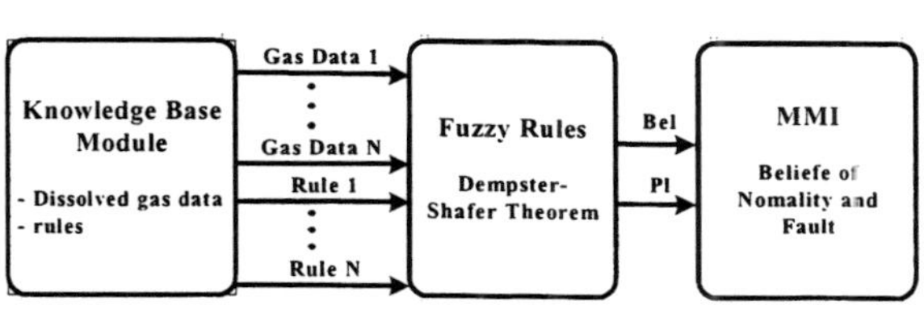

Fig. 2. Inference engine module

3.3.1 Settlement of the Uncertainty of the Boarder

For settling the uncertainty of the boarder of dissolved gas threshold level, the fuzzy rules were used. For setting up the effective membership degree, the formula suggested by Dombi was used, and it is Eq.(1). This method has the advantage of being able to set the wanted membership degree through changing the value of coefficients (LEE, *et al.*, 1998).

$$\mu(x) = \frac{(1-\upsilon)^{\lambda-1}(x-a)^{\lambda}}{(1-\upsilon)^{\lambda-1}(x-a)^{\lambda} + \upsilon^{\lambda-1}(b-x)^{\lambda}} \quad (1)$$

where the a is the value of the lowest limit, b is the value of the highest limit, λ is change rate and υ is bending point.
The membership degree was obtained using the Eq.(1) between a and b. The membership degree became 0.01 under a and 0.99 over b for removing the error of computation. The table 3 represents parameters(a, b, λ, υ) of key gas components.

Table 3 Parameters of key gas components

Parameter	H_2	CH_4	C_2H_6	C_2H_4	C_2H_2	CO	CO_2
a	160	80	24	24	0.8	2,400	24,000
b	240	120	36	36	1.2	3,600	36,000
λ	4 for all						
ν	0.6 for all						

Based on Table 3, the membership degree μ(x) of key gas components is drawn like Fig. 3.

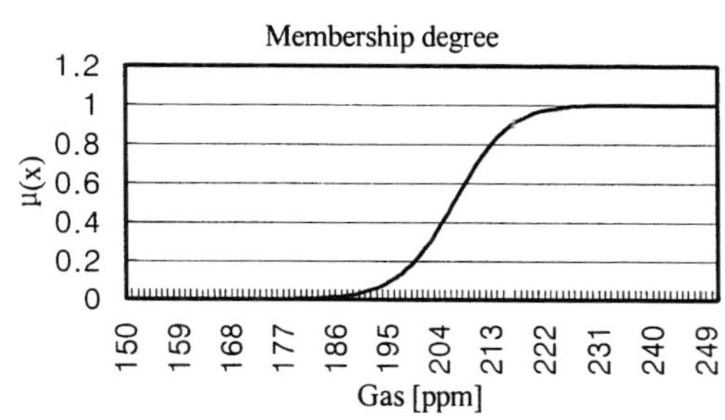

(a) H_2

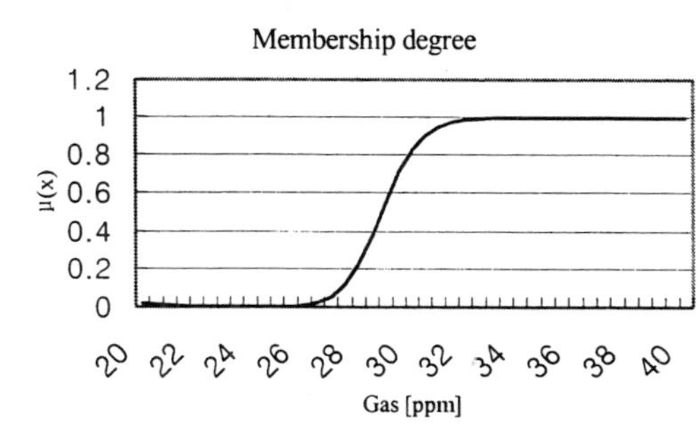

(b) C_2H_2

Fig. 3. Membership degree of each gas

3.3.2 Settlement of the Uncertainty of the Rules

The knowledge contains the uncertainty inevitably. Thus the method dealing with truth and lie only like mathematical logic can't represent the uncertainty occurred practically. Early the Bayesian probability was used for dealing with the uncertainty, but because of the difficulty of the dealing with ignorance, Dempster suggested the new notion of the upper and lower probability in 1967. Shafer improved the Dempster's theory and changed the upper and lower to belief and plausibility for giving the meaning.
The function, f, representing the fuzzy measure is

equal to Eq.(2).

$$f : P(X) \rightarrow [0,1] \tag{2}$$

condition :

$$m(\phi) = 0 \quad \text{and} \quad \sum_{A \in P(X)} m(A) = 1$$

where m(A) is the evidence of set A and is said to a basic probability assignment. The belief measure and plausibility measure can be defined like Eq.(3) and (4) based on a basic probability assignment, the other way, m(A) is obtained using Bel, Pl like Eq.(5).

$$Bel(A) = \sum_{B \subseteq A} m(B) \tag{3}$$

$$Pl(A) = \sum_{B \cap A \neq 0} m(B) \tag{4}$$

$$m(A) = \sum_{B \subseteq A} (-1)^{|A-B|} Bel(B) \tag{5}$$

3.4 Human-Machine Interface Module

The HMI module has the function connecting the user to the system. In this paper, HMI module was constructed to GUI environment, if the amount of each gas are given, then the belief measure of the normality and fault is inferred and displayed.

Fig. 4 and Fig. 5 are the input windows of the gas data and the output windows of inferred results respectively.

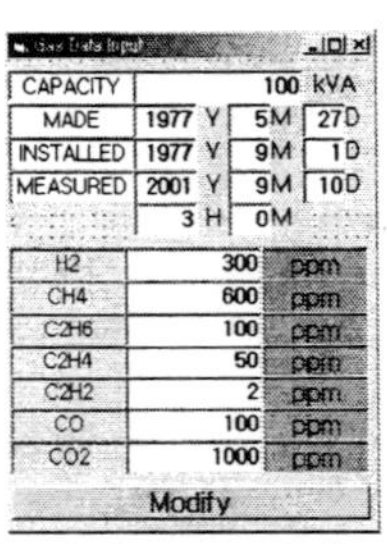

Fig. 4. Input-windows of gas data

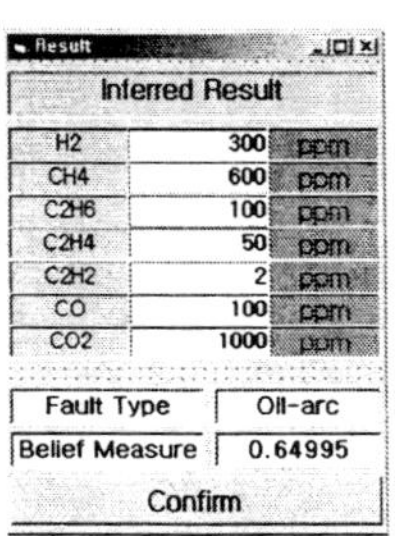

Fig. 5. Result of inference

4. DATABASE

Because the DGA is affected easily by the establishment time, construction, capacity and oil quantity of transformers and the measured value has the different value according to the measured methods, the history of each transformer must be managed for effective diagnosis. Also the data needed in monitoring and diagnosing transformers will increase very much as time goes. Especially, the effective management of data is the difficult circumstances because of the overlap of data. Thus the construction of database is necessary for the effective management of data and the fast query of needed data (Moon, *et al.*, 2000; Korth).

In this paper, thus, as the proposed expert system was connected to database, the history management and synthetic management of transformers became possible.

First, the characteristics of entity and attribute were summarized (Table 4), entity-relationship model was expressed (Fig. 6) based on entity-attribute and database was developed (Fig. 7).

Table 4 Characteristics of entity and attribute

Entity	Attribute
Transformer	Capacity(KVA), Computerized Number(Pno), Made YM(Make YM), Installed YM(Set YM)
Connection	Transformer Identification(TR ID), Computerized Number(Pno)
Input Data	Measured YMDYMS(GetYMDHMS), Gas ppm
Output Data	Measured YMDYMS(GetYMDHMS), Belief(Bel), Plausibility(Pl), Action(act)

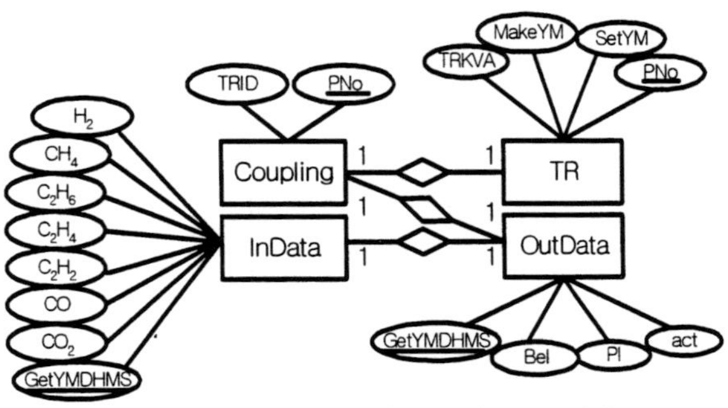

Fig. 6. DB entity-relationship model

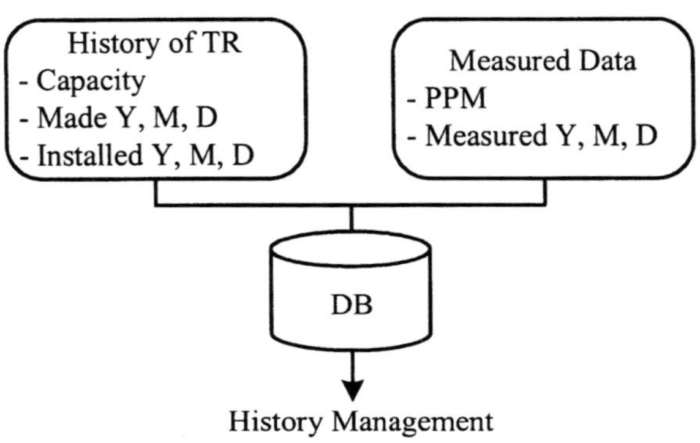

Fig. 7. Development of database

5. CASE STUDIES

Using the four data represented at IEEE Std. and the measured one data, the expert system was verified. The first data is normal, the second is overheated fault of insulation oil, the third is overheated fault of solid insulation material, the fourth is arc fault of insulation oil and the fifth is arc fault of solid insulation material. Table 5, 6, 7, 8 and 9 is the case studies of each data.

Table 5 Case study 1

Gas	H_2	CH_4	C_2H_6	C_2H_4	C_2H_2	CO
ppm	10	20	5	5	0	100
IEEE	Normal					
Expert System	Arc(Solid Insulation Material) Belief Measure of fault : 0.1115					

At Table 5, because the gas exceeding threshold level doesn't exist, IEEE Std. judged this case normality and the proposed expert system judged this case normality like that belief measure of fault was inferred to 0.1115, too.

Table 6 Case study 2

Gas	H_2	CH_4	C_2H_6	C_2H_4	C_2H_2	CO
ppm	150	20	50	50	0	4500
IEEE	Overheat(Insulation Oil)					
Expert System	Overheat(Insulation Oil) Belief Measure of fault : 0.9036					

Table 7 Case study 3

Gas	H_2	CH_4	C_2H_6	C_2H_4	C_2H_2	CO
ppm	250	200	0	0	0	4500
IEEE	Overheat(solid insulation material)					
Expert System	Overheat(solid insulation material) Belief Measure of fault : 0.9234					

At Table 6 and 7, like viewing CO, it is sure that this phenomenon is overheated fault, and the quantity of C_2H_6 and C_2H_4 divide this fault to solid insulation material and insulation oil. That is, when the overheated fault is sure, if the quantity of C_2H_6 and C_2H_4 exists, then it is the fault for insulation oil, else it is the fault for solid insulation material. In this case, proposed expert system divided this fault to overheated fault of oil and solid material exactly and gave the belief measure 0.9036 and 0.9234, respectively.

Table 8 Case study 4

Gas	H_2	CH_4	C_2H_6	C_2H_4	C_2H_2	CO
ppm	300	600	100	50	2	100
IEEE	Arc(Insulation Oil)					
Expert System	Arc(Insulation Oil) Belief Measure of fault : 0.6500					

At Table 8, based on the quantity of H_2 and CH_4, it is sure that this phenomenon is arc fault, and this case was judged to arc fault of oil because the quantity of CO is very small. Expert system represented the same result and gave the belief measure 0.6500, too.

Table 9 Case study 5

Gas	H_2	CH_4	C_2H_6	C_2H_4	C_2H_2	CO
ppm	5480	2790	39	308	448	3190
IEEE	Arc(Solid insulation material)					
Expert System	Arc(Solid insulation material) Belief Measure of fault : 0.8539					

At Table 9, In the end, all data exceeds the threshold level. but based on its ratio, IEEE std. judged this fault to arc fault of solid material. Expert system judges this case same fault, too. And belief measure of fault is inferred to 0.8539.

6. CONCLUSIONS

In this paper, the expert system was developed that can diagnose the silicone oil-filled transformers effectively, using DGA that used generally widely because reliability is high among the state diagnosis methods.

The developed expert system consisted of knowledge based module, inference engine module and HMI module. The knowledge base consisted of the knowledge using the rule and transformed the descriptive knowledge to numerical value. The inference engine module infers the kind of fault using inputted data. Especially, for solving the uncertainty of the boundary of danger level, fuzzy rules and Dombi's equation was used. Also, for solving the uncertainty of the rules, Dempster-Shafer's theory was applied and belief measure of fault was calculated. The HMI module was programmed to GUI environments by using the BASIC language. Also, expert system was connected to database, for diagnosing the transformers more effectively by managing the history of it.

In case studies, the expert system was verified, as the results of expert system were same to real results. After this, for more effective diagnosis, various diagnosis methods must be added. Also if the developed expert system is complemented through accumulation of many real gas data and exact management of the history of transformer, the more effective diagnosis will be possible.

This work was supported by the Soongsil University Research Fund

REFERENCES

Chatterton, W. J. and Goudie, J. L. (2000). An Update on Silicone Transformer Fluid: Manufacturing, Maintenance and End of Life Options. *Conference record of the IEEE International Simposium on Electrical Insulation*. pp. 412-416.

Dombi J. (1990). Membership Function as an Evaluation. *Fuzzy Set and System*, 35, pp. 1-21.

Giarratano, J. (1989). *Expert Systems*, PWS-KENT Publishing Company.

IEEE. (1999) *Trial-Use Guide for the Interpretation of Gases Generated in Silicone-Immersed Transformer Draft-9*. IEEE. Std. P-1258

Jeon, Young-Jae (1996). A Fault Diagnostic Expert System for Transformers using Dissolved Gas Analysis. *M. S. Thesis of Soongsil University*. Korea.

Kim Jae-Chul (1998). *et al*, A Kohonen Network for Fault Diagnosis for Power Transformers using Dissolved Gas Analysis. *Trans. on KIEE*. **Vol. 47, No. 6.**

Korth F. H (2001). *Database Systems*. McGraw-Hill.

Lee, Jae-Kyu, et al (1998)., *principal and development of expert system*. JiYoung-Sa. Korea.

Moon Jong-Fil (2000). The Monitoring Program and Database Construction for Diagnosing Distribution Transformer in On-line. *The Conference of KIEE*, pp. 177-179.

Toskey, G. A. *et al* (1995) Manufacturing Considerations for Silicone Fluid Filled Transformer. *Conference on Electrical Electronics Insulation*. pp.275-279.

ELSEVIER
IFAC
PUBLICATIONS
www.elsevier.com/locate/ifac

OPTIMAL DESIGN OF RESISTIVE TYPE HTS-FCL USING MONTE CARLO METHOD

Yoon, Jae-Young Kim, Jong-Yul Choi, Heung-Kwan
Korea Electrotechnology Research Institute

Abstract: Nowadays, one of the serious problems in KEPCO system is much higher fault current than the SCC(Short Circuit Capacity) of circuit breaker. As the superconductivity technology has developed, the HTS-FCL(High Temperature Superconductor-Fault Current Limiter) can be one of the attractive alternatives to solve the fault current problem. But the parameters of HTS-FCL should be designed optimally to have the best performance. Under this background, this paper presents the optimal design method of parameters for resistive type HTS-FCL using Monte Carlo technique.

1. INTRODUCTION

As the power system has developed, the fault current increases, then the short circuit capacity can be exceeded in KEPCO transmission system. This comes up to a real serious problem of system planning and operation viewpoints.[1] Various countermeasures such as separation of busbars and lines, enhancement of circuit breaker's short circuit capacity, application of fault current limiter and BTB(Back-to-Back) HVDC system can be alternatives.[2] All of these alternatives have drawbacks of viewpoint on system stability or effectiveness of fault current reduction and etc. But, as the HTS technology has developed, the HTS-FCL can be the most effective alternative to reduce the fault current as compared with other countermeasures in terms of system stability, cost effectiveness and fault current reduction.[3-6]

In order to apply the HTS-FCL to practical power system, the simulation results of transient behavior including steady state, quenching and recovery state is needed. Also it is necessary for the dynamic model to represent overall phenomena, in order to obtain the simulation results, and the optimal design of HTS-FCL parameter should be performed to maximize the effect of fault current reduction and maintain the stability of power system when fault occurs. Therefore, it is presented the optimal design method of parameters for resistive type HTS-FCL using Monte Carlo technique and the typical case study is performed to confirm the effectiveness of this method with typical power system model which is similar to the practical system under the various conditions in this paper.

2. PARAMETER OF RESISTIVE TYPE HTS-FCL

HTS-FCL has a zero resistance under the static conditions. But, if fault current exceeds the critical value, called quenching state, the resistance of HTS-FCL increases and the fault current is limited to a certain specified value. The fault current reduces below the critical value after HTS-FCL operates and fault clears. After that, the characteristics of HTS-FCL are recovered by reducing the resistance as a zero.

The important parameters to represent the dynamic characteristics of HTS-FCL are described in Table 2.1. Among parameters in Table 2.1, the operation characteristics of HTS-FCL are dependent on design parameters such as R_{FIN}(Final resistance value), I_{OP}(Initial operating current), T_{FCL}(time constants). These parameters have the reverse characteristics to determine the magnitude of fault current. For example, as the R_{FIN} is larger, I_{OP} is smaller or T_{FCL} is shorter than others, the effectiveness of HTS-FCL to reduce the fault current is higher and higher. But, even if these three parameters are well-designed by prescribed value, the overall effect of HTS-FCL is dependent on the stochastic characteristics of mutual effect of each parameter. Therefore, it is necessary to consider the stochastic characteristics of HTS-FCL parameters for optimal design.

Table 2.1 General definitions of HTS-FCL parameter

Parameter	Definitions	Remarks
CB_{SCC}	SCC of circuit breaker (kA)	Short Circuit Capacity
I_{FCL}	fault current flowing into HTS-FCL	
I_{op}	Initial operating current of HTS-FCL (kA)	upper limit : (SCC - Margin) lower limit : (Max. load current +Margin)
R_{FIN}	final resistance value in quenching state(Ω)	dependent on CB_{SCC} I_{op}
T_{FCL}	time constants of HTS-FCL(msec)	
T_{REC}	recovery time of HTS-FCL(msec)	related to reclosing

3. EMTDC MODEL OF HTS-FCL

This paper presents the transient model of HTS-FCL considering static and quenching state by Fig. 3.1. When fault occurs, the fault current of HTS-FCL model in Fig. 3.1 is dependent on various parameters described in Table 3.1. As a consequence, the resultant resistance value of HTS-FCL is influenced on the nonlinear physical factors such as fault current, final resistance value, initial operating current and temperature.

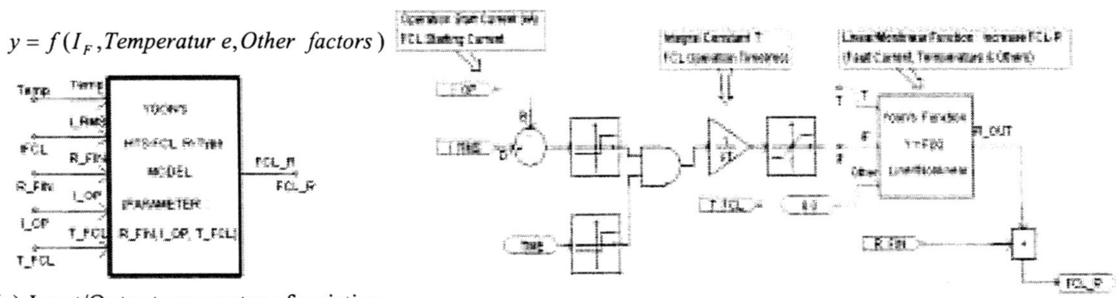

(a) Input/Output parameter of resistive HTS-FCL model

(b) Detailed contents of resistive HTS-FCL model

Fig. 3.1 Transient model of resistive HTS-FCL model

4. DESIGN METHOD OF OPTIMAL PARAMETER

Monte Carlo Simulation is one of the optimal design method applied to the case that input parameters have stochastic characteristics and mutual dependency. So, Monte Carlo simulation is a probabilistic method considering the input parameter as a random variable and can be used by effectiveness methodology to design optimal parameter of HTS-FCL. In this paper, R_{FIN} and I_{op} are considered as random variables and the design procedures are as follows.

(step_1) Setting up the upper and lower limit of random variables and perform the case study for time constants of HTS-FCL(T_{FCL})

○ Final resistance value as a random variable : $10 \leq R_{FIN} \leq 60\ (\Omega)$

○ Initial operating current as a random variable : $5 \leq I_{op} \leq 10$ (kA)

○ Time constants for case study : $0.001 \leq T_{FCL} \leq 0.01$ (sec)

(step_2) Determine input random parameters for each case using random generator within upper and lower limit.

(step_3) Simulation of model system using random input parameters determined in step_2 for each case

(step_4) Analysis of study results using optimization function of Eq. (2) : This optimization function is to minimize the steady state continuous fault current flowing into HTS-FCL when fault occurs.

$$Min_x[f(x)] = Min\,|\,I_{FCL}\,| \qquad (2)$$

where, I_{FCL} : Steady state continuous fault current flowing into HTS-FCL(kArms)

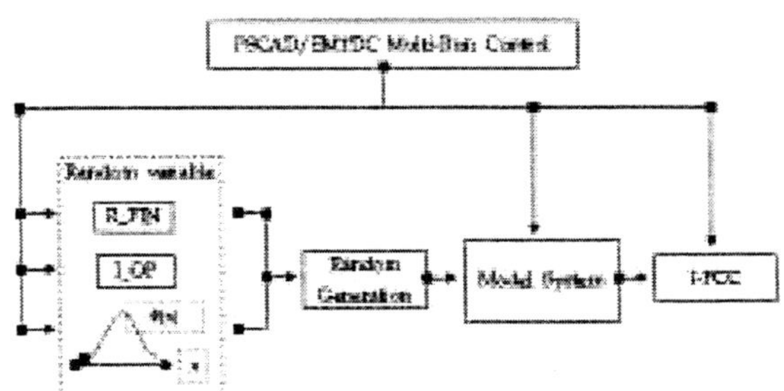

Fig 4.2 Overview of Probabilistic analysis

5. CASE STUDY RESULTS

5.1. Overview of analysis

5.1.1. Model system

Model system of Fig 5.1 is presented to confirm the effectiveness of this method. This model system represents the similar characteristics to KEPCO 154kV system, which have similar average short circuit capacity, overhead line, cable configurations and etc.

5.1.2. Model system data

Basic data to analyze the model system with HTS-FCL are as follows.

○ Equivalent source of sending and receiving end : $1.02\angle 10°$(PU) and $1.0\angle 0°$

○ Source impedance : $2.223(\Omega)\angle 85.0°$ (in sending and receiving end)

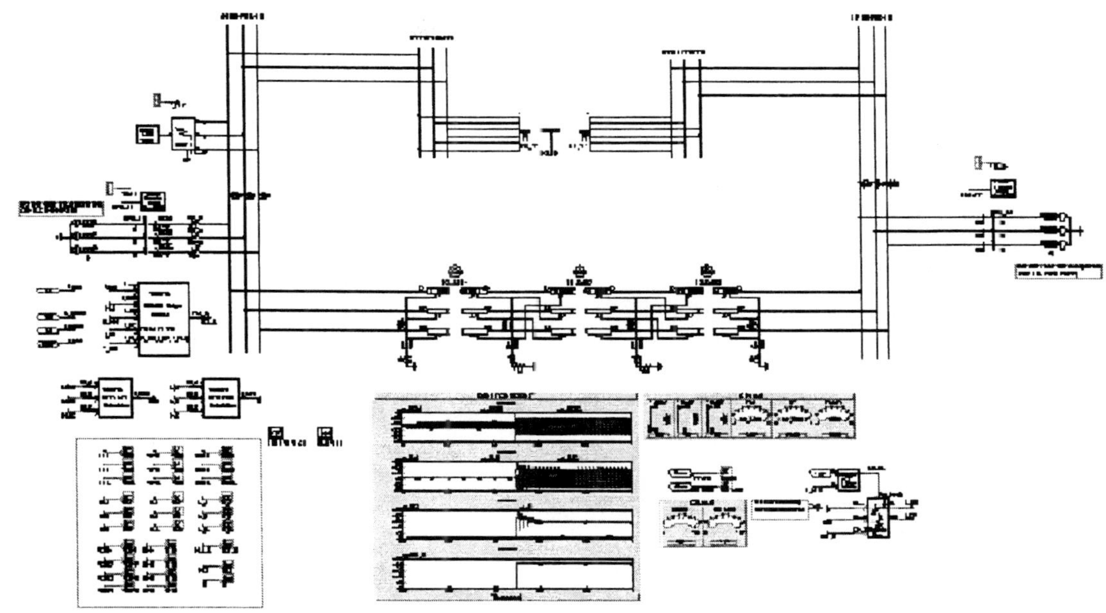

Fig 5.1 Model system for KEPCO 154kV system with resistive HTS-FCL

- ○ Overhead line : ACSR 410[mm^2]x2B, frequency independent, non-transposed model (10.0km)
- ○ Cable line : F 2000[mm^2], frequency independent, non-transposed model (10.0km)
- ○ Random variable (R_{FIN} & I_{op}) : with limits specified in step_1

5.2. Case study results

5.2.1. Base case

Fault study in model system is performed as the following upper and lower limits of random variables.

- ○ $10 \le R_{FIN} \le 60$ (Ω)
- ○ $5 \le I_{op} \le 10$ (kA)

The input parameter of these random variables varies based on three different methods of 100 cases.

- ○ Sequential : input parameter varies step by step for each case
- ○ Normal : input parameter varies by normal distribution
- ○ Random : input parameter varies by random generator

When T_{FCL} is 0.002, the study results of this base case are described in Table 5.1. It means that the steady state continuous fault current in the case of sending bus 3-phase fault has a minimum value of R_{FIN}(57.3Ω) and I_{op} (5.0kA) when random generator is used to determine the input parameter. When HTS-FCL doesn't exist, the total steady state continuous fault current is 50.0kA. These results confirm the effectiveness of HTS-FCL reducing fault current.

Generally speaking, it is expected to have a minimum value when the R_{FIN} is 60.0(Ω) and I_{op} is 5.0(kA). But, it proved to be inaccurate via sequential simulation. This is caused by the complex and mutual effect of random variables. If R_{FIN} is 60.0(Ω), IFCL(fault current flowing into HTS-FCL) goes down more rapidly under the critical value(I_{op}= 5.0) as compared with other specific cases, HTS-FCL is no longer operate as a countermeasure of fault current reduction in this case.

Table 5.1 Study results of basic case

CASE	Optimal parameter	Results IF(kA)	IFCL(kA)	FCL_R	Remarks
No HTS-FCL		50.0			Without HTS-FCL
T_{FCL}= 0.002 (sec)	R_{FIN} = 57.0(Ω) I_{op} = 5.0(kA)	24.3	3.31	27.1(Ω)	Sequential
T_{FCL}= 0.002 (sec)	R_{FIN} = 56.1(Ω) I_{op} = 5.1(kA)	24.3	3.37	26.6(Ω)	Normal distribution
T_{FCL}= 0.002 (sec)	R_{FIN} = 57.3(Ω) I_{op} = 5.0(kA)	24.3	3.30	27.2(Ω)	Random

*) IF : Total fault current, IFCL : fault current flowing into HTS-FCL

FCL_R : Final fault resistance

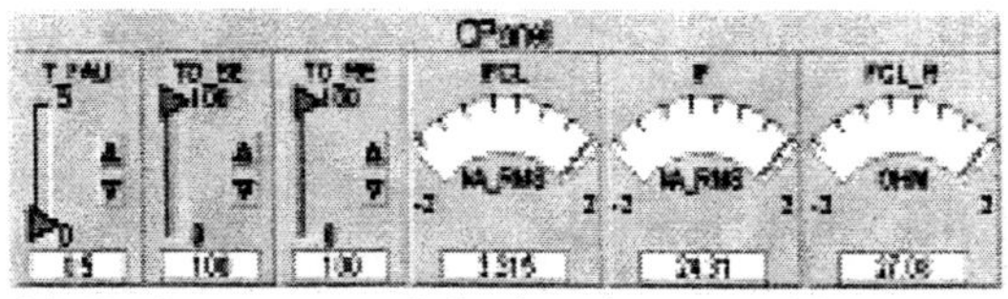

(a) Fault current and final resistance with HTS-FCL (Sequential case)

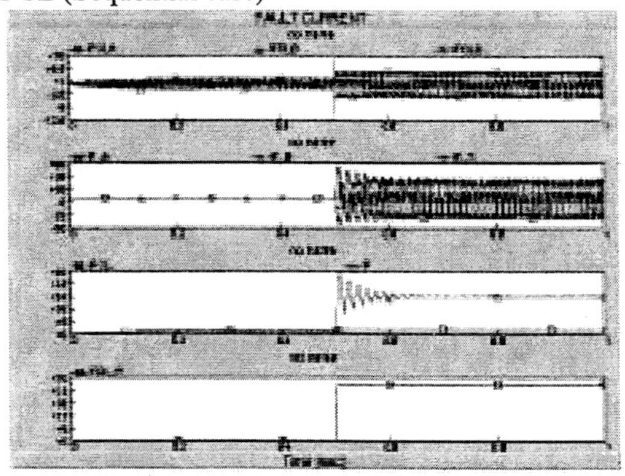

(b) Fault current and final resistance with HTS-FCL (Sequential case)

Fig 2.5 Basic study results with HTS-FCL (fault occurs T=0.5sec)

5.2.2. Case study

In this paper, some cases with different time constants of HTS-FCL are analyzed. The overall study results are described in Table 5.2 and 5.3. Same as the above study results, the optimal parameter is dependent on the complex and mutual effect of random variables.

Table 5.2 Case study results with various T_{FCL}(input parameter varies sequentially)

CASE	Optimal parameter	Study results IF(kA)	IFCL(kA)	FCL_R	Remarks
T_{FCL}= 0.001 (sec)	R_{FIN} = 59.0(Ω) I_{op} = 5(kA)	24.18	3.05	29.5(Ω)	
T_{FCL}= 0.003 (sec)	R_{FIN} = 56.0(Ω) I_{op} = 5(kA)	24.44	3.56	25.2(Ω)	
T_{FCL}= 0.005 (sec)	R_{FIN} = 59.0(Ω) I_{op} = 5(kA)	24.61	3.89	23.0(Ω)	
T_{FCL}= 0.01 (sec)	R_{FIN} = 60.0(Ω) I_{op} = 5(kA)	24.89	4.38	20.4(Ω)	

Table 5.3 Case study results with various T_{FCL}(input parameter varies randomly)

CASE	Optimal parameter	Study results IF(kA)	IFCL(kA)	FCL_R	Remarks
T_{FCL}= 0.001 (sec)	R_{FIN} = 59.9(Ω) I_{op} = 5.0(kA)	24.16	3.00	29.96(Ω)	
T_{FCL}= 0.0015 (sec)	R_{FIN} = 51.4(Ω) I_{op} = 5.25(kA)	24.29	3.27	27.4(Ω)	

6. CONCLUSION

This paper presents the optimal design method of resistive type HTS-FCL using Monte Carlo technique. The overall study results show that the optimal parameter is dependent on the complex and mutual effect of random variables. It is confirmed the effectiveness of HTS-FCL in model system with similar characteristics of KEPCO 154kV system. As a consequence, it is expected that resistive HTS-FCL can be one of the most effective alternative to reduce the fault current. The detailed study for the comparisons with other alternatives countermeasures in terms of system stability, cost effectiveness and fault current reduction are needed.

Acknowledgment

The authors would like to express their great gratitude to the Centre of Applied Superconductivity Technology in Korea for their financial support and cooperation.

REFERENCES

1. KEPCO, *A report on the reduction of fault current,* 1995.11
2. KERI, *A basic study for the development of HVDC transmission system,* 2002.6
3. KERI, *Comparisons of HTS-FCL,* 2002.6
4. M. Noe, B. R. Oswald, *Technical and economical benefits of superconducting fault current limiters in power systems,* IEEE Transactions on applied superconductivity, 1999.6
5. J. N. Nielsen, J. J. Ostergaard, *Applications of HTS fault current limiters in the Danish utility network.*
6. H. Kameda, *Setting method of specific parameters of a superconducting fault current limiter considering the operation of power system protection"* IEEE Transactions on applied superconductivity, 1999.6
7. Charles A. Gross, *Power System Analysis,*
8. EMTDC Manual, Custom Model, 1986

ELSEVIER
IFAC
PUBLICATIONS
www.elsevier.com/locate/ifac

RISK MANAGEMENT AND DERIVATIVES IN KOREA ELECTRICITY MARKETS

Ki Wook Park

Team Manager, KOSPO
167, Samsung-Dong, Kangnam-Gu, Seoul 135-791, Korea

Abstract: Deregulated electricity markets in Korea introduce new risks to generation company("Gencos") and require sophisticated and analytical risk management. This paper describes typical methods to calculate Value-at-Risk(VaR) and derivatives to hedge the risks. Electricity derivatives help Gencos manage and hedge price risks in a competitive electricity market. Weather derivatives can be used to hedge volume risks on which the weather has a big influence, such as the risk of declining sales in the power sector as a result of a change in the weather. Finally, the risk management strategy which MOCIE[1] makes at early stage of restructuring in Korea is reviewed in terms of efficiency and effectiveness. *Copyright © 2003 IFAC*

Keywords: Risk, Covariance matrices, Distributions, Monte Carlo simulation

1. Restructuring and Deregulation

It's a kind of trend that an increasing number of power exchanges(or power pools), are in the process of being organized worldwide. These exchanges are more or less established in a different manner, and a variety of experiences have been met along the way. Some implication can be gained with regard to which elements of the different market designs are reasonable and efficient. Trading through power exchanges, with associated spot prices, is a new phenomenon. Traditionally inter-utility trade was encouraged to make ends meet with supply rather than market based terms. Utilities pursued a backup source of supply in case of unforeseen events and higher than expected loads. With the advent of power exchanges, utilities can rely on the spot market whenever in-house generation is insufficient or too expensive compared to the spot prices. Market prices are secured for purchases as well as sales.

1.1 England and Wales.

The E&W pool was designed to let the National Grid Company (NGC) execute central system planning. System operations and market operations are both coordinated and run by the NGC through a poolco model. All physical power generation in E&W has to be traded through the pool in other words only generators submit bids. Bids are submitted in the Day-Ahead market for next day deliveries in the 48 half-hour periods. NGC forecasts the demand on the day before delivery. Based on this, System Marginal Prices are calculated using a pricing algorithm. Generators receive the so-called Pool Purchase Price (PPP), which is the sum of the System Marginal Price and the Capacity Payments. The PPP takes into consideration of the possible lost load and the value of lost load. Prices are made available to pool members the day before delivery, meaning that prices are set on an ex-ante basis.
The E&W pool was evaluated by the national regulator OFGEM(Office of Gas & Electricity Markets). OFGEM expressed that the market design should be replaced by an exchange for trade in physical electricity. NETA(New Electricity Trading Arrangements), a redesigned E&W spot market, was commenced in March 2001. The new electricity trading arrangements (NETA) for the UK suggested a radical but not unprecedented departure from the use of an auction market for trading electricity. Under the new arrangements outlined by the industry regulator, a spot market or power exchange has replaced the previous day-ahead Pool and allows participants to

[1] Ministry of Commerce, Industry and Energy

trade electricity close to the time of their physical requirements.
The new spot market enables participants to trade power either to meet their full requirements or to supplement contracts written under longer-term agreements. The spot market allows trading up to 1 hour ahead of dispatch (delivery) of the electricity contract.

1.2 Scandinavia : Nord Pool

Even if Nord Pool is owned by the national system operators of the participating countries, market operations are kept separate from system operations. The Nordic spot market allows for bilateral physical trade, meaning that market participants may freely enter bilateral contracts. Trade at the exchange is voluntary, and both supply and demand bids are allowed. Nord Pool operates several markets, of which the Day Ahead (for all countries) and Hour Ahead (for Sweden and Finland only) represents the physical spot markets. The market clearing price (MCP) is calculated for the Nordic system as one, and is often referred to as the system price. Price zones are established in case of congestion, with each of the countries representing separate zones. Nord Pool, on behalf of the system operators, settles for possible congestion. This is the reason why no competitive physical spot markets may be established in Scandinavia. Nord Pool furthermore offers Forwards and Futures, both physical(Elspot and Elbas) and financial. These products are settled using the system price as the reference price. Separate financial accounting is used for the Future markets. In addition, the system operator, Statnett, operates the regulating market for real time balancing of the system.

1.3 U.S

New England Power Pool; The New England Power Pool, formed in 1971, is a voluntary association of electric utilities in New England who established a single regional network to direct the operations of the major generating and transmission (bulk power system) facilities in the region. NEPOOL built a state-of-the-art Control Center to centrally dispatch the bulk power system using the most economic generating and transmission equipment available at any given time to match the electric load of the region. This approach netted millions of dollars in savings for NEPOOL utilities and their customers, while increasing the overall reliability of the bulk power system. NEPOOL will continue to exist as the entity representing not only traditional electric utilities but companies that will participate in the emerging competitive wholesale electricity marketplace. ISO New England has a services contract with NEPOOL to operate the bulk power system and to administer the wholesale marketplace.

PJM; The PJM Interchange Energy Market is a real-time balance market, providing for market operations as well as system operations. Generators and loads bid to the PJM Day Ahead market and/or submit Day Ahead schedules. Based on this, the hourly Locational Marginal Prices (LMPs) of approximately 1,600 nodes are calculated through an optimal dispatch algorithm. Hourly schedules and LMPs resulting from the Day Ahead calculations are binding for the market participants. In real-time balancing operations the PJM calculates LMPs through optimal power flow calculations every 5 minutes. The prices that market participants get are the hourly weighted average Ex-Post prices for each node calculated from the 12 five-minute LMPs occurring over the hour in question. Bilateral trade is allowed in this model, meaning that participation is voluntary. However, schedules have to be submitted

1.4 Restructuring of Electricity in Korea

On January 21, 1999, the Ministry of Commerce, Industry and Energy (MOCIE) published a Plan for restructuring the electricity industry in Korea (the Restructuring Plan). The overall objectives of the Restructuring Plan are to introduce competition and thereby increase efficiency in Korean electricity industry, ensure a long-term, intensive and stable supply and promote consumer convenience through the expansion of consumer choice. The Restructuring Plan sets forth the Government's broad policy directions and calls for flexible implementation over the next 10 years or more.
It is expected to be processing from 2004 to 2009, the distribution subsidiaries would purchase power directly from the generation companies through the competitive bidding pool system described above. In the role as the Transmission Company, KEPCO would charge the distribution companies a transmission fee. Because KEPCO going to be intended to let it act as the sole electricity Transmission Company in Korea. During this phase, each distribution company will still be permitted to operate only in a specific geographical area. The electricity industry will be regarded to assure competition, cost effectiveness and sufficient supply.

2. Risk Management in Electricity Industry

Deregulated electricity markets and emergence of centralized physical markets in electric power introduce different kind of risks to markets participants in both operations and financial ways. The risks encountered by Gencos in Korea can be managed and hedged by various risk management products such as forwards/futures, options and swaps. This paper will cover market price and financial risks that require new techniques and mathematics to deal with them. Some of the methodologies have similarities in the electric power ; some in the financial world, and some are truly new incorporating synthesis of both.

2.1 Risk management

Value-at-Risk(VaR) answers s fundamental question facing all risk managers in Gencos: The concept of VaR is a very appealing one- it can be developed for any kind of portfolio and importantly can be aggregated across portfolios containing different kinds of instruments. But there are numerous ways in which VaR can be calculated and each method gives a different answer. In this paper, we compare and contrast the three main methods which are currently used for computing Var. These are variance-covariance, historical and Monte Carlo simulation.

Variance- Covariance; The variance-covariance approach also known as delta-normal VaR. It assumes that the returns for all the market variables which are the spot, forward and futures prices are normally distributed. All information can therefore be captured by the variance-covariance matrix of the returns- the means are typically assumes to be zero. The assumption of normally distributed returns and the delta approximation leads to the portfolio returns can be expressed in terms of the standard deviations of the underlying returns via a simple matrix equation. The variance-covariance approach is widely used and accepted as the basic method for evaluating VaR for portfolios without a large options component. Its advantages are that it is simple to understand and computationally efficient. But the method has drawbacks. Firstly the assumption of normally distributed returns is not realistic for electricity markets, which are characterized by fat-tail return distributions. Secondly, it is very difficult to improve the model while retaining the simple delta-normal VaR calculations because the simplicity relies on the normality assumption . Also, the method is not really appropriate for derivatives portfolio, since the delta equivalent approximation cannot consider all the options information, which can lead to a major underestimation of the VaR.

Monte Carlo; The Monte Carlo VaR method is based on the statistical simulation of the joint behavior of all relevant market variables and uses this simulation to generate future possible values of the portfolio. The model assumed by many market participants is typically the same as for the delta-normal VaR methods - the returns are normally distributed. But the advantages of this approach are that it allows for the accurate modeling of market behavior such as the addition of jumps, market gapping, and incorporation of the knowledge of futures events. It also leads to a more accurate representation of the dependence of the prices of options on the market variables via the use of the trading models. The problems with this approach are that it is relatively complex to understand and implement, and can be computationally demanding because of the need to revalue the portfolio many times. However, it differs in that the dependence of non-linear derivatives on the market variables is more accurate, because option pricing models are used to compute the changes in the option prices for the each of the simulated states of the underlying forward curves.

Historical Simulation; This method uses historical time series of market variables to obtain the distribution of portfolio returns. It therefore differs from Monte Carlo simulation, which uses a model of the market variables and a set of random numbers to generate a portfolio distribution. Under the assumption that this past history of changes is a good forecast for the range of changes the portfolio could have in the future, we can compute a VaR estimate at our chosen confidence level. The advantages of this approach are that it is simple to understand, intuitive and relatively straightforward to implement. It can be applied to all instruments and all market risk types- all that is required is sufficient history of the relevant market variables. It allows the use of the pricing models used for trading to revalue option positions. The approach is particularly valuable if the user does not believe that the distributions of market returns can be described by normal distributions. The disadvantages of historical simulation are that it is computationally demanding in the same way as Monte Carlo simulation. Furthermore, the portfolio distribution, and thus the VaR estimate is completely determined by the distribution of the underlying market variables over the time periods selected.

3. Derivatives in Electricity Market

3.1 Electricity Derivatives

Increased competition in wholesale and retail electricity markets is likely to lower electricity prices, but will also result in greater price volatility as the industry moves away from administratively determined, cost-based rates and towards market-driven prices. Price volatility introduces new risks for Gencos. In a competitive environment, some Gencos will sell their power in potentially volatile spot markets and will be at risk if spot prices are insufficient to cover generation costs. Consumers will face greater seasonal, daily, and hourly price variability and, for commercial businesses, this uncertainty could make it more difficult to assess their long-term financial position. Electricity futures and other derivatives help Gencos manage, or hedge, price risks in a competitive electricity market.

Electricity Futures/Forward; Futures contracts can be used to hedge or to speculate. An entity with a long (short) position in the electricity market can hedge by selling (buying) a future. A speculator, in contrast, takes an outright long or short position in expectation of a price move. Someone with a long futures position (i.e., has purchased futures) profits when prices increase and loses when prices decline. Someone with a short futures position (i.e., has sold futures) profits when prices decline and loses when prices increase. Hedgers mitigate risk by taking opposite positions in the physical and futures markets. The fact that the hedging positions oppose the physical positions shows that hedgers are insulated from price changes because gains in the physical position are offset by losses in the futures position, and vice versa. With a perfect hedge, the magnitude of the corresponding gains and losses in the physical and futures positions will be exactly the same.

Electricity Swaps; A price swap is a negotiated agreement between two parties to exchange or "swap" specific price risk exposures over a predetermined period of time. Swaps are widely used in natural gas and oil markets and were introduced in electricity markets in 1995. Price swaps, which are traded in the "over-the-counter" (OTC) markets rather than on an exchange, serve the same economic function as futures contracts. One party agrees to pay a fixed payment stream, while the other agrees to pay a variable payment stream. The buyer of the swap makes the fixed payment while the seller of the swap makes the variable payment. When the swap transaction is initiated, the two parties must agree on the following: the fixed price, the determinant of the variable price, the time period covered by the swap, and the notional size of the swap. Just like the buyer of a future, the buyer of a swap profits when prices increase and loses when prices decrease relative to the fixed payment level.

Electricity Options; In 1996, NYMEX also introduced options for electricity. There are two types of options: a put option and a call option. The buyer of an electricity put option (also called "floors") pays a premium for the right, but not the obligation, to sell electricity at a specified price, the strike or exercise price, at a specified point in time. End users use call options (also called "caps") to place a maximum ceiling price (relative to an indexed price) that they will pay for the commodity at a specified point in time. Gencos and end users can use combinations of calls and puts to ensure a particular price range. Gencos use put options (also called "floors") to guarantee a minimum price for their electricity (relative to an indexed price) in conjunction with the physical sale of their commodity. Gencos would still benefit from increases in commodity prices but would avoid the risk of lower prices.

3.2 Weather Derivatives

Weather derivatives differ from conventional derivatives in that there is no original, negotiable underlying or price of an underlying, which normally forms the basis of any derivative. For example, financial derivatives are based on stocks, stock indices, bonds, exchange rates or currencies – all of which are themselves negotiable objects, unlike the weather. The underlying of weather derivatives is based on weather conditions data, such as on the temperature, which influence the trading volume of some goods. Weather derivatives cannot hedge the price of the underlying, as it is impossible to put a price on the weather. Consequently, weather derivatives hedge other risks with which the weather has a close correlation, such as the risk of declining sales in the energy and power sector as a result of a change in the weather and the price changes that could result from this. The idea behind a weather volume hedge is that the results of weather-sensitive sectors can be subject to great volatility – even if prices remain unchanged – due to a change in demand or volume.

Weather Options; Most weather derivatives have concentrated in the field of options designed to protect against fluctuating temperatures. Options of this kind are based on heating degree days (HDDs) and cooling degree days (CDDs). These indices represent the average temperature for a given period of time. The degree days are calculated as the absolute difference between a reference temperature (for example 18°C) and the actual average temperature. The degree days determined in this way are added over the underlying period of the contract. If the average temperature is above 18°C on one day, it is referred to as a cooling degree day; if the temperature is lower than that, it is known as a heating degree day. HDDs measure the average temperature for the winter half-year, while CDDs usually measure that for the summer half-year. While HDD options can be used to obtain protection against excessively warm winters, CDD options afford a safeguard against excessively cool summers. The option strategies that are available are HDD/CDD calls and puts, as well as combination strategies, such as collars, strangles, spreads and straddles. The strike value determines the number of degree days upwards of which the option can be exercised. In the case of a call, the strike value for the relevant period must be below the total number of actual degree days in order to make a profit from the option; whereas a strike value above the total number of actual degree days is the target for a put.

Weather Swaps; Swaps are privately negotiated financial contracts in which two parties agree to exchange, or "swap", specific price risk exposure over a predetermined period of time. They are OTC instruments that can be customized to meet a particular set of needs. Swaps allow for strategies designed to protect against market price fluctuations. There is no "cost" for swap. When used in connection with floating price energy contracts, swaps afford a buyer and seller protection from adverse price movements, in exchange for giving up the ability to capitalize on benefit price movements. There are no standardized swap transactions. Nonetheless, most transactions involve an exchange of periodic payments between two parties, with one side paying a fixed price and the other side paying a variable price. Specific terms of swap agreements - including the fixed of the weather index and its floating price reference, the terms of the contract, and the quantity to be hedged - are established by the two parties involved, and can vary, subject to their specific needs and objectives. A swap contract is settled in cash, usually against as agreed-upon market price index, and is customized as to volume, timing, location, seasonality, and swing

4. Risk Factors and Risk Management Objectives at Early Stage of Competition in Korea

4.1 Risk Factors

Risk management strategy which is being developed in Korea is intended to mitigate some of the key risks associated with the start of the two-way bidding pool in Korea in 2004. The main risks are spot price and revenue stability risks, availability risks, market power risks and risks associated with new entry. Spot price and revenue stability risks are mitigated by fixing the prices between the retailer and the generators for a large portion of the market. Availability risks are mitigated by setting up a multilateral financial hedge "insurance" scheme called the Horizontal Hedges. Market power risks are mitigated by use of the contracting, hedging obligations on Genco who has incentives to use market power and giving the retailers some physical dispatch rights over certain plant. Risks associated with new entry are mitigated by providing for reserve capacity contracts when required to give signals to encourage new entry.

4.2 The Objectives

Market Power and Competition; Risk management strategy will introduce a set of pro-competitive measures to the Korean market. The KHNP[2] hedging obligation will require KHNP to contract for a high proportion of their output and hence mitigate any potential market power. The retailer's Physical Dispatch Rights will perform a similar role for the peaking portion of the market. The Horizontal Hedge contracts which effectively insure generators against failures and Force Majeure will improve liquidity for all generators.

Revenue Stability; Risk management strategy will promote enhanced revenue stability for generators. A high proportion of two-way contracts will be used initially, falling gradually over time to be replaced with market driven contracts. Horizontal Hedges will provide protection from Force Majeure and other outages. Physical Dispatch Rights provide fixed payments for the use of at least some peaking plant, which removes some market risk for these plant.

Support for New Investment; Prices in the Two Way Bidding Pool (TWBP) are the primary signal for new investment. It is intended that the risk management strategy contracts will provide revenue stability and will progressively transition from "shaped" financial outcomes to "market-based" financial outcomes through the enhanced proportion of market-based contracts over time.

[2] Korea Hydro & Nuclear Power Company. In 2003, KHNP's market share is over 40% of total electricity generated.

Furthermore, the financial impact of the risk management strategy is intended to minimise price shocks for retail customers as the market transitions to a pricing level consistent with the long run marginal cost of new entry into Korea, on average.

4.3 Risk Management Contracts(Derivatives)

Vesting contract; Vesting contracts are deemed to be any form of contract between generators and retailers imposed on the counter-parties at the time of market start. The contract conditions and financial parameters are typically tailored to meet specific market transition objectives. Vesting contracts are normally part of a comprehensive set of financial arrangements that define the commercial relationships and key business risk allocations between the restructured electricity industry businesses in the period immediately following structural and market reform.

Vesting contracts may be simple one-way or two-way financial hedge contracts, but an overall vesting contract proposal can also include highly customized contracts which link contract price and volume to external factors in the market such as some aspect of market volume, fuel price movements, consumer price movements, force majeure events affecting either of the contract counter-parties, generator plant availability and/or dispatch level; etc.

Vesting contracts can be used to cover from 0 to more than 100% of a participant's exposure to volatile pool prices. Vesting contract cover can be structured to very closely match hedging requirements from year to year, whatever level that may be.

They also allow much of a participant's pool price exposure to be hedged while retaining the right economic signal in the form of the spot market clearing price to drive the participant's operating decisions at the margin essentially in real time.

PDR; Physical dispatch rights establish a regulatory obligation for wholesale purchase to enter into and hold a minimum level of physical contracts with nominated participants(i.e. "approved" generators) and for Gencos to sell a minimum level of physical contracts to be eligible for dispatch. Physical dispatch right enforces forward contract cover to reduce retailer exposure to volatile spot market prices at the time of dispatch. It minimizes trading volumes in the TWBP with consequential reductions in TWBP prudential requirements. It possibly assists in underwriting new generation projects depending upon the timing of when wholesale purchasers seek the contract cover required to meet their expected obligation.

Horizontal hedging contract; Horizontal hedging contracts (contracts between Gencos or between retailers) can be used to support the traditional vertical vesting contracts to enable certain risks to be shared more widely within a sector, reducing the concern that any single entity will be forced to bear a disproportionate share of the cost of a particular

adverse market outcome. For example, a horizontal hedging contract could be structured to provide an option to one Genco to call upon one or more other Gencos in the event of a plant failure that leaves it exposed to higher spot market prices.
The financial structure of a horizontal hedging contract resembles a conventional one-way vesting contract. However, instead of being a firm contract whenever the pool price exceeds the contract strike price, difference payments under the contract only apply when the plant availabilities of the counter-parties satisfy certain conditions defined in the contract. In exchange for the more limited application of the contract compared with a more conventional one-way hedge, the cost of the insurance provided in the form of the option fees payable are correspondingly reduced.

Hedging obligation;. A broadly based hedging obligation would enforce forward contract cover to reduce retailer exposure to volatile spot market prices at the time of dispatch, and could also be used to underwrite new investment in generation. A hedging obligation imposed on a dominant participant can quite effectively mitigate its market power. Schemes of this type would require the dominant participant to sell a defined amount of contracts in such a way that the contract prices are set by a market and are not controlled by the dominant participant. Once the contracts are in place, the effective size of the dominant participant operating in the spot market is equivalent to its un-contracted capacity that has been reduced to an acceptable level. A broadly based hedging obligation being imposed on wholesale participants ensures that the market is largely hedged against volatile pool prices at all times while still preserving the appropriate economic price signal at the margin for all market participants.

5. Conclusion

This paper has discussed the methodology of risk management and a variety of derivatives. In electricity market, historical price return distributions are fat-tailed, leading to fat tails in the portfolio return distribution. It should be clear from these examples that delta-normal VaR is not appropriate for electricity markets, and that historical VaR is far the best at capturing the fat-tails of the portfolio return distribution. However, historical VaR only gives an accurate picture if history repeats itself. However, with Monte Carlo VaR, knowledge of future changes in the market can be incorporated in the simulations, which can be especially useful in the nascent power markets. Therefore, we need to consider that both historical and Monte Carlo VaR should be used together to help get a more accurate picture of the portfolio risk. Also, electricity/weather derivatives can now protect earnings from anticipated drops in price and demand. Predictability and stability of the cash flow are of crucial importance for Gencos. Because electricity/weather hedging can stabilize earnings volatility, Gencos can use them to produce the stable and predictable results investors and commercial lenders appreciate. Financial market innovation in recent years has been unusually fast and extensive. There has been a remarkable growth in the use of derivative instruments. Gencos in Korea can install derivative instruments to protect themselves against adverse changes in various external factors in deregulated markets. MOCIE is also structuring risk management strategy to stabilize the market. But MOCIE dose not consider structural stapling which can bring Gencos positive effects. It can accomplish similar hedging benefits compared to the vesting contracts which need to devise specific contracts. Also, it could reduce the need for financial liquidity in the market since retail/Gencos would only be financially exposed to pool prices to the extent that they are "net" retailers or "net" generators.

REFERENCES

Helle Granli (1999). *A Comparison and Evaluation of Different Power Exchange, Issue No. 4.* CALPOL Market Report.
John C. Hull (1997). *Options, Futures, and Other Derivatives, third Edition.* Prentice Hall.
Mark Garman, Carlos Blanco and Robert Ericson (2000). *Weather Deiivatives: Instruments and Pricing Issues.* Financial Engineering Associates.
Robert Jarrow and Stuart Turnbull (2000). *Derivative Securities,* 2nd edition. South-Western College Publishing.
Eltermine, Eloption (1998). Nord Pool.
The Elspot Market. No 2 (1999). Nord Pool.
The Manual for the Personnel of Electricity Market Operation (1999). KEPCO.
The Proposed Pool Price Stabilization & Risk Management Framework (2002). KEPCO.

ELSEVIER
IFAC
PUBLICATIONS
www.elsevier.com/locate/ifac

RESERVE PROCUREMENT & PRICING VIA JOINT DISPATCH IN RESTRUCTURED ELECTRICITY INDUSTRY

M. Rashidinejad
Department of Elec. Engineering
Shahid Bahonar University
Kerman, Iran
mrashidinejad@yahoo.co.uk

M. Fotuhi-Firuzabad
Department of Elec. Engineering
Sharif University of Technology
Tehran, Iran
fotuhi@sina.sharif.edu

Y. H. Song
Department of Elec. Engineering
Brunel University
London, UK

Abstract--Procurement and pricing of electric power and ancillary services is one of the most important issues of electricity market manager under new restructured environment. Ancillary services are necessary to provide security and reliability constraints applied to the dispatching and scheduling problem while they are essential for commercial functions. In a market payments should be based on energy and ancillary services capacities. Spinning Reserve (SR) is an ancillary service that is coupled with energy capacity, should be procured and priced via a joint economic dispatch mechanism to provide a right signal for all electricity customers. This paper presents a new approach for joint economic dispatch as an optimal criterion for spinning reserve allocation and pricing. Case study shows the significance of the proposed methodology. *Copyright © 2003 IFAC*

Keywords: Power system, Procurement, Pricing, ancillary services, Spinning Reserve, Restructured electricity industry.

1. INTRODUCTION

The increasingly competitive electricity industry towards a deregulated power market is resulting in greater interest in sufficient and efficient utilisation of generation and ancillary services. In most of electricity industry around the world, the System Operators (SO) may face serious problems in the acquisition, pricing and cost of ancillary services [Hirst and Kirby, 1997]. The problem is raised because the electricity industry as a whole may not yet recognise the importance of ancillary services either technically or economically. Traditionally, units were just dispatched to meet energy demand.

Ancillary services are necessary to provide security and reliability constraints applied to the dispatching and scheduling problem as well as they are essential for commercial functions [Cheung, et. al., 1999]. In a market where payment is based on energy supplied, the generators prefer to have their units on-line and loaded near the high economic dispatch limit. Because most of ancillary services are provided by the same pieces of equipment that produce energy, energy markets and ancillary services markets are tightly coupled. Any problem in one market may cause problem in other markets.

Since bulk-power systems must maintain a continuous and near-instantaneous balance between generation and load, operating reserve is a backup capability of those systems to prevent such unwanted imbalances. Furthermore, ancillary services are required to maintain system reliability and to affect commercial transaction. A real-time market may provide a balance in energy and ancillary services between generation and consumption. Spinning Reserve (SR) is an ancillary service that may maintain this balance when a major generation or transmission line unexpectedly fails [Flynn, et. al. 2001, Rashidi-Nejad, et. al., 2001].
This paper focuses on procurement spinning reserve in conjunction with energy and shows the impact of this procurement on the price of energy as well.

2. EXTENDED JOINT DISPATCH PROBLEM

The ordinary economic dispatch (ED) problem is stated by specifying the loads in MW to be supplied at certain nodes, with the objective of minimum generation costs. In a comprehensive dispatch problem, many other constraints needed to be considered including generation limits, voltage conditions, security limits as well as reliability limits. The extended ED (EED) is a complementary ordinary ED considering additional reliability and security constraints. In this section an extended joint ED (EJED) is introduced to include spinning reserve capacity as well as generating reserve capability.

2.1. Mathematical Formulation and Optimal Solution

The general ED mathematical formulation [Wood and Wollenberg, 1996, Xie, et. al. 2000] is summarised as follows:

$$MinF(x)$$

$$S.t.\begin{cases} g(x) = 0 \\ h(x) \le 0 \end{cases}$$

Where; $x^T = (x_1^T, x_2^T, x_3^T)$ is a variable vector including the following sub-vectors.

x_1 : all real power generator outputs variables P_i^g, x_2 : all spinning reserve capacity of generating units SR_i and x_3 : all voltage magnitudes and all voltage angle variables V_i, θ_i.

The equality set of constraint can be split into the subsets as follow:

$$g^T = (g_1^T, g_2^T, g_3^T)$$

Where; g_1 : demand-supply equation for real power equilibrium, g_2 : demand-supply equation for real power equilibrium and g_3 : spinning reserve capacity of each generating unit, which for each generating unit can be written as:

$$SR_i = \min[(P_{i-\max}^g - P_i^g), (srf_i * P_{i-\max}^g)],$$

srf_i is the spinning reserve factor as a contribution factor for each generating unit.
The optimality conditions, for the above equations can be derived applying Lagrangian function:

$$L = F(x) + \lambda^T g(x) + \mu^T h(x).$$

Applying the Kuhn-Tucker conditions to the *F(x)* satisfies all constraints at the same time and the following equation can be obtained:

$$\frac{\partial L}{\partial x} = \frac{\partial}{\partial x}(F(x) + \lambda^T g(x) + \mu^T h(x)_{|x^*, \lambda^*, \mu^*} = 0$$

$$\frac{\partial L}{\partial \lambda} = g(x)_{|x^*} = 0$$

$$diag\{\mu\}\frac{\partial L}{\partial \mu} = diag\{\mu\} h(x)_{|x^*, \mu^*} = 0$$

$$\mu \ge 0$$

2.2. Model Description

Generally, classical ED involves minimising generation cost functions with regard to the output limits of generating units with or without considering network losses [Wood and Wollenberg, 1996]. In the EJED model, the objective function includes cost functions of generating units and spinning reserve cost functions of all contributed units. The constraints include minimum and maximum generation limit for each generating unit, minimum system reserve requirements and spinning reserve capacity limits for the generating units. The unit's spinning reserve potential capacity is considered as a percentage of maximum generation capacity or the difference between maximum generation and actual generation output. The model also employs a total spinning reserve need as a percentage of the maximum load demand. EJED model is based on the objective function of the classical economic dispatch [Wood and Wollenberg, 1996, Ma and Sun, 1998], and it is extended to include spinning reserve cost function. The quadratic cost curve expression of real power is adopted in this paper.

$$Min\{F = \sum_{i=1}^{m}[f_i(P_i^g) + SRC_i(SR_i)]\},$$

Where;

$$f_i(P_i^g) = A_i^p * (P_i^g)^2 + B_i^p * (P_i^g) + C_i^p$$

is the production cost function of *ith* generating unit.

The spinning reserve cost function contains two parts. The cost of reserve capacity provision (COR) for *ith* unit, which is: $COR_i = RCC_i * SR_i$. Ramping cost (RC) that may be identified for compensating ramping-up and/or ramping-down as required by the system. This part of spinning reserve cost can be presented by quadratic cost curve by the following equation,

$RC_i = CRR_i * (\Delta P_i)^2$ [Flynn, et. al., 1999],

where; ΔP_i is the difference between two generation stages. If this difference can be taken into account as required spinning reserve, which *ith* unit must deliver to the system, therefore the following equation can be derived:

$RC_i = CRR_i * (SR_i)^2$ and

then, $SRC_i(SR_i) = CRR_i * (SR_i)^2 + RCC_i * (SR_i)$

where

F : objective function,

$f_i(P_i^g)$: active power generation cost function for *ith* unit,

$SRC_i(SR_i)$: spinning reserve cost function for *ith* unit,

A_i^P, B_i^P, C_i^P : active power cost function coefficients for *ith* unit and

CRR_i, RCC_i : spinning reserve cost function coefficients for *ith* unit, while m : number of generating units in power system.

Supply-demand constraint is an equality constraint that shows the equilibrium between real power generation and consumption including network losses as well. Therefore, total real power generation needs to meet total demand plus network losses,

$$\sum_{i=1}^{m} P_i^g = \sum_{j=1}^{n} P_j^d + P_{loss} \quad ,$$

where; P_i^g : Real power generation of *ith* unit ,

P_j^d : real power demand of *jth* bus and

P_{loss} : total real power system losses. Generally, real power output limits of *ith* generator is:

$$P_{i-\min}^g \le P \le P_{i-\max}^g .$$

But in this model, real power output limits can be split to two distinct limits, of which

$$P_i^g + SR_i \le k_i P_{i-\max}^g , \; k_i P_{i-\min}^g \le P_i^g ,$$

where;

SR_i : spinning reserve of ith unit ,

$P_{i-\max} g$: maximum real power generation of *ith* unit,

$P_{i-\min}^g$: minimum real power generation of *ith* unit and k_i : a stochastic binary variable, which is 0 when the generator of *ith* bus is out of service and is 1 when generator at *ith* bus is in service.

The spinning reserve capacity for *ith* generating unit can be either the difference between the maximum generation capacity and the real output of the unit, $P_{i-\max}^g - P_i^g$, or a percentage of the installed capacity $srf_i * P_{i-\max}^g$ [Wood and Wollenberg, 1996, Ongsakul and Tippayachai, 2000]. It can be written in the form of a conditional mathematical function as:

$$SR_i = \begin{cases} P_{i-\max}^g - P_i^g \, ...if.....P_i^g \ge (1 - srf_i) P_{i-\max}^g \\ srf_i * P_{i-\max}^g ....if......P_i^g \le (1 - srf_i) P_{i-\max}^g \end{cases}$$

This equation can be converted to another mathematical form such as:

$$SR_i = \min[(P_{i-\max}^g - P_i^g), (sraf_i * P_{i-\max}^g)] k_i$$

and

$$SR_{i-\min} < \sum_{i=1}^{m} SR_i < SR_{i-\max} ,$$

where; $SR_{i-\max}$: maximum system reserve,

srf_i : spinning reserve allocation's factor for *ith* unit ,

$SR_{i-\min}$: minimum system reserve requirements

To achieve the true optimal dispatch, transmission losses must be taken into account. In this paper, the B-matrix loss technique [Wood and Wollenberg, 1996] is applied to calculate the system losses,

$$P_{loss} = \sum_{i=1}^{m} \sum_{j=1}^{m} P_i B_{ij} P_j + \sum_{i=1}^{m} P_i B_{j0} + B_{00} \quad ,$$

where;

B_{ij} : the i*j*th element of the loss coefficient square matrix coefficient,

B_{j0} : the *i*th element of the loss coefficient vector and

B_{00} : the loss coefficient constant.

3. EJED MDOEL AND SOLUTION PROCEDURE

A solution process of the proposed method is illustrated in Figure 1. As the flow diagram Figure 1 shows, initial data are entered to the model, in which system reserves requirement are also included. Solving a pre-dispatch without considering reserve the real power outputs can be derived and the system loss will be calculated. Then, solving the EJED results of the optimal values of real power outputs and reserve can be obtained. If power flow is converged, the corresponding Lagrangian multipliers obtained are the prices of real power and spinning reserve. If power flow is not converged then it is needed to decrease the SRR, and try to run again until getting convergence. The results include the real power output of generating units, spinning reserve contributions of the units, real power spot price and spinning reserve price for capacity reservations.

3.1.EJED Case Study and Results Analysis

In this section, an application of quadratic programming for extended joint ED (EJED) is implemented. To extend the ED, spinning reserve cost function is added to the overall objective function of the ED. Spinning reserve constraints are also added to the set of ED constraints.

In fact, spinning reserve limitations include two different types of constraints: generating unit limits for reserve and system reserve requirements [Zhang, et. al 2000, Proctor, et. al. 2000].

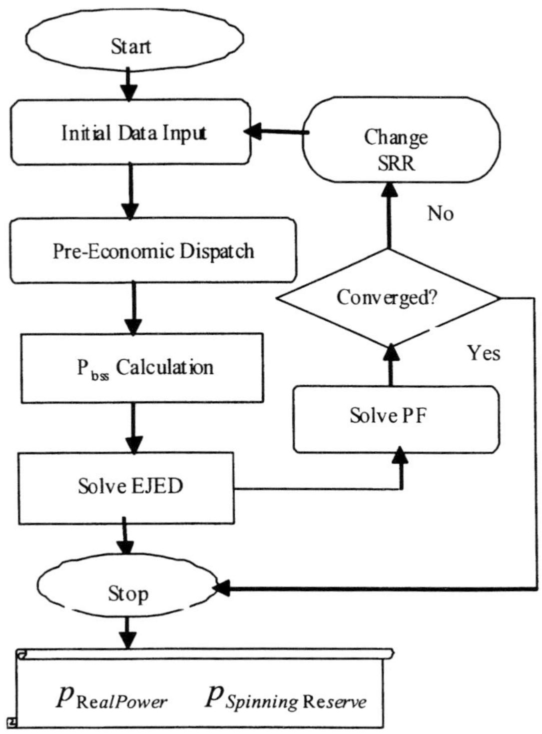

Figure 1- EJED Flow Diagram.

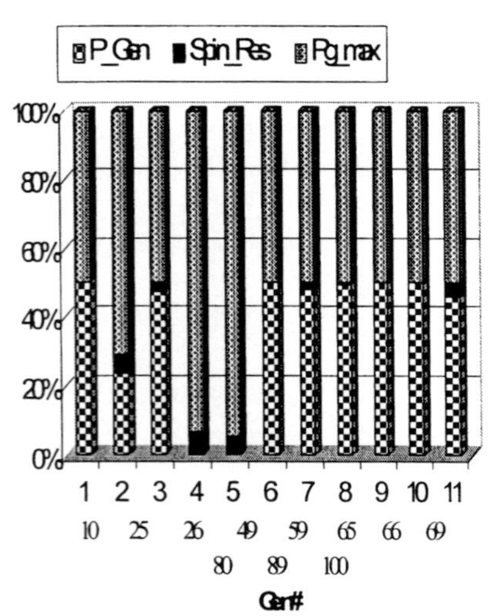

Figure 2- Real power generations and spinning reserve contributions of IEEE 118-bus.

Generating unit limitations includes unit reserve capacity, treated as equality constraint and unit ramp-rate, treated as inequality constraint. System reserve requirement, which should not be less than the total reserve capacity in a power system, is treated as inequality constraints [Singh, 1999, Bobo and Mauzy, 1994].

The EJED is implemented to an IEEE 118-bus system through MATPOWER [Zemerman and Dequiang, et. al. 1997], which is a power system optimisation software. The outcomes of this case study in terms of unit reserve contributions as well as the effect of providing system reserve requirements are derived in this section. For a specified level of demand and system reserve requirement, only eleven of generating unit (among 54 existing generators in the system) are automatically selected to meet the real power demand, while among them six generating unit provide reserves. The contribution generating units for real power and spinning reserve are illustrated in Figure 2. As it is shown in Figure 2, for these generating units, some part of their capacity is allocated for real power and some part for spinning reserve.

In Figure 3 variations of spot price at each bus are illustrated. The difference between the prices of real power, in case of considering spinning reserve provision and without considering spinning reserve, can be seen from Figure 3. It may be discussed that, the difference between two corresponding prices at each bus, which is the impact of spinning reserve provision on the real power price and the basis real power price, can be declared as spinning reserve price at that bus.

Figure 4 shows the hourly price of real power and spinning reserve for different level of real power demand. It is shown that by increasing demand the hourly price of peal power as well as spinning reserve price increases.

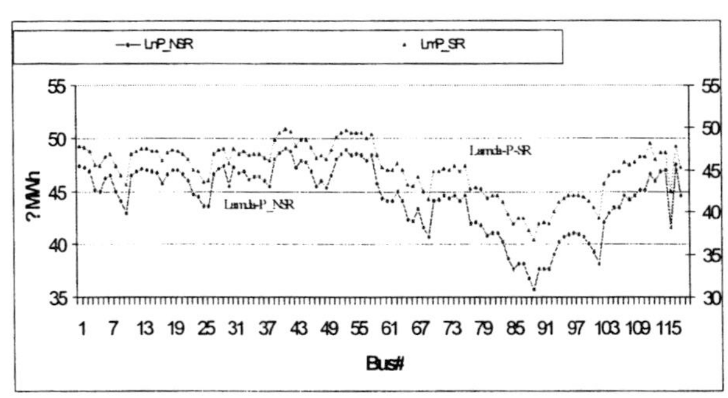

Figure3- Variations of real power prices with and without spinning reserve for IEEE 118-bus system.

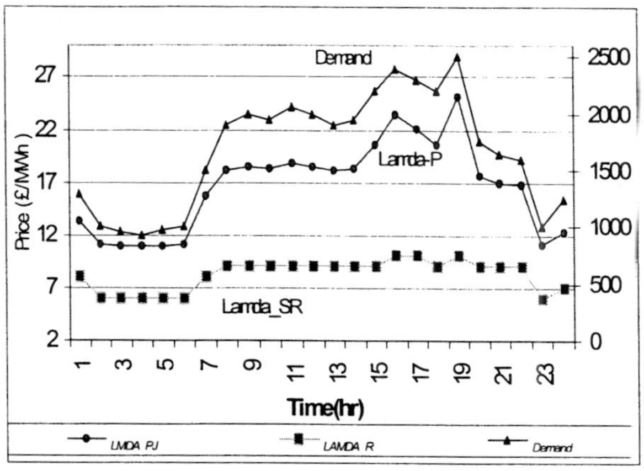

Figure 4- Variations of demand, real power price and spinning reserve price for 24 hour.

4. CONCLUDING REMARKS

In this paper, a new model for calculating optimal price of electric energy and spinning reserve via Sequential Quadratic Programming (SQP) is suggested. This model includes the contribution of each unit to provide the reserve for maintaining system reserve requirement (SRR). For spinning reserve allocation a flexible reserve percentage algorithm is suggested, which it is shown that by applying this criterion the optimum value of spinning reserve can be derived. It illustrates that the proposed extended joint economic dispatch (EJED) models is an efficient methodology in terms of considering reliability as well as optimality of power system utilisation under deregulated environment. The model also considers pinning reserve costs and unserved energy cost as well as energy production costs. The results show that the price of energy and spinning reserve derived from the proposed model is higher than just a simple price without considering interrelationship between actual energy and reserves. This model can be extended for large power system to take acceptable results in the context of cost-based pricing mechanism and spinning reserve allocation in deregulated electricity markets.

5. Acknowledgement

The first author would like to acknowledge financial support provided by the Power System Regional Company of Kerman, Iran.

6. REFERENCES

D. R. Bobo and D. M. Mauzy, "Economic Dispatch with Responsive Spinning Reserve Constraints", IEEE Trans. On Power System, Vol. 9, No.1, Feb. 1994.

K. W. Cheung, P.Shamsollahi, and S. Asteriadis, "Functional Requirements of Energy and Ancillary Service Dispatch for the Interim ISO New England Electricity Market", IEEE Power Engineering Society Winter Meeting, Vol.1, 1999, pp.269-73, Piscataway, NJ, USA.

M. Flynn, W. P. Sheridan, J. D. Dillon, and M.J. O'Malley, "Reliability and Reserve in Competitive Electricity Market Scheduling", IEEE Trans. on Power Systems, Vol. 16, No. 1,pp.241-249, Feb. 2001.

M. E. Flynn, M. J. O'Malley, P.Brown and M.Power "Balance Between Generating Plant Costs And System Load Following Requirements", CIGRE, 21, rue d'Artois F-75008 Paris, p. 400-03 London Symposium 1999.E. Hirst and B. Kirby, "Generating Competitive Markets for Ancillary Services", Oak Ridge National Laboratory, Energy Division, Tennessee, Oct. 1997.

X. Ma and D. Sun, "Energy and Ancillary Service Dispatch in a Competitive Pool", IEEE Power Engineering Review, p54, January 1998.

W. Ongsakul and J. Tippayachai, "Constrained Economic Dispatch by Micro Genetic Algorithm Based on Migration and Merit Order Load Solution", International Conference on Electricity Utility Deregulation and Restructuring and Power Technologies 2000, p. 510-17, 4-7 April 2000, City University, London.

R. Proctor, B. Fox, and D. Flynn, "Reserve Trading Within Power Purchase Agreement", International Conference on Electricity Utility Deregulation and Restructuring and Power Technologies 2000, pp. 137-141, 4-7 April 2000, City University, London.

M. Rashidi-Nejad, Y. H. Song, M. H. Javidi-Dasht-Bayaz, "Contingency Reserve Pricing via a Joint Energy and Reserve Dispatching approach", Energy Conversion & Management, Elsevier, Accepted 27 Jan. 2001.

H. Singh, "Auction for Ancillary Services", Decision Support Systems, Vol. 24 p183-191, 1999.

A. J. Wood and B. F. Wollenberg, "Power Generation Operation & Control", John Wiley & Sons, New York, 1996.

K. Xie, Y. H. Song, E. Yu and G. Y. Liu, "Decomposition Model and Interior Point Method for Optimal Spot Pricing of Electricity in Deregulated Environments", IEEE Trans. on Power System, Vol. 15, No. 1, pp.39-50, Feb. 2000.

D. R. Zemerman and G. Dequiang, "MATPOWE, A Power System Simulation Package", Version 2. School of Electrical Engineering, Cornell University, Ithaca, December 24, 1997,

http://www.pserc.cornel.edu//matpower/matpower.html.

X. P. Zhang, C. Y. Chen, Z. Zheng, G. Strbac, J. Kubokawa, "Allocation of Frequency Regulating Services", International Conference on Electricity Utility Deregulation and Restructuring and Power Technologies 2000, pp. 349-354, 4-7 April 2000, City University, London.

ELSEVIER
IFAC
PUBLICATIONS
www.elsevier.com/locate/ifac

APPLICATION OF BAYESIAN-COURNOT NASH GAME MODEL IN ESTABLISHING BIDDING STRATEGY WITH INCOMPLETE INFORMATION AMONG NON-COOPERATIVE GENERATION COMPANIES

Byeong-Gook Kwon, Seung-Chul Lee, Dong-Soo Choi , Jong-Hwan Kim

School of Electrical & Electronic Engineering
Chung-Ang University in Seoul Korea

Abstract : In this paper, a technique that can apply Bayesian-Cournot Nash(BCN) equilibrium model in establishing electric power bidding strategies for generation companies in competitive electric power market is proposed. Each generation company must compete with others in order to maximize its profit. Accurate informations on the operation costs and payoffs of the generation companies are assumed be unavailable. Consequently, the competitive market among the generation companies is modeled as a non-cooperative game with incomplete information. Under the incomplete information environment in the power market, the operation cost function of the opponent generation company is estimated using the Bayesian conditional probability through continuous transactions game. A Cournot Analysis Model is applied for Nash Equilibrium as the best response strategy of the each generation company. *Copyright © 2003 IFAC.*

Keywords : game theory, Bayesian, Cournot, Nash equilibrium, noncooperative game, incomplete information, electric power market

1. Introduction

The electric power industries in many countries wordwide are now in the process of transition from their traditional vertically integrated forms into deregulated and competitive ones. In competitive environments, each generation company is concerned with its profit maximization rather than system-wide cost minimization[1][2]. Three basic models for the electric power market can be the pool model, the bilateral model and the mixed model. The pool model is considered to be the intuitive mechanism to implement in most countries. In the Pool market, each generation company establishes its own bidding strategy for generation amount to sell, and the market sets the clearing price of the power transacted[2][3].

Studies have been performed recently in applying on the game theory to solve power system problems in competitive electricity environments. Normally, the Nash game theory is applied to simulate and analyze the economic behavior of the generation companies in the pool market[4][5]. Each generation company is introduced as a player, economic benefits constitute payoffs and players' strategies are composed of the bidding price and the bidding generation.

The Nash game equilibrium of the electric power transactions has been analyzed by economic models, such as Cournot, Bertrand, Stackelberg and Supply Function Method. Cournot analysis method, which is applied to determine equilibrium production and consumption levels of market, has the productions as strategy[6]. Bertrand analysis has strategy that select the price instead of the productions[7]. Supply Function Method is a transaction model that bids by the supply function instead the productions or the price[9]. Stackelberg method is an analysis of the follower's decisions atfer the leader picks up a preferential strategy in market[8]. A Stackelberg price leadership model for simulating deregulated electricity market was also presented in [10].

In particular, several works have used the Cournot method to analyse the production levels for generation companies[11]-[15]. However, they are based on the hypothesis that power market information is complete. In this paper, we model the power transactions activities in the power market as a non-cooperative game with incomplete information. Each generation company does his best to maximize the profits. In such environment, the operation cost functions of the opponent generation companies are estimated using a Bayesian conditional probability[2] through continuous transaction games. The estimated transaction game is a game with complete but incomplete information. Using the Cournot analysis according to the KT(Kuhn-Tucker) point[15] with the cost function estimated by the Bayesian approach, the Nash Equilibrium is found as the power transactions solution between two generation companies.

2. Analysis of the Power Transactions

2.1 Generation Costs and Benefits

In this paper, the proposed power transactions game assumes that ;

1) The bidding generation amount of each generation company is less than the total system demand in a specific power market.

2) The transmission losses are ignored.

3) All information such as cost function and payoff of each generation company is not revealed in the market.

A single market in which N generation companies offer a electric power to a large number of consumer is considered. Suppose that each generation company has currently u_i generation units committed, where $i = 1,2,\cdots,N$. Then the total power output of the i generation company is $P_i = \sum_{k=1}^{u_i} P_k$. The total power output of all companies except the i company becomes $P_j = \sum_{m=1}^{M} P_m - 1$.

Suppose that the generation cost function of the i generation company for the current output P_i can be represented by a second order polynomial as the following form ;

$$C_i = a_i + b_i P_i + c_i P_i^2 \tag{1}$$

where, C_i : cost function of company i

P_i : generation output

a_i, b_i, c_i : cost coefficients

2.2 Formulation Using a Bayesian Theory

Each generation company in the competitive power Pool establishes bidding strategies to maximize its economic benefits. However, since the cost function of the opponent generation are invisible, each generation company can not estimate the profits to be returned. In this paper the analysis of the power transaction game is based on the assumption that generation companies use Bayesian approaches in dealing with incomplete information. In the estimation process of the probability distribution of the unknown cost functions of the opponent companies, the expectation of the payoff is maximized. Each generation company defines the probability distribution of the game using informations that the ISO (Independent System Operator) provides to all participants, like the known contracts for fuel prices, the availability of transmission lines, etc.

The Bayesian approach proposed in this paper transforms the game with incomplete information into the game with complete information. Hence, Nash equilibrium is applied to the transformed game, and each generation company's optimal bidding strategies are derived for the equilibrium condition.

We consider N generation company participant Pool market in which company i and $j(j = 1,2,\cdots,i-1,i+1,\cdots,N)$ compete to supply the Pool's load.

Company i knows his type $t_i^x (x = 1,2,\cdots,X)$ in the game but does not knows his opponent's type $t_j^y (x = 1,2,\cdots,Y)$; conversely, company j knows which type t_j^y represents him in the game but does not know company i's type t_i^x. We assume each generation company's type depends on fuel kinds. For instance, companies may define different F scenarios for each fuel kind. A probability distribution $\phi(f)$ is used to model the uncertainty on fuel kind such that ; $\sum_{f=1}^{F} \phi(f) = 1$. Let D_i^f, D_j^f be the probability for company i, j to take fuel of type t_i^x, t_j^y. The basic probability distribution of the game form a probability matrix Π_{ij}, such that element π_{xy} corresponds to the probability that company i is type x and company j is type y ; hence π_{xy} is calculated as ;

$$\pi_{xy} = \sum_{f=1}^{F} \left[\phi(f) D_i^f(x) D_j^f(y) \right] \tag{2}$$

The Bayesian conditional probability is the probability that company i is playing against company j's type t_j^y, subject to the condition that company i is type t_i^x.

$$\Pr_i^f(y) = \Pr(t_j^y \mid t_i^x) = \frac{\pi_{xy}}{\sum_{y=1}^{Y} \pi_{xy}} \tag{3}$$

Likewise,

$$\Pr_j^f(x) = \Pr(t_i^x \mid t_j^y) = \frac{\pi_{xy}}{\sum_{x=1}^{X} \pi_{xy}} \tag{4}$$

A generation company's strategy depends on his type. Let s_i^x be the company i's strategies with type x. Company's benefit depend not only on his strategy but also of his opponent's strategies. The benefits B_i are referred to as conditional benefits. In order to maximize B_i, generation company i should know his opponent's type. Company i tries to estimate his opponent j's the expected costs.

Suppose that the types of each generation company are the independent event. From Eq.(3) and (4), the company i 's generation cost function can be obtained as follows ;

$$C'_j = \sum_{f=1}^{F}[\phi(f)\Pr_j^f a_j^f(y)] + \sum_{f=1}^{F}[\phi(f)\Pr_j^f b_j^f(y)]P_j + \sum_{f=1}^{F}[\phi(f)\Pr_j^f c_j^f(y)]P_j^2 \quad (5)$$

In Eq.(5), C_j can be expressed with

$$C'_j = a'_j + b'_j P_j + c'_j P_j^2 \quad (6)$$

Likewise,

$$C'_i = a'_i + b'_i P_i + c'_i P_i^2 \quad (7)$$

In Eq.(6) and (7) generation companies are considering all potential types of opponents. The estimated game is a game with complete but incomplete information. Using the Cournot model with estimated cost function by the Bayesian approach, we analyze the Nash Equilibrium as the power transactions solution among each generation companies.

2.3 Cournot-Nash Equilibrium

The Cournot model has become a widely used model in-cooperative oligopoly theory until now[7]. In this model, each generation company takes actions independently and each one maximizes its profit by choosing an output quantity. A Cournot-Nash Equilibrium state is a set of supply power outputs of the companies obtained by each company maximizing an individual payoff function. Cournot model has the following assumptions :

1) companies make decisions simultaneously
2) market price is depend on the market quantities
3) companies compete only with quantities
4) total supply is equal to total demand

The supply function generally take the continuous curve shown in Figure 1. (a). The market price ρ (spot price) is decided by the inverse demand function shown in Figure 1. (b), $\rho = f_d(d)$.

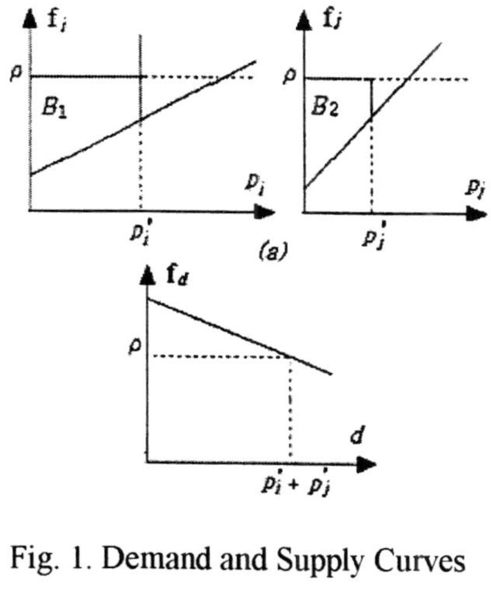

Fig. 1. Demand and Supply Curves

By differentiating the generation cost function $C_j = a_j + b_j P_j + c_j P_j^2$ with respect to the output P_i. The marginal cost function becomes

$$f_i(p_i) = \frac{\partial C_i}{\partial P_i} = b_i + 2c_i P_i \quad (8)$$

The inverse demand function can be formulated as

$$f_d(d) = d_0 - \omega(P_i + P_j) \quad (9)$$

where, d_0 : maximum load

ω : price elasticity

Since generation company i knows the estimating of the j 's cost function, generation company i will choose an output to maximize its benefit. In the power market, the economic benefits of generation company i is expressed as ;

$$\begin{aligned} B_i &= \rho P_i - C_i(P_i) \\ &= f_d(d)P_i - C_i(P_i) \\ &= -(\omega + c_i)P_i^2 + (d_0 - b_i - \omega P_j)P_i - a_i \end{aligned} \quad (10)$$

A Cournot Nash equilibrium is set of supply power outputs of the companies obtained by each company maximizing an individual benefit function. A Counot model about generation companies under the condition of technical constraints with complete information is established as follows ;

$$\max_{P_i} B_i = -(\omega + c_i)P_i^2 + (d_0 - b_i - \omega P_j)P_i - a_i$$

$$\text{subject to} \quad P_i^{\min} \le P_i \le P_i^{\max} \quad (11)$$

The price ρ in Eq. (10) is the uniform ρ. Since the benefit function B_i of the company i in Eq. (11) is concave and quadratic, the benefit maximizing condition is given by setting the partial derivative equal to zero. KT(Kuhn-Tucker) conditions are not only necessary conditions but also sufficient ones for the existence of extremum about nonlinear programming.

Differentiating the benefit function with respect to the i company output, we obtain

$$\frac{\partial B_i}{\partial P_i} = -2(\omega + c_i)P_i + d_0 - b - \omega P_j = 0 \quad (12)$$

KT points should be satisfied with the conditions:

$$\begin{cases} \dfrac{\partial B_i}{\partial P_i} - \lambda_1 + \lambda_2 = 2 \\ \lambda_1(P_i - P_i^{\max}) = 0 \\ \lambda_2(P_i^{\max} - P_i) = 0 \\ \lambda_1 \ge 0, \lambda_2 \ge 0 \end{cases} \quad (13)$$

where λ_1 and λ_2 are the generalized Lagrange multipliers. The KT points can be obtained using (12) and (13) as follows :

if $\lambda_1 = -2(\omega + c_i)P_i^{\min} + d_0 - b_i - \omega P_j \geq 0$

and $\lambda_2 = 0$, the KT point is given by

$$P_i = P_i^{\min} \tag{14}$$

if $\lambda_1 = 0$ and

$\lambda_2 = 2(\omega + c_i)P_i^{\max} - d_0 + b_i + \omega P_j \geq 0$

it is given by

$$P_i = P_i^{\max} \tag{15}$$

if $\lambda_1 = 0$, $\lambda_2 = 0$, it is given by

$$P_i = \frac{d_0 - b_i - \omega P_j}{2(\omega + c_i)} \tag{16}$$

Obviously, (14) and (15) are the KT points with constraints, while (16) is just the KT point without any constraint. The optimal point of the nonlinear programming must be one of (14), (15) or (16). The steps to obtain the Cournot-Nash equilibrium state with technical constraints are as follows :

Step 1 : The linear equations of the companies i and j can be obtained from (12). The partial derivative B_i with respect to the quantity P_i is

$$P_i^* = \frac{d_0 - b_i - \omega P_j}{2(\omega + c_i)} \tag{17}$$

This function P_i^* is called the company i 's best response to the company j 's quantity. Likewise,

$$P_j^* = \frac{d_0 - b_j - \omega P_i}{2(\omega + c_j)} \tag{18}$$

The two simultaneous equations P_i^* and P_j^* that define the Cournot best response functions. By substituting P_j in Eq.(17) with P_j^* of Eq.(18)

$$P_i = \frac{(d_0 + b_j - 2b_i)\omega + (d_0 - b_i)2c_j}{3\omega^2 + 2(2c_i + 2c_j)\omega + 4c_i c_j}$$

$$P_j = \frac{(d_0 + b_i - 2b_j)\omega + (d_0 - b_j)2c_i}{3\omega^2 + 2(2c_i + 2c_j)\omega + 4c_i c_j} \tag{19}$$

The outcome of Cournot game, i.e the pair of (P_i^*, P_j^*) is a Cournot-Nash equilibrium.

Step 2 : Checking the results obtained at step 1. If it is the Kuhn-Tucker points, the Cournot-Nash equilibrium state is obtained. Otherwise, adjust the results according the Kuhn-Tucker points (14), (15). The quantity P_j^* with respect to the quantity $P_j^{\max}$ or $P_j^{\min}$ is

$$P_j^* = \begin{cases} \dfrac{d_0 - b_i - \omega P_i^{\max}}{2\omega + 2c_j} \\ \dfrac{d_0 - b_i - \omega P_i^{\min}}{2\omega + 2c_j} \end{cases} \tag{20}$$

3. A Case Study

Suppose the Bayesian-Cournot Nash(BCN) equilibrium game model is used with a linear demand function and quadratic cost functions of each generation companies. In these examples, there are 3 generation companies, whose cost coefficients are listed in Table I.

We use F=2 scenarios on fuel kind with probability ϕ (1)=0.6 and ϕ (2)=0.4 repectively. In the first scenario, we assume that the probabilities for each company to be type 1 and type 2 are :

$$D_1^1 = [0.7 \quad 0.3]$$

$$D_2^1 = [0.2 \quad 0.8]$$

$$D_3^1 = [0.4 \quad 0.6]$$

For the second fuel scenario we use :

$$D_1^2 = [0.4 \quad 0.6]$$

$$D_2^2 = [0.5 \quad 0.5]$$

$$D_3^2 = [0.2 \quad 0.8]$$

Using the Bayesian conditional probability in Eq.(2), the matrix Π_{12} and Π_{13} in the company 1 and 2, 3 can be obtained as

$$\Pi_{12} = \begin{bmatrix} 0.164 & 0.416 \\ 0.156 & 0.264 \end{bmatrix}$$

$$\Pi_{13} = \begin{bmatrix} 0.200 & 0.380 \\ 0.120 & 0.300 \end{bmatrix}$$

Using the matrix Π_{12} and Π_{13} in Eq.(3) and (4), the conditional probability vectors between the company i and j are :

$\Pr_i^1 = [0.314 \quad 0.686]$ $\Pr_i^2 = [0.329 \quad 0.671]$

$\Pr_j^1 = [0.569 \quad 0.431]$ $\Pr_j^2 = [0.585 \quad 0.415]$

In Eq.(5)~(7), the parameters of each company's marginal cost function can be obtained as

$$C_i = 2.077P_i + 0.05P_i^2$$

$$C_j = 2.648P_j + 0.05P_j^2$$

Table I. The cost coefficients of the three generation companies

Coefficient & Fuel kind \ Company No. & Type		i		j			
		1		2		3	
		type 1	type 2	type 1	type 2	type 1	type 2
b	f=1	2.0	2.2	1.8	2.0	3.0	3.2
($/Mwh)	f=2	2.8	3.0	2.0	3.2	3.2	3.4
c	f=1	0.03	0.04	0.02	0.06	0.02	0.08
($/Mw^{2}h)	f=2	0.08	0.12	0.04	0.08	0.04	0.10

The coefficients of the demand function can be obtained from ISO and are assumed as d_0 : 20, ω : 0.04. We assume the power output constriants of company i are equal, $P_i^{\min}$ =35, $P_i^{\max}$ =90. From Eq.(19), the Cournot-Nash equilibrium P_i and P_j can be computed as

$$P_i = 82.21 \qquad P_j = 78.13$$

Each generation company can be obtained the equilibrium point of a pure strategy shown in Fig. 2.

Finally, the spot price ρ in the power market is calculated as

$$\rho = d_0 - \omega(P_i + P_j) = 13.59$$

From Eq.(10), the maximum benefit of each company can be obtained as

$$B_i = P_i(\rho - b_i - 2c_iP_i) = 270.64$$

$$B_j = P_j(\rho - b_j - 2c_jP_j) = 244.46$$

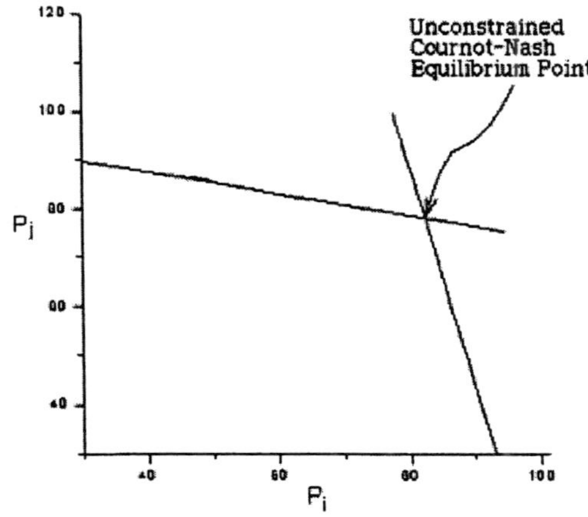

Fig. 2. Cournot-Nash Equilibrium point

4. Conclusion and Future Work

In this paper, the Bayesian-Cournot Nash(BCN) equilibrium game model is proposed to analyze the bidding strategies for generation companies. Opponent's cost and benefit functions are estimated with Bayesian conditional probabilities. Then the Cournot-Nash equilibrium is found using the estimated cost functions. The case study shows a pure Cournot-Nash equilibrium without constraints.

Future work may include the constraints such as the line losses, the wheeling and the congestion costs, etc. applications of the mixed Nash equilibrium strategies, and expansions to n-player and dynamic games.

References

[1] F. Nishimura, R. D. Tabors, "Benefit optimization of centralized and decentralized power systems in a multi-utility environment", IEEE Tranc. on Power Systems, Vol. 8, No. 3, pp. 1180-1186, Aug. 1993.

[2] R. W. Ferrero, J. F. Rivera, and S. M. Shahidehpour, "Application of Game with Incomplete Information for Pricing Electricity in Deregulated Power Pools", IEEE Trans. on Power Systems, Vol. 13, No. 1, pp. 184-189, Feb. 1998.

[3] J. D. Weber, T. J. Overbye, "A Two-Level Optimization Problem for Analysis of Market Bidding Strategies", IEEE PES Summer Meeting, Vol. 2, pp. 682-687, 1999.

[4] Xiaomin Bai, S. M. Shahidehpour, V. C. Ramesh, and Erkeng Yu, "Transmission Analysis by Nash Game Method", IEEE Trans. on Power Systems, Vol. 12, No. 3, pp. 1046-1052, August 1997.

[5] R. W. Ferrero, J. F. Rivera, and V. C. Ramesh, "Transaction Anaysis in Deregulated Power Systems Using Game Theory", IEEE Trans. on Power Systems, Vol. 12, No. 3, pp. 1340-1347, August 1997.

[6] H. R. Varian, Intermediate Microeconomics, New York : Norton, 1996, pp. 259-310

[7] B. Allen and M. Hellwing, "Pricing-Setting Firms and the Oligopolistic Foundations of perfect Competition", American Economic Review, Vol. 76, No. 2, 1986.

[8] A. J. Robson, "Stackelberg and Marshall", American Economic Review, Vol. 8, No. 1, 1990.

[9] P. D. Klemplerer and M. A. Meyer, "Supply Function Equilibrium in Oligopoly Under Uncertainly", Journal of Economics, Vol. 57, No. 6, pp. 1243-1277, 1989.

[10] Morin, T.L.; Nderitu, G.; Sparrow, F.T.; Yu, Z.; "A Stackelberg price leadership model with application to deregulated electricity markets", Power Engineering Society Winter Meeting, 2000. IEEE , Vol. 3, pp. 1814 - 1819, Jan. 2000.

[11] Z. Younes, M. Ilic, "Generation strategies for gaming transmission constraints, will the dergulated electric power market be an oligopoly ?", Decision Support System, Vol. 24, pp. 207-222, Jan. 1999.

[12] S. Borenstein, J. Bushnell, "The competitive effevts of transmission capacity in a deregulated electricity industry", RAND J. Econ., Vol. 31, pp. 294-325, summer 2000.

[13] Benjamin F. Hobbs, "Linear Complementarity Models of Nash-Cournot Competition in Bilateral and POOLCO Power Markets", IEEE Trans. on Power Systems, Vol. 16, No. 2, pp. 192-202, May. 2001.

[14] L. B. Cunningham, R. Baldick, "An Empirical Study of Applied Game Theory : Transmission Contrained Cournot Behavior", IEEE Trans. on Power Systems, Vol. 17, No. 1, Feb. 2002.

[15] Y. B. Zhang, X. J. Luo, and J. Y. Xue, "Constrained Competitive Cournot Equilibrium with Incomplete Information in Electricity Market", Proceedings of the 2002 International Conference on Power Systems Technology, Vol. 4. V.4, Oct. 2002.

ELSEVIER
IFAC
PUBLICATIONS
www.elsevier.com/locate/ifac

STUDY OF IDENTIFICATION FOR THE INERTIA CONSTANT IN A SMALL POWER SYSTEM FOR LFC

Y.Goto, S.Matsushita, K.Yukita
K.Mizuno,K.Ichiyanagi
Aichi Institute of Technology

H.Aoki, Y.Mizutani
Tokai University

Abstract: This paper has tried the identification for the inertia constant in a small power system, for the betterment of the LFC control characteristic. The power demand estimation method was used for the method of identifying an inertia constant power system. Until now, the method that used as an output set-point control by the power demand estimation is improved to the identification method. The power demand estimation method is using the frequency variation, the tie line power flow deviation, the output deviation of power plant, the inertia coefficient and damping coefficient of power system. In this paper, the actual electric power demand, the output of power plant, etc. was observed and the inertia coefficient is calculated using this estimation method. Then, the inertia constant and the damping constant are not observed.

In this paper, the inertia coefficient of the system of 2000kW small-scale system was calculated. *Copyright © 2003 IFAC*

Keywords: Power distribution control, Power system control, Power system simulation, Power systems, Identification

1. INTRODUCTION

The electric power supply, with which the electric power system is the stability and the high reliability, is required more and more by upgrading and diversification of social life and industry. In power system, the participation of In-dependent Power Producer (IPP), Distributed Generator such as wind power generation, solar photovoltaic generation, etc. seem to increase with deregulation and liberalization of power market recently. And, electricity trading which is the region monopoly until now, becomes also possible such as electricity trading and electric power consignment which exceeded the region. Therefore, sophisticated control technology is more required. And, the control method of electric power in the large-scale electric power system is complicated.

At present, Load Frequency Control (LFC) of control method, which maintains the frequency and the tie line power flow of the power system center in the standard value, is carried out. The tie line power flow bias control (TBC), for the purpose of maintaining frequency and interconnection tidal current in standard value, and the flat frequency control (FFC) , for the purpose of maintaining frequency is carried out in present LFC, respectively.(BERGEN,VITAL 2000)(IliC Zaborsky,2000)(Eric Hirst, Brendan Kirby 2000)(Jayant Kumar, kah-Hoe Ng, Gerald Sheble, 1997)

In previous deregulation system, FFC is carried out by the large-capacity system linked with the system, and TBC has been carried out in the small-capacity system, in Japan. However, all interconnection power systems began to carry out TBC, under the deregulation environment, for the congestion of the transmission line, etc.. Therefore, the sophisticated technology of LFC is more required, and the research is widely made. Until now, LFC using optimum control theories, fuzzy theory, H infinity control theory, etc. is proposed in great numbers in order to improve the control performance of frequency control. And, system constant identification technique using spectral analyses, etc. is also proposed in order to grasp the system constant with the aim of the improvement in LFC technology. (Hamaba, etal.,1991)(Hayashi , etal., 1992)(Toshio Inoue, etal, 1997)(K.Yukita, etal 1997)This paper has tried the identification for the inertia constant in a small power system, for the betterment of the LFC control characteristic. The power demand estimation method was used for the method of identifying an inertia constant power system. Until now, the method that used as an output set-point control by the power demand estimation is improved to the identification method. The power demand estimation method is using the frequency variation, the tie line power flow deviation, the output deviation of power plant, the inertia coefficient and damping coefficient of power system.

In this paper, the actual electric power demand, the output of power plant, etc. was observed and the inertia coefficient is calculated using this estimation method. Then, the inertia constant and the damping constant are not observed.

In this paper, the inertia coefficient of the system of 2000kW small-scale system was calculated.

2. POWER DEMAND ESTIMATION

The power demand estimation method estimates the power load fluctuation quantity of each area. And the estimated value is given the TBC-type output setting control as output set value of the power plant for the adjustment in each area. In this method, the power demand estimation method uses in the identification of the system constant. Then, this inertia constant uses the general value. However, this inertia value is not an optimum value in each case. Therefore, it may occur for this error for the power estimation value at the error. In this study, by doing observed value of electric power demand value, frequency, output deviation of power plant in actual system, the inertia coefficient was calculated using the power demand estimation.

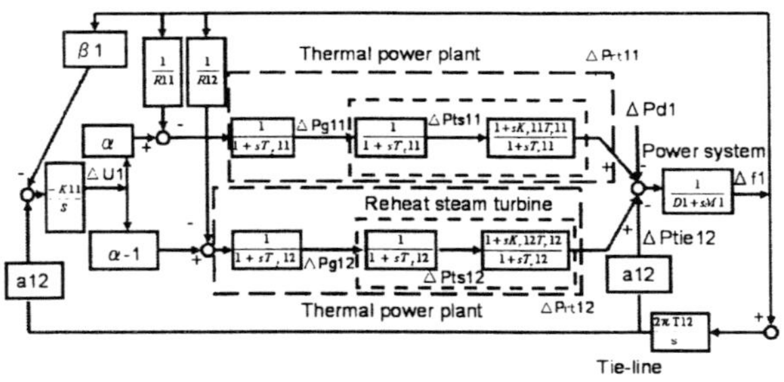

Fig.1 LFC model

2.1 Demand Estimation

In classical LFC model system shown in Fig1(IIiC Zaborsky2000)(Hadi Saadat, 1990), the following equation using the power plant output deviations ⊿Prt1, ⊿Prt2, the tie line power flow deviation ⊿Ptie12 and the frequency variation of each area ⊿f1 is used in order to estimate the load disturbance ⊿Pd1.

$$\Delta\tilde{P}_{d1}(k) = \frac{1}{2}\{-\frac{M_1}{T_{dk}}[\Delta f_1(k) - \Delta f_1(k-1)] - D_1\Delta f_1(k-1) + \Delta P_{rt1}(k-1) + \Delta P_{rt2}(k-1) + \Delta P_{tie12}(k-1)\} + \frac{1}{2}\Delta P_{d1}(k-1) \quad (1)$$

where M_1: Inertia coefficient T_{dk}: Sampling time.
D_1: Damping coefficient.

The load fluctuation quantity ⊿Pd1 calculated in the equation (1) is used by the arithmetic mean with the value before the 1 sample. By this fact, the prediction accuracy is improved. Then, in literature (K.Yukita, Y.Goto, Y.mizutani, 1997), it is used using this estimated value as output set value of each power plant to improve the characteristic of LFC.

2.2 Identification technique of the system constant using Demand Estimation.

In Fig.1, the electric power balance equation is arranged on ⊿Pd.. Until now, in some reports, the representative value has been used the inertia coefficient and the damping coefficient of the system, since to grasp the accurate value is very difficult. Equation (1) is transformed here like the following equation, if the power load fluctuation quantity ⊿Pd is observable, and it seems to be able to calculate the inertia coefficient M of the system.

Then, the identification of inertia coefficient of the system was tried using following equation in this paper.

$$M_1(k) = \{T_{dk}[\Delta P_{rt1}(k-1) + \Delta P_{rt2}(k-1) + \Delta P_{tie12}(k-1) - D_1\Delta f_1(k-1) + \Delta P_{d1}(k-1) - 2\Delta P_{d1}(k)]\} / \{\Delta f_1(k) - \Delta f_1(k-1)\} \quad (2)$$

In this paper, by obtaining mean value like following equation in order to calculate the inertia coefficient M one after another in the equation (2) in the k time, it is adopted.

$$M = \sum_{k=1}^{n} M(k)/n \quad (3)$$

Where k is the sampling time, n is the finish time.

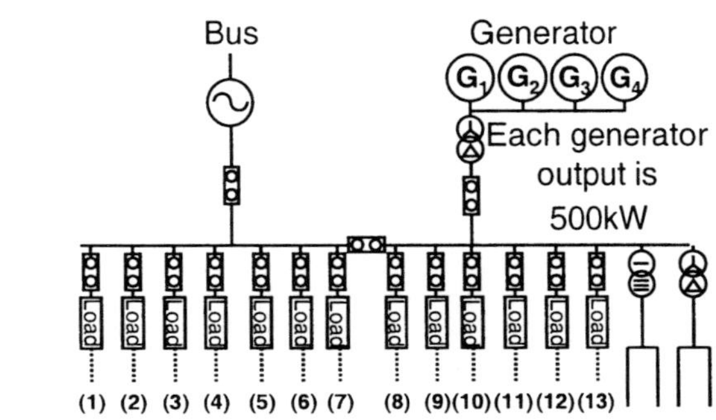

(a) Single wire circuit diagram

(b) General drawing

Fig.2 Single wire circuit diagram of the small-scale system

Table 1. Capacity of each load

Load number	Capacity	Load number	Capacity
1	–	8	800kVA
2	Cpare	9	1000kVA
3	1100kVA	10	955kVA
4	1130kVA	11	1050kVA
5	1350kVA	12	750kVA
6	1600kVA	13	Cpare
7	575kVA		

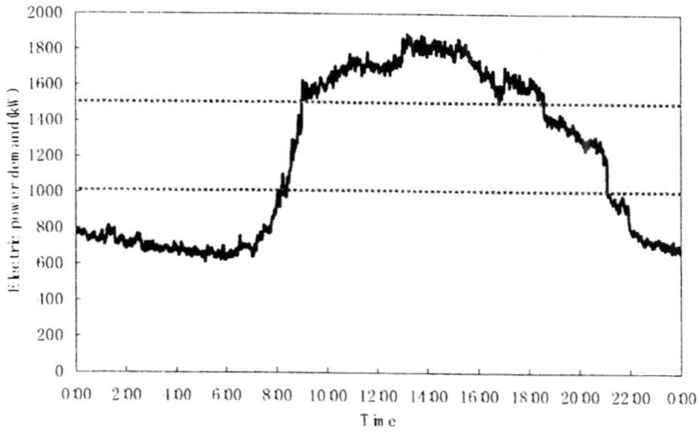

Fig.3 Daily load curve. (2002/2/8)

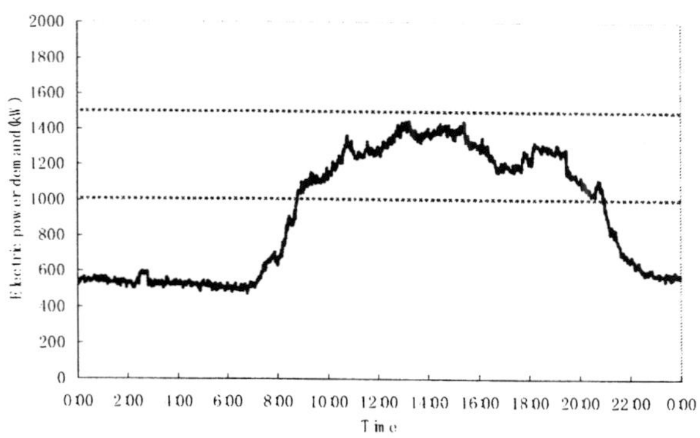

Fig.4 Daily load curve. (2002/5/16)

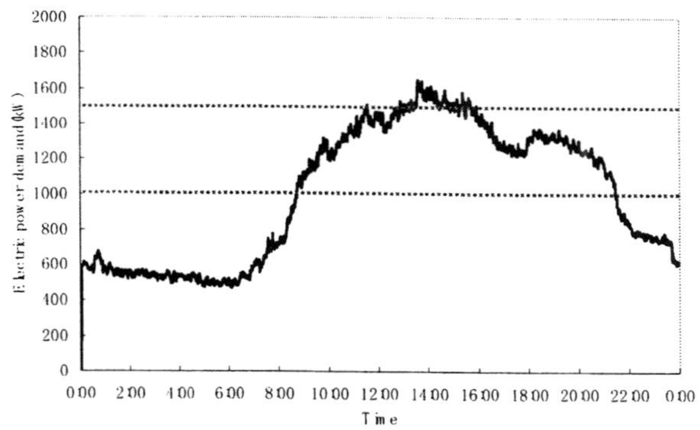

Fig.5 Daily load curve. (2002/6/4)

Table 2. Characteristic capacity of the daily load

2002/2/8		2002/5/16		2002/6/4	
Average	60.00168647	Average	59.99409257	Average	60.00751677
Standard error	0.000862189	Standard error	0.000993178	Standard error	0.000870151
Median value	60.0012	Median value	59.99565	Median value	60.005
Mode	59.9782	Mode	60.00365	Mode	60.02827
Standard deviation	0.032432525	Standard deviation	0.037688444	Standard deviation	0.033019916
Dispersion	0.001051869	Dispersion	0.001420419	Dispersion	0.001090315
Kurtosis	0.478310486	Kurtosis	0.792300041	Kurtosis	1.10720545
Strain	-0.01544862	Strain	-0.202035265	Strain	0.339280222
Range	0.253	Range	0.29745	Range	0.23586
Minimum	59.8697	Minimum	59.82615	Minimum	59.90895
Maximum	60.1227	Maximum	60.1236	Maximum	60.14481
Total	84902.38635	Total	86391.4933	Total	86410.82415
Sample number	1415	Sample number	1440	Sample number	1440
Maximum (1)	60.1227	Maximum (1)	60.1236	Maximum (1)	60.14481
Minimum (1)	59.8697	Minimum (1)	59.82615	Minimum (1)	59.90895
Confidence interva	0.001691308	Confidence interva	0.001948234	Confidence interva	0.001706903

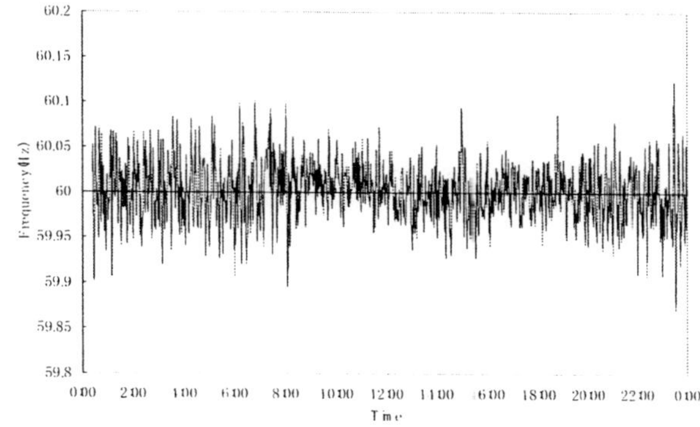

Fig.6 Frequency Characteristic. (2002/2/8)

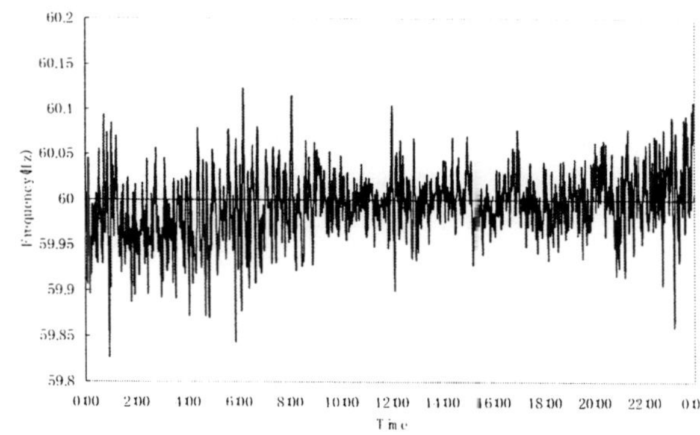

Fig.7 Frequency Characteristic. (2002/5/16)

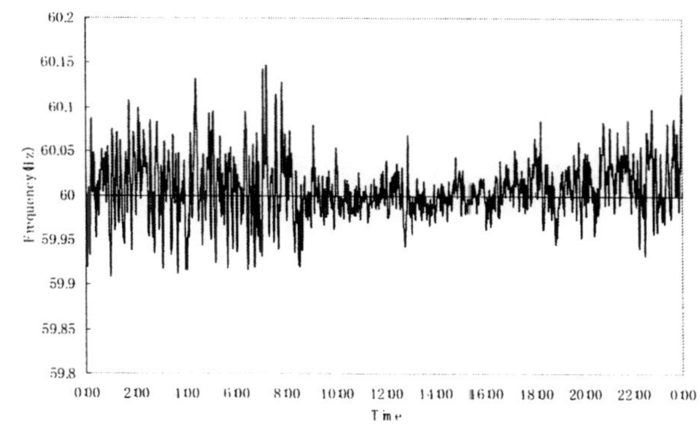

Fig.8 Frequency Characteristic. (2002/6/4)

Table 3. Characteristic capacity of the frequency

2002/2/8		2002/5/16		2002/6/4	
Average	1194.588735	Average	925.1105	Average	1006.423691
Standard error	12.02042496	Standard error	9.181508	Standard error	10.21607286
Median value	1259.251749	Median value	1036.371	Median value	1122.915794
Mode	687.2925436	Mode	510.2188	Mode	523.6169747
Standard deviation	456.1430557	Standard deviation	348.4137	Standard deviation	387.6727079
Dispersion	208066.4873	Dispersion	121392.1	Dispersion	150290.1284
Kurtosis	-1.736577828	Kurtosis	-1.75333	Kurtosis	-1.612129067
Strain	0.06067035	Strain	0.003065	Strain	-0.068366538
Range	1278.262739	Range	974.449	Range	1651.602366
Minimum	608.3516719	Minimum	470.0241	Minimum	0
Maximum	1886.614411	Maximum	1444.473	Maximum	1651.602366
Total	1720207.778	Total	1332159	Total	1449250.116
Sample number	1440	Sample number	1440	Sample number	1440
Maximum (1)	1886.614411	Maximum (1)	1444.473	Maximum (1)	1651.602366
Minimum (1)	608.3516719	Minimum (1)	470.0241	Minimum (1)	0
Confidence interva	23.57946228	Confidence interva	18.0106	Confidence interva	20.04001569

3. SMALL-SCALE SYSTEM

The single wire circuit diagram of the small-scale system used in this paper is shown in Fig.2. Electric power consumption capacity of this system is about 2000kW, and it possesses the private power generator of 4-machines. Each generator is the 500kW rated capacity. And, in the time of which the consumed electric power of the system is over 1000kW, the first power generation starts. Then, the electric power supply of 500kW is changed by this generator. Afterwards, the generator starts one by one at every 500kW in the electric power demand. Shortage minute of the electric power is supplied from the infinite bus.

The generator which worked this time does the fixed output operation (500kW fixed operation), and it has done the GF operation. However, the electric power conditioning is carried out by the infinite bus, when the generator does rated operation. The power distribution to the load is carried out at 6600V. The load capacity in this system is shown here in Table 1. Fig.3-Fig.5 is a daily load curve in this system, and it shows the total of the electricity consumption at each load No. 1-No.12 at Fig.2. And, Fig.6 - Fig.8 shows the first frequency fluctuation. Fig.3 and Fig.6 are February 8th, 2002. Fig.4 and Fig.7 are May 16th, 2002. And, Fig.5 and Fig.8 are June 4th, 2002.

As a characteristic of the load in this system, there is no decline of a demand in the noon observed generally, as it is shown at Fig.3-Fig.5. And, a demand begins to tend to temporarily increase in around 17 o'clock. It is proven that this system tends to be peculiar for this. Mean value and standard deviation of this load characteristic are shown at Table 2. The Table II is shown that the standard deviation on February eighth is 456kW, May 16th is 348kW and June 4th is 387kW. It is proven to be under the capacity of a generator machine in this system. Fig.6 - Fig.8 frequency characteristics are compared. Those many are fluctuation within ±0.05Hz. Then, the fluctuation over ±0.10Hz is only to appear in a moment. In this frequency characteristics, the average is about 60.0Hz, and it is proven to be very stable. The characteristic of the frequency is shown at Table 3.

4. Identification result and examination

Using Fig.3-Fig.8 characteristic of the pre-chapter, it tries to use the equation (2) of the identification of the inertia coefficient of Fig.1. However, the calculation was tried in this paper as D=0 in order to calculate damping coefficient D of the system, since it is difficult. And, the inertia coefficient put in the instant was calculated, since Fig.3-Fig.8 sampling period is 1 minute. It seems that the calculation in the 1 dividing stage is appropriate here for the control period of LFC, since it is within seconds or within a few minutes. The mean value was calculated in each day, and in respect of this calculated inertia coefficient, mean value of the value required further was calculated. The result is shown at Table 4. The inertia coefficient is 2.48 on February 8th, it is 1.12 on May 16th, and it is 2.47 on June 4th. It becomes a near value here in February 8th and June 4th. It is proven to become half value than other two on May 16th about. That the load on May 16th is small for this seems to influence. On the effectiveness of the value of calculated inertia coefficient, the electric power demand on June 15th 2002, was estimated. The result is shown in Fig.9 and Fig.10. At Fig.9, the following are shown: Electric power demand value which took the sum to Load No.1- Load No.12 and estimate by the Demand Estimation

Table 4. Inertia coefficient of system

Day	M (p.u.kW/Hz)
2/8	2.48
5/16	1.12
6/4	2.47
Average	2.02

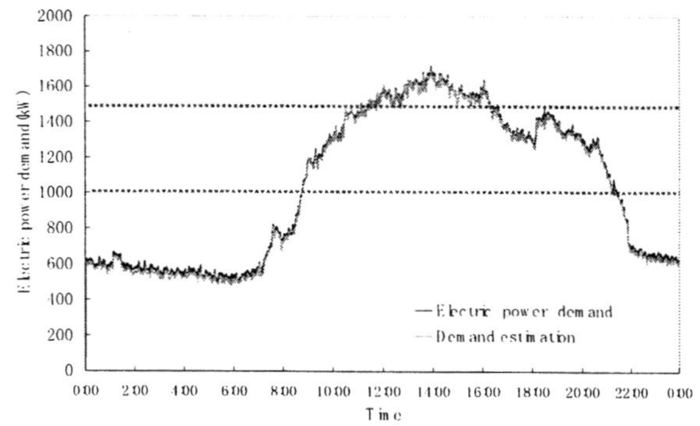

Fig.9 Daily load curve and estimate. (2002/6/5)

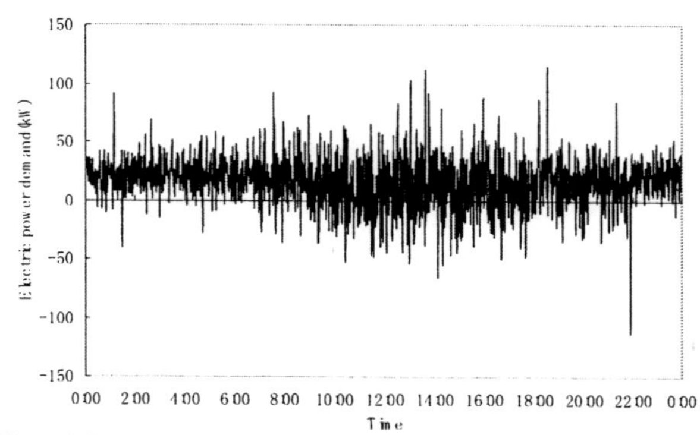

Fig.10 Estimated error

The error of electric power demand value and estimate is shown in Fig.10 in order to examine the error at Fig.9. The estimated error is 16.2kW in the average, as it is shown in Fig.10, and it is proven to be about 120kW even in the maximum. It is proven to become within about 6%, in respect of this, when the 2000kW system capacity is considered.

5. CONCLUSION

In this paper, the effectiveness by actual small-scale system was examined in respect of Demand Estimation which authors had proposed until now for control performance improvement of load frequency control. In addition, the identification of inertia coefficient M of the system in the LFC model was tried.

It became a value which was near for 3.3 that as the result, inertia coefficient M becomes 2.02, and that it has reported the convention. And, it was possible to estimate DE at the error within about 6% (within 120kW) of the system capacity.

6. REFERENCE

BERGEN,VITAL "POWER SYSTEM ANALYSIS" PRENTICE HALL, (2000)

IliC Zaborsky, "Dynamics and Control of Large Electric Power Systems", Wiley-interscience, (2000)

Jayant Kumar, kah-Hoe Ng, Gerald Sheble, "AGC SIMULATOR PROICEBASED OPERATION PART 1:A MODEL", IEEE Transactions on Power Systems, Vol.12, No.2, May, (1997)

Jayant Kumar, kah-Hoe Ng, Gerald Sheble, "AGC SIMULATOR PROICEBASED OPERATION PART 2:CASE STUDY RESULTS", IEEE Transactions on Power Systems, Vol.12, No.2, May, (1997)

Eric Hirst, Brendan Kirby "Time-Averaging Period for Regulation Service", IEEE Power Engineering REVIW, Vol20, No.7, (2000)

Hadi Saadat, "Power System Analysis McGrawHill", (1990)

Hamaba, Yokoyama, Kato, Takabayashsi, "System constant estimation method using the normal period operation data", Inst. of Electrical Engineers of Japan paper magazine B, Vol.111, No.6, (1991)

Hayashi, Kawada, Nagasaki, Yokoyama "The estimation method by spectral analysis of electric power-frequency characteristic constants of electric power system", Inst. of Electrical Engineers of Japan paper magazine B, Vol.112, No.11, (1992)

Toshio Inoue, Haruhito Taniguchi, Yasuyuki Ikeguchi, Kiyoshi Yoshida, "Estimation of Power System Ineria Constant and Capacity of Spinningreserve Support Generators Using Measured Frequency Transients", IEEE Transactions on Power Systems, Vol.12, No.1, (1997)

K.Yukita, Y.Goto, Y.mizutani, "The load frequency control which used fuzzy control considering regulatory effect of control factor", Inst. of Electrical Engineers of Japan paper magazine B, Vol.116-B, No.1, (1997)

ELSEVIER
IFAC
PUBLICATIONS
www.elsevier.com/locate/ifac

FAST ALGORITHMS FOR SOLVING LOAD FLOW EQUATIONS[1]

Fenghong Yang[§], Yun Tang[§†], Ping Luo[‡], Ming Rao[£]

[§] *Department of Mathematical Sciences, Tsinghua University, Beijing 100084, China;* [‡] *Department of Computer Science and Technology, Tsinghua University, Beijing 100084, China;* [£] *CCUC, University of Alberta, Edmonton, T6G 2G6, Canada.*

[†] *ytang@math.tsinghua.edu.cn*

Abstract: This paper presents two new and fast inexact Newton methods, called Newton–CAV method and simplified Newton method, to solve load flow equations of power systems. The new algorithms are inherently parallel. Tests of the simplified Newton-CAV method are performed on the IEEE 118-bus system and the IEEE 662-bus system under Matlab 6.1 environment. The numerical results show that it is feasible and fast. Moreover the approach of choosing relaxation parameters in CAV method is obtained. *Copyright © 2003 IFAC*

Keywords: load flow solution; nonlinear equation; inexact Newton method; iterative method; parallel computation

1. INTRODUCTION

Calculation of load flows is one of the most fundamental power system operation and planning problem. The work presented in the paper is to introduce new algorithms for solving load flow equations.

The load flow problem can be represented through a system of nonlinear algebraic equations

$$\boldsymbol{F}(\boldsymbol{x}) = \boldsymbol{0}, \tag{1}$$

where $\boldsymbol{F} : \boldsymbol{R}^n \rightarrow \boldsymbol{R}^n$ is a differential mapping. The most attractive method for solving (1) is the classical Newton method due to its local quadratic convergence. The Newton update equation is formulated as:

$$\boldsymbol{J}(\boldsymbol{x}) \bigtriangleup \boldsymbol{x} = -\boldsymbol{F}(\boldsymbol{x}), \tag{2}$$

where $\boldsymbol{J}(\boldsymbol{x})$ is the Jacobian matrix of $\boldsymbol{F}$ at $\boldsymbol{x}$. A single iteration of Newton process involves solving linear equation (2) for the state update $\bigtriangleup \boldsymbol{x}$ and then updating the state vector $\boldsymbol{x}$ by $\boldsymbol{x}+ \bigtriangleup \boldsymbol{x}$.

Such linear equations are typically solved by two classes of methods: direct methods and iterative methods. Many direct methods such

[1] This work is supported by National Key Basic Research Special Fund (No. G1998020309).

as Gaussian elimination, LU factorization etc., have been introduced to the load flow computation. In addition to the direct methods, there are iterative approaches to solve the linear system which leads to a class of methods, named inexact Newton method (Brown, 1987; Demdo et al., 1982; Eisenstat and Walker, 1996), as a variation of the classical Newton method. That is, we can use an iterative method to solve (1) only approximately. More precisely, an inexact Newton method is the method that for given initial guess $\boldsymbol{x}_0$ generates a sequence $\{\boldsymbol{x}_k\}$ of approximating to the solution $\boldsymbol{x}^*$ of (1) according to the following steps (Demdo et al., 1982):

For $k = 0$ step 1 until converges Do
Find some step $\boldsymbol{s}_k$ which satisfies

$$\boldsymbol{J}(\boldsymbol{x}_k)\boldsymbol{s}_k = -\boldsymbol{F}(\boldsymbol{x}_k) + \boldsymbol{r}_k, \text{where} \frac{\|\boldsymbol{r}_k\|}{\|\boldsymbol{F}(\boldsymbol{x}_k)\|} \leq \eta_k,$$

Set $\boldsymbol{x}_{k+1} = \boldsymbol{x}_k + \boldsymbol{s}_k$.

Here the forcing term η_k may depend on $\boldsymbol{x}_k$. Taking $\eta_k \equiv 0$ gives Newton method. The Generalized Minimal Residual (GMRES) method (Saad and Schultz, 1986) is a promising technique which has received great attention recently from researchers in the field of load flow computation. A significant research effort then has been expended to examine the Newton-GMRES method and parallel Newton-GMRES method (Flueck and Chiang, 1998; Tu and Flueck, 2002) in the calculation of power systems.

On the other hand, as a new iterative parallel technique suitable for large and sparse unstructured systems of linear equations, Y. Censor et al presented firstly the component averaging (CAV) algorithm in (Censor et al., 2001). The algorithm achieved good performance on image reconstruction problems of the Herman head phantom. The convergence of the algorithm was given completely in (Censor and Elfving, 2002). In this article, the CAV method is combined with the Newton method, which yields a kind of inexact Newton methods, called Newton-CAV method and simplified Newton-CAV method.

The organization of the paper is as follows. The component averaging method is introduced in section 2. The Newton-CAV algorithm and simplified Newton-CAV algorithm are presented in section 3. The latter is tested on the IEEE 118-bus and the IEEE 662-bus systems under Matlab 6.1 environment in section 4. Conclusions and future work are presented in the last section.

2. COMPONENT AVERAGING ALGORITHM

In this section, we introduce the component averaging (CAV) algorithm (Censor et al., 2001). CAV method is a parallel iterative method suitable for large and sparse unstructured linear equations.

Consider the linear equations

$$\boldsymbol{A}\boldsymbol{x} = \boldsymbol{b}, \tag{3}$$

where $\boldsymbol{A}$ is an $m \times n$ sparse matrix, $\boldsymbol{x} \in \boldsymbol{R}^n$ and $\boldsymbol{b} \in \boldsymbol{R}^m$.

In the following discussion, let $\boldsymbol{a}^i$ be the ith row of $\boldsymbol{A}$, s_l the number of nonzero elements in the jth column of $\boldsymbol{A}$ and $\|\boldsymbol{x}\|_G^2 =< \boldsymbol{x}, \boldsymbol{G}\boldsymbol{x} >$ the associated ellipsoidal norm, where $\boldsymbol{G}$ is an $n \times n$ symmetrically positive definite matrix.

CAV algorithm (Censor et al., 2001):
Initialization: $\boldsymbol{x}_0 \in \boldsymbol{R}^n$ is arbitrary.
Iterative step: Given $\boldsymbol{x}^k$, compute $\boldsymbol{x}^{k+1}$ by using, for $j = 1, 2, \cdots, n$, the formula:

$$x_j^{k+1} = x_j^k + \lambda_k \sum_{i=1}^{m} \frac{b_i - < \boldsymbol{a}^i, \boldsymbol{x}^k >}{\sum_{l=1}^{n} s_l (a_l^i)^2} a_j^i \tag{4}$$

where $\{\lambda_k\}_{k \geq 0}$ are relaxation parameters, commonly confined to the interval $\varepsilon \leq \lambda_k \leq 2 - \varepsilon$ for some fixed but arbitrarily small $\varepsilon \geq 0$ and $\{s_l\}_{l=1}^n$ are defined as above. Let $\boldsymbol{S} = \text{Diag}(s_1, s_2, \cdots, s_n)$ (we assume $s_l \neq 0, l = 1, 2, \cdots, n$) and $\boldsymbol{D}_s = \text{Diag}(\frac{1}{\|\boldsymbol{a}^1\|_S^2}, \frac{1}{\|\boldsymbol{a}^2\|_S^2}, \cdots, \frac{1}{\|\boldsymbol{a}^n\|_S^2})$, then (4) can be rewritten as:

$$\boldsymbol{x}^{k+1} = \boldsymbol{x}^k + \lambda_k \boldsymbol{A}^T \boldsymbol{D}_s (\boldsymbol{b} - \boldsymbol{A}\boldsymbol{x}^k). \tag{5}$$

Obviously, (5) is similar to the forms of Jacobian method and Gauss-Seidel method.

Theorem 2.1 (Censor and Elfving, 2002) Assume that $H = \{\boldsymbol{x} \in \mathbf{R}^n | \boldsymbol{A}\boldsymbol{x} = \mathbf{b}\} \neq \emptyset$ and that the relaxation parameters are restricted to

$$0 < \varepsilon \leq \lambda_k \leq \frac{2-\varepsilon}{\rho(\boldsymbol{A}^T\boldsymbol{D}_s\boldsymbol{A})} \text{ for all } k \geq 0,$$

where ε is an arbitrarily small but fixed constant and $\rho(\boldsymbol{A}^T\boldsymbol{D}_s\boldsymbol{A})$ is the spectral radius of matrix $\boldsymbol{A}^T\boldsymbol{D}_s\boldsymbol{A}$ with $\boldsymbol{D}_s$ defined above. Then any sequence $\{\boldsymbol{x}_k\}_{k\geq 0}$, generated by CAV algorithm, converges to a solution of the system (3). If, in addition, $\boldsymbol{x}_0 \in range(\boldsymbol{A}^T)$, then $\{\boldsymbol{x}_k\}_{k\geq 0}$ converges to the solution of (3), which has the minimal Euclidean norm.

Remark: It may play an important role for the selection of the relaxation parameter λ in the convergence of the algorithm.

3. NEWTON-CAV ALGORITHMS

Now let us turn to Eq.(1). Combining the classical Newton method with CAV method mentioned above, we obtain the following two inexact Newton–CAV methods.

Algorithm 1 (Newton–CAV method):

Let $\boldsymbol{x}_0$ be an initial vector, $\varepsilon_1 > 0$ small enough, $k = 0$.

While $\|\boldsymbol{F}(\boldsymbol{x}_k)\| \geq \varepsilon_1$ Do

Compute $\boldsymbol{A} = \boldsymbol{J}(\boldsymbol{x}_k), \boldsymbol{b} = -\boldsymbol{F}(\boldsymbol{x}_k)$.

Solve $\boldsymbol{A}\boldsymbol{s}_k = \boldsymbol{b}$ by the CAV method (noting it may be implemented in parallel).

$\boldsymbol{x}_{k+1} = \boldsymbol{x}_k + \boldsymbol{s}_k$.

$k = k + 1$.

End while

In the above algorithm we have to compute the Jacobian matrix $\boldsymbol{J}(\boldsymbol{x})$ in each Newton process. If assuming $\boldsymbol{J}(\boldsymbol{x}_k) \equiv \boldsymbol{J}(\boldsymbol{x}_0)$ fixed for all $k > 0$, we obtain another algorithm in the following.

Algorithm 2 (Simplified Newton–CAV method):

Let $\boldsymbol{x}_0$ be an initial vector, $\varepsilon_1 > 0$ small enough, $k = 0$.

Calculate $\boldsymbol{A} = \boldsymbol{J}(\boldsymbol{x}_0)$.

While $\|\boldsymbol{F}(\boldsymbol{x}_k)\| \geq \varepsilon_1$ Do

Compute $\boldsymbol{b} = -\boldsymbol{F}(\boldsymbol{x}_k)$.

Solve $\boldsymbol{A}\boldsymbol{s}_k = \boldsymbol{b}$ by the CAV method (noting it may be implemented in parallel).

$\boldsymbol{x}_{k+1} = \boldsymbol{x}_k + \boldsymbol{s}_k$.

$k = k + 1$.

End while

By using the classical theory (Brown, 1987; Demdo et al., 1982), convergence of the above two algorithms are trivial. And they are parallelizable inexact Newton methods.

4. NUMERICAL TESTS

In the section, we test the algorithm 2 on the IEEE 118-bus and the IEEE 662-bus systems under Matlab 6.1 environment and give the results in comparison with Newton method and simplified Newton method. In calculation, we set $\varepsilon_1 = 10^{-3}$ and the approximate start is used as an initial vector.

Firstly, we consider the influence of the relaxation parameter λ on the convergence.

In the following tables, out-iter denote the numbers of outer iteration and t_{CPU} the total sums of Newton iteration and CAV procedure in seconds. Table 1 and Table 2 show that relaxation parameter λ impact evidently on the outer iterative number and CPU time, as well as the closer λ are to the upper bound, the faster the algorithm converges. So λ can be chosen based on this rule.

Table 1. Numerical results for IEEE 118-bus power system

λ	out-iter	t_{CPU}(s)
1.0	39	1.668
1.5	26	1.23
2.0	20	1.01
2.5	16	0.875
3.0	13	0.781
3.5	12	0.75
3.8	11	0.703
3.9	*	*
4.0	*	*

Table 2. Numerical results for IEEE 662-bus power system

λ	out-iter	t_{CPU}(s)
1.0	43	25.311
1.5	29	20.780
2.0	22	17.514
2.5	18	16.484
3.0	15	16.952
3.5	13	14.874
3.8	12	13.890
3.9	*	*
4.0	*	*

In the two tables above, $*$ denote the algorithm isn't convergent.

Now we give comparison results. Herein we set the inner iterative number equal to 100 and $\lambda = 3.8$.

Table 3. Comparison of numerical result of simplified Newton-CAV method with those of Newton method and simplified Newton method (IEEE 118-bus power system)

algorithms	out-iter	t_{CPU}(s)	ratio e
N	2	0.61	1.1525
SN	2	0.33	2.1303
SN-CAV	11	0.703	1

Note: In the table let N denote the Newton method, SN the simplified Newton method and SN-CAV the simplified Newton-CAV method. The ratio $e = \frac{t_3}{t_i}$, where t_i are the CPU time of the Newton method, the simplified Newton method and the simplified Newton-CAV method for $i = 1, 2, 3$, respectively. The notations in Table 4 are similar.

Table 4. Comparison of numerical result of simplified Newton-CAV method with those of Newton method and simplified Newton method (IEEE 662-bus power system)

algorithms	out-iter	t_{CPU}(s)	ratio e
N	3	26.186	0.5304
SN	3	9.577	1.4503
SN-CAV	12	13.89	1

Table 3 and Table 4 show that the inexact Newton methods presented in the paper are potential to solving the load flow equations. It follows from the change of the ratio e that the trend is obvious in larger power systems.

5. CONCLUSION AND DISCUSSION

In the paper we introduce the component averaging (CAV) algorithm to compute load flows for power systems and present new inexact Newton methods, called Newton-CAV method and simplified Newton-CAV method. It has the similar form with inexact Newton methods by using Jacobian or Gauss-Seidel algorithm to solve the Newton process. But it has the advantage of avoiding computing the inverse matrix in iteration. So the new algorithm is convenient and fast in real application. Under Matlab 6.1 environment, we test the algorithm 2 on the IEEE 118-bus and the IEEE 662-bus power systems. The numerical results show that it is potential to solve load flows for large power systems. At the same time, we find an approach of choosing relaxation parameter λ, that is, the closer to the upper bound, the faster the algorithm converges.

To understand performance of the new algorithms, however, it is necessary to test them on other nonlinear problems. Their parallel implement are ongoing based on a message-passing distributed memory multiprocessor architecture, such as a cluster of workstations.

REFERENCES

Brown, P. N. (1987). A local convergence theory for combined inexact-newton/finite-difference projection methods. *SIAM Journal on Numerical Analysis*, **24**:407–434.

Censor, Y. and Elfving, T. (2002). Block-iterative algorithms with diagonally scaled oblique projections for the linear feasibility problems. *SIAM Journal on Matrix Analysis and Applications*, **24**:40–58.

Censor, Y., Gordon, D., and Gordon, R. (2001). Component averaging: An efficient iterative parallel algorithm for large and sparse unstructured problems. *Parallel computing*, **27**:777–808.

Demdo, R. S., Eisenstat, S. C., and Steihaug, T. (1982). Inexact newton methods. *SIAM Journal on Numerical Analysis*, **19**:400–408.

Eisenstat, S. C. and Walker, H. F. (1996). Choosing the forcing terms in an inexact newton methods. *SIAM journal on scientific computing*, **17**:17–32.

Flueck, A. and Chiang, H. (1998). Solving the nonlinear power flow equations with an inexact newton method using gmres. *IEEE TRANS on power systems*, **13**(2):267–273.

Saad, Y. and Schultz, M. H. (1986). Gmres: A generalized minimal residual algorithm for solving nonsymmetric linear systems. *SIAM Journal of Scientific and Statistical Computing*, **7**(3):856–869.

Tu, F. and Flueck, A. J. (2002). A message-passing distributed-memory newton-gmres parallel power flow algorithm. *IEEE PES Winter Meeting*, pages 1477–1482.

ELSEVIER
IFAC
PUBLICATIONS
www.elsevier.com/locate/ifac

A NEW FAULT DETECTIOR FOR THREE-PHASE HALF-WAVE CONVERTER USING WAVELET TRANSFORM

Jong-Min Cheon, Choon-Kyung Kim, Jong-Moo Lee, Seog-Ju Kim and Soonman Kwon

Korea Electrotechnology Research Institute, Changwon, Gyungsangnamdo, Korea
Tel : 82-55-1449 ; Fax : 82-55-280-1476 ; E-mail : jmcheon@keri.re.kr

Abstract: In this paper, we introduce a new method of detecting faults in a three-phase half-wave converter used to transmit DC power into Control Rod Drive Mechanisms (CRDMs) in a Nuclear Power Plant (NPP). The failures in a three-phase half-wave converter can have a bad influence upon the operation of CRDM. This situation makes the system generate the urgent alarm and stop the movement of control rods. Existing systems use complex analogue circuits, but we adopt simple and efficient digital technology, in accordance with a current tendency. The digital system needs a new algorithm for fault detection and the wavelet transform is used to design the new fault detector and improve the ripple detector of Westinghouse type NPP.

Keywords: Nuclear power plant, Control system, Power circuits and Fault detection

1. INTRODUCTION[1]

The primary functions of the Control Rod Control System (CRCS) in a Nuclear Power Plant (NPP) are to transmit the power to Control Rod Drive Mechanisms (CRDMs) according to the control signal and to enable the Control Rod Assemblies (CRAs) attached into CRDMs to move vertically. The reactor power is controlled by this vertical movement of the control rods composed of neutron absorber. In an emergency, CRCS interrupts the power offered to CRDM and causes control rods to drop into the core to shut down safely (K.K.Jeong, 1991; K.S.Park, 1997).

The NPPs built long ago have aging devices and the shortage of components and want to replace their existing complex analogue systems with more efficient and simpler systems in instrumentation and control equipment. The latest NPPs also trend toward digital system (P.H.Sung, *et al.* ,2001). Because of this current tendency and advantages of digital system, we got to design a new CRCS based on digital technology (S.Kwon, *et al.* ,2001). CRCS is composed of Main Control Unit (MCU) and Power Control and Monitoring Unit (PCAMU). Our digitalized system adopts Programmable Logic Controllers (PLCs) for MCU and Digital Signal Processors (DSPs) for PCAMU, respectively. MCU gets general orders of rod speed & direction for the realization of the rod movement. PCAMU convert the three-phase MG set power into sequenced, amplitude-controlled DC current pulses in obedience to MCU's orders. The converted power is transmitted to selected CRDM coils for moving control rods. Considering currents and voltages from CRDM coils, PCAMU also monitors the operations of CRDM (J.M.Cheon, *et al.* ,2001).

Among many faults and errors in this system, especially, we focus on the failures in three-phase half- wave converters of the Power Converter Module (PCM) embedded in PCAMU. Because these faults can exert a harmful influence on CRDM's normal operations, it is important to detect the occurrence of the failures quickly and correctly. In the new system based on digital technique, DSP takes charge of monitoring and detecting faults. A new algorithm for detecting the faults of interest needs to be devised and programmed as software to design DSP instead of the existing analogue ripple detector in Westinghouse model.

We establish the failures in three-phase half-wave converters as the targets to be detected by the newly devised detection method. Wavelet transform is used to design the new detector and the superiority of that method is investigated from the results of simulations.

2. TARGET FAULTS FOR DETECTION

2.1 Three-phase half-wave converter

Three thyristors are connected to each phase of the AC power supply in a three-phase half-wave converter. The reason why thyristors are used is to achieve three fixed levels of current and to arrive at these current levels as fast as possible. To obtain various levels of current, the thyristor has to be able to turn on at different times during the waveform. This turn-on is referred to as the firing angle. The

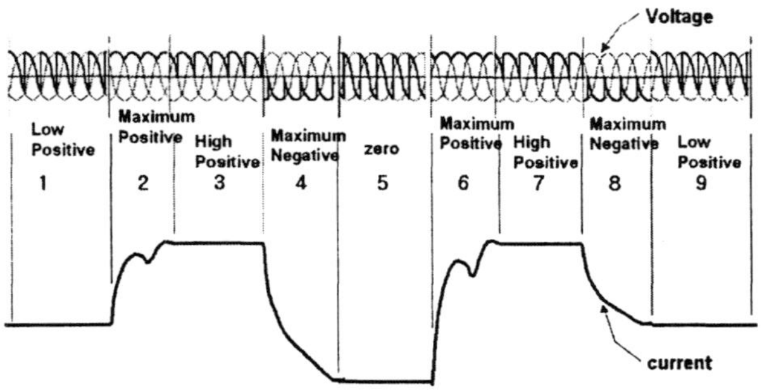

Fig. 1. The voltage and current waveforms of a stationary gripper coil

firing angles vary automatically in response to the current command signals from MCU.

Figure 1 depicts the current waveform for a stationary gripper coil and its voltage waveform that produces the current. The stationary gripper coil is one of three coils in CRDM. The voltage is measured by the Hall potential transformer at the three-phase half-wave converter and the current is measured by the Hall current transformer at each coil.

When MCU calls for the full current, the thyristors are fired at a point that forces the current to rise exponentially at the maximum rate. As the current approaches the full current level, the firing points are gradually retarded until the average coil voltage is enough to maintain the full level. If MCU wants lower level than the present level, the firing point is further retarded to another point where the average voltage is maximum negative. As stated above, the firing point is controlled by the current level that MCU requires (K.K.Jeong, 1991).

2.2 Phase Failure

The faults occurring in a three-phase half-wave converter are called the phase failure. The phase failure has two kinds of failures, the missing phase failure and the diode action failure. The former is caused by a thyritor without conduction and the latter is due to a thyristor which has lost gate control. When these faults occur, the alarm circuit generates the urgent alarm and the movement of control rods is prohibited. The missing phase failure is considered in the positive forcing like the period 2 and 6 in Figure 1 and the diode action failure is related to the negative forcing such as the period 4 and 8 in Figure 1 (K.K.Jeong, 1991). The existing system does not detect the phase failure except the positive and negative forcing periods, but our new fault detector watches the occurrence of the phase failure through every period.

2.3 Phase Failure

We can see the ripple voltage waveforms caused by the missing phase failure and the diode action failure in Figure 2. From this figure, you can see the difference of the ripple ratio between the normal value and the abnormal value from a failure. Aiming at this point, the existing Westinghouse model detects the occurrence of failures.

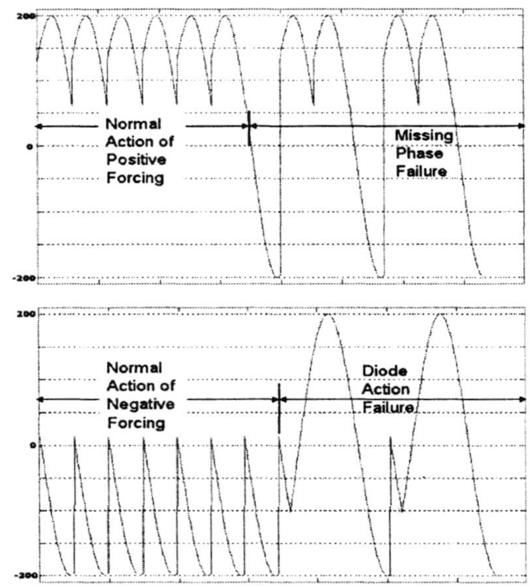

Fig. 2. The voltage and current waveforms of a stationary gripper coil

We can see the ripple voltage waveforms caused by the missing phase failure and the diode action failure in Figure 2. From this figure, you can see the difference of the ripple ratio between the normal value and the abnormal value from a failure. Aiming at this point, the existing Westinghouse model detects the occurrence of failures. In the case of the missing phase failure, the ripple ratio of the positive forcing state to the fault state is 1 to 4. The ripple ratio of the negative forcing state to the diode action failure is 1 to 2.

Figure 3 shows the Westinghouse model ripple detector which takes out the ripple part from the voltage waveform (K.K.Jeong, 1991). The output voltage at ⓔ is proportional to the height of the ripple in the input coil voltage and inputs to the phase failure section of the failure detector card. At the time when ⓔ voltage exceeds +8.2V, the phase detector informs the alarm card of some fault occurrence. To separate the normal and abnormal state effectively, we need to modify their ripple ratios. In Figure 3, the ripple of the negative forcing part is large enough to make the ripple detector voltage be excess of +8.2V. From this reason, the ripple detector possesses a Zener diode as the negative clipper. Including this negative clipper, the ripple detector can modify the ripple ratio of the positive forcing normally to the missing phase failure into 1 to 3 and the ripple ratio of the negative forcing normally to the diode action failure also into 1 to 3. Figure 4 displays the modified ripple ratios.

The ripple detector output of the missing phase failure occurrence at 0.05sec can be seen in Figure 5. We can find the existing ripple detector's ability to detect a failure but we also cannot miss the weakness

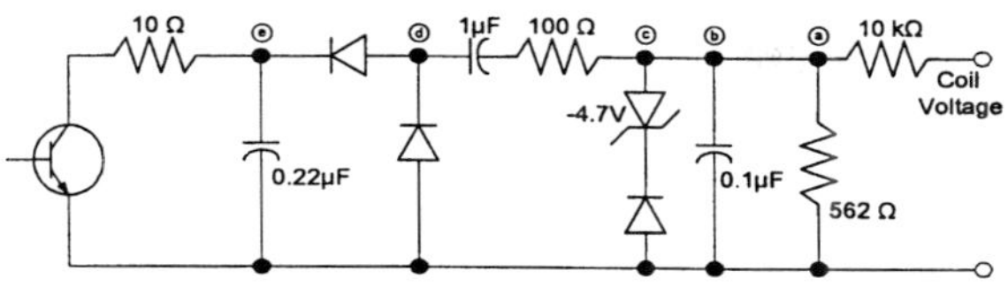

Fig. 3. The ripple detector in Westinghouse model NPP

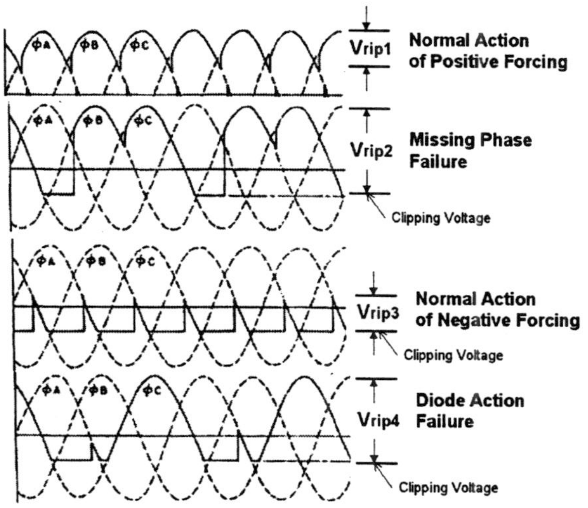

Fig. 4. The modified voltage ripple ratios

of this system. It does detect the phase failure only during the positive and negative forcing periods but the period during which the control rods stop must be watched.

If the stationary gripper coil current for the control rod stop is abnormal, the control rods under that current condition may drop. That situation means an urgent alarm and it is necessary to watch the voltages of other periods like the rod stop period. When we design a new fault detector for the phase failure, we consider the necessity of watching every period of voltage and make existing analogue systems with complex devices digitalized. A digital system is organized with hardware and software. To program the software, we need a new algorithm that detects the phase failure. Until now, many measures have been adopted to detect faults. Some representative methods are Neural Network (J.H.Lee, *et al.*, 2000; H.Bae, *et al.*, 2002), Neuro-Fuzzy system (J.K.Kang, *et al.*, 1999; S.H.Kim, et al., 1999), Fourier Transform (C.H.Jeong, *et al.*, 2002; J.H.Oh, *et al.*, 2002), Wavelet Transform (D.S.Hong, *et al.*, 2001; G.Y.Park, *et al.*, 2002; W.Y.Yim, 2002) and Kalman Filter (Y.S.Lee, 2002).

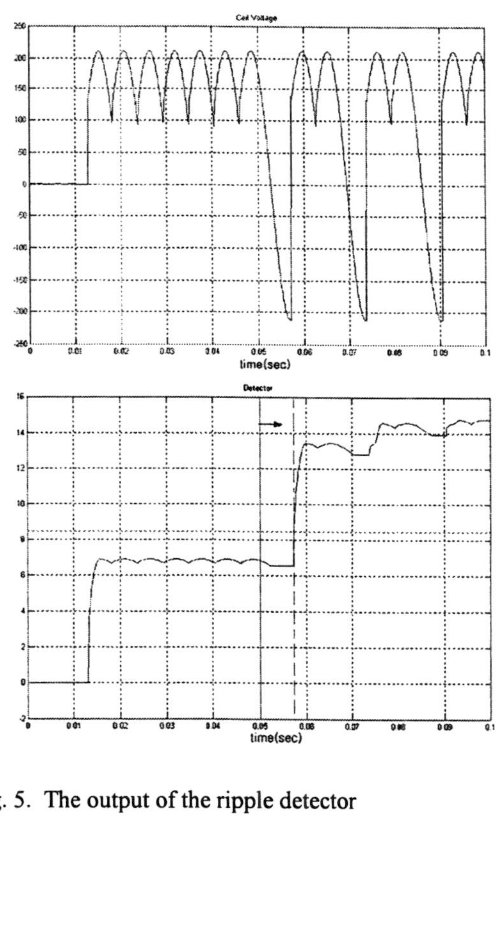

Fig. 5. The output of the ripple detector

In this paper we use Wavelet Transform to devise a new fault detection method and the reason why this transform is used is presented by comparing the performances of Fourier transform and showing the solution of problems in the existing ripple detector in the next chapters.

3. DESIGN A NEW FAULT DETECTOR

3.1 Using Fourier Transform

Fourier transform is the most frequently used method in the signal processing and analysis (G.Y.Park, *et al.*, 2002). Especially, when we want to know the values of some frequencies included in a signal, this works well. We thought that the ripple frequencies would change at the fault occurrence and Fourier transform could find out the ripple frequencies of normal state and abnormal state.

The ripple frequency of normal operation is three times as many as the frequency of the AC power supply. If the power with 60 Hz is supplied, the ripple frequency of the coil voltage is 180 Hz. Figure 6 shows that the dominant ripple frequency of the normal state is 180 Hz and 60 Hz is the dominant ripple frequency of the failure state. From these results, we can confirm Fourier transform's ability to catch a failure occurrence. However, what we know

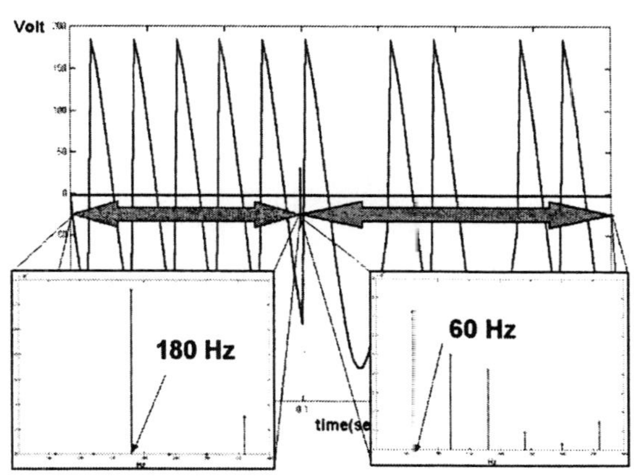

Fig. 6. The frequency responses of Fourier transform

is not when the failure happens but the fact of the failure occurrence. This shortcoming makes us feel something wanting and find other methods for designing the new fault detector. When we investigate the performances of the wavelet transform, the ability of local analysis gives us a hint about finding the time when a failure occurs (M. Misiti, *et al.*,1996; C.S.Burrus, *et al.*, 1998).

3.2 Using Wavelet Transform

When an arbitrary signal *s(t)* is selected, Equation 1 represents the continuous wavelet transform (CWT) of *s(t)*.

$$C_{CWT}(a,b) = \frac{1}{\sqrt{a}} \int_{-\infty}^{\infty} s(t)\psi\left(\frac{t-b}{a}\right)dt \qquad (1)$$

where Ψ is the wavelet function, '*a*' is the scale factor and '*b*' is the position factor. The scale factor '*a*' varies the scale of the wavelet function and the position factor '*b*' means the time position of the wavelet function. In calculating wavelet coefficient '*C*', every possible scale factor gives us a fair amount of work and an awful lot of data. Choosing factors based on powers of two (so-called dyadic scale and position factors), we can analyse much more efficiently and accurately. These measures generate the discrete wavelet transform (DWT). The DWT equation is as follows.

$$C_{DWT}(m,k) = \frac{1}{\sqrt{a_0^m}} \sum_k s(k)\psi\left(\frac{t-ka_0^m}{a_0^m}\right) \qquad (2)$$

DWT shows a basic process of filtering. That process is the decomposition of lowpass filters and highpass filters. The lowpass filters are called 'Approximations' and the highpass filters are 'Details'. Figure 7 shows the decomposition process of s(t) and this is called the wavelet decomposition tree.

$$s = cD_1 + cD_2 + cD_3 + \cdots cD_n + cA_n \qquad (3)$$

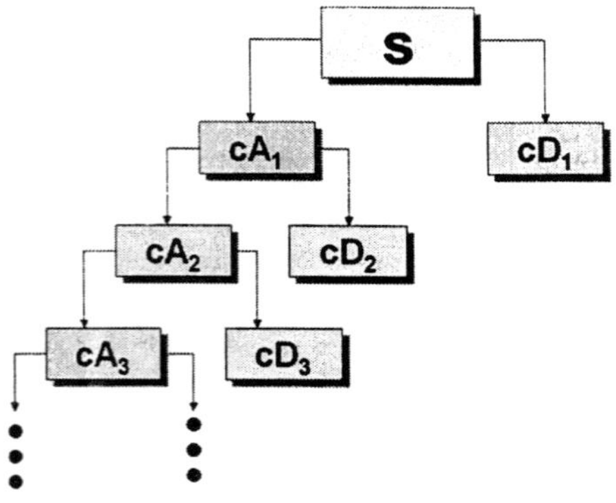

Fig. 7. The wavelet decomposition tree

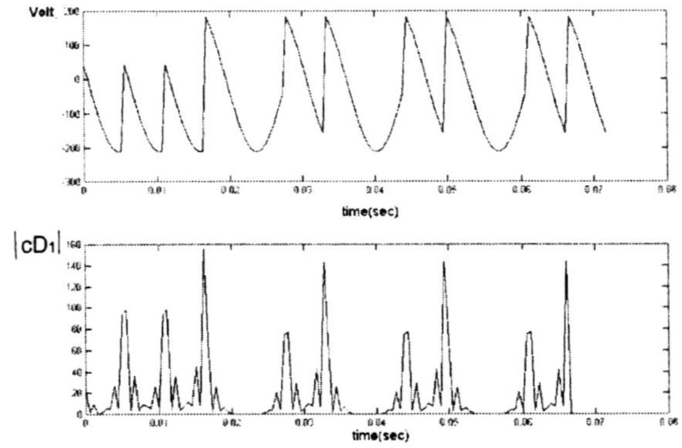

Fig. 8. The result of DWT for the case of the missing phase failure during the negative forcing period

Using DWT, we adopt Daubechies DB4 as the mother wavelet and focus on level-1 Detail component cD1 to know when a fault occurs.

As we can see in Figure 8, the remarkably big values of cD1 at each point where the signal changes its trend. In the normal state, the interval of the peak points is constant as about 5.6msec. But the interval of the peak points exceeds the normal state's in the failure state. Aiming this point, we devise a new fault detection algorithm informing the time at which the fault happens.

3. SIMULATIONS

In performing simulations the sampling time is fixed at about 0.56msec. The width of the time window has become settled at 16.8msec(30 points).

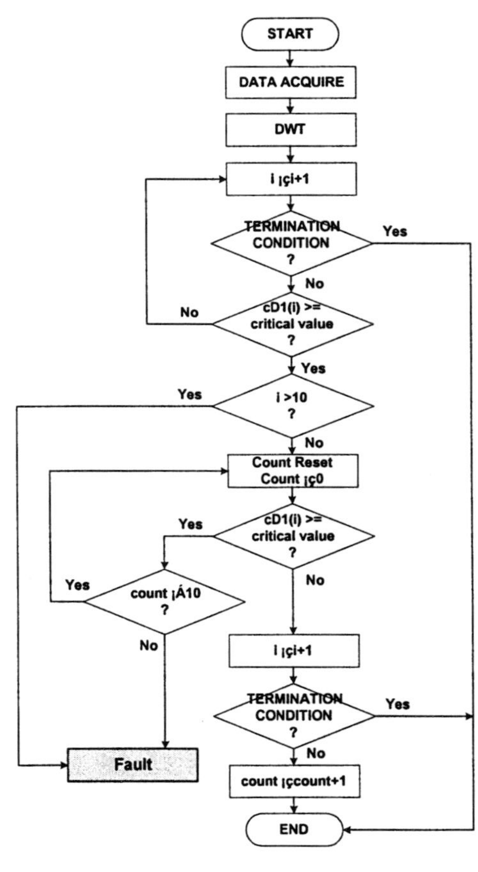

Fig. 9. The flowchart of the proposed fault detection method

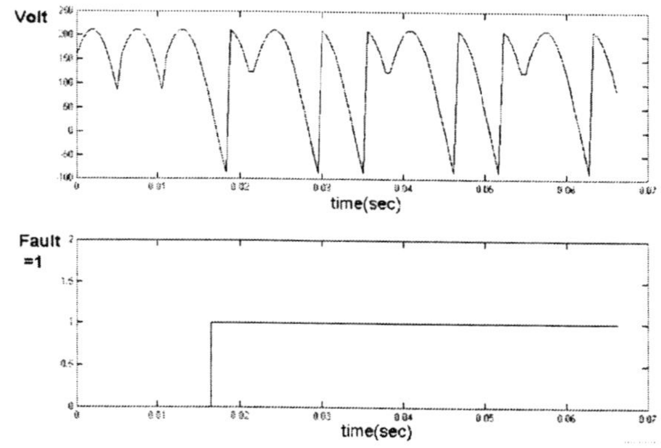

Fig. 10. The diode action failure during Positive Forcing

The flowchart for detecting the time of a failure occurrence is shown in Figure 9. First of all, the acquired data is transformed by DB4-DWT and the interval of the peak points is examined. The examined interval is more or less than the normal state interval means the fault occurrence.

Figure 10 shows the case of the diode action failure during the positive forcing. When the firing points are accelerated for the positive forcing, the diode action failure will not be caught generally. Because the ripple detector works only at the point of the change of the current level, only the failures of the positive or negative forcing are caught. This problem prohibits the existing ripple detector from detecting the diode action failure and passes this overlooked abnormal situation until the next state. Our detector has improved that problem and the result of Figure 10 demonstrates the benefits. At about 0.016sec, the proposed detector informs us the failure occurrence.

Figure 11 shows the case of the missing phase failure during the negative forcing. In this case of the retarded firing points in which the average voltage is maximum negative, the missing phase failure cannot be detected. The existing detector which detects the failure only in the forcing period misses and passes this abnormal situation to the next state without any actions. However, the proposed detector can catch the missing phase failure at about 0.022sec.

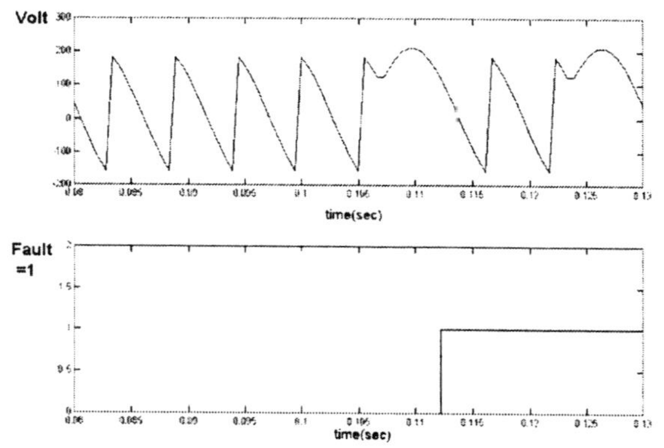

Fig. 12. The diode action failure during the rods keeping still

In Figure 12, the diode action failure during the rods keeping still can be detected at about 0.112sec and the urgent alarm is generated.

In the newly designed CRCS, the double holding is adopted and the movable grippers replace the role of the troubled stationary grippers holding the drive rod.

Figure 13 shows the case in which the missing phase failure occurs during the state of rods keeping still. The phase missing failure can be detected at about 0.105sec.

3. CONCLUSIONS

We design a new CRCS based on digital technology and hope that this digitalized system solves the problems in the existing analogue CRCS system. In this paper, we devise a new algorithm that detects the phase failure for designing DSP. Fourier transform also detects the phase failure by its ability to analysing the ripple frequency but it cannot give the correct time information. The wavelet transform was utilized for the new detector and its local analysis performance catches the correct time when a fault occurs. The new detector is able to watch the failures over the whole current states. The results of our simulation showed the proposed detector's strong points stated above.

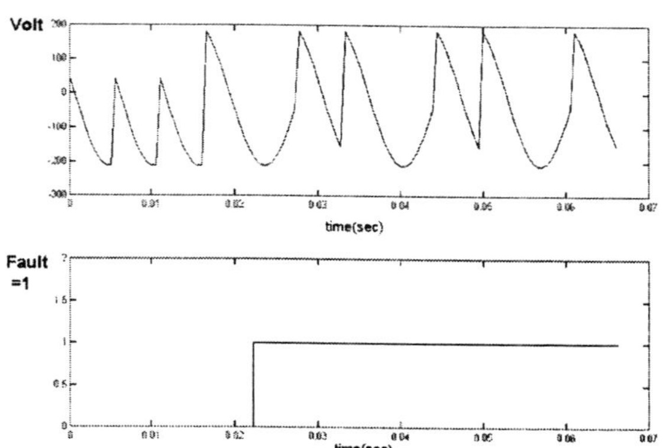

Fig. 11. The missing phase failure during Negative Forcing

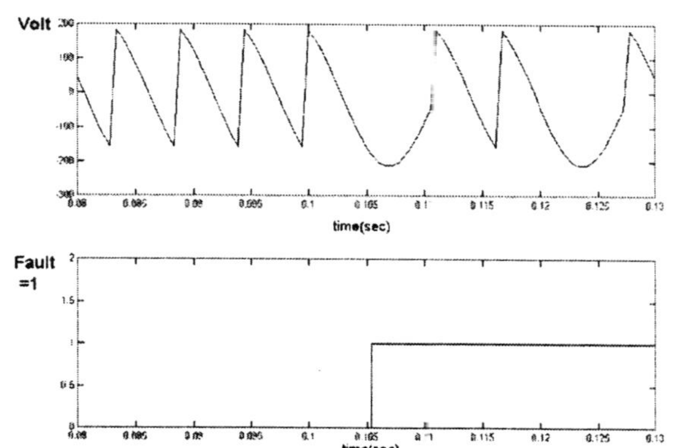

Fig. 13. The missing phase failure during the rods keeping still

REFERENCES

C.H.Jeong, M.H.Song, E.S.Kang and K.M.Kim (2002), "Efficient FFT-Based Fault Detection Using Mean Absolute Difference", *Proc. of ICCAS,* pp. 7-10, Oct..

C.Sidney Burrus, Ramesh A., Gopinath and Haitao Guo (1998), *Introduction to Wavelets and Wavelet Transform*, Prentice-Hall Inc.

D.S.Hong, C.W.Ryu and W.Y.Yim (2001), "A Study on High Impedance Fault Detection using Wavelet Transform and Neural Network", *Trans. of KIEE*, **Vol. 50A**, **No. 3**, pp. 105-110, March.

H.Bae, S.I.Kwon, J.K.Lee, C.K.Song and S.Kim (2002), "The Fault Diagnosis using Two-state Neural Networks for Nuclear Power Plants", *Trans. of KFIS*, **Vol. 12, No. 2**, pp. 129-134.

G.Y.Park, B.S.Park, J.S.Yoon, D.H.Hong and J.H.Jin (2002), "Development of Fault Monitoring Technique for Agitator Driving System", *Proc. of ICCAS,* pp. 1-6, Oct..

J.H.Lee, S.W.Park and B.H.Seo (2000), "Fault Diagnosis for a System Using Classified Pattern and Neural Networks", *Trans. of KIEE*, **Vol. 49D, No. 12**, pp. 643-650, Dec..

J.K.Kang, S.H.Kim and S.R.Lee (1999), "Neuro-Fuzzy based Fault Diagnosis System for Closed-loop Control System", *Proc. of the 14th KACC*, pp. C124-C127, Oct..

J.M.Cheon, J.B.Ahn, C.K.Kim, J.M.Lee and S.Kwon (2001), "Design of a Common Control Rod Control System(CRCS)", *Proc. of the KIEE Summer Annual Conference*, pp. 2331-2333, July .

J.H.Oh, S.Y.Yun and J.C.Kim (2002), "Feature Extraction of Fault Current using Fourier Transform in the Multi-Shot Reclosing Scheme", *Trans. of KIEE*, Vol. 49A, No. 2, pp. 50-55, Feb..

K.K.Jeong (1991), *Control Rod Control System,* Nuclear Power Training Institute, Korea Hydro & Nuclear Power Co.,Ltd.

K.S.Park, D.H.Hwang, M.S.Kim, Y.C.Lim, B.M.Kim and S.Y.Sung (1997), "A New Power System for CRDM", *Proc. of the I&C, Automation and Robotics Research Meeting joint announcement*, pp. 221-224.

M. Misiti, *et al.* (1996), *Wavelet Toolbox for Use with MATLAB*[B]

P.H.Sung, et al. (2001), *Development of Quantitative Reliability Evaluation Methodologies of KNGR Digital I&C Components*, KAERI.

S.H.Kim, J.S.Kim, T.Park, J.R.Lee and G.T.Park (1999), "Design of Fault Diagnostic System based on Neuro-Fuzzy Scheme", *Trans. of KIEE*, **Vol. 48A**, No. 10, pp. 1272-1278, Oct..

S.Kwon, et al. (2001), *Development of CEDM and Generator Control System,* KERI.

Y.S.Lee, et al. (2002), *Modeling CRCS for Fault Diagnosis*, KERI and Kunsan National Univ.

W.Y.Yim (2002), "Identification of Operated Protective Device in Distribution Power System using Wavelet Transform and ANN", *Proc. of ICCAS,* pp. 307-312, Oct..

ELSEVIER
IFAC
PUBLICATIONS
www.elsevier.com/locate/ifac

DEVELOPMENT OF REMOTE CORROSION MONITORING AND CONTROL SYSTEM FOR OIL TANKS BY USING A HIGH EFFICIENCY CP RECTIFIER

Jeong-Hyo Bae *, Tae-Hyun Ha, Hyun-Goo Lee,
Dae-Kyeong Kim, Yoon-Cheol Ha

Korea Electrotechnology Research Institute
28-1, Seongju-dong, Changwon-city, 641-120, Korea

Abstract: Recently, advanced countries have used the corrosion monitoring system in chemical plants and oil tanks in order to protect them from corrosion because corrosion causes large accidents, soil pollution, and money loss. Generally, the owners of such facilities adopt Cathodic Protection (CP) systems for corrosion protection.
However, Korea has adopted neither the CP system nor the corrosion monitoring system for oil tanks yet.
This study presents a system we developed. It is a system that can not only automatically and remotely monitor corrosion but also remotely control corrosion by using a high efficiency CP rectifier. We hope that the results of this study will be used for the effective and economic operation of CP system. It can also extend the life of an oil tank. *Copyright © 2003 IFAC*

Keywords: rectifiers, remote control, monitoring elements, prediction methods

1. INTRODUCTION

Direct and indirect loss from corrosion in Korea has reached the mark of 20 trillion won (4% of GNP: as of 2000), but the researches to reduce this loss have not been active. It has significant ramification as accidents caused by oil tank corrosion at chemical plants or oil tanks can result in loss of human lives and environmental pollution. (D.K. Kim etl 7, Hakryeul Lee etl 5)

For many years, developed countries have commercialized and used electrochemical corrosion protection system as well as corrosion monitoring system to protect oil tanks from corrosion. In 1994, the U.S. Environmental Protection Agency (EPA) adopted new regulations that require all underground storage tanks to be equipped with electrochemical corrosion protection facility and other corrosion control system by the end of 1998. As a result, oil tanks are required to use cathodic protection and the good working of corrosion protection facilities, crack monitoring, steam monitoring, underground monitoring, statistical inventory monitoring, and inside monitoring on oil tank are required to have monthly and annually. (John Morgan, NACE Standard)
In 1999 in Korea, National Institute of Environmental Research reported in its "Study on the restoration of polluted soil and system development" that electrochemical corrosion protection needs to be used for oil tanks. But in reality, there is no gas station that has electrochemical corrosion protection facilities or corrosion monitoring systems for its oil tanks. Only when environmental laws related to this problem are enacted, measures to deal with this problem will be adopted. And also there are no skilful corrosion engineers in this field. So, we need not only cathodic protection systems, but also automatic corrosion monitoring system and corrosion control system in order to protect against large accidents, soil pollution, and money loss.
Accordingly, this study introduces a remote corrosion control system that safely maintains and controls facilities through proper corrosion protection by automatically and remotely monitoring the corrosion status of an oil tank as well as remotely controlling a high efficiency CP rectifier.

2. DOMESTIC AND INTERNATIONAL STATUS

2.1 Domestic Status

Experts and the press have warned about the danger and harmful effects of oil leak many times, but the researches on corrosion protection technology for underground tanks have been neglected.
In 1996, a research team at Industrial Chemistry Dept. of Konkuk University reported that they tested 529 oil tanks at 175 gas stations nationwide for oil leak and 31.9% or 191 tanks were leaking. And the amount of oil leak was 109,300 liters per month, which is equivalent to 5,797 five-gallon kegs. When

this amount of oil leak is converted to 10,110 gas stations, the number of gas stations in Korea as of June 2000 as shown in Table 1, the amount will be enormous.

Table 1 The number of gas stations in Korea (as of June 2000)

Company Province	SK	LG-Caltax	Inchon Oil	S-Oil	Hyndai Oil	Total
Seoul	311	250	37	100	120	818
Gyeonggi	790	628	95	230	432	2175
Busan	153	116	6	58	78	411
Gangwon	264	163	-	72	129	628
Chungbuk	208	137	1	91	136	573
Chungnam	370	263	1	149	315	1098
Jeonbuk	287	230	-	102	154	773
Jeonnam	327	299	-	164	194	984
Gyeongbuk	546	334	-	263	261	1404
Gyeongnam	417	280	18	133	266	1114
Jeju	54	49	-	29	-	132
TOTAL	3727	2749	158	1391	2085	10110

The Korean government has started studying and investigating oil tank corrosion accidents recently. And in 1999, National Institute of Environmental Research reported that electrochemical corrosion protection need to be used for oil tanks in its "Study on the restoration of polluted soil and system development." Soil Conservation Division of Ministry of Environment has given an advance notice of the enactment of Article 23 of Environment Conservation Law, which prescribes "The source of soil pollution shall be liable for the damage and clean up," and has been working for the enforcement at the end of 2002. However, there is no gas station that is equipped with electrochemical corrosion protection facilities or corrosion monitoring systems for its oil tanks. Only when environmental laws related to this problem are enacted, measures to deal with this problem will be adopted.

2.2 *International Status*

Developed countries have researched on the corrosion protection technology for underground tanks for many years. One of the methods to protect oil tanks from corrosion is to coat the surface of an oil tank with polyethylene having excellent insulation property. Another method is to use non-corrosive material, such as ceramic. However, asphalt coating has been preferred for most oil tanks at gas stations for economic reason. As time passes, these tanks cause environmental pollution due to oil leak and active researches have been carried out on this problem in recent years. The researches carried out abroad are introduced below. (Kuhn, R.J, Australian Standard, ASTM Standard, W. v. Baeckmann, W. Schwenk)

In 1990, researchers in Japan including Tsunoda pointed out that the bottom of an oil tank buried deep in asphalt mortar or oil contaminated soil was more susceptible to corrosion than the top part even though cathodic protection was provided and proposed a new method called Finite Element Method.

In the U.S. in 1992, the use of ring shape cable type anodes having bigger diameter than regular cathodic protection for the bottom of oil tank was proposed and its design and application was reported in NACE conference.

In 1995, ARCO evaluated methods to prevent corrosion of aboveground storage tanks constructed along the Trans Alaska Pipeline System (TAPS). It carried out experiments to improve cathodic protection of the bottom of tanks, discussed related problems and presented solutions.

In 1981, a system developed by the U.S. Steel Research Center that integrated sacrificial anodes, protective coating and insulation bushing for oil tanks was reported in ASTM STP 741 as a method to reduce the corrosion of carbon steel used for underground facilities.

In 1984 in Hungary, an evaluation was carried out to highlight the direct loss including the cost of tank repair and replacement caused by underground steel tank corrosion and the indirect loss from interrupted operation and environmental effects and to assess the economic aspect of applying the electrochemical theory and corrosion technology for corrosion protection of fuel storage tanks.

In 1994, a U.S. researcher, J.A. Lehmann published an article in Materials Performance Journal. In it, he proposed that the corrosion protection of underground storage tanks should be a very important national consideration with the establishment of the EPA regulations. And cathodic protection and other corrosion control methods should be applied to the facilities needing corrosion protection according to the new regulations and enforcement regulations.

3. REMOTE CORROSION MONITORING AND CONTROL SYSTEM

3.1 *Development Direction*

Before developing a remote corrosion monitoring and control system, an analysis of user requirements was carried out to decide which functions and specifications are necessary and to develop the system more effectively. And the results are shown below.

1) The cost should be as low as possible.
2) The system should be free of malfunction.
3) The distance for remote monitoring should be 50m at the minimum.
4) It should be equipped with MMI (Man Machine Interface) so that a novice user can operate it easily.
5) The system should be able to summarize, process statistically and log the CP potential data.
6) It should be able to store the input voltage of ±10[Vdc] or under.
7) The measurement resolution should be less than 10[mV].
8) It should be designed as outdoor type and durable.
9) The minimum sampling time of potential data should be 1 sec.

10) It should be able to check the potential data of a random position real-time from a computer terminal in the main control room.
11) The system should be programmed so that corrosion status of a random position can be judged automatically when it is measured real-time.
12) The potential data of corrosion protection facility should be organized into a D/B and stored long term. The information should be controlled with the information on corrosion protection facility and utilized as basic data for future maintenance and inspection of the facility.
13) The remote corrosion monitoring device, corrosion control device and CP rectifier should be formed into one unit if possible to reduce the size and cost. It should be designed so that local corrosion protection control is possible.

3.2 System Overview

As shown in Figure 1, the oil tank corrosion monitoring and control terminal consists of the Central Processing Unit that controls the whole system, the Analog to Digital (A/D) Converter that converts the analog signal of potential difference between the underground oil tank structure and the reference potential into a digital signal and makes an input, the memory that stores digitalized data, the CP rectifier that is an output device for electrochemical corrosion protection of oil tank and communication interface that transmits data to a remote server.

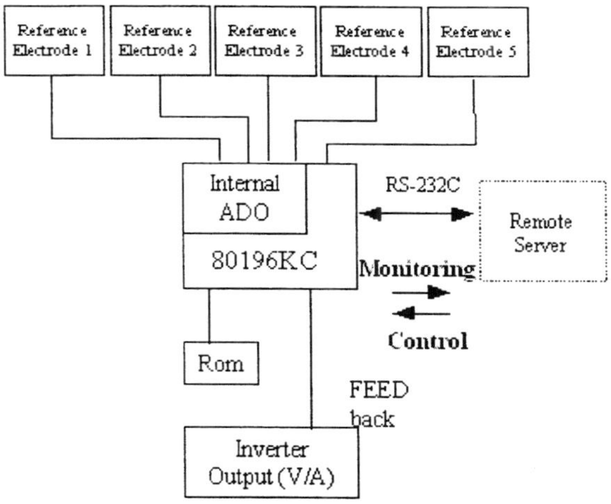

Fig. 1 Diagram of oil tank corrosion monitoring and control system concept.

The operation principle can be summarized as follows. The analog signal of the oil tank potential measured against the reference potential is entered into the A/D converter. The A/D converter digitalizes the signal into 10-bit resolution. And this data is transmitted to a PC through communication interface by the control of CPU and this data is used as a feedback signal for the rectifier to protect the oil tank automatically according to the set corrosion protection potential. The measured CP potential data is sent to the remote server through the communication interface (RS-232) of corrosion monitoring terminal and the remote server sends a rectifier output control signal to the communication interface of corrosion control monitoring terminal to maintain the optimum CP potential by comparing the measured data with the preset CP potential. In general, each gas station in Korea has 5 tanks, so the number of input channels is set to 5, and 1 channel is designed to monitor the drainage point. The specifications of oil tank corrosion monitoring and control system can be summarized as in Table 2.

Table 2 Basic specifications of oil tank corrosion monitoring and control system.

No	Item	Specifiation
1	Number of input channel	5
2	Input channel voltage range	-9999~9999
3	Resolution	10 mV
4	Sampling time	0.1 sec
5	Alarm	No
6	Display	LED
7	Operating temperature	-10℃~40℃
8	Dimension	420(W)×450 (H)×260(D)
9	Switching speed	70 kHz
10	Output range	60 V/20 A

Figure 2 shows the main board of remote corrosion monitoring and control system and Figure 3 shows the external view of the system.

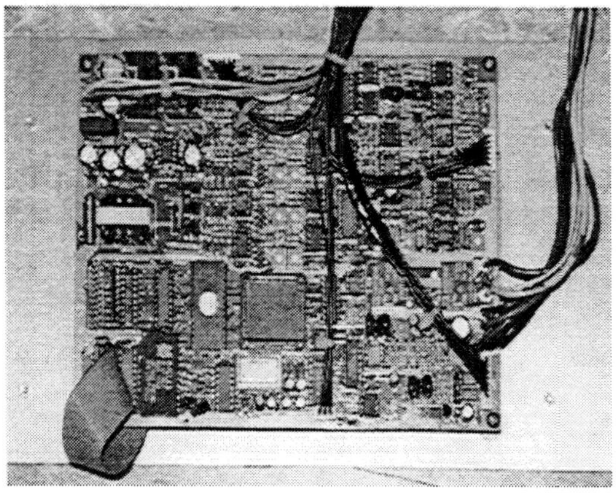

Fig. 2 Picture of the main board for remote corrosion monitoring and control system.

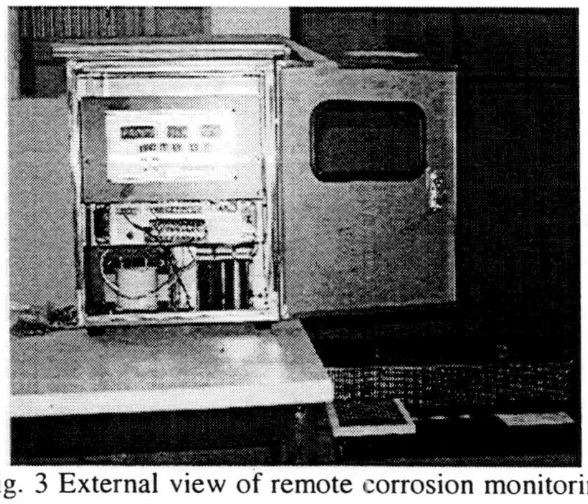

Fig. 3 External view of remote corrosion monitoring and control system.

4. APPLICATION PROGRAM

4.1 Overview

The application program consists of the system program that runs on the PC and controls the oil tank corrosion monitoring and control terminal and the firmware that runs on the corrosion monitoring and control terminal.

The firmware program collects necessary data through the digital rectifier and controls its output in accordance. Necessary information is communicated to the system program that controls oil tank corrosion monitoring and control terminal through the PC and RS-232. When the operation is controlled remotely, the set value is received from the PC and the digital rectifier is controlled. In the case of local mode, previously saved set value is received from the front control panel and the digital rectifier is controlled.

The system program running on the PC has oil tank information, anode information, and installation environment information. It monitors the status real-time by using the data transmitted from the corrosion monitoring and control terminal and controls the corrosion monitoring and control terminal in the remote mode. It also enables easy management by analyzing the transmitted data.

4.2 Firmware Program for Corrosion Monitoring and Control Terminal

Functions of the firmware program for corrosion monitoring and control terminal:

1) System Initialization: When power is supplied to the corrosion control terminal, a command to initialize the status of each hardware is executed. Each hardware port and internal memory area are initialized. And the command to initialize the real time clock module that is used to control system time is executed.
2) Data Collection: The analog value needed at the channel specified by the A/D module is converted to a digital value. And the converted value is stored in a specific memory area and transmitted to the monitoring server as necessary. Several necessary analog values are displayed in the 7-segment LED display area of the front panel.
3) Communication with the PC: The program is linked to the PC via RS-232 communication and exchanges data with the PC. It transmits the data from the oil tank corrosion monitoring and control terminal to the PC and receives a suitable control signal from the system program of the PC.
4) Control of Digital Rectifier: The digital rectifier is controlled by the input entered from the front panel. It is controlled according to the output voltage or the CP potential of each mode.

Flow of Firmware Program

The flow chart of the firmware program of corrosion monitoring and control terminal is as follows (Figure 4). The overall principle of the flow is as follows. When power is applied, the I/O, memory and Real Time Clock (RTC) of corrosion control terminal are initialized. And whether there is a transmission request from the higher system program is checked and the potential of oil tank and the output of rectifier are transmitted to the PC accordingly. And when the rectifier control command is received, the rectifier output is controlled to the set value by using the Pulse With Modulation (PWM) method. And local button input is checked and the corresponding value is displayed with the LED.

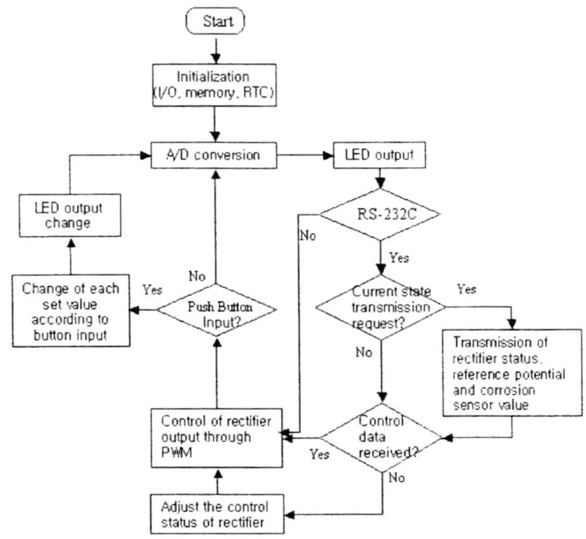

Figure 4. Flow chart of the firmware program of corrosion monitoring and control terminal

The functions of application program can be summarized as in Table 3 and the main screen of application program is shown in Figure 5.

Table 3 Functions of application software

Menu	Sub-menu	Details
1. Main	Main	Main screen of system
2. System Manage	User registration /deletion	User password registration/deletion Edit allowed to registered user
	Gas station history	View, edit or print the general history of gas stations
	Communication port property	View or edit the property of communication port
	System variables	View or edit the system variables
	Time synchronization	Time synchronization between terminal and monitoring server
	System closing	System closing
3. Anode /TB	Anode history	View or edit the history of anode
	Test box history	View or edit the history of test box
	CP potential monitoring	View or print the transmitted CP potential
	CP potential variation	View or print the variation of CP potential for a given period.
	CP potential alarm log	View the CP potential on overprotection or under protection.
4. Oil Tank	Oil tank history	View or edit the history of oil tanks
	Corrosion coupon history	View or edit the history of corrosion coupon
5. Rectifier	Rectifier history	View or edit the history of rectifier
	Output status monitoring	View or print the transmitted information on output status
	Output status variation	View or print the variation of output status
	Output status alarm log	View the CP potential on poor output
6. Help	Help index	Guide for use of main functions
	System information	Brief information on the system

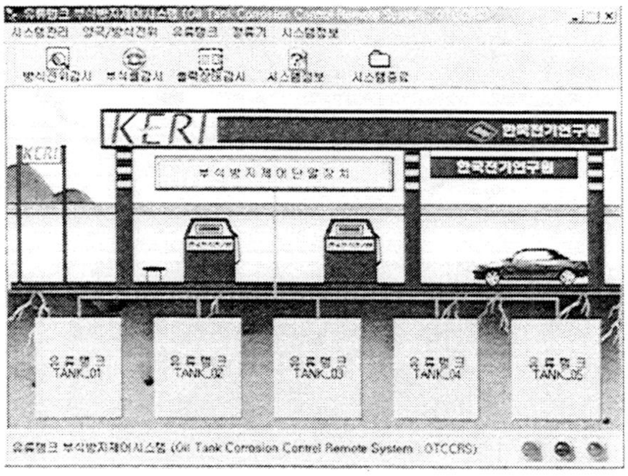

Figure 5. Main screen of application program

5. CONCLUSIONS

We have developed a system that can provide optimum corrosion protection for facilities by remotely controlling a high efficiency CP rectifier as well as monitoring oil tank corrosion. The system will not only bring import replacement effects but also prevent environment pollution due to oil leak. It will be possible to manage facilities more economically and systematically.

We think the results of this study will be very useful for the oil tank owners.

REFERENCE

D.K. Kim etl 7, (1995) *Studies on the Current and Potential Distribution of Cathodic Protection System,* KOGAS.

Hakryeul Lee etl 5, (1998) *Corrosion Protection Technology,* Vol 2, MOCT

John Morgan, (1993) *Cathodic Protection,* Second Edition NACE

NACE Standard, (1993) *External Cathodic Protection of On-Grade Metallic Storage Tank Bottoms,* NACE RP0193-93

Kuhn, R.J. (1953) *Criteria for Steel and Cast Iron*, Proceedings of the American Petroleum Institute, Vol 14, P153

Australian Standard, (1993) *Galvanic(sacrificial) Anodes for Cathodic Protection*, AS 2239

NACE Standard, (1993) *Cathodic Protection Design*, Recommended Practice RP B 401

ASTM Standard, (1994) *Annual Book of ASTM Standard, Wear and Erosion: Metal Corrosion*, ASTM Volume 03.02

W. v. Baeckmann, W. Schwenk, (1975) *Handbook of Cathodic Protection The Theory and Practice of Electrochemical Corrosion Protection Techniques,* BSI Code of Practice for Cathodic Protection, Portcullis Press LTD

ELSEVIER
IFAC
PUBLICATIONS
www.elsevier.com/locate/ifac

3-D ASSESSMENT OF HUMAN HAZARD BY OUTDOOR ELECTRICAL FACILITIES IN SUBMERGED AREA

**Tae-Hyun Ha*, Hyun-Goo Lee, Dae-Kyeong Kim,
Jeong-Hyo Bae, Ji-In Lee, Yoon-Cheol Ha**

*Korea Electrotechnology Research Institute
28-1, Seongju-dong, Changwon-city, 641-120, Korea*

Abstract: In this paper, we show three-dimensional distribution of voltages resulted from the leakage current originated from outdoor electrical facilities in a submerged area. In case these facilities are grounded by the neutral line multiple grounding method, the existence of ungrounded electrical facilities can cause a disastrous effect on near-by passengers. In order to investigate this situation, we installed a real-scale test field for the experiment type I (for the leakage current path between a enclosure grounded electrical facility and another enclosure grounded one), and that for the experiment type II (for the leakage current path between a enclosure grounded electrical facility and another ungrounded one). For both cases, we carried out three-dimensional monitoring of the voltage distribution while varying additional conditions such as the exposure of the underground cables and the finishing of cable connection part. The result shows that a disastrous effect on human safety can arise from the leakage current without a pertinent measure for the construction and maintenance of outdoor electrical facilities.

Keywords: distribution feeders, leakage currents, voltage distribution, safety analysis

1. INTRODUCTION

As the urbanization grows more active, a variety of outdoor electrical facilities such as streetlights, signal lights, billboards, air conditioner outdoor units, etc., become more ubiquitous. Submersion of these facilities, commonly connected to the neutral line of 22.9 kV distribution feeders, can make a harmful effect on near-by human beings, since the leakage current from the facilities causes a sharp increase in the potential around them.

Numerous studies have been carried out on the hazards of electric shock (IEEE, 2000). Theoretically, the voltage distribution in such a situation can be easily simulated by computer-assisted numerical analysis (SES, 2000; Ground-it.com, 2001). To our knowledge, however, no real-scale investigation for the situation herein has been reported yet.

In this study, we analyze the effect of the leakage current on human safety by monitoring the voltage distribution for the cases of the experiment type I: formation of the electrical pass between an enclosure grounded electrical facility (streetlight) and another enclosure grounded one (cabinet panel) and the experiment type II: formation of the electrical pass between the enclosure grounded one (streetlight) and another ungrounded one (air conditioner outdoor unit).

2. REAL-SCALE TEST FIELD

2.1 Configurations

In order to analyze the effect of the leakage current on human safety and monitor the voltage distribution when outdoor electrical facilities are submerged, we designed and constructed a simplified real-scale test field on the basis of the preliminary investigation of the actual condition.

① The size of the test field is 12 m × 6 m × 1.5 m (width×length×height) and the distance between the grounded outdoor electrical facility ("test facility") and the secondary facility, grounded or ungrounded, is 6 m.

② The inner wall of the test field is made of wood in order for insulation and wrapped by a tent for waterproofing.

③ The bottom of the test field is sectionalized into two parts, i.e., an asphalt pavement and a footpath, to reproduce the actual condition.

④ The copper sulphate reference electrodes (CSE) for potential measurement are fixed on a lattice frame at intervals of one meter. We designed the frame to be lifted up to 1.5 m from the bottom of the test field in order to monitor the vertical distribution of the voltage attendant upon the submersion water level.

⑤ The frame is flexible enough to be in contact with the bottom when measuring the voltages of the uneven bottom, where there is difference in height between the road and the footpath, while it makes the end of all electrodes same-heightened during the measurement in water.

⑥ In order to produce the leakage current, several current sources are intentionally provided, i.e., exposure of the ELP 30 mm underground conduit tube (exposed area: 3 cm × 3 cm) and 600 V CV 1×5.5 mm^2 underground cable (exposed length: 1 cm) as well as the cable connection part of the test facility. The latter is carried out using the vinyl tape for insulation and using the terminal board for complete exposure. For the terminal board, 60 A standard is used.

⑦ All the connections in the circuit are designed to be controlled from outside of the test field and completely insulated from the submerged area except the parts in ⑥ for the intentional exposure experiment.

⑧ Grounding in the test field was performed at two points for the test facility and another one at one point, 20 m apart from the test field, for neutral wire grounding.

Figure 1 is the floor plan of the test field showing the points of voltage measurement for type I and type II.

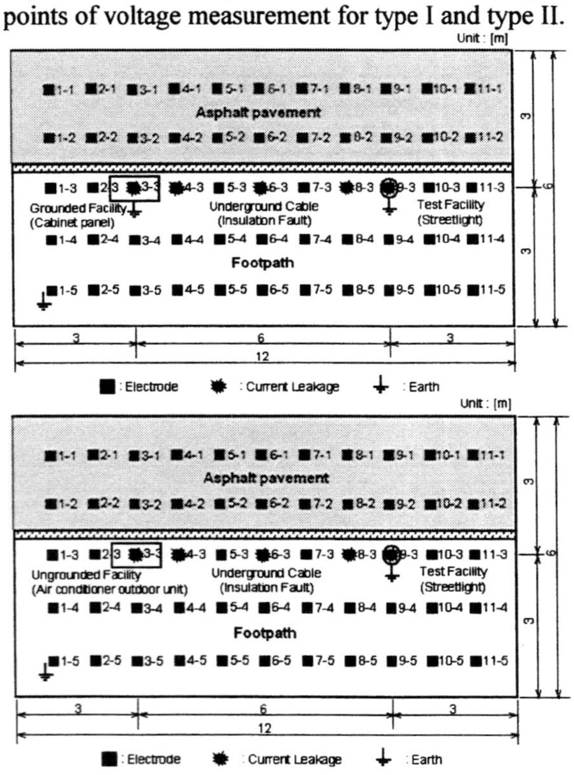

Fig. 1. Floor plan drawings of the test fields

2.2 *Soil Resistivity in Test Field*

Table 1 shows the measured soil resistivity after excavating the bottom of the test field. The depth of the excavation is about 2 m.

Table 1 The measured soil resistivity at the bottom of the test field

Pin Interval [m]	Measured Resistance [Ω]	Soil Resistivity [Ω·m]
1	17.00	106.76
2	11.50	144.44
3	8.85	166.73
4	6.90	173.32

2.3 *Grounding Resistance in Test Field*

The measured electrical resistivity of water after filling the test field up to the submersion position (about 150 cm high from the footpath) was 170 Ω·m. Table 2 shows the measured grounding resistance at each point leaving the water filled.

Table 2 The measured grounding resistance

Facility	Grounding Resistance [Ω]
Test Facility	21.1
Grounded Cabinet	18.0
Neutral Line Grounding	44.5

3. EXPERIMENTALS

3.1 *Circuit Configuration*

Figure 2 shows the circuit configurations for type I and type II. Only the test facility can be the current source for the former case, while the ungrounded facility as well as the test facility can be the current source for the latter case. Because the cabinet panel in upper figure is well insulated and propped sufficiently high with a concrete foundation, it does not play a role as a leakage current source when submerged.

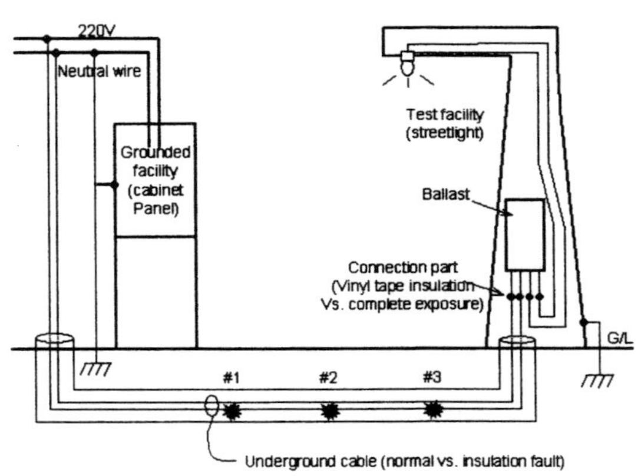

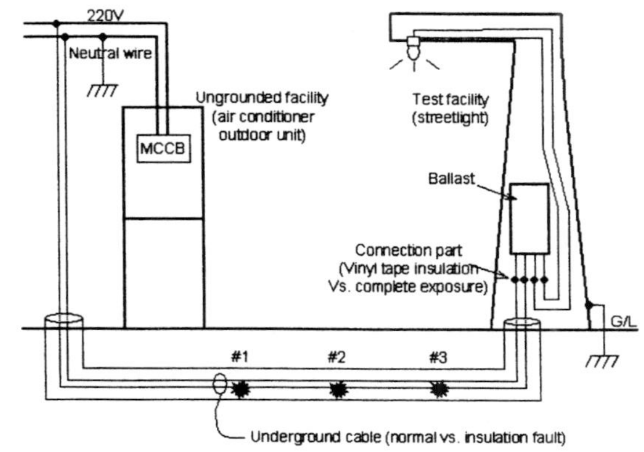

Fig. 2. Circuit configurations for the test fields

3.2 Experimental Variables

Measurement of potential at each position is carried out under the following experimental variables:

① Grounding: the enclosure of the cabinet panel used for experiment type I is grounded by connecting to the neutral wire while that of the air conditioner outdoor unit used for experiment type II is not grounded.
② Exposure of underground cable: none vs. fault at each position #1(4-3), #2(6-3) and #3(8-3).
③ Electrical connection part: vinyl insulation taping vs. complete exposure.
④ Submersion water level: bottom level vs. submersion of the secondary facility.
⑤ Measurement height: bottom level → 30 cm → 60 cm → 90 cm → 120 cm → water level.

3.3 Measurement Equipment

The main experimental equipments used for measuring the leakage current and the potential are a digital oscilloscope (TDS 3032, Tektronix) and multi-meter/switch system (2750, Keithley). The simultaneous recording of the 55 potential values is performed using two sets of multi-channel modules (7702, 40 channels per module) of the multi-meter/switch system.
Insulating transformer is used as power supply equipment for the test field, which facilitates the power grounding from the secondary. Excessive current can be interrupted through the use of the digital panel-meter and electromagnetic switches.
The voltage values in this paper are the potential differences between the potential measured at each location in Figure 1 and the potential at location numbered 1-5. All the potential measurements are carried out with respect to the copper sulphate reference electrode (CSE).

4. RESULTS AND DISCUSSION

4.1 Voltage Distribution

The voltage distribution shows that in case of the experimental condition of type I, where the test facility is the leakage current source and the other is an enclosure grounded one, the voltages near the test facility are higher than those near the grounded facility. The maximum voltage near the test facility becomes much higher when the electrical connection part of the test facility is completely exposed to water than insulated by vinyl tape.
Table 3 shows the maximum voltage and maximum voltage gradient at the given experimental variables (section 3.2). The maximum voltage occurs near the test facility, where the leakage current passes into the ground via the steel enclosure of the streetlight. As a result the voltage at the bottom is the highest. Note all the maximum voltage gradients in the table are lesser than 2.5 V/m.

Table 3 The measured maximum voltage and maximum voltage gradient in experiment type I

Experimental variables			Max. voltage [V]	Max. voltage gradient [V/m]
Connection part	**Position of insulation fault**	**Measuring height [cm]**		
Insulated by vinyl taping	none	bottom	0.6144	0.3503
		30	0.5022	0.233
		60	0.5055	0.2424
		90	0.5271	0.2878
		120	0.4665	0.2002
		water level	0.4544	0.1884
	#1	bottom	0.8067	0.3451
		30	0.741	0.2978
		60	0.6809	0.2725
		90	0.6542	0.2585
		120	0.6696	0.2485
		water level	0.6243	0.208
	#2	bottom	1.079	0.208
		30	0.9073	0.4456
		60	0.9042	0.2547
		90	0.8772	0.2755
		120	0.8486	0.2634
		water level	0.8204	0.233
	#3	bottom	1.3912	0.5618
		30	1.2242	0.3705
		60	1.221	0.3823
		90	1.1764	0.3534
		120	1.1726	0.3201
		water level	1.1429	0.2919
Completely exposed to water	none	bottom	4.3181	2.2992
		30	3.8356	1.7319
		60	3.9117	1.8706
		90	3.8535	1.8115
		120	3.751	1.7104
		water level	3.5724	1.4846
	#1	bottom	4.3128	2.0389
		30	3.9794	1.7212
		60	3.9547	1.759
		90	3.8889	1.7126
		120	3.7783	1.5515
		water level	3.6782	1.4398
	#2	bottom	3.21	1.1958
		30	3.1571	1.1999

	60	3.0729	1.1177
	90	3.0272	1.0492
	120	2.9576	0.9404
	water level	2.8841	0.8385
#3	bottom	3.7479	1.5645
	30	3.4754	1.2955
	60	3.4623	1.1906
	90	3.3804	1.1472
	120	3.2358	0.9852
	water level	3.1626	0.8632

Figures 3 and 4 show the 3-D potential distribution profiles for the experiments marked as shaded rows in Table 3 in order to compare the effect of the electrical insulation of the electrical connection part of the test facility, using Kriging method (Davis, 1986).

As for the experimental condition of type II, where the ungrounded outdoor facility is the leakage current source, the voltages near the test facility are lower than those near the current source. The maximum voltage near the ungrounded facility and the test facility is much higher when the electrical connection part of the test facility is completely exposed than when insulated by vinyl tape. Table 4 shows the maximum voltage and maximum voltage gradient at the given experimental variables. The maximum voltage occurs near the ungrounded facility, where the leakage current passes through water and concrete because the facility is propped about 60 cm above the bottom with the concrete foundation. As a result, the maximum voltage appears at a height of about 60 cm from the bottom.

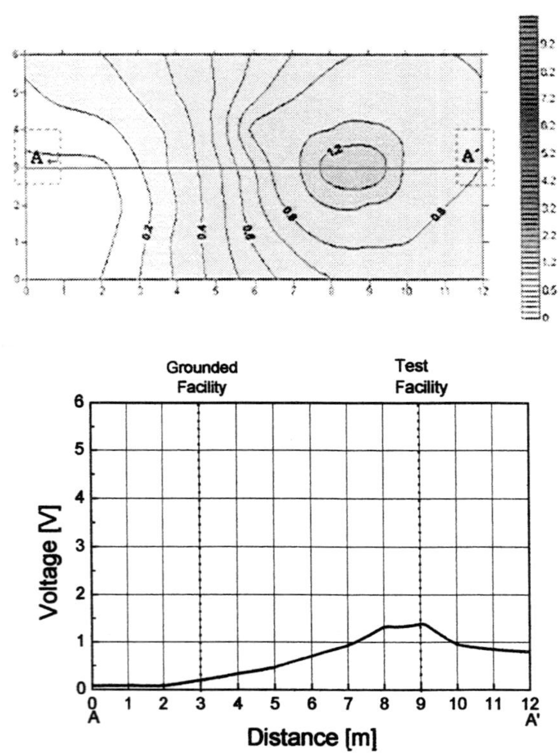

Fig. 3. 3-D voltage distribution in experiment type I with the connection part insulated by vinyl taping

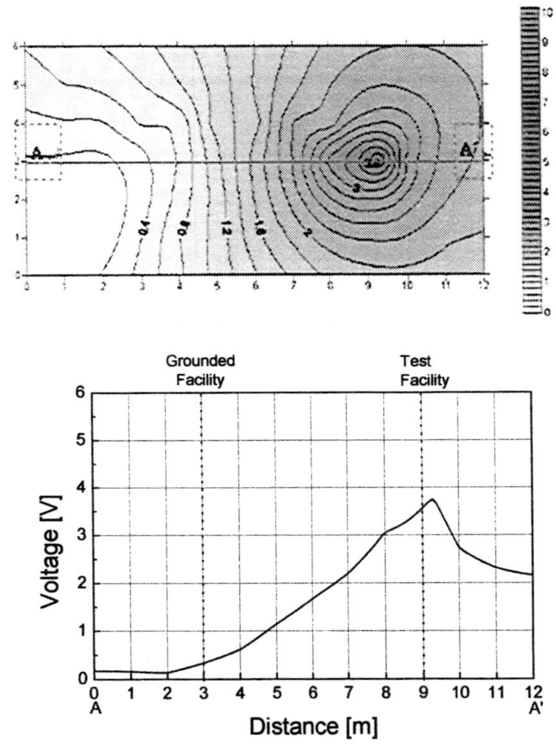

Fig. 4. 3-D voltage distribution in experiment type I with the connection part completely exposed

Table 4 The measured maximum voltage and maximum voltage gradient in experiment type II

Experimental variables			Max. voltage [V]	Max. voltage gradient [V/m]
Connection part	Position of insulation fault	Measuring height [cm]		
	none	bottom	2.9228	1.7623
		30	4.4906	3.4789
		60	4.2169	2.8586
		90	3.5721	1.8484
		120	3.333	1.4719
		water level	3.2566	1.375
Insulated by vinyl taping	#1	bottom	3.0333	2.0066
		30	4.4322	3.427
		60	4.2244	2.8429
		90	3.5651	1.8212
		120	4.2244	2.8429
		water level	3.308	1.4078
	#2	bottom	2.9069	1.6216
		30	4.4598	3.3887
		60	4.1617	2.779
		90	3.6074	1.8395
		120	3.3783	1.5019
		water level	3.359	1.456
	#3	bottom	2.9413	1.6457
		30	4.4166	3.3717
		60	4.2304	2.8755
		90	3.6427	1.8557
		120	3.3722	1.4834
		water level	3.3938	1.466

		bottom	2.775	1.7435
		30	5.4727	4.8143
	none	60	5.7863	2.4305
		90	3.2256	1.7122
		120	3.0469	1.4653
		water level	3.0926	1.4971
		bottom	2.6971	1.65
		30	5.3044	4.6067
	#1	60	4.0595	2.8375
		90	3.1124	1.6371
Completely exposed to water		120	2.9251	1.3867
		water level	2.9182	1.3623
		bottom	5.7879	5.7831
		30	4.6021	4.4155
	#2	60	3.6652	3.0899
		90	2.6374	1.7228
		120	2.4828	1.5202
		water level	2.5315	1.7227
		bottom	5.8169	5.6282
		30	5.1849	4.4279
	#3	60	4.1814	2.9868
		90	3.2303	1.6951
		120	3.0934	1.5066
		water level	3.2303	1.6951

Figures 5 and 6 show the 3-D potential distribution profile for the experiments marked as shaded rows in Table 4, of which the condition is to vary the electrical insulation of the electrical connection part of the test facility while other variables are fixed.

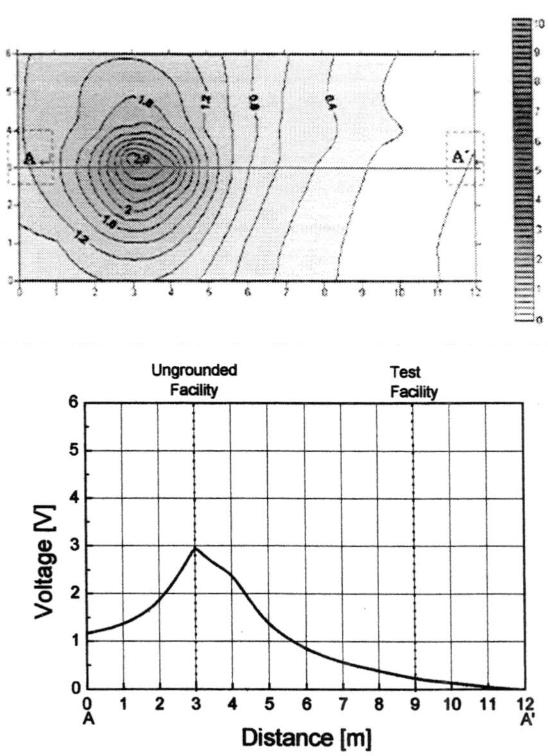

Fig. 5. 3-D voltage distribution in experiment type II with the connection part insulated by vinyl taping

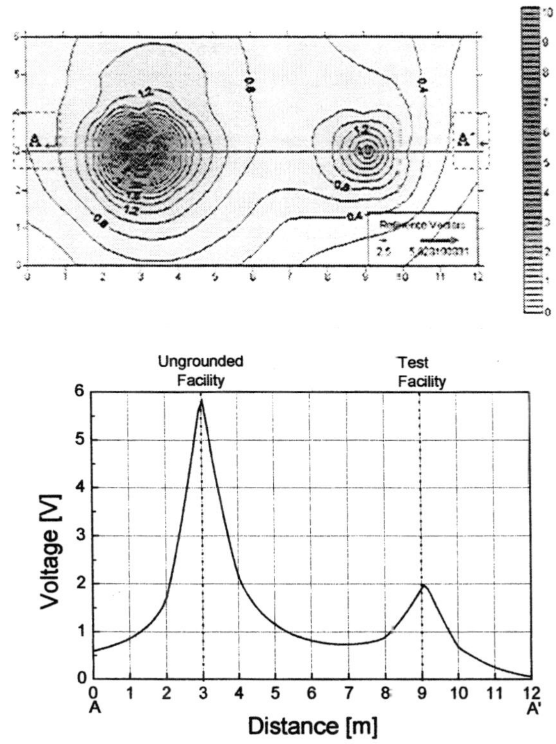

Fig. 6. 3-D voltage distribution in experiment type II with the connection part completely exposed

4.2 Discussion

As for the allowable limit of the touch voltage regarding the human safety, the guide from Japan Electric Association (JEA) is commonly used, where it is 2.5 V in water (JEA, 1971). In practice, this means 2.5 V within about 1 m, considering the location of human body in water, which can be also regarded as the voltage gradient.

For all the conditions in experiment type I, where the test facility is the current source, it can be concluded that there is no harmful effect on human body in water unless one touches directly the current leaking facility. This is easily readable from the maximum voltage gradients in Table 3.

On the other hand, in case of the situation in experiment II, the leakage current can cause a disastrous effect on human beings either near the ungrounded current source or in contact with that facility. As shown in Table 4, the catastrophic effect grows remarkably when another nearby facility or cable becomes additional current source owing to the inadequate insulation. The vectors shown in Figure 6 shows the voltage gradients higher than 2.5 V/m and the corresponding area can be thought as a danger zone.

5. CONCLUSIONS

In the present experimental condition, the effect of the leakage current on human safety was affected by the exposure of underground cable, the insulation of the electrical connection part and the enclosure grounding of the outdoor electrical facilities.

Since the multiple grounding of neural in 22.9 kV distribution system is the most widely used method at the present, the formation of current path between the leakage current source and grounding electrodes is inevitable in submerged area. The voltage increase near the current source, however, can be mitigated if we take a pertinent measure for the outdoor electrical facilities to meet the submersion situation, including the better insulation of the electrical connection part, the proper maintenance of underground cables as well as the ensuring of enclosure grounding of outdoors electrical facilities. In addition the installation of residual current protective devices will prevent such a human hazard substantially.

REFERENCES

Davis, J.C. (1986). *Statistics and Data Analysis in Geology*, 2nd Ed., John Wiley & Sons.

Ground-it.com (2001). *KWIKGRID®, TACLINK®, CONIND™*, Computer Softwares, Ground-it.com Consulting Ltd.

IEEE (2000). *IEEE STD 80-2000, IEEE Guide for Safety in AC Substation Grounding*, IEEE.

JEA (1971). *JEAG8101-1971*

SES (2000). *Resistivity,Grounding, Cable & Line Parameters, Fault Current Distribution, Electrical Interference and Electromagnetic Fields Analysis Software CDEGS™*, Safe Engineering Services & Technologies Ltd.

ELSEVIER
IFAC
PUBLICATIONS
www.elsevier.com/locate/ifac

HARMONIC ANALYSIS AND ITS COUNTERMEASURE ON ELECTRIC RAILWAY IN KOREA

Young Hoon Ahn, Chang Mu Lee

Mr.Ahn working for KNR, Mr.Lee working for KRRI

Abstract: These days, KNR has many construction projects of electrification. There are harmonic problems in construction of AC 25kV-60Hz electric railway. To solve them, we have analyzed a harmonic characteristic of AT power supply system. We have found out harmonic countermeasure. This paper presents three kinds of model for harmonic analysis and three cases of harmonic countermeasure for electrified conventional line. *Copyright © 2003 IFAC*

Keywords: Power supply system, harmonics, harmonic analysis

1. INTRODUCTION

There are many electrification construction projects that replace diesel fuel with electricity fuel as power source of railway system in KNR(Korean National Railroad). Processing electrification of railway, harmonic current generating from operating electric rolling stock causes a disturbance of railway signal system and an induced electricity trouble of communication system near by railroad track. Especially, when the harmonic current flows into a power source of AT power supply system in electric railway, it rises overheating and vibration of power condenser, disordering operation of protective relay. So that, it is very important thing in a side view of power quality and reduction of construction cost to reflect the results of analysis and estimation through harmonic simulation on construction process. This paper describes simulation evaluation cases for harmonic current and adopted harmonic analysis models to construction process of electric railway in south Korea.

2. MODEL FOR HARMONIC ANALYSIS OF AT POWER SUPPLY SYSTEM

AT power supply system of electric railway is like fig-1. Harmonic current generating from electric rolling stock running on track makes a reproduction on an enlarged scale of harmonic current resonating at specific harmonic according to impedance characteristic of AT power supply system.

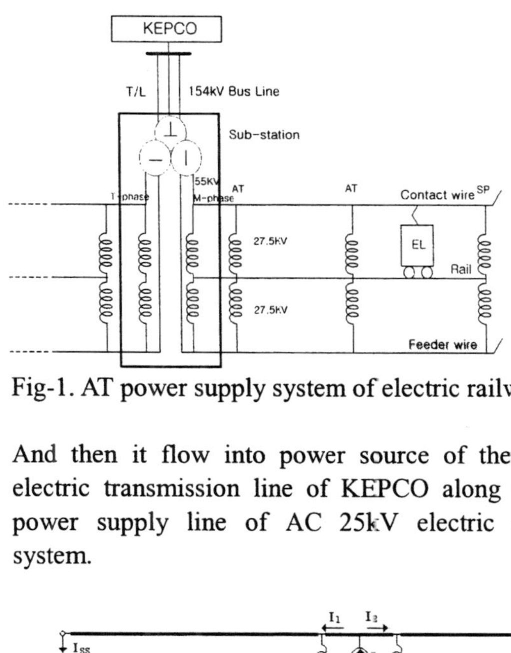

Fig-1. AT power supply system of electric railway

And then it flow into power source of the power electric transmission line of KEPCO along the AT power supply line of AC 25kV electric railway system.

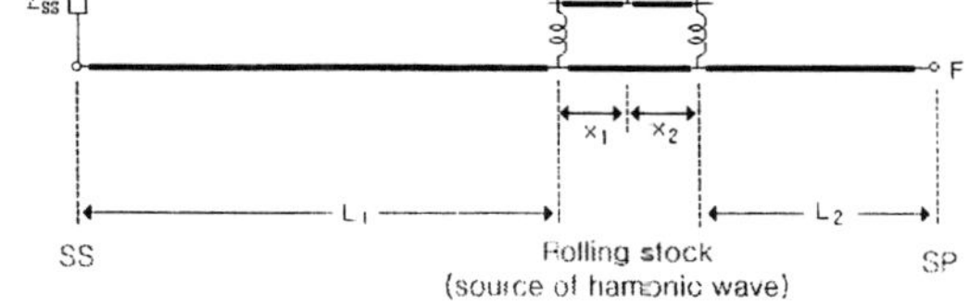

Fig-2. Harmonic equivalent circuit of M-phase(or T-phase)

To define harmonic circuit of AT power supply system, Zss is impedance of power source side at M-

Phase(or T-Phase) on Substation and electric rolling stock is harmonic source. So then, harmonic equivalent circuit is like fig-2.

A heavy solid line mark the distributed constant circuit, X1 and X2 is distance between vehicle and AT, L1 presents distance between substation and AT, and L1 means distance between AT and sectioning post.

2.1 Model using the Distributed Constant Circuit

The Model using the Distributed constant circuit has studied by RTRI(Railway Technical Research Institute) in Japan to calculate the magnification factor and resonance of harmonic on AT power supply system.

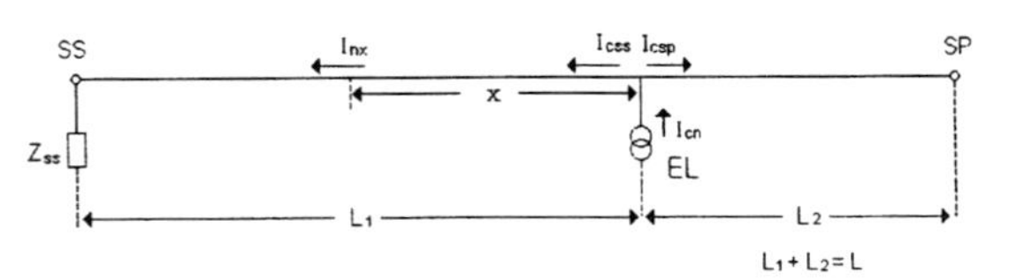

Fig-3. A close shot of harmonic equivalent circuit for AT power supply system

The magnification factor of harmonic current flowing into AT power supply system from the harmonic source can be calculated with the model replaced a AT power supply system like fig-2. with a close shot of harmonic equivalent circuit like fig-3. This method use to analysis with calculation values of equivalent impedance at a substation side and a terminal of AT power supply system. Harmonic magnification factor K means INX/ICN. This value is like following function. (Joachim Holtz and Heinz-jurgen Klein., 1989; Yoshifumi Mochinaga,1994; A. Bonner, et.al., 1996a, 1996b)

$$K = \frac{\cosh\gamma l_2 (Z_{ss}\sinh\gamma(l_1-\chi) + Z_0\cosh\gamma(l_1-\chi))}{Z_{ss}\sinh\gamma l + Z_0\cosh\gamma l}(p.u.) \quad (1)$$

in here, $Z_0 = \sqrt{\frac{Z}{Y}}$

Z_0: characteristic impedance of Feeder line side

$\gamma = \sqrt{Z \cdot Y}$: propagation constant of Feeder line side

Z_{SS} : Addition of main transformer`s impedance and power source`s impedance(Ω)

Using the harmonic impedance of power source and flowed harmonic current into substation, harmonic voltage is directly calculated at Incoming side of SS and Bus-line from KEPCO. Therefore, THD(Total Harmonic Distortion) at these points can be find by the following function.

$$THD = \frac{1}{V_1}\sqrt{(V_2^2 + V_3^2 + V_4^2 + V_5^2 + \cdots + V_n^2)} \times 100(\%) \quad (2)$$

2.2 Model using the Four- Port Network

Harmonic current source connect to secondary side of both Auto-Transformer(T-R) and primary side of both Auto-Transformer(T-F) connect through Feeder line in the model of fig-1. Also there is admittance by capacitance among three conductor(C, R, F) in catenary system. Therefore, the Network is not exactly modeled with Four-terminal constant. Therefore, KRRI(Korea Railroad Research Institute) has developed model using Four-Port Network for exact analysis of harmonic characteristics. The model considering self-impedance, mutual-impedance and admittance of AT power supply system has structured as Transfer Matrix Function including entire circuit elements(Impedance of power source, AT of Substation, AT aof Sectioning post, AT in the middle of line, Feeder line and electric rolling stock).

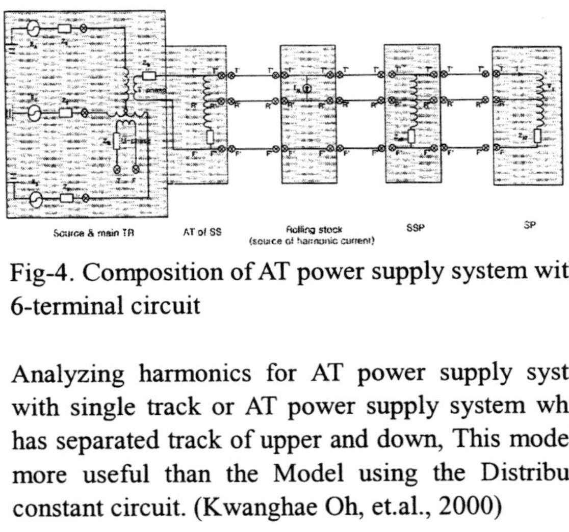

Fig-4. Composition of AT power supply system with 6-terminal circuit

Analyzing harmonics for AT power supply system with single track or AT power supply system which has separated track of upper and down, This model is more useful than the Model using the Distributed constant circuit. (Kwanghae Oh, et.al., 2000)

2.3 Model using the 8-Port Network

The Gyoungbu high speed railway and a constructing railroad in recent take a AT power supply system with the common earth method which is commonly connected to both earth circuit and conductor groups of upper and down track. It is not adopt to analyze by the conventional method of harmonic analysis. So that, another suitable harmonic analysis model based on the Theory of Circuit Network has developed by KRRI for the AT power supply system with the common earth method. (Kwanghae Oh, hanmin Lee, Shanghoon Chang and Junghoon Kim, 2002)

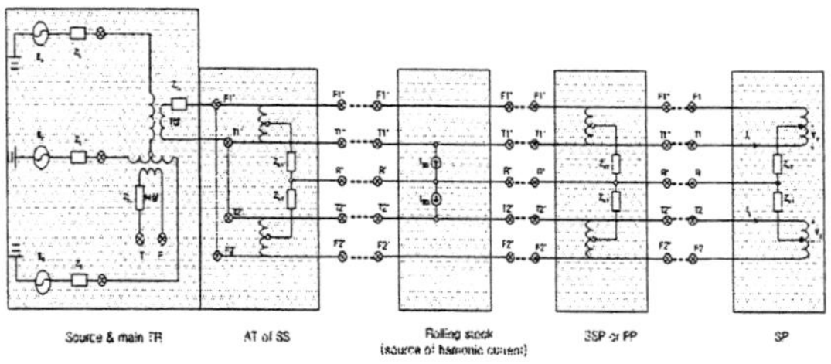

Fig-5. Composition of AT power supply system with 10-terminal circuit

2.4 Resonance characteristic of AT power supply system

As results of examination for harmonic characteristics at the AT power supply line by the analysis models, a general tendency of the harmonic magnification ratio and the resonance characteristics are like the following things.

If a length of AT power supply line is setting regularly, a resonance point(resonance frequency) is fixed without relation to a position of harmonic source.

If a length of AT power supply line is setting, a harmonic magnification ratio at a resonance point is highly enlarged when electric rolling stock is located faraway from a substation.

A harmonic resonance is generated at low order, and its scale is decreased a little when a distance of AT power supply line is long. (Kwanghae Oh, et al., 2000; Changmu Lee, Moonseub Han and Younghoon Ahn.,2002; Kwanghae Oh, hanmin Lee, Shanghoon Chang and Junghoon Kim, 2002)

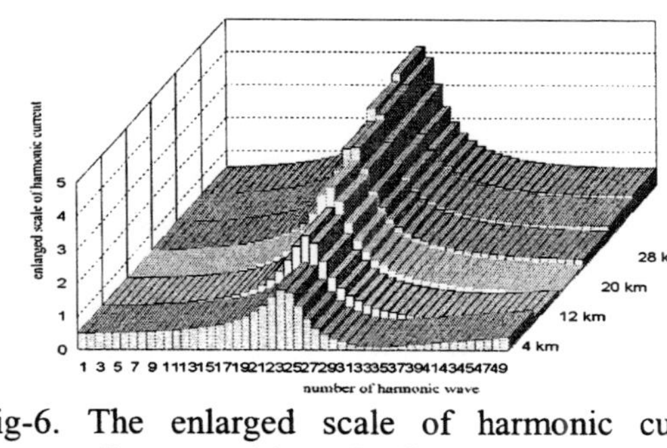

Fig-6. The enlarged scale of harmonic current according to position of rolling stock

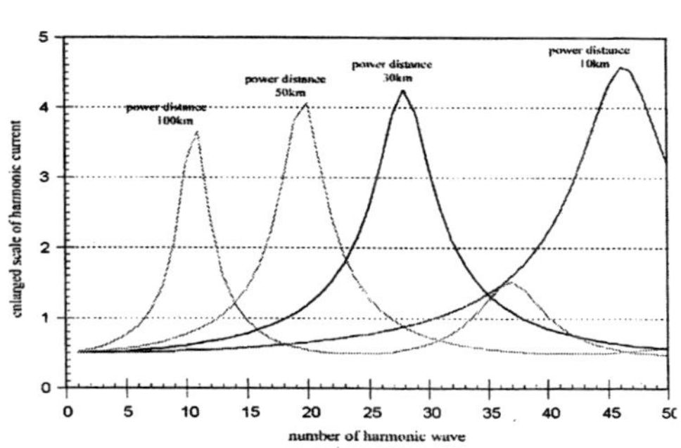

Fig-7. the mutual relation of distributed power distance and harmonic resonance

3. Harmonic analysis and its countermeasure of AT power supply system

3.1 Harmonic characteristic according to class of electric locomotive

We have investigated the results of harmonic current characteristics for the four kinds of locomotives operating or testing in Korea.

First, the electric locomotive(EL8000 type) has initially imported from Europe to Korea in 1972, and this train operating on the electrified conventional lines(Jungang Line, Youngdong Line, Taebaek Line) has used a control method by Thyristor.

Secondly, to replace EL8000 type with EL8100 type in the electrified conventional lines, a electric locomotive which was newly developed has used a control method by PWM.

Thirdly, it is use a control method by Thyristor that the high speed train(KTX) will be operated in Gyoungbu high speed railway Line in 2004 by a maximum speed of 300km/h.

Lastly, high speed train of the Korean type(G7 train) with a maximum speed of 350km/h has developed by KRRI, and this train using a control method by PWM has been testing on the test track of high speed railway. (Kwanghae Oh, et al., 2000; Changmu Lee, Moonseub Han and Younghoon Ahn.,2002; Kwanghae Oh, hanmin Lee, Shanghoon Chang and Junghoon Kim, 2002)

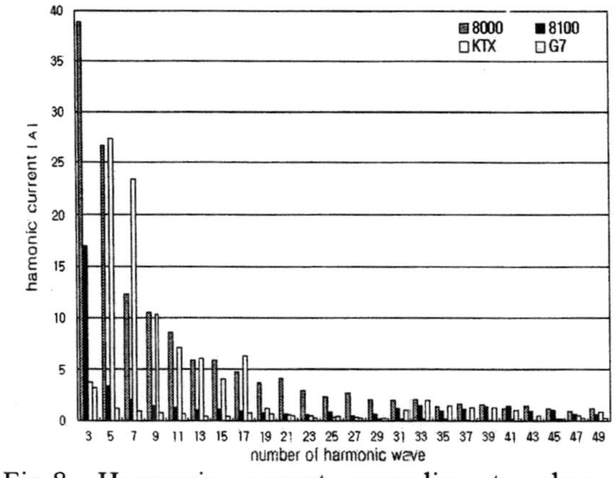

Fig-8. Harmonic current according to class of electric locomotive.

The upper picture presents a measurement result of harmonic current according to the class of electric locomotives. As the measured result of harmonic current generating from EL8000 type, we knew the result is the rise of generating current quantity on harmonic order according to a increase of traction load current, and the drop of generated current quantity according to a increase of harmonic order. This result is because of the usage of Thyristor phase-control method on EL8000 type.

The harmonic current is especially generated much quantity on the low harmonic order. Therefore, we have installed the LC Filter in substation to restrain a harmonic on the low order, also the RC Bank in sectioning post has been sued to refrain a harmonic current on high order operating from a resonance of AT power supply line. As a measured result of harmonic current generating from EL8100 type, we have taken very little a variation of harmonic current according to a variation of traction load current and very small a scale of harmonic current. It is because of the usage of PWM control method on EL8100 type. A pattern of harmonic generating current from the KTX is similar to it from the EL8100 type because of the usage of Thyristor-Phase control method. Therefore, the countermeasure equipment for harmonics should be installed at the substation in AT power supply system for the KTX to remove the harmonics on low order. In case of the G7 train, the generated quantity of harmonic current is little. It is because of the usage of PWM control method in a similar way to the EL8100 type. In patterns of the generated harmonic from the vehicle using Thyristor-Phase control method, the scale of harmonic current changes according to the scale of load current for a traction force of train. The occurrence of harmonic current decreases according to the increase of harmonic order. In patterns of the generated harmonic from the vehicle using PCM control method, the harmonic current is constant without regard to the scale of load current for a traction force of train and tends to increase.

3.2 Cases of Harmonic countermeasure in electrified conventional line

The electrification of Chungbuk-Line

The harmonic of electrified Chungbuk-Line was predicted with the analysis method based on the distributed constant circuit has studied by RTRI. The table-1 shows a result of evaluation for distorted factor of voltage at out-point of power source transmitting from KEPCO to Jeungpyoung-SS.

Table-1. Distorted ratio of voltage by condition of rolling stock operation

Condition of rolling stock operation	Distorted ratio of voltage			
	None	RC Bank	AC Filter	RC+AC
EL8000 (only)	3.81%	2.23%	1.40%	0.87%
EL8100 (only)	0.73%	0.40%	0.37%	0.19%
Mixing operation	2.30%	1.33%	0.86%	0.54%

The criterion for distorted factor of voltage is satisfied with a installation of AC Filter(LC Filter removing harmonics of 3,5,7 and 9 order) or AC Filter and RC Bank as countermeasure, the countermeasure equipment is not needed when the EL8000-Type operate only on track. In mixing operation, the criterion is almost satisfied with a installation of the AC Filter.

The electrification of Gyeongbu-Line

We have a schedule of mixing operation both The KTX and the EL8100 type in Electrified Cyeongbu-Line. Especially, a schedule of high speed train operation is planned as a headway of 16min between Dongdaegu and Busan by The KTX train with large scale load power of 15,486kw. The harmonic of Electrified Gyoungbu-Line was predicted with the analysis method base on the distributed constant circuit has studied by RTRI. To evaluate harmonic up to 19 order by simulation, the EL(harmonic source) is placed each 4 or 6 train unit at a certain position on the distributed power supply side of M-phase and T-phase in Substation.

The table-2 shows a predictive result of harmonic current at Gyoungsan-SS. a specification of RC Bank is R=750Ω, C=0.18μF. In this result, we found the criterion is pleased when RC Bank is installed at sectioning post.

Table-2. Distorted ratio of voltage by position of measurement

Position of measurement	Distorted ratio of voltage		
	None	RC Bank	AC Filter
Incoming side of SS	13.63%	1.51%	1.37%
Bus-line from KEPCO	12.62%	1.39%	1.27%

Design	Report of practice design	Review of practice design
Harmonic number	19 harmonics	49 harmonics
Condition of operation	The fixed position of Rolling stock without position	Running position of rolling stock according to planning of train operation
Harmonic pattern	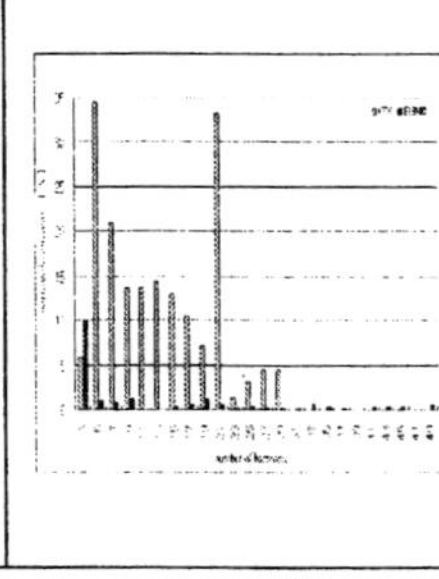	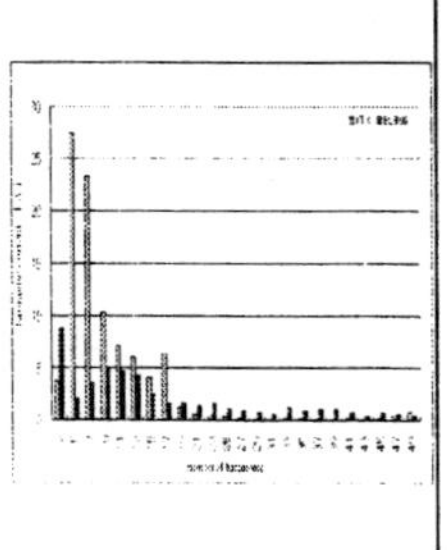

Fig-9. The comparison of harmonic pattern

We have reviewed a practice design to estimate a precision harmonic in construction process, used a method with Four-Port Network of KRRI, and continuously simulated with an actual operation plan of the KTX and EL8100 type instead of a fixed rolling stock position. In harmonic analysis, The most important element which is the harmonic current generating from vehicle was taken from result of actual measurement. The fig-9 resents a mutual comparison of harmonic pattern and condition of train operation in practice design and review of practice design. The table-3 which is a result of design review shows a predictive result of voltage distorted ratio according to installation of harmonic countermeasure facility as review of practice design when the vehicle scheduled to 3 levels with plan of train operation goes. Taking a criterion as maximum predictive load in future, have to install RC Bank(r=550Ω, C=0.21μF) in sectioning post and AC Filter(LC Filter removing harmonics of 3,5,7 and 9 order) in substation to satisfy the criterion.

Table-3. Distorted ratio of voltage by plan of train operation

Plan of train operation	Distorted ratio of voltage			
	None	RC Bank	AC Filter	RC+AC
2004 year	5.05%	1.97%	1.33%	1.02%
2009 year	7.67%	3.61%	1.61%	1.32%
Maximum load	7.51%	3.34%	1.67%	1.37%

The electrification of Honam-Line

KNR has a mixing operation schedule of three type trains(EL8100, KTX, G7 is consist of 10 coach) by 4 steps in the electrified Honam-Line. We have adopted a analysis method with Four-Port Network developed by KRRI for harmonic prediction, gotten harmonic data streaming into substation when electric rolling stocks running on section of AT power supply of substation according to the mixing train operation schedule, estimated voltage distorted ratio at out-point of KEPCO with this data. We have selected R=750Ω and C=0.18μF as parameter of RC Bank, AC Filter is 3,5,7 and 9 order for LC filter as removing filter . The table-4 shows a predict result of voltage distorted ratio when three types of train by 4 levels of planning of train operation perform with the mixed train operation in AT power supply line of Dugye substation. According to this prediction result, we have to install RC Bank and AC Filter as harmonic countermeasure equipment to be satisfied with the criterion of voltage distorted ratio in 2004 year.

Table-4. Distorted ratio of voltage by plan of train operation

Plan of train operation	Distorted ratio of voltage		
	None	RC Bank	RC+AC
Level 1 (2004Year)	6.83%	4.15%	0.62%
Level 2 (2014Year)	13.55%	7.51%	0.89%
Level 3 (2024Year)	20.90%	9.45%	1.08%
Level 4 (2034Year)	20.91%	9.41%	1.11%

In the electrified Honam-Line, it was reexamined harmonics of report of practice design because of change of conditions like the quantity of generated harmonic current in locomotives. Fig-10 presents a comparison of simulation condition in report of practice design and review of practice design. Especially, the parameters of RC Bank and AC Filter were changed new parameters and the harmonic current generating from vehicle changed a recent measured result which is the most important thing in simulation of harmonic analysis.

Design	Report of practice design	Review of Practice design
RC Bank	R=750Ω, C=0.18μF	R=550Ω, C=0.21μF
AC Filter	LC filter reducing harmonics(3,5,7,9 level)	LC filter reducing harmonics(3,5,7 level)
Harmonic pattren	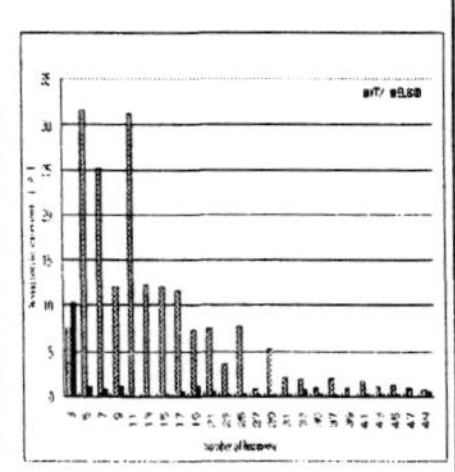	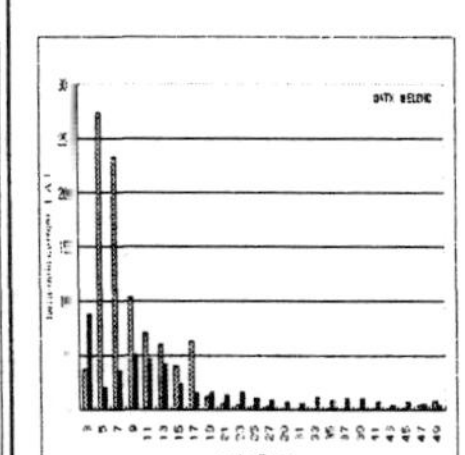

Fig-10. Comparison of harmonic pattern

The table-5 shows a predictive result of the reviewed estimation for voltage distorted ratio by installation of harmonic countermeasure facility according to 4-steps of train operation plan.

Table-5. Distorted ratio of voltage by plan of train operation

Plan of train operation	Distorted ratio of voltage			
	None	RC Bank	AC Filter	RC+AC
Level 1 (2004Year)	0.78%	0.38%	0.30%	0.26%
Level 2 (2014Year)	1.84%	0.78%	1.41%	0.61%
Level 3 (2024Year)	1.85%	0.80%	1.35%	0.63%
Level 4 (2034Year)	2.17%	0.92%	1.48%	0.73%

The criterion is satisfied with a value of voltage distorted ratio without harmonic countermeasure when train run according to train operation plan in 2004 year. But RC Bank in sectioning post has to be installed as harmonic countermeasure equipment to be satisfied with the criterion of voltage distorted ratio when train operates according to train schedule after 2014 year .

4. CONCLUSION

The harmonic of electric railway system can be exactly analyzed by the model of distributed power supply system including operation pattern and characteristic of electric rolling stock. The exact harmonic analysis take a estimation a scale of harmonic countermeasure equipment securing a better electric power in the construction process of electrified conventional railway systems, and make a calculation the optimal construction cost preventing the excess or little of design. Therefore, We have to consider following particulars to reflect exact analysis and countermeasure of harmonics in construction planning of electric railway by stages.

The application of harmonics analysis model that has as characteristic as distributed power supply system of electric railway.

The precise measurement and peculiarity comprehension for the generated harmonic current from types of electric rolling stocks.

The operation of harmonic simulation should be reflected running condition of electric rolling stock according to schedule plan of train.

The design of harmonic countermeasure facility has to consider synthetically the characteristic of distributed power supply system and electric rolling stock, the train operation pattern.

REFERENCES

A. Bonner, et.al., (1996a). Modeling and simulation of the propagation of harmonics in electric power networks PART Ⅰ. IEEE Transactions on Power Delivery, Vol.11, No.1, pp.452-465.

A. Bonner, et.al., (1996b). Modeling and simulation of the propagation of harmonics in electric power networks PART Ⅱ. IEEE Transactions on Power Delivery, Vol.11, No.1, pp.466-474.

Changmu Lee, Moonseub Han and Younghoon Ahn. (2002). Investigation of harmonic evaluating factor in electric railroad. KIEE Proceedings, pp.1251-1253.

Gilsoo Jang, Hwachang Song and Byongjun Lee. (2001). On-line harmonic distortion index for power quality assessment, KIEE Transactions on Electric Power Systems, Vol.50, No.8, pp.373-378.

Joachim Holtz and Heinz-jurgen Klein. (1989). The Propagation of Harmonic Currents Generated by inverter-Fed Locomotives in the Distributed Overhead Supply System. IEEE Transactions on Power Electronics, Vol. 4, No.2, pp168-174.

Kwanghae Oh, et.al., (2000). Harmonic analysis based on four-port representation for traction power supply system. ICEE Proceedings, pp.165-168.

Kwanghae Oh, hanmin Lee, Shanghoon Chang and Junghoon Kim. (2002). 8-port network model for harmonic analysis on the test track in Seoul-Pusan high-speed railway. KIEE Transactions on Electric Machinery, Vol.51, No.3 pp.99-106.

Korean National Railroad. (1998). Report of practice design for the electrification of Chungbuk-Line(Jochwon-woumseong) with double track.

Korean National Railroad. (1999). Report of practice design for the facilities of transmission & transformer on Gyoungbu-Line(Dongdaegu-Busan, Daejeon, Daegu and Busan).

Korean National Railroad. (2001). Report of practice design for the facilities of transmission & transformer and the buildings of substation on Honam-Line

Korea Railroad Research Institute. (1999). Evaluation of Power Quality and Its Countermeasure In Seoul-Daegeon High-speed Railway.

Sanghoon Chang, et.al., (2002). Harmonic analysis for traction power supply system using Four-port Network Model. KIEE Transactions on Electric Power Systems, Vol.51, No.6, pp.255-261.

Yoshifumi Mochinaga. (1994). Higher harmonics Resonance on AT Feeding Circuit and Countermeasures to Suppress it. T.IEE Japan, Vol.114-D, No.10, pp.978-986.

ELSEVIER
IFAC
PUBLICATIONS
www.elsevier.com/locate/ifac

MODELING AND CONTROL OF A FUEL CELL SYSTEM IN TRANSPORT APPLICATIONS

J.Lachaize*, M.Fadel*, S.Caux*, P.Shott, L.Nicod*****
**Laboratoire d'Electrotechnique et d'électronique industrielle, UMR, LEEI INPT ENSEEIHT, CNRS*
2 rue Camichel 31071 Toulouse cedex 7 France.
Tel : +33(0)5 61 58 82 08
lachaize@leei.enseeiht.fr ;fadel@leei.enseeiht.fr ; caux@leei.enseeiht.fr
***Laboratoire Hydrogène et Pile à Combustible, LHPAC,CEA/DTA/DEM/SPCM-17 rue des Martyrs*
38054 Grenoble cedex 9 France.
Tel: +33(0)4 38 78 48 40
pascal.schott@cea.fr
****ALSTOM Transport SA, 50 rue du docteur guinier BP 4- 65600 Séméac France.*
Tel : +33(0)5 62 53 48 57
laurent.nicod@transport.alstom.com

Abstract: This paper presents a Modeling approach to control the PEM fuel cell auxiliaries and the design of control laws. In a second time, we present the structure specification of an electrical power train with fuel cell for high power transport applications. This work shows that we can characterize and adjust the system control on an operating cycle with the simulation tools. *Copyright © 2003 IFAC*

Keywords: Modeling, Simulation, Control Methods, Real-Time Operation of Power Plants, non-conventional energy sources, distributed energy generation.

1- INTRODUCTION.

The use of fuel cell is now envisaged to supply electric energy in high power rail transport applications. So we realize a study to control the structure of the fuel cell system and the electrical chopper which composed the electric power train.

In a first time, we present the fuel cell model and its auxiliaries used to design the controllers. Then we explain the control strategy of these auxiliaries, only for air compressor and valve pressure.

In a second time, we detail the power train structure as well as the control scheme of each electrical chopper and first results of energy management.

2- GLOSSARY.

Fi	Molar flow.
N	Number of cells bystack.
F	Faradays constant 96485 C/mol.
I	Current in A.
λ_{H_20}	Hydration rate of the membrane.
X_{0_2}	Molar fraction of 0_2 in air =21%
StO_2	Stochiometry rate of oxygen.
R	Molar gas constant. 8.13 J/K/mol.
V_{cath}	Cathodic Volume in m^3
T_{fc}	Temperature in Kelvin.
Pi	Pressure in Pascal.
Cyl	Compressor capacity in m^3.
ω	Rotation speed in rad/s.
γ	Polytropic exponent, for air γ=1.4.
M	Molar masse of air.

3- FUEL CELL SYSTEM.

The fuel cell system is composed by the fuel cell core associated with all the necessary auxiliaries for an embedded fuel cell system.

This schema represents all the functions present in a fuel cell system:

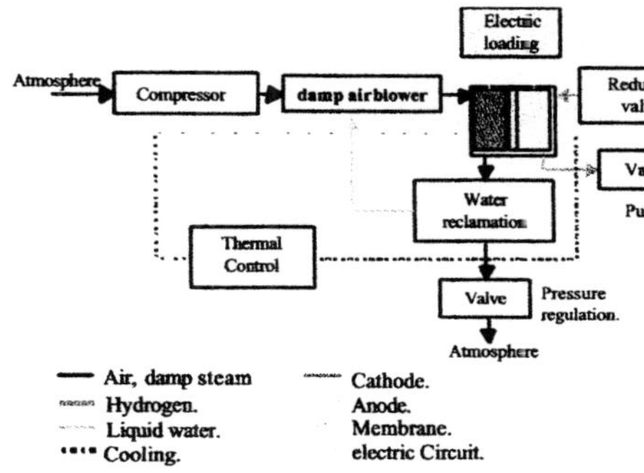

Fig 1. : Fuel cell system.

The Modeling process consists in writing the variations of the species concentrations, then in calculating the pressure in each compartment. After this step we use a quasi-static model to establish the voltage variation of the fuel cell [1] [2].

The nominal point of the system is:

$$P_{outlet} = 1\ 10^5 Pa\,;\ P_{inlet} = 1.6\ 10^5 Pa\,;\ T_{fc} = 353.15°K\,;$$

$$StO_2 = 1.6\,;\ Power\_nom = 400kW\,;\ Unom_{min} = 375V$$

4- FUEL CELL MODELING.

4.1- Cathodic species schedule.

These equations describe the variations of the gaseous species concentrations in the cathodic compartment.

The balance of oxygen and nitrogen allows us to write the following equation:

$$\begin{cases} \dfrac{dnO_2}{dt} = 0.21F_{in} - F_{O_2c} - X_{0_2}F_{valve} \\ \dfrac{dnN_2}{dt} = 0.79F_{in} - X_{N_2}F_{valve} \\ \dfrac{dn_{vap}}{dt} = \dfrac{X_{vap}}{1-X_{vap}}F_{in} - X_{vap}F_{valve} \end{cases} \quad (1)$$

Where: $F_{O_2c} = \dfrac{NI_{fc}}{4F}$ $\quad X_{vap} = \dfrac{P_{sat}(T_{fc}°)}{P_{cath}}$ $\quad X_i = \dfrac{P_i}{P_{cath}}$

With :

- F_{in} inlet molar flow of the compressor.
- F_{valve} outlet molar flow of the valve.
- Xi molar fraction of i species .
- F_{O2c} oxygen molar flow consumed by the fuel cell.

4.2- Anodic species schedule.

In the anodic compartment, we can write the same type of balance.

$$\frac{dnH_2}{dt}=F_{H_2in}-F_{H_2c}-F_{H_2valve} \qquad F_{H_2c}=\frac{NI_{fc}}{2F} \tag{2}$$

With :

- F_{H2in} inlet molar flow of the Reducing valve.
- F_{H2c} hydrogen molar flow consumed by the fuel cell.
- $F_{H2valve}$ outlet molar flow of the valve.

4.3- Pressure calculation.

We use the perfect law gases hypothesis to calculate the pressure:

$$P_{cath}=\frac{RT_{fc}}{V_{cath}}(n_{O_2}+n_{H_2}+n_{vap}) \tag{3}$$

The dynamic model of the pressure variation is:

$$P_{cath}=\frac{RT_{fc}}{V_{cath}}\frac{1}{p}\left((1+\frac{X_{vap}(P_{cath},T_{fc})}{1-X_{vap}(P_{cath},T_{fc})})F_{comp}-F_d\right) \tag{4}$$

With: $F_d=F_{O2\,cons}+F_{valve}$

We can also calculate the partial pressures:

$$P_{O_2}=\frac{RT_{fc}}{V_{cath}}n_{O_2} \qquad P_{H_2}=\frac{RT_{fc}}{V_{cath}}n_{H_2} \tag{5}$$

4.4- Voltage calculation.

The voltage variations of the fuel cell are computed with a quasi-static model, because the chemical reaction dynamic is faster than the system dynamics.

This voltage depends on the current density in the fuel cell, the partial pressures (of hydrogen and oxygen), the temperature of the reaction, and the hydration of the membrane. $Ufc=f(I,PO_2,PH_2,T_{fc},\lambda_{H_20})$

Hypothesis of the model:

- Insignificant Anodic activation voltage.
- Uniform current density.
- Uniform temperature.

The voltage expression is:

$$Ufc=E_{Nernst}-\frac{RT_{fc}}{2\alpha F}\ln\left(\frac{j}{j_o}\right)-R_{mem}\,j-j\left(c\frac{j}{j_{max}}\right)^2 \tag{6}$$

With :

- α= symmetry factor (0,5).
- j_o = current exchange.j_o=[10^{-7}A/cm^2;10^{-5}A/cm^2].
- R_{mem}=f(T_{fc},λ_{H2O}) membrane resistance (0,162 Ω/cm^2).
- j_{max}=2.5A/cm^2 maximum current.
- c=0.471 coefficient due to simplification.

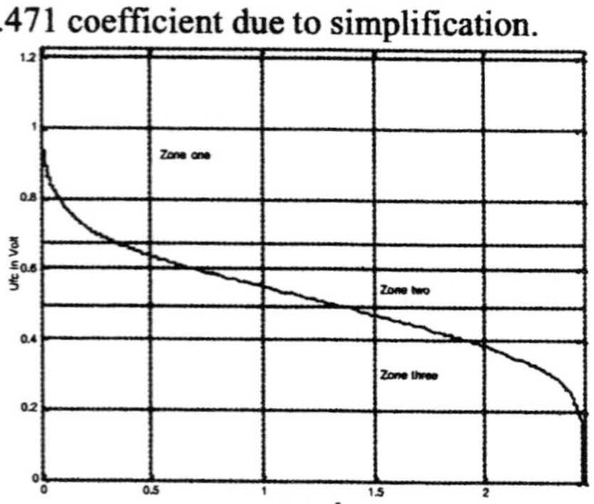

Fig 2. : Polarisation curve.

In the polarisation curve, we can see three zones:

- Cathodic activation (zone one): $\frac{RT_{fc}}{2\alpha F}\ln\left(\frac{j}{j_o}\right)$.
- Ohmic loss (zone two) :$R_{mem}j$.
- Diffusion limitation (zone three): $j(c\,j/j_{max})^2$.

The Nernst law is valid on all the voltage range. This law represents the voltage of the oxidation-reduction reaction :

$$E_{Nernst}=\frac{\Delta G^0}{2F}+\frac{RT_{fc}}{2F}\ln\left(\sqrt{\frac{P_{O_2}}{10^5}}\frac{P_{H_2}}{10^5}\right) \tag{7}$$

5- COMPRESSOR MODELING.

After Modeling the PEM core, we must study the air supply process to obtain the dynamics of the voltage developed by the fuel cell.

The air supply system of the fuel cell is composed by a screw type compressor. We choose this type of compressor to avoid the flow oscillations and the lubrication. The rotation is made by a synchronous drives with permanent magnets.

NB: we work here on a low-pressure fuel cell system.

5.1- Outlet flow compressor.

The outlet flow is detailed in [3] and is:

$$q_{molar}=\frac{P_{outlet}}{RT_{outlet}}Cyl\frac{\omega}{2\pi}\eta_v \tag{8}$$

The volumetric efficiency η_v allows us to include several phenomena in the model:

- The total mechanical clearances between rotor and compressor casing .
- Compression rate.
- Rotor peripheral speed.
- Flow density.

The volumetric efficiency depends on the rotor speed and compressor rate ($tx=\frac{P_{outlet}}{P_{inlet}}$) as seen on fig 3.

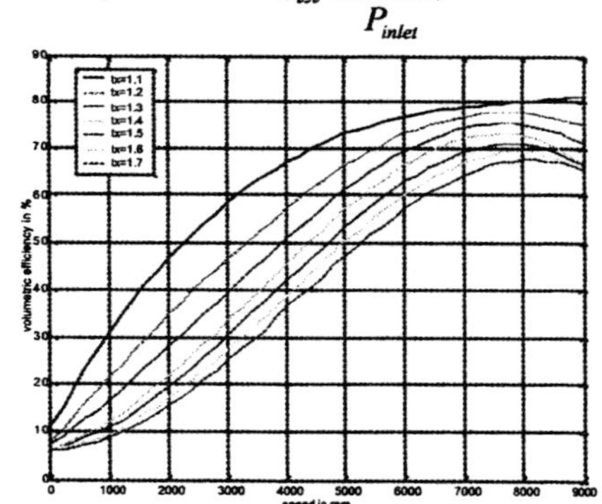

Fig 3. : Volumetric efficiency.

5.2- Flow simplification modeling.

The association of the synchronous machine (with speed control) and the compressor can be represented by the following diagram:

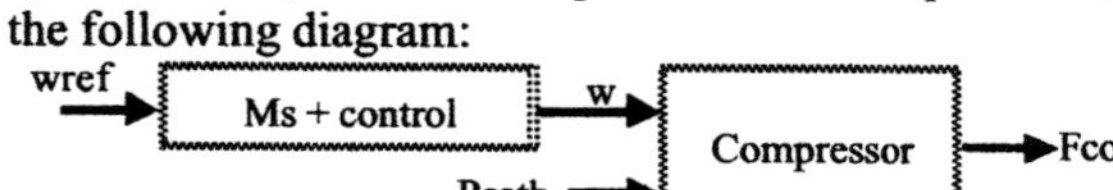

Fig 4. : Synchronous machine and compressor.

From this functional diagram we can write the evolution law of the compressing flow *'Fcomp'*. We place upstream speed reference an inverse model of the compressor in order to have a linear relation between the variable of entry *'e'* and the flow of the compressor *'Fcomp'*.

The inverse model is representing by the following maps on Matlab:

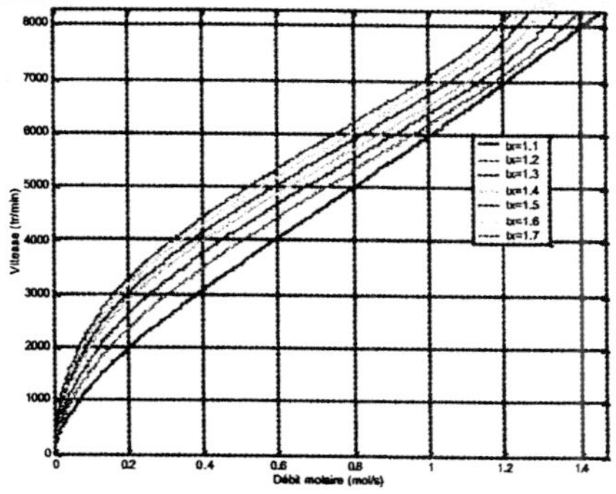

Fig 5. : compressor inverse model.

Therefore, we can write the following diagram:

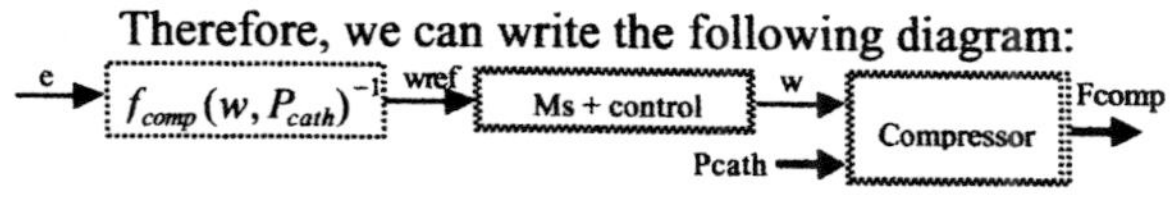

Fig 6. : flow modeling.

The synchronous engine and its controller are equivalent to a first order system. So the exit flow of the compressor is represented by:

$$F_{comp} = \frac{1}{p}\left(\frac{1}{\tau}e - \frac{1}{\tau}F_{comp}\right) \qquad (9)$$

6- VALVES MODELING.

The molar flow of the valves is expressed in function of the opening section, the temperature, and the outlet and inlet pressures [4]:

$$F_{valve} = S_{open}\sqrt{\frac{2P_{inlet}^{\ 2}}{MRT_{inlet}}\frac{\gamma}{\gamma-1}\left[\left(\frac{P_{outlet}}{P_{inlet}}\right)^{\frac{2}{\gamma}} - \left(\frac{P_{outlet}}{P_{inlet}}\right)^{\frac{\gamma+1}{\gamma}}\right]} \qquad (10)$$

This valve models is linearized around the nominal point. So the new model is: $F_{valve} = S_{open}K_{valve}$

7- AUXILIARY CONTROL STRATEGY

7.1- Steady state modeling.

Therefore, the modelisation of the fuel cell system result in a two states model [2][5]where:

$$X = [\mathrm{Pcath\ Fcomp}]^t$$

$$\dot{X} = \begin{bmatrix} 0 & (1+\frac{X_{vap}(P_{cath},T_{fc})}{1-X_{vap}(P_{cath},T_{fc})})\frac{RT_{fc}10^{-5}}{V_{cath}} \\ 0 & -\frac{1}{\tau} \end{bmatrix} X + \begin{bmatrix} -\frac{RT_{fc}10^{-5}}{V_{cath}} & 0 \\ 0 & -\frac{1}{\tau} \end{bmatrix} \begin{bmatrix} Fd \\ e \end{bmatrix}$$

In this modelisation the pressure is in bar.

The nonlinear model is linearized around the nominal point. So we can write :

$$\dot{X} = \begin{bmatrix} 0 & (1+\frac{X_{vap_n}}{1-X_{vap_n}})\frac{RT_n10^{-5}}{V_{cath}} \\ 0 & -\frac{1}{\tau} \end{bmatrix} X + \begin{bmatrix} -\frac{RT_n10^{-5}}{V_{cath}} & 0 \\ 0 & -\frac{1}{\tau} \end{bmatrix} \begin{bmatrix} Fd \\ e \end{bmatrix} \qquad (11)$$

7.2- Feedback Control Design.

The system controller is implemented in digital so we numerize the steady state. The sampling frequency is equal to: $fe = 166\,Hz$

The new steady state is:

$X(k+1) = A_d X(k) + B_d u(k)$ and $Y(k) = C_d X(k)$

The control structure schema is:

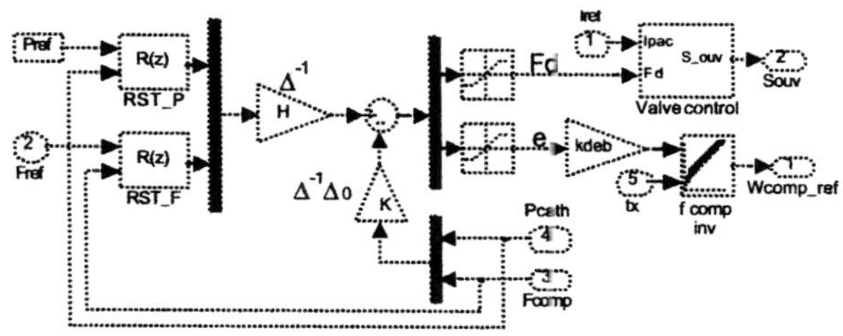

Fig 7. : Pressure and flow control.

In order to create uncoupling of the two states variables we write:

$$u(k) = \Delta^{-1}v(k) - \Delta^{-1}\Delta_0 X(k) \qquad (12)$$

With: $\Delta = B_d$ and $\Delta_0 = A_d$

with this steady state feedback the new system is simplified as: $P_{cath} = z^{-1}v_1$ *and* $F_{comp} = z^{-1}v_2$.

After uncoupling the system we implement RST regulator in order to control the two variables with different dynamics.

The regulator structure is:

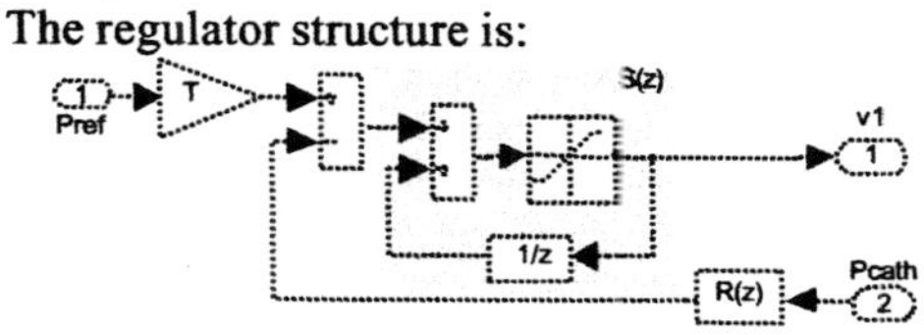

Fig 8. : Pressure and flow regulator.

Where:

$$S(z^{-1}) = 1 - z^{-1}; \quad R(z^{-1}) = r_0 + r_1 z^{-1}; \quad T = R(1) \qquad (13)$$

The polynomial regulator coefficients are calculated by the pole placement. The chosen characteristic polynomial of the close loop is.

$$P(z^{-1}) = 1 + p_1 z^{-1} + p_2\ z^{-2} = (1 - z^{-1}\exp(-\omega Td))^2 \qquad (14)$$

with:

$\omega = 15\,rad/s$ for pressure loop.

$\omega = 50\,rad/s$ for flow compressor loop.

We obtain:

$$r_0 = p_1 + 1 \ and\ r_1 = p_2 \qquad (15)$$

The flow reference is calculated with the Faradays law:

$$F_{ref} = \frac{N\,I\,St_{O2}}{4F\,X_{O_2}} \qquad (16)$$

7.3- Simulation results.

This simulation represents the evolution of pressure and flow when power is requested on fuel cell. It is realized with a very complete model on simulink.

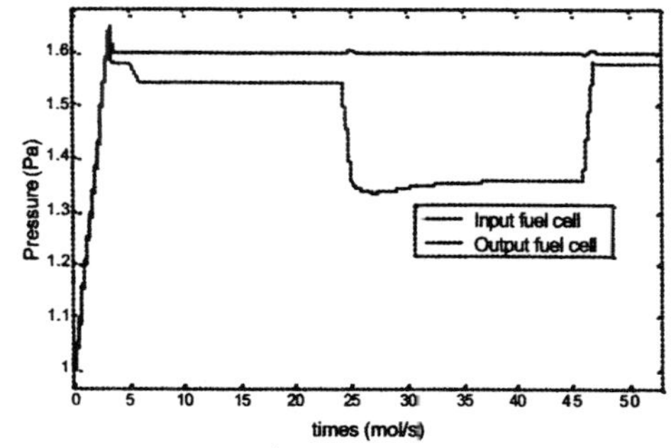

Fig 9. : Cathodic pressure.

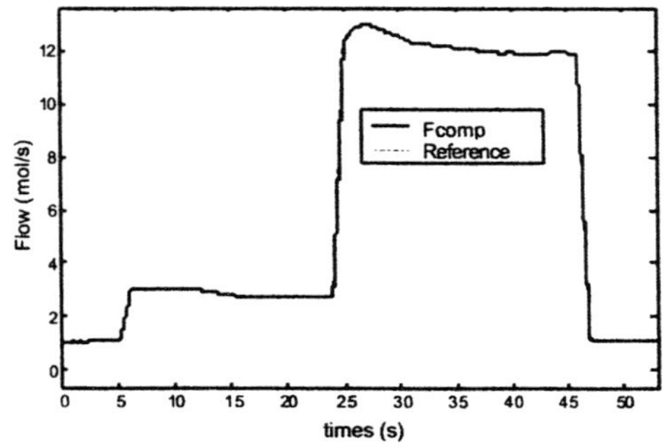

Fig 10. : Compressor flow

As we can see on this simulation the pressure is remained constant despite the compressor flow variation. We can also see that the compressor flow tracks the flow reference.

8- TRACTION CHAIN.

8.1- *Architecture.*

The fuel cell is now, connected to the DC link with an electronic chopper. The existing fuel cell sub-system has to be designed for high power and does not exist for the moment. That is why an Energy Storage System (ESS) must be associated with the fuel cell. In addition the ESS is able to absorb the braking energy of the vehicle. The ESS technology chosen is supercaps. This ESS is connected to the DC link by another electronic chopper.

Both sources must give the power demand by the propulsion motor and the auxiliaries of the vehicle. The electrical choppers are connected by parallel structure as shown on fig 11.

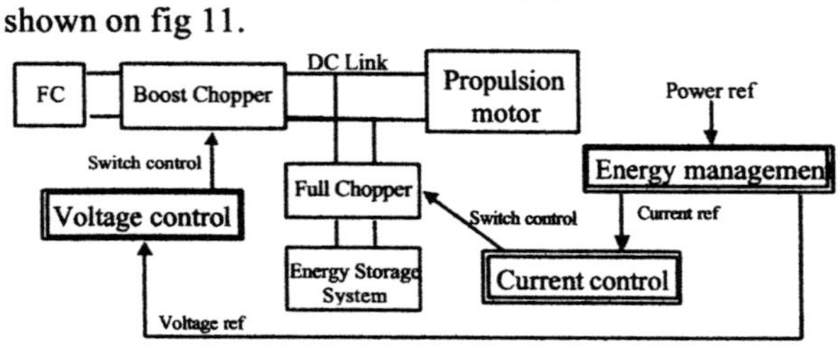

Fig 11. : Power Train and control structure.

The fuel cell chopper is a Boost because in the high power rail transport applications the voltage of the Dc link is high. This chopper is controlled for the DC link voltage regulation. So a loop voltage regulation is used.

The supercaps full chopper is a Buck/Boost because this chopper must be reversible in current. This chopper is in current regulation control mode to manage the charge and discharge of the ESS. The current reference is computed in function of the ESS power reference which is obtained by an energy strategy.

8.2- *Boost Modeling and Control*

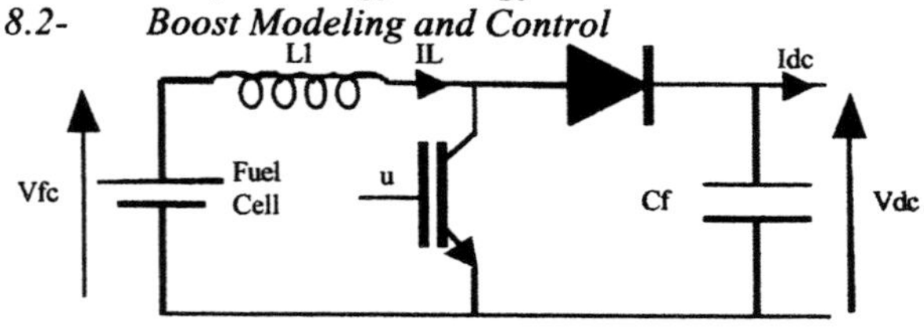

Fig 12. : Boost principle schema.

$\{L1 = 0.95mH \ and \ Cf = 3.2mF\}$

The power switch is control by PWM with α duty cycle. The constant frequency is fd=1/Td .With the PWM control. Chopper is described by the following average model.

$$\begin{cases} Vfc = L_1 \dfrac{dIL}{dt} + (1-\alpha)Vdc \\ IL(1-\alpha) = Cf \dfrac{dVdc}{dt} + Idc \end{cases} \quad (17)$$

This chopper is controlled for the DC link voltage regulation. In the aim to have good control of the voltage we have realized two regulation loops : one on the current and an other one on the voltage[6]. To obtain suitable transfer function for control purpose, we realized a linearization by an inverse model. So we write:

$$\alpha = \frac{VL' - Vfc}{Vdc} + 1 \quad (18)$$

With this linearization the current model is:

$$T_1(p) = \frac{Il(p)}{VL'(p)} = \frac{1}{L_1 p} \quad (19)$$

To implement a digital controller we numerize the transfer function, the sampling frequency is equal to the switch frequency.

$$H(z^{-1}) = \frac{B(z^{-1})}{A(z^{-1})} = \frac{Td}{L_1} \frac{z^{-1}}{(1-z^{-1})} \quad (20)$$

Finally we establish a classical sampled regulation structure (Fig. 13) with an RST regulator with these polynomials corresponding to proportional and integrator controller:

$$S(z^{-1}) = 1 - z^{-1}; \quad R(z^{-1}) = r_0 + r_1 z^{-1}; \quad T = R(1) \quad (21)$$

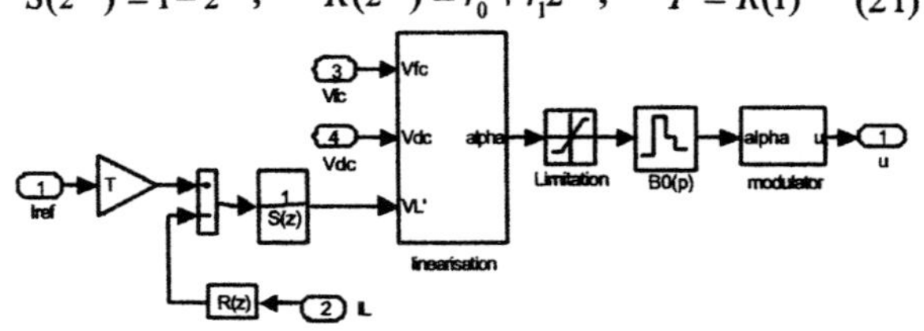

Fig 13. : Boost current loop.

The regulation coefficients are calculates by the pole placement. The characteristic polynomials of the closed loop is.

$$P(z^{-1}) = 1 + p_1 z^{-1} + p_2 \ z^{-2} = (1 - z^{-1} \exp(-\omega Td))^2 \quad (22)$$

with: $\omega = 2\pi fd/10 = 3142 \ rad/s$.

We obtain:

$$\begin{cases} r_0 = \dfrac{L_1}{Td}(p_1 + 2) \\ r_1 = \dfrac{L_1}{Td}(p_2 - 1) \end{cases} \quad (23)$$

In order to avoid windup we have implemented an anti-windup structure in the regulation loop.

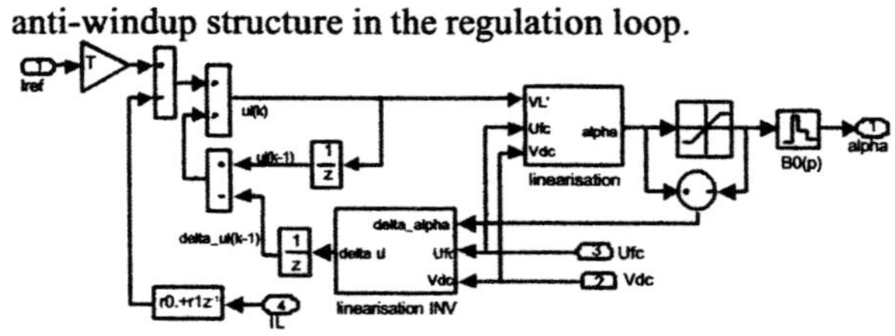

Fig 14. : Boost current loop implementation.

In the same way, to add a voltage control loop, we realised a linearisation by an inverse model with:

$$Iref = \frac{Vdc}{Vfc}(Iref' + Idc) \quad (24)$$

So we can write the voltage model:

$$T_2(p) = \frac{Vdc(p)}{Iref'(p)} = \frac{1}{Cf \, p} \quad (25)$$

We have here the same type of transfer than in the voltage loop so we made the same digital controller with also an anti-windup.

The structure of this controller is show in fig 15:

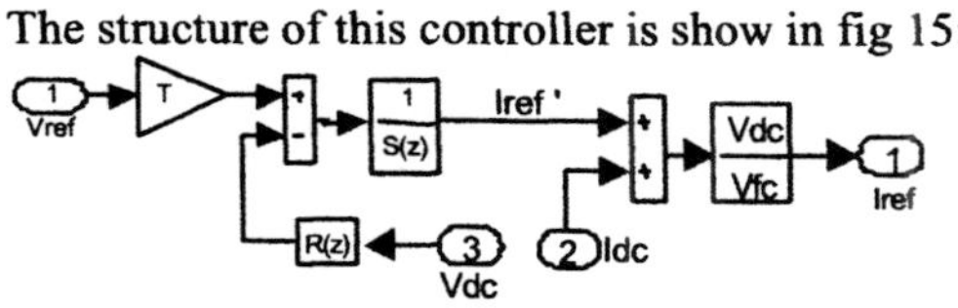

Fig 15. : Boost voltage loop.

The coefficient of the regulation are.

$$\begin{cases} r_0 = \dfrac{Cf}{Te}(p_1 + 2) \\ r_1 = \dfrac{Cf}{Te}(p_2 - 1) \end{cases} \qquad (26)$$

Here we have the same characteristic polynomials with: $\omega = 2\pi fd/100 = 314\,rad/s$.

8.3- Buck/Boost Modeling and Control.

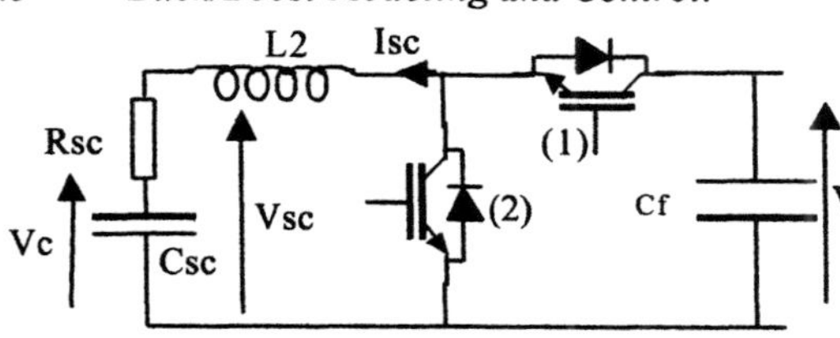

Fig 16. : Buck/Boost principle schema.

$\{L2 = 0.94mH\,;Cf = 3.2mF\,;rsc = 55m\Omega\,;Csc = 47F\}$

This chopper is described by the following average model which allows us to establish the equation in the two working phases[7].

Differentials equations in Buck case:

$$\frac{disc}{dt} = \frac{1}{L_2}Vdc\,\alpha_1 - \frac{Rsc}{L_2}isc - \frac{Vc}{L_2} \text{ and } \frac{dVc}{dt} = \frac{isc}{Csc} \qquad (27)$$

Differentials equations in Boost case:

$$\frac{disc}{dt} = \frac{1}{L_2}Vdc\,(1-\alpha_2) - \frac{Rsc}{L_2}isc - \frac{Vc}{L_2} \text{ and } \frac{dVc}{dt} = \frac{isc}{Csc} \qquad (28)$$

In order to simulate this chopper we write:

$$Vpwm = \alpha_1\,Vdc\;Buck/Boost + (1-\alpha_2)\,Vdc\;\overline{Buck/Boost} \qquad (29)$$

Where Buck/Boost binary variable represents the type of the chopper operation. This chopper control is of course a current regulation to manage the charge and discharge of the ESS. In order to have one model and one regulator we write that the control in the boost mode is the complement of the control in the Buck mode. In this case the Boost model can be wrote as a Buck model. But in the control loop we must compute the complement value of the rate cycle when the Boost operation is activated. The current transfer function is:

$$T(p) = \frac{Isc}{\alpha} = \frac{Csc\,Vdc\;p}{1 + Rsc\,Csc\;p + L_2\,Csc\;p^2} \qquad (30)$$

The DC link voltage is constant because this voltage is control by the Boost chopper.

To implement a digital controller we numerize the transfer function, the sampling frequency is equal to the switch frequency.

$$H(z^{-1}) = \frac{B(z^{-1})}{A(z^{-1})} = \frac{b_1 z^{-1}(1-z^{-1})}{1 + a_1 z^{-1} + a_2 z^{-2}} \qquad (31)$$

We establish a sampled regulation structure as on fig 17. We compute an RST regulator with these polynomials:

$$S(z^{-1}) = 1 - z^{-1}\,;\; R(z^{-1}) = r_0 + r_1 z^{-1}\,;\; T = \frac{A(1) + b_1 R(1)}{b_1} \qquad (32)$$

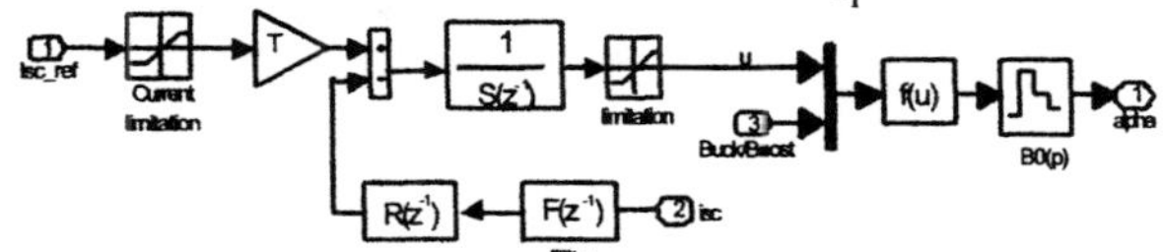

Fig 17. : Buck/Boost current loop.

The function in the f(u) bloc is the implemented function to calculate the complement value of the α rate cycle when the Boost operation is activated. So we write:

$$alpha = u\;Buck/Boost + (1-u)\overline{Buck/Boost} \qquad (33)$$

The type of the chopper operation variable (Buck/Boost) is calculated by:

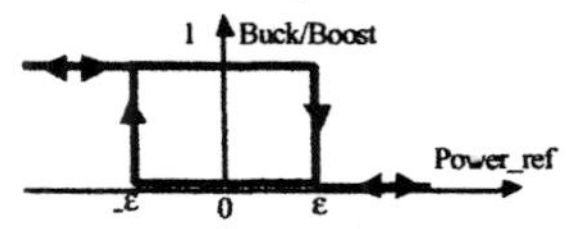

Fig 18. : Buck/Boost operation.

This is simulated by a relay because the zeros value of the current must be in the two modes because the chopper can make a current with a zero average value. So we have:

Buck mode when Buck/Boost=1.

Boost mode when Buck/Boost=0.

The balance point of this converter is for the average current equal to zero. When the chopper is in one mode we have an intermittent flow for little value of the current. The first solution in order to control the chopper is to hold the mode to the last active mode and to filter the current. The second solution is to switch between the both mode when we have zero current. We have decided to implement the first solution so we filter the supercaps current by a numerical Butterworth (5 orders) filter $F(z^{-1})$. The regulation coefficient are calculated by the pole placement.

$$\begin{cases} r_0 = (p_1 - a_1)/b_1 \\ r_1 = (p_2 - a_2)/b_1 \end{cases} \qquad (34)$$

The characteristic polynomials is:

$$P(z^{-1}) = 1 + p_1 z^{-1} + p_2\,z^{-2} = (1 - z^{-1}\exp(-\omega Td))^2 \qquad (35)$$

Where: $\omega = \omega_{fil}/10 = 1500\,rad/s$ and ω_{fil} is the angular frequency cutting of the filter.

Due to current overshoot when the chopper operation changed, we must initialised the duty cycle at each changed to set the current to zero.

The initialised value is: $\alpha_{init} = Vsc/Vdc$

The implementation structure of the loop regulation with initialisation and anti-windup is:

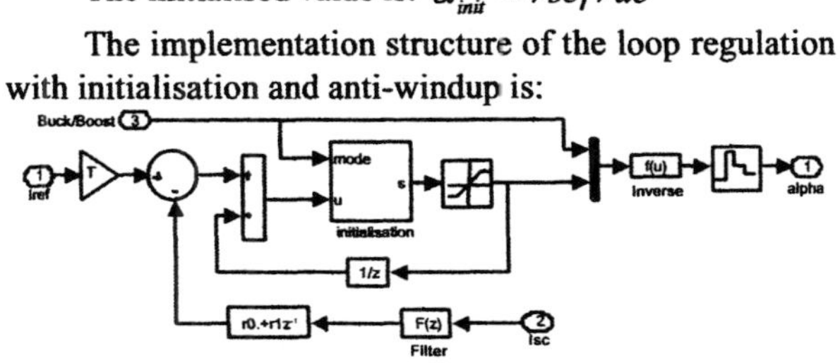

Fig 19. : Current regulation implementation.

The initialization is computed by the structure seen on fig 16:

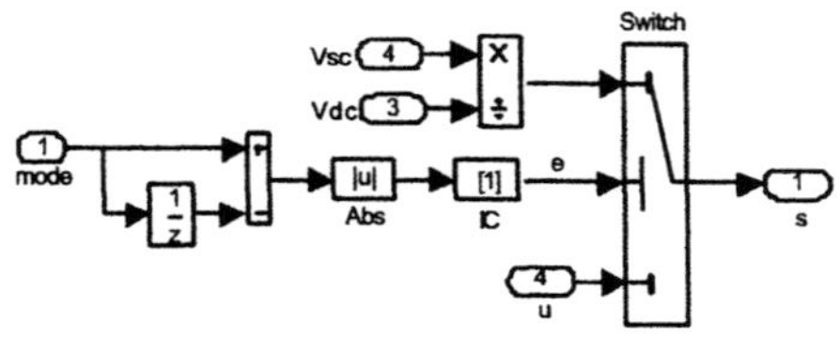

Fig 20. : Initialization computed.

The initialization structure detect the mode modification and put the control variable to the init value. The current reference is calculated in function of the ESS power reference and the loses estimation of the chopper. The power reference is provided by an energy strategy.

In Buck case :

$$ESS\ power\ ref(k) = Power\ ref(k) - Losses(k-1) \quad (36)$$

In Boost case :

$$ESS\ power\ ref(k) = Power\ ref(k) + Losses(k-1) \quad (37)$$

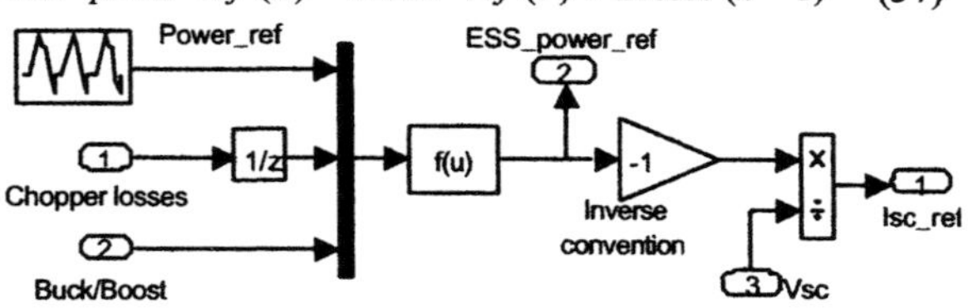

Fig 21. : Current reference.

The current reference is also managed by a logical function in order to limit the supercaps voltage (in max and min value). The losses of the chopper are estimated in average value by a value plant[8][9].

8.4- Simulation results.

The aim of this simulation in average value is to view some responses of the main system variables. Data used for the power reference are taken from ADVISOR (NREL software) simulation.

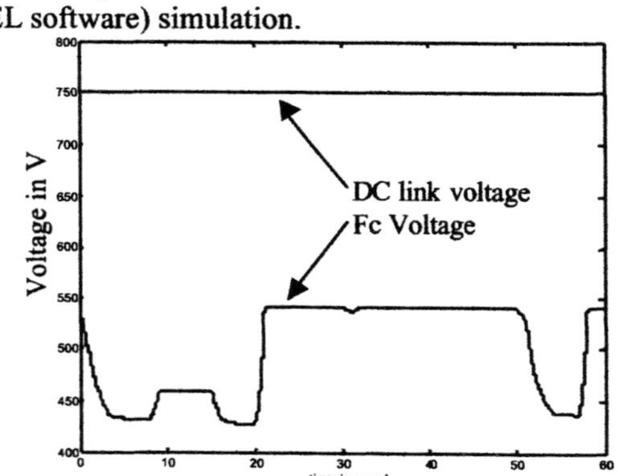

Fig 22. : DC link and Fc voltage

We can see that the first converter has a voltage regulation and this voltage remained constant despite the power request on the DC link.

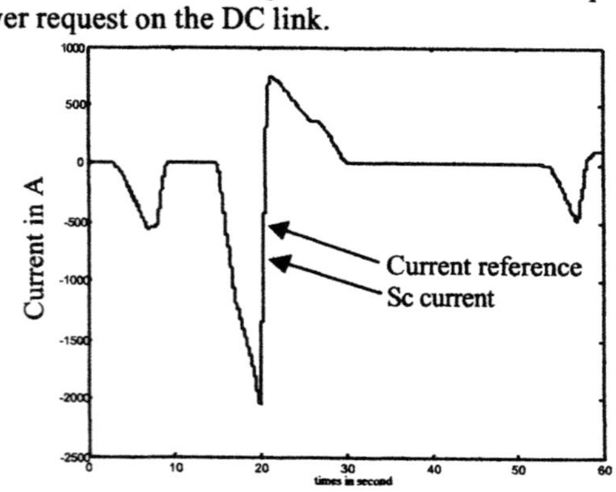

Fig 23. : Desired and real current

The simulation shows also that the second chopper has a current regulation and the current in the supercaps tracks the current reference. We can also see that this chopper is reversible in current.

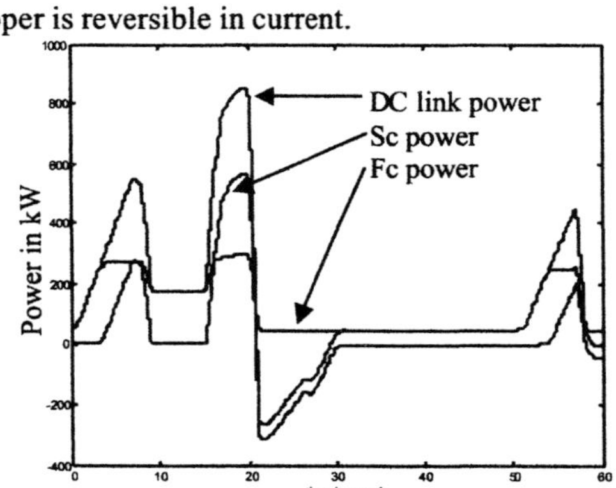

Fig 24. : power traction on an operating cycle.

As we can see on this simulation, the power of the propulsion motor and auxiliaries is separated between the fuel cell and the supercaps following the energy management chosen. The quasi-static model seems to be sufficient for the power train simulation but, we are now testing if we must develop a dynamic model of the fuel cell.

9- CONCLUSION.

In order to control the fuel cell system we develop the fuel cell model, the air compressor model, and the valves model. We have only developed in this paper the control structure of the air supply system and pressure valves. But in the fuel cell system, there are other auxiliaries to control, like the cooling system. So we must continue the study on the control structure.

10- REFERENCES.

[1] W Friede, S Raël, B Davat: "PEM fuel cell models for supply of an electric load", proc of Electrimacs 2002, August 18-19.

[2] Jay T.Pukrushpan, Anna G.Stefanopoulou, Huei Peng: "Modeling and Contol for PEM Fuel Cell Stack System", proc of the American Contol Conference – Anchorage, Michigan - May 8-10-2002.

[3] VARAIX corporation documentation: "Compressor efficiency definitions" and " Evaluation of vairex air system technology for automotive fuel cell power systems (fcps) ": website www.vairex.com

[4] Bird, Stewart, light foot: book: "Transport Phenomena" p481, Wiley international edition, ISBN: 0-471-07395-4

[5] Wiartalla Andreas, Lang Olivier, Pischinger Stefan, Schönfelder Carter: "Efficiency Boosting of Fuel Cell System Considering Part Load Operation", proc of FISITA 2001 World Automotive Congress -Helsinki, Finland – June 2-7-2002

[6] M.Fadel :"Lois de commande pour une alimentation AC/DC à absorption de courant sinusoïdal", proc of 3EI 2000.

[7] B.J.Arnet, L.P.Haines: "High Power DC-to-DC Converter For Supercapacitors", proc of IEMDC 2001 – Boston, Massachusetts.

[8] International Rectifier documentation: "Application characterisation of IGBT's" : website: www.irf.com/technical-info/an990/an-990.htm

[9] EUPEC corporation data-sheet: " EUPEC :FZ 1600 R 17 KF6 B2" : website www.eupec.com

ELSEVIER
IFAC
PUBLICATIONS
www.elsevier.com/locate/ifac

AN INTERFACE CONSTRUCTION OF FUEL CELL AND BATTERY HYBRID SYSTEM

Soo-Bin Han, Suck-In Park, Hak-Geun Jeoung, Bong-Man Jeoung, Won-Young Lee

Korea Institute of Energy Research,
71-2, Jang-Dong, Yusong-Gu, Taejon 305-343, Korea
Phone: +82-42-860-3121
Fax: +82-42-860-3102
E-mail: sbhan@kier.re.kr

Abstract: This paper describes an interface method to utilize the hybrid system with fuel cell and battery. The hybrid system has unique advantage to manage energy state between high energy system (fuel cell) and high power system (battery) according to various type of load. First, the hybrid system is completely modelled and simulated by using Simplorer. The models are sufficiently correct by considering many parameters. Interface method is realized by constructing a DC-DC converter, and experimented under various load conditions. *Copyright © 2003 IFAC*

Keywords: Fuel cell, Battery, Hybrid system, DC-DC converter, Current control

1. INTRODUCTION

Recently, successable development of new energy device such as fuel cell and super-capacitor naturally lead to new area research of the hybrid system (Maggiore, 1997; Spyker, et al, 1996). Conventional energy storage system only based on the battery has many problems, and is not sufficient for a variety of applications in performances such as life, size, response time, power and energy capability. While battery is a storage device with medium energy but high power, fuel cell is a high energy generation device.

Fuel cell power system can be used without additional battery for large power generation plant, but hybrid system with battery is general approach for electric vehicle and moveable power supply where require fast response and additional standby power for controller. So hybrid energy system is composed of different energy systems, especially battery/fuel cell system, is new approach for optimizing the energy system in various applications.

This paper describes interface method between fuel cell and battery. First, modelling and simulation of hybrid energy system components such as fuel cell, battery and interface are performed using Simplorer, which is a new simulation tool suitable for complex systems including circuit, control and mechanical system. The models are sufficiently correct by considering many parameters. Interface methods between different energy system are also presented and really constructed. Then, various test of hybrid system are performed and its characteristics are discussed.

2. SYSTEM MODELING AND SIMULATION

2.1 Fuel cell review and modelling

Fuel cell is similar to battery in that both produce a DC current by using an electrochemical process. Two electrodes, an anode and a cathode, are separated by an electrolyte. Like batteries, fuel cells are combined into groups, called stacks, to obtain a usable voltage and power output. Unlike battery, however, fuel cell doesn't release energy stored in the cell or run down when the energy is gone. Instead, it convert the energy in a hydrogen-rich fuel directly into electricity and operate as long as it is supplied with fuel. Fuel cell emits almost none of the sulfur and nitrogen confounds released by conventional generating methods, and can utilize a wide variety of fuels: natural gas, coal-derived gas, landfill gas, biogas, or alcohol.

Fig. 1 shows that an I-V characteristic curve where the fuel cell voltage decreases with increasing current from 100% in the no load condition to about 40 ~ 50% in the full load condition. Normally, the peak power occurs near the rating current. A discharging load is needed in the fuel cell system to avoid the high voltage state produced by the electrochemical reaction. Some fuel cell system uses resistance load directly between two fuel cell electrodes as discharging load. For hybrid vehicle applications, battery charging from fuel cell is operated as a kind of discharging load. So the operating load range is located between a discharging load and a maximum load which is normally above a full load.

Fuel cells system is composed of a variety of different complex systems including mechanical system, electrochemical system, and chemical systems. So its complete model is almost impossible, but following modeling method can be used.

- Language description such as C, FORTRAN or Matlab script language.
- LUT(Look up table) method based on real measured I-V characteristics
- Time varying voltage source

Each method has its own advantage, but LUT method as shown in Fig. 2 is more useful than others in a view of practice for electrical engineer.

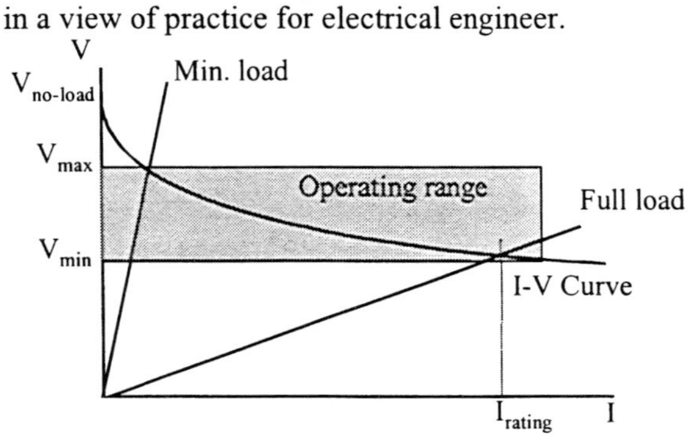

Fig. 1. I-V characteristic of fuel cell

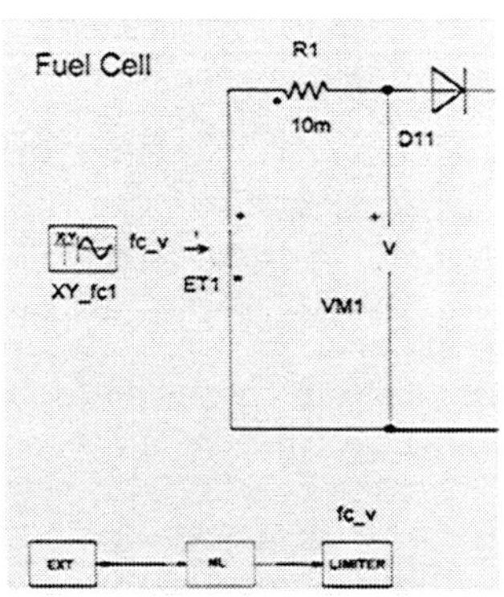

Fig. 2. Fuel cell model by LUT approach

2.2 Battery modelling

The possible battery models are various. Among them, most commonly used model is the one which consists of an ideal battery with open circuit voltage and a constant equivalent internal series resistance. The improvement of this model is possible by accounting for the non-linear characteristics of both the open circuit voltage and internal resistance represented by the state of charge (SOC).

Used Simplorer model in this paper is shown in Fig. 3. The model is based on the consideration of battery SOC which is calculated by integrating the battery current. Fig. 4 shows a model test results, where SOC decreases with battery charge current and battery voltage also decreases with SOC.

2.3 Interface method and model

Interface between different energy system is possible by a variety of methods. Simple method uses only passive element as resistor or inductor although not efficient and unpractical approach. The output voltage of fuel cell decreases severely as the current increases. In the worst case, the output voltage at rated current is below half of one at no load. It is uneconomical to design an inverter to meet a very wide range of input DC voltage and constant AC output. This would result in a large equivalent capacity and poor performance. To avoid this, a chopper or DC-DC converter is connected as an interface to convert the broad range of input voltage to a constant output voltage or current, which is to be fed to the inverter.

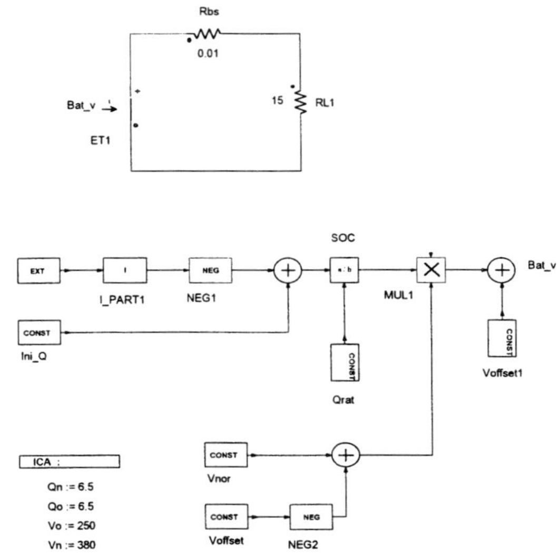

Fig. 3. Battery model by Simplorer

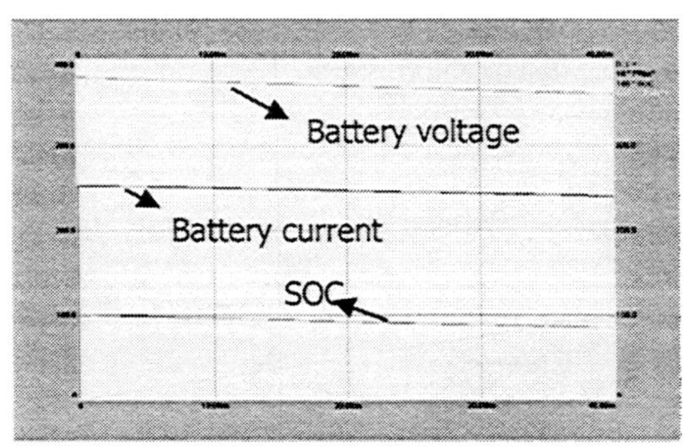

Fig. 4. Battery model test

However, the total system cost including converter and inverter might be higher, and the system efficiency might be lower compared to that of single inverter unit; hence, the system design should be determined from the comparison of merits and demerits of these systems. Until now, it is a major tendency to use power conditioning system include both converter and inverter for AC commercial output, and use only converter for DC applications.

There are various topologies of converters each exhibiting different characteristics. Simple inductor coupled converters may provide a means of lifting (Boost converter) or reducing (Buck converter) the voltage from the fuel cell to interface to a common DC link. The output of these converters is not isolated from the input, however if needed, isolated type topology based on the bridge or push pull topology is used (Severns, et al, 1985). So interface converter can be varied to system spec. and applications.

In this paper, current mode controlled (Redl, et al, 1985) boost converter is used as shown in Fig. 5. A test result is shown in Fig. 6, where although input voltage is time varying source, inductor current is following to current command and output voltage is regulated as a constant value. In hybrid energy system, input voltage source and output voltage source can be any energy system such as battery, fuel cell and super-capacitor according to applications. Fig. 7 shows how hybrid energy storage system can be modeled uniformly by using submodule. Fig. 8 shows simulation results of hybrid system with input-side fuel cell and output-side battery. When current command and load varies, various operation of fuel cell and battery can be simulated.

3. INTERFACE CONSTRUSTION AND TEST

Interface converter for hybrid system is composed as shown in Fig. 9, where boost type DC-DC converter of Fig. 7 is used in power stage. Power stage is controlled with IGBT gate drive signal given by analog type current controller for real time operation.

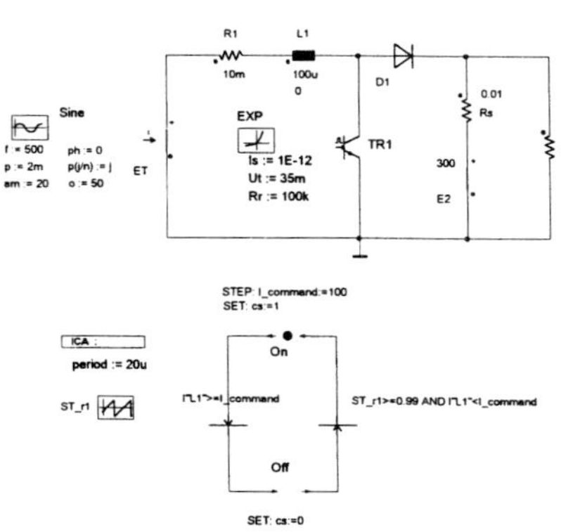

Fig. 5. Current mode controlled Boost converter as an interface converter

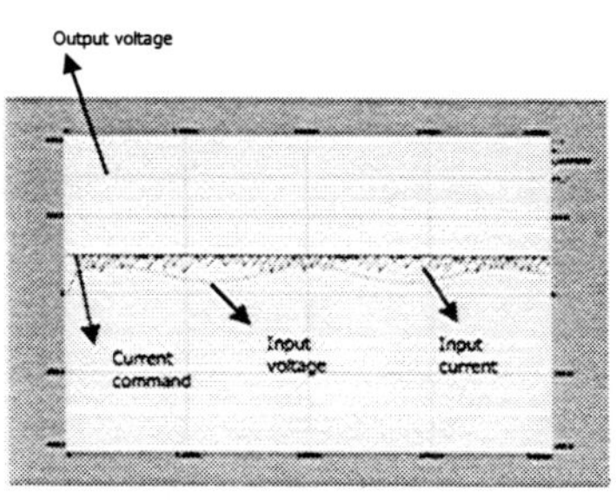

Fig. 6. Current mode control test

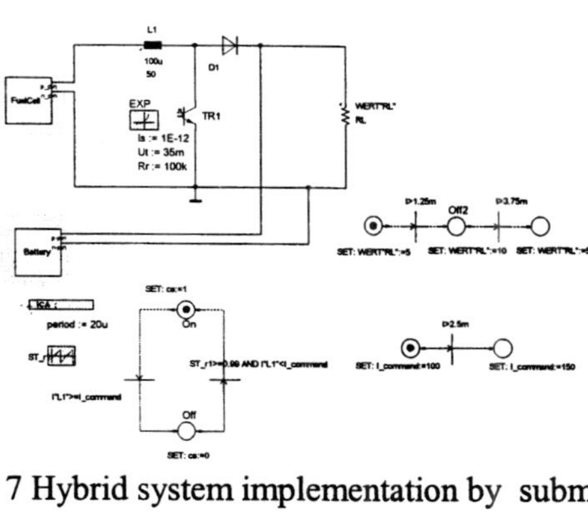

Fig. 7 Hybrid system implementation by submodule

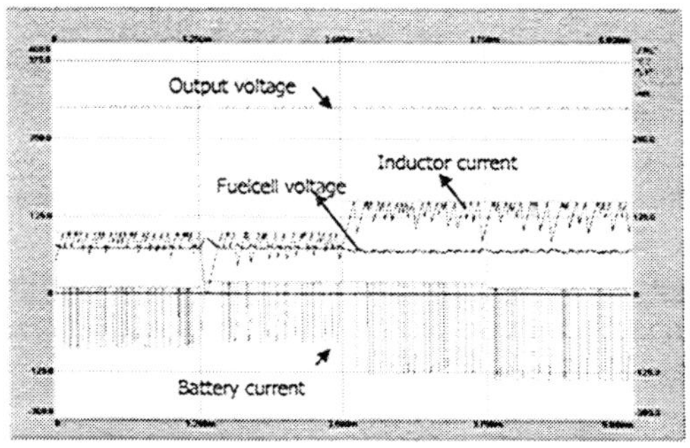

Fig. 8 Fuel cell - Battery system simulation

Current command is generated by digital controller where necessary reference current is synthesized based on the current, voltage and power reference given by power control unit. A 1102 control board manufactured by dSPACE is used as a power control unit in PC. In control concept, current control loop is located in a inner loop and voltage control loop is in a outer loop as shown in Fig. 10 because current controller must be operated in a faster sampling time than the voltage controller.

For real system operation, we construct that a fuel cell simulated power supply like ABC 150 is used for realizing the static characteristic curve of the PEM(Polymer Exchange Membrane) type fuel cell and the interface output is connected to electronic load and lead-acid type battery bank as shown in Fig. 11. Interface converter covers the input range of 30-150V and 0-300A for the output range of 100-350V with maximum 25kW power rating. The experiment view is shown in Fig. 12.

The response time of fuel cell stack is actually fast because it is several msec order. But additional mechanical and chemical systems such as fuel supply system make total fuel cell system very sluggish. So fuel cell is normally operated as base power. In this case, fuel cell system should be used with battery for compensating transient load condition. When abrupt load changes, the battery follows the change until the fuel cell power get to the requirement to the load. Fig. 13 shows an operation test when load is varied in a stepped form. As load changes, load current changes abruptly and battery current follows the load changes in the beginning. However converter output shows constant current.

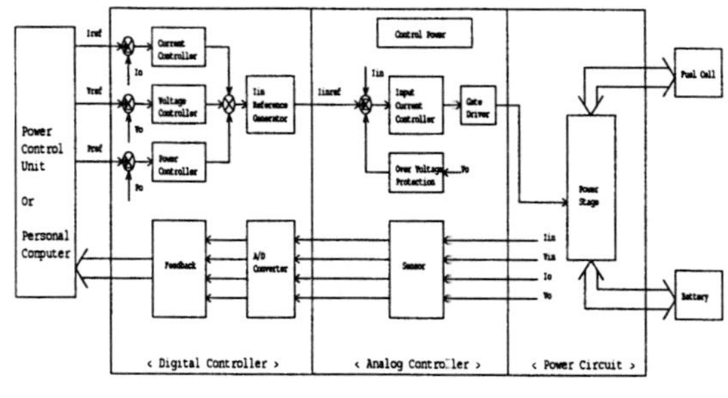

Fig. 9. Interface converter structure

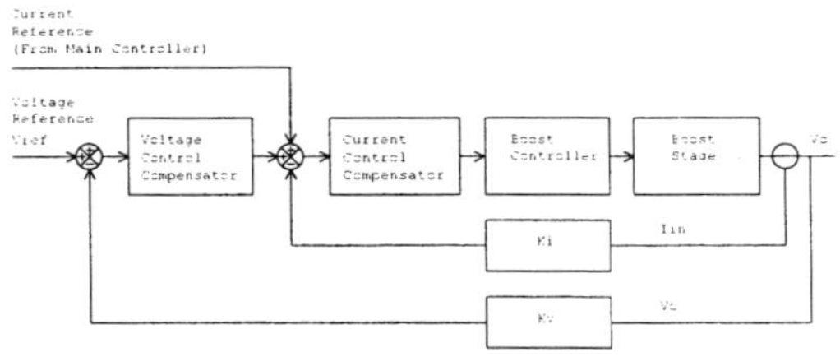

Fig. 10. Used control loop

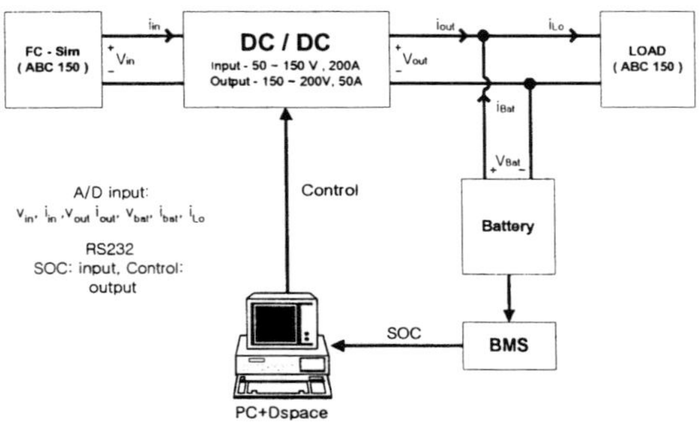

Fig. 11. Whole block of system experiment

An interface converter characteristic according to the input current is shown in Fig. 14. As input current increases, input power and output power also increase. Under high current above 150A, output power is not generated exactly proportional to the input current due to increased internal loss. So the efficiency of the interface converter is saturated to about 93% under high current as shown in Fig. 15.

4. CONCULSION

This paper describes an interface method to utilize the hybrid system with fuel cell and battery hybrid energy storage system. System component is sufficiently modelled and simulated by Simplorer for design of real system. An experiment is performed in case that fuel cell system is operated as a base power with long time constant and battery power is used as transient load following. Optimum system design and power managing control is actually different to various load types. The authors will research the design and control criteria for different applications in a next step.

Fig. 12. Photo of experiment system

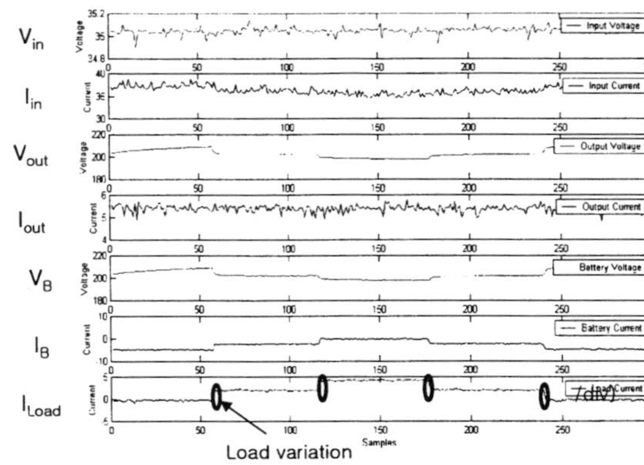

Fig. 13. Test operation for load variation

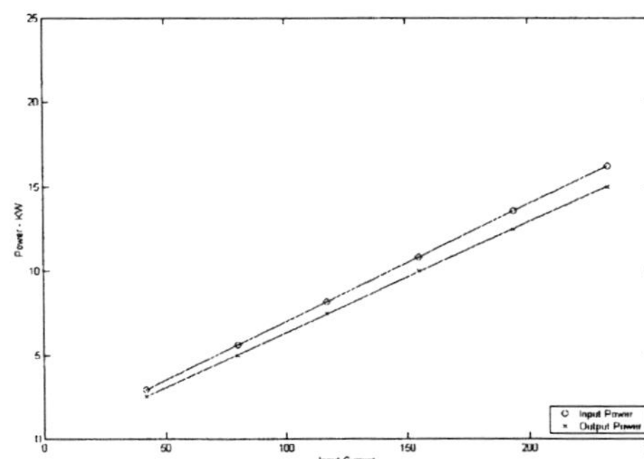

Fig. 14. Input and output power characteristics (X-axis: current, Y-axis: power)

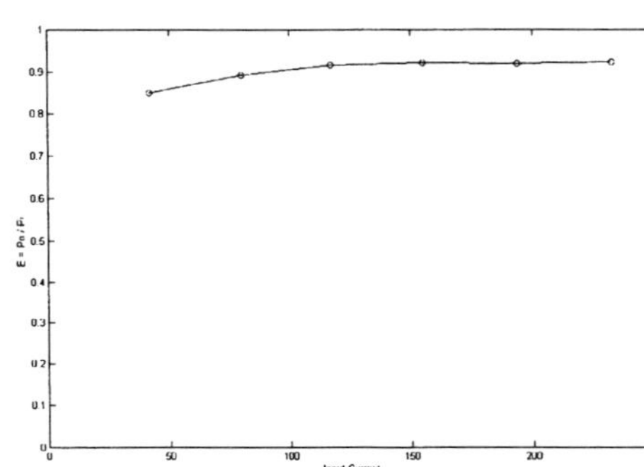

Fig. 15. Efficiency characteristics as current change (X-axis: current, Y-axis: efficiency)

REFERENCES

Spyker, R.L., et al (1996). Evaluation of Double layer Capacitors for Power Electronic. *IEEE, IAS Conference Record*, pp. 725-730.

Maggiore, A. (1997). Solid Polymer Fuel Cells of Industrial Interest. *Intertech Conference Records of Commercializing Fuel Cell Vehicles 97*.

Severns, R.P., and G. Bloom (1985). Modern DC to DC Switchmode Power Converter Circuit. VNR.

Redl, R. and N. Sokal (1985). Current Mode Control, Five Different Types, Used with the Three Basic Classes of Power Converters. *Power Electronics Specialists Conference Record,* pp. 771-785

ELSEVIER
IFAC
PUBLICATIONS
www.elsevier.com/locate/ifac

DEVELOPMENT OF DISTANCE RELAY MODELS FOR REAL TIME DIGITAL SIMULATOR

*Lee Joo-Hun, **Yoon Yong-Beum, **Cha Seung-Tae, *Lee Jin, *Choe Jong-Woong

* *LG Industrial Systems Co., Ltd.* ** *Korea Electric Power Research Institute of KEPCO*

Abstract: This paper discusses how to develop a software model of the digital relay that can be added to real time digital simulator(RTDS) model library and is then subjected to the same outputs as the actual relay. The software model is stand-alone and can be used with actual relays. It is also possible to conduct interactive real-time tests when the system effects of the relay action need to be investigated. The characteristics of the mho type and the quadrilateral type, which are commonly used in recently developed relays, are modelled in this paper. Single circuit line and double circuit line system are used for model verification. The transmission lines are each 100 km in length and are modelled as distributed parameter lines but not frequency dependent. The transmission lines in the single circuit system are modelled as ideally transposed line. The mutual coupling data with the parallel line was taken account in the transmission lines for the double circuit system. CTs and PTs are included and operated in their linear region during the tests. For the purpose of testing the relay model accuracy the faults have been applied at various points on the protected line. Its accuracy is assessed against theoretical values. *Copyright © 2003 IFAC*

Keywords: Digital relays, Distance relay model, Real Time Digital Simulator, Capacitive Voltage Transformer

1. INTRODUCTION

In recent years, strong interest in applying protective relaying system to power system simulation has been appeared on the scene from numerous technical articles because the need to increase the transmission capability of existing and new electrical networks is forcing the relay engineer to examine the performance of present and future relays than has been the case in the past. Especially the need to observe the transient response of the power system network adopting the high-speed relays is imminent.
The use of simulation software such as EMTP, EMTDC, PSS/E and EUROSTAG, which is commonly used to observe the response of a power system network, would be a satisfactory way to simulate a power system network if only the action of primary protection is considered. But if the system stability or the subsequent action of other relays on a parallel line is considered, then the effect of the primary relay trip must be fed back into the simulation. This requires other devices that are not widely available. Therefore it is desirable to have a model of the relay that can be included in the simulation. The simulation software mentioned above contains a group of relay models in its model library for its users. Those models are intended to represent the general principles of relays and are not intended to exactly represent any particular product of any relay manufacturer. Rather, they can model the effects of actual relays given properly specified data and proper interpretation by the engineer. Many utilities already have complete EMTP models of their power system that have been verified against actual field data. These models have been used for performance testing of new and existing protection schemes, studies on high impedance fault(HIF) in transmission line, the effects of DC-Offset and harmonics, optimal zone setting and so on. However RTDS has no relay models in its model library because RTDS is originally invented for the test of

the actual relay. Therefore the need of relay models gradually increases among the users.

This paper describes the development and testing method for a software model of the digital relay that can be run on RTDS.

2. DEVELOPMENT OF RELAY MODEL

Most of protective relays were the electromechanical type or solid-state type, and these are still in wide use, but digital type designs are now in common. Although this trend continues, it may be a long time before all devices are completely replaced. In developing a software model for the electromechanical type relay and the solid-state type relay, it is very difficult to construct an accurate model. The primary reason is the lack of design data on which the model to be based. The second reason is the difficulty involved in fitting a digital model to a non-linear analogue device. On the contrary, in case of the digital relay, it is relatively easy to develop an accurate model because the protective relaying algorithms remain basically the same in spite of varying with different manufacturers and constantly changing. According to the trend and the above-mentioned reasons, digital type relays have been developed by user defined component(UDC) method for RTDS. The UDC method provides a tool that allows end users to develop their own component models and run them on the RTDS just as another components. All the details related to linking them to the existing RTDS software are handled automatically by RTDS compiler. The software model of the relay developed by UDC method can be used interactively with the simulation running on after the relay action takes place.

2.1 Typical architecture of digital relays

The typical architecture of digital relays is shown in Fig, 1. Generally, Digital relays are made up with three stages as follows.

① Analog-digital converting stage
② Calculation and decision stage
③ Digital inputs and outputs stage

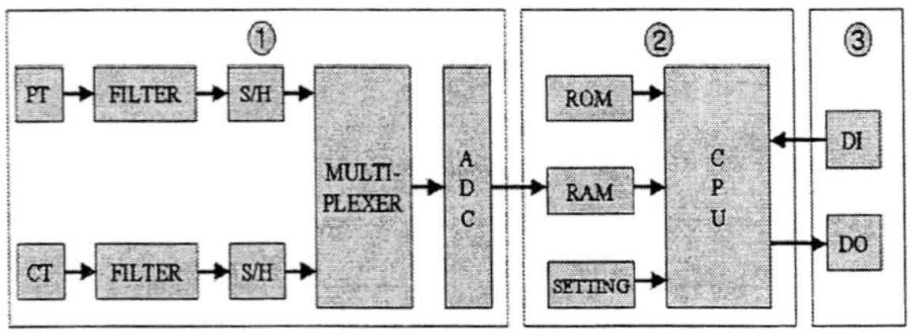

S/H : sampling holder
ADC : analog-digital converter
SETTING : setting interface
CPU : central processing unit
DI, DO : digital input and output interface

Fig. 1. Typical architecture of digital relays

On the stage of analog-digital conversion(①) in Fig.1, The sampling holders, multi-plexer and A/D converter have not been modelled because all the input signals are already in digital on RTDS. The auxiliary PTs and CTs have been regarded as operate in their linear region. The voltages and currents are taken through a low pass filter and a DC-Offset removal filter, then are processed in Discrete Fourier Transformation(DFT) to give the fundamental components in terms of sin and cosine components for passing on to next stage.

2.2 Impedance Calculation of relay models

The calculation and decision stage(②) in Fig.1 plays the most important role in the software model for the relay. This stage can be divided into two parts. The first part is responsible for the calculation of the impedance from the relay to the fault point using the fundamental components of voltages and currents. it has been modelled as the following. Fig. 2.

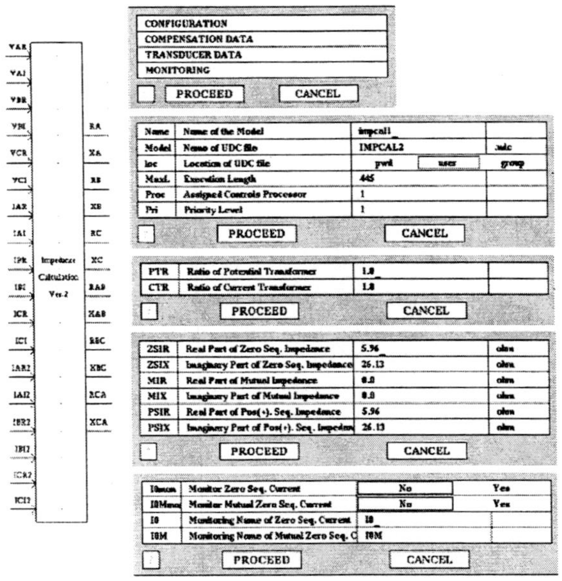

(a) icon (b) parameter input window
Fig. 2. Impedance calculation model for RTDS

In order to correctly calculate the impedance to the fault point, the current input to the earth fault calculating elements is compensated by the residual current(3Io) of the protected line in the case of a single circuit and by the residual current of the protected line and residual current of the parallel line in the case of a double circuit line. The equations for ground fault are as follows.

$$I_{or} = I_{ar} + I_{br} + I_{cr} \tag{1}$$

$$I_{oi} = I_{ai} + I_{bi} + I_{ci} \tag{2}$$

$$I_{omr} = I_{a2r} + I_{b2r} + I_{c2r} \tag{3}$$

$$I_{omi} = I_{a2i} + I_{b2i} + I_{c2i} \tag{4}$$

$$K_r^{'} = \frac{K_r}{3} = \frac{Z_0 - Z_1}{3Z_1} \tag{5}$$

$$K_m^{'} = \frac{K_m}{3} = \frac{Z_m}{3Z_1} \tag{6}$$

$$I'_{ar} = I_{ar} + K'_r * I_{or} + K'_m * I_{omr} \quad (7)$$

$$I'_{ai} = I_{ai} + K'_r * I_{oi} + K'_m * I_{omi} \quad (8)$$

$$D_a = (I'_{ar})^2 + (I'_{ai})^2 \quad (9)$$

$$R_a = \frac{V_{ar} * I'_{ar} + V_{ai} * I'_{ai}}{D_a} \quad (10)$$

$$X_a = \frac{V_{ai} * I'_{ar} - V_{ar} * I'_{ai}}{D_a} \quad (11)$$

$$Z_a = \frac{V_a}{I_a + K'_r * I_o + K'_m * I_{om}} \quad (12)$$

Where,

V_a : phase 'a' voltage

I_a, I_b, I_c : phase currents of protected line

I_{a2}, I_{b2}, I_{c2} : phase currents of parallel line

I_o : zero-sequence current of protected line

I_{om} : zero-sequence current of parallel line

K_r, K'_r, K_m, K'_m : compensation factor

Z_o : zero-sequence impedance

Z_1 : positive-sequence impedance

Z_m : mutual impedance

$*_r$: real part

$*_x$: imaginary part

$*_m$: mutual part

Zb and Zc are calculated in the exactly same way.

The following equations are used for impedance calculation of phase-phase fault.

$$V_{abr} = V_{ar} - V_{br} \quad (13)$$

$$V_{abi} = V_{ai} - V_{bi} \quad (14)$$

$$I_{abr} = I_{ar} - I_{br} \quad (15)$$

$$I_{abi} = I_{ai} - I_{bi} \quad (16)$$

$$D_{ab} = I_{abr}^2 + I_{abi}^2 \quad (17)$$

$$R_{ab} = \frac{V_{abr} * I_{abr} + V_{abi} * I_{abi}}{D_{ab}} \quad (18)$$

$$X_{ab} = \frac{V_{abi} * I_{abr} - V_{abr} * I_{abi}}{D_{ab}} \quad (19)$$

$$Z_{ab} = R_{ab} + jX_{ab} = \frac{V_{ab}}{I_a - I_b} \quad (20)$$

Where,

V_{ab} : line-line voltage between phase 'a' and 'b'

I_{ab} : line-line current between phase 'a' and 'b'

Zbc and Zca are calculated in the exactly same way.

2.3 Zone characteristics of relay models

The other part is in charge of the decision as to whether the relay operates or not. If a fault occurs, the impedance moves into the operating zones of the relay and then the tripping logic outputs a trip command. The operating zones of the relay represented on the R-X plane have a characteristic such as impedance, reactance, mho, quadrilateral and so on. The characteristics of the quadrilateral type and the mho type are commonly used in recently developed relays, so those have been modelled as the following Fig. 3 and 4. Most of model's input parameters can be obtained directly from KEPCO's relay setting rule. The rest of input parameter are fixed or recommended by manufacturers. Their parameter input windows are represented in the similar way shown in Fig. 2. The quadrilateral type and the mho type model can be represented a variety of operation zone shape as their setting parameters are changed. So they can cover most of digital relay widely used in Korea such as DLP, M-DAR, MDT-A2, MDT-F and MXL1E.

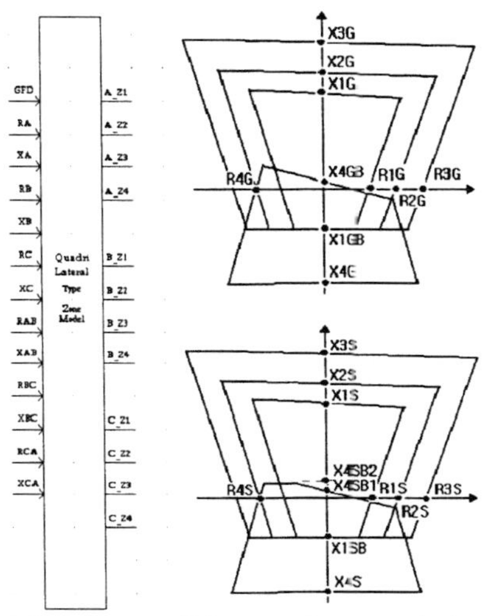

(a) icon (b) zone characteristics

Fig. 3. Quadrilateral model for RTDS

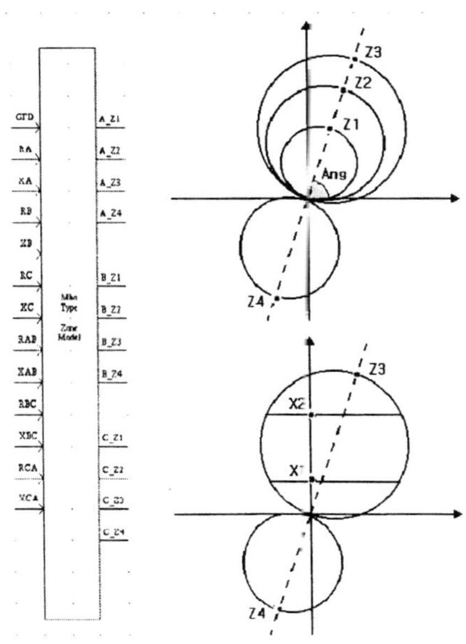

(a) icon (b) zone characteristics

Fig. 4. Mho model for RTDS

The last stage(③) is composed of digital input and digital output. This stage was not modelled because most of components needed on this stage have already been made as the library model of RTDS.

3. SOFTWARE MODEL TESTS

One rack of RTDS is used for the software model tests. The rack consists of 10 3PC cards, 1 IRC card and 1 WIC card. The one line diagrams and data for the software model tests are shown in Fig. 5 and 6.

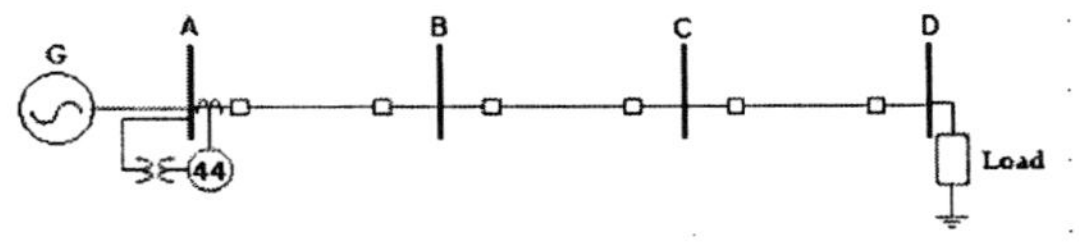

- *Gen. data :*
V = 154 0 kV, Z1 = 0.001 [ohm], Z0 = 0.0 [ohm]
- *Line data for each section :*
Z1=Z2=1.85+j37.67[ohm], Z0=36.18+j122.77[ohm]
- *Load data :*
120+j30[ohm]

Fig. 5. One line diagram of test system(single circuit)

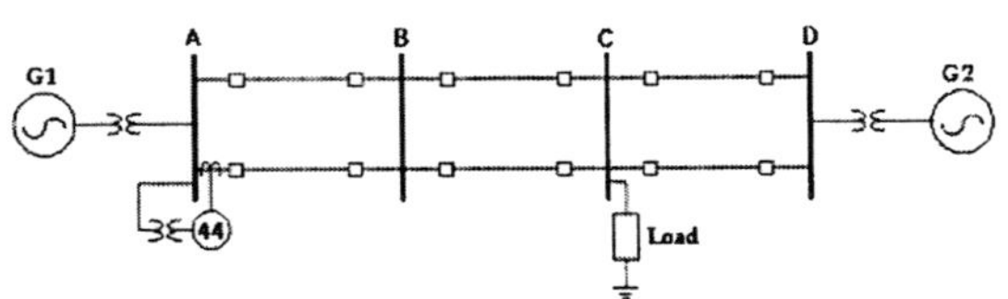

- *Gen1. data :*
V = 154 9.6 kV,Z1 = 0.001 [ohm],Z0 = 0.0 [ohm]
- *Gen2. data :*
V = 1540.0 kV ,Z1 = 0.001 [ohm],Z0 = 0.0 [ohm]
- *Line data for each section :*
Z1=Z2=4.25+j28.97[ohm] ,Z0=21.38+j98.78[ohm], Zm = 17.12+j57.21[ohm]
- *Load Data :*
120+j30[ohm]

Fig.6.One line diagram of test system(double circuit)

Those systems are the 154 kV system. The transmission lines are each 100 km in length and are modelled as distributed parameter lines but not frequency dependent. The transmission lines in the single circuit system are modelled as ideally transposed line. The mutual coupling data with the parallel line was taken account in the transmission lines for the double circuit system. CTs and PTs are included and operated in their linear region during the tests. Currents and voltages are low pass filtered and then are DC-Offset removal filtered. Thereafter the signals are subjected to a DFT whose sampling rate is 32 samples per 60 Hz cycle. The phase of DFT is controlled by the frequency tracking routine, which is based on phase locked loop(PLL), that alters phase to maintain synchronism between the sampling rate and the system frequency. Then the results are used in various relay calculating functions such as impedance elements where Z=V/I type calculations are performed. The calculated values for Z are compared with theoretical values and are checked to verify whether they lie inside preset impedance zones and trip signals are issued appropriately. The preset impedance of zone 1, 2, 3 and 4 is 85, 150, 250, 150 % of the protected line impedance respectively. For the purpose of testing the relay model accuracy the symmetrical and unsymmetrical faults have been applied at 50, 75, 90, 150, 170, 250, 270 % of the protected line length from the relay location.

4. RESULTS OF MODEL TEST

With the fault points being changed, the variety of faults have been applied for the software model test.

4.1 Results of impedance calculation

The accuracy of impedance calculation is measured by means of comparison between the software model's outputs and theoretical values. The results of the comparison are summarized in Table 1 and 2. The maximum error((theoretical value-output value/ theoretical value)) is below about 1 %.

Table 1 . Summary of single circuit tests (unit : %)

Fault Point	*Phase-phase fault Impedance Error*	*Ground fault Impedance Error*
50	0.53	0.01
75	0.86	0.14
90	0.60	0.44
150	0.71	0.09
170	0.50	0.50
250	0.84	0.01
270	0.97	0.01

Table 2 . Summary of double circuit tests (unit : %)

Fault Point	*Phase-phase fault Impedance Error*	*Ground fault Impedance Error*
50	0.30	0.40
75	0.37	0.09
90	0.40	0.40
150	0.17	0.13
170	0.30	0.60
250	0.23	0.05
270	0.41	0.35

All the trip signals from the software model can be monitored in runtime. Figure 7 shows the impedance trajectories for phase ‘a’ in RTDS/PSCAD run time window, when a ground fault is applied at phase ‘a’, 85 % of the protected line length. X(Y) axis denotes real(imaginary) part of impedance in Fig. 7.

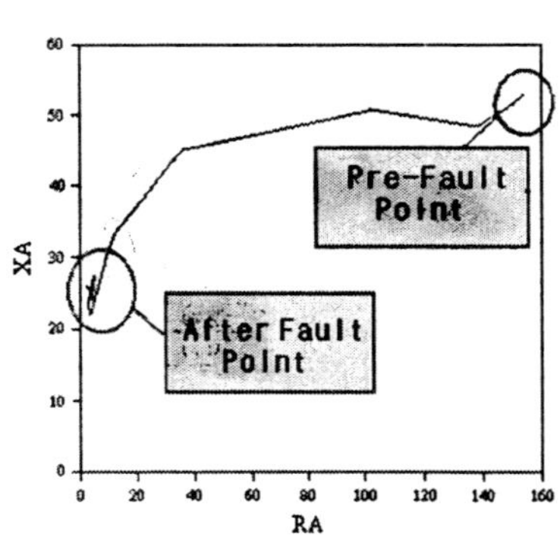

Fig.7. Impedance trajectories (phase ‘a’ fault, 85%)

4.2 Results of zone characteristics tests

It is rather difficult to decide whether trip signals are issued or not, when the impedance trajectories move around the boundaries of each zone of relays. The quadrilateral type model and the mho type model shown in Fig. 3 and 4 are tested and verified under the conditions that the impedance inputs are very close to their zone boundaries. Figure 8 and 9 show pick-up points on R-X plane for zone 1 of the quadrilateral and the mho type model respectively. All the zones are verified in the same way. The models can operate properly unless the differences between the input values and the values on their zone boundaries are less than 1.0e-4.

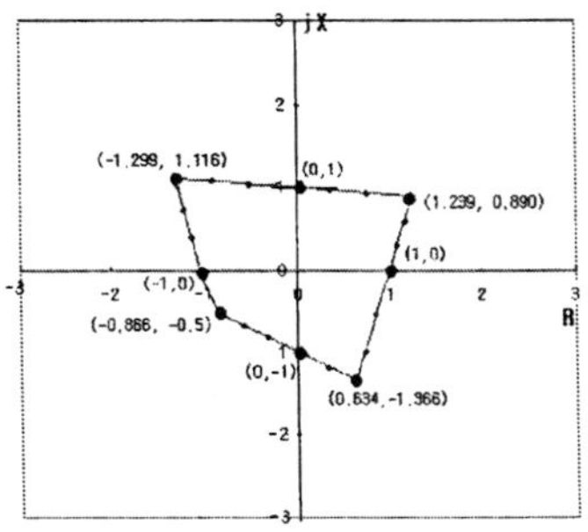

Forward & backward reactance angle = 5 [deg]
Forward right & left blinder angle = 75 [deg]
Backward right & left blinder angle = 75 [deg]
Zone 1 resistance = 1.0 [ohm]
Zone 1 reactance = 1.0 [ohm]
Zone 1 backward reactance = 1.0 [ohm]

Fig.8. Pick-up points on R-X plane(quadrilateral)

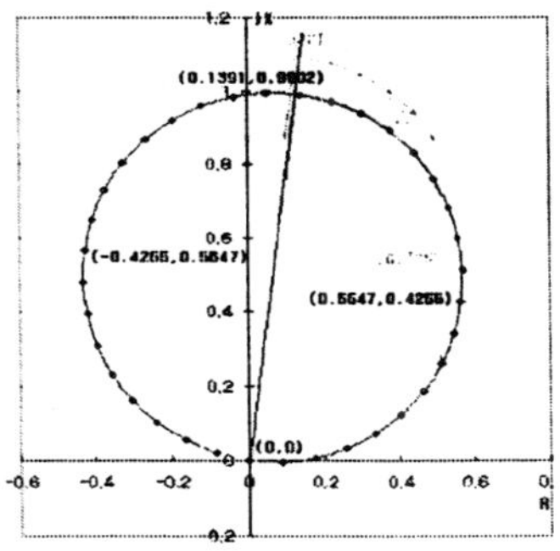

Characteristic angle : 82 [deg]
Zone 1 Imp = 1.0 [ohm]

Fig.9. Pick-up points on R-X plane(mho)

With the fault phases, types and applying orders being changed at regular intervals, the variety of faults have been applied for the software model test and then checked to verify whether trip signals are issued from the models appropriately or not, which are connected with the fault-related phases. Figure 10 shows trip signals which are issued by quadrilateral type model, when ground faults are applied on phase 'a', 'b', 'c' of single circuit line shown in Fig.5 in turn at 25 msec intervals. Each fault lasts 100 msec. It takes about 5.0e-2 msec for the software model to issue trip signals, when the faults are applied. The results of mho type model are almost same.

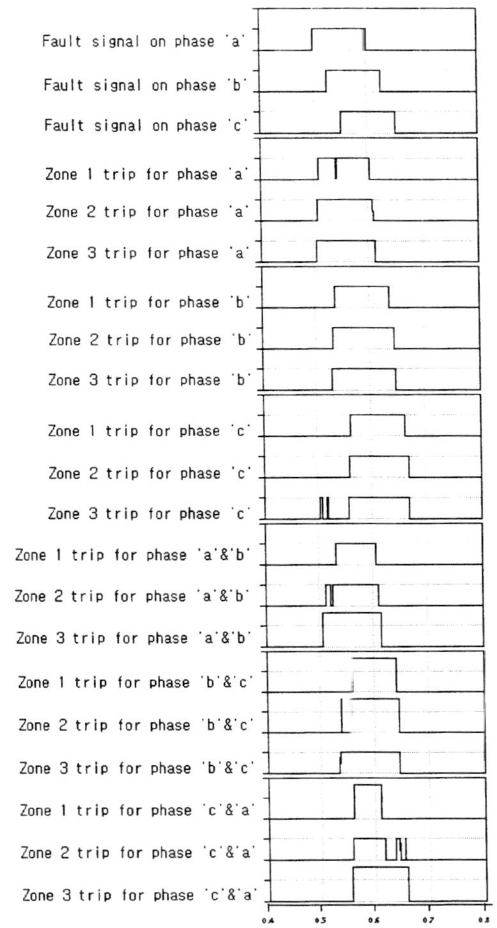

Fig.10. Fault applying signals and trip signals

5. CASE STUDY

5.1 Relay performance during CVT transients.

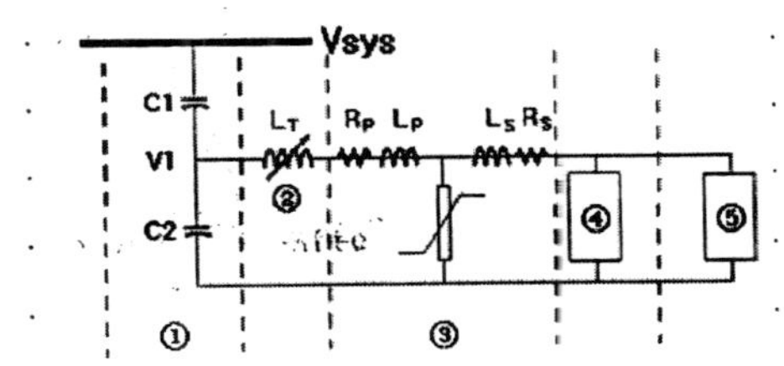

① Capacitive divider stack
② Tuning reactor
③ Step-down Transformer
④ Ferroresonance-suppression Filter
⑤ Burden

Fig.11. Generic CVT structure

Capacitive Voltage Transformers(CVT) as shown in Fig. 11 are commonly used in high-voltage transmission line. CVT has been worked satisfactorily with conventional relays for many years. However, as the requirement for faster protective relays grows, so does the concern over the poor transient response of some CVT for certain system conditions. Poor CVT transient response causes the distance relay to overreach. The burden and ferroresonance-suppression circuit of CVT also affect the performance of distance relaying.

The simple system as shown in Fig.12 is built with CVT and distance relay model to determine the performance of proposed distance relay model during CVT transients. These tests are performed under different conditions because the transient responses of CVT are affected by system impedance ratios, fault locations and fault resistance.

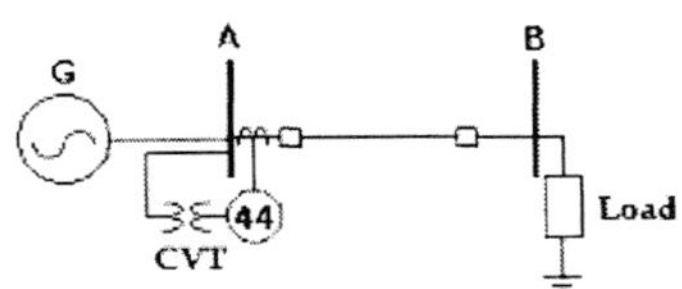

- Gen. data :

V = 154 0 kV, Z1 = 0.001 [ohm], Z0 = 0.0 [ohm]

- Line data :

$C_{grading}$ =695[pF], $C_{bus\text{-}ground}$ =1200[pF]

- Load data :

50[ohm]

- CVT data :

C1 = 0.013[μ F], C2=0.264[μ F]

Step-down transformer ratio =(8000:200)

Burden = 100[ohm]

Fig.12. One line diagram of study system

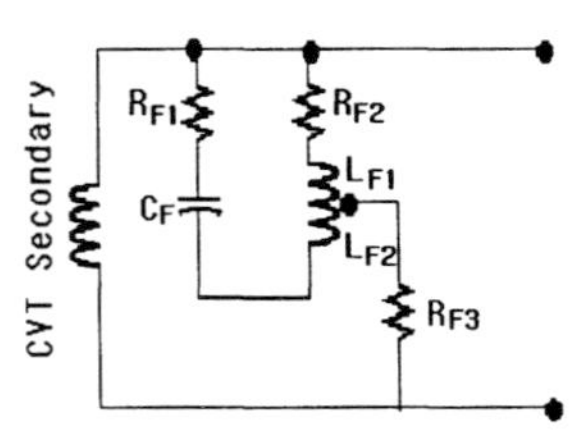

R_{F1}=1.06[ohm], R_{F2}=4.24[ohm], R_F=40.0[ohm]

L_{F1}=0.01[H], L_{F2}=0.394[H], C_F=8.0[μ F]

Fig.13. Ferroresonance-suppression filter

These kinds of tests are very useful to the relay engineer but it is very hard and dangerous to test with real devices because of the ferroresonance.

5.2 Test results of relay performance

Figure 14 shows the calculated impedance of the distance relay model during CVT transients. The active ferroresonance-suppression circuit is adopted. The circuit acts like a band-pass filter and thereby introduces extra time delay in the CVT secondary output. When a fault is applied at the end of the radial line, relays can not pick up on time due to CVT transient errors.

Figure 15 shows the distance relay performance as a function of CVT burden. The CVT transient characteristics are influenced by the magnitude and angle of the connected burden. Solid-state and digital relays have very small burden. The burden of these relays is nearly resistive. When using a CVT, the relay engineers need to calculate the total burden of all devices connected to the CVT and make sure the burden is not excessive to assure proper distance relay protection.

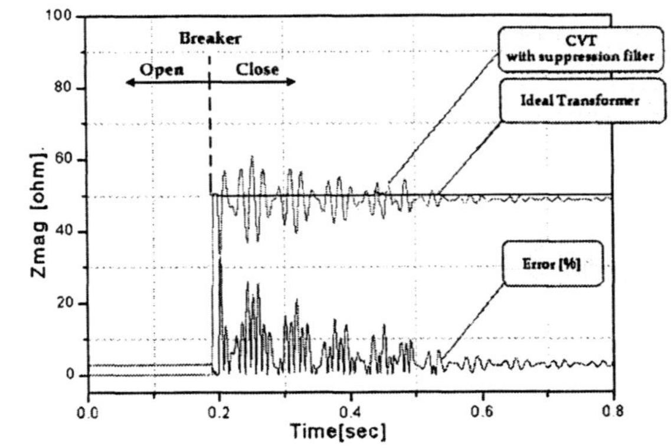

Fig.14. Magnitude of Impedance during transients

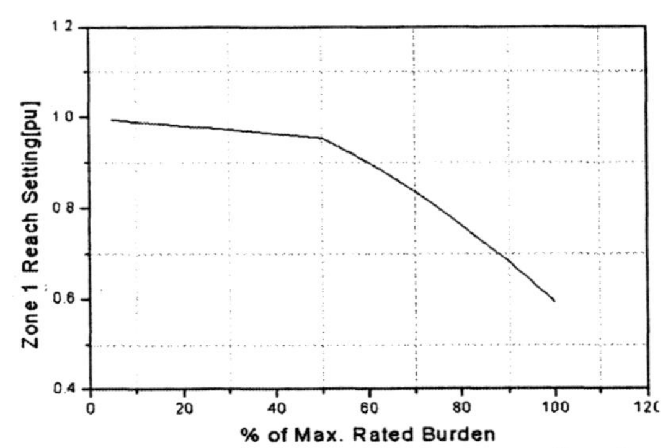

Fig.15. Distance relay performance as a function for CVT burdens

6. CONCLUSION

This paper describes how to implement and verify the software distance relay model that can be running on RTDS(Real Time Digital Simulator). They can operate properly around their zone boundaries. The maximum error of impedance calculation is below about 1 % and the total operation time is about 5.0e-2 msec. It is possible to represent protection system effectively on RTDS because the software model is stand-alone and can be used with real relays. It is also possible to conduct interactive real-time tests when the system effects of the relay action need to be investigated.

REFERENCES

McLauren P. G. (1992). A Real Time Digital Simulator for Testing Relay . *IEEE*. Vol .7, No. 1, pp207-213.

McLauren P. G., Swift. G. S., Dirks, E. N., Jayasinghe, R.P. and Pernando, I (1995). An accurate software model for off-line assessment of a digital relay. Stockholm Power Tech.

McLauren P. G. (1995). Using Real Time Digital Simulator to Develop an Accurate Model of a Digital Relay. *ICDS*.

Tutorial on Electromagnetic Transient Program Applications to Power System Protection, *IEEE PES*, 01TP150

ELSEVIER
IFAC
PUBLICATIONS
www.elsevier.com/locate/ifac

SHORT-TERM REACTIVE POWER LOAD FORECASTING USING MULTIPLE TIME-SERIES MODEL

Jong-Man Cho, Jin-Hack Kim
Woo-Hyun Park, Yun-ho Lee, Jin-O Kim

Korea Power Exchange, Seoul, KOREA
Dept. of Electrical Engineering, Hanyang University, Seoul, KOREA

Abstract: This paper shows that the loads of active and reactive power have a significant positive relationship and there exist two types of relationship between them. In investigating the cross plots at every hour, it is found out that their relationship is linear from 0 to 8 hours, while they have two piece-wise linear curves from 9 to 23-rd hours. Also, reactive power load is estimated and forecasted using active power load as an explanatory variable with OLS (Ordinary Least Squares) regression methods. MAPE (Mean Absolute Percentage Error) for each model is calculated for the evaluation of one-hour ahead forecasting. *Copyright © 2003 IFAC*

Keywords: short-term reactive power load forecasting, OLS, threshold, MAPE

1. INTRODUCTION

Lots of studies and researches have been made in various fields for developing reliability and stability of power supply with the methods for decreasing cost and forecasting models for active power loads. However, forecasting has been mainly focused on active power load. In spite of the fact that reactive power load plays an important role because of the strong relationship between system voltage and reactive power loads, studies and researches on the forecasting of reactive power load and its relationship with active power have been seldom made. Therefore, in this paper, reactive power load is forecasted using the relationship between active and reactive power loads.

In the first part of this paper, we introduce the methods of calculating active and reactive power loads and figure out their relationship with scatter plot. Also, we apply the regression results of reactive power load on the calculations of the power factor, and the effective amount of fixed and switched capacitance. In the second part of this paper, we estimate and forecast reactive power load using active power load as the explanatory variable. Finally, we calculate MAPE (Mean Absolute Percentage Error) with out-of-sample data at each hour to evaluate the effectiveness of reactive power forecasting.

The test data used in the case study was supplied by Shin In-Chon Substation in KEPCO system from January 2002 through August 2002.

2. RELATIONSHIPS BETWEEN ACTIVE AND REACTIVE POWER

2.1 Calculation of Reactive Power Load

To investigate the correlation of active and reactive power, first it is necessary to calculate the reactive power at load side, i.e., pure reactive power load. Active and reactive power measured from EMS are only the amount installed into secondary winding of the transformer (most likely 22.9 kV side), which includes reactive power supplied from static condenser at 22.9 kV side. Therefore, pure reactive power load (Q_L) before compensation is the value subtracted compensation amount of reactive power from metered reactive power. Fig.1 shows the structure of 154 kV substations, which describes the relation of measured and pure reactive power load.

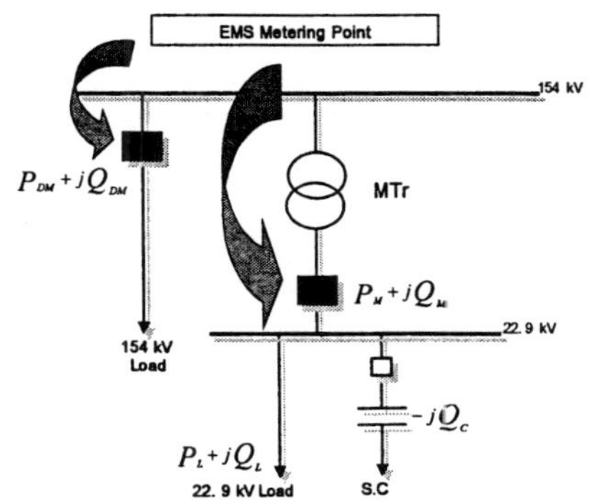

Fig. 1. Structure of 154 kV substations

Total active and reactive powers at the bus of 154 kV are expressed as follows:

$$P_T = \sum P_{DM} + \sum P_M = \sum P_{DM} + \sum P_L \tag{1}$$

$$Q_T = \sum Q_{DM} + \sum Q_M = \sum Q_{DM} + \sum Q_L + \sum Q_C \tag{2}$$

Therefore, total pure reactive power load can be calculated as;

$$Q_L = \Sigma Q_{DM} + \Sigma Q_M - \Sigma Q_C \quad (3)$$

2.2 Regression between Reactive and Active power

To investigate the correlation of active and reactive power, first we need to show the relationship between total active power (P_T) and total reactive power after compensation (Q_T), and between P_T and pure reactive power load before compensation (Q_L).

Observation of data for two variables on the scatter plot shows the correlation of them. There is positive linear relationship between P_T and Q_T in Fig. 2(a), and their relationship can be verified in Fig. 2(b), where the trend of P_T and Q_T have similar patterns from January to February 2002.

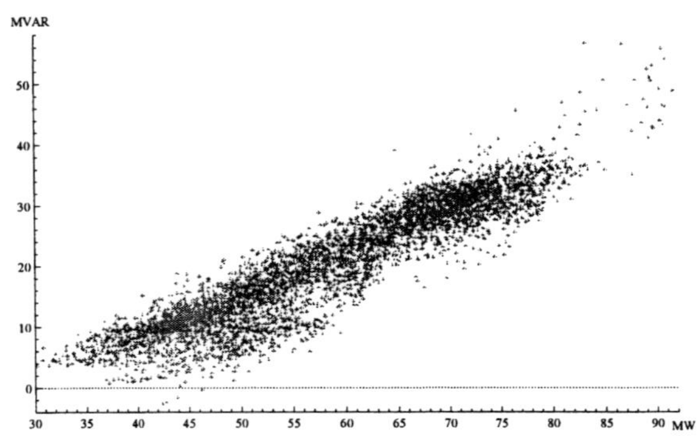

(a) Scatter Plot of P_T and Q_T

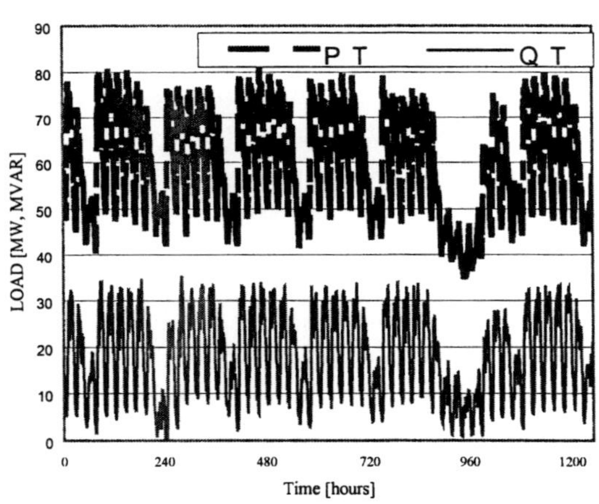

(b) Comparison of P_T and Q_T (Jan. 1, to Feb. 28, 2002.)

Fig. 2. Correlation of P_T and Q_T

In order to examine which regression model is proper for the forecasting performance, several test statistics were used in this paper such as Coefficient of Determination (denoted in R^2), Durbin-Watson statistic (DW), standard error (SE), and pooled standard error (Pooled SE).

R^2 is used for measuring the proportion of the variance of the dependent variable about its mean, which is explained by the independent variables. It can vary between 0 and 1, and the regression model is said to be proper as its R^2 value is higher. DW is most popular to test the hypothesis which residuals are white noise, i.e. $e_t \sim$ i.i.d $(0, \sigma^2)$. It is known that residuals are independent each other as DW is close to 2 in a regression model. Pooled SE is obtained by using the weighted average of SE according to degree of freedom for each model.

For a close examination, regression in linear model is used and the result is as follows.

$$Q_T = 0.58846 P_T - 11.606 \quad (4)$$

In this case, R^2 =0.52, DW=0.733, and the SEs corresponding to the coefficients of 0.58846 and -11.606 are 0.8413 and 0.0119, respectively.

Fig. 3 is the scatter plot of between P_T and Q_L. The relationship between P_T and Q_L has non-linear relationship, which is divided into upper and lower regimes based on critical threshold value. Comparing Fig. 2(a) with Fig. 3, it can be shown that compensation of reactive power is executed only in the upper regime.

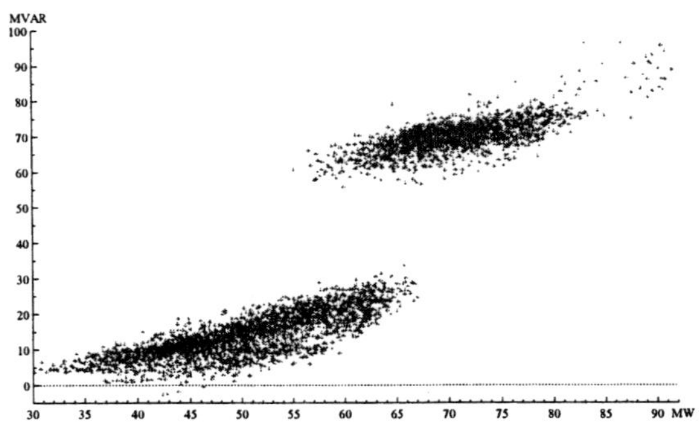

Fig. 3. Scatter Plot of P_T and Q_L

Again, a linear model is used for regression of upper regime and the result is as follows

$$Q_T = 0.5941 P_T + 27.924 \quad (5)$$

where R^2 =0.5127, DW=0.726, and the SEs corresponding to the coefficients of 0.5941 and 27.924 are 0.0122 and 0.86302, respectively.

Let us denote the active power and reactive power load at time *t*, by x_t and y_t^*, respectively. Under the assumption of conformality, the following holds

$$y_t^* = \tan(\theta) x_t \quad (6)$$

where $\cos(\theta)$ is the power factor of the composite feeder loads. The net reactive power at time *t*, y_t, can be expressed as follows:

$$\begin{aligned} y_t &= y_t^* - (C_0 + C_t) \\ &= \tan(\theta) x_t - (C_0 + C_t) \quad (7) \end{aligned}$$

where

C_o: total fixed reactive power, which is not varying as active power load changes such as that of the fixed capacitors.

C_t: total capacitive reactive power due to the switched capacitors at time t.

The term C_t can be expressed in terms of each individual capacitor output in the following way,

$$C_t = \sum_{i=1}^{n_s} I_{ct}^{0} \beta_{t,i} \;\; ; \;\; \beta_{t,i} = \{0 \; or \; 1\} \tag{8}$$

where

I_{ct}^{o}: reactive power rating of the i-th capacitor on the feeder

$\beta_{t,i}$: on/off status of i-th switched capacitor at time t

n_s: number of switched capacitors.

Consequently, y_t can be expressed as follows:

$$y_t = \begin{cases} \tan(\theta) x_t - C_o & if \; \beta_{t,i} = 0 \\ \tan(\theta) x_t - (C_o + C_t) & if \; \beta_{t,i} = 1 \end{cases} \tag{9}$$

A linear regression analysis between y_t and x_t can afford the value of variable $\tan(\theta)$ and calculate such as power factor of the feeder loads, the total amount of the fixed and switched reactive power components. The results of linear regression analyses of the feeder's data in Table 1 are given by Fig. 2(a) and Fig. 3.

Table 1. The regression result of Fig. 2(a) and Fig. 3

	$\tan(\theta)$	C_o	$C_o + C_t$	P.F
Figure 2	0.58846		-11.606	0.862
Figure 3	0.59405	27.924		0.859

F.C (Fixed Capacitor) ≈27.924 MVAR
S.C (Switched Capacitor) ≈ 39.53 MVAR

The sign of C_o is an indication of the reactive magnitudes of inductive to capacitive reactive power existing on feeder.

2.3 Forecasting procedures

Before we perform the load forecasting using multiple time-series model as shown in eqs. (11) and (12), to obtain the confidential forecasting results, there remain the problem how we determine the lag lengths (in other words, how we determine N_d and N_h). In order to solve this problem, this paper uses the t-value test statistics. The t-value of each coefficient can be used for the selection of significant input variables. Input variables whose t-values are larger than the critical t-value at a specified significance level have significant effect on the endogenous variable. The t-value of each regression coefficient is defined as

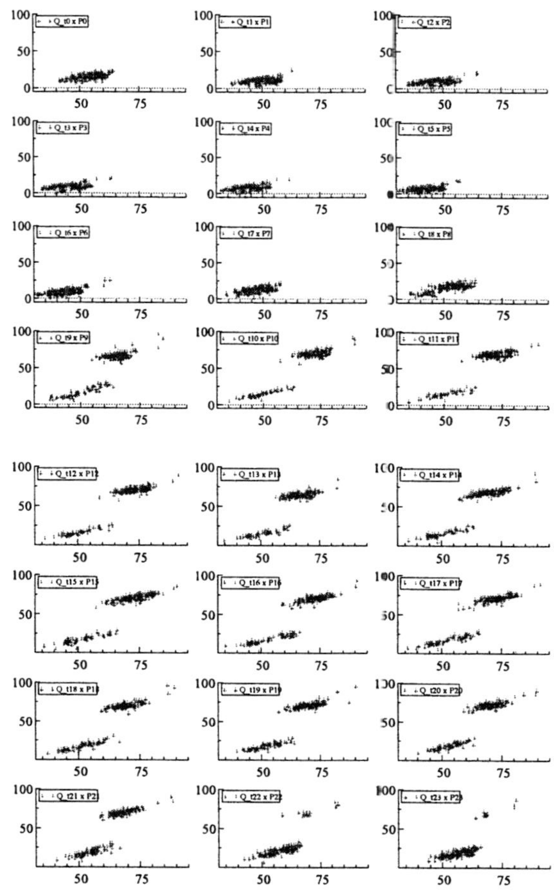

Fig. 4. P_T And Q_L in the scatter plot at each hour

$$t-value_{l\in N_K} = \frac{\hat{\beta}_l}{\sigma\sqrt{d_{ll}}} \tag{10}$$

where $\hat{\beta}_l$ is the estimate of the l-th regression coefficient by OLS algorithm, d_{ll} is the l-th diagonal element of, $(X^T X)^{-1}$ and σ is the standard error of the regression. The critical t-values for some significance levels are described in Table 2.

Table 2. Critical t-value for each significance level

Significance Level [%]	5	10	20
Critical t-value	1.96	1.645	1.28

To forecast Q_L, the relationship between P_T and Q_L is shown in the scatter plot in each hour. In Fig. 4, you can conform there are two forms of scatter plot. One form has liner relationship between two variables from 0 to 8-th hour. The other form has non-linear relationship from 9 to 23-rd hour, which has upper regime and lower regime based on critical threshold value. For those reasons, the form from 0 to 8 hour was forecasted by using linear regression. In addition, used P_T for explanatory variables to increase the prediction. The form from 9 to 23-rd hour found out the threshold value, which has the lowest value of errors. After that, we found the each lower and upper regimes' linear function, then used to two-step estimation for Q_L 's prediction. Due to this reason, two variables can be predicted by threshold value, which has non-linear relationship. In other words, if P_T can be predicted by using this then Q_L can be predicted, too.

From the Fig. 5, this predicts process can be divided into two. Method 1 is showing prediction of P_T and Method 2 showing two variables has either linear relationship or non-linear relationship. From this process, if the output comes out as linear relationship then use linear model to forecast. In other hand, if the output comes out as non-linear relationship then find out threshold value and use P_T from method 1 to predict Q_L.

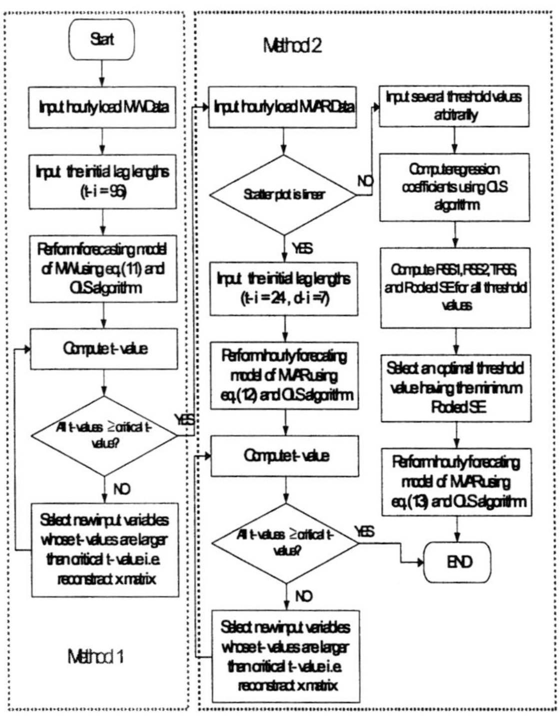

Fig. 5. Flowchart for forecasting reactive power load

As shown in Fig. 2(a), P_T and Q_L have linear relationship. As a result, P_T forecasting model can be made as follow.

$$P_T = f(P_{T(t-N_h)}, Q_{T(t-N_h)}) = a_o + \sum_{i=1}^{N_h} b_i P_{T(t-i)} + \sum_{j=1}^{N_h} c_j Q_{T(t-j)} \quad (11)$$

where the following notation is used: h is the hour of the forecast, $P_{T(t-i)}$ and $Q_{T(t-j)}$ is historical load data, a_0 , b_i and c_j are regression coefficients for each variable, N_h is hourly lag length. The input variables of the proposed model consist of the historical hourly load data. The initial numbers of the hourly lag lengths (N_h) is 96. Fig. 5 shows the flow of hourly load forecasting process at a specified significance level. In this flow, forecasting can be performed in *t*-static. When all *t*-values of the coefficients are larger than the critical *t*-value of particular significance level, forecasting performances stop.

The initial numbers of the daily and hourly lag lengths, N_d and N_h have 7 and 24, respectively. Forecasting model for Q_L can be made as below.

[Forecasting form for 0 to 8-th hour]

$$Q_L = f(\hat{P}_T, P_{T(t-N_h)}, P_{T(d-N_d)}, Q_{L(t-N_h)}, Q_{L(d-N_d)}) = a_o + a_1 \hat{P}_T + \sum_{l=1}^{N_h} b_l P_{T(t-l)} + \sum_{m=1}^{N_d} c_m P_{T(d-m)} + \sum_{n=1}^{N_h} d_n Q_{L(t-n)} + \sum_{j=1}^{N_d} e_j Q_{L(d-j)} \quad (12)$$

where $P_{T(d-N_d)}$ and $Q_{L(d-N_d)}$ is historical load data, $\hat{P}_T$ is forecasting P_T , a_0 , a_1 , b_l , c_m , d_n and e_j are regression coefficients for each variable, N_d is daily lag length.

[Forecasting form for 9 to 23-rd hour]

$$Q_L = \begin{cases} a\hat{P}_T + b & \text{if } \hat{P}_T > A \\ c\hat{P}_T + d & \text{if } \hat{P}_T \le A \end{cases} \quad (13)$$

where

A : Threshold value

$\hat{P}_T$: Threshold value and forecasting value

a, b, c, d : Regression coefficients

Fig. 6 shows the process of the selection of optimal threshold active power. To obtain the optimal threshold value, we arbitrarily choose several values of P_T . For each arbitrary threshold value, Q_L were regressed on P_T for lower and upper regimes and RSS of each regime was obtained, which are called RSS1 and RSS2 in this paper, respectively. Pooled SE can be obtained with total residual sum of squares (TRSS). Next, Pooled SE for each threshold value was compared with each other and optimal threshold value that has the minimum Pooled SE was chosen.

$$\textit{Pooled SE} = (\textit{TRSS} / \textit{Degess of freedom})^{0.5} \quad (14)$$

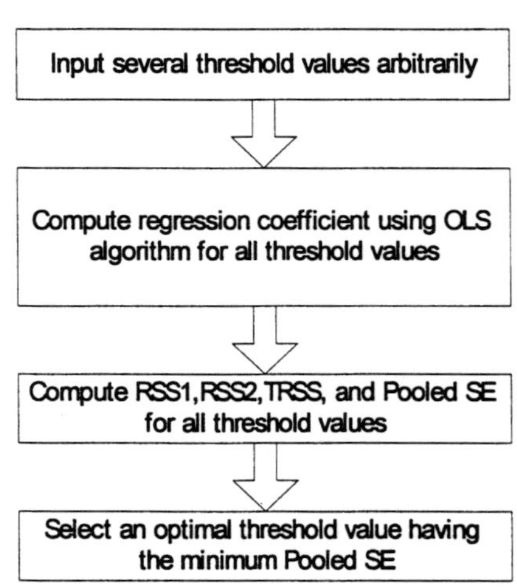

Fig. 6. Procedure of the selection of optimal threshold value

The forecasting accuracy of the proposed model was evaluated by the average of absolute percentage errors of the hours in the forecasting period, and the mean absolute percentage error (MAPE) is computed by

$$MAPE[\%] = \frac{1}{N}\sum_{n=1}^{N}\frac{|Actual(n) - Forecast(n)|}{Actual(n)} \times 100 \qquad (15)$$

where Actual and Forecast mean the actual and the forecast reactive power in the forecasting period, respectively, and N is the number of hours in the forecasting period.

2.4 *Output of Regression Result*

P_T output of regression analysis can be shown Table 3. It is obtained from application of the proposed multiple time series mode of P_T using the Q_T as the explanatory variable.

Table 3. The regression results of P_T

R^2	DW	Sigma	MAPE [%]
0.9894	2.00	1.252	1.6411

The Table 3 shows that a proposed multiple time series model can give a very accurate forecast. In this case, MAPE is 1.64%

Table 4 summarizes the results of Pooled SE for each arbitrary threshold value. The range of the threshold value of Q_L is 56 to 67.

Table 4. Selection of optimal threshold value at each hour

Time	Threshold value (MW)	RSS1	RSS2	TRSS	Pooled SE
9	57	99.120	234.221	333.341	4.083
10	65	35.880	161.420	197.301	3.141
11	64	20.543	76.807	97.350	2.206
12	65	41.912	79.715	121.627	2.466
13	58	30.014	86.667	116.681	2.415
14	56	33.813	92.018	125.832	2.508
15	60	26.082	67.115	93.932	2.167
16	65	25.528	90.979	116.507	2.414
17	64	11.695	84.848	96.543	2.197
18	63	27.964	55.852	83.816	2.047
19	63	36.362	74.632	110.994	2.356
20	63	24.419	97.451	121.870	2.468
21	62	32.479	185.173	217.651	3.299
22	67	150.396	7.593	157.999	2.811
23	67	187.977	3.070	191.047	3.091

Output of Q_L from regression analysis can be shown in Table 5. It is obtained from application of the proposed multiple time series mode and non-linear model of Q_L using P_T as the explanatory variable. In this paper, all MAPEs value have forecasted outcome, which has 10% of total observation data.

Table 5. Forecasting errors of Q_L at each hour

Time	Minimum Error	Maximum Error	MAPE[%]
0	0.026	1.153	5.214
1	0.089	0.609	5.348
2	0.044	1.225	7.049
3	0.037	0.482	2.551
4	0.006	0.460	2.906
5	0.024	0.510	2.938
6	0.008	1.513	7.212
7	0.040	1.043	3.964
8	0.025	0.980	3.404
9	0.215	7.795	8.384
10	0.285	5.276	7.555
11	0.048	4.298	4.572
12	0.018	4.572	5.607
13	0.735	4.213	5.448
14	0.020	4.757	5.870
15	0.143	4.077	4.346
16	0.126	6.660	4.219
17	0.346	5.399	4.045
18	0.152	4.336	4.121
19	0.078	3.847	5.342
20	0.074	4.944	4.248
21	0.096	9.436	5.836
22	0.662	5.338	10.126
23	0.301	5.246	13.490
Average	0.149	3.674	5.575

3. CONCLUSION

In preparing the first research on relationship between active and reactive power load we found out that there are two types of relationships; from 0 to 8 hour, they are linear, during 9-th to 23-rd hour, they are two piece-wise linear. Thresholds of the latter are ranged about 56 to 67 values of active power loads. We estimated and forecasted reactive power by using active power as explanatory variable with OLS method. We calculated MAPE with out-of-sample data and the resulted average of MAPE was 5.575[%]. In the process of forecasting reactive power, our results can give strong implication to future studies about forecasting active power loads if they include reactive power data as an explanatory variable to forecast active power loads, they will get better results. Our further study will be focused on more detailed specifications for forecasting reactive power, which includes other explanatory variables such as temperature and humidity, etc.

ACKNOWLEDGEMENT

This work was supported by the research fund of Korea Power Exchange

REFERENCES

Bruce L. Bowerman and Richard T. O'Connell, 'Time Series Forecasting: Unified concepts and Computer Implementation", Duxbury Press, Boston, 1987

Joseph F. Hair, Rolph E. Anderson, and Ronald L. Tatham, William C. Black, "Multivariate Data Analysis", Prentice Hall, Inc., New Jersey, 1998

Korea Power Exchange (KPX), "A Study on the Voltage Operation and Determination Methods of Load Power Factor for the Power Transmission System", July 2002

Kyung-Hun Lee, Yunho Lee, Jin-O Kim, "TAR (Threshold Autoregressive) Model for short-Term Load Forecasting Using Nonlinearity of Temperature and Load.", KIEE. Trans. Vol. 50A, No. 9, September.2001

M. T Hagan and S. M. Behr, "The Time Series Approach to Short-Term Load Forecasting", IEEE Trans. On Power Systems, Vol. 2, No. 3, pp.785~791,1987

S. Civanlar and J.J. Grainger, "Forecasting Distribution Feeder Loads: Modelling and Application to Volt/Var Control.", IEEE Trans. On Power Delivery, Vol. 3, No. 1, January 1988

S. Civanlar and J.J. Grainger, "Volt/Var Control on Distribution Systems with Lateral Branches Using Switched Capacitors and Voltage Regulators, Part 1: The Problem", IEEE Tran. On Power Apparatus and Systems, November 1985, pp.3278-3283

S. Civanlar and J.J. Grainger, "Volt/Var Control on Distribution Systems with Lateral Branches Using Switched Capacitors and Voltage Regulators, Part 2: The Problem", IEEE Tran. On Power Apparatus and Systems, November 1985, pp.3284-3290

S. Civanlar and J.J. Grainger, "Volt/Var Control on Distribution Systems with Lateral Branches Using Switched Capacitors and Voltage Regulators, Part 3: The Problem", IEEE Tran. On Power Apparatus and Systems, November 1985, pp.3291-3297

ELSEVIER
IFAC
PUBLICATIONS
www.elsevier.com/locate/ifac

IMPLEMENTATION OF INTELLIGENT SMART MODULE FOR PREDICTION OF THE REMAINED BATTERY CAPACITY

Byung-Woo Jung, Sung-Sik Kim, Jung-Mo Yang, Hyun-Chan Cho, and Yong-Taek Oh,

School of Information Technology, Korea University of Technology and Education.

3-304 ho, CIS Lab, First Engineering Building, KUT,
Byung-chun myun , Chun-an city, Chung-nam do, 300-708, South, Korea.
Phone : +82-41-560-1171, Fax : +82-41-564-3261.
jungbw92@empal.com

Abstract --- In this paper, We propose a implementation of an intelligent smart module to predict a decrease in a capacity and a remain capacity by a cycle life of Li-ion battery. In Li-ion battery it is hard to predict the remain capacity because characteristic of output voltage is not linear and the capacity for charge is decreasing as charge and discharge. Therefore this intelligent smart module predicts the remain capacity with high accuracy in Li-ion battery by an algorithm for the remain capacity using fuzzy logic and a data acquisition board for measuring voltage, current and temperature. And this paper compares this results using fuzzy logic with the charge-discharge experimental results. *Copyright © 2003 IFAC*

1. INTRODUCTION

There are several Li-ion battery characteristics such as cycle life, charge-discharge characteristic, thermal characteristic, self-discharge characteristic (charge retention), and recovered capacity, etc. In these characteristics the cycle life by charge-discharge reduces full charge capacity (FCC) of battery and the discharge characteristic is non-linear. A remain capacity in Li-ion battery can be measured by converting voltage into frequency and monitoring voltage with frequency. Or it can be measured by numerical operation with counting the number of charge and discharge and considering current. But it is hard to monitor the exact remain capacity with these methods because they do not consider non-linear discharge characteristic, self-discharge characteristic, and difference between theoretical value and real decrease in the capacity by cycle life.[1]

This paper suggests the intelligent smart module using fuzzy logic control to solve these problems and shows this usefulness with the experimental results. The experiment in this paper was done as below.

At first the data acquisition board was made for measuring exact voltage, current and temperature of Li-ion battery. Using this board the charge-discharge experimental data were collected.

And second non-linear discharge characteristic of Li-ion battery was solved by fuzzy logic.

At last the decrease in the battery capacity by cycle life, temperature, etc. were calculated using fuzzy estimator. And the experimental results and results by fuzzy logic were compared.

2. CHARACTERISTICS OF LI-ION BATTERY

This chapter shows the charge-discharge characteristics, thermal characteristic, cycle life characteristic, self-discharge (charge retention), and recovered capacity, etc. of Li-ion battery ICR18650.

The charge was done by constant current (CC) at first and then constant voltage (CV).

Discharge voltage was not settled at fixed value. Figure.1 shows the charge and discharge characteristics of Li-ion battery. [1][2]

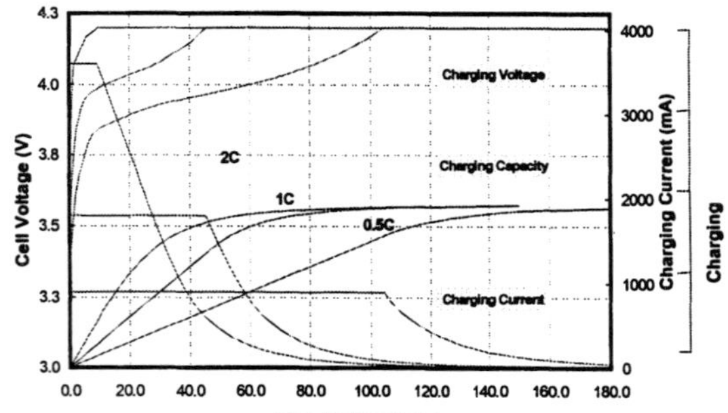

(a) Charge characteristic

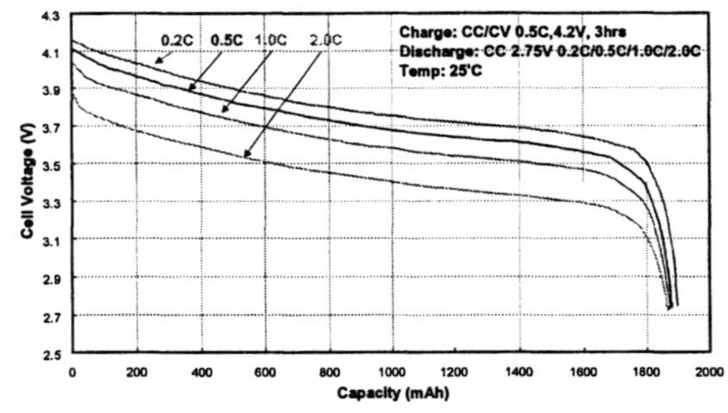

(b) Discharge characteristic

Figure. 1. The charge and discharge characteristics of Li-ion battery

In general the battery capacity is increasing as temperature is increasing, but the cycle life is rapidly decreasing. The capacity of rechargeable batteries is decreasing as the charge-discharge is being repeated. The cycle life is defined as the number of charge-discharge until the battery capacity decrease up to 80% of the initial capacity. Figure.2 shows thermal characteristic and cycle life of Li-ion battery.

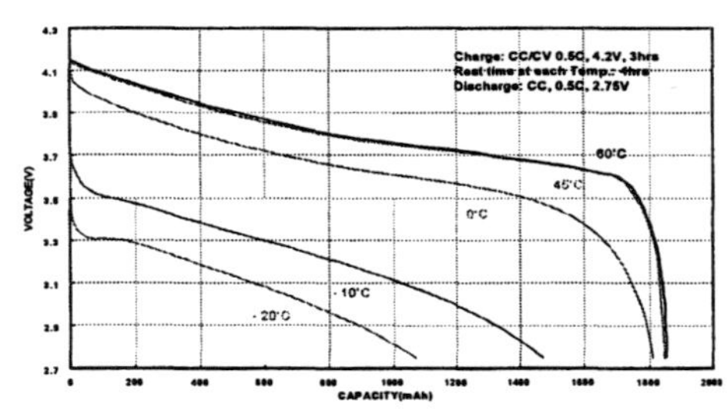

(a) Thermal characteristic [0.5C]

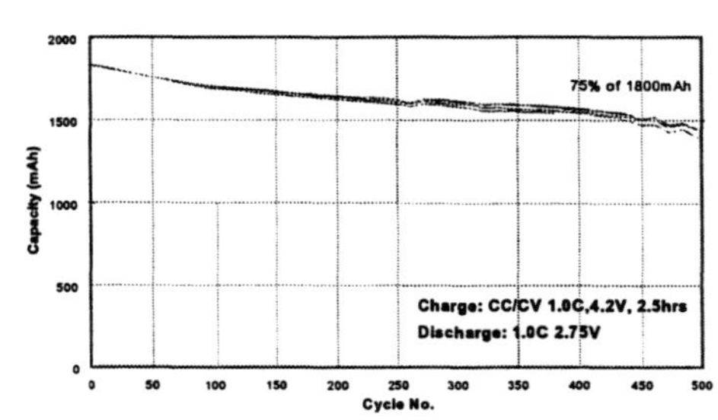

(b) Cycle life

Figure. 2. Thermal characteristic and cycle life of Li-ion battery

The self-discharge is defined as a phenomenon that the battery capacity decreases by Li-ion battery chemical reaction. This can be evaluated as "Retention Capacity (or Charge Retention)" and "Recovered Capacity". These terms show how much capacity are remained and decrease after being left for long time. Figure.3 shows the self-discharge characteristic curve after full charge and a recovery characteristic curve that shows a rate of the valid battery capacity to the initial capacity when the battery is recharged. [2]

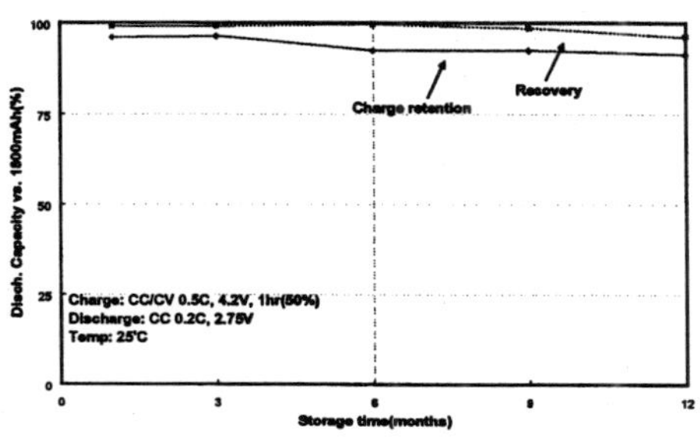

Figure. 3. Charge retention and recovered capacity

3. DESIGN OF AN INTELLIGENT SMART MODULE

3. 1 Data acquisition board and experimental equipments

In this paper the data acquisition board was made and used. We had to detect very small voltage such as 4.2×1/10,000[V] for the Li-ion battery used in this paper. But there was no rapid voltage change. So we selected low speed but high resolution AD converter. To get a data about FCC we saved the voltage value in the PC at real time while charging a new battery and an old battery using constant current. And we got the decrease in Li-ion battery tank capacity using fuzzy estimator with the data in PC. Figure.4 is the data acquisition board and charge-discharge experimental equipment. [8]

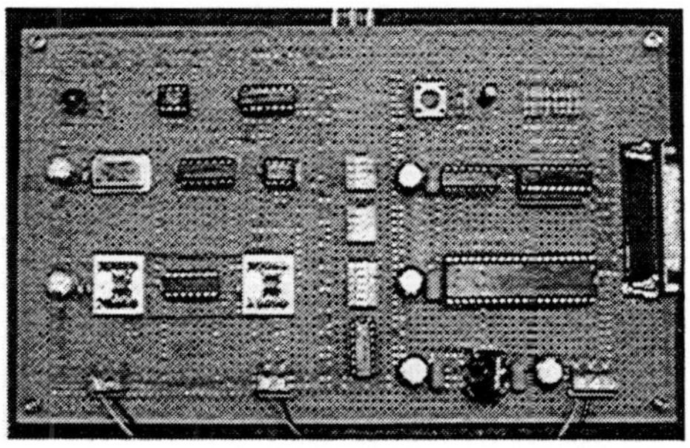

(a) Data acquisition board

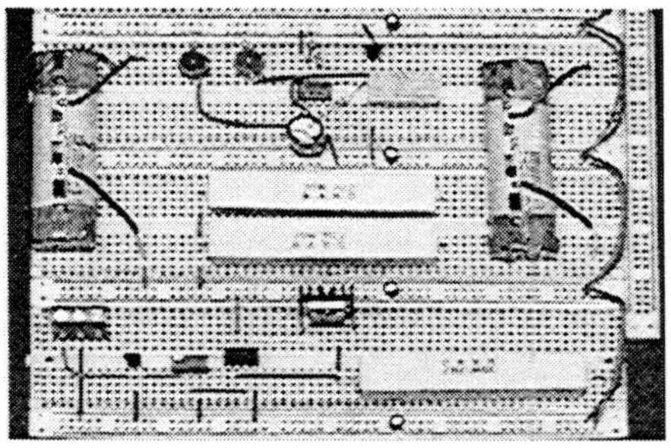

(b) Charge-discharge experimental equipment

Figure. 4. Data acquisition board and experimental equipment

3.2 Intelligent smart module system

Figure.5 is a block diagram of Intelligent smart module including fuzzy algorithm to measure the remain capacity of Li-ion battery.

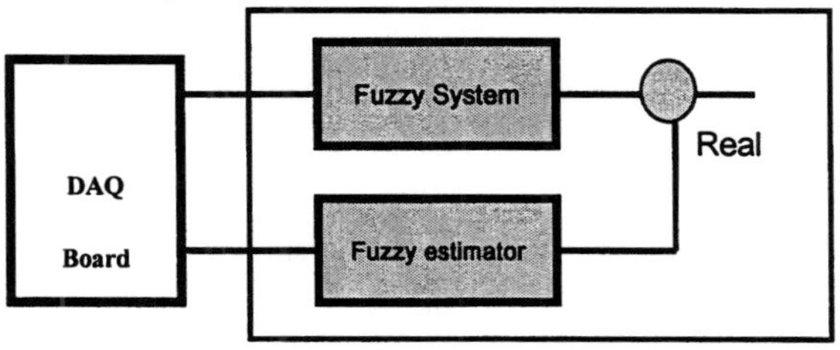

Figure. 5.
Block diagram of Intelligent smart module

The algorithm is as follows for the Fuzzy System.

Step 1. Fuzzifier

To apply fuzzy algorithm we changed the crisp value to proper linguistic variable. There are 2 inputs and 1 output. V and dV are inputs and RC is output. [3] The membership function of each input and output is Figure.6, Figure.7, and Fugire.8 And the discharge characteristic of Li-ion battery is so non-linear that the membership function regions isn't regulating. For example, the discharge voltage of Li-ion battery rapidly decrease from 3.3V to 2.5V.

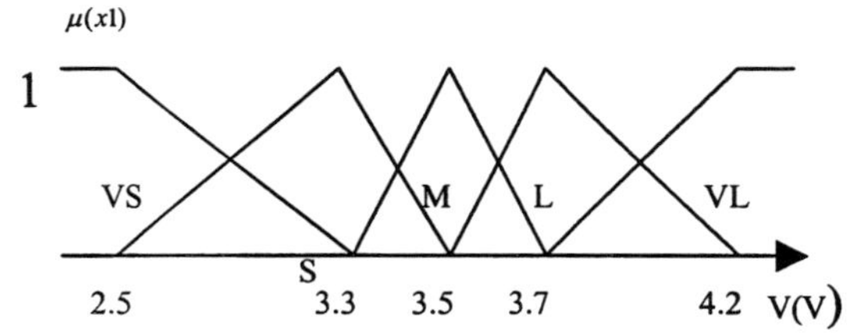

Figure. 6. Membership function of V

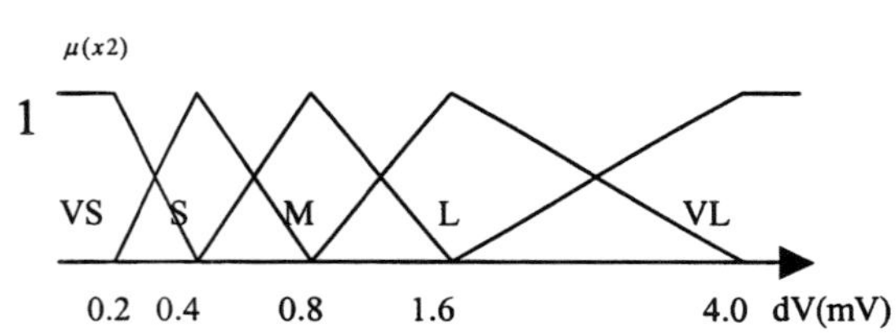

Figure. 7. Membership function of dV

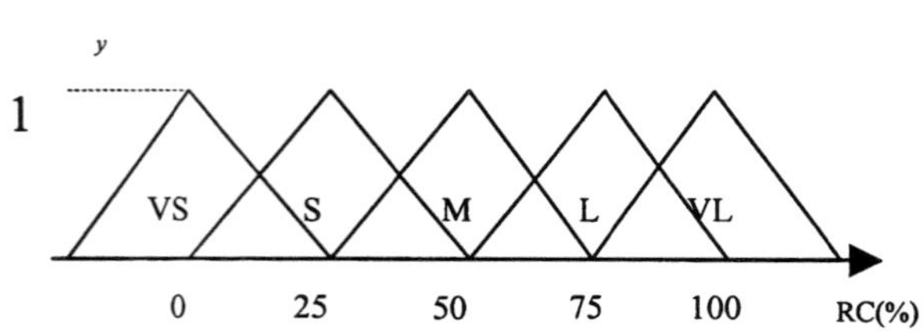

Figure. 8. Membership function of RC

The linguistic variables of Figure-6, Figure-7, and Figure-8 are defined in Table-1

Table-1 Definition of the linguistic variables used in fuzzy logic

Variables	VS	S	M	L	VL
Definition	Very Small	Small	Medium	Large	Very Large

Step 2. Fuzzy rules and fuzzy inference

Table-2 shows fuzzy rules for output RC. Mamdani's max-min compositional rule is used and fuzzy logic algorithm is applied for fuzzy approximate reasoning. [4][5]

Table-2 Fuzzy rules

RC		V				
		VS	S	M	L	VL
dV	VS	S	M	L	VL	VL
	S	VS	S	M	L	VL
	M	VS	S	M	L	L
	L	VS	S	M	L	L
	VL	VS	S	S	M	L

Step 3. Defuzzifier and fuzzy estimator

Simplified center of gravity such like equation (1) is used for getting the remain capacity output using defuzzifier. [6]

$$y^{*COG} = \frac{\sum_{i=1}^{5} \mu(y_i) \times y_i}{\sum_{i=1}^{5} \mu(y_i)} \qquad (1)$$

Fuzzy estimator is for evaluating the oldness of Li-ion battery. Fuzzy estimator using PGU calculates the cycle life and tank capacity of battery, which is changed because of the self-discharge characteristic. Figure-9 shows a graph of the derivative of voltage versus current, and the fuzzy estimator. [7]

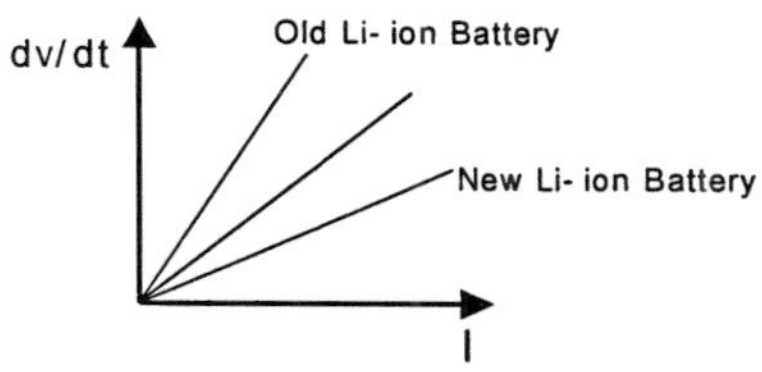

(a) Graph of the derivative of voltage versus current

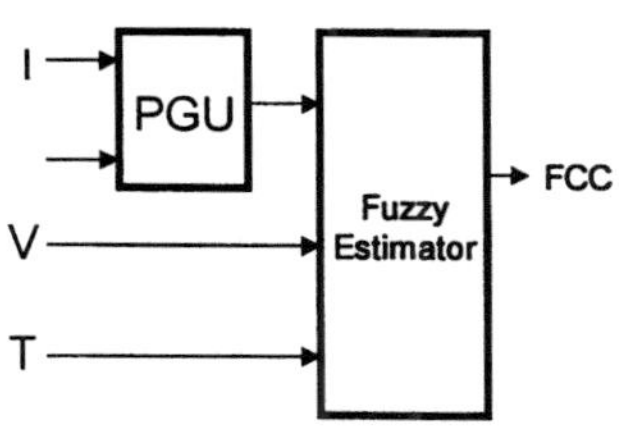

(b) Fuzzy estimator

Figure. 9. Graph of the derivative of voltage versus current, and fuzzy estimator

4. EXPERIMENTAL RESULT

We saved the charge-discharge experimental data into a personal computer. We compared that data in PC with the results from fuzzy system, and the results from fuzzy system with fuzzy estimator to show the usefulness of this system for predicting the remain capacity of Li-ion battery.

4. 1 Results by fuzzy algorithm

When we applied only fuzzy algorithm without fuzzy estimator to this experiment, we got the results as below. For the new battery the results from fuzzy algorithm have 2% error from the charge-discharge experimental results. For the old battery the results from fuzzy algorithm have 3.1% error from the charge-discharge experimental results. But the tank capacity of the old battery is reduced by 4% from the new battery. When we compare the old battery to the new one with considering this decrease in the capacity, we can say there is 7.1% error. This 7.1% error comes from the decrease in the tank capacity and fuzzy system error. As the battery is older, this error becomes bigger. Figure.11 shows the results from fuzzy system .

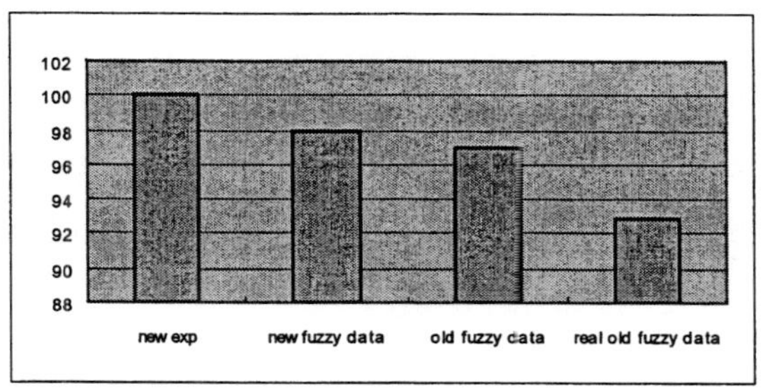

Figure. 11. Results from fuzzy system

4.2 The results from fuzzy system with fuzzy estimator

When we added fuzzy estimator to this experiment, we got the results as below. There is no battery aging for the new battery, so FCC is 1. The results from fuzzy system with fuzzy estimator have small error of 1.1% from the charge-discharge experimental results. For the old battery we got the FCC value as 0.963. The results from fuzzy system with fuzzy estimator have 1.5% error from the charge-discharge experimental results. Therefore when we use fuzzy system with fuzzy estimator, we can have small error because the decrease in the tank capacity is considered. Without fuzzy estimator error was high because we had no FCC value.

Figure.12 shows the results from fuzzy system with fuzzy estimator.

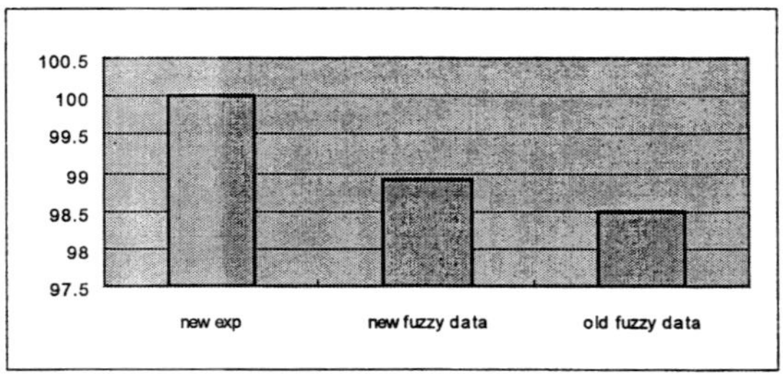

Figure. 12. Results from fuzzy algorithm with fuzzy estimator

5. CONCLUSION

It is hard to measure the remain capacity of Li-ion battery because Li-ion battery has non-linear discharge characteristic and the capacity for charge is reduced by the repeated charge-discharge and the self-discharge characteristic. For this problem we made the data acquisition board with high resolution and designed the intelligent smart module using fuzzy system with fuzzy estimator. When only fuzzy system without fuzzy estimator was applied to the intelligent smart module there were good results for the new battery. In the case of the old battery there was an error and this problem was solved by using fuzzy system with fuzzy estimator. In this paper we experimented using constant current load. So we are experimenting using not constant current load but the real variable load and we are trying to reduce an error for the high accuracy system. And we are researching for the concrete application

ACKNOWLEDMENTS

This work was support by Grant No.R05-2000-000250-0(2002) from the Basic Research Program of the Korea Science and Engineering Foundation.

.

REFERENCES

[1] A.Nishimoto, M. Watanabe, Y.Ikeda, S Kohjiya, Electrochimica Acta (1998)

[2] Samsung SDI, Performances of the Cylindrical Li-ion ICR18650-18 (2000)

[3] Lee, C. C., "Fuzzy logic in control system: Fuzzy logic controller, Parts I and II.", IEEE Trans. Syst. Cybern., Vol 20, No. 2, pp 404-435, 1990.

[4] Wasserman, P.D., Neural Computing: Theory and Practice, Van Nostrand Reinhold, NY.

[5] Rao, V.B. and Rao, H. V., "C++ Neural Networks and Fuzzy Logic", 2nd ed. MIS press, NY, 1995.

[6] S. Chae, Y. S. Oh, "Fuzzy theorem and control", Chenongmoongak Press, 1995.

[7] Minh Ta-Cao, Hoang Le-Huy, Model Reference Adaptive Fuzzy Controller and Fuzzy Estimator for High Performance Induction Motor Drivers

[8] Albert D. Helfrick, William D. Cooper, Modern Electronic Instrumentation and Measurement Techniques

ELSEVIER
IFAC
PUBLICATIONS
www.elsevier.com/locate/ifac

POWER SYSTEM STABILITY PREDICTION USING LINE POWER FLOW

Yoon-Sung Cho*, Gilsoo Jang, Byongjun Lee, Sae-Hyuk Kwon

School of Electrical Engineering, Korea University, Seoul 136-701, South Korea

Abstract: Power system stability is correlated with system structure, disturbances and operating conditions, and power flows on transmission lines are closely related with those conditions. This paper proposes a methodology to identify correlative power flows for power system transient and small-signal stability prediction. In transient stability sense, the Critical Clearing Time is used to select some dominant contingencies, and Transient Stability Prediction index is proposed for the quantitative comparison among the contingencies. Also, this paper discusses a methodology to identify crucial transmission lines of which flows are related with small-signal stability by introducing a sensitivity factor based on eigenvalue sensitivity technique. On-line monitoring of the selected lines enables to predict transient stability in real-time. Also, a procedure to make a priority list of monitored transmission lines is proposed. The procedure is applied to a test system, and it shows capabilities of the proposed method. *Copyright © 2003 IFAC*

Keywords: Transient stability, Small-signal stability, TSP, CCT, Eigenvalue, Sensitivity

1. INTRODUCTION

Electric power systems have become more complex and the operating characteristics of many power networks around the world have been changing considerably. It makes an increasing need for on-line stability monitoring. Recent research works deal with the analysis of electro-mechanical characteristics of synchronous generators to a certain fault as well as stability estimation for on-line stability control. Power system stability is correlated with system structure, disturbances and operating conditions, and power flows on transmission lines are closely related with those conditions. If correlated transmission line flows can be identified, monitoring of the lines with Energy Measurement System enables system stability to be enhanced.

This paper proposes a methodology to identify correlative power flows for power system transient and small-signal stability prediction. In transient stability sense, the Critical Clearing Time (CCT) is used to select some dominant contingencies, and TSP (Transient Stability Prediction) index is proposed for the quantitative comparison among the contingencies. The proposed procedure is to be applied to the KEPCO system in the year 2003. For small-signal stability prediction purpose, eigenvalue sensitivity with respect to power flows is used to make the priority list. The proposed procedure is to be applied to a simple two-area system. A procedure incorporating above two approaches is proposed to identify crucial transmission lines for the system stability in this paper. The proposed procedure is to be applied to a 6-generator 20-bus test system, and simulation results show capabilities of the proposed method.

2. IDENTIFICATION OF CORRELATIVE TRANSMISSION LINES

An algorithm to identify correlative transmission lines for on-line stability diagnosis is proposed in this paper. The procedure is described in Figure 1.

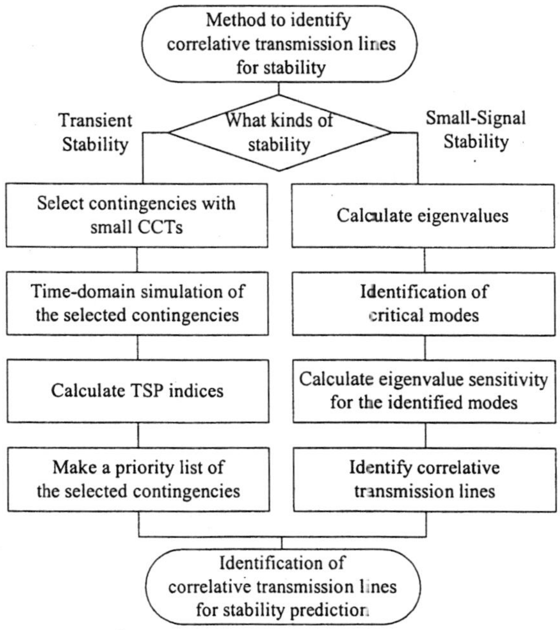

Fig 1. Identification procedure

2.1 Transient stability

CCT is a good index to evaluate transient stability in power system. It is defined as the time for which a fault can last without losing synchronism. Power transfers in transmission lines at CCT can determine if generators will lose synchronism, and an index is proposed based on this assumption. The proposed index is Transient Stability Prediction (TSP) index, which is the summation of normalized power variation rate. Transmission lines which are sensitive to faults can be determined by comparing indices.

$$TSP = \sum_{t=t_0}^{t_1} \frac{|P_2 - P_1|/|P_1|}{\Delta t} \tag{1}$$

t_0 = fault initiating time
t_1 = fault clearing time
P_1 = power flow
P_2 = power flow at the next time-step

The comparison of TSP indices has a meaning at same fault case since the indices are dependent on the fault. The fault cases used in this paper are selected using contingency studies. Each step is as follows:

① Calculate CCTs for the all buses. CCT is calculated by the gap criterion after time-domain simulation.
② Select contingencies with small CCTs.
③ Calculate TSP indices of the selected cases.
④ Make a priority list of the selected contingencies

The above procedure is implemented as an automatic calculation module in a time domain power system simulation program, PSS/E. Hence, the proposed algorithm can be performed automatically whenever system configuration changes. The priority list of the transmission lines is provided as a result of this procedure.

2.2 Small-signal stability

The identification method in terms of small-signal stability is based on eigenvalue sensitivities with respect to power flows of T/Ls.

2.2.1 Eigenvalue sensitivity

System dynamics are represented as equation (2).

$$\dot{x} = Ax \tag{2}$$

Equation (3) shows the solution of equation (2).

$$x(t) = \sum_{i=1}^{N} u_i v_i^T x(0) e^{\lambda_i t} \tag{3}$$

In equation (2), x is the $N \times 1$ state vector of the system, A is $N \times N$ matrix with distinct eigenvalues $\lambda_i (i = 1,2,\cdots,N)$: in (3), $u_i (i = 1,2,\cdots N)$ are the linearly independent eigenvectors of A which satisfy :

$$Au_i = \lambda_i u_i \tag{4}$$

and $v_j (j = 1,2,\cdots,N)$ are corresponding eigenvectors of A^T which satisfy :

$$v_j^T A = \lambda_j v_j^T \tag{5}$$

u_i and v_j satisfy the equation :

$$u_i^T v_j = v_j^T u_i \{ \begin{matrix} = 0 (i \neq j) \\ \neq 0 (i = j) \end{matrix} \quad (i, j = 1,2,\cdots N) \tag{6}$$

Eigenvalue sensitivity with respect to line power flow, P_{ab}, is given in equation (7).

$$\frac{\partial \lambda_i}{\partial P_{ab}} = \frac{v_i^T \frac{\partial A}{\partial P_{ab}} u_i}{v_i^T u_i} \tag{7}$$

where,

$$\frac{\partial A}{\partial P_{ab}} = \frac{\partial A}{\partial x_i} \cdot \frac{\partial x_i}{\partial P_{Gj}} \cdot \frac{\partial P_{Gj}}{\partial P_{ab}} \tag{8}$$

v_i = left eigenvector
u_i = right eigenvector.

2.2.2 Generator models

Generators are modeled by the classical model in this approach, and state equations of the model are as follows.

$$M_i \dot{\varpi}_i = P_i - P_{ei} \tag{9}$$
$$\dot{\delta}_i = \varpi_i - \varpi_s \quad i = 1,2,\cdots,n$$

where,

$$P_i = P_{mi} - E_i^2 G_{ii}$$
$$P_{ei} = \sum_{j=1, j\neq i}^{n} [C_{ij} \sin(\delta_i - \delta_j) + D_{ij} \cos(\delta_i - \delta_j)]$$
$$C_{ij} = E_i E_j B_{ij}$$
$$D_{ij} = E_i E_j G_{ij}$$

and

E_i : internal bus voltage of generator i
M_i: inertia constant of generator i
P_{mi} : mechanical power input of generator i
G_{ii}: driving point conductance of node i
ω_i: rotor speed of generator i
ωs : synchronous speed
δ_i : rotor angle of generator i

This model is used to represent the generators without excitation control in this work. System equation is developed using equation (9), and it is given in equation (10).

$$\begin{bmatrix} \dot{x}_1 \\ \vdots \\ \dot{x}_{n-1} \\ -- \\ \dot{x}_n \\ \vdots \\ \dot{x}_{2n-2} \end{bmatrix} = \left[\begin{array}{ccc|c} & 0 & & I \\ \hline A_{11} & \cdots & A_{1,n-1} & \\ \vdots & \ddots & \vdots & 0 \\ A_{n-1,1} & \cdots & A_{n-1,n-1} & \end{array}\right] \begin{bmatrix} x_1 \\ \vdots \\ x_{n-1} \\ -- \\ x_n \\ \vdots \\ x_{2n-2} \end{bmatrix}$$

or

$$\begin{bmatrix} \dot{x}_1 \\ \dot{x}_2 \end{bmatrix} = \begin{bmatrix} 0 & U \\ A_\omega & 0 \end{bmatrix} \begin{bmatrix} x_1 \\ x_2 \end{bmatrix} \qquad (10)$$

where,

U = The identity matrix

x_1 = The n-1 vector of the angle changes δ_Δ

x_2 = The n-1 vector of the speed angles $d\delta_\Delta / dt$

$$[A_\omega]_{ii} = -\frac{1}{M_i} \sum_{j=1, j\neq i}^{n-1} F_{ij} - \frac{1}{M_i} F_{in} - \frac{1}{M_n} F_{ni}$$

$$[A_\omega]_{ij} = \frac{1}{M_i} F_{ij} - \frac{1}{M_n} F_{nj}$$

2.2.3 Relation of line active power and rotor angle

Electrical power of *i*-th generator is given in equation (11).

$$P_{ei} = \sum_{j=1}^{N} V_i V_j |Y_{ij}| \cos(\theta_{ij} - \delta_i + \delta_j) \qquad (11)$$

Equation (12) shows $\partial X_i / \partial P_{Gj}$ which is required to calculate eigenvalue sensitivity.

$$\begin{vmatrix} \Delta P_{e1} \\ \vdots \\ \Delta P_{e(n-1)} \end{vmatrix} = \begin{vmatrix} \sum_{j=1, i\neq j}^{N} F_{1j} & \cdots & -F_{1(n-1)} \\ \vdots & \ddots & \vdots \\ -F_{(n-1)1} & \cdots & \sum_{j=1, i\neq j}^{N} F_{ij} \end{vmatrix} \begin{vmatrix} \Delta x_1 \\ \vdots \\ \Delta x_{(n-1)} \end{vmatrix}$$

or

$$\Delta P = J_{px} \Delta X \qquad (12)$$

where,

$$F_{ij} = C_{ij} \cos(\delta_i - \delta_j) - D_{ij} \sin(\delta_i - \delta_j)$$

Hence, in (12)

$$\Delta X = (J_{px})^{-1} \Delta P = J_{xp} \Delta P \qquad (13)$$

Therefore

$$\Delta X_i = [J_{xp}]_{ij} \Delta P_j \quad (\Delta P_{ej} \neq 0) \qquad (14)$$

2.2.4 Relation of generator active power and line active power

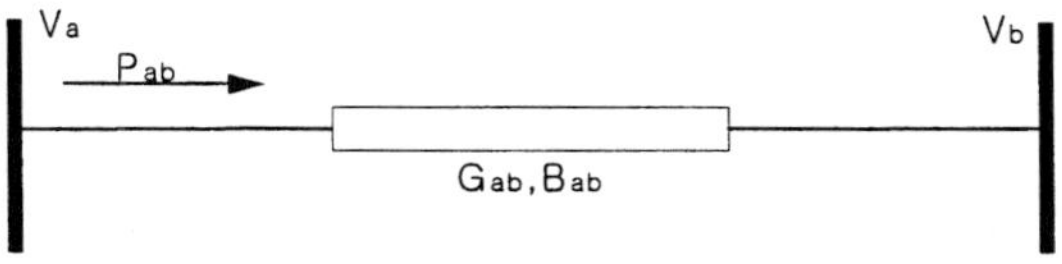

Fig. 2. Active power between two buses

The active power between two buses can be represented with the line admittance, $G_{ab}+jB_{ab}$. $\partial P_{Gj}/\partial P_{ab}$ is obtained by substituting P_{ab} into equation (9), since P_{ab} can be represented with respect to generator rotor angles.

2.2.5 Correlative lines identification procedure

① Identification of critical modes:
- Calculate eigenvalues from system matrix
- Select less damped modes with low frequencies

② Calculate eigenvalue sensitivity with respect to line active power for the identified modes

③ Identify correlative transmission lines

3. APPLICATION OF THE PROPOSED PROCEDURE

3.1 Transient stability

3.1.1 A 6-generator 20-bus test system

Table 1 Priority list w.r.t. transient stability

From	To	Bus 152	Bus 154	Bus 3005
151	101	4277.32	1967.44	975.78
151	102	4277.32	1957.44	975.78
151	152	.	1447.63	641.19
151	201	8386.71	6703.05	7253.62
152	153	.	953.87	577.75
152	202	.	17718.46	6037.99
152	3004	.	9090.21	49902.91
153	154	4926.28	.	586.35
153	3006	12451.01	5918.89	9582.51
154	203	3604.81	.	475.08
154	205	1951.34	.	3184.26
154	3008	10303.78	.	12742.68
201	202	981.72	2729.67	605.33
201	204	3545.88	5666.06	415.88
202	203	4235.21	6312.19	343.93
203	205	4957.12	6565.18	716.72
204	205	3069.72	6079.01	424.75
205	206	3221.23	6192.94	651.88
3001	3002	12722.48	5331.17	9951.33
3001	3003	6490.08	5750.78	7515.23
3002	3004	12722.48	5331.17	9951.33
3003	3005	6464.92	5750.21	.
3004	3005	4773.92	1441.57	.
3005	3006	10447.11	5744.75	.
3005	3007	3080.24	2021.52	.
3005	3008	5180.26	15673.39	.
3007	3008	4651.57	25801.96	5621.25

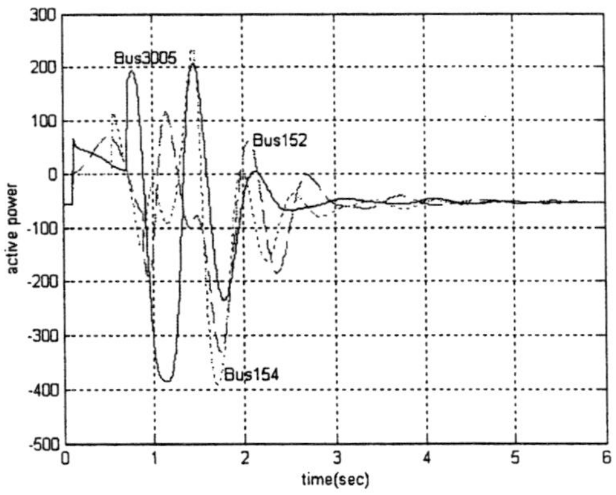

Fig. 3. Active power between #154~3008(before CCT)

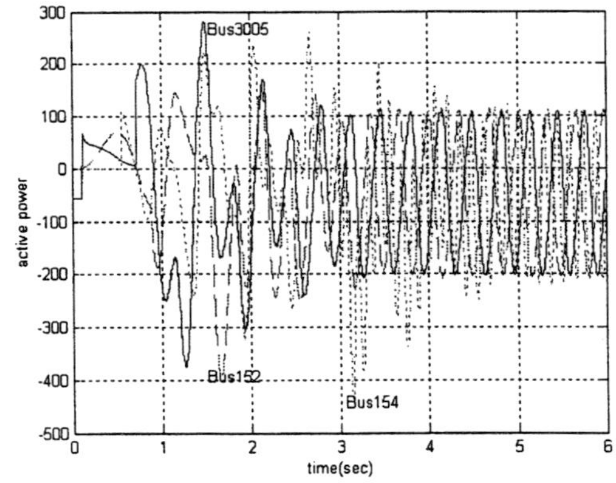

Fig. 4. Active power between #154~3008(at CCT)

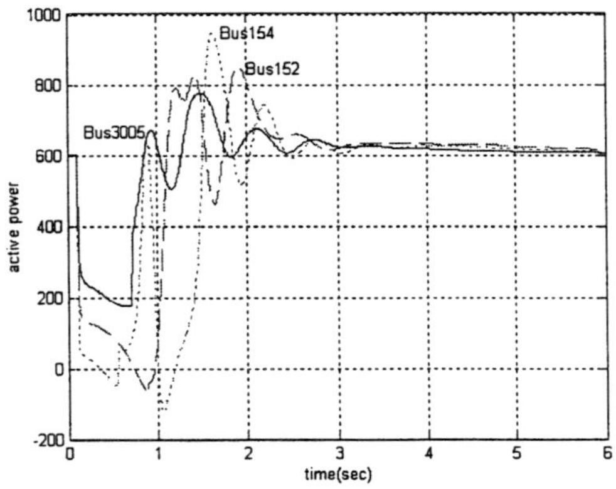

Fig. 5. Active power between #204~205(before CCT)

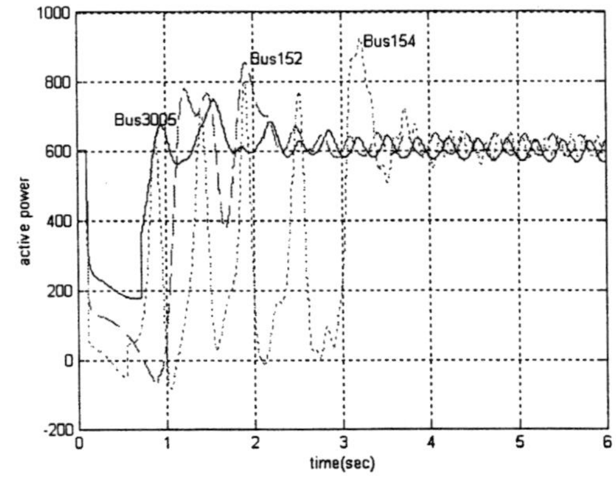

Fig. 6. Active power between #204~205(at CCT)

The proposed method is applied to a test system which has 6 generators and 20 buses. The fault at bus is considered in this study. PSS/E is used for nonlinear time simulation program. The time-step used is 0.0083 sec (0.5 cycle). Some critical ones are given here.

① CCT of bus # 152 = 0.7913 sec
② CCT of bus # 154 = 0.3494 sec
③ CCT of bus #3002 = 0.6166 sec

Table 1 shows TSP indices of transmission lines, and the priority list based on the indices is proposed. The results say that lines between buses 151 and 201, buses 154 and 3008, and buses 3001 and 3002, and buses 3002 and 3004 are related to the faults which are not on their buses. Time domain simulation of those contingencies are performed to verify identification method, and Figures 3, 4, 5, and 6 show the line flow between buses #154 and #3008 has more significant variations with fault clearing time than other ones.

3.1.2 The KEPCO system

The test system in this work is based on the KEPCO system in the year 2003. The system consists of 230 generators, 964 buses and 1955 AC branches. Its total generation is 48598.9 MW and 13217.2 MVR and 753 loads have 24847.2 MW and 7793.7 MVR. The fault at 2 line 3 phase line-to-ground is considered in this study. CCTs are calculated by the developed module, and some critical ones are given here.

Table 2 Simulation Conditions

Case No.	Fault location from-to	CCT (sec(cycle))
1	7150~7600	0.0620(3.72)
2	5150~5600	0.1024(6.14)
3	10300~10401	0.1196(7.18)
4	6100~6300	0.1326(7.96)
5	6600~7101	0.1349(8.10)

Table 3 shows TSP indices of transmission lines, and the priority list is proposed based on the indices. The results say that lines between buses 1700 and 2500, buses 4400 and 6951, buses 4900 and 6600, and buses 6450 and 7155, buses 7150 and 7600, and buses 8300 and 8800 are related to the faults which are not on their buses. Time domain simulation of those contingencies are performed to verify identification method, and Figures 7, 8, 9, and 10 show the line flow between buses 4400 and 6951 has more significant variations with fault clearing time than other ones.

Table 3 Priority list w.r.t. transient stability

Fault Location		CCT				
from	To	Case No.1	Case No.2	Case No.3	Case No.4	Case No.5
1700	2500	14588.1	4565.7	17422.4	27763.6	12808.7
2400	2600	6444.4	6352.3	6669.3	18041.1	4467.6
2400	3600	2194.1	6499.3	823.6	1935.6	495.5
3350	3800	2243.2	5721.1	2893.3	7419.4	2195.6
4100	4700	4946.1	6296.2	8642.8	42348.4	`6173.1
4400	4450	5089.8	3753.7	7322.5	7927.6	5053.4
4400	4600	3211.9	9307.8	3577.4	40537.4	2113.3
4400	6951	19411.1	20092.6	19475.1	.	12782.9
4900	6600	39443.7	17336.3	46964.8	18007.3	.
5150	5155	2588.6	.	4215.1	7510.4	4380.1
5155	5500	2853.8	.	2711.8	7881.9	6394.9
5600	5700	4034.1	.	4792.4	7536.9	4104.6
6020	6030	6357.9	17892.5	5439.1	5439.1	3822.3
6100	6900	4147.8	9892.1	3413.1	182816.2	2443.7
6150	6950	4129.8	6782.9	3897.5	7075.9	2636.5
6300	6900	3523.4	6892.5	2832.8	.	1973.2
6300	6950	6837.6	13626.6	6930.3	.	4941.2
6450	7155	.	17757.1	44517.1	21859.5	23759.4
7150	7155	.	622.6	4610.7	1621.1	2339.4
7150	7600	.	12247.5	35525.9	11564.4	21623.4
7250	7900	1792.2	1163.6	12825.1	1001.9	1455.1
8150	8155	3231.2	4733.2	5931.2	1955.1	5585.5
8156	8250	1792.4	8527.4	5709.6	2667.1	12001.8
8300	8800	3104.1	3101.4	9638.1	2157.8	3118.7
8600	8800	17046.6	7539.5	37815.3	5342.1	34925.6
10300	10350	7803.3	7803.3	.	2979.6	6577.5
10301	10700	4922.4	4922.4	.	5386.8	18986.9

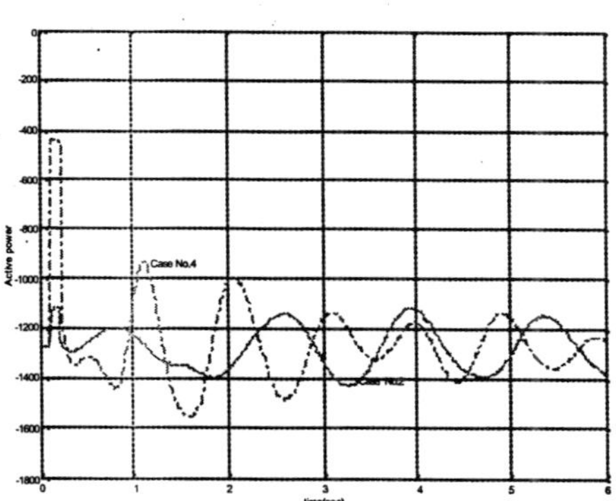

Fig. 7. Active power between #4400~6951 (before CCT)

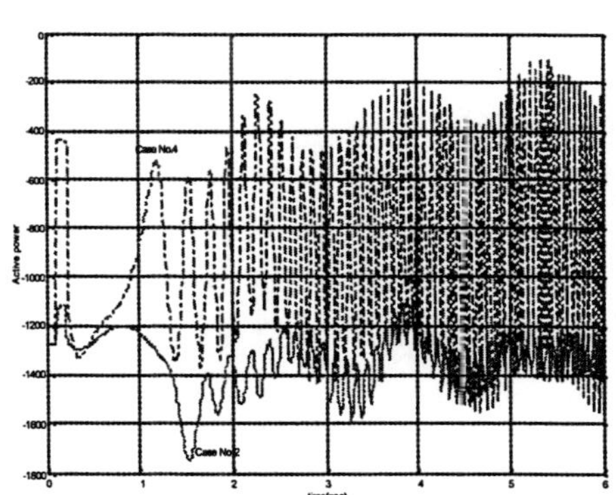

Fig. 8. Active power between #4400~6951 (at CCT)

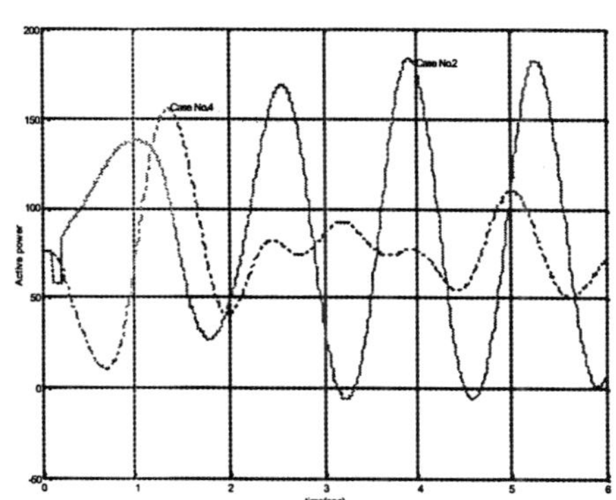

Fig. 9. Active power between #8300~8800 (before CCT)

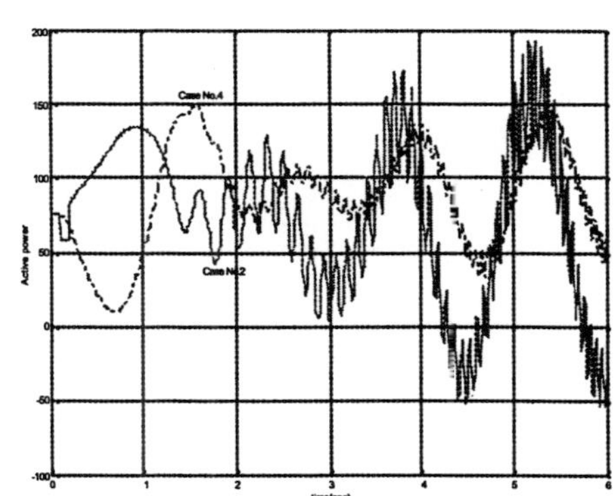

Fig. 10. Active power between #8300~8800 (at CCT)

3.2 Small-signal stability

3.2.1 A simple two-area system

The proposed method is applied to a simple two-area system which has 4 generators and 9 buses. Damping of the modes isn't represented since classical model is used for the generator modeling for the derivation of the eigenvalue sensitivity w.r.t. line flow. Table 4 shows the modes of a simple two-area system, and table 5 shows $\partial\lambda/\partial P_{ab}$ of the modes. The eigenvalues have different variations from line flows respectively. Therefore, it is possible to determine correlated lines with the sensitivity values.

Table 4. Modes of a two-area system

	$\lambda_{1,2}$	$\lambda_{3,4}$	$\lambda_{5,6}$
200MW	±j3.05377	±j8.87407	±j8.94562
300MW	±j2.83321	±j8.90799	±j8.97002
400MW	±j2.35730	±j8.95629	±j9.00335

Table 5. Eigenvalue sensitivity of bus #7 and 8

	$\partial\lambda_{1,2}/\partial P_{ab}$	$\partial\lambda_{3,4}/\partial P_{ab}$	$\partial\lambda_{5,6}/\partial P_{ab}$
200MW	$\mp$j0.16105	±j0.03263	±j0.03195
300MW	±j0.36540	±j0.08670	±j0.02006
400MW	±j0.95970	±j0.21240	±j0.02122

3.2.2 A 6-generator 20-bus test system

Damping of the modes isn't represented since classical model is used for the generator modeling for the derivation of the eigenvalue sensitivity w.r.t. line flow. Hence, critical modes are selected only by observing the frequencies. Table 6 shows the modes of the test system, and the modes 3 and 4 are identified as the critical modes.

Table 6. Modes of the test system

Eigenvalue		ω[rad/s]	F[Hz] (T[s])
$\lambda_{1,2}$	±j1.645E+02	±1.645E+02	26.19 (0.038)
$\lambda_{3,4}$	±j0.348E+02	±0.348E+02	5.535 (0.181)
$\lambda_{5,6}$	±j1.274E+02	±1.274E+02	20.27 (0.049)
$\lambda_{7,8}$	±j1.089E+02	±1.089E+02	17.33 (0.058)
$\lambda_{9,10}$	±j1.089E+02	±1.089E+02	17.96 (0.056)

Table 7. Eigenvalue sensitivity of modes 3 & 4

	$\partial\lambda_{3,4}/\partial P_{ab}$
Bus 154~3008	±j2.103E-06
Bus 204~205	±j2.498E-07
Bus 201~202	±j5.576E-07

Table 7 shows $\partial\lambda/\partial P_{ab}$ of the critical modes ($\lambda_{3,4}$), and the eigenvalues have different variations from line flows respectively. Therefore, it is possible to determine correlated lines with the sensitivity values. Sensitivity derivation with detail generator modeling is under way in order to include effects on damping of modes.

4. CONCLUSIONS

This paper deals with an identification method for stability monitoring. Transmission lines which are correlated with power system transient and small-signal stability are identified using the proposed method, and time-domain simulation results support the method. Also, the TSP index introduced in this work identified line flows with large variation to various disturbances. Eigenvalue sensitivity with respect to line power flows shows the possibility for the use in identifying crucial transmission lines to the critical modes.

Measurement systems with GPS (Global Positioning System) can be put into the selected transmission lines in order to monitor power flows in real-time, and it could enhance power system transient stability with proper prediction methods.

ACKNOWLEDGMENT

This work was supported by the Advanced Power System Research Center designated by the Ministry of Commerce, Industry & Energy.

REFERENCES

A.A Fouad and Vijay Vittal, " Power System Transient Stability Analysis Using The Transient Energy Function Method", Englewood Cliffs, New Jersey:Prentice-Hall, 1992

Jang, G. "Nonlinear control design for stressd power systems using normal forms of vector fields", Ph.D. Thesis, Iowa State University, Ames, Iowa,1997

P.M. Anderson and A.A. Fouad, "Power System Control and Stability", IEEE Press, 1994

P. Kunder, "Power System Stability and Control", McGraw-Hill, 1994

PSS/E Application Manual, Power Technologies, Inc., 1995

P.W. Sauer and M.A. Pai,"Power System Dynamics and Stability", Prentice-Hall, 1998

Ota H, Kitayama Y, Ito H, Fukushima N, Omata K, Morita K, Kokai Y, "Development of transient stability control system (TSC system) based on on-line stability calculation", Power Systems, IEEE Transactions on , Volume: 11 Issue: 3 , Aug. 1996

Taylor C.W., "The future in on-line security assessment and wide-area stability control", Power Engineering Society Winter Meeting, 2000. IEEE , Volume: 1 , 2000

ELSEVIER
IFAC
PUBLICATIONS
www.elsevier.com/locate/ifac

A STUDY ON THE ALGORITHM AND ITS PRACTICAL USE FOR IMPROVING THE RELIABILITY OF AUTOMATIC POWER RESTORATION SYSTEM

J. B. Choo, I. J. Song, N. H. Lee

Korea Electric Power Research Institute

Abstract: Recently, according to a rapid increase of the power demand, the 154kV lines of Korea Electric Power Corporation (KEPCO) are continuously expanded and become to the more complex loop system. As a result of increase of the fault capacity and the unbalance of power load flow, the circuit breaker have been exceeding the capacity limit of those in the loop power system. To solve this problem, this paper presents algorithm and application of Automatic Power Restoration System (hereafter called "APRS")

Keywords: Power System, Restoration, Transmission & Distribution

1. INTRODUCTION

The Korea power systems as infrastructure are based on 765kV – 345kV – 154kV. Due to the rapid growth of power demand and for the power supply reliability, The Korea Electric Power Corporation has built 345kV & 154kV transmission lines of a considerable complex loop-type. Such a complex loop-type system has the advantage of maintaining the higher power supply reliability than that of radial systems. However, it causes to increase the number of transmission line exceeded the interrupt capacity on substations has been continuously increased while the KEPCO system is under fault. This has led to the 154kV radial operation with bus & line isolation operations and the loss of the power supply reliability has been ignored.

The APRS which has been installed in the 154kV radial system, is the first large-scale system in Korea and has been designed to deal with the fault capacity limits on transmission lines and the power overflow and improve the power supply reliability (The minimization of the unserviced area and the fast automatic restoration) under fault.

2. ALGORITHM OF THE APRS

The APRS which has been designed to do suitable actions or tasks which used to need human intervention for 154kV network restoration in the case of a local disturbance. With the APRS , the fault section is identified, isolated and restored back to the normal state within around 10 seconds

The APRS system consists of one main system and extend to eight substations. For acquiring data and output control commands, the fiber optic communication cables are used between the main system and the local terminal units. The schematic of the APRS is shown in figure 1.

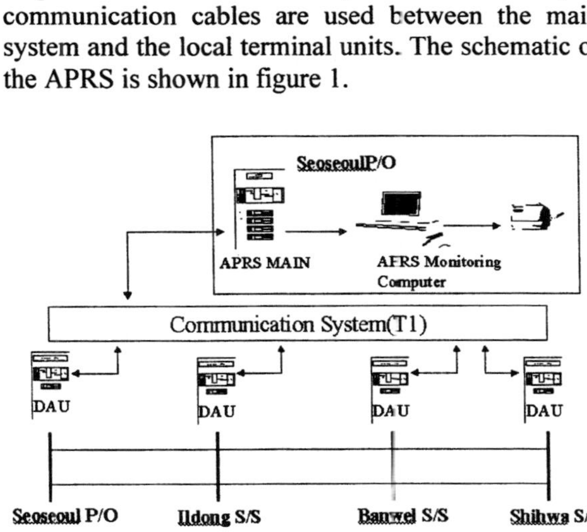

Fig.1. The schematic of the APRS

Figure 1 illustrates the type of system information acquired from the 4 substations (CB & ES & DS ON/OFF status, protective relay operation, T/L power flow, M.Tr MW, 154kV bus Voltage) in real time and transmitted to the main system. For the successful restoration the APRS keeps track of the information mentioned above and furthermore. APRS detects due to the current conditions include the protective relay operations. And change in the status of the circuit breakers. If the line fault is recognized and the blackout area is restored by the restoration procedures as described in figure 2.

This paper presents the automatic restoration algorithm of the APRS which consists of identification of the current system operation, diagnosis of fault area, fault area isolation and blackout area restoration mentioned below, and Section 3 finally deals with a case of the APRS operation in real KEPCO 154kV radial system

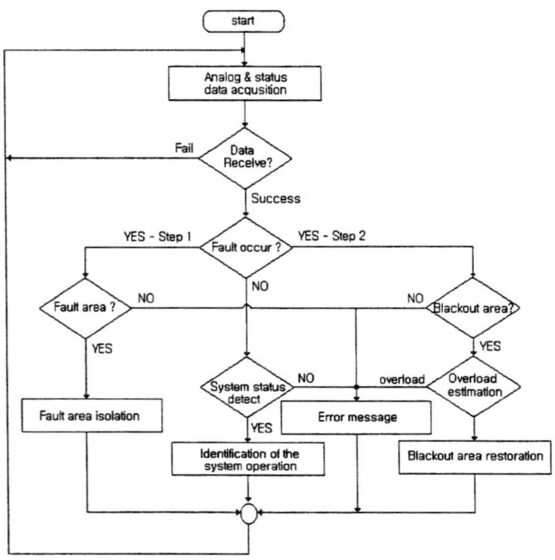

Fig.2. The restoration procedure of the APRS

2.1 *Identification of the current system operation*

154kV power system to be installed APRS shown in figure 1 has two network regions that consists of the overhead transmission lines between each substations. This area has the ability to provide greater power transfer capability in the case of a blackout of one network region and indeed has the higher possibility of such a blackout occurrence that it is known that two circuits are open simultaneously. In view of this, the APRS must be able to identify the status of the current system operation changing as network conditions and power flow in order to completely process the restoration procedures when fault occur. Specially, the APRS has been considered to recognize system operation as a normal status, in case that one line of two circuits is not in service, using the status data of circuit breakers and earth switches. Figure 3 shows this procedures.

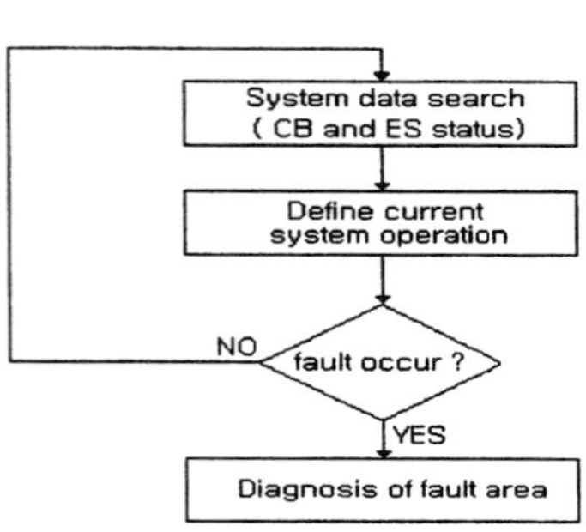

Fig.3. Identification of the current system operation

2.2 *Diagnosis of fault area*

The APRS finds fault area using CB status data and protective relay information. If circuit breakers are detected to be changed, It means that the system may has a fault or system operator may change the CB for any purposes and the APRS will detect relevant relay information. If relays operate, the APRS defines if the system is bus or line fault. And for the higher reliability, relative transmission lines power flow, M.TR load and bus voltage data should be used. In case that the status of CB are changed and the operation of primary and back-up protective relays are detected, the APRS finds fault area according to the protection zone of the primary and back-up relays. In case of bus fault, the APRS finds where the fault area is, on the basis of the CB status data and relay information such as a differential relay and under voltage relay. Figure 4 describes the diagnosis procedures of fault area.

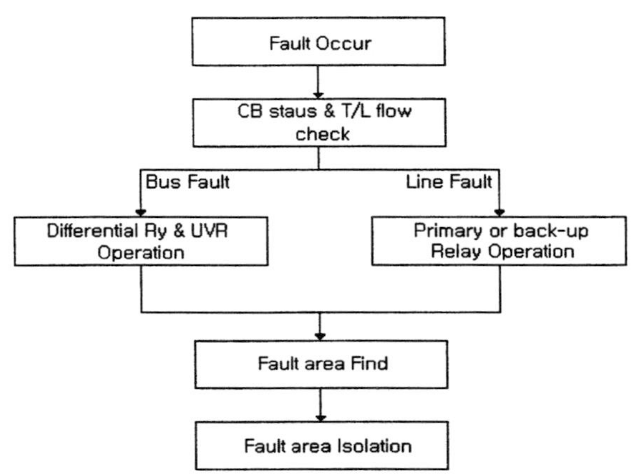

Fig.4. The diagnosis of fault area

2.3. *Fault area isolation*

When the APRS supply power to blackout area from other network lines that have a sufficient power transfer capability, this step is to isolate fault area from normal area in order to prevent supplying power to the fault area.

That is, the APRS commands to trip the CB of fault area detected by diagnosis of fault area mentioned above and the fault area will be isolated fault area from blackout area

2.4. *Blackout area restoration*

After isolating fault area, the APRS process the restoration procedures as shown in figure 5. If the blackout area is found, the APRS will detect power-supplying routes for the restoration of the blackout area. And then the blackout area will be completely restored by closing CB to transfer power. In this case, the APRS must have a function to estimate overload if the power supplying lines provide sufficient power transfer capability to the blackout area before closing CB. The APRS has been saving the power flow data of individual transmission lines until fault occurs.

Before closing CB for restoration, it estimates if the capability limit of the power supplying lines large than the current power flow of this lines added to pre-faulted MW data of the blackout area. As a result of the overload estimation, it will make a decision to close CB. Figure 5 describes the procedures of the blackout area restoration.

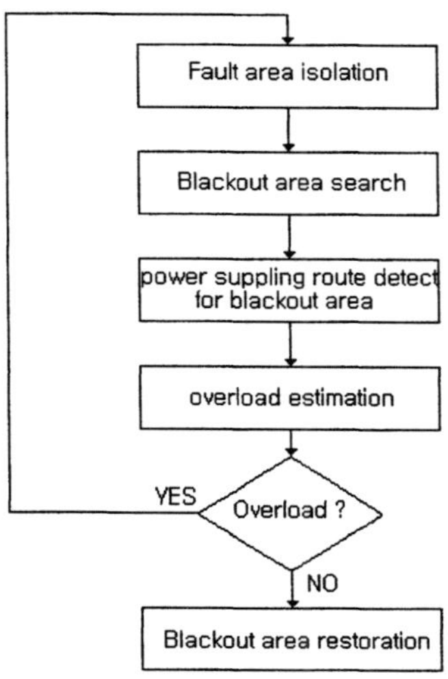

Fig.5. Blackout area restoration

3. IMPLEMENTATION AND OPERATION OF APRS

3.1 *Installation and implementation of APRS*

The APRS system consists thus of one main system and four DAU as mentioned above. The main system being installed in the 'Seoseoul'P/O has the specifications described as follows:

- Logic and Operation CPU Board : MVME162 Board
- Communication and Data Temporary Storage Memory Board * 4EA
- Communication Modem(CSU) * 4EA
- Monitor
- Printer

This main system can display CB & DS operations, T/L power flow, M.TR load and as well as bus voltage onto the monitoring screen in real time, and provide the report-quality printouts stating the on/off status of circuit breakers & disconnecting switches and the relevant protective relay operations in real time. The data acquisition unit being installed in a substation is composed of as follows:

- Communication and I/O Control CPU Board
- Analog Input Board
- Digital Input Board
- Digital Output Board
- Communication Modem (Channel Service Unit)

KEPRI installed the main system at 'Seoseoul'P/O and DAU at its 'Ildong', 'Shihwa', 'Seoseoul', and 'Banwel' substations. This two systems are connected via T1(1.544Mbps) optical transmission reliability. This area consists of the overhead transmission lines between 'Sinsiheung' substation and 'Seoseoul'P/O. As shown in figure 6 , it is configured to be operated as line isolation where the 154kV 'Shiwel' #1,2 T/L CB in 'Banwel'S/S remains open and the 154kV 'Shihwa'S/S remains closed. The 'Ildong'S/S is supplying 'Banwel'S/S using the 154kV 'Banwel' #1,2 T/L and the 'Seoseoul'S/S is supplying 'Ildong'S/S using the 154kV 'Ildong'#1,2 T/L.

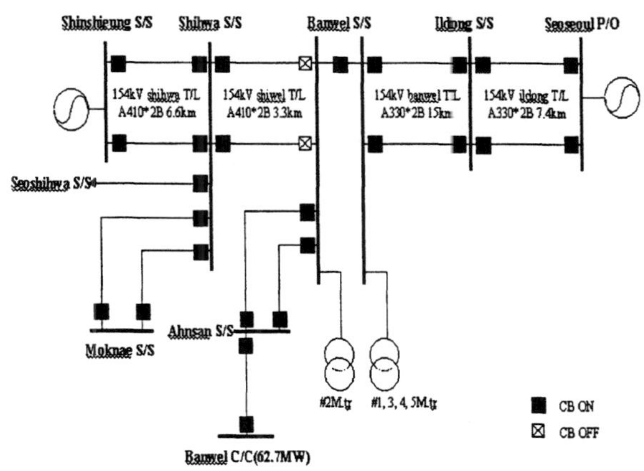

Fig.6. 154kV KEPCO network operated as line isolation

3.2 *A case study of APRS*

A double circuit fault of 'Ildong'S/S and 'Banwel'S/S makes the end substation 'Banwel'S/S blackout. If such a blackout does occur, the APRS detects and decides where the fault is on the basis of restoration procedures stating in this paper. If the fault area turns out to be on the transmission lines between 'Ildong'S/S and 'Banwel'S/S, 'Banwel'S/S has the 'Banwel'#1,2T/L CB Off to isolate a fault section shown in figure 7.

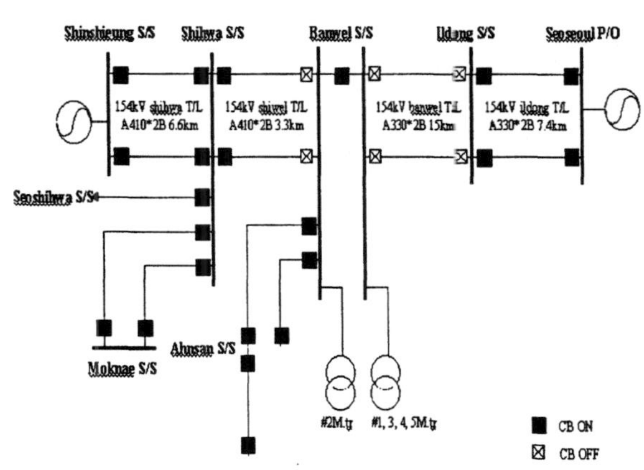

Fig.7. Isolation of a fault section between 'Ildong'S/S and 'Banwel'S/S

After isolating the fault section, the APRS estimates overload If 'Shiwel'#1,2T/L can supply sufficient power to the outage area of 'Banwel'S/S. If it turns out to be able to supply power, the APRS will restore power service in the outage area while closing the 'Shiwel'#1,2T/L CB in 'Banwel'S/S shown in figure 8. Such a blackout could be restored automatically within 10 seconds after the 'Banwel'#1,2T/L CB of 'Ildong'S/S was tripped.

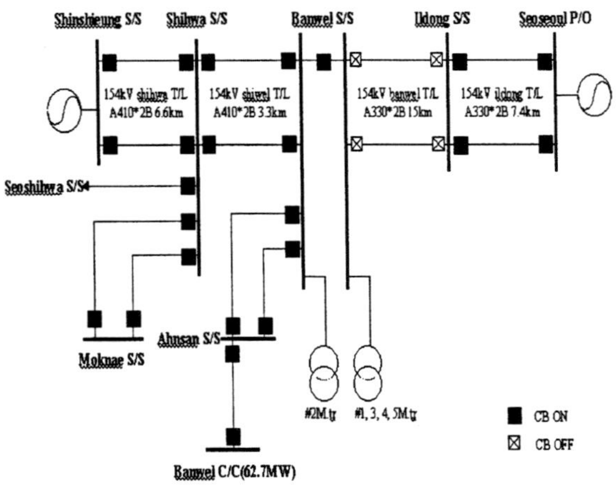

Fig.7. Restoration of the outage area of 'Banwel'S/S

4. CONCLUSIONS

Experiments with the APRS system installed in KEPCO system have shown that is capable of performing much speedy restoration of blackout areas such as 'Ildong', 'Banwel' and 'Ansan' areas. This APRS also contribute to the improvement of power supply reliability in the areas mentioned above. It is expected that the sounding results of the actual restoration of power systems would further increase its use and flexibility.

Due to the increase of the fault capacity limits on 154kV transmission lines, the numbers of bus and line isolation substations are growing continuously. This trend is only likely to continue in KEPCO power system and the APRS will be used widely in other areas.

REFERENCES

G. Morin(1987), "Service Restoration Following a Major Failure on the hydroquebec Power System", IEEE Trans. PWRD-2, NO2, 454

M. M. Adibi(1987), " Power System Restoration – The second Task Force Report" IEEE Trans.PWRS-2,NO2,927

Development and Testing of an Automatic Power Restoration System: Application of a Real–time Power System Simulator, International Electric Research Exchange workshop 99.9

H. J. Yoo, S. C. Hong, and T. W. Kim(2000), "Improving the 154kV Radial Transmission System Reliability by Automatic Power Restoration System" Korea Electric Power Research Institute, Technical report(TR.98PJ06.J2000.287

ELSEVIER
IFAC
PUBLICATIONS
www.elsevier.com/locate/ifac

APPLICATION EFFECTS OF SFCL IN KOREAN POWER SYSTEM

Jong-Yul Kim, Heung-Kwan Choi, Jae-Young Yoon, Hak-Man Kim

Korea Electrotechnology Research Institute, Changwon, 641-600, Korea

Abstract: As power systems grow more complex and power demands increase, the fault current tends to gradually increase. In the near future, the fault current will exceed a circuit breaker rating for some substations, which is an especially important issue in the Seoul metropolitan area. In this paper, the resistive and inductive SFCLs are applied to reduce the fault current in Korean power system and their technical and economic impacts are evaluated. *Copyright © 2003 IFAC*

Keywords: Superconductor, Fault Current Limiter, Technical and Economic Impact.

1. INTRODUCTION

The transmission network in Korean power system is protected by 40kA rate of a circuit breaker for 345kV systems and 31.5kA and 50kA rate of a circuit breaker for 154kV systems. As power system s grow more complex and electrical demands increase, the fault current has the tendency of increasing gradually.

Unless an appropriate countermeasure is applied, the fault current will soon exceed the circuit breaker rating for some substations. To solve this problem, several methods are implemented on the power system; changing the system configuration from a loop to a radial system; splitting the bus; and upgrading lower rated equipment to higher rated equipment.

Superconductors are utilized for fault current limiter applications because they can transform into a normal state in a few milliseconds, have high resistivity in the normal state and return to the superconducting state once fault conditions are removed. Superconducting fault current limiter (SFCL) reduces the fault current and allows the use of lower rated circuit breakers. Therefore, we can obtain cost savings by avoiding the need to upgrade lower rated equipment in existing installations. In these days, companies world-wide have conducted research on SFCL: ABB (Switzerland), GEC-Alsthom (France), Tokyo Electric (Japan), General Atomics (USA), and Siemens (Germany) [1,2,3]. They planned to develop 154kV voltage class SFCL at the research centre and university in Korea. In this paper, resistive type and inductive type SFCL are applied to reduce the fault current in Korean power systems and their technical and economic impacts are evaluated.

2. INTRODUCTION TO SFCL

■ *Resistive Type SFCL*

The simplest superconducting limiter concept, the series resistive limiter, exploits the non-linear resistance of superconductors in a direct way. For a full-load current of I_{FL}, the superconductor would be designed to have a critical current of $2I_{FL}$ or $3I_{FL}$.

During a fault, the fault current pushes the superconductor into a resistive state and resistance R appears in the circuit.

The superconductor in its resistive state can also be used as a trigger coil, pushing the bulk of the fault current through a resistor or inductor. The advantage of this configuration, shown in Fig. 1, is that it limits the energy that must be absorbed by the superconductor.

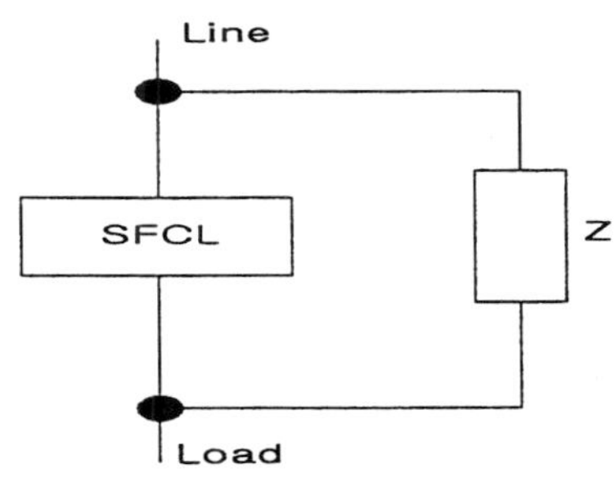

Fig.1. Resistance Type SFCL

The fault-current limiter FCL normally is a short across the copper inductive or resistive element Z. During a fault, the resistance developed in the limiter shunts the current through Z, which absorbs most of the fault energy.

The trigger coil approach is appropriate for transmission line applications, where tens of megawatt-seconds would be absorbed in a series resistive limiter[4].

■ *Inductive Type SFCL*

Another concept uses a resistive limiter on a transformer secondary, with the primary in series in the circuit. This concept illustrated in Fig. 2, yields a limiter suitable for large-current and is coupled to an High Temperature Superconducting(HTS) winding W_{HTS}.

During normal operation, zero impedance is reflected to the primary. Resistance developed in the HTS winding during a fault is reflected to the primary and limits the fault.

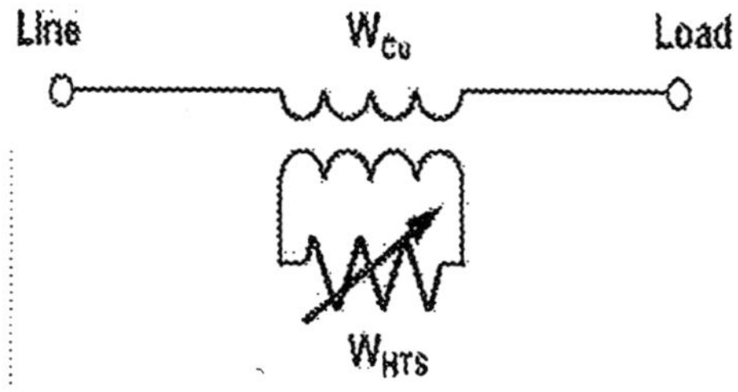

Fig.2. Inductive Type SFCL

The inductive limiter can be modelled as a transformer. The impedance of this limiter in the steady state is nearly zero, since the zero impedance of the secondary HTS winding is reflected to the primary. In the event of a fault, the large current in the circuit induces a large current into secondary and the winding losses superconductivity. The resistance in the secondary is reflected into the circuit and limits the fault[4].

3. RESULT OF FAULT CURRENT ANALYSIS

We investigated the fault current of 154 kV buses in the Seoul metropolitan area in 2004, 2006, and 2010. For the fault current analysis, we have been using the PSS/E program and peak data of Korea Electric Power Company (KEPCO) in 2004,2006,2010. Figs. 3-5 show the results of fault current analysis in Seoul metropolitan area.

In the figures, the horizontal axis represents the bus in the Seoul metropolitan area and the vertical axis represents the fault current of the bus. The fault current, shown in Figs.3-5, is about 30-40kA in 2004 and increases gradually as time goes by.

In 2010, it exceeds 50 kA that is maximum rating of a 154 kV circuit breaker. Unless a proper countermeasure is applied, the power system will be damaged by the fault current.

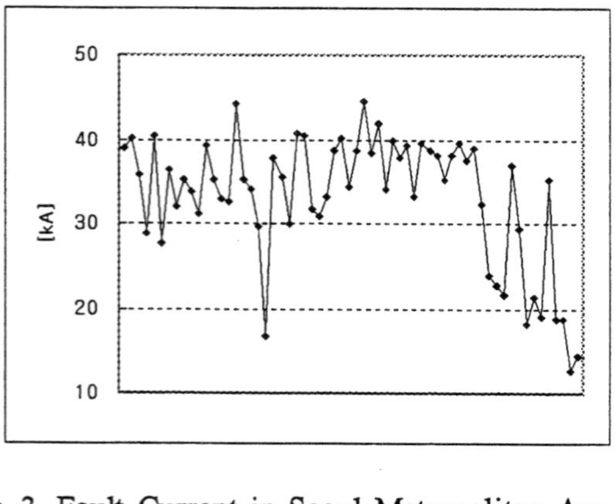

Fig. 3. Fault Current in Seoul Metropolitan Area in 2004

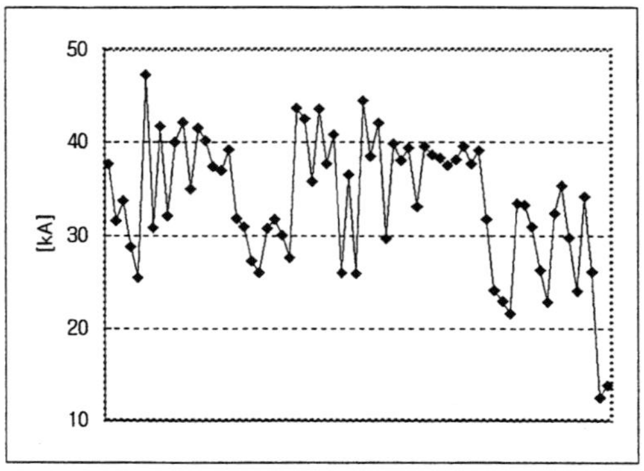

Fig. 4. Fault Current in Seoul Metropolitan Area in 2006

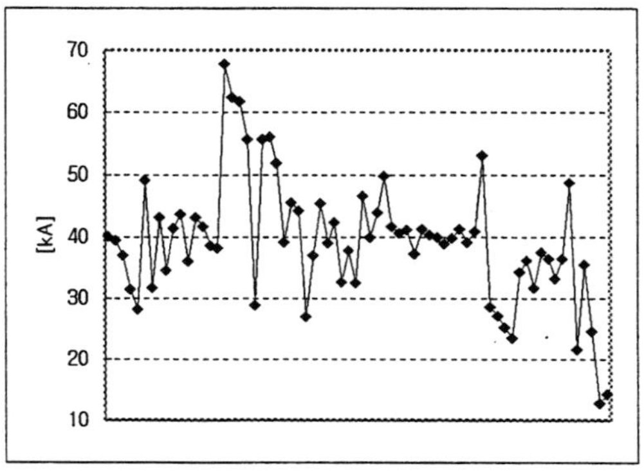

Fig. 5. Fault Current in Seoul Metropolitan Area in 2010

4. APPLICATION OF SFCL IN POWER SYSTEM

4.1 Test Power System

We have studied many cases. But in this paper, we introduce the southern Seoul metropolitan area case. Because the southern Seoul metropolitan area has a very high load density, the possibility that the fault current exceeds the circuit breaker rating is higher than any other area in Korea. We used KEPCO peak data for 2010 and the three-phase fault was considered. Assuming a 50 kA rating of the 154 kV circuit breaker, the fault currents of eight buses exceed the rating of the circuit breaker as shown in Table 1.

Table 1 Bus and Fault Current

Bus	Fault current [kA]
2510	67.9
2520	62.5
2525	61.7
2530	55.6
2540	55.6
2570	56.1
2580	52.0
2845	53.2

4.2 Procedure for Applying SFCL

We proposed a procedure for applying SFCL to power systems to resolve the fault current problem as shown in Fig. 6. First, investigate the contribution of the fault current and select a candidate location at each bus where the fault current exceeds the circuit breaker rating. The selected candidate locations are evaluated for the application effect. Finally, according to the application effect, determine the most effective location to limit the fault current.

Until the fault current problem is resolved, install the additional SFCL at next effective location.

1) Select a Candidate Location

Though the SFCL is installed at the same bus, the fault current reduction effect is different according to the location's relation to the neighbor bus. By using the fault current contribution, we select a candidate location at each bus where the fault current exceeds the circuit breaker rating. In the case of bus 2510, the fault current from neighbor bus 2500 is larger than other neighbor buses. Therefore, the candidate location of bus 2510 is position ①, as shown in Fig. 7. After all, we select a total of eight candidate locations as shown in Table 2.

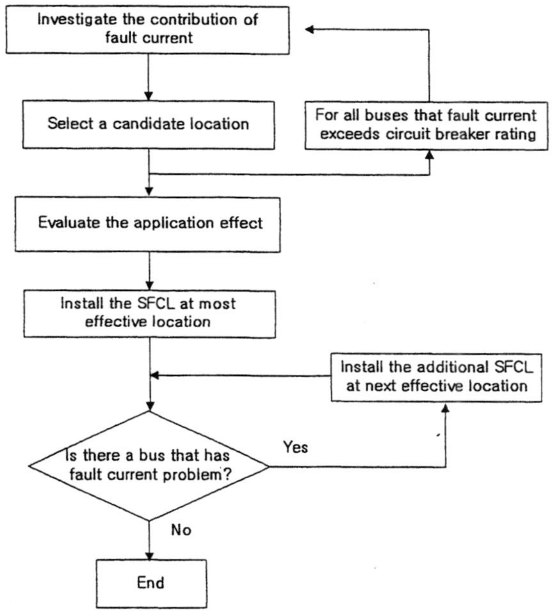

Fig. 6. Procedure of Applying SFCL in Power System

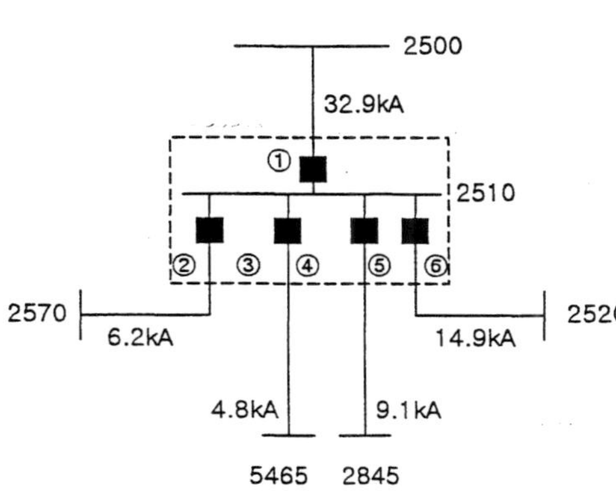

Fig. 7. Contribution of Fault Current at Bus 2510

Table 2 Candidate Location

Installation bus	Neighbor bus
2510	2500
2520	2510
2525	2520
2530	2520
2540	2530
2570	2510
2580	2570
2845	2510

2) Evaluate the Fault Current Reduction Effect

The resistive type SFCL and the inductive type SFCL are installed in the trunk transmission line to evaluate the fault current reduction effect shown in Fig. 8. The impedance of the SFCL after quenching is assumed to be 0.05 [pu] (based on 100[MVA]).

In case the SFCL is installed at bus 2525, the fault currents of seven out of eight buses are reduced below 50 kA as shown in Table 3. Therefore, bus 2525 can be considered the most effective location among candidate locations.

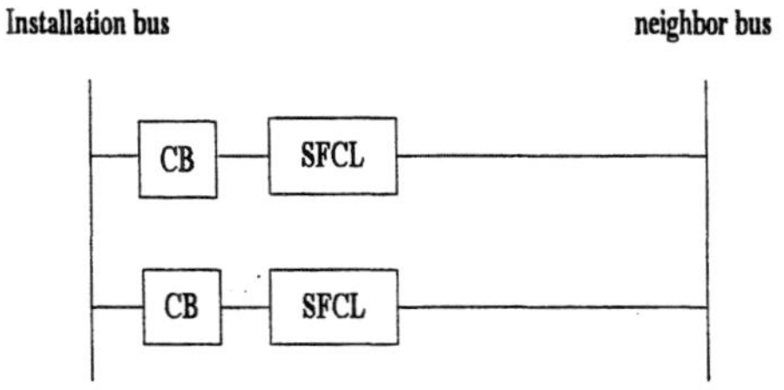

Fig. 8. Installation of SFCL in Transmission Line

Table 3 Application Effect of Candidate Location

Installation bus	Neighbor bus	Resistive type SFCL	Inductive type SFCL
2510	2500	5	5
2520	2510	0	0
2525	2520	7	7
2530	2520	4	4
2540	2530	4	4
2570	2510	4	4
2580	2570	1	1
2845	2510	1	1

4.3 Results of Case Study

According to the proposed procedure shown in Fig. 6, the SFCL is installed at bus 2525 and bus 2510 (seen in Table 4). The resistive type SFCL and the inductive type SFCL are considered and Table 5 shows the results of the case study

As seen in Table 5, the SFCL at bus 2525 and bus 2510 reduce to 50 kA the fault current of eight buses that have the fault current problem.

Table 4 Installation of SFCL

Installation bus	Neighbor bus	CKT
2525	2520	2
2510	2500	4

Table 5 Results of Case Study

Bus	Without SFCL [kA]	Resistance type SFCL [kA]	Inductive type SFCL [kA]
2510	67.9	43.5	39.6
2520	62.5	39.8	36.4
2525	61.7	31.7	35.3
2530	55.6	37.9	34.3
2540	55.6	38.2	34.5
2570	56.1	38.5	34.8
2580	52.0	36.6	33.1
2845	53.2	39.8	35.9

5. ECONOMIC BENEFIT OF SFCL

5.1 Cost o f upgrading Circuit Breakers

In a case study, the fault currents of eight buses exceed the circuit breaker rating. The typical conventional solution is the replacement of substation circuit breakers at an estimated cost of about $8,000,000.

By applying the SFCL, the fault current is reduced below the circuit breaker rating, eliminating the need to upgrade the substation circuit breakers and saving.

Table 6 summarizes the number of substation circuit breakers and the estimated upgrading cost.

Table 6 Cost of Upgrading Circuit Breakers(CB)

Bus	No. of CBs	Estimated Price of CB	Upgrading Cost
2510	16	$125,000	$2,000,000
2520	8	$125,000	$1,000,000
2525	6	$125,000	$750,000
2530	6	$125,000	$750,000
2540	8	$125,000	$1,000,000
2570	8	$125,000	$1,000,000
2580	6	$125,000	$750,000
2845	6	$125,000	$750,000
Total	64	-	$8,000,000

In this calculation, the upgrading cost is based only on circuit breakers. However, other electrical equipment, including DS, may need to be replaced in some buses. Thus, the total cost for upgrading electrical equipment is expected to be more than $8,000,000.

5.2 The Prospect Of Price Of SFCL

At this point, forecasting the cost of SFCL is difficult. The cost of the HTS wire and the refrigeration system in SFCL are keys to commercial success. In this paper, we not only review the present HTS wire cost and refrigeration system cost, but also project these costs for the future[5].

■ *High Temperature Superconducting(HTS) Wire*

The cost of HTS wire is generally described by a figure-of-merit measured in dollars per killoamp-meter($/kA-m), which is dependent on two parameters: first, the maximum amount of current the HTS wire will conduct; and second, the manufacturing cost per meter of wire. Both parameters are expected to improve as a result of advances in manufacturing techniques. Fig. 9 shows the prospective cost of HTS wire ($/kA-m) from 2005 to 2025. In this figure, the asymptotic value is about $20/kA-m.

■ *Refrigeration System*

The impact of refrigeration on the future competitiveness of SFCL is critical. The 1999 benchmark cost of a medium-sized refrigeration unit was about $60,000/kW at 77 K. The manufacturers assert that the cost of refrigeration will decrease as demand increases and more units are produced.

Therefore, the cost will drop to less than $20,000/kW.

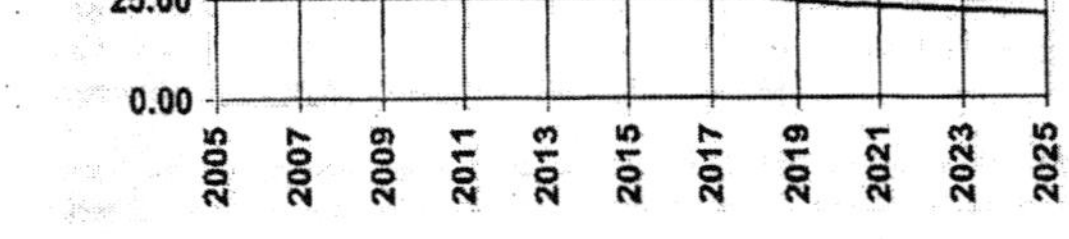

Fig. 9. Prospect of HTS Wire Cost

As a result of advance in manufacturing technique, economic benefit from applying SFCL is expected to be enormous in the future.

6. CONCLUSION

In this paper, we have carried out the technical and economic effects of SFCL in power system. In the case study, the SFCL is installed at two different buses, the most effective locations to reduce the fault current. By applying the SFCL, the fault current is reduced below circuit breaker rating, preventing the need to replace the circuit breakers and any other equipment and thus yielding a cost-savings of, more than $8,000,000.

As the cost of HTS wire and refrigeration drop, we can expect an enormous the economical benefit in the future.

ACKNOWLEDGMENT

The authors would like to express their gratitude with special thanks to the Center for Applied Superconductivity Technology in Korea for their financial support and cooperation.

REFERENCES

[1] L.R.Lawrence.Jr, C.Cox and D.Broman, "High temperature superconductivity: The products and their benefits," Bob Lawrence & Associates, Inc, 2000.

[2] M. Sjostrom and D. Politano, "Technical and Economical Impacts on a Power System by Introducing an HTS FCL," *IEEE Trans on Applied Superconductivity Conference, 2000.*

[3] R. Mikkonen, "Highlights of SC Power Applications in Europe," *IEEE Trans on Applied Superconductivity,* Vol.12, No.1, pp 782-787, 2002.

[4] Robert S. Sokolowski, "Power Applications of Superconductivity in Japan and Germany," WTEC panel report, 1997.

[5] J.Mulholland, T.P.Sheahen and B.Mcconnell, "Analysis of future prices and markets for high temperature superconductors," U.S Department of Energy, 2001.

ELSEVIER
IFAC
PUBLICATIONS
www.elsevier.com/locate/ifac

REIGNITION SURGE EFFECT IN A VACUUM CIRCUIT BREAKER ON INDUCTION MOTOR USING THE PSCAD/EMTDC

W.H. LEE, JIN LEE
Power System Team, R&D CENTER, LG Industrial Systems, Cheongju, KOREA

Abstract: When switching off an induction motor by Vacuum Circuit Breaker, it would damage the motor insulation by switching reignition surges. In order to investigate switching surge effects on motor insulation the simulations as parameters of cable length, motor size, with/without arrester around the Vacuum Circuit Breaker are performed by a Vacuum Circuit Breaker model that was developed by authors using the PSCAD/EMTDC. And the switching reignition surge profiles are obtained. Based on the surge profiles, the proper locations of the surge arrester for the protection of induction motor in distribution system are proposed. The results would be useful to analysis the reignition surges effects and to select adequately the location of arrester in the distribution system. *Copyright © 2003 IFAC*

Keywords: Vacuum Circuit Breaker, switching surge, induction motor, PSCAD/EMTDC

1. INTRODUCTION

Vacuum breakers have many advantages over other types of breakers for applications in distribution systems such as lower initial cost and reduced maintenance requirements. There have been concerns about using Vacuum breakers for switching induction motors, since the breaker operations may cause switching over-voltage waves with steep fronts, and motors are generally more vulnerable to impulse over-voltages compared with other apparatus like transformers. However, with proper protection by surge suppressors, arresters, the over-voltages associated with breakers can be controlled to a harmless level. So, typical switching surge voltage profiles of the Vacuum breaker for the induction motor are requested so that proper protection devices are set-up in the motor driving distribution system in accordance with the profiles.

In this paper, we provide reignition surge simulations of the Vacuum breaker Energization and De-energization using the VCB model developed by author in the EMTDC and qualities of surge over-voltages in an example system. The surge profiles produced by the simulations are compared with the IEEE Guide 522-1992. It is clarified that most cases are not meet the IEEE Guide unless the surge arrester

In such cases, we recommended the proper locations of arrester to be installed to reduce the harmful surges. The VCB model developed and the details of these over-voltage control measures are included in next sections.

2. VACUUM CIRCUIT BREAKER MODEL

A considerable amount of research has been done to develop models for studying the reignition of circuit breakers. Most of these models have been implemented into EMTP. But we developed the VCB model by EMTDC in which has a same calculation engine and further graphic environments than EMTP. The breaker model is a re-ignition simulation model. So the re-ignition process in the VCB model is controlled by the dielectric withstand capability and high frequency quenching capability characteristics. That is, the model controls repeatedly the switches(open/close) in EMTDC for implementation of the reignition surges profiles depend on whether the voltage across the VCB contacts is higher or less than the dielectric capability and the high frequency current at nearby current zero through the VCB is greater or less the high frequency clearing ability. The process in the simulation was calculated by the control program in EMTDC. The control program is characterized by the properties of the

- The ability of the circuit breaker to chop the current before its natural zero ("60Hz normal current chopping level").
- The Dielectric Withstand Capability(DWC) between the vacuum contacts when opening.
- High Frequency Interruption Capability(HFIC) of the vacuum discharge.

2.1 Normal chopping

Current chopping is the interruption of an AC current shortly before natural (sinusoidal) current zero. As a result, current will be not interrupted at $i=0$ but $i=I_0$ with I_0 the chopping current. The normal chopping current sets the initial conditions for the following transients. Principally, the prediction of the 60Hz chopping current is very complex. An explicit expressions are found in some papers[1,2]. But its validity has been investigated only for a small region. Nowadays, the mostly used CuCr contacts produce chopping current values in order of 3 ~ 5 A[3]. We chose 5A as chopping current for the simple simulations.

2.2 Dielectric Withstand Capability

The withstand capability is measured by the breakdown voltages taken at the different times after contact separation during opening. The time after contact separation is of course related to the contact travel and the contact material. These breakdown voltages are statistical in nature as illustrated in the paper[3]. In order to quantify the statistical trend many data points are required, typically anywhere between few hundred and few thousand points per graph. Using standard statistical regression analysis a best-fit distribution can be found. For the simulations, dielectric withstand voltages characteristics is approximately expressed by the following equation[4] using the statistical analysis.

$$V = A\,[T]^{B} \qquad (1)$$

Where

V : VCB dielectrical recovery voltage [V]

T : time from interruption [s]

A and B : constants

$A = 6.6 \times 10^6$ and $B = 0.9$ were used as A phase's dielectric recovery voltage characteristics. B, C phase voltage characteristics are same as A phase's.

2.3 High Frequency Interruption Capability

This is the third of the critical features that determine the unique behaviour of the vacuum device during transients. Current interruption in vacuum contact is controlled by complex process of arc plasma existing between the contacts at the current zero. The interruption capability of vacuum switches is measured by the rate of decay of the current prior to current zero, dI/dt. Since during electrical transients the surge currents are typically of high frequency interruption capability that should be represented in the mathematical model. In the paper[4], an interruption capability is represent as a straight line by which illustrates the best fit of the measured data points of high frequency dI/dt as in the case with the DWC. For simple simulations, we use a fixed value of 270A / μs as The High Frequency Interruption Capability.

2.4 The VCB modelling in EMTDC

Fig. 1 shows the block diagram of the VCB model developed by author using EMTDC Custom component by which the user can 'introduce' into the main algorithm of the program his/her own piece of code. In many cases this code represents models of components that are not available in PSCAD standard library.

A vacuum interrupt model should treat the behaviour of the vacuum gap in physically correct way, without knowing about the electrical circuit. This means that reasonable criteria must be chosen to switch between the different states. So, the VCB model obtained the voltages(U_s, U_L) between contacts, current(I_s) through VCB contacts, and old VCB switch states. And then the VCB states determined by the control program are delivered to VCB contacts open/close control module. The module (EMTDC S/W model) finally was operated by the decision. So, the switch module is open when the current nearly zero is smaller than the HFIC and close when the voltage between contacts is higher than the DWC of the breaker.

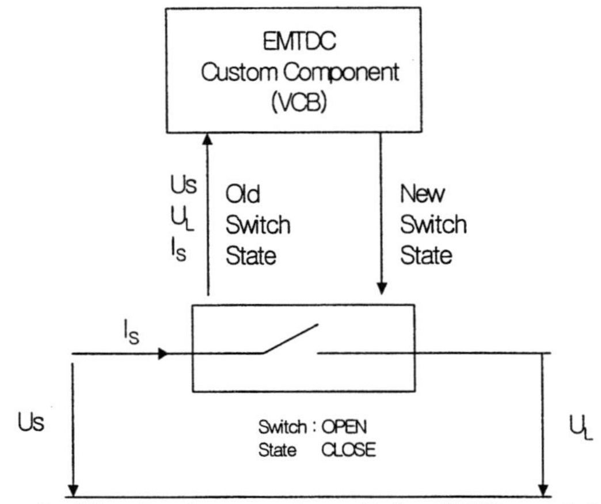

Fig. 1. The implementation of the VCB model in EMTDC.

3. SWITCHING SURGE SIMULATIONS WITH THE VCB MODEL

In Fig. 2, it is shown a network to be simulated. The time step used is 3μs, the simulation time is 0.2 second and the plot time is 2μs. First, we simulated the starting(closing) surge analysis of VCB on motor terminal in the network. Second, re-ignition surge analysis of the VCB when the VCB opened the line in which connected to the induction motor is performed based on some conditions. The surge analysis profiles obtained are compared to the IEEE Guide for "Testing Turn-to-Turn Insulation on Form-Wound Stator Coils for Alternating-Current Rotating Electric Machines" – Standard 522-1992(6.5p.u.)[6]. Under all circumstances, the surges at the motor terminal should be lower than the motor impulse withstand capability(6.5 p.u.). So if the simulation cases in which the surge amplitude is larger than 6.5 p.u. are founded, we recommended the location, rating of the surge arrester to be installed for protection of the induction motor in the distribution system.

3.1 DESCRIPTION OF THE NETWORK

A. Voltage Source

The voltage source is assumed to be at 13.8kV (1.05 p.u.). The source impedances include a short-circuit impedance representing the transformer impedance, an equivalent transformer and breaker capacitance. Source positive impedance is accounted that corresponds to the 1000MVA maximum short circuit capacity (about 41kA, 0.19 ohms for 13.8kV system).

B. Cables

The short line model (T model) was used to represent the cables that connect the induction motor. The used values were R = 0.424ohms/mile, L= 9.77Mh/mile, C = 0.937μF/mile from Westinghouse T&D Reference Book[5] for 750MCM Cables. The cable length is being from 200ft to 2000ft.

C. MOTOR

A motor used in simulation was modelled by the three-phase R-L load in parallel with line-to-ground capacitance. The three-phase motor models include a terminal to ground capacitance, and the equivalent winding impedance under the locked rotor condition. A starting current six times of the rated current is assumed to obtain the impedance. The parameters of the motor models used are as follows [5].

For 1000 HP motor,
R=6.96ohms, L=85mH, C=0.016μF
For 5000 HP motor,
R=1.392ohms, L=17mH, C=0.066μF
For 8800 HP motor,
R=0.79ohms, L=9.65mH, C=0.094μF

D. SURGE ARRESTER

The arrester is already modelled in EMTDC as nonlinear resistor. The electrical characteristics of the arrester used are found in the PDV/PVR arrester manual book[7] from OHIO Brass Company. The PDV/PVR arresters comply with ANSI and IEC standards for distribution arresters. The arrester ratings are selected according to the fact that the systems are ungrounded. 15KV(MCOV=12.7KV) PDV-100 arrester is used.

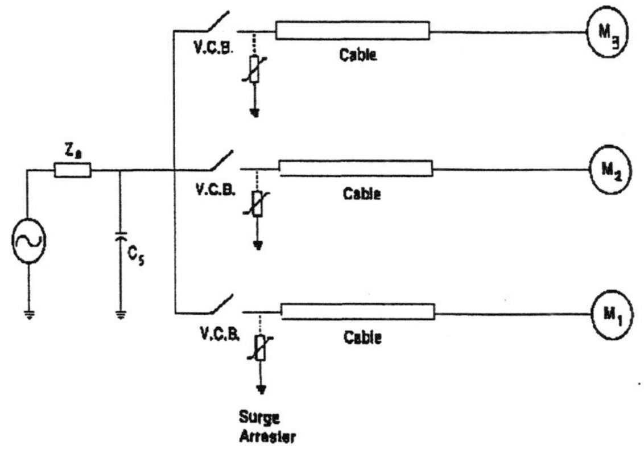

Fig. 2. The test network

3.2 SURGE SIMULATIONS

Operation conditions considered as variables in the simulations include 4 variables as follows:

- Motor size
  (1000HP, 5000HP, 8800HP)
- Cable length
  (200ft, 300ft, 500ft, 1000ft, 1500ft, 2000ft)
- Switching operation
  (Energization, De-energization)
- Presence of arrester
  (No/With arrester)

A. STARTING SURGE SIMULATIONS

The energization is the starting operation of the induction motor by the VCB switching (VCB open -> closing). The simulation results on starting surge profiles obtained are given table 1. All cases of starting surge profile are not founded to exceed the limit of IEEE Standard 522-1992(6.5p.u.).

Table 1. The starting surge profiles on induction motor(Y: below the IEEE Guide)

System Voltage	Motor size(HP)	Cable length	Peak Voltage KV	Peak Voltage P.U.	IEEE Std. 522-1992 OK?
13.8Kv	1000HP	200ft	19.0KV	1.76p.u.	Y
		300ft	23.0KV	2.13p.u.	Y
		500ft	23.0KV	2.13p.u.	Y
		1000ft	22.5KV	2.08p.u.	Y
		1500ft	22.5KV	2.08p.u.	Y
		2000ft	21.7KV	2.01p.u.	Y
	5000HP	200ft	22.0KV	2.04p.u.	Y
		300ft	23.1KV	2.14p.u.	Y
		500ft	24.4KV	2.26p.u.	Y
		1000ft	21.7KV	2.01p.u.	Y
		1500ft	19.3KV	1.79p.u.	Y
		2000ft	20.13KV	1.86p.u.	Y
	8800HP	200ft	21.0KV	1.95p.u.	Y
		300ft	22.0KV	2.04p.u.	Y
		500ft	22.38KV	2.07p.u.	Y
		1000ft	22.17KV	2.05p.u.	Y
		1500ft	18.15KV	1.68p.u.	Y
		2000ft	17.0KV	1.57p.u.	Y

B. REIGNTION SURGE SIMULATIONS

First of all, the important concern is the amplitude of the surge voltage at the motor terminal at which should be kept below the IEEE Guide.

The results of Re-ignition surges simulation in Table 2 show clearly that the limit can be exceeded without the proper protection device for over-voltages. That is, if cable length is longer than or equal 500ft when no arresters are used then re-ignition surges at the motor terminal exceeded the withstand capability limit (6.5 p.u.). See the cases (A500R, A1000R, A1500R, A2000R, B500R, B1000R, B1500R, B2000R, C500R, C1000R, C1500R) in the table 2.

Surge arresters at the VCB are sufficient for limiting the surge over-voltages magnitude at the motor terminals in table 2. But if cable length connected to the motors (8800, 5000HP) is equal or greater than 1000ft, the surge over-voltages are exceeded the IEEE Std. 522-1992 limit (6.5 p.u.) – see A1500R, A2000R, B1000R, B1500R in the table 2.

On the other hand, with surge arrester at the motor terminal any surges over-voltages are not founded to exceed the IEEE Std. 522-1992 limit in the table 2. That is, surges can be reduced below the withstand capability limit by placing the 15KV rated arrester at the motor terminal directly. So, according to the simulation results we recommended that a surge arrester should be installed at the motor terminal if the cable connected is greater than 1000ft.

In table 2, the C2000R is not exceeded the limit in spite of no arrester. If cable length is longer than 2000ft, the amplitude of surges are slowly decreased because the capacitances of the cable are increased and surge impedances are decreased [4].

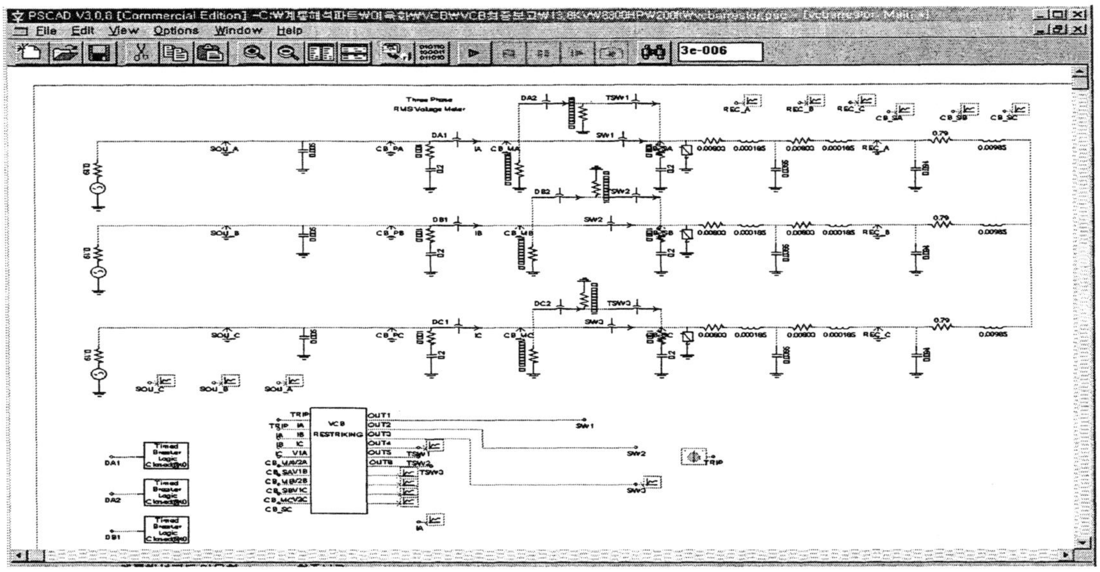

Fig.3 The test system implemented on the PSCAD/EMTDC.

Fig. 3 shows the simulation network on PSCAD/EMRDC. In Fig. 3, a box named "VCB RESTRIKING" is the EMTDC Custom Component developed by the authors. Inputs for the model are trip signal, three phase current values through VCB contacts, three phase voltages across the contacts. The outputs of the model are decision values (open : 1, close : 0) by which the switches(TSW1, SW1) are controlled to present the reignition surges.

Table 2. The re-ignition surge profiles on induction motor (Y: below the IEEE Guide, N: exceeded the Guide)

Rated Motor (HP)	Cable Length	Surge Voltages (NO Arrester)		IEEE Std. 522-1992 OK?	Surge Voltages (Arrester at motor terminal)		IEEE Std. 522-1992 OK?	Surge Voltages (Arrester at VCB)		IEEE Std. 522-1992 OK?	Case #
		Peak KV	p.u		Peak KV	p.u		Peak KV	p.u		
6600HP (Reactor Coolant Pump)	200ft	57	5.26	Y	39	3.62	Y	51	4.6	Y	A200R
	300ft	60	5.57	Y	41	3.80	Y	52	4.8	Y	A300R
	500ft	70	6.51	N	42	3.89	Y	60	5.5	Y	A500R
	1000ft	105	9.74	N	43	3.98.	Y	67.7	6.3	Y	A1000R
	1500ft	95	8.85	N	42	3.89	Y	84.1	7.8	N	A1500R
	2000ft	110	10.2	N	41	3.80.	Y	74	6.8	N	A2000R
5000HP	200ft	60	5.56	Y	37	3.43	Y	50	4.6	Y	B200R
	300ft	63.3	5.87	Y	38.5	3.57.	Y	59	5.4	Y	B300R
	500ft	120	11.1	N	41.0	3.80	Y	70	6.4	Y	B500R
	1000ft	114	10.6	N	42.0	3.89	Y	75.7	7.0	N	B1000R
	1500ft	153	14.2	N	41.0	3.10	Y	71.9	6.6	N	B1500R
	2000ft	80	7.42	N	44.0	4.06	Y	65	5.6	Y	B2000R
1000HP	200ft	61	5.66	Y	28	2.59	Y	48	4.4	Y	C200R
	300ft	70	6.5	Y	32	2.96	Y	53	4.9	Y	C300R
	500ft	74.5	6.90	N	34.8	3.22	Y	55.9	5.2	Y	C500R
	1000ft	76.5	7.10	N	39	3.62	Y	52	4.8	Y	C1000R
	1500ft	70.1	6.51	N	39	3.62	Y	50	4.6	Y	C1500R
	2000ft	55	5.1	Y	39	3.62	Y	50	4.6	Y	C2000R

Fig. 4 shows the switching surge simulation results according to the location of surge arrester. Fig. 4 (a) shows the reignition surge voltage profiles on the induction motor by the VCB without arrester. Fig. 4 (b) shows the reignition surge voltage profiles with arrester at motor terminal. Fig. 4 (c) shows the reignition surges with arrester at VCB. Summary, if arrester is properly located according to cable length, the surges when switching off can be lower than the motor withstand capability limits.

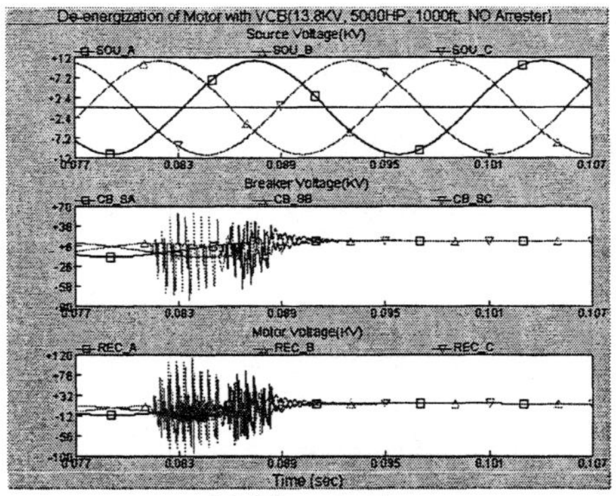

(a) No arrester

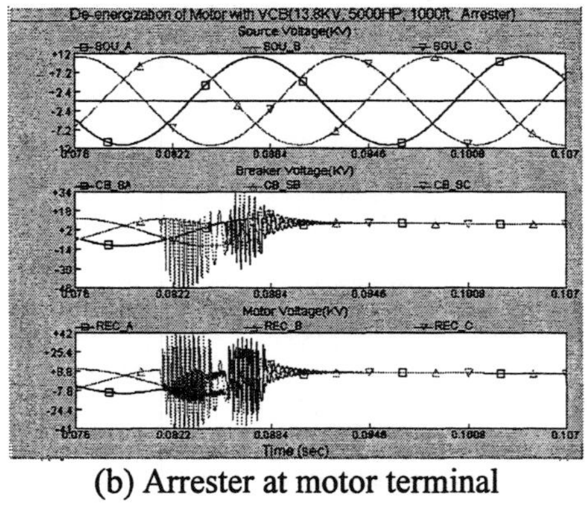

(b) Arrester at motor terminal

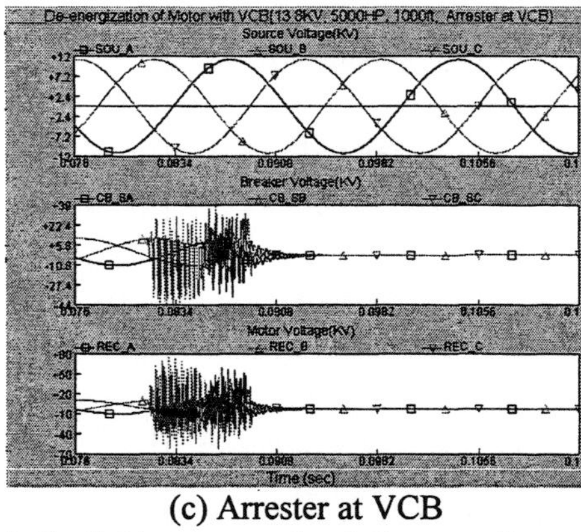

(c) Arrester at VCB

Fig. 4. Switching surge simulation results according to the location of surge arrester. (B1000R case)

4. CONCLUSIONS

Using the PSCAD/EMTDC a VCB model for surge analysis was developed. The breaker model is a re-ignition simulation model in which the re-ignition process is controlled by normal chopping, the dielectric withstand capability and high frequency interruption capability characteristics.

Simulations with the VCB model for switching surge analysis on the induction motor are performed on the test network. The starting surges are not constituted a threat to motor insulation. Some of reignition surges exceeded the motor IEEE Standard 522-1992. Surge arresters installed at VCB are sufficient for limiting the over voltage surges at motor terminal, provided that the cable length is less than 1000ft. For cables greater 1000ft, a surge arrester should be installed at nearly motor terminal for protection of the induction motor insulation.

The VCB model proposed in this paper would be useful to analysis the reignition surges effects of the VCB and to select adequately the location of arrester for reducing the surges in the distribution system.

REFERENCES

[1] J. Helmer, M.Lindmayer(1996), Mathematical Modelling of the High Frequency Behaviour of Vacuum Interrupters and comparison with Measured Transients in Power Systems, XVIIth International Symposium on Discharges and Electrical Insulation in Vacuum

[2] Janko Kosmac, et al(1995), A Statistical Vacuum Circuit Breaker Model for Simulation of Transient Overvoltages, IEEE Transaction on Power Delivery, Vol. 10, No. 1, January.

[3] Glinkowski, et al (1999). IEEE TUTORIAL ON THE VACUUM SWITCHGEAR sponsored by The Power Engineering Committee.

[4] K. Yokokura, et. al. (1981), Multiple restriking voltage effect in a vacuum circuit breaker on motor insulation, IEEE TRANSACTIONS ON PAS, Vol. PAS-100, No.4, April

[5] Westinghouse T&D Reference Book

[6] IEEE(1992), Testing Turn-to-Turn Insulation on Form-Wound Stator Coils for Alternating-Current Rotating Electric Machines, IEEE Standard 522-1992.

[7] PDV/PVR arrester manual Book (1996) from OHIO Brass Company.

BIOGRAPHIES

Wook-Hwa Lee was born in Kyungnam, South Korea, on March 27, 1969. He received B.S., M.S., Ph. D., degrees from the Electrical engineering of the KonKuk University, Seoul, Korea, 1991, 1993, 1998. Since 1999, Dr. Lee is a research engineer with Power System Team, R&D CENTER, LG Industrial Systems, Cheongju, 361-720, South of KOREA. His special fields of interest included Surge analysis, transients and Power Quality analysis in the distribution power systems.

Jin Lee was born in South Korea on Jan. 12, 1963. He received his B.S. degree from Seoul National University, Seoul, Korea in 1985. He joined LG Industrial Systems as a research engineer in 1985. He had worked with measuring devices, protective relays and distribution SCADA. He is presently working as a senior engineer with power system research team of R&D Center, LG Industrial Systems.

ELSEVIER
IFAC
PUBLICATIONS
www.elsevier.com/locate/ifac

SECURITY-CONSTRAINED OPTIMAL POWER FLOW USING FIRST-ORDER CONTINGENCY SENSITIVITY MATRIX

Jong-Ryul Won, Kil Choi

Dept. of Electrical & Electronic Eng., Anyang University, South Korea

Abstract: This paper proposes a MATLAB program for solving security-constrained optimal power flow using linear programming. Security-constrained optimal power flow is capable of finding an optimal generation satisfying bus voltage limits, line flow limits, reactive generation limits, even if unexpected contingency may occur. First-order sensitivity matrixes are obtained based on base case power flow solutions with and without contingency, respectively. This program is tested for a modified IEEE 14-bus system with 5 generators. Results show a good ability of finding feasible optimal solutions in case of a single line contingency. *Copyright © 2003 IFAC*

Keywords : Security-constrained optimal power flow, Linear programming, Contingency

1. INTRODUCTION

Operation of a power system is required to satisfy desirable economical benefits and reliability. However, these two requirements are incompatible with each other. Two much requirement of economical benefits can reduce power system stability and reliability, and vice versa. Therefore, effective method is requested to meet these two goals properly. Optimal power flow(OPF) is an alternative measure for this method. It has been also applied to power system analysis and operation in time of drastic change of electric power industry due to its restructuring. Especially security-constrained optimal power flow(SCOPF) is essential to economical and reliable power system operation. It can obtain an optimal solution which maintains transmission and voltage limits, while minimizing generation costs in case of N-n line contingency(A. J. Wood *et al* and H. Harsan *et al*). A great variety of solution methods have been applied to the OPF problems. Generally linear programming-based newton methods, successive linear programming, and interior point methods have been applied to OPF problem.

MATLAB provides a variety of effective tools for the calculation of matrix algebra and optimization problems etc. It has been used for many scientific and engineering parts(Hadi Saadat). Linear programming(LP) has been applied to many optimization problems due to its simplicity and convenience(B. Stott *et al*, O. Alsac *et al*, and M. Olofsson *et al*).

This paper presents a MATLAB program for solving SCOPF using linear programming. Two kinds of first-order sensitivity matrixes are obtained by base case AC power flow solutions with and without contingency in advance. Unlike conventional λ-iteration method, this program utilizes directly sensitivity matrix with respect to control variables(generations). Transmission line overloading and bus voltage limits are all satisfied in this SCOPF solutions in case of no outage or post-outage state. Contingency sensitivity matrixes are calculated based on the power flow solution obtained after simulating contingency from the base case generation state. Therefore post-contingency constraints are all satisfied exactly. Even in the multi-contingency case(N-n contingency), these matrixes are calculated only one time. These sensitivity matrixes can be used in many other power system analyses. This program is applied to a modified IEEE 14-bus system.

2. SECURITY-CONSTRAINED OPTIMAL POWER FLOW

The optimal power flow is a valuable tool for minimizing the cost of operating an electric power system. Security-constrained optimal power flow is aimed at determining a feasible, minimum cost operating point, such that in the event of a contingency the post-contingency state will remain feasible. This is a kind of preventive mode in the security analysis. Its general formulation is shown as the following equations.

$$Min.\ f = C^T \cdot x \tag{1}$$

$$A \cdot x \leq b \tag{2}$$

$$A_c \cdot x \leq b_c \tag{3}$$

$$x_{\min} \le x \le x_{\max} \tag{4}$$

,where x represents the control variables(generation ΔP_G). C is the cost vector, and eq.(2) operation constraints (bus voltage limit and transmission line overload limits) if any contingency does not happen, and eq.(3) operation constraints to be maintained even if any contingency happens in the optimal state. Eq.(4) shows the limit of control variables. Sensitivity matrixes A and A_c are obtained by base-case AC power flow solutions directly.
Detailed formulation is represented in the following equations. First, slack generation is derived by the other generations, and this is included in the cost function.

$$Min.\sum_{k=1}^{N_G} C_k \cdot \Delta P_{Gk} = Min.\left(\sum_{k=2}^{N_G} C_k\right) + C_1 \cdot \Delta P_1$$
$$= Min.\sum_{k=2}^{N_G}\left(C_k + C_1 S_k\right)\cdot \Delta P_{Gk} = Min.\sum_{k=2}^{N_G} C_k^{'} \Delta P_{Gk} \tag{5}$$

Constraints are also derived based on control variables(ΔP_G). In this paper, equality constraints are not needed since it is already included in the sensitivity matrix induction eq.(6).

s.t.

$$\Delta P_D = 0,\ \Delta Q_D = 0 \tag{6}$$

$$\begin{bmatrix} \theta_{D,\min} - \theta_{D,0} \\ V_{\min} - V_0 \end{bmatrix} \le A_1 \cdot \Delta P_G \le \begin{bmatrix} \theta_{D,\max} - \theta_{D,0} \\ V_{\max} - V_0 \end{bmatrix} \tag{7}$$

$$Q_{G\min} - Q_0 \le A_2 \cdot \Delta P_G \le Q_{G\max} - Q_0 \tag{8}$$

$$F_{\min} - F_{line0} \le T \cdot \Delta P_G \le F_{\max} - F_{line0} \tag{9}$$

$$P_{G\min} - P_{Go} \le \Delta P_G \le P_{G\max} - P_{Go} \tag{10}$$

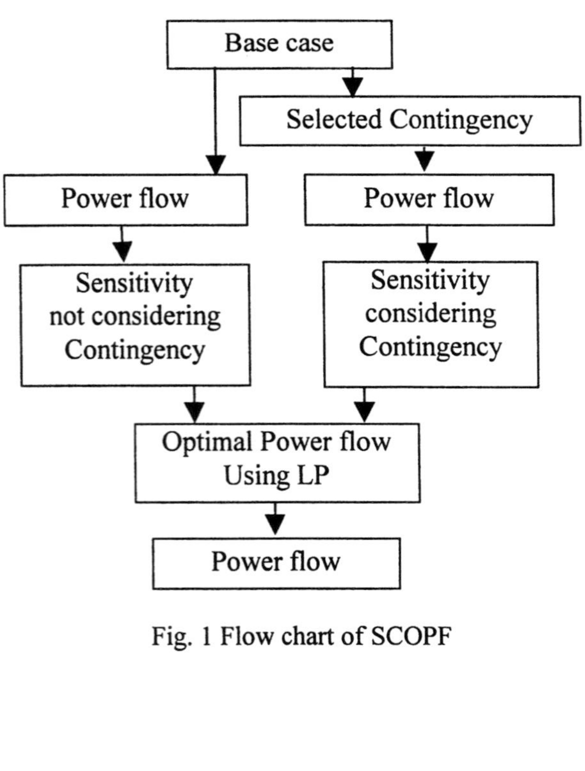

Fig. 1 Flow chart of SCOPF

Contingency constraints are derived by the following formulations.

$$\begin{bmatrix} \theta_{D,\min} - \theta_{D,0} \\ V_{\min} - V_0 \end{bmatrix} \le A_{C1} \cdot \Delta P_G \le \begin{bmatrix} \theta_{D,\max} - \theta_{D,0} \\ V_{\max} - V_0 \end{bmatrix} \tag{11}$$

$$Q_{G\min} - Q_0 \le A_{C2} \cdot \Delta P_G \le Q_{G\max} - Q_0 \tag{12}$$

$$F_{\min} - F_{line0} \le T_C \cdot \Delta P_G \le F_{\max} - F_{line0} \tag{13}$$

The derivation of above first-order sensitivity matrixes A_1, A_2, T, and A_{C1}, A_{C2}, T_C are explained next section. Detailed flow chart of SCOPF is shown in Fig. 1.

3. CALCULATION OF SENSITIVITY MATRIX

Sensitivity matrix of voltage variation, reactive power generation and transmission flows with respect to control variables(Originally ΔP_G , ΔV_G , ΔP_L , ΔQ_L , however in reality ΔP_G is a major control variable) is explained. Power balance condition is simply represented by the conditions $\Delta P_D = 0$, $\Delta Q_D = 0$.

Generally, bus active and reactive powers are represented by the equations of bus voltages and line admittances.

$$P_i = \sum_{k=1}^{n} V_i V_k \left[G_{ik}\cos(\theta_i - \theta_k) + B_{ik}\sin(\theta_i - \theta_k)\right] \tag{14}$$

$$Q_i = \sum_{k=1}^{n} V_i V_k \left[G_{ik}\sin(\theta_i - \theta_k) - B_{ik}\cos(\theta_i - \theta_k)\right] \tag{15}$$

Jacobian function with respect to voltage magnitude and angles is represented as the following equation. In this jacobian calculation, slack bus is excluded.

$$\begin{bmatrix} \Delta P_G \\ \Delta Q_G \\ \Delta P_D \\ \Delta Q_D \end{bmatrix} = \begin{bmatrix} \frac{\partial P_G}{\partial \theta_G} & \frac{\partial P_G}{\partial V_G} & \frac{\partial P_G}{\partial \theta_D} & \frac{\partial P_G}{\partial V_D} \\ \frac{\partial Q_G}{\partial \theta_G} & \frac{\partial Q_G}{\partial V_G} & \frac{\partial Q_G}{\partial \theta_D} & \frac{\partial Q_G}{\partial V_D} \\ \frac{\partial P_D}{\partial \theta_G} & \frac{\partial P_D}{\partial V_G} & \frac{\partial P_D}{\partial \theta_D} & \frac{\partial P_D}{\partial V_D} \\ \frac{\partial P_D}{\partial \theta_G} & \frac{\partial P_D}{\partial V_G} & \frac{\partial P_D}{\partial \theta_D} & \frac{\partial P_D}{\partial V_D} \end{bmatrix} \begin{bmatrix} \Delta\theta_G \\ \Delta V_G \\ \Delta\theta_D \\ \Delta V_D \end{bmatrix}$$

$$= \begin{bmatrix} A_{11} & A_{22} & B_{11} & B_{22} \\ A_{33} & A_{44} & B_{33} & B_{44} \\ C_{11} & C_{22} & D_{11} & D_{22} \\ C_{33} & C_{44} & D_{33} & D_{44} \end{bmatrix} \begin{bmatrix} \Delta\theta_G \\ \Delta V_G \\ \Delta\theta_D \\ \Delta V_D \end{bmatrix} \tag{16}$$

In this matrix, sensitivity matrixes with respect to control variables are derived one by one with respect to control variables.

$$\begin{bmatrix}\Delta\theta_D\\ \Delta V_D\end{bmatrix}=\begin{bmatrix}D_{11} & D_{22}\\ D_{33} & D_{44}\end{bmatrix}^{-1}\begin{bmatrix}\Delta P_D\\ \Delta Q_D\end{bmatrix}-\begin{bmatrix}D_{11} & D_{22}\\ D_{33} & D_{44}\end{bmatrix}^{-1}\bullet\begin{bmatrix}C_{11} & C_{22}\\ C_{33} & C_{44}\end{bmatrix}\begin{bmatrix}\Delta\theta_G\\ \Delta V_G\end{bmatrix} \quad (17)$$

$$\Delta P_G=\begin{bmatrix}A_{11} & A_{22}\end{bmatrix}\begin{bmatrix}\Delta\theta_G\\ \Delta V_G\end{bmatrix}+\begin{bmatrix}B_{11} & B_{22}\end{bmatrix}\begin{bmatrix}\Delta\theta_D\\ \Delta V_D\end{bmatrix} \quad (18)$$

ΔP_G is re-formulated by substitution of eq.(17) into eq.(18).

$$\Delta P_G=\begin{bmatrix}M_1 M_2\end{bmatrix}\begin{bmatrix}\Delta\theta_G\\ \Delta V_G\end{bmatrix}+\begin{bmatrix}N_1 N_2\end{bmatrix}\begin{bmatrix}\Delta P_D\\ \Delta Q_D\end{bmatrix} \quad (19)$$

Therefore, $\Delta\theta_G$ with respect to control variables can be derived by eq.(19).

$$\Delta\theta_G=M_1^{-1}\Delta P_G-M_1^{-1}M_2\Delta V_G-M_1^{-1}\begin{bmatrix}N_1 N_2\end{bmatrix}\begin{bmatrix}\Delta P_D\\ \Delta Q_D\end{bmatrix}$$

$$=\begin{bmatrix}M_1^{-1} & -M_1^{-1}M_2 & -M_1^{-1}N_1 & -M_1^{-1}N_2\end{bmatrix}\begin{bmatrix}\Delta P_G\\ \Delta V_G\\ \Delta P_D\\ \Delta Q_D\end{bmatrix} \quad (20)$$

and, $\Delta\theta_G$, ΔV_G variation with respect to control variables is capable of being expressed as the following equation.

$$\begin{bmatrix}\Delta\theta_G\\ \Delta V_G\end{bmatrix}=\begin{bmatrix}M_1^{-1} & -M_1^{-1}M_2 & -M_1^{-1}N_1 & -M_1^{-1}N_2\\ 0 & 1 & 0 & 0\end{bmatrix}\begin{bmatrix}\Delta P_G\\ \Delta V_G\\ \Delta P_D\\ \Delta Q_D\end{bmatrix}$$

$$=\begin{bmatrix}\Omega_1 & \Omega_2 & \Lambda_1 & \Lambda_2\\ \Omega_3 & \Omega_4 & \Lambda_3 & \Lambda_4\end{bmatrix}\begin{bmatrix}\Delta P_G\\ \Delta V_G\\ \Delta P_D\\ \Delta Q_D\end{bmatrix} \quad (21)$$

Also load bus voltage variation with respect to control variables is expressed by substitution of eq.(21) into eq.(17).

$$\begin{bmatrix}\Delta\theta_D\\ \Delta V_D\end{bmatrix}=\begin{bmatrix}D_{11} & D_{22}\\ D_{33} & D_{44}\end{bmatrix}^{-1}\begin{bmatrix}\Delta P_D\\ \Delta Q_D\end{bmatrix}-\begin{bmatrix}D_{11} & D_{22}\\ D_{33} & D_{44}\end{bmatrix}^{-1}\begin{bmatrix}C_{11} & C_{22}\\ C_{33} & C_{44}\end{bmatrix}*$$

$$\left(\begin{bmatrix}-M_1^{-1} & -M_1^{-1}M_2\\ 0 & E\end{bmatrix}\begin{bmatrix}\Delta P_G\\ \Delta V_G\end{bmatrix}+\begin{bmatrix}-M_1^{-1}N_1 & -M_1^{-1}N_2\\ 0 & 0\end{bmatrix}\begin{bmatrix}\Delta P_D\\ \Delta Q_D\end{bmatrix}\right)$$

$$=\left(\begin{bmatrix}D_{11} & D_{22}\\ D_{33} & D_{44}\end{bmatrix}^{-1}-\begin{bmatrix}D_{11} & D_{22}\\ D_{33} & D_{44}\end{bmatrix}^{-1}\begin{bmatrix}C_{11} & C_{22}\\ C_{33} & C_{44}\end{bmatrix}\begin{bmatrix}-M_1^{-1}N_1 & -M_1^{-1}N_2\\ 0 & 0\end{bmatrix}\right)*$$

$$\begin{bmatrix}\Delta P_D\\ \Delta Q_D\end{bmatrix}-\left(\begin{bmatrix}D_{11} & D_{22}\\ D_{33} & D_{44}\end{bmatrix}^{-1}\begin{bmatrix}C_{11} & C_{22}\\ C_{33} & C_{44}\end{bmatrix}\begin{bmatrix}-M_1^{-1} & -M_1^{-1}M_2\\ 0 & E\end{bmatrix}\begin{bmatrix}\Delta P_G\\ \Delta V_G\end{bmatrix}\right)$$

$$=\begin{bmatrix}U_1 & U_2\\ U_3 & U_4\end{bmatrix}\begin{bmatrix}\Delta P_D\\ \Delta Q_D\end{bmatrix}+\begin{bmatrix}W_1 & W_2\\ W_3 & W_4\end{bmatrix}\begin{bmatrix}\Delta P_G\\ \Delta V_G\end{bmatrix}$$

$$=\begin{bmatrix}W_1 & W_2 & U_1 & U_2\\ W_3 & W_4 & U_3 & U_4\end{bmatrix}\begin{bmatrix}\Delta P_G\\ \Delta V_G\\ \Delta P_D\\ \Delta Q_D\end{bmatrix} \quad (22)$$

ΔQ_G with respect to control variables is also derived by eq.(16), (21) and (22).

$$\Delta Q_G=\begin{bmatrix}A_{33} & A_{44}\end{bmatrix}\begin{bmatrix}\Delta\theta_G\\ \Delta V_G\end{bmatrix}+\begin{bmatrix}B_{33} & B_{44}\end{bmatrix}\begin{bmatrix}\Delta\theta_D\\ \Delta V_D\end{bmatrix}$$

$$=\left(\begin{bmatrix}A_{33} & A_{44}\end{bmatrix}\begin{bmatrix}\Omega_1 & \Omega_2 & \Lambda_1 & \Lambda_2\\ \Omega_3 & \Omega_4 & \Lambda_3 & \Lambda_4\end{bmatrix}+\begin{bmatrix}B_{33} & B_{44}\end{bmatrix}\begin{bmatrix}W_1 & W_2 & U_1 & U_2\\ W_3 & W_4 & U_3 & U_4\end{bmatrix}\right)\begin{bmatrix}\Delta P_G\\ \Delta V_G\\ \Delta P_D\\ \Delta Q_D\end{bmatrix}$$

$$=\begin{bmatrix}H_1 & H_2 & H_3 & H_4\end{bmatrix}\begin{bmatrix}\Delta P_G\\ \Delta V_G\\ \Delta P_D\\ \Delta V_D\end{bmatrix} \quad (23)$$

Line MW flow ΔF_l variation with respect to control variables is represented as follows. First line flow connecting bus i and bus k is formulated as following equation.

$$F_{ik}=-G_{ik}V_l^2+V_iV_k\left[G_{ik}\cos(\theta_i-\theta_k)+B_{ik}\sin(\theta_i-\theta_k)\right] \quad (24)$$

Therefore its variation is derived as follows.

$$\Delta F_l=\begin{bmatrix}\dfrac{\partial F_l}{\partial\theta_G} & \dfrac{\partial F_l}{\partial V_G}\end{bmatrix}\begin{bmatrix}\Delta\theta_G\\ \Delta V_G\end{bmatrix}+\begin{bmatrix}\dfrac{\partial F_l}{\partial\theta_D} & \dfrac{\partial F_l}{\partial V_D}\end{bmatrix}\begin{bmatrix}\Delta\theta_D\\ \Delta V_D\end{bmatrix} \quad (25)$$

By substitution of eq.(21) and (22) into eq.(25) its variation with respect to control variables is expressed.

$$\Delta F_l = \begin{bmatrix} \frac{\partial F_l}{\partial \theta_G} & \frac{\partial F_l}{\partial V_G} \end{bmatrix} \begin{bmatrix} \Omega_1 & \Omega_2 & \Lambda_1 & \Lambda_2 \\ \Omega_3 & \Omega_4 & \Lambda_3 & \Lambda_4 \end{bmatrix} \begin{bmatrix} \Delta P_G \\ \Delta V_G \\ \Delta P_D \\ \Delta Q_D \end{bmatrix} +$$

$$\begin{bmatrix} \frac{\partial F_l}{\partial \theta_D} & \frac{\partial F_l}{\partial V_D} \end{bmatrix} \begin{bmatrix} W_1 & W_2 & U_1 & U_2 \\ W_3 & W_4 & U_3 & U_4 \end{bmatrix} \begin{bmatrix} \Delta P_G \\ \Delta V_G \\ \Delta P_D \\ \Delta Q_D \end{bmatrix}$$

$$= \begin{bmatrix} T_1 & T_2 & T_3 & T_4 \end{bmatrix} \begin{bmatrix} \Delta P_G \\ \Delta V_G \\ \Delta P_D \\ \Delta Q_D \end{bmatrix} \qquad (26)$$

Finally, slack generation variation with respect to control variables is derived.

$$\Delta P_1 = \begin{bmatrix} \frac{\partial P_1}{\partial \theta_G} & \frac{\partial P_1}{\partial V_G} \end{bmatrix} \begin{bmatrix} \Delta \theta_G \\ \Delta V_G \end{bmatrix} + \begin{bmatrix} \frac{\partial P_1}{\partial \theta_D} & \frac{\partial P_1}{\partial V_D} \end{bmatrix} \begin{bmatrix} \Delta \theta_D \\ \Delta V_D \end{bmatrix}$$

$$= \begin{bmatrix} \frac{\partial P_1}{\partial \theta_G} & \frac{\partial P_1}{\partial V_G} \end{bmatrix} \begin{bmatrix} \Omega_1 & \Omega_2 & \Lambda_1 & \Lambda_2 \\ \Omega_3 & \Omega_4 & \Lambda_3 & \Lambda_4 \end{bmatrix} \begin{bmatrix} \Delta P_G \\ \Delta V_G \\ \Delta P_D \\ \Delta Q_D \end{bmatrix}$$

$$+ \begin{bmatrix} \frac{\partial P_1}{\partial \theta_D} & \frac{\partial P_1}{\partial V_D} \end{bmatrix} \begin{bmatrix} W_1 & W_2 & U_1 & U_2 \\ W_3 & W_4 & U_3 & U_4 \end{bmatrix} \begin{bmatrix} \Delta P_G \\ \Delta V_G \\ \Delta P_D \\ \Delta Q_D \end{bmatrix}$$

$$= \begin{bmatrix} S_1 & S_2 & S_3 & S_4 \end{bmatrix} \begin{bmatrix} \Delta P_G \\ \Delta V_G \\ \Delta P_D \\ \Delta Q_D \end{bmatrix} \qquad (27)$$

In this problem, by power balance condition, $\Delta P_D = 0$ and $\Delta Q_D = 0$ are assumed. And generator voltage variation cannot affect generation cost. Therefore only active generation ΔP_G is considered. Only sensitivity matrix Ω_1, W_1, W_3, H_1, T_1, S_1 are used.
Therefore sensitivity matrix is defined as follows.

$$A_1 = \begin{bmatrix} W_1 \\ W_3 \end{bmatrix}, \quad A_2 = [H_1], \quad T = [T_1]. \qquad (28)$$

Contingency constraints sensitivity matrixes A_{C1}, A_{C2}, T_C are derived in the same way, but They are calculated based on the power flow solution obtained after simulating contingency from the base case generation state. Therefore post-contingency constraints can be satisfied exactly.

4. CASE STUDY

4.1 Input Data

Proposed SCOPF algorithm is programmed by using MATLAB. Especially *LINPROG* function is used for LP. This algorithm is tested in the modified IEEE 14-bus system. Additional constraints, such as generation P & Q limits, voltage profiles' limit, and line flow limits, are properly assigned. Data is shown below tables. Bus diagram is also added in the below figure. Specially, voltage magnitude limit is assigned to 0.95~1.05 [p.u.] in the steady state, and 0.93~1.07 [p.u.] in outage case. Slack bus is selected to bus number 1. Cost function is assumed to be linear function of generation.

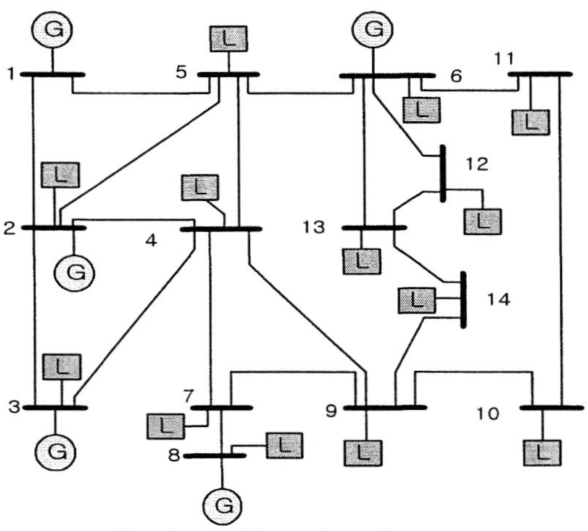

Fig.2 IEEE 14-bus diagram

Table 1 IEEE 14-bus data

Bus No.	P_{Gmax}	P_{Gmin}	Q_{Gmax}	Q_{Gmin}	P_L	Q_L	COST
1	200.0	10.0	200.0	-200.0	0.0	0.0	0.6
2	50.0	20.0	200.0	-200.0	11.7	12.7	3.5
3	80.0	15.0	200.0	-200.0	64.2	14.0	12.4
4	0.0	0.0	0.0	0.0	47.8	-3.9	0.0
5	0.0	0.0	0.0	0.0	7.6	1.6	0.0
6	120.0	10.0	200.0	-200.0	21.2	7.5	1.6
7	0.0	0.0	0.0	0.0	0.0	0.0	0.0
8	45.0	10.0	200.0	-200.0	35.0	5.0	5.0
9	0.0	0.0	0.0	0.0	29.5	16.6	0.0
10	0.0	0.0	0.0	0.0	9.0	5.8	0.0
11	0.0	0.0	0.0	0.0	3.5	1.8	0.0
12	0.0	0.0	0.0	0.0	6.1	1.6	0.0
13	0.0	0.0	0.0	0.0	13.5	5.8	0.0
14	0.0	0.0	0.0	0.0	14.9	5.0	0.0

4.2 Analysis of Results

Table 2 Result of OPF and SCOPF(Generation)

Bus	Initial P_G	OPF Result	SCOPF Result	COST Coeff.	P_{Gmax}	P_{Gmin}
1	92.22	99.90	97.25	0.6	200.0	10.0
2	50.00	20.00	50.00	3.5	50.0	20.0
3	70.00	15.00	19.34	12.4	80.0	15.0
6	30.00	94.41	60.10	1.6	120.0	10.0
8	25.00	37.99	40.31	5.0	5.0	10.0

Simulation results are summarized at the Table 2,3,4 and 5. As shown in the Table 3, base case line flow has overloading on the 1-5 line and 4-9 line. These overlodings are removed after the OPF simulation. However, even at the optimal point after OPF, line overloading also happens, when 5-4 line has an outage. This problem has solved by proposed SCOPF performance as shown in the Table 3. Cost variation is shown in the Table 4. SCOPF result raises cost value compared with OPF since it has more constraints.

Table 3 Comparison of line flows' variation

Line No.	Flow Limit	Initial Flow	OPF Result (1)	In (1) 5-4 line outage	SCOPF Result (2)	In (2) 5-4 line outage
1-2	240.0	10.79	39.90	74.46	24.45	55.72
3-2	48.0	-9.94	-32.18	-48.36	-33.15	-47.88
2-4	84.0	26.15	15.39	48.87	21.90	52.37
1-5	72.0	81.45	60.00	28.27	72.08	44.10
2-5	48.0	12.92	-0.21	-16.79	7.03	-8.01
3-4	36.0	15.74	-17.02	-0.84	-11.71	3.02
5-4	96.0	49.90	65.22	-	59.19	-
5-6	60.0	35.89	-13.51	3.62	12.31	28.19
4-7	48.0	17.01	1.55	-2.53	1.47	-2.28
8-7	72.0	-10.00	2.99	2.99	5.31	5.31
4-9	24.0	26.15	13.34	1.33	19.28	8.24
7-9	60.0	6.85	4.51	0.40	6.75	2.95
9-10	24.0	-1.57	-11.00	-20.97	-5.91	-15.04
6-11	30.0	14.19	23.83	34.18	18.62	28.00
6-12	24.0	9.25	10.55	12.06	9.83	11.19
6-13	36.0	20.50	25.24	30.59	22.65	27.47
9-14	36.0	5.07	-0.65	-6.80	2.45	-3.26
10-11	24.0	-10.57	-20.00	-29.97	-14.91	-24.04
12-13	24.0	3.15	4.45	5.96	3.73	5.09
13-14	24.0	10.14	16.17	23.02	12.86	19.03

Table 4 Comparison of costs

	Initial Cost	OPF Result	SCOPF Result
COST	12.71	6.57	7.71

Table 5 Comparison of voltage magnitudes variation

Bus	Initial Voltage	OPF Result (1)	In (1) 5-4 line outage	SCOPF Result (2)	In (2) 5-4 line outage
1	1.0000	1.0000	1.0000	1.0000	1.0000
2	1.0000	0.9994	1.0000	0.9999	1.0000
3	1.0000	0.9996	1.0000	0.9999	1.0000
4	0.9890	0.9894	0.9831	0.9895	0.9827
5	0.9947	0.9944	0.9984	0.9947	0.9988
6	1.0000	1.0006	1.0000	1.0010	1.0000
7	0.9940	0.9939	0.9922	0.9942	0.9920
8	1.0000	0.9994	1.0000	0.9999	1.0000
9	0.9650	0.9605	0.9504	0.9635	0.9543
10	0.9702	0.9644	0.9543	0.9677	0.9591
11	0.9885	0.9889	0.9867	0.9886	0.9873
12	0.9924	0.9950	0.9960	0.9935	0.9949
13	0.9870	0.9900	0.9910	0.9883	0.9898
14	0.9558	0.9509	0.9416	0.9537	0.9458

Voltage magnitudes maintain all limited values after OPF and SCOPF as shown in the Table 5. However, in the SCOPF result, more stable results are obtained.

5. CONCLUSIONS

This paper presented a MATLAB program solution for SCOPF using linear programming. Two kinds of first-order sensitivity matrixes were obtained by base case AC power flow solutions with and without contingency in advance. This program was applied to a modified IEEE 14-bus system, and the results showed its effectiveness and usefulness.

REFERENCES

A. J. Wood, B. F. Wollenberg.(1996). *Power Generation Operation and Control.* John Wiley & Sons Inc. 1996.

Hadi Saadat (1999). *Power System Analysis.* McGraw Hill.

B. Stott, J. L. Marinho (1979). Linear Programming For Power-system Network Security Applications. *IEEE Trans. on Power Apparatus and System.* **Vol.PAS-98. No.3.** pp.1093-1097

O. Alsac, J. Bright, M. Prais, B. Stott (1990). Further Developments in LP-Based Optimal Power Flow. *IEEE Transactions on Power Systems.* **Vol. 5. No. 3.** pp.697-711.

M. Olofsson, G. Andersson, L. Soder (1995). Linear Programming based Optimal Power Flow Using Second Order Sensitivities. *IEEE Transactions on Power Systems.* **Vol.10. No.3.** pp.1691-1697.

H. Harsan, N. Hadjsaid, P. Pruvot (1997). Cyclic Security Analysis for Security Constrained Optimal Power Flow. *IEEE Transaction on Power Systems.* **Vol. 12. No.2.** pp.948-953.

ELSEVIER
IFAC
PUBLICATIONS
www.elsevier.com/locate/ifac

CONTROL ANALYSIS AND STABILITY FOR STATIC VAR GENERATOR

Hanxiang CHENG[1,2], Xianggen YIN[1], Zhihua WANG[1], Shaohua CHEN[2]

1. Faculty of Electrical and Electronic Engineering
Huazhong Univ. of Science and Technology, 430074, Wuhan, Hubei, China
2. Faculty of Automation
Guangdong Univ. of Technology, 510090, Guangzhou, Guangdong, China

Abstract: Usually SVG plays a role of reactive power-source in the power distribution and transmission systems. As the control of SVG is dependent of system parameters and control parameters, the dynamic performance and stability becomes a very important research field for SVG control. According to the simple power system topology, the relationship of control characteristic among equivalent load, system transmission impedance, damping inductance, adjustable reactive power of SVG, control angle δ, and the DC-link voltage of inverter has been analyzed in this paper. Especially stabilization of SVG and the accepted parameter-range of control parameters under stabilization condition have been descript in detail also. The processing method used for SVG stabilization discussed in this paper would be applied to any configuration of SVG. Through complex calculation, real simulation, and by the selected control parameters in the descript range, some good results are obtained.

Key Words: SVG, Stabilization, Power Electronics

1. INTRODUCTION

Applications of power electronics in the field of power distribution and transmission systems is attracting wide attention, especially, the FACTS (Flexible AC Transmission Systems) technology is developed rapidly, of which, SVG (Static Var Generator) is more likely applicable in power system because of its anticipated large-scale application in future, and flexible control strategy. SVG is of many advantages over traditional static reactive-power compensators (SVC). It can continuously adjust followed in and/or followed out reactive power, control AC bus voltage, improve system stability, enhance dynamic stabilization of system, and damp system-surge. The demands that are made on the power system because of unbundling of present usage to allow open access to third parties for transmission of power will force application of advanced controllers, and improve transmission quality of power system. Nevertheless, the stability and dynamic characteristic are related to the load characteristic, distributed impedance of transmission line to user, AC current damping- inductance between connected point of power bus and SVG, and control parameters in the SVG control. There are rare publications [1] about the stability of SVG, so how to control SVG, achieve optimized effective of power system, and enlarge control angle δ under stable condition, are the main problem to be solved in this paper.

2. REACTIVE-POWER COMPENSATION DESCRIPTION OF SVG

Normally speaking, SVG is consisted of power electronics, capacitance, and distributed impedance. In other words, if the distributed resistance and the loss of power electronics are ignored, SVG produces neither active power, nor absorbs active power. So the SVG could be considered a reactive power source. This reactive-power gets excitation energy not from outside but from power system itself. On the assumption that system power U_s is supplied through transmission line-impedance $Z_s = r_s + jx_s$ to user, at the node point; SVG is connected to the bus. So the simple system topology could be gotten as in Fig.1.

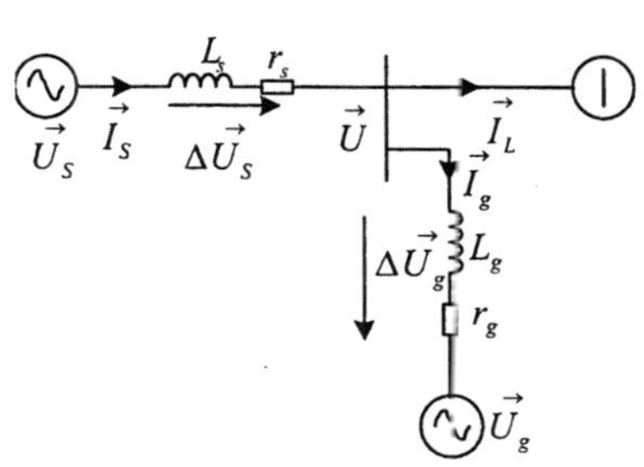

Fig.1 The simple topology of power system

Where $\vec{I}_L$ is equivalent load current representing all load in the system; r_g and L_g are the equivalent resistance and current-damping inductance respectively; $\vec{U}$ represents node-voltage; $\vec{U}_S$ represents system power voltage or applied power source. From above topology the node voltage $\vec{U}$ can be described as:

$$\vec{U} = \frac{\vec{U}_s + (Z_S / Z_G)\cdot \vec{U}_g - \vec{I}_L \cdot Z_S}{1 + Z_S / Z_G} \qquad (1)$$

Because SVG is a reactive-power source and if the loss of branch SVG is ignored, the added constrained condition would be:

$$\vec{U} = k\vec{I}_g \angle \pm 90^0 \qquad (2)$$

Where k is impedance constant which include system impedance, equivalent load, and other parameters. Supposing that the angle between node-voltage vector and system-voltage vector is α_u, as illustrated in Fig.2, so the relationship between $\vec{I}_g$ and $\vec{U}$ can be described as

$$\vec{U} = U_s - \vec{I}_s(r_s + jx_s) = U_s - (\vec{I}_g + \vec{I}_L)\cdot(r_s + jx_s) \text{ or}$$

$$\begin{aligned} U\angle\alpha_u &= U_s - \{ [I_{g\alpha} + I_{L\alpha} + j(I_{g\beta} + I_{L\beta})]\cdot(r_s + jx_s)\} \\ &= U_s - (I_{g\alpha} + I_{L\alpha})r_s - (I_{g\beta} + I_{L\beta})x_s \\ &\quad - j[(I_{g\alpha} + I_{L\alpha})x_s + (I_{g\beta} + I_{L\beta})r_s] \end{aligned}$$

The dotted line in Fig.2 represents producing reactive-power while the real line represents absorbing reactive-power to SVG. Considering the vector relationship of dark real line in Fig.2, the α_u could be calculated as:

$$tg(\alpha_u) = \frac{-x_s A_1 - r_s A_2}{U_s - r_s A_1 - x_s A_2} \qquad (3)$$

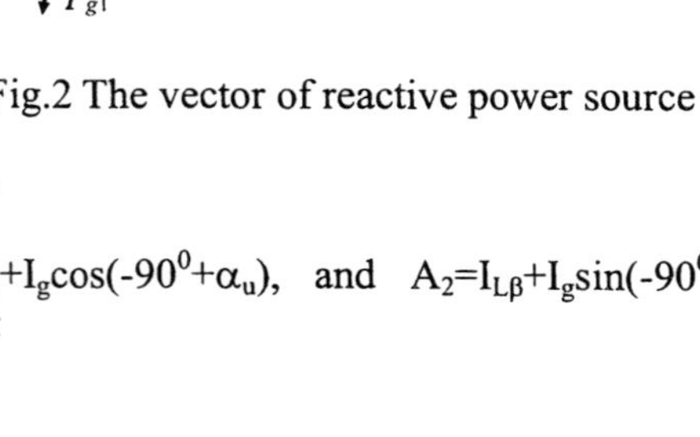

Fig.2 The vector of reactive power source

Where

$A_1 = I_{L\alpha} + I_g\cos(-90^0 + \alpha_u)$, and $A_2 = I_{L\beta} + I_g\sin(-90^0 + \alpha_u)$. Hence:

$$I_g = \frac{tg\alpha_u(U_s - I_{L\alpha}r_s - I_{L\beta}x_s) - x_s I_{L\alpha} - r_s I_{L\beta}}{x_s\cos(-90^0 + \alpha_u) - r_s\sin(-90^0 + \alpha_u) - A\cdot tg\alpha_u} \qquad (4)$$

$A = x_s\sin(-90^0 + \alpha_u) + r_s\cos(-90^0 + \alpha_u)$,
hence

$$\begin{aligned} \vec{U} &= U_s - \vec{I}_s(r_s + jx_s) \\ &= U_s - (I_g\angle(\alpha_u \pm 90^0) + I_L\angle\varphi_L \end{aligned}$$

Here, if I_g is positive, the current direction of SVG branch departs the connected bus towards SVG, at the same time, the symbol takes negative 90^0; if I_g= "-"; taking positive symbol before 90^0, the current direction is reversed as reference direction as in Fig.1.

In fact, the angle α_u represents the angle difference between the vector of SVG equivalent voltage-source $\vec{U}_g$ and the vector of system voltage source $\vec{U}_s$, i.e. the control angle δ under the assumption of angle crossover of vector $\vec{U}$ and vector $\vec{I}_g$ as shown in Fig.2. Supposed that U=380V, r_s=0.023Ω, transmission line inductance x_s= 0.0178Ω, equivalent resistance of SVG branch r_g=0.03Ω, the damping inductance together with filter of SVG branch x_g=0.059Ω, equivalent load current I_L=200A, and the load power factorcosφ_L=cos50^0, the relationship of control angle δ and other electrical quantity is drawn in Fig.3 according the equations (3) and (4).

With the same manner above, we can get the similar figures when some system parameters changing (for instance, x_s, x_g, I_L), and the conclusions can be made for the purpose of correct SVG operation:

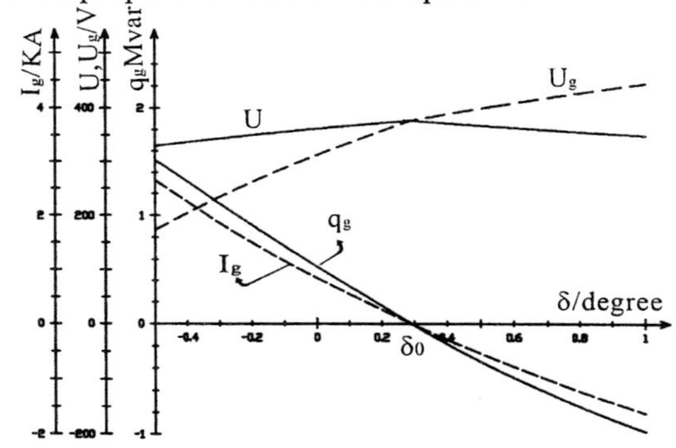

Fig.3 The relationship of δ and other variables

1) The magnitude of equivalent SVG voltage source should be higher than that of node voltage when SVG sent out reactive power; and the equivalent voltage source of SVG would be lower than that of the node voltage when SVG absorbs reactive power.
2) The range of SVG control-angle δ is not large, otherwise, the node-voltage and/or banch current of SVG would be acutely decreased and/or increased.
3) Controlable range of δdepends closely upon

system transimission impedance Z_s. The greater the Z_s is, the wider the range of δ. Furthemore, the base point δ depends also upon system impedence Z_s, and the central control point δ_0 shifts right as Z_s increases.

4) The range of control angle δ will be decreased when damping inductance L_g increases. Fig.4 presents the SVG characteristic of larger L_g which is 10 times larger as that in Fig.3 when keeping other parameters constant. It is clear that δ can shift only in a narrow range when node-voltage drop allows only 10%, and adjustable reactive power is smaller. Generally, a larger damping-inductance is expected for the reason of power-polution and commutation, but it conflicts with control range of δ.

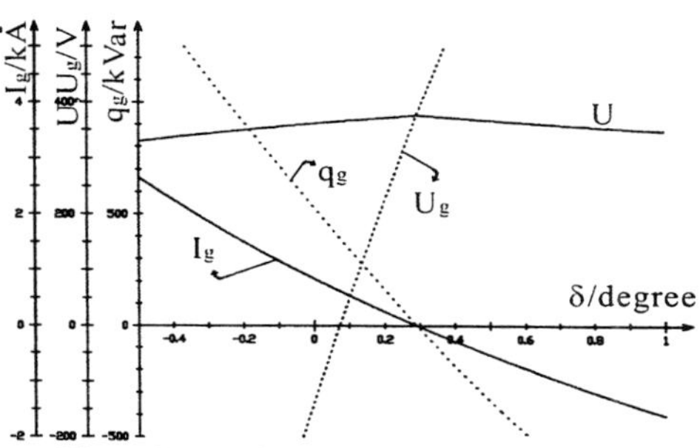

Fig.4 SVG characteristic of larger L_g

5) Keeping all other parameters constant, decreasing transmission impedance could enlarge the range of δ while node voltage-drops is within allowable level, so the reactive-power and node-voltage can be controlled in a larger range.

Ensuring power quality, such as reactive power, node-voltage, and et al, is the basic idea of SVG control system through output triggering-pulses of power electronics which makes the pulses shift forwards or bachwards with the reference of basic control point.

3. MATHEMATIC MODE OF SVG CONTROL SYSTEM

As described in above section the SVG is regarded as an ideal reactive-power source, but real SVG is consisted of a number of nonlinear elements, such as power electronics, capacitance, some distributed resistance and inductance in fact. Normally speaking, SVG includes an inverter, an optional rectifier or some other voltage/current regulators, which depend on the material configuration of SVG in order to get steady DC-link voltage/current. It may be a simple three-phase bridge inverter with rectifier, or some other voltage/current supposition inverter for the adjustment of high reactive power. Because of the reason of volume and weight, DC-link capacitor is normally adapted in most inverter; this is so called voltage-type SVG as shown in Fig.5. In this type of SVG, the capacitance voltage, i.e. DC-link voltage, is a very important quantity, it corresponds to equivalent voltage-source amplitude U_g of reactive power as shown in Fig.1. The mathematic mode of this type of voltage-source SVG will be described in this section.

For simplicity, supposing the control object is reactive power over the transmission line and node voltage at connected bus, so the corresponding control system would be depicted in Fig5. In the Figure, U represents node-voltage; I_s system output current, U_d the DC-link voltage, U_g the input voltage of SVG, r_s and Ls transmission resistance and inductance respectively, r_g and L_g equivalent resistance and inductance of SVG branch.

The controller is mainly consisted of a piece of digital signal processing (DSP) microcomputer, other logic components and measuring interfaces. The control process includes reactive power calculation, PI control, add/minus and the calculation of control angle δ based on synchronous rotated α, β coordinate. Since $\vec{U}_s = U_{s\alpha}+jU_{s\beta}=U_s$, $\vec{I}_s = I_{s\alpha}+jI_{s\beta}$, and $\vec{I}_L = I_{L\alpha}+jI_{L\beta}$, the equations can be written as following for this feed-back control system:

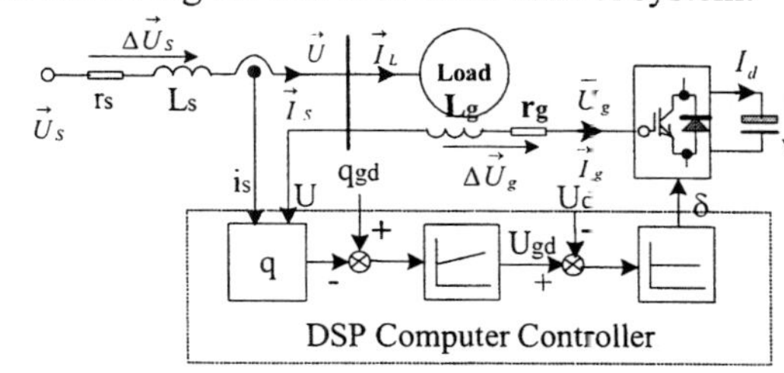

Fig.5 Feed-back control system of SVG

$$q = -\frac{3}{2}\left(u_{s\alpha} i_{s\beta}\right)$$

$$u_{gd} = k_{qp}\left(q_{gd} - q\right) + k_{qi}\int\left(q_{gd} - q\right)dt;$$

$$\delta = k_{up}\left(u_{gd} - u_d\right)$$

Here q_{gd} is the desired reactive power, which normally takes 0, k_{qp} is proportional coefficient of q regulator, k_{qi} is integral coefficient, and k_{up} is proportional coefficient of voltage regulator. So the corresponding differential equations will be derived respectively:

$$\dot{\delta} = k_{up}\dot{u}_{gd} - k_{up}\dot{u}_d \tag{5}$$

$$\dot{u}_{gd} = \frac{3}{2}k_{qp}\left(\dot{U}_s i_{s\beta} + U_s \dot{i}_{s\beta}\right) + \frac{3}{2}k_{qi}U_s i_{s\beta} + k_{qi}q_{gd}$$

$$= \frac{3}{2}k_{qp}U_s \dot{i}_{s\beta} + \frac{3}{2}k_{qi}U_s i_{s\beta} + k_{qi}q_{gd}$$

$$= \frac{3}{2}k_{qp}U_s(\dot{i}_{g\beta} + \dot{i}_{L\beta}) + \frac{3}{2}k_{qi}U_s(i_{g\beta} + i_{L\beta}) + k_{qi}q_{gd} \tag{6}$$

Where $\vec{(\ )}$ and $\dot{(\ )}$ represent vector and differential symbol. For inverter, the relationship between DC-link voltage U_d and node-voltage $\vec{U}$ can be presented as:

$$\vec{U}_g = \vec{U} - \Delta\vec{U}_g = \vec{H}U_d \quad (7)$$

In the expression above, $\Delta\vec{U}_g$ is voltage-drop vector of SVG branch; $\vec{U}$ is a transfer vector from AC to DC, which can be defined as [2]:

$$\vec{H} = He^{j(\gamma_H+\delta)};\ \gamma_H = \int \omega_s dt \quad (8)$$

Here, ω_s is frequency of power system; δ is trigger angle of SVG control referenced to power supplier; H is scalar quantity, for a given configuration of SVG, it is a constant. So the input voltage expression of SVG could be written as:

$$U_{g\alpha} = H\cdot U_d\cos\delta \quad (9)$$

$$U_{g\beta} = H\cdot U_d\sin\delta \quad (10)$$

From the expressions above, $\Delta\vec{U}_g$ and $\Delta\vec{U}_s$ can be written as following:

$$\Delta\vec{U}_g = \vec{I}_g r_g + L_g\frac{d\vec{I}_g}{dt}$$

$$= r_g I_{g\alpha} + L_g\dot{I}_{g\alpha} + j(r_g I_{g\beta} + L_g\dot{I}_{g\beta})$$

$$\Delta U_{g\alpha} = r_g I_{g\alpha} + L_g\dot{I}_{g\alpha} \quad (11)$$

$$\Delta U_{g\beta} = r_g I_{g\beta} + L_g\dot{I}_{g\beta} \quad (12)$$

$$\Delta\vec{U}_S = \vec{I}_S r_s + L_S\frac{d\vec{I}_S}{dt} = r_s I_{s\alpha} + L_s\dot{I}_{s\alpha} + j(r_s I_{s\beta} + L_s\dot{I}_{s\beta})$$

Transferring this express to α, β coordinator:

$$\Delta U_{s\alpha} = r_s I_{s\alpha} + L_s\dot{I}_{s\alpha} \quad (13)$$

$$\Delta U_{s\beta} = r_s I_{s\beta} + L_s\dot{I}_{s\beta} \quad (14)$$

Since $i_{s\beta}=i_{g\beta}+i_{L\beta}$, $i_{s\alpha}=i_{g\alpha}+i_{L\alpha}$, so node voltage could be expressed after simplification:

$$u_\alpha = i_{g\alpha}r_g + L_g\dot{i}_{g\alpha} + Hu_d\cos\delta\ ;$$

$$u_\beta = i_{g\beta} + L_g\dot{i}_{g\beta} - Hu_d\sin\delta,\ \text{i.e.}$$

$$u_\alpha = u_{s\alpha} - r_s i_{s\alpha} - L_s\dot{i}_{s\alpha}$$

$$= U_s - r_s(i_{g\alpha} + i_{L\alpha}) - L_s(\dot{i}_{g\alpha} + \dot{i}_{L\alpha})$$

$$u_\beta = u_{s\beta} - r_s i_{s\beta} - L_s\dot{i}_{s\beta}$$

$$= -r_s(i_{g\beta} + i_{L\beta}) - L_s(\dot{i}_{g\beta} + \dot{i}_{L\beta})$$

Considering all the representations above, the SVG branch current can be written as:

$$\dot{i}_{g\beta} = \frac{1}{L_g}u_\beta - \frac{r_g}{L_g}i_{g\beta} + Hu_d\sin\delta$$

$$-\frac{1}{L_g}\left[r_s(i_{g\beta} + i_{L\beta}) + L_s(\dot{i}_{g\beta} + \dot{i}_{L\beta})\right] - \frac{r_g}{L_g}i_{g\beta} + Hu_d\sin\delta$$

i.e.,

$$\dot{i}_{g\beta} = -\frac{r_s + r_g}{L_s + L_g}i_{g\beta} + \frac{L_g Hu_d\sin\delta}{L_s + L_g} - \frac{L_s\dot{i}_{L\beta}}{L_s + L_g} - \frac{r_s i_{L\beta}}{L_s + L_g}$$

With the similar deriving process, α part of SVG branch current is expressed as:

$$\dot{i}_{g\alpha} = -\frac{r_s + r_g}{L_s + L_g}i_{g\beta} - \frac{L_g Hu_d\cos\delta}{L_s + L_g} + \frac{u_s}{L_s + L_g} - \frac{L_s\dot{i}_{L\alpha}}{L_s + L_g} - \frac{r_s i_{L\alpha}}{L_s + L_g}$$

So expression (6) can be written again as follow:

$$\dot{u}_{gd} = \frac{3k_{qi}U_s(L_s + L_g) - 3k_{qp}U_s(r_s + r_g)}{2(L_s + L_g)}i_{g\beta} + \frac{3k_{qp}U_s L_g Hu_d}{2(L_s + L_g)}\delta$$

$$+\frac{3k_{qp}U_s(\dot{i}_{L\beta}(L_s + L_g) - L_s\dot{i}_{L\beta} - r_s i_{L\beta})}{2(L_s + L_g)} + \frac{3k_{qi}U_s}{2}i_{L\beta} + k_{qi}q_{gd}$$

(15)

Considering the relationship of DC-link voltage, the DC-link current can be represented as (16)

$$I_d = \frac{3}{2}\text{Re}\left\{\vec{H} * \vec{I}_g^{*}\right\} = \frac{3H}{2}(\cos\delta\cdot I_{g\alpha} + \sin\delta\cdot I_{g\beta}) \quad (16)$$

Here, "Re" is real part of correspond complex; the star symbol " * " represents conjugate complex. So the differential equation of DC-link voltage could be written as:

$$\dot{u}_d = \frac{1}{C}I_d = \frac{3H}{2C}(\cos\delta\cdot I_{g\alpha} + \sin\delta\cdot I_{g\beta}) \quad (17)$$

Taking (17) into (5), the differential equation of control angleδ can be written as another former:

$$\dot{\delta} = k_{up}\dot{u}_{gd} - k_{up}\dot{u}_d = k_{up}\dot{u}_{gd} - \frac{3Hk_{up}}{2C}(i_{g\alpha}\cos\delta + i_{g\beta}\sin\delta)$$

In the discussion of section 2, it has been mentioned that the shifting range of δ is very narrow at some conditions, and perhaps is a little larger at some other conditions, but the absolute value of δ is normally small during adjustment of reactive power or the node voltage, it depends on practical system. So in the time domain it may be supposed that the varying range of control angle δ not wide under steady condition in order to get a set of first-order differential equations in the n state variables, it could be considered that $\sin\delta \approx \sin\delta_0$, or $\sin\delta \approx \delta$, $\cos\delta \approx 1$. The δ_0 is an average value of δ. Furthermore the DC-link voltage u_d is usually also steady, which varies in a limited range, so $u_d\sin\delta$ could be approximatively expressed as δU_{d0}, i.e. DC-link voltage takes the steady value. This happens especially in the superposition configuration of SVG because of more adjustable points in a period, which would produce more stable DC-link voltage.

Taking $L=L_s+L_g$ and $r=r_s+r_g$, the differential equations above would be written as following:

$$\dot{i}_{g\beta} = \frac{L_g HU_{d0}}{L}\delta - \frac{r}{L}i_{g\beta} - \left(\frac{r_s i_{L\beta} + L_s\dot{i}_{L\beta}}{L}\right)$$

$$\dot{i}_{g\alpha} = -\frac{r_g}{L}i_{g\beta} - \frac{L_g H}{L}u_d + \frac{U_s}{L} - \frac{L_s\dot{i}_{L\alpha}}{L} - \frac{r_s i_{L\alpha}}{L}$$

$$\dot{\delta} = \frac{3k_{up}k_{qp}L_g HU_{d0}U_s}{2L}\delta - \frac{3k_{up}H}{2C}i_{g\alpha} + a_{14}i_{g\beta} + V_1$$

Here,

$$a_{14} = \frac{3k_{up}U_s(k_{qi}L - k_{qp}r)}{2L} - \frac{3k_{up}H\sin\delta_0}{2C};$$

$$V_1 = k_{up}\left[\frac{3k_{qp}U_s(L_g\dot{i}_{L\beta} - r_s i_{L\beta})}{2L} + \frac{3k_{qi}U_s}{2}i_{L\beta} + k_{qi}q_{gd}\right]$$

In the representations above load current i_L and its differential corresponding to its α and β part are known quantities; U_s is an effective value of power system. Hence, all the representations above would be expressed as a matrix form:

$$\begin{bmatrix}\dot{\delta}\\ \dot{u}_d\\ \dot{i}_{g\alpha}\\ \dot{i}_{g\beta}\end{bmatrix} = \begin{bmatrix}\frac{3k_{up}k_{qp}U_sL_gHU_{d0}}{2L} & 0 & -\frac{3k_{up}H}{2C} & a_{14}\\ 0 & 0 & \frac{3H}{2C} & \frac{3H\sin\delta_0}{2C}\\ 0 & \frac{-L_gH}{L} & -\frac{r}{L} & 0\\ \frac{L_gHU_{d0}}{L} & 0 & 0 & -\frac{r}{L}\end{bmatrix} \cdot \begin{bmatrix}\delta\\ u_d\\ i_{g\alpha}\\ i_{g\beta}\end{bmatrix} + \begin{bmatrix}V_1\\ 0\\ \frac{U_s - L_s\dot{i}_{L\alpha} - r_s i_{L\alpha}}{L}\\ \frac{L_s\dot{i}_{L\beta} + r_s i_{L\beta}}{L}\end{bmatrix}$$

The form can be compactly written in brief as:

$$\dot{X} = AX + V \qquad (18)$$

Here, X is state vector, V is known equivalent power vector, and A is the characteristic matrix.

4. SYSTEM STABILITY ANALYSIS

Expression (18) is in fact reduced to a linear, time-invariant state matrix equation. So using all linear system methods to analyzing system stability can solve the differential equations. There are many different ways to solve this problem. For linear time-invariant systems, all common stability definitions are equivalent; each implies the others, and each holds if and only if all of the system's characteristic roots are in the left half of the complex plane (LHP). For first- and second-order systems, stability is determined by inspection of the characteristic polynomial, and the system is said stable if all the polynomial coefficients have the same algebraic sign. For higher-order polynomials, representing higher-order systems, the algebraic signs of the polynomial coefficients may or may not yield information as to stability. Furthermore no coefficient may be zero in a system with LHP roots because there are no minus signs involved and thus no way for a coefficient to be canceled. In this practical circumstance the characteristic polynomial is the 4th-order polynomial, it can be written as:

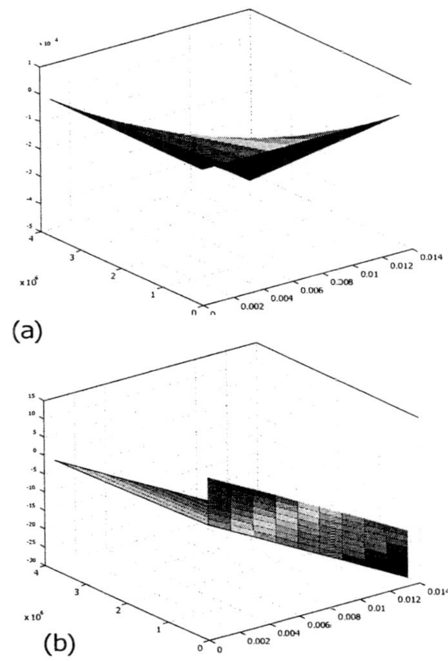

Fig.6 the value-field of control parameters

$$|sI-A| = a_4s^4 + a_3s^3 + a_2s^2 + a_1s + a_0 = 0 \qquad (19)$$

Where, $a_4=1$ and other coefficients are respectively as following:

$$a_3 = \frac{2r}{L} - \frac{3k_{up}k_{qp}L_gU_sHU_{d0}}{2L}$$

$$a_2 = \frac{3L_gH^2(1 + k_{up}U_{d0}\sin\delta_0)}{2LC} + \frac{r^2}{L^2} - \frac{3k_{up}k_{qp}U_sL_gHU_{d0}r}{2L^2} - \frac{3k_{up}k_{qp}L_gHU_{d0}U_s}{2L}$$

$$a_1 = \frac{3H^2L_gr - 9k_{up}k_{qp}L_g^2H^3U_{d0}U_s + 3k_{up}L_gH^2U_{d0}r\sin\delta_0}{2L^2C} - \frac{3k_{up}k_{qi}L_gHU_{d0}U_sr}{2L^2}$$

$$a_0 = \frac{3k_{up}L_g^2H^3U_{d0}U_s}{2L^3C}(4k_{qp}r - 3k_{qi}L) \qquad (20)$$

If the coefficients $a_0 \sim a_4 > 0$, the system under discussion may be stable. In normal control system, control parameters, such as k_{qp}, k_{qi}, k_{up} in this system, are changeable. If system parameters are known, the control parameter can be determined for a stable system. Assuming that U_s=400V, $r_s=9\times10^{-3}\Omega$, $L_s=0.1\times10^{-3}$H, $r_g=14\times10^{-3}\Omega$, $L_g=0.35\times10^{-3}$H, $C=1.1\times10^{-3}$F, H=0.74219, U_{d0}=537V and $\delta=-5^0$, the values range of k_{qp}, k_{qi}, and k_{up} will be derived by taking these known conditions into expression (20), and after simplification the inequalities are:

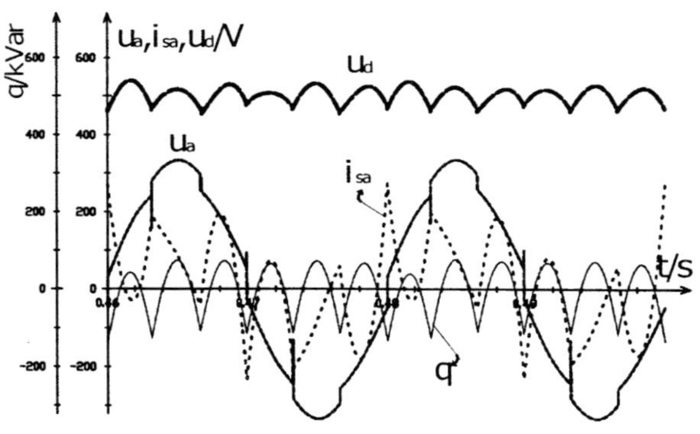

Fig.7 Stable output curves of SVG

$$0.00135k_{qi} - 0.092k_{qp} + k_{up}k_{qp} < 5.495\times10^{4}$$
$$k_{qi} < (5.495\times10^{-4} + 0.092k_{qp} - k_{up}k_{qp})/0.00135$$
$$k_{qi} < (2612/k_{up} - 9.5063\times10^{6}k_{qp} - 27.344)/0.00135$$
$$k_{qi} < (29860/k_{up} - 2.39\times10^{8}k_{qp} - 1.537\times10^{6})/9.5063\times10^{6} \tag{21}$$

The common solutions of the 4 inequalities above are the values that the control parameters k_{qp}, k_{up} and k_{qi} should be taken. Taking k_{qp}, k_{up}, k_{qi} as x, y, and z-axis in 3 dimensions respectively, so the value-field of k_{qp}, k_{up} and k_{qi} is depicted in Fig.6. From Fig.6 it is implied that the k_{qp} and k_{qi} are very small (less than 10^{-3}), and k_{up} could be a little larger. How to get an optimal set of the parameters is dependent on stability tolerance.

Taking the control parameters k_{qi}=1.2E-2, k_{qp}=1.9E-4, k_{up}=3.2E-5, and other system parameters are same as in inequality (21), which are clearly coincident with inequality (21) also, the output simulation curves of simple three-phase bridge SVG with three-phase rectifier for the purpose of a stable DC-link voltage are shown in Fig. 7 according to the control strategy of Fig. 5

However with the inspection of coefficients sign may get a stable system, but at some other conditions the system may not be stable. For this occasion the Routh-Hurwitz criterion is usually used as the criterion of higher order characteristic polynomial. The detail information about the Routh-Hurwitz criterion is described in reference [3].

5. RESULTS ANALYZING AND ILLUSTRATION

In the section above we have discussed stabilization method to SVG, and obtained parameters set by inspection of characteristic polynomial sign or with Routh-Hurwitz criterion as precise set of parameter selection in order to get a stable system of SVG, and verified a stable operation through real simulation by the selected parameters above. In the deriving process, however, the used simplification conditions $\cos\delta\approx1$, $\sin\delta\approx\delta$ and $\sin\delta u_d \approx \delta U_{d0}$ may involve some offset. The curve u_d is nearly balanced with a certain amount of 6th harmonic under stable condition, the u_d offset amplitude of maximum and minimum is about 80V, and the average value is about 510V, so the simplification of $u_d\approx U_{d0}$ could not bring larger error. But the fluttering scope of control angle δ is not in a narrow range as expected, the fluttering range is about $0\sim18^0$ from real simulation Fig.8. So the $\sin\delta\approx\delta$ may bring some error in the determination of stability with equation (20), this case could be much better improved for a SVG configuration of multi-level voltage source, because there is more control in a single output period.

All in all, the SVG stability determination using inequality (20) could be suitable for SVG control under most conditions. If some unstable errors happen, the Routh-Hurwitz criterion could perhaps be a good solution for stability problem.

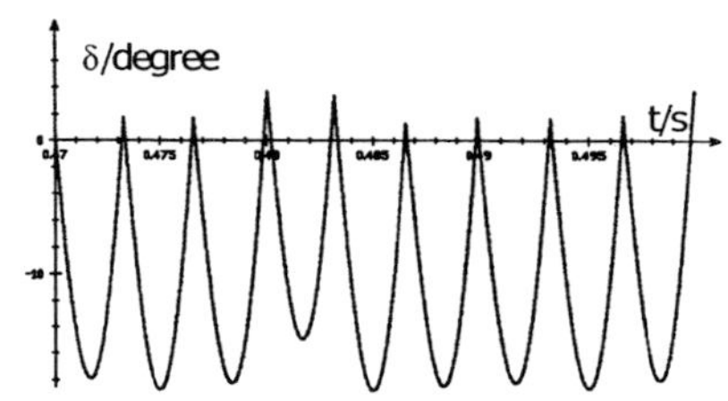

Fig.8 The stable δ curve corresponding to Fig.6

REFERENCE

[1] Yuanzhang Sun, Mu Zhao, Xiong Ni. Stability analysis of power system in STATCOM control, Automation of electric power systems. Vol.25, №11 2001

[2] Michael Weinhold. Drephasiger Pulsstromrichter zur Speisung von Gleichspannungszwischenkreisen ohne Amplitudensteuerung de metzseitigen Stromrichterspannungen. Dissertation. Fakultät für Elektrotechnik an der Ruhr-Universität Bochum. Germany 1993 pp3-4

[3] Gene H. Hostetter, Clement J. Savant, Jr., Raymond T. Stefani "DESIGN OF FEEDBACK CONTROL SYSTEMS", 1982 CBS College Publishing, New York, N.Y. USA

ELSEVIER
IFAC
PUBLICATIONS
www.elsevier.com/locate/ifac

ADAPTIVE FUZZY CONTROL FOR VARIABLE SPEED VARIABLE PITCH WIND TURBINES

ZHANG Xin-fang, XU Da-ping

*Department of Automation,
North China Electric Power University 102206 P. R. China*

Abstract: Automatic control is crucial for the efficiency and reliability of wind power generation systems. Wind energy conversion systems have strong nonlinear multi variable with many uncertain factors and disturbances. A new method for estimating the optimal rotating speed at low wind speed is proposed using fuzzy logic inference system. The method avoids the inaccuracy owing to measuring the wind speed. Adaptive fuzzy controllers based on variable universe are employed to improve wind energy capture for variable speed, variable pitch wind turbines. The fuzzy algorithm keeps the form of rules and varies universe of discourse according to the control error, fine control accuracy can be achieved. The maximum power coefficient is realized by generator torque control at low wind speed, and also pitch angle is adjusted to keep rated output power at high wind speed. Simulation results verify the effectiveness of the proposed methods. *Copyright © 2003 IFAC*

Keywords: wind turbines, fuzzy system, adaptive fuzzy control, variable universe, pitch control, torque control

1. INTRODUCTION

The development of renewable natural energy has attracted considerable interest in recent years primarily due to concerns about the environment pollution caused by the burning of fossils, and its continually diminishing reserves. Wind turbine generator system provides an environmentally friendly, economically competitive and socially beneficial means of electricity generation. Variable speed (VS), horizontal axis grid integration wind turbine (WT) provides a better efficiency than fixed speed wind turbine, especially in low wind speed range. The mechanical stresses are reduced, gusts of wind can be absorbed, i.e., energy is stored in the mechanical inertia of the turbine, creating an "elasticity" that reduces torque pulsations (Muller, *et al.*, 2002).

Variable speed wind turbine (VSWT) can be controlled by two speed controllers. One is for electric torque control of generator at low wind speed and the other one is for blade pitch controlling at high wind speed. The objectives of the controllers are to achieve maximum power coefficient Cp at low wind speed and keep the rated output power at high wind speed. The paper proposes a new method of estimating the optimal wind turbine speed by using fuzzy logic inference system. The method avoids the inaccuracy owning to measuring the wind speed.

The VSWT system has strong nonlinear characteristics with many uncertain factors, such as meteorological conditions and continuously varying ac system load. Therefore, traditional linear control techniques based on the complete mathematical model cannot gain better performance. Intelligent control based on fuzzy logic is an efficient means for wind turbines and wind energy engineers pay much attention to it (Chen, *et al.*, 2000; Prats. *et al.*, 2000; Prats. *et al.*, 2002). Fuzzy control is useful when there are not accurate models and when only qualitative information is available. But, it is very difficult for fuzzy control to get fine control accuracy. Adaptive fuzzy control based on variable universe (Li, 1999) is applied to the VSWT system. This fuzzy control algorithm not only has merits of traditional fuzzy control, but also has fine accuracy and better dynamic performance. This paper describes the principles of the control scheme and reports the simulation results. A Simplified system model of a 600 kW VSWT is taken as an example to demonstrate the proposed fuzzy control strategy.

2. A SIMPLIFIED MODEL OF VSWT

A simplified structure of a variable speed wind turbine generating system is shown in figure 1. The rotor of wind turbine extracts energy from the wind and converts it into mechanical power and a generator converts mechanical power into electrical power. In most systems, the rotor shaft and the generator shaft are coupled through a gearbox. The generator is coupled to the grid through a transformer and a power electronic converter making the generator output to match the frequency and voltage of the grid. Control system and protection system are important part of modern WT.

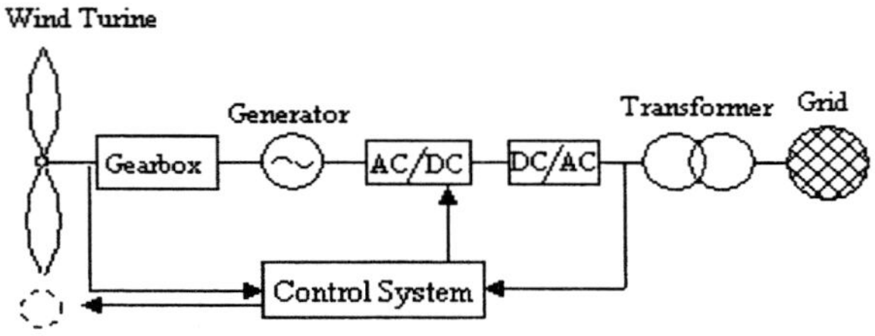

Fig. 1. Structure of Modern Wind Turbine

2.1 Rotor Equations

The power and torque extracted from the wind can be expressed as the following equations

$$P_r = \frac{1}{2} Cp(\beta,\lambda)\rho\pi R^2 V_w^3 \quad (1)$$

$$T_r = \frac{P_r}{w_r} = \frac{1}{2} Cp(\beta,\lambda)\rho\pi R^2 \frac{V_w^3}{w_r} \quad (2)$$

Where P_r is power extracted from the wind [W], T_r is aerodynamic torque of the rotor [Nm], ρ is air density [kg/m^3], R is rotor radius [m], Cp is power coefficient (maximum value Betz's limit 59.3%), β is pitch angle of rotor blades [deg], λ is tip speed ratio, the ratio between blade tip speed and wind speed upstream the rotor.

$$\lambda = \frac{w_r R}{V_w} \quad (3)$$

Where w_r is rotor speed [rad/s], V_w is wind speed [m/s].

Power coefficient $Cp(\beta,\lambda)$ is a function of the tip speed ratio λ and the pitch angle β. The calculation of it requires knowledge of aerodynamics and blade element theory. As the computations are rather complicated, numerical approximations have been developed (Heier, 1998). Here the following function will be used (Slootweg, *et al.*, 2001).

$$Cp(\beta,\lambda) = 0.22(\frac{116}{\lambda_i} - 0.4\beta - 5)e^{\frac{-12.5}{\lambda_i}} \quad (4)$$

$$\frac{1}{\lambda_i} = \frac{1}{\lambda + 0.08\beta} - \frac{0.035}{\beta^3 + 1} \quad (5)$$

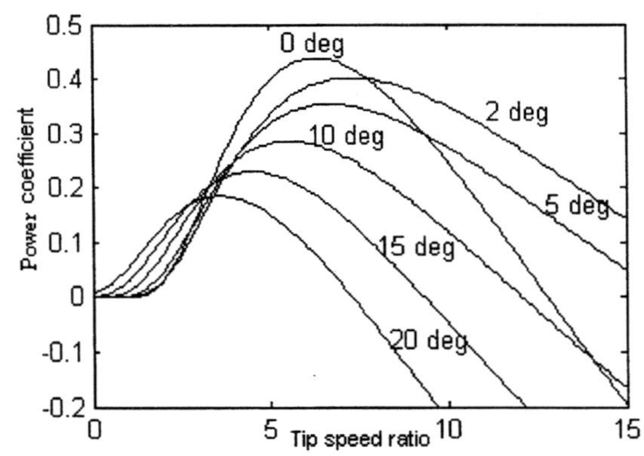

Fig. 2. Power Coefficient Cp.

Figure 2 shows the Cp curve with various values of λ and β. For a determined pitch angle β, there are corresponding optimal tip speed ratio λ_{opt} and maximum power coefficient $Cp_{\max}$. Clearly the turbine speed should be changed with wind speed so that optimum tip speed ratio is maintained. The following equation provides the expression of maximum aerodynamic torque of the wind turbine.

$$T_{ropt} = \frac{Cp_{\max}(\beta,\lambda_{opt})\rho\pi R^5 w_{ropt}^2}{2\lambda_{opt}^3} = K w_{ropt}^2 \quad (6)$$

Where constant $K = \frac{Cp_{\max}(\beta,\lambda_{opt})\rho\pi R^5}{2\lambda_{opt}^3}$

2.2 Motion Equation of Generator and WT

The modeling of the wind turbine and generator may be simplified into a single motion equation (Chen, *et al.*, 2000).

$$(\frac{J_r}{G^2} + J_g)\frac{dw_g}{dt} = (\frac{T_r}{G} - T_G) - (\frac{D_r}{G^2} + D_g)w_g \quad (7)$$

Where w_g is generator rotor speed [rad/s], J_g is generator rotor inertia [kg m^2], J_r is turbine rotor inertia, G is gearbox ratio, D_r is turbine friction coefficient [Nm/rad], D_g is generator friction coefficient [Nm/rad], T_G is electromagnetic torque applied on generator shaft [Nm].

A frequency converter (AC/DC and DC/AC in figure 1) decouples the generator from the grid, allowing the rotor speed to vary by controlling the generator reaction torque. Since mechanic response is slower than electric response, the dynamic characteristics of the generator are not included in the wind turbine model. The electric time constant is negligible compared to the mechanic time constant. The load torque command of the wind turbine is instantaneously transmitted to the shaft.

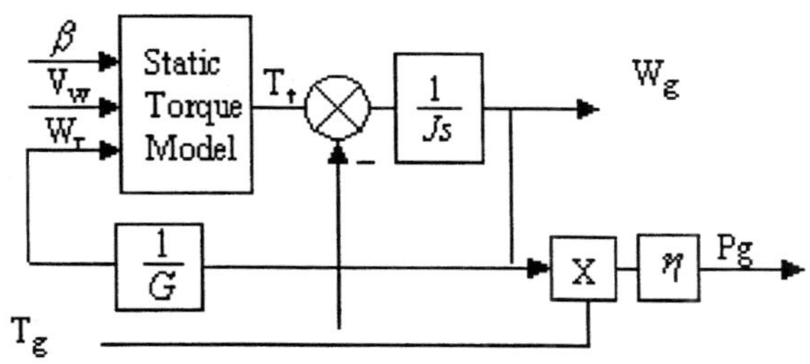

Fig. 3. The Simplified Model of WT

Figure 3 presents the simplified model of a VSWT. The input variables of the model are wind speed V_w, pitch angle β and generator torque T_g. The output variables are generator speed w_g and electric power P_g. Where η is the generator efficiency, $T_t = T_r / G$, $J = (J_r / G^2 + J_g)$.

3. REFERENCE ROTATING SPEED

The generator torque demand can be set to Kw_{ropt}^2 for open loop variable speed control of a wind turbine with w_{ropt} set to the measured rotor speed. This ensures that in the steady state the turbine will work with an optimum tip speed ratio. A close loop speed controller has been used in this paper. In many control schemes the blade rotating speed is regulated according to the measured wind speed. This is inaccurate as the wind speed is generally different at every point on the surface covered by the blades. Reference speed w_{ref} cannot be obtained by equation (8) because the aerodynamic torque acting on the rotor blades (T_{ra}) cannot be measured directly, as a part of it accelerates the rotor itself and is therefore not acting on the shaft.

$$w_{ref} = w_{ropt} = \sqrt{\frac{T_{ra}}{K}} \tag{8}$$

In this paper, a new method is proposed to get optimal rotating speed by fuzzy logic inference system. The input to the employed algorithm consists of the produced active power, the rotor rotating speed and their delayed values as shown in Figure 4.

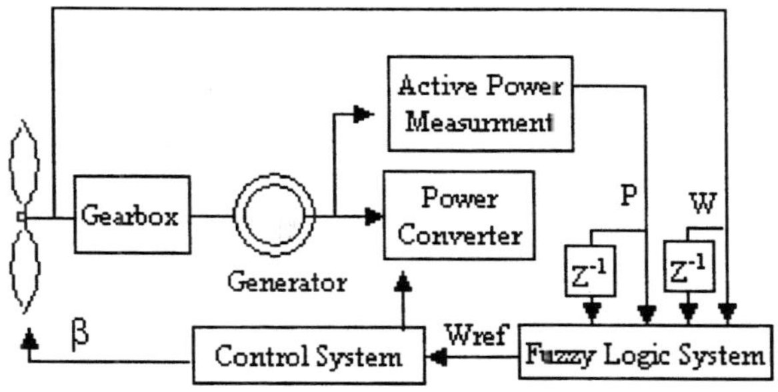

Fig. 4. Reference Rotating Speed Fuzzy System

$$\Delta P(k-1) = P(k-1) - P(k-2) \tag{9}$$

$$\Delta W(k-1) = W(k-1) - W(k-2) \tag{10}$$

Where $\Delta P(k-1)$ is the change of the power during $(k-1)$ time internal, $\Delta W(k-1)$ is the change of the rotating speed during $(k-1)$ time internal. The output $\Delta W_{ref}(k)$ is the reference rotating speed increment at low wind speed. Kanellos and Hatziargyriou (2002) assume that the aerodynamic power change equals the sum of the produced active power change and the change of the blades kinetic power. And the following rule is employed so that the operating point will change properly and a new rotating speed closer to the optimal is achieved.

$$\Delta W_{ref}(k) = \partial \cdot sign(\Delta W(k-1)) \cdot \Delta P(k-1) \tag{11}$$

Where ∂ is a parameter. The choice of appropriate values for it is mandatory as it affects controller's response speed and steady state error. But the determination of ∂ is very difficult. Fuzzy logic system provides a systematic and efficient way to incorporate expert knowledge. It can produce the nonlinear mapping between inputs and outputs. A fuzzy system generally includes four parts such as fuzzification, knowledge base, fuzzy inference engine and defuzzification. All the inputs and the outputs are normalized and partitioned to seven fuzzy sets: Negative Big (NB), Negative Medium (NM), Negative Small (NS), Zero (ZE), Positive Small (PS), Positive Medium (PM) and Positive Big (PB). Knowledge base employs 49 rules, as shown in table 1.

Table 1 Fuzzy Logic Rules

$\Delta W \backslash \Delta P$	NL	NM	NS	ZE	PS	PM	PL
NL	PL	PL	PM	NS	NM	NL	NL
NM	PL	PM	PS	NS	NS	NM	NL
NS	PM	PS	PS	ZE	NS	NS	NL
ZE	NL	NM	NS	ZE	PS	PM	PL
PS	NM	NS	NS	ZE	PS	PS	PM
PM	NL	NM	NS	PS	PS	PM	PL
PL	NL	NL	NM	PS	PM	PL	PL

Fuzzy system calculates the reference rotating speed for speed controller to realize maximum power coefficient at given wind speed. But large change of reference speed will cause the oscillation in the rotating speed at the optimal operating point. This condition can be avoided by tuning the normalization factors, fuzzy membership functions and fuzzy partitions.

4. ADAPTIVE FUZZY CONTROL FOR VSWT

4.1 Variable Universe and Fuzzy Algorithm

Although fuzzy control is superior to traditional linear control techniques in many ways, it is not an efficient method to gain fine control accuracy. It is possible to solve the problem by combining fuzzy control with PID control. But, oscillation arises from switching of different control strategy, and it is difficult to tune the control parameters. Adaptive fuzzy control based on variable universe (Li 1999) keeps the form of rules and varies universe of discourse according to the control error. When the error is changing smaller, the universe of fuzzy control is compressing accordingly, vice versa. The universe compression is equivalent to the addition of rules, thus fine control accuracy can be gained.

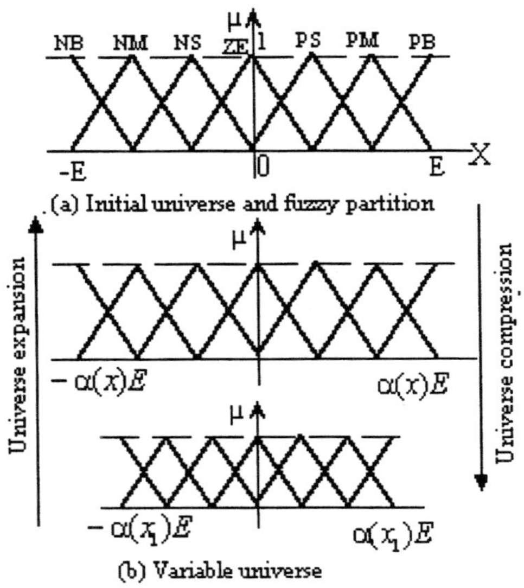

Fig. 5. Expansion and Compression of Universe

The universes of discourse of input variables $x_i(i=1,2,\Lambda,n)$ are $X_i=[-E,E](i=1,2,\Lambda,n)$. The universe of output variable y is $Y=[-U,U]$. $\{A_{ij}\}(1\le j\le m)$ and $\{B_j\}(1\le j\le n)$ are fuzzy partitions of universes X_i and Y respectively. A fuzzy control can be represented as an n-ary piecewise interpolating function.

$$y(x_1,x_2,\Lambda,x_n)=F(x_1,x_2,\Lambda,x_n)=\sum_{j=1}^{m}\prod_{i=1}^{n}A_{ij}(x_i)y_j \quad (12)$$

The so-called variable universe means that some universes, such as X_i and Y, can change with changing variables x_i and y, denoted by

$$X_i(x_i)=[-\partial_i(x_i)E_i,\partial_i(x_i)E_i] \quad (13)$$

$$Y(y)=[-\beta(y)U,\beta(y)U] \quad (14)$$

Where $\partial_i(x_i)(i=1,2,\Lambda,n)$ and $\beta(y)$ are adaptive factors which have many formations. The situation of variable universe changing is shown as figure 5. The paper employs the following adaptive factors.

$$\partial(x)=1-\lambda\exp(-kx^2),\lambda\in(0,1),k>0 \quad (15)$$

$$\beta(t)=K_I\sum_{i=1}^{n}p_i\int_0^t x_i(\tau)d\tau+\beta(0) \quad (16)$$

Where $p_i(i=1,\Lambda,n)$ is constant parameters, K_I is proportional constant, for it is regarded as a design parameter, the initial value $\beta(0)$ is viewed as a design parameter to be adjusted according to practical situation, here $\beta(0)=1$. The algorithm of variable universe adaptive fuzzy control evolves as the following form (Li. *et al.*, 2002)

$$u(t)=\left(K_I\sum_{i=1}^{n}p_i\int_0^t e_i(\tau)d\tau+\beta(0)\right)U\sum_{j=1}^{m}\prod_{i=1}^{n}A_{ij}\left(\frac{x_i(t)}{\partial_i(x_i(t))}\right)y_j \quad (17)$$

4.2 Control System Design of VSWT

The proposed control system of VSWT has integrated a torque controller and a pitch controller, as shown in figure 6.

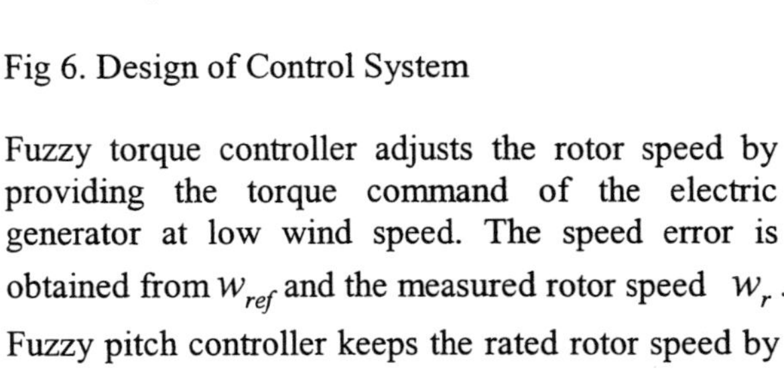

Fig 6. Design of Control System

Fuzzy torque controller adjusts the rotor speed by providing the torque command of the electric generator at low wind speed. The speed error is obtained from w_{ref} and the measured rotor speed w_r. Fuzzy pitch controller keeps the rated rotor speed by adjusting the reference pitch angle β_{ref} at high wind speed.

5. SIMULATION RESULTS

The considered 600 kW VSWT is equipped with a three blades and blade pitch control. Wind turbines start to provide power from 3~4 m/s wind speed. Rated power is given for nearly wind speed to 12.5

m/s. Wind speed higher than 25 m/s is recommended to brake wind turbines. Its parameters are shown in appendix A. Pitch angle is limited as 0~25 deg. Pitch angle rate is 6 deg/s. The response to two wind sequences (average wind speed: 8 m/s, 15 m/s) is simulated in time domain (s). At low wind speed pitch angle keeps zero deg. In figure 7 the wind speed, the rotor speed, the pitch angle, the output power are depicted. In all graphs, the straight line corresponds to the high wind speed, the dotted line to the low wind speed.

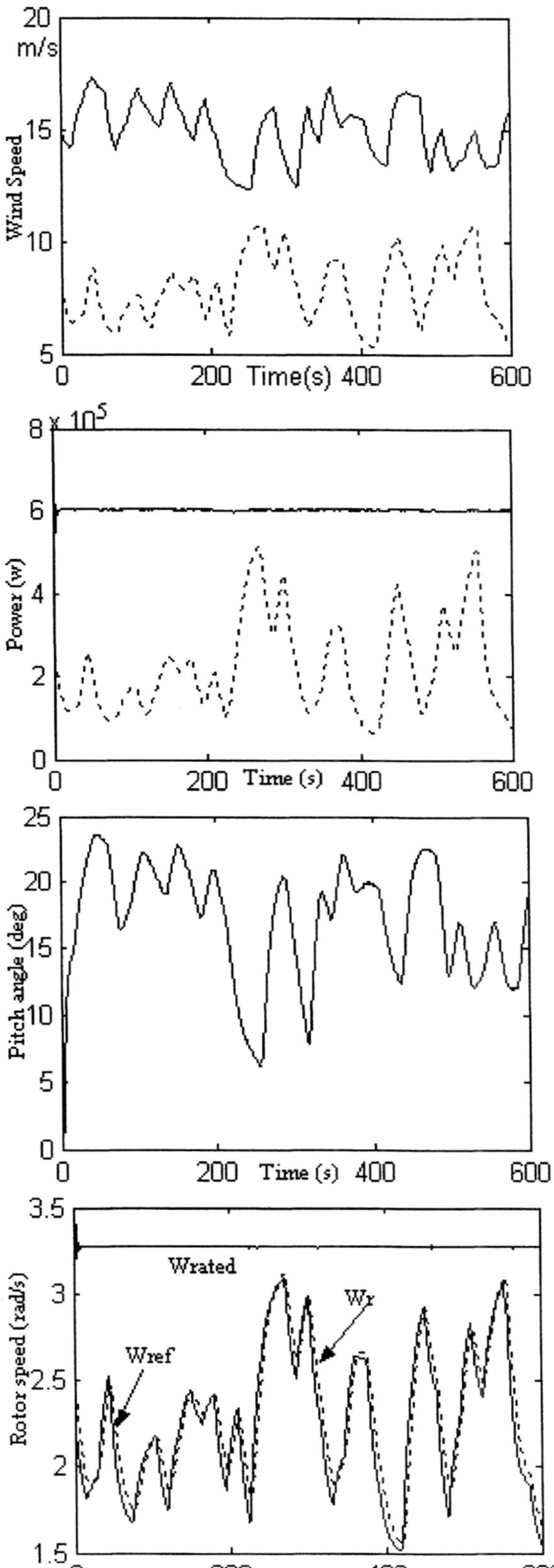

Fig.7. Starting from above: wind speed, power, pitch angle and rotor speed.

5. CONCLUSIONS

In this contribution, Adaptive fuzzy control based on variable universe has been designed for improving captured energy of VSWT. Advantages of this fuzzy algorithm include not only nonlinear control actions, less dependence on mathematical models and good robustness, but also fine accuracy and adaptive ability. A new method of estimating optimal rotating speed is proposed using fuzzy logic inference system. It avoids the inaccuracy of measuring wind speed. Torque fuzzy control at low wind speed and blade pitch fuzzy control at high wind speed have been used in the designed model. Simulation results demonstrate the effectiveness of the proposed control strategy.

REFERENCES

Chen, Z. Gomez, S.A. and McCormick, M (2000). A Fuzzy Logic Controlled Power Electronic System for Variable Speed Wind Energy Conversion Systems. *IEEE Power Electronics and Variable Speed Drives Conference*, pp. 114-119.

Kanellos F.D. and Hatziargyriou N.D. (2002). A New Control Scheme for Variable Speed Wind Turbines Using Neural Networks. *IEEE Power Engineering Society Winter Meeting*, **Vol:1**, pp. 360 –365.

Li Hongxing (1999). Adaptive Fuzzy Controllers Based on Variable Universe. *Science in China (Series E).*, **Vol: 42,** pp. 10-20.

Li Hongxing, Miao Zhihong and Wang Jiayin (2002). Variable Universe Stable Adaptive Fuzzy Control of Nonlinear System. *Science in China (Series E).* **Vol: 45**, pp. 65-75.

Muller, S. Deicke, M. etc (2002). Doubly Fed Induction Generator Systems for Wind Turbines. *Industry Applications Magazine,* IEEE, **Vol: 8,** pp. 26 –33.

Prats, M.A.M., Carrasco, J.M. etc. (2000). Improving Transition between Power Optimization and Power Limitation of Variable Speed, Variable Pitch Wind Turbines Using Fuzzy Control. *26th Annual Confjerence of the IEEE,* **Vol : 3**, pp. 1497-1502.

Prats, M.A.M., Carrasco, J.M. etc.(2002). A New Fuzzy Logic Controller to Improve the Captured Wind Energy in A Real 800KW Variable Speed Variable Pitch Wind Turbine. *Power Electronics Specialists Conference of IEEE,* **Vol:1**, pp. 101-105.

S.Heier (1998). *Grid Integration of Wind Energy Conversion System,* Chicester, UK:John wiley& Sons Ltd.

Slootweg, J.G. Kling, W.L and Polinder, H. (2001). Dynamic Modelling of a Wind Turbine with Doubly Fed Induction Generator. *Power Engineering Society Summer Meeting, IEEE* , **Vol: 1,** pp. 644 –649.

APPENDIX A

Diameter: 45 m
Number of blades: 3
Rotor inertia: 500000 kg m^2
Gearbox ratio: 1: 50
Rated wind speed: 12.5 m/s
Rated power: 600 KW
Air density: 1.225 kg/m^3

ELSEVIER
IFAC
PUBLICATIONS
www.elsevier.com/locate/ifac

ANALYSIS ON DETERMINING THE ALLOCATION RATIO OF TRANSMISSION SERVICE COSTS USING THE ARBITRATION GAME

Koo-Hyung Chung, Chong-Il Yoo*, Dong-Joo Kang, Balho H. Kim**

Dept. of Electric Engineering, Hongik University, Seoul, Korea
**Korea Electric Power Corporation, Seoul, Korea*
***Korea Electrotechnology Research Institute, Seoul, Korea*

Abstract: In many parts of the world, the electricity industry is undergoing unprecedented changes. Hence, in order to restructure the electricity industry readily and efficiently and minimize the confusion related to restructuring, it is required the systematic studies on transmission pricing and the allocation of transmission service costs between generators and distributors. However, the allocation ratio of transmission service costs have been determined by regulator arbitrarily under the common practice since there is not any criterion with respect to allocating transmission service costs even now. This paper demonstrates that determining the allocation ratio of transmission service costs between generators and distributors is regulator's own authority using game theory. For the analysis, the competition for generators and distributors to determine own allocation ratio is modeled as the arbitration game. *Copyright © 2003 IFAC*

Keywords: Transmission service cost, allocation ratio, game theory, final-offer arbitration, Nash equilibrium

1. INTRODUCTION

The electricity industry worldwide is experiencing unprecedented restructuring. As restructuring has been advanced, traditional vertically-integrated utilities are separated to generation, transmission and distribution respectively. The restructuring aims to increase efficiency in electricity industry. Transmission remains as a natural monopoly and is centrally controlled due to economy of scale while competition is introduced in generation and distribution (and/or retail). Therefore, the systematic studies on transmission pricing and the allocation of transmission service costs are required to restructure the electricity industry readily and efficiently and minimize confusion related to restructuring.

In most electricity market, since there is not any criterion with respect to allocating transmission costs even now, the regulator determines the allocation ratio of transmission service costs between generators and distributors arbitrarily under common practice. Although KEMA proposes the allocation ratio of 50:50 between generators and distributors in Korea, it is not based on any theoretical criterion.

This paper demonstrates that determining the allocation ratio of transmission service costs between generators and distributors is regulator's own authority using game theory. For this analysis, the competition for generators and distributors to determine own allocation ratio is modeled as the arbitration game.

2. DECISION OF TRANSMISSION COST ALLOCATION RATE BETWEEN GENERATORS AND DISTRIBUTORS

2.1 The Arbitration Game

The right to strike of a labor union is generally restricted by the legislation when a dispute is caused in public utility such as railway, electricity etc. Instead, an arbitration committee, as third party, mediates between contending parties and the decision of the arbitration committee takes a legal effect.

There are many procedures of the arbitration. One is that an arbitration committee settles the dispute arbitrarily. Another procedure is as follows: first, an arbitration committee requests that each party should submit own final offer concerned with the dispute. Then, the arbitration selects one between submitted offers. This procedure is called 'final-offer arbitration'.

This paper analyzes on Nash equilibrium strategy to determine the allocation ratio of transmission service costs between generators and distributors through final-offer arbitration. That is, it is assumed that each market participant negotiates with the opponents to determine transmission cost allocation rate. If all participants cannot reach a consensus on allocation of transmission service costs, the regulator selects one of all participants' final offers. Then, each participant's allocation ratio of transmission service costs is set by this regulator's decision.

2.2 Basic Assumptions

This paper introduces several assumptions for applying the arbitration game to determination procedure of each market participant's allocation ratio of transmission service costs. The assumptions are as follows:

- It is assumed that one generator representing all generators and one distributor representing all distributors participate in this arbitration game. Since determining the allocation ratio of transmission service costs is the competition between generation part and distribution part, this assumption is reasonable.
- All participants submit own final offer to regulator while they do not know the opponent's. Consequently, this arbitration game is modeled as simultaneous move game.
- The regulator has an optimum allocation ratio of transmission service costs that considers this ratio rational for public benefit. Hence, the regulator selects closer one to own optimum ratio between two submitted final offers when all participants cannot reach a consensus on allocation of transmission service costs.
- It is assumed that a generator and a distributor forecast the same regulator's optimum allocation ratio of transmission service costs.

2.3 The Regulator's Arbitration Procedure

Let α_g be a generator's allocation ratio of transmission service costs finally submitted by the generator and α_d be a generator's allocation ratio finally submitted by the distributor. Moreover, let α_s be a generator's optimum allocation ratio considered by the regulator. Each α_x has any value between 0 and 1 and a distributor's allocation ratio of transmission service costs is 1-α_g. In arbitration game, therefore, each participant uses the generator's allocation ratio of transmission service costs as own strategy.

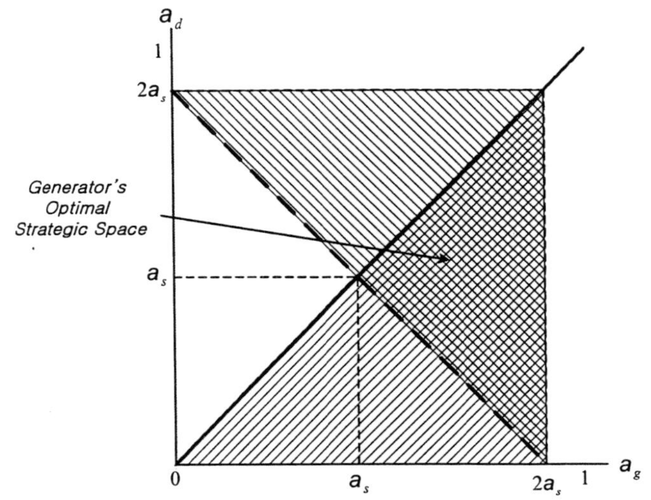

Fig. 1. The generator's optimal strategic space

In this game, since each participant can choose any value between 0 and 1 as own final offer, their strategic space is $S_x = [0,1]$ respectively. The regulator's final-offer arbitration procedure is described as follows:

- If $\alpha_g = \alpha_d = \alpha$, then the generator's allocation ratio of transmission service costs is determined as α without regulator's arbitration.
- When $\alpha_g \neq \alpha_d$, the regulator chooses one of following alternatives as the generator's allocation ratio finally.
  - If $|\alpha_s - \alpha_g| > |\alpha_s - \alpha_d|$, then the regulator chooses α_d as the final arbitration alternative.
  - If $|\alpha_s - \alpha_g| < |\alpha_s - \alpha_d|$, then the regulator chooses α_g as the final arbitration alternative.
  - If $|\alpha_s - \alpha_g| = |\alpha_s - \alpha_d|$, then the regulator chooses own optimum allocation ratio, α_s, as the final arbitration alternative.

2.4 The Generator's Optimal Strategy

In competitive electricity market, if the generation part absorbs transmission service costs, each generator intends to reflect these costs in own energy bidding price. However, since such reflecting costs in energy bidding price may cause weakening the generator's competitiveness in generation market, all generators ought to absorb a portion of transmission service costs. Therefore, each generator desires to absorb transmission service costs as less as possible. For absorbing less cost, the generator must submit final offer lower than the distributor's, α_d.

$$0 \leq \alpha_g \leq \alpha_d \tag{1}$$

Moreover, this generator's offer must be closer to the regulator's optimum allocation ratio than the distributor's.

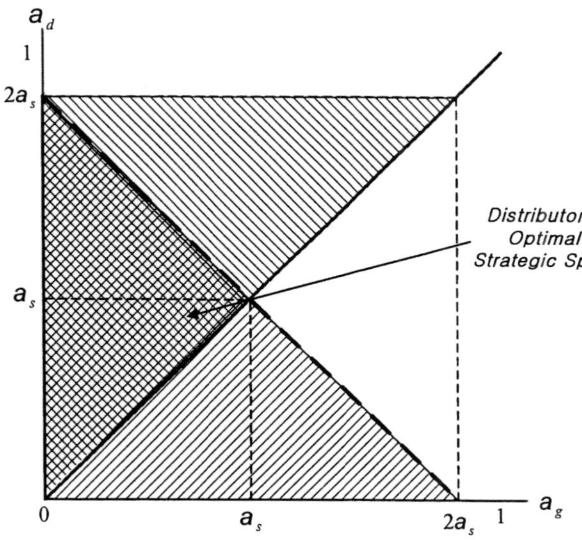

Fig. 2. The distributor's optimal strategic space

$$|\alpha_s - \alpha_g| < |\alpha_s - \alpha_d|$$

Arranging this equation, it is derived as follows:

$$\alpha_s < \frac{\alpha_g + \alpha_d}{2} \quad (2)$$

From (2), the generator ought to choose such final offer that the average value of own alternative and the distributor's expected alternative is larger than the regulator's optimum value.

By (1) and (2), the generator's optimal strategy must satisfy the following condition, (3).

$$2\alpha_s - \alpha_d < \alpha_g \leq \alpha_d \quad (3)$$

This condition is modified to $0 < \alpha_g \leq \alpha_d$ when $2\alpha_s - \alpha_d \leq 0$.

As the result, the generator's optimal strategic space is the same as shown in Fig.1.

2.5 The Distribution Entity's Optimal Strategy

If the distributor part absorbs a large portion of transmission service cost, each distributor transfers this costs to end-users. However, since electricity rate paid by end-user is a regulation case, government restricts to increasing the end-users' electricity rate. As the result, if the distributor cannot reflect this costs to the end-users' electricity rate properly, then it causes to decrease the distributor's profits. Therefore, the distributor also desires to absorb transmission service costs as less as possible. For absorbing less cost, the distributor must submit final offer higher than the generator's, α_g

$$\alpha_d \geq \alpha_g \quad (4)$$

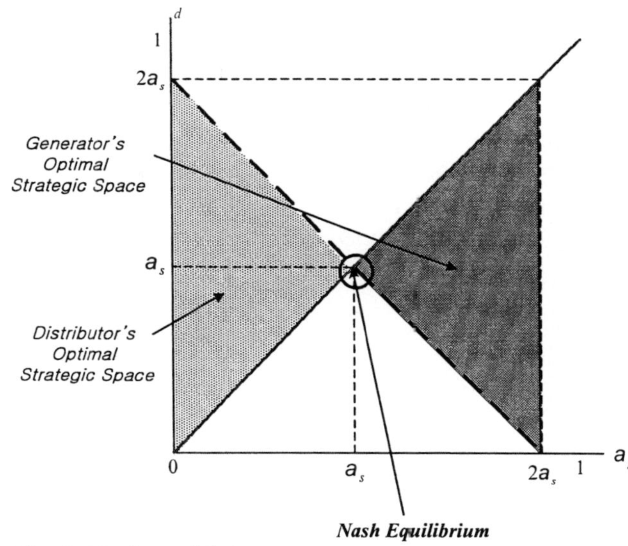

Fig. 3. Nash equilibrium strategy

Moreover, this distributor's offer must be closer to the regulator's optimum allocation ratio than the generator's.

$$|\alpha_s - \alpha_d| < |\alpha_s - \alpha_g|$$

Arranging this equation, it is derived as follows:

$$\alpha_s > \frac{\alpha_g + \alpha_d}{2} \quad (5)$$

From (5), the distribution ought to choose such final offer that the average value of the generator's expected alternative and own alternative is larger than the regulator's optimum value.

By (4) and (5), the distributor's optimal strategy must satisfy the following condition, (6).

$$\alpha_g \leq \alpha_d \leq 2\alpha_s - \alpha_g \quad (6)$$

This condition is modified to $\alpha_g \leq \alpha_g < 1$, if $2\alpha_s - \alpha_g > 1$. Fig.2 shows the distributor's optimal strategic space.

2.6 Nash Equilibrium Strategy

From each entity's optimal strategic space described in (3) and (6), Nash equilibrium strategy can be derived. In Nash equilibrium, there is no incentive to change own strategy when opponent's strategy is given. This means that Nash equilibrium must exist on the intersection between the generator's and distributor's optimal strategic space. As the result, Nash equilibrium strategy is determined as regulator's optimum allocation ratio of transmission service costs, α_s, as shown in Fig.3. That is, Nash equilibrium strategy is that both the generator and distributor submit own final offer as the regulator's optimum allocation ratio.

Consequently, the decision on allocation ratio of transmission service costs between the generator and the distributor is the regulator's own authority. In other words, although allocation ratio of transmission service costs is determined by negotiation between the generator and the distributor to reflect on competitive electricity marketplace situation, final decision needs the regulator's arbitration when one participant's profit is contrary to the opponent's.

3. CONCLUSION

This paper analyzes the determination procedure on allocation ratio of transmission service costs between the generation part and the distribution part using arbitration game model. For this analysis, final-offer arbitration model is introduced as the procedure of this game. In final-offer arbitration model, the regulator requests to each participant submitting own final offer concerned with dispute and then chooses one between two submitted offers. Consequently, Nash equilibrium is determined as the regulator's optimal alternative when the generator and the distributor cannot reach a consensus on allocation of transmission service costs. In facts, consensus on transmission cost allocation is so difficult since one participant's transmission cost allocation amount is contrary to the opponent's. Therefore, the determination on allocation ratio of transmission service costs between the generator and the distributor needs the regulator's arbitration.

Although introduced assumptions are impractical, it is expected that this paper can provide basic information on transmission cost allocation through arbitration game model. Moreover, it is expected that the additional study on derivation of the regulator's optimal allocation ratio of transmission service costs should be performed based on the result of this paper.

REFERENCES

Balho H. Kim and Jong-Bea Park (2001), *Transmission pricing,*

EESRI (2002), *A study on the allocation of transmission service costs*, KEPCO

Prahit K. Dutta (1999), *Strategies and games*, MIT Press

ACKNOWLEDGEMENT

This work has been supported by EESRI (02-전-01), which is funded by MOCIE (Ministry of commerce, industry and energy).

ELSEVIER
IFAC
PUBLICATIONS
www.elsevier.com/locate/ifac

A STUDY ON RECLOSING OPERATION CONSIDERING FAULT CURRENT LIMITER

Kazuto Yukita*, Junya Fujita*, Daisuke Kojima*,
Yasuyuki Goto* Katsuhiro Ichiyanagi* Yasuhito Tabata** Shigeaki Ogawa**

*Aichi Institute of Technology , **Chubu Electric Powe Co.Inc.

Abstract : This paper proposes the effective reclosing operation method in the current limiter introduction power system. From a angular velocity, a field system voltage, a phase of the generator, the proposed method calculates the optimum reclosing time domain. The current limiter used reactance type and resistive type. The examination was carried out using the Inst. of Electrical Engineers of Japan one-machine infinite bus system model in order to demonstrate the effectiveness of the proposal method. In the simulation, Impact ver2.3 was used.

Keywords - Power systems, Power system control, Voltage stabilizers, Voltage control, Power transmission

1 .INTRODUCTION

In order to build earth environment preservation and an eco-friendly electric power system, it is in the tendency, which environmental harmony type power generation systems, such as photovoltaic equipment, wind power system, and fuel cell, will increase into the system. Since such a power generation system has many comparatively small-scale things, it is dotted with small-scale plant in a system, and it is in the tendency of complication more. In the system in which this small-scale plant exists, the fault current when fault occurs on a transmission line is more increasing. In order to solve this problem, the introduction to the system of FCL (Fault Current Limiter) is considered and studied (IEEJ 2001b). In the meantime, the reclosing operation that it releases the transmission line under trouble, since whether the temporary trouble is the trouble or whether it is an eternal trouble are judged, and closes circuit must be done, when the trouble arose in the transmission line. Automatic reclosing operation is carried out, as for this reclosing operation, the execution time changes with voltage class, kinds of fault, etc. However, the present condition is set up in consideration of the generating time of a secondary arc etc (IEEJ 2001a,IEEJ 1999a). Authors have so far reported that there is a time [the improvement effect in system stability to be by execution time] domain about this reclosing operation (Yukita, et al, 2002,Kojima, et al, 2001). Then, this paper considered reclosing operation in a FCL introduction system further. It was the one machine infinite bus system model which is an IEEJ standard model (IEEJ 1999b) which took into consideration the governor (GOV) and automatic voltage adjustment machine (AVR) of a Generator, and FCL was installed in the Generator near end and eternal ground fault was assumed at the near end here. The FCL uses R type FCL that is resistive, and L type FCL that is the reactance type here. The simulation in this paper was performed using Impact Ver2.3(TEPC).

2 .MODEL SYSTEM

2.1 One machine infinite bus system model

In this paper, Fig.1 shows the one-machine infinite bus systems model that is the Inst. of Electrical Engineers of Japan (IEEJ) standard model system is used. The Generator is connected to an infinite bus through two transmission line. Moreover, FCL was installed in the Generator near the end of a transmission line. Fig.2 and Fig.3 show the generator considers AVR and GOV. In this paper, the eternal three-line ground fault, which seemed to be the severest in the trouble, was assumed

2.2 FCL model

In this paper the FCL model was used two type, the R-type FCL. FCL. Fig.5 shows the operating characteristic of FCL used in this paper. This FCL shall operate in about 2ms after fault.

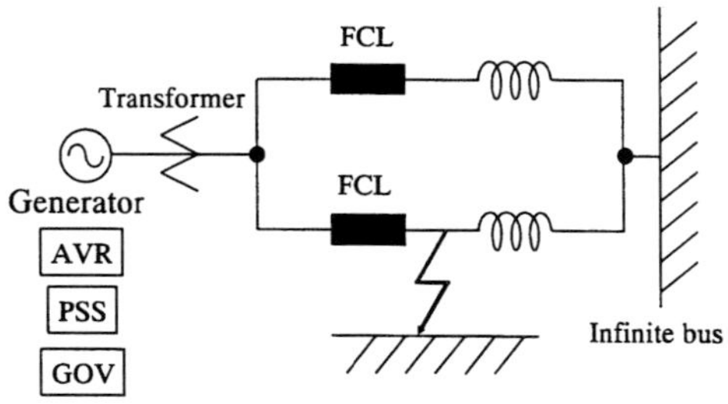

Fig. 1. One machine infinite bus system.

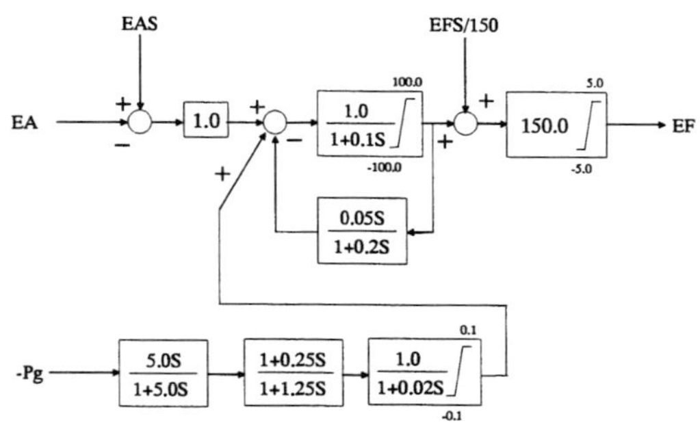

Fig. 2. Block diagram of AVR.

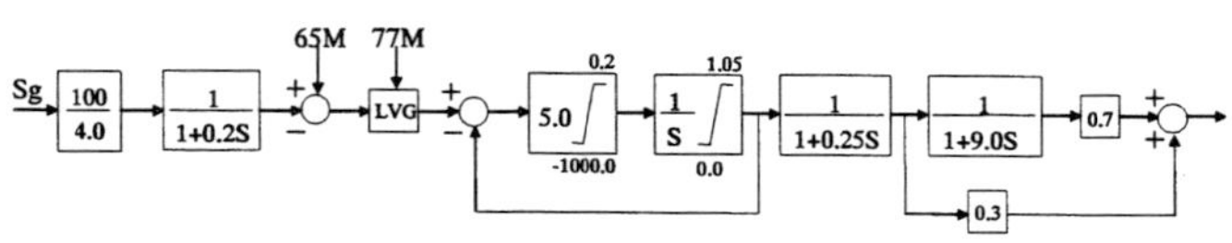

Fig. 3. Block diagram of GOV.

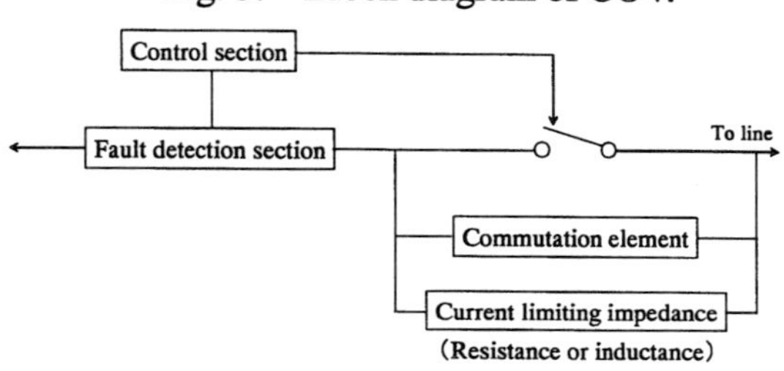

Fig. 4. Basic configuration of FCL.

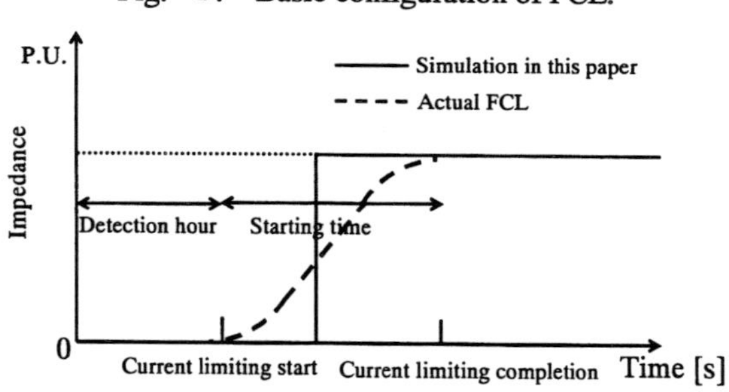

Fig. 5. Operating characteristic of FCL

3.RECLOSING

3.1. The system that FCL has not been introduced

The reclosing operation used in this paper is a method in reference(Yukita, et al, 2002), and it does generator phase difference δ, generator angular velocity ω and generator field voltage ed' of the generator before the fault occurrence with the standard, and each deviation ($\Delta\delta$, $\Delta\omega$, Δed') after the fault occurrence is obtained. Then, reclosing operation time with the system stability improvement effect was calculated.

The computed conditions are shown as follow.

$$\Delta\delta < 0 \quad (1)$$

$$\Delta\omega < 0 \quad (2)$$

$$\Delta ed' < 0 \quad (3)$$

If reclosing operation is carried out in this domain, I will think that the improvement effect in stability is expectable. Fig.6 shows time sequence at this time.

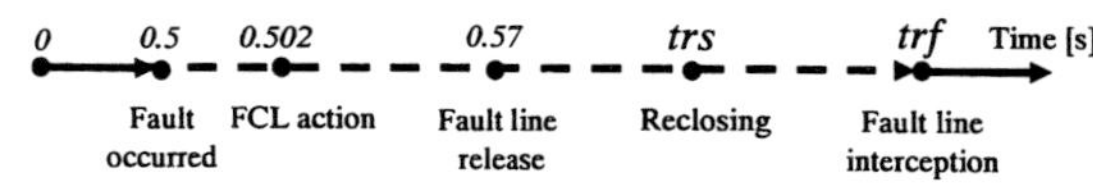

Fig. 6. Time sequence.

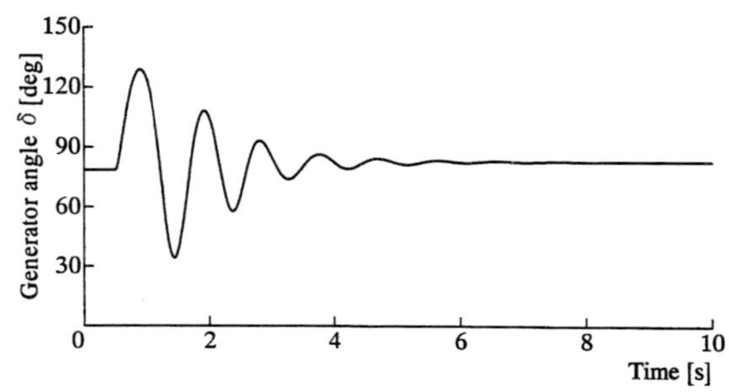

Fig. 7. In case of no reclosing operation.

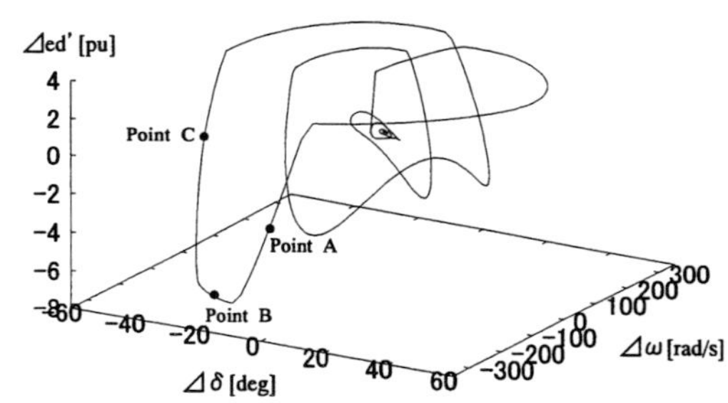

Fig. 8. Characteristic in $\Delta\delta - \Delta\omega - \Delta$ed'.

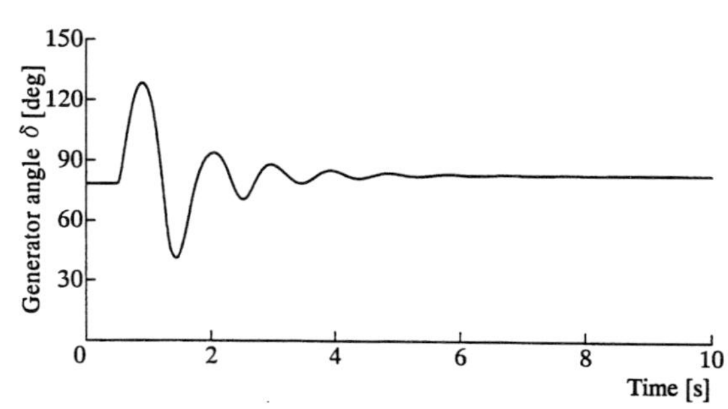

Fig. 9. In case of reclosing operation. (trs=1.3s)

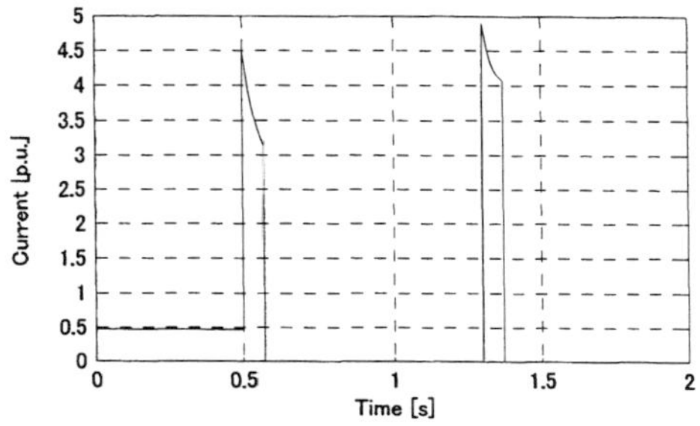

Fig. 10. Line current of fault line.

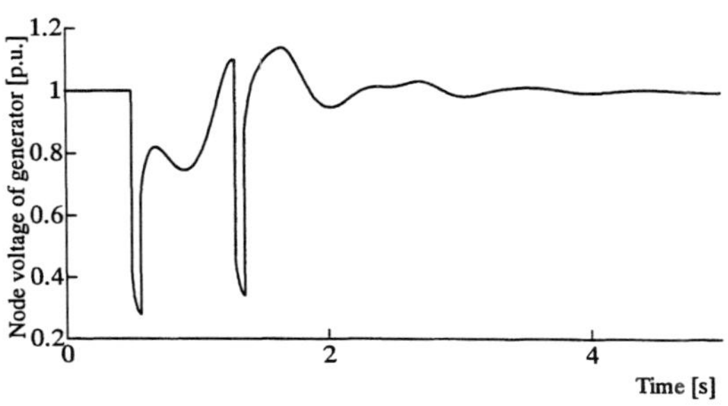

Fig. 11. Node voltage of generator.

Fig.7 shows the generator phase oscillation characteristic in the case of the reclosing operation. Fig.8 shows $\Delta\delta$, $\Delta\omega$, Δed'characteristic in this time. The region where the stability improvement effect is expected is shown for a same figure the point A to the point C. Fig.9 shows the phase oscillation characteristic in the case of the reclosing operation in a point B which is center in this region almost. In comparison with Fig.7 and Fig.9, the phase oscillation is suppressed, and it proves effect.

Fig.10 shows execution value characteristic of this time fault current. It is proven that fault occurrence after current increases.

3.2 The system that R type FCL is introduced

In the system , the R type FCL, the resistance was used for the commutation element, was introduced. It tries to examine the reclosing operation.Fig.12 shows the generator phase oscillation characteristic in the system, that the R type FCL was introduced in order to calculate the reclosing operation time.Fig.13 shows $\Delta\omega$-$\Delta\delta$-Δed' characteristic. Fig.14, Fig.15 and Fig.16 show the characteristic by the reclosing operation in a point A – point C of Fig.13. The R-type FCL in this paper was supposed to operate three-phase after about 2ms after the fault occurrence simultaneously. Fig.17 and Fig.18 show the current characteristic in the case of reclosing operation in a point B of Fig.13. Fig.19 shows the node voltage of the generator. Fig.20 shows the generator phase oscillation in the case of the reclosing operation in 50 cycles, and Fig.21 shows the case of reclosing operation in the 60 cycles, which are about 1 second after the trouble line release that are the report of literature (IEEJ 2001b). It is proven that in the case of the reclosing operation in a point B, which is a center in the region in Fig.13 almost, it suppresses the oscillation most, and that it is a good characteristic.

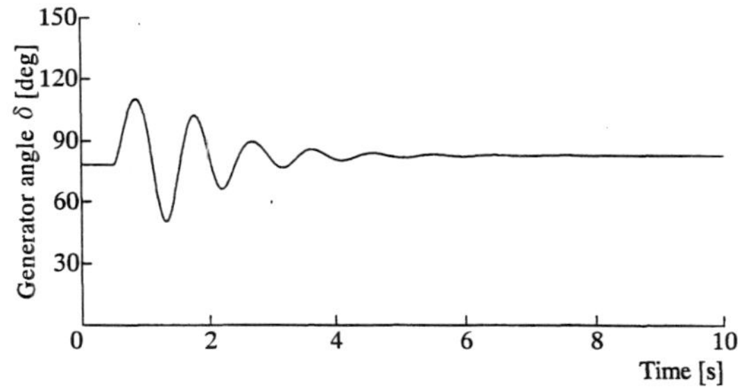

Fig. 12. In case of no reclosing operation.

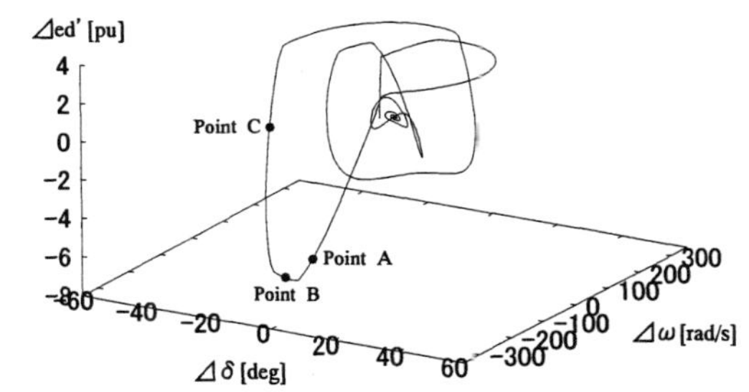

Fig. 13. Characteristic in $\Delta\delta-\Delta\omega-\Delta$ed'.

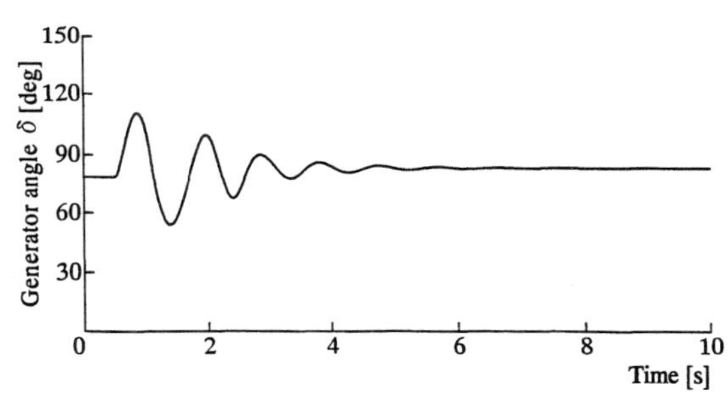

Fig. 14. In case of reclosing operation.
(point A, trs=1.13s)

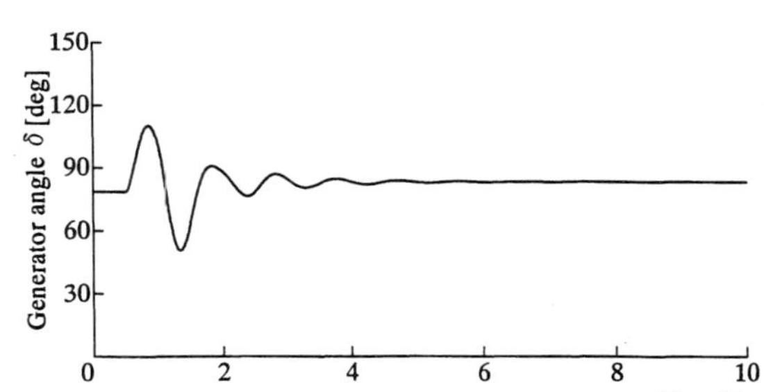

Fig. 15. In case of reclosing operation.
(point B, trs=1.2s)

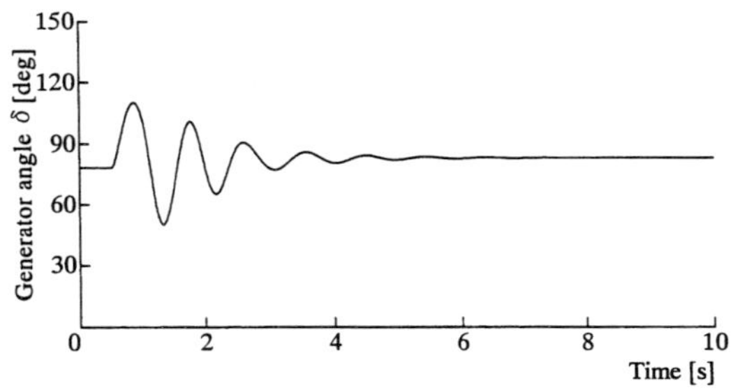

Fig. 16. In case of reclosing operation.
(point C, trs=1.28s)

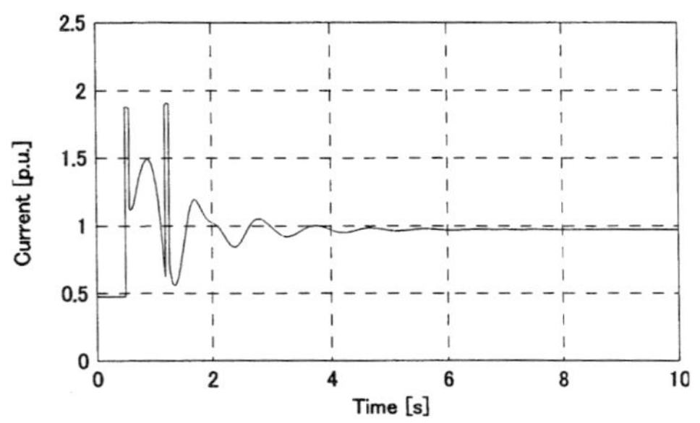

Fig. 17. Line current of soundness line.

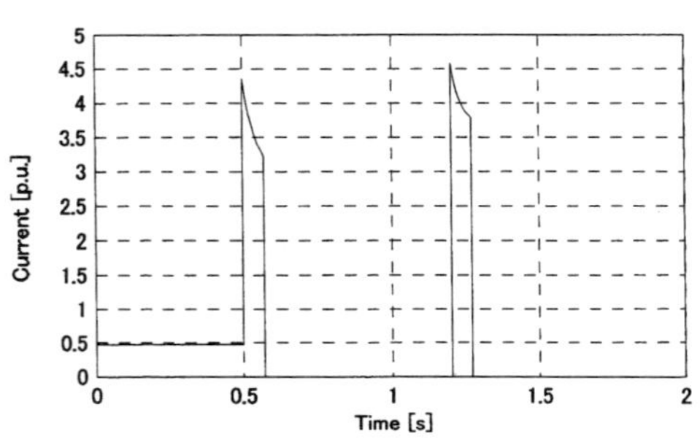

Fig. 18. Line current of fault line.

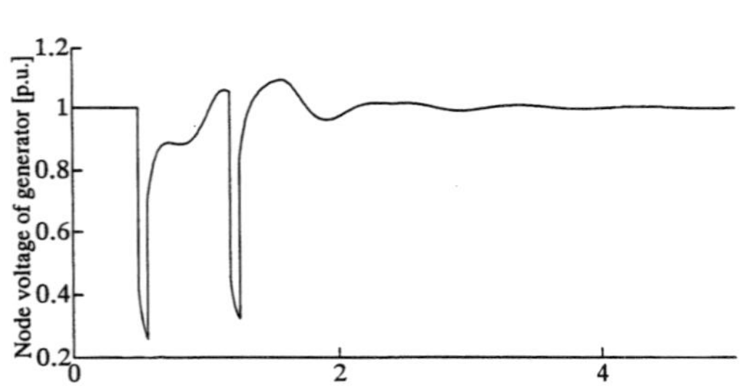

Fig. 19. Node voltage of generator.

3.3 The system that L type FCL is introduced

In the system , the L-type FCL of which the reactance was used for the commutation element was introduced, it tries to examine the reclosing operation. Fig.22 shows the generator phase oscillation characteristic in the system . The L type FCL was introduced in order to calculate the reclosing operation time. Fig.23 shows $\Delta\omega$-$\Delta\delta$-Δed' characteristic. The L type FCL was supposed to operate

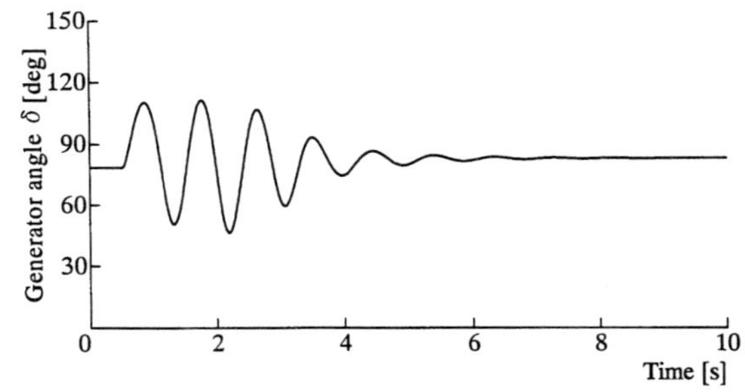

Fig. 20. In case of reclosing operation.
(50cycle, trs=1.4s)

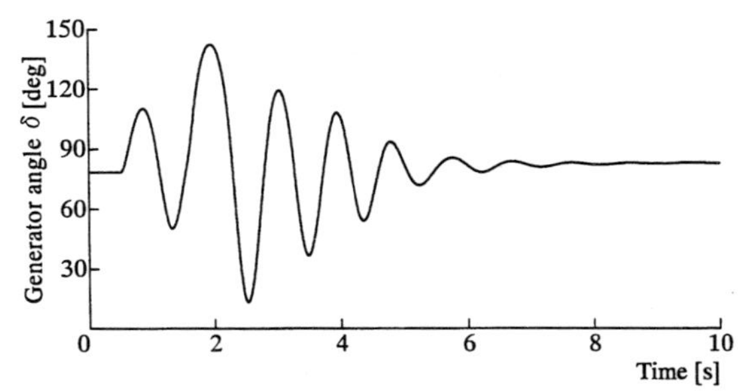

Fig. 21. In case of reclosing operation.
(60cycle, trs=1.57s)

three-phase after about 2ms after the fault occurrence as well as R type FCL simultaneously.Fig.24, Fig.25 and Fig.26 show the characteristic by the reclosing operation in a point A to the point C of Fig.23. Fig.27 and Fig.28 show current characteristic in the case of reclosing operation in a point B of Fig.23. In the fault line, it is proven that L type FCL also suppresses the increase in the fault current similarly regarded as R type FCL after fault occurrence. Fig.29 shows the node voltage of the generator .It is proven here that L type FCL suppresses the increase in the fault current in comparison with Fig.18 and Fig.28 further than R type FCL. And, it is proven that the lowering of the voltage is suppressed of L-type FCL in comparison with Fig.19 and Fig.29 further than R-type FCL. Fig.30 shows generator phase oscillation in the case of the reclosing operation in 50 cycles, and Fig.31 shows the case of reclosing operation in the 60 cycles, which are about 1 second after the trouble line release that are the report of literature (IEEJ 2001a). It is proven that in the case of the reclosing operation in a point B, which is a center in the region in Fig.23 almost, it suppresses the oscillation most, and that it is a good characteristic.

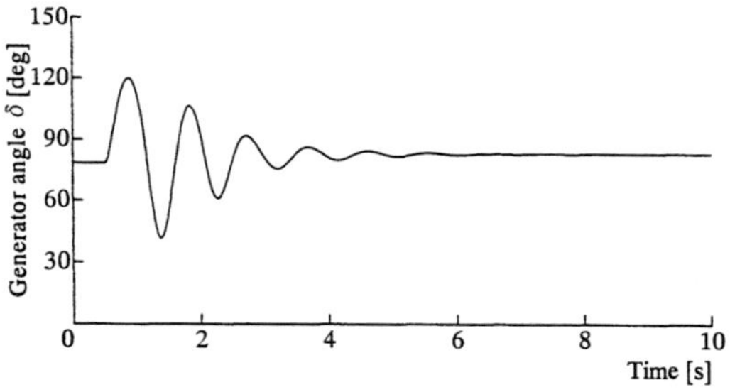

Fig. 22. In case of no reclosing operation.

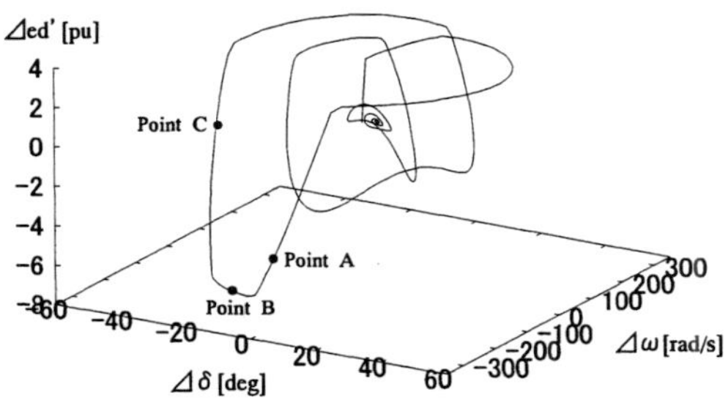

Fig. 23. Characteristic in $\Delta\,\delta - \Delta\,\omega - \Delta\,ed'$.

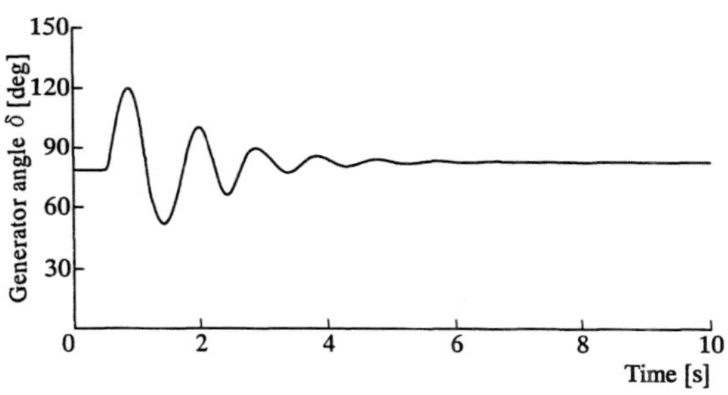

Fig. 24. In case of reclosing operation. (point A, trs=1.17s)

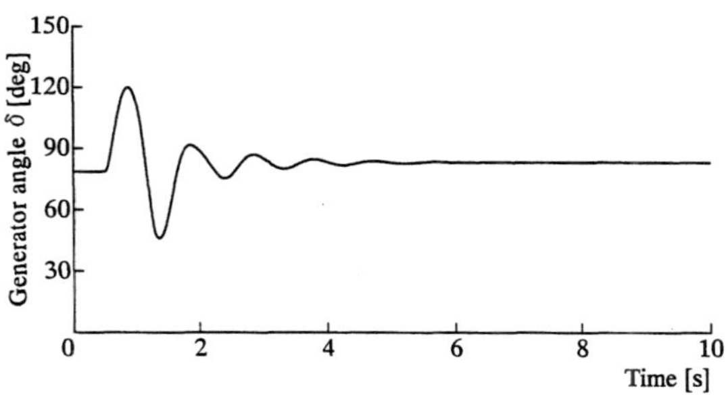

Fig. 25. In case of reclosing operation. (point B, trs=1.25s)

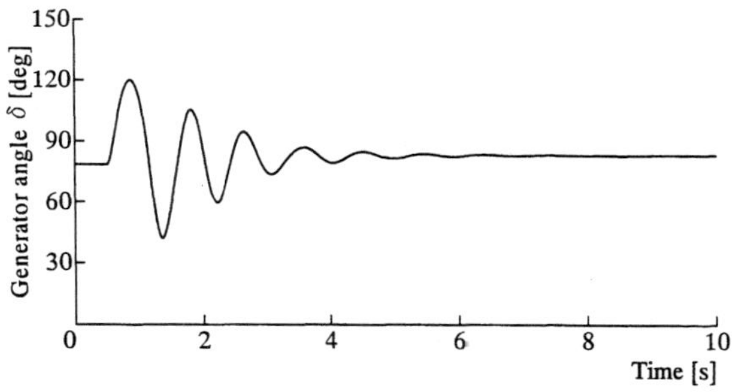

Fig. 26. In case of reclosing operation. (point C, trs=1.33s)

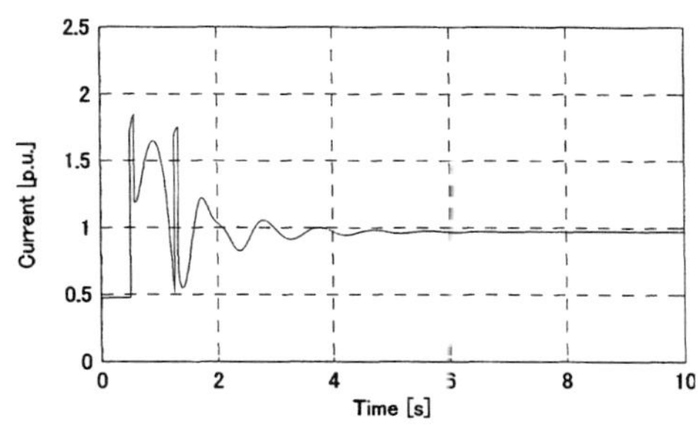

Fig. 27. Line current of soundness line.

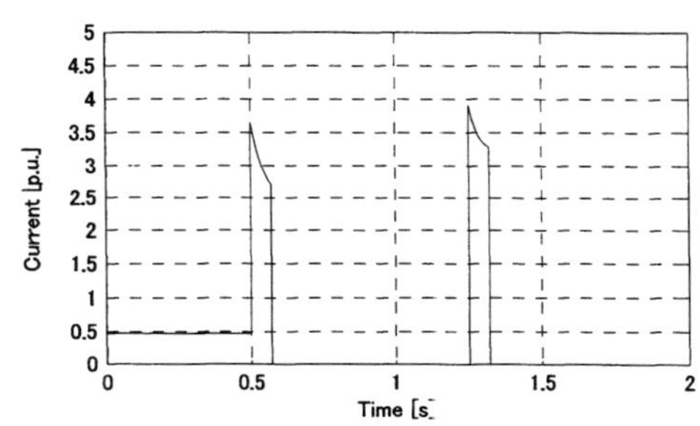

Fig. 28. Line current of fault line.

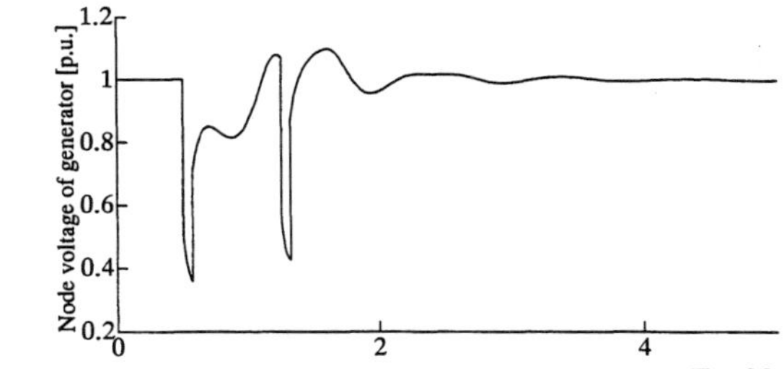

Fig. 29. Node voltage of generator.

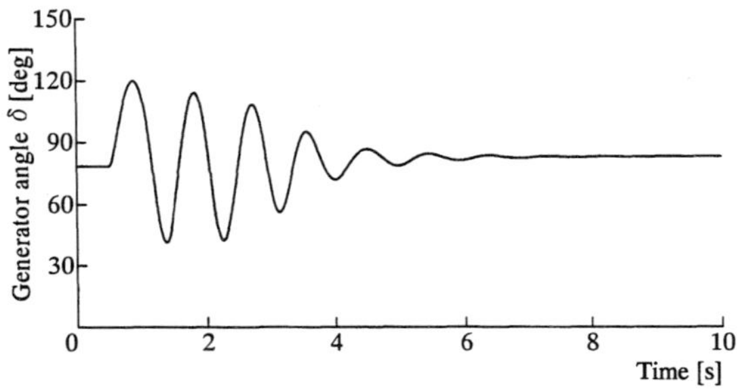

Fig. 30. In case of reclosing operation.
(50cycle, trs=1.4s)

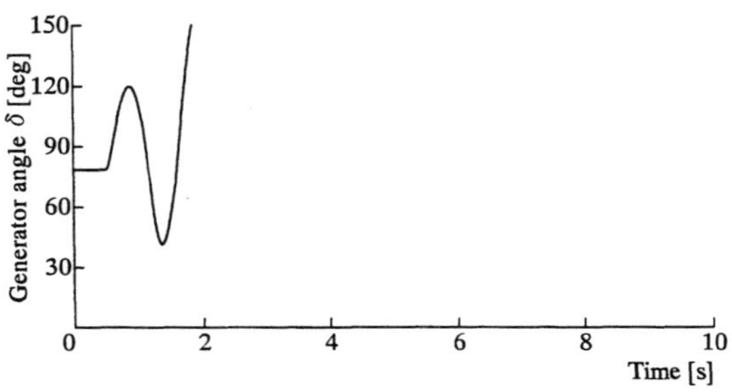

Fig. 31. In case of reclosing operation.
(60cycle, trs=1.57s)

4. CONCLUSION

In this paper, the reclosing operation was examined using the system that FCL was introduced for the fault current control in electric power system in which the complication advances. The reclosing operation was carried out in the region that in this paper, it noticed the time in generator phase, generator angular velocity and generator field voltage, and that they expect the stability improvement effect. It was examined using a machine infinite system model, which was the IEEJ standard model system on the effectiveness. As the result, it was shown that the system stability improvement effect could be expected by carrying out the reclosing operation in the region.

REFERENCE

"Development of the system analysis synthesis environment (Impact) by PC" The Tokyo Electric Power Co., Inc.

Kojima, Kato, Yamada, Yukita, Goto, Ichiyanagi, Matumura : "Examination of the optimal reclosing which used the curve recently in an electric power system", (2001)

The Institute of Electrical of Japan : "A power system step out and accident influencing prevention relay technology", Institute of Electrical Engineers of Japan technical report, No801, (2001a)

The Institute of Electrical of Japan: "Application technology of the FCL equipment for fault current control, and analysis evaluation", Institute of Electrical Engineers of Japan technical report, No 850, (2001b)

The Institute of Electrical of Japan : "It is restoration operation in case of the accident of an electric power system.", Institute of Electrical Engineers of Japan technical report, No354

The Institute of Electrical of Japan:"Systematization and engineer training" of protection relay system engineering , Institute of Electrical Engineers of Japan technical report, No717, (1999a)

The Institute of Electrical of Japan : "The standard model of an electric power system", Institute of Electrical Engineers of Japan technical report No.754 (1999b) http://www.iee.or.jp/pes

Yasuda, Yokoyama, Tada : "A power transmission network low research" about the influence which an identification [next] model does to a multi-opportunity system simulation , Electric power technology · Electric power system technical joint study group, PE-02-119, PSE-02-129, (2002)

Yukita , Kojima , Yamada , Mizuno , Goto , Ichiyanagi : "High-speed reclosing operation for the improvement in system stability at the time of electric power system fault", Power · Energy section convention, (2002)

ELSEVIER
IFAC
PUBLICATIONS
www.elsevier.com/locate/ifac

A STUDY OF NOVEL POWER SYSTEM STABILIZER

Wei Wang **Long Wu** **Heping Xu** **Weiqun Liu** **Xianming Chen**

Nanjing Automation Research Institute
P.O.BOX 323,Nanjing 210003 PRC Email:ecti@nari-china.com

Abstract: The use of power system stabilizer (PSS) to control generator excitation system is most cost-effective method to damp low-frequency oscillation. However, incremental active power (ΔPe) based PSS suffers from inter-area oscillations, which is restricted to use because of neglecting increment of angle speed of system. A novel power angle (δ) based PSS, which eliminate the noise brought from using differentiation of δ, is proposed to overcome the disadvantages of ΔPe based stabilizer. On-line identifier for inductance (X_Σ) between local unit and equivalent infinite bus, has been introduced to mitigate low-frequency upset. The test results of the new PSS is also given. In comparison with ΔPe based PSS, the novel PSS shows its advantages. *Copyright © 2003 IFAC*

Keywords: PSS, on-line identifier, δ-based, inter-area oscillation, excitation control

1. INTRODUCTION

In order to improve power system stability, fast excitation control of alternator has been widely used to electrical power industry since the mid 1960s (P.L. Dandeno, *et al.*, 1968). Especially, PSS based on small signal analysis theory is adopted to excitation control, which is proved a good approach to mitigate low frequency oscillations.

Due to insufficiency of accuracy of small signal theory, many methods and theories, such as fuzzy control, adaptive control (M. and OP, 1999), non linear control (Qiang and Yuanzhang , 1989), etc. theories, have been introduced to overcome disadvantages of small signal method. These methods may work well, if it is able to obtain equivalent system parameters properly. In additional, it is difficult for field operators to tune parameters concerned correctly that limits their application.

There are two major advantages for PSS: robust parameters tuning and facility in control. On the other hand, there has been little incentive to date to consider an adaptive control system (PRABHA KUNDUR,1994).

However, conventional PSS suffers from inter-area mode oscillations (PRABHA KUNDUR,1994). Since the effectiveness of existing PSS, as far as the inter-area mode is concerned, drops significantly. It is necessary for us to improve conventional ΔPe based PSS. Improved incremental angle speed ($\Delta\omega_r$) based PSS overcomes some disadvantages of ΔPe based PSS by integrating increment of mechanical power (ΔPm), which faces to deal with noise in measurement of rotor speed (ω_r).

There are three items discussed in this paper:

-To estimate Pm and incremental angle speed of the whole system ($\Delta\omega_s$) by synthesizing generator terminal electric quantities.

-To establish equivalent system model and run-time identifier for X_Σ

-To propose δ based PSS design method

Test results of comparison between the new PSS and the conventional PSS are present, in this paper.

2. EQUIVALENT SYSTEM MODEL

To study inter-area mode oscillation, two-area model is an ideal choice. As well known, it is impossible to estimate angle speed change of equivalent area by power flow calculation with local generator terminal electrical quantities. Since requirement for extra equivalent system arguments is extravagance.

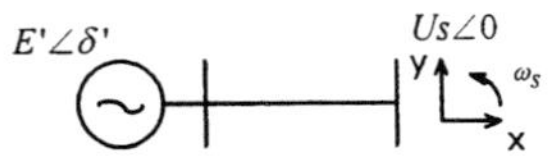

Fig. 1 Single unit infinite bus with changeable ω_s

2.1 Single unit infinite bus with changeable ω_s

PSS designed according to single machine infinite bus, is able to deal with inter-area mode oscillation if neglecting $\Delta\omega_s$. Here a changeable ω based single machine infinite bus system shown in Fig.1, will be discussed. From ordinary point of view of single machine infinite bus, ω_s is always considered as a constant. But from point of view of two-machine system, it trend to be assumed that the remote equivalent machine has a large moment of inertia, which damp angle speed change. Hence, inter-area oscillations could be hidden.
In order to overcome disadvantages mentioned above, X_Σ is estimated dynamically in system described by Fig.1. In practice, in the course of low-frequency oscillation, change of voltage of infinite bus (Us) is sensitive to upset. So that, change of Us needs to be considered.

2.2 Equivalent system model

Referring to Fig.1, it is easy to write following equations for system:

$$\begin{cases} \dot{U}_t = \dot{E}' - jx'_d \cdot \dot{i}_t \\ T_j \dfrac{d\omega_r}{dt} = P_m - P_e - D(\omega_r - \omega_s) \\ \dfrac{d\delta}{dt} = \omega_r - \omega_s \end{cases} \qquad (1)$$

In equations (1), derivation of ω_s in generator rotor equation is neglected, since equivalent infinite bus has mighty moment of inertia. Similarly, change of ω_s caused by local unit also is neglected in respective equation. For calculating ω_s, assuming that :
(I) X_Σ is constant in the course of oscillation
(II) neglecting saliency
(III) neglecting resistance of generator stator and line
From equation (1), we can conclude that ω_r and power angle δ will change when system angle speed ω_s changes. Consequently, change of ω_s drives or damps rotor movement. This process is naturally regulated by the whole system itself. In this case, as a auxiliary control, PSS ought not to work.

3. δ BASED PSS DESIGN

Maybe someone has a suspicion that measure of δ is possible. Is it possible to measure ω_s since noise is great ? It will be discussed how to measure ω_s and δ , how to estimate X_Σ later. At first, let's review the phasor diagram. In Fig.2, solid line denotes phasors in steady state of system. Dashed line denotes ones in dynamic states of system.

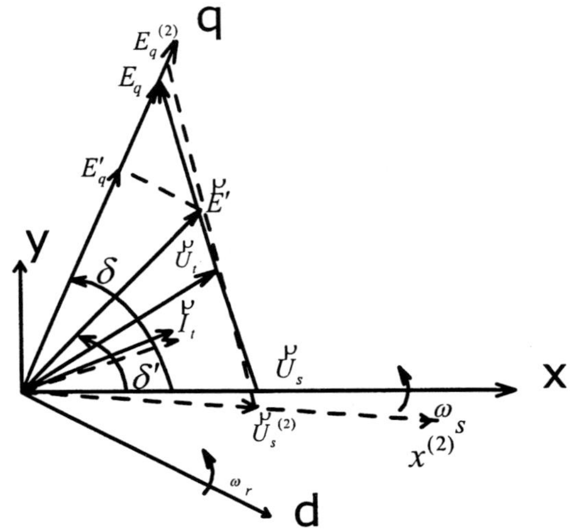

Fig.2 Phasor diagram of single machine infinite bus equivalent system.

According to assumption (I), X_Σ can be identified under steady state of system. The system voltage $\dot{U}_s^{(2)}$ can also be obtained by synthesis of X_Σ and $\dot{I}_t^{(2)}$. The phase angle between $\dot{U}_s^{(2)}$ and $\dot{U}_s$ should ascribes to relative motion between local generator and system. So $\Delta\omega_s$ can be calculated with the phase angle.

3.1 Measure and deduction of δ

Firstly, three phase angles are defined as follows:
δ_{et} represents phase angle between generator transient potential (E') and generator terminal voltage ($\dot{U}_t$).
δ_{rt} -phase angle between E_q and $\dot{U}_t$.
δ_{ts} -phase angle between $\dot{U}_t$ and $\dot{U}_s$.
Obviously,

$$\delta = \delta_{rt} + \delta_{ts} \qquad (2)$$

Secondly, according to Fig.1 and equation (1), equation (3) describes the relation between $\dot{U}_t$ and $\dot{U}_s$.

$$\dot{U}_t = \dot{U}_s + jx_\Sigma \cdot \dot{i}_t = U_s + jx_\Sigma \cdot i_{tr} - i_{ti} \qquad (3)$$

where: i_{tr} represents real part of terminal current i_t
i_{ti} - image part of i_t
Therefore

$$\delta_{ts} = a\tan\left(\frac{x_\Sigma \cdot i_{tr}}{U_s - i_{ti}}\right) \qquad (4)$$

To measure δ_{rt}, two methods are given as follows:
The one method is based on vector rotational principle. In a symmetrical three phase AC system, we can get equation (5) and (6).

$$j\sqrt{3}\dot{U}_a = \dot{U}_c - \dot{U}_b = \dot{U}_{cb} \qquad (5)$$

$$-j(\dot{U}_a + jx_q \cdot \dot{I}_a) = x_q \cdot \dot{I}_a + \dot{U}_{cb} \qquad (6)$$

δ_{rt} can be measured easily according to equation (6).
The other method is based on park transformation. In the coordinate rotating with terminal angle speed ω_e ,

expression of $\dot{U}_t$ and $\dot{I}_t$ can be obtained.Calculation of δ_{rt} can be implemented by digital automation voltage regulator (AVR) with some simple operations.

3.2 *Identifier of* X_Σ

Because X_Σ will vary with change of center point of equivalent system, X_Σ needs to be identified dynamically. Electrical active power (Pe) and reactive power (Qe) of generator could be represented by equation (7) and (8).

$$P_e = E'U_s \sin(\delta')/(x'_d + x_\Sigma) \quad (7)$$

$$Q_e = (U_s^2 - E'U_s \cos(\delta'))/(x'_d + x_\Sigma) \quad (8)$$

Two equations above are of the constraint conditions of solving X_Σ and U_s. When excitation voltage of local unit has a significant change, X_Σ can not be modified according to assumption (I). When local unit regulates its active power, the whole power system will enter a new frequency adjustment state. X_Σ will be identified a new value as soon as the equivalent system enters into steady, will be identified newly. Thus, fluctuation of X_Σ reflects influence of local unit to the whole system. If X_Σ is stiff without any change, that is to say local unit is insignificant to the system. In other words, infinite bus equivalent system is proper. Before the new PSS enter into work status, X_Σ must be estimated appropriately. Otherwise, the PSS won't work at all.

3.3 *Computation of* $\Delta\omega_s$

When the active power of local generator arises or falls, ω_s will also arise or fall. If the PSS does not enter into work state, AVR will keep terminal voltage as reference voltage. Based on small signal theory, equations concerned can be listed as follows:

$$\begin{aligned}\Delta P_e &= E'U_s \cos(\delta')\Delta\delta'/(x'_d + x_\Sigma) \\ &+ E'\sin(\delta')\Delta U_s/(x'_d + x_\Sigma) \\ &+ U_s \sin(\delta')\Delta E'/(x'_d + x_\Sigma)\end{aligned} \quad (9)$$

$$\Delta\delta_{es} = \delta' + \Delta\delta' - \delta_{es}^{(2)} \quad (10)$$

$\delta_{es}^{(2)}$ denotes phase angle between E' and $\dot{U}_t$, which superscript denotes for calculation dynamical state (dashed in Fig.2).

$\Delta\delta_{es}$ is the phase angle produced by $\Delta\omega_s$.

$$\Delta\omega_s = \frac{d(\Delta\delta_{es})}{dt} \quad (11)$$

Oscillations in one part of a system can excite unit oscillation in another part of the system due to resonance (M. Klein, *et al.*, 1992). While phase angle between oscillation source and oscillation sink is larger than 90 degree, local unit might fail to damp this upset by using conventional PSS. $\Delta\omega_s$ can be used to explain the conclusion (M. Klein, *et al.*, 1992) which shows the importance of using appropriate load models in the design of PSS for enhancement of the stability of an inter-area mode.

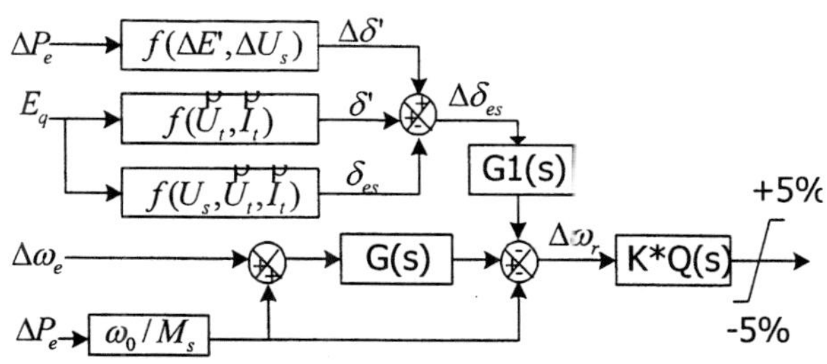

Fig.3 Control block diagram of δ based PSS.

3.4 *Design of new PSS*

Design of new PSS needs to avoid using differentiation of many quantities, since differentiation is source of noise. According to equation (1), $\Delta\omega_r$ can be obtained by integrating (ΔPm-ΔPe). But the $\Delta\omega_r$ includes component of $\Delta\omega_s$. For effect of $\Delta\omega_s$ to local unit, the camp winding of local unit could react rapidly. Output of the PSS should not to include component of $\Delta\omega_s$, since it is impossible to obtain the phase angle between system fluctuation and local unit oscillation. Fig.3 shows control block diagram of the new PSS. In Fig.3:

$$\Delta P_e = \frac{1}{s}\cdot\frac{s5.02}{1+s5.02}\cdot\frac{s2.0}{1+s2.0}P_e \quad (12)$$

$$\Delta\omega_e = 2H\cdot\frac{s5.02}{1+s5.02}\cdot\frac{s2.0}{1+s2.0}\omega_e \quad (13)$$

$$G(s) = \frac{1}{(1+s0.06)^4} \quad (14)$$

$$G_1(s) = s\cdot\frac{0.396}{s^2 + s0.889 + 0.396}\cdot\frac{s5}{1+s5} \quad (15)$$

$$K*Q(s) = K_{pss}\cdot\frac{1+sT_1}{1+sT_2}\cdot\frac{1+sT_3}{1+sT_4} \quad (16)$$

where :

Kpss, T1, T2, T3 and T4 all are variable which need to be decided by field test. Generally, equation (16) is phase lead transfer function. According test results, it is necessary to appended low pass filter for eliminating noise of differentiation in equation (15).

4. TEST RESULTS

4.1 *Introduction to test system and parameters*

Table 1 parameters of test system

	New PSS	existing PSS		New	Existing
x'_d	0.324p.u.		T2	0.04s	3.62s
xq	1.56 p.u.		T3	0.1s	0.07s
xd	1.56 p.u.		T4	0.04s	0.06s
T1	0.1s	0.7s	Kpss	0.5	2.5

A novel real time simulator of synchronous generator for closed loop test of AVR, which is introduced in detail in the reference (Wang , *et al.*, 2001) is used to test the new PSS. The new PSS is developed by Nanjing Automation Research Institute (NARI), State Power Corp of PRC, which is embedded in SAVR2000, a full digital excitation controller. So, the new PSS is a subroutine of software of AVR. In the test, parameters adopted are listed in Table.1

4.2 Test results

By comparison between test results of the new PSS and those of existing PSS, it can be concluded that the new PSS plays a better role in mitigating low-frequency upset than that of the conventional PSS. Number of oscillation of response using new PSS is less than that of existing PSS. Moreover, periods of new stabilizer regulation is shorter than period of existing one. However, there is not apparent difference between use of the new PSS and that of existing one in the ground fault test.

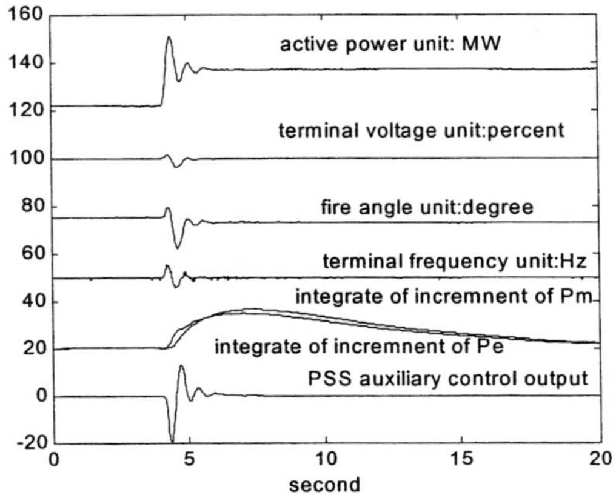

Fig.3a Response using new PSS with mechanical power 0.03p.u. step test

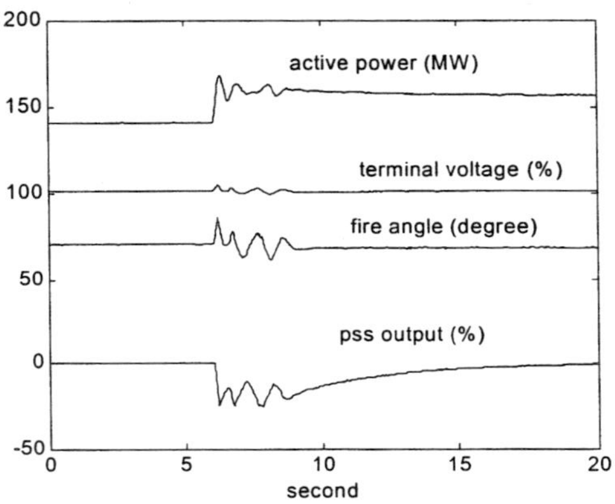

Fig.3b Response using conventional PSS with mechanical power 0.03p.u.step test

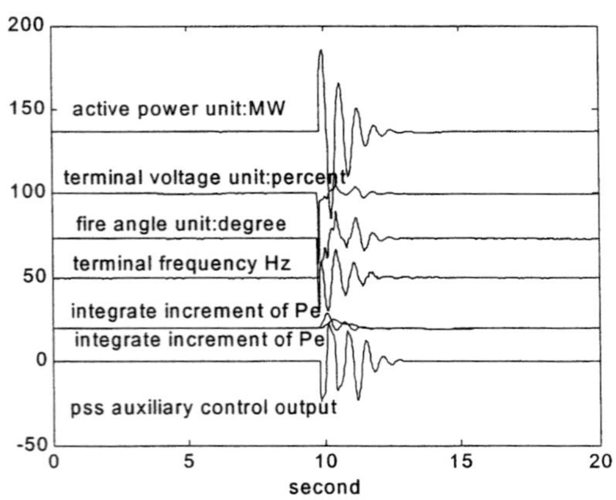

Fig.4a Response using new PSS with three phase ground fault cleared in 0.08s

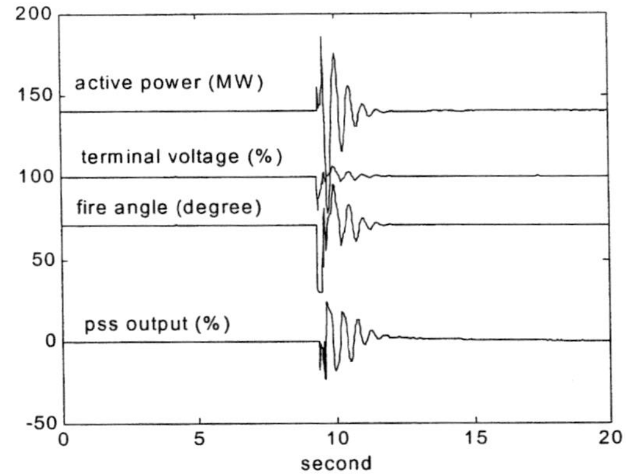

Fig.4b Response of conventional PSS with three phase ground fault cleared in 0.08s

5. CONCLUSION

Results of experiment show that the new PSS is effective to mitigate inter-area low-frequency mode oscillations. The new PSS is also capable of dealing with local oscillations which include both power plant internal oscillations and two-area oscillations. Almost all input quantities can be obtained by measurement and deduction of generator terminal electrical quantities. So that the new PSS can avoid noise brought from measurement. In addition, it is benefit to realize the new PSS.

However, use of the new PSS is also restricted when upset source is closed to local unit or when $\Delta\omega_s$ is very great. So, there is a lot of work for us to do further to mitigate fully low-frequency oscillations.

REFERENCES

P.L. Dandeno, A.N. Karas, K.R. McClymont, and W. Waston (1968). Effect of High-Speed Rectifier Excitation System on Generator Stability Limits. In: *IEEE Trans* **Vol. PAS-87,** pp.190-201.

Qiang Lu and Yuanzhang Sun. (1989). *Nonlinear Stabilizing Control Multimachines System*, In: *IEEE PES* **Vol. 4 No.1,** pp.1357-1362.

PRABHA KUNDUR (1994). Power System Stability and Control. McGraw-Hill, Companies, Inc.

Wang Wei , Zeng JiLun and Xu HePing (2001). Realization of Real-Time Synchronous Generator Simulation Based on DSPs for AVR Closed-Loop Tests. In: *ICEE 2001* **Vol. 2,** pp.224-227.

M. Sanaye-Pasand and OP Malik (1999). A Fuzzy Logic Based PSS Using a Standardized Rule Table. In: *Electric Machines and Power Systems* **Vol. 27 No.3,** pp.295.

M. Klein, G.J. Rogers, S. Moorty and P. Kundur (1992), Analytical Investigation of Factors Influencing Power System Stabilizers Performance. In: IEEE Trans on EC, **Vol. 7 No.3,** pp382-390.

D.C. Lee, R.E. Beaulieu, and J.R.R Service (1981), A Power System Stabilizer Using Speed and Electrical Power Inputs – Design and Field Experience. In: IEEE Trans on PAS, **Vol.PAS-100 No.9,** pp4151-4155.

ELSEVIER
IFAC
PUBLICATIONS
www.elsevier.com/locate/ifac

INFLUENCE OF HV LINES ON PARALLEL PIPELINES

Hyun-Goo Lee, Tae-Hyun Ha, Jeong-Hyo Bae, Yoon-Cheol Ha, Dae-Kyeong Kim

Underground Systems Group
Korea Electrotechnology Research Institute

Abstract: Because of the continuous growth of energy consumption and the tendency to site power lines and pipelines along the same route, the close proximity of power lines and buried metallic pipelines has become more and more frequent. Therefore, there has been and still is a growing concern about possible hazards resulting from the influence of power lines on metallic pipelines.
Underground pipelines that run parallel to or in close proximity to power lines are subjected to induced voltages caused by the time-varying magnetic fields produced by the power line currents. The induced electro-motive force causes currents circulation in the pipeline and voltages between the pipeline and surrounding earth.
This paper analyses the induced voltage on the gas pipelines buried in parallel with overhead HV lines. *Copyright © 2003 IFAC*

Keywords: Power lines, Pipelines, Parallel, Induced voltage

1. INTRODUCTION

With the expansion of industrialization and urbanization, the sections where the power line and underground metallic pipeline run in parallel in close proximity have increased rapidly in Korea. This causes induced voltages on underground metallic pipelines due to the power line currents. The currents flow through the pipeline because of the voltage difference induced on the pipeline and flow out through the damaged areas of pipeline coating. This could cause AC corrosion in the pipeline, which could in turn lead to disastrous accidents, such as gas explosion or oil leakage (Gummow, 1999).
In North America, research has been carried out to protect buried pipelines since late 1970s (Dabkowski and Taflove, 1978). However, the status of research in this area is still in its infant stage in Korea.
This paper analyses the induced voltage on a gas pipeline when an overhead power transmission line runs parallel to it. The amplitude of induced voltage is determined by three parameters; first, the power line currents according to the operating conditions; second, the distance between power line and pipeline and the length that parallels between them, and third, the phase arrangement of power line and cable type. When the length in parallel is long, it is not possible to analyse the induced voltage using numerical analysis method, such as the finite element methods and the boundary element methods. Therefore, the induced voltage is analysed by using nodal network analysis that models the parameters of equivalent circuit for the parallel section. Through this, it is possible to know the inductive influence of power line on a near metallic pipeline.

2. NODAL NETWORK ANALYSIS

2.1 Overview

Two different types of analysis methods are used to analyze the voltage induced on the buried pipeline due to the power line currents. One is the numerical analysis method, which includes finite element methods and boundary element methods. The other is nodal network analysis.
In the case of using the numerical analysis method, if the length that parallels becomes long, calculation could take a long time depending on the computer capacity and accurate analysis is difficult. On the

other hand, nodal network analysis uses π type concentrated equivalent circuit impedance matrix and produces more reliable results. Therefore, it is widely used for the calculation of induced voltage on the pipeline due to the power line (Southey and Dawalibi, 1998).

2.2 Theoretical Background

The formula used for the calculation of induced voltage on the pipeline due to a power line is as follows

$$\frac{dV(x)}{dx} + ZI(x) - E = 0 \quad (1)$$

$$\frac{dI(x)}{dx} + YV(x) = 0 \quad (2)$$

where Z is the impedance per unit length of the circuit pipeline and earth, Y is the admittance per unit length of the circuit pipeline and earth, and E is the induced voltage on the pipeline per unit length (CIGRE, 1995).

Figure 1 is the diagram of circuits that shows the influence of the power line on a parallel pipeline.

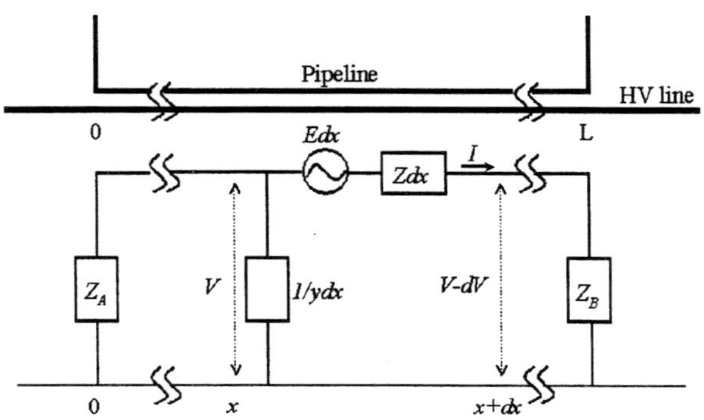

Fig. 1. Influence of a HV line on a parallel pipeline

2.3 Analysis Program

CONIND is a sophisticated pipeline electrical interference analysis tool. It is used to analyse the electrical effects of induced voltage on pipelines from parallel power lines operating under normal steady state load conditions as well as during power system faults.

CONIND represents the pipeline network in a segmented format consisting of discrete lengths of pipeline, for which the power line coupling effects are constant. The results of the calculations contain all pipeline voltages and currents that flow through the pipeline network.

The program calculation is based on an iterative development of the Thevenin equivalents of the pipeline segments. It does not use a matrix-based solution method. It requires data entry to describe each pipeline segment and the coupling between that segment and each current carrying parallel power line conductor (Simpson, 2000).

The calculation of induced voltage on the pipeline is illustrated in Figure 2. For every pipeline segment in a pipeline model, it is possible to develop the equivalent of Figure 2. The Thevenin impedance with voltage V_θ and impedance Z_θ is calculated looking into the pipeline from the left side. The new Thevenin equivalent becomes the V_L and Z_L for the next pipeline segment. It may need to be combined with other Thevenins due to other pipeline segments and ground electrodes. The pipeline diameter, wall thickness, coating resistance, soil resistivity and other parameters are represented by the characteristic impedance Z_0 and propagation constant γ (Dawalibi, 1989).

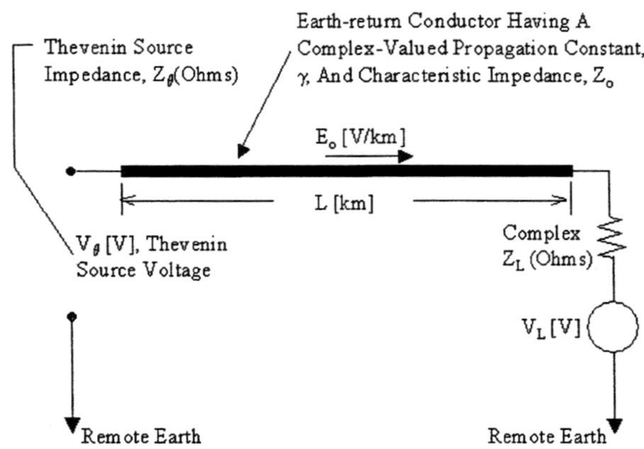

Fig. 2. Pipeline Thevenin Calculation

3. ANALYSIS CONDITIONS

3.1 Parameters used for Analysis

The material of the gas pipeline and soil resistivity used for the analysis of the induced voltage on the gas pipeline running parallel to the power line is shown in Table 1. The gas pipeline is manufactured according to the manufacturing specifications for commonly used 30" main supply pipelines. For the soil resistivity, the value most frequently used in Korea for grounding and the anode design for cathodic protection is used, And the current for each phase flowing through the power line is shown in Table 2. Here, the allowable currents for each power line are used as the current for each phase.

Table 1. Specification of gas pipeline and soil resistivity

Item		Specification
GAS Pipeline	Material	Steel
	Diameter	508[mm]
	Thickness	9.5[mm]
	Resistivity	0.159[μΩ·m]
	Relative Permeability	232
PE Coating	Insulation Resistance	70[kΩ·m^2]
	Thickness	1.75[mm]
	Dielectric Constant	2.3
Soil Resistivity		100[Ω·m]

Table 2. Phase currents in the power line

Cable Type	Allowable currents [A]
ACSR-OC160mm^2	455
ACSR-OC95mm^2	275
ACSR-OC58mm^2	206

3.2 Analysis Model

The cross sectional view of the analysis model used for the analysis of the induced voltage on the gas pipeline according to the length in parallel and the distance is shown in Figure 3.
The height of the power line is assumed to be 14[m] above the earth's surface. And for the distance between each phase, the actual value used for the installation of the steel cross arm of the tower is applied. The analysis is done for a case where the buried gas pipeline runs parallel to the power line along a certain distance without any separate grounding device and where 3 phases are in balance. For the distance between the pipeline and the power line, the distance from the tower to the pipeline is analysed while keeping the height of power line constant.
Also, a 13[km] section where the gas pipeline runs parallel to the 22.9kV power line is selected. Then it is divided into 40 segments through the examination of diagram and field survey for the analysis.

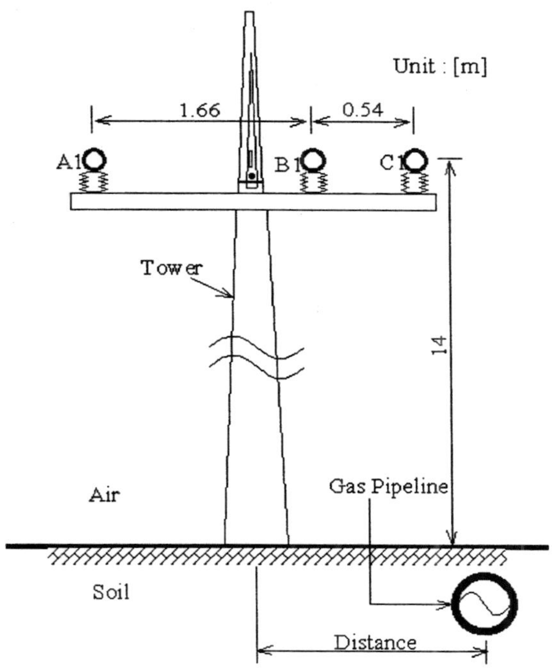

Fig. 3. Cross sectional view of analysis model

4. ANALYSIS RESULTS

4.1 Induced voltage according to the length in parallel

The results of the analysis of induced voltage on the gas pipeline running parallel to the 22.9kV power line according to the length is shown in Figure 4. As the length in parallel increased from 0 to 1,000[m], the maximum induced voltage increased from 0 to 0.392[V].

Table 3. Field Survey for 13[km] parallel Section

No	22.9kV D/L	Parallel Length [m]	Remark
1	160 ㎟ 2 Circuit	660	
2	160 ㎟ 1 Circuit	660	
3	160 ㎟ 2 Circuit	108	
4	160 ㎟ 1 Circuit	340	
5	-	145	No Parallel
6	160 ㎟ 1 Circuit	400	
7	-	410	No Parallel
8	160 ㎟ 1 Circuit	295	
9	160 ㎟ 1 Circuit	600	
10	160 ㎟ 1 Circuit	590	
11	160 ㎟ 2 Circuit	280	
12	160 ㎟ 1 Circuit	160	
13	160 ㎟ 1 Circuit 58 ㎟ 1 Circuit	140	
14	160 ㎟ 2 Circuit	310	
15	160 ㎟ 2 Circuit 58 ㎟ 1 Circuit	160	
16	160 ㎟ 2 Circuit	230	
17	160 ㎟ 2 Circuit 95 ㎟ 1 Circuit	250	
18	160 ㎟ 2 Circuit	920	
19	160 ㎟ 2 Circuit 58 ㎟ 1 Circuit	220	
20	160 ㎟ 2 Circuit	1,020	
21	160 ㎟ 2 Circuit 58 ㎟ 1 Circuit	200	
22	160 ㎟ 2 Circuit	95	
23	160 ㎟ 2 Circuit 58 ㎟ 1 Circuit	130	
24	-	140	No Parallel
25	160 ㎟ 2 Circuit 58 ㎟ 1 Circuit	100	
26	160 ㎟ 2 Circuit	260	
27	160 ㎟ 2 Circuit	550	
28	160 ㎟ 3 Circuit	230	
29	160 ㎟ 2 Circuit	360	
30	160 ㎟ 2 Circuit	240	
31	160 ㎟ 2 Circuit 95 ㎟ 1 Circuit	400	
32	160 ㎟ 1 Circuit 95 ㎟ 1 Circuit	100	
33	95 ㎟ 1 Circuit 58 ㎟ 1 Circuit	170	
34	160 ㎟ 1 Circuit 95 ㎟ 1 Circuit	160	
35	160 ㎟ 2 Circuit	130	
36	160 ㎟ 3 Circuit	100	
37	160 ㎟ 4 Circuit	180	
38	-	180	No Parallel
39	160 ㎟ 4 Circuit	350	
40	160 ㎟ 4 Circuit	1,000	

4.2 Induced voltage according to the distance between the pipeline and the power line

The results of the analysis of induced voltage on the gas pipeline running parallel to the 22.9kV power line according to the distance between them is shown in Figure 5. As the distance increased from 0 to 500[m], the maximum induced voltage decreased in reverse from 0.479 to 0.029[V].

4.3 Induced voltage in 13[km] parallel section

Inductive induced voltage under normal condition for the real parallel field is analyzed using the CONIND.

Table 3 is the results of field survey for gas pipeline 13[km] running parallel to the 22.9kV distribution lines.

The allowable current for the power line is used for the current flowing in the power line under normal conditions in consideration of the worst case. And the analysis is done for the normal steady state conditions.

The analysis results are shown in Figure 6. The maximum induced voltage is about 2.5[V] at a terminal without grounding.

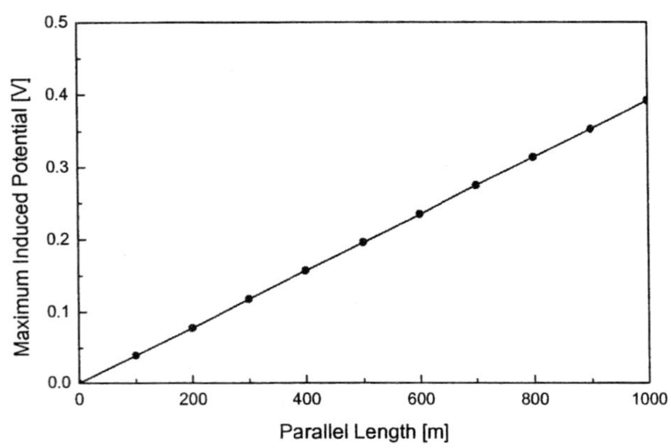

Fig. 4. Induced voltage according to the length in parallel

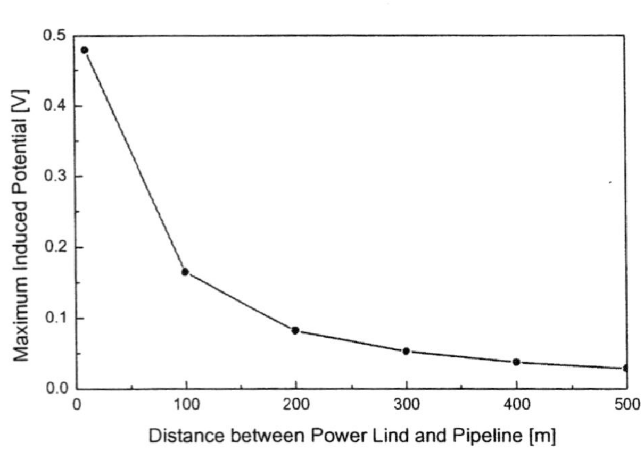

Fig. 5. Induced voltage according to the distance

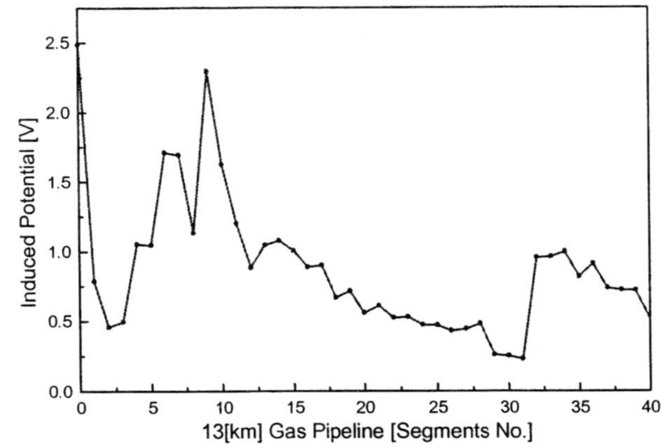

Fig. 6. Induced voltage for 13[km] parallel section

5. CONCLUSION

This paper analyze the induced voltage on the gas pipeline when an overhead power line runs parallel to a gas pipeline by using nodal network analysis that models the parameters of equivalent circuit for the parallel section. The amplitude of induced voltage is determined by three parameters; first, the power line currents according to operation conditions; second, the distance between power line and pipeline and the length in parallel, and third, the phase arrangement of power line and cable type. The analysis shows that the amplitude of induced voltage is in direct proportion to the length in parallel while it is in inverse proportion to the distance. And the induced voltage on the 13[km] section gas pipeline running parallel to the 22.9kV power line is analyzed, and the maximum induced voltage is about 2.5[V] at a terminal without grounding.

This paper contributes to find out the influence of power line on a near metallic pipeline.

REFERENCE

CIGRE (1995), *Guide on the Influence of High Voltage AC Power Systems on Metallic Pipelines*. Working Group 36.02

Dabkowski J. and Taflove A. (1978), *Mutual Design Considerations for Overhead AC Transmission Lines and Gas Transmission Pipelines*, PRC/AGA Contract PR-132-80

Dawalibi F.P. (1989), *Analysis of Electrical Interference from Power Lines to Gas Pipelines*, IEEE Transactions on Power Delivery, **Vol. 4, No. 3**, p.415-421

Gummow R.A. (1999), *Cathodic Protection Considerations for Pipelines with AC Mitigation Facilities*, PRC Contract PR-262-9809

Simpson Ian B.K. (2000), *CONIND Reference Manual*, Ground-it.com Consulting Ltd.

Southey R.D. and Dawalibi F.P. (1998), *Computer Modelling of AC Interference Problems for the Most Cost-Effective Solutions*, NACE International, **Corrosion98, Paper No.564**.

ELSEVIER
IFAC
PUBLICATIONS
www.elsevier.com/locate/ifac

DYNAMIC SIMULATION OF VOLTAGE INSTABILITY WITH PARTICULAR REFERENCE TO PROTECTION STRATEGIES USING EMTP

J. Y. Heo[1], C. H. Kim[1], R. K. Aggarwal[2], N. O. Park[3]

[1]*Information & Computer Engineering, Sungkyunkwan University Suwon, 440-746, Korea and NPT Center*
[2]*Electronic and Electrical Engineering, Bath University Bath BA2 7AY, UK*
[3]*KERI, 665, Naeson 2-dong, Uiwang-si, Gyeonggi-do, Korea*

Abstract: The simulation of voltage instability and voltage collapse can be basically classified into static simulation and dynamic simulation; initially, the static simulation was the one used. The static method requires less CPU resource, provides the voltage stability indices and gives much insight into the voltage and the power problem. However, this is a less accurate method than the dynamic simulation performed in the time domain. When it becomes difficult to secure the voltage stability margin via the static stability simulation approach, it becomes necessary to perform the dynamic simulation; in this case, it is required to model the dynamic components of the power system such as a generator and a load. The dynamic simulation provides the accurate results of the voltage instability. However, this method also has some drawbacks in that, it cannot provide the sensitivity information and the margin of the stability, is time consuming and needs much CPU resource. Nonetheless, system transient stability studies based on dynamic simulation are important in order to ascertain the protection / critical fault clearance time strategies that need to be adopted.
This paper uses EMTP MODELS to perform a dynamic simulation of voltage instability and voltage collapse. The load model is implemented with EMTP MODELS and this model is then interfaced with a test power system. The test results clearly show that we can simulate the voltage instability and the voltage collapse dynamically using EMTP MODELS.

Keywords: Voltage stability, Voltage collapse, Dynamic models, Load modelling, Electric power systems

1. INTRODUCTION

Voltage stability has been studied for some decades now. There is, however, no generally accepted definition of voltage stability or voltage collapse because of the complex phenomenon and the variety of ways in which it can manifest itself in a power system (Karlsson, 2000). Both IEEE and CIGRE have their respective definitions (Karlsson, 2000; Delfino, *et al.*, 1996). Voltage instability has been studied in the areas of analysis, simulation, countermeasure, etc.
There are several power system components known to contribute to voltage instability. The voltage instability simulation requires the modelling of the power system components, especially voltage control device and loads (Balanathan, *et al.*, 2002). The simulation of voltage instability can be basically classified into static simulation and dynamic simulation (Nagao, *et al.*, 1997). The static simulation is carried out based on the power flow solutions in the steady state; this method requires much less CPU and it can be used to provide much insight into voltage/reactive power problem. But, the static simulation cannot give an accurate result in the time domain (Morison, *et al.*, 1993). On the other hand, the dynamic simulation provides the most accurate replication of the power system during voltage instability. However, this method is time-

consuming and also does not readily provide sensitivity information or stability margin of voltage instability (Morison, *et al.*, 1993), but is nevertheless important for ascertaining protection strategies which are largely dictated by the transient stability constraints associated with a power system.
This paper presents a technique whereby the load model is implemented and the dynamic simulation of voltage instability is carried out using EMTP MODELS.

2. VOLTAGE INSTABILITY

2.1 Load Model

Load dynamic response plays a key role in a power system voltage instability problem. Load characteristics are an expression giving the active or the reactive power consumed by the load as a function of voltage, load demand. The general forms of load characteristics are (Custsem and Vournas, 1998):

$$P = P(z, V) \tag{1}$$

$$Q = Q(z, V) \tag{2}$$

Where, P is a power consumed by the load, V is a load voltage and z is a load demand.
The well-known exponential load is :

$$P = P_O(\frac{V}{V_0})^{\alpha} \tag{3}$$

Where, P_0 is the real power at nominal voltage, V_0 is a nominal voltage, α=0 : constant power load, α=1 : constant current load, α=2 : constant impedance load. When power consumed by load is a function of load state variable, it is called transient load characteristics. The transient load characteristics are given as (Custsem and Vournas, 1998):

$$P = P_t(z, V, x) \tag{4}$$

$$Q = Q_t(z, V, x) \tag{5}$$

Where, P is a power consumed by the load, P_t is transient state characteristics for active power, Q_t is transient state characteristics for reactive power, V is a load voltage, z is a load demand and x is load state variable.
There are several models of the loads such as an induction motor, load tap changer (LTC), thermostatic load, generic aggregate load, etc. The load dynamics of multiplicative generic load model is given as (Custsem and Vournas, 1998):

$$P = z_p P_0(\frac{V}{V_0})^{\alpha_T} \tag{6}$$

$$P_0 T_p \frac{dz_p}{dt} = P_s - z_p P_t = P_0(\frac{V}{V_0})^{\alpha_S} - z_p P_0(\frac{V}{V_0})^{\alpha_T} \tag{7}$$

Where, P is power consumed by load, P_s is the steady state characteristic, P_t is the transient state characteristic, T_p is time constant, z_p is state variable, α_S is the steady state load exponent and α_T is the transient load exponent.

2.2 Simulation Method of Voltage Instability

As mentioned before, the simulation of voltage instability can be classified into static and dynamic simulations (Nagao, *et al.*, 1997).
The static simulation is based on the power flow solution of the power system. The most common technique for power flow includes some type of Newton iteration. However, these algorithms become ill-conditioned as the system nears the point of voltage collapse. These have particular difficulties in power flow convergence of stress cases close to the maximum power transfer (Kim, *et al.*, 2002). Thus, several newer algorithms have modifications for more stable solutions (Kundur, 1999). The static simulation provides the fast analysis result of voltage instability, much insight into voltage/reactive problem and the voltage stability margin.
However, when it appears difficult to determine the voltage stability margin via the static simulation, it is necessary to perform the dynamic simulation for the more detailed analysis (Balanathan, *et al.*, 2002). The dynamic method is based on a technique of transient simulation in the time domain. This method requires the dynamic models of generators, loads, etc., and provides the most accurate replication of the time response of the power system. However, time domain simulations are time consuming and more effort is required for analysis of results (Nagao, *et al.*, 1997). It should be mentioned that although the QSS(Quasi-Steady-State) simulation offers a good compromise between the static simulation and the dynamic simulation (Kundur, 1999), i.e. it has the advantages of both static and dynamic simulations, from a point of view of ascertaining the protection / critical fault clearance strategies for a power system network, it is important and sufficient to perform system stability studies purely based on the dynamic(or transient) simulation method and hence the methodology and result's presented herein are based on this approach.

3. DYNAMIC SIMULATION OF VOLTAGE INSTABILITY USING EMTP MODELS

MODELS is a symbolic language interpreter for the EMTP that has recently gained popularity for the electro-magnetic transients phenomenon modelling. It provides the monitoring and controllability of power systems as well as some other algebraic and relational operations for programming. Importantly, MODELS is able to change dynamically the states of power system in response to the simulated results of EMTP power system (Kim, *et al.*, 2000; Dube, 1996; Martinez, 1993; Wilson and Nordstorm, 1993, Funabashi, *et al.*, 1998). MODELS that has the interaction with the EMTP power system (Peterson and Wall, 1991; Chaudhary, *et al.*, 1994) can be used to implement the dynamic load model. By using MODELS the various load models such as the exponential load, polynomial load model, etc can be implemented flexibly.

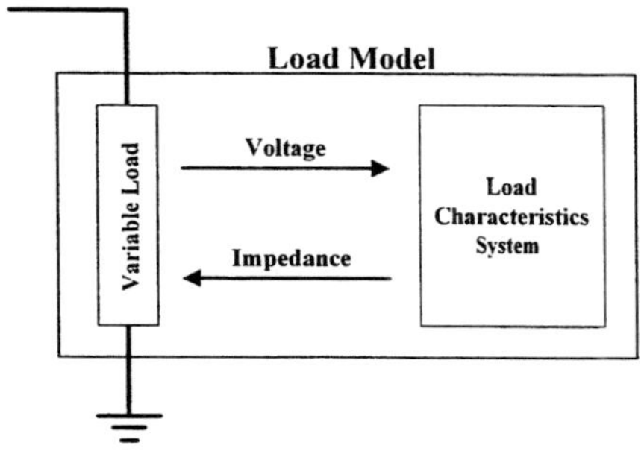

Fig. 1. Structure of load model

Fig. 1 shows the system diagram of the load model that interacts with the power system. The impedance of the variable load can be changed by load characteristics system including the load characteristic equation. The load power is determined by the load characteristics equation that is a function of the load voltage and the load state variable, and then the load impedance is calculated using the load power and the load voltage.

Fig. 2 shows a block diagram of the load model using EMTP MODELS. Tyep-60 slave source used in EMTP is the voltage/current source that can be controlled by TACS or MODELS (Dube, 1996). The current value of the source is set as shown in section (a) according to $\overline{Z}$ and V. In a serial RLC circuit, the voltage in terms of the current is given as :

$$v(t) = Ri(t) + L\frac{di(t)}{dt} + \frac{1}{C}\int_{t_0}^{t} i(t)\,dt \qquad (8)$$

In the s-domain, this equation becomes :

$$V(s) = RI(s) + sLI(s) + \frac{1}{sC}I(s) \qquad (9)$$

For a lagging load, eqn 9 is simplified as :

$$V(s) = RI(s) + sLI(s) \qquad (10)$$

$$I(s) = \frac{V(s)}{R + sL} \qquad (11)$$

The combination of section (a) and type-60 slave source is operated as the variable impedance which follows the input value $\overline{Z}$.

Section (c) is the load characteristic which gives the complex power $\overline{S}$ consumed by the load as a function of V_{rms} and state variable.

The complex power of section (c) in Fig. 2 and rms values of voltage are used in section (b) to determine the load impedance using the equation given as :

$$\overline{Z} = \frac{V_{rms}^2}{|\overline{S}|^2}(P + jQ) \qquad (12)$$

Where, $\overline{Z}$ is the load impedance, Vrms is the rms value of the load voltage and $\overline{S}$ is the complex power.

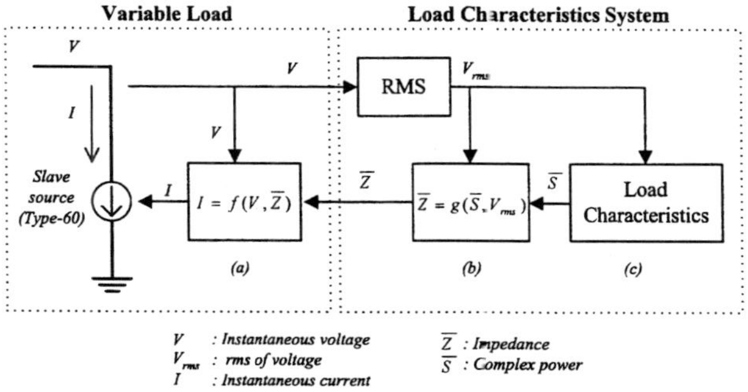

Fig. 2. Block diagram of load model using EMTP MODELS

The EMTP MODELS code which is forced to realize the block diagram of Fig. 2 is shown in Fig. 3. The voltage values of type-60 slave source are sent to the input of the load model and the output current values of load model are sent to the type-60 slave source. The computation of exponential and differential equation is carried out in the load characteristics. These operations are included in the library function of the MODELS. The rms values of voltage, the power consumed by the load, the load impedance and the load current are calculated by the MODELS.

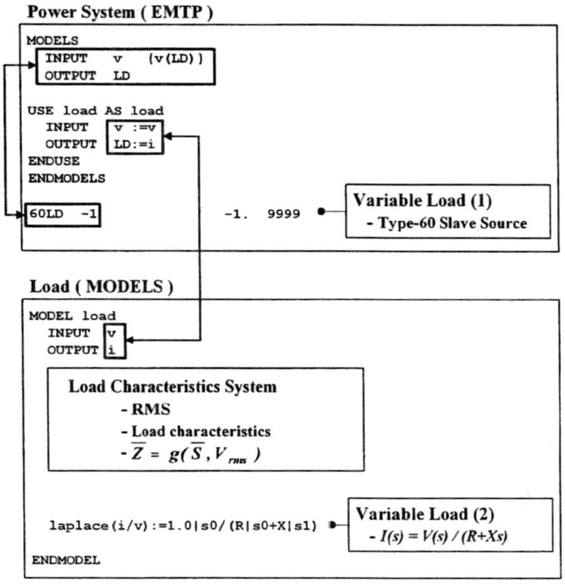

Fig. 3. EMTP MODELS code of load model

4. SIMULTATION RESULTS

4.1 Simulation Method

A set of simulation tests were carried in the test model of a power system shown in Fig. 4 which is interfaced with the model of a load implemented with EMTP MODELS. The transmission system modelled comprises of a total line length of 500km.

A multiplicative generic dynamic load model is connected to bus 2, and constant active power load is applied whereby α_S is set to 0 and α_T is set to 1. A capacitor is inserted in parallel with the load model to prevent non-convergence during the simulation.

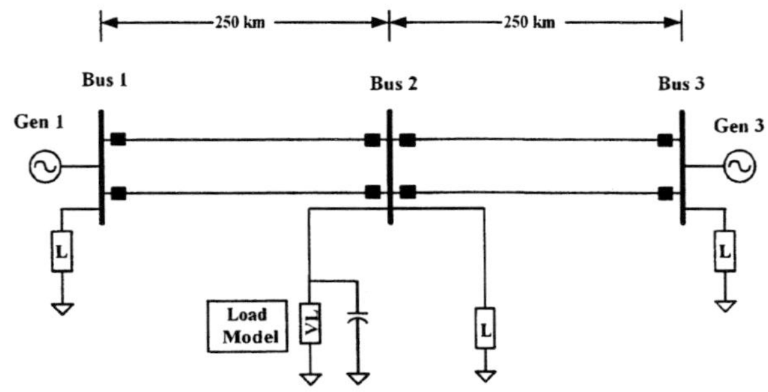

Fig. 4. Model of power system

4.2 PV curves

When a disturbance consists of a loss of a transmission line between buss 2 and 3 leading to a reduced power transfer to the load, static PV curves for bus 2 are calculated for pre- and post-disturbance conditions using EMTP and are shown in Fig. 5a and Fig. 5b, respectively. The maximum power that can be consumed by the load is 7.4 p.u in Fig. 5a. After the loss of one transmission line, the maximum power is reduced to 5.7 p.u like Fig. 5b. This effectively means that voltage collapse can occur due to loss of a transmission line such as the trip of the circuit breaker when the consumed power is above 5.7 p.u by the load on bus 2.

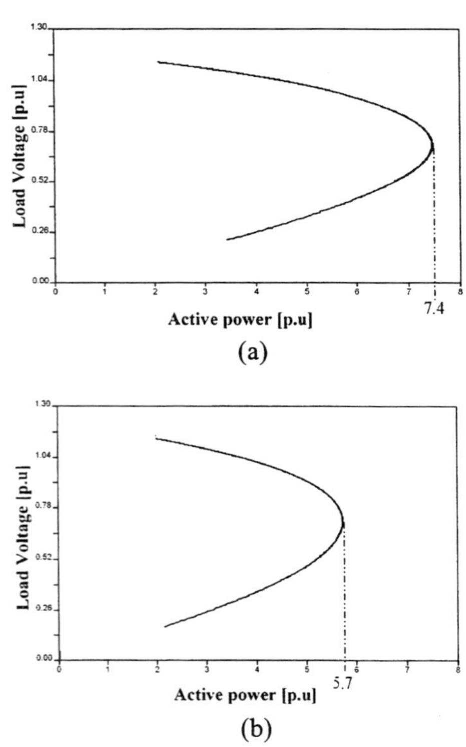

Fig. 5. PV curves as seen from bus 2 in case of no loss of lines and loss of one transmission line between bus 2 and bus 3. (a) No loss of lines. (b) Loss of a line between bus 2 and bus 3.

4.3 Voltage Instability and Voltage Collapse

When P_0 of load model is set to 6.0 p.u and no disturbance occurs, the load voltage and the active power consumed by the load model in the time domain are shown in Fig. 6a, and Fig. 6b shows them in the PV plane. P_0 is a smaller power than the one at nose point in Fig. 5a. Thus, the power and the voltage stay at the intersection point of PV curve as in Fig. 5a and the constant power load characteristics with P_0=6.0 p.u. In Fig. 6b, the start-up transient contributes to the trajectory before the steady state point.

When P_0 of load model is set to 6.0 p.u and one transmission line between bus 2 and bus 3 is disconnected 2s after the start of the simulation, the load voltage and the active power consumed by the load model in the time domain is shown in Fig. 7a, and Fig. 7b shows them in the PV plane.

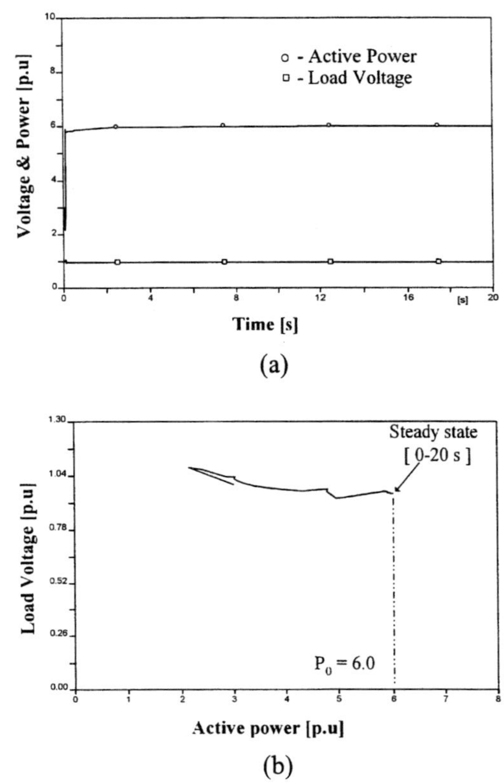

Fig. 6. Result of simulation when P_0 is set to 6.0 p.u and no disturbance occurs. (a) Voltage and active power in the time domain. (b) Voltage and active power in the PV plane.

Before 2s, the PV curve is the same to as in Fig. 5a, but the PV curve shrinks to one shown in Fig. 5b after 2s. After the disturbance, the load demand becomes larger than the maximum power that is can be transmitted to the load, and the PV curve (Fig. 5b) of the post-disturbance no longer intersects the characteristics curve of the load. Before the voltage and the power pass the nose point in the PV plane, the higher demand causes some voltage drop but results in a higher load power. However, after passing the nose point, the larger demand causes a decrease in both the voltage and the load power. This decreasing of the voltage and the power continues until the voltage collapses 14s after the disturbance.

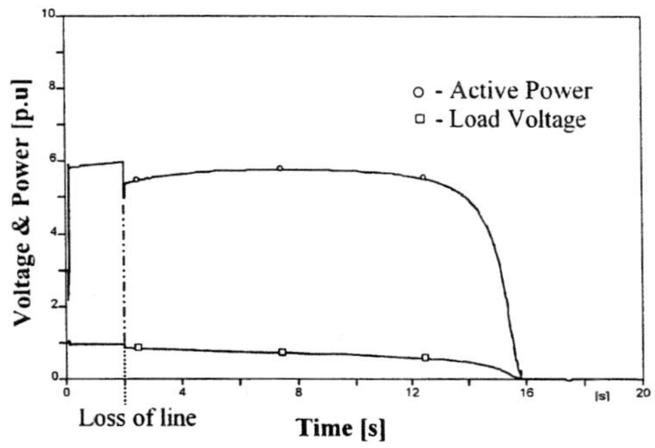

(a)

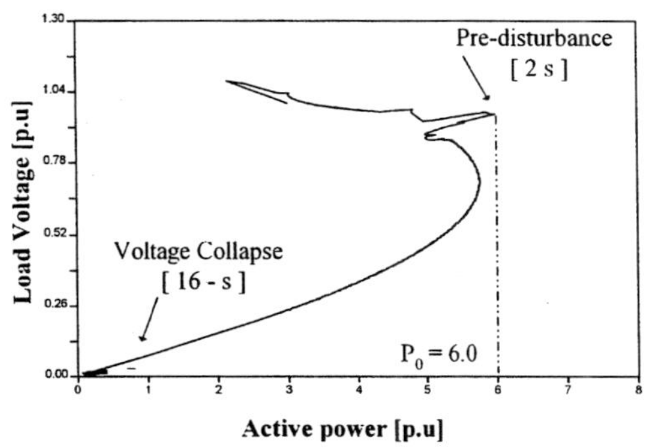

(b)

Fig. 7. Result of simulation when P_0 is set to 6.0 p.u and one line between bus 2 and bus 3 is disconnected 2s after the start of the simulation. (a) Voltage and active power in the time domain. (b) Voltage and active power in the PV plane.

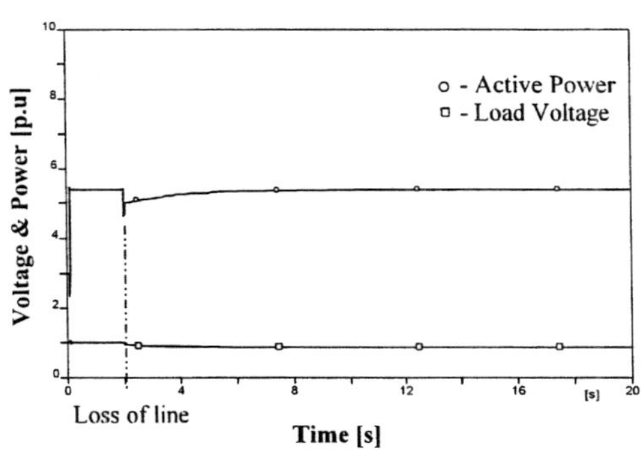

(a)

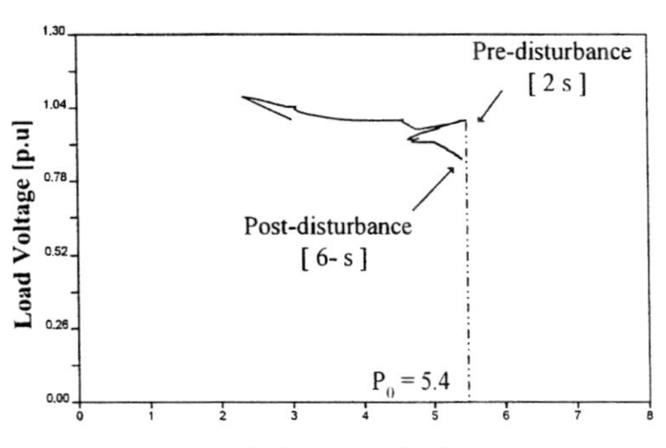

(b)

Fig. 8. Result of simulation when P_0 is set to 5.4 p.u and one line between bus 2 and bus 3 is disconnected 2s after the start of the simulation. (a) Voltage and active power in the time domain. (b) Voltage and active power in the PV plane.

When P_0 of load model is set to 5.4 p.u and one transmission line between bus 2 and bus 3 is disconnected 2s after the start of the simulation, the load voltage and the active power consumed by the load model is shown in Fig. 8a, and Fig. 8b shows them in the PV plane. Here again, before 2s, the PV curve is the same as in Fig. 5a, but the PV curve shrinks to one shown in Fig. 5b after 2s. Although the PV curve shrinks, this shrinkage does not result in a loss of intersection. The higher demand causes some voltage drop but results in a higher load power. This is the expected operation for protection against the voltage collapse. The voltage and the power decrease drastically at the moment of the disturbance, but the power is restored to the pre-disturbance level, 4s after the disturbance.

4.4. Load Shedding (Aggarwal, et al., 1993; Ahn, et al., 2001)

When P_0 of the load model is set to 6.0 p.u, one transmission line between bus 2 and bus 3 is disconnected, 2s after the start of the simulation and the load is shed at 8.3s; the load voltage and the active power consumed by the load model in the time domain are shown in Fig. 9a, and Fig. 9b shows them in the PV plane. After the disturbance, the voltage starts to decrease at bus 2. The voltage reaches 0.7 p.u and the load at bus 2 is disconnected. This load shedding makes the PV curve to expand, and results in a new intersection point of the PV curve and the characteristic curve.

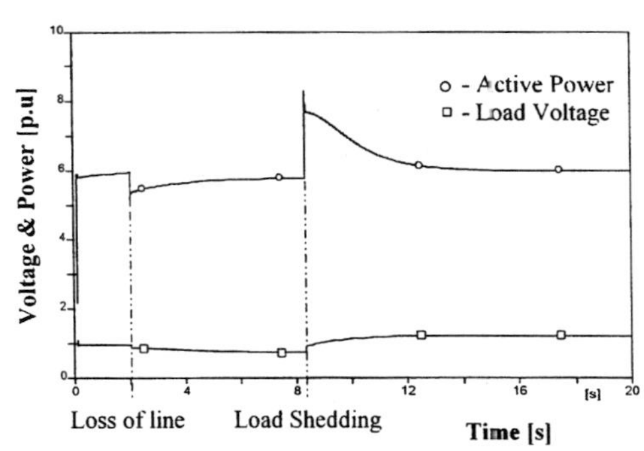

(a)

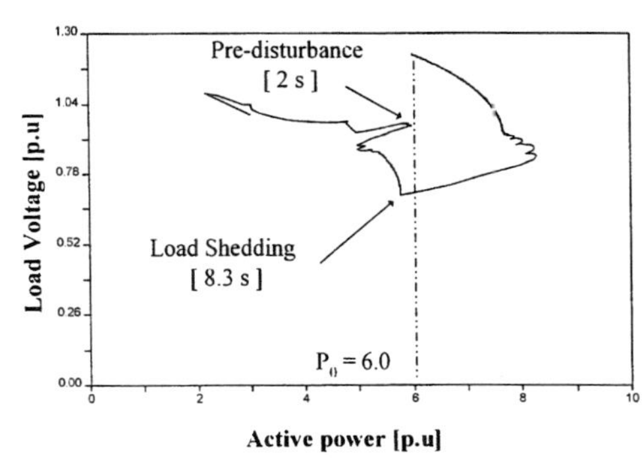

(b)

Fig. 9. Result of simulation when P_0 is set to 6.0 p.u, one line between bus 2 and bus 3 is disconnected 2s after the start of the simulation and the load is shed at 8.3s. (a) Voltage and active power in the time domain. (b) Voltage and active power in the PV plane.

5. CONCLUSION

This paper clearly demonstrates that EMTP MODELS can be used to simulate the voltage instability and the voltage collapse, and the load model can be implemented by EMTP MODELS. The load model is interfaced with the test power system, and the changes in the voltage and the power are simulated, following a disturbance such as loss of a transmission line.

A load model with a variable impedance using the current source has problems of non-convergence close to voltage collapse, and thus careful usage of the load model is necessary. Much time is required to simulate voltage instability because computation of exponential and differential equation in the load model are performed at each time step, and repeated simulations are also required because a single result of the simulation cannot give sufficient information for voltage instability. As mentioned before, although these are some of the drawbacks of the dynamic simulation approach, this methodology is necessary for ascertaining the protection / fault clearance strategies that need to be adopted for a particular power system network. The simulation results presented here show the detailed process leading to voltage instability and voltage collapse, and this method can be used for dynamic studies requiring coordination of controls and protections. Finally, it should be emphasised that to the best of authors' knowledge, hitherto, no work has been reported on stability simulation by EMTP, in particular EMTP MODELS; this allows a true interaction between a power system and dynamic load models which can have a significant bearing on the system stability and hence the technology presented here is novel.

6. ACKNOWLEDGEMENT

The authors would like to thank Korea Ministry of Science and Technology and Korea Science and Engineering Foundation for their support through ERC program.

REFERENCES

Aggarwal R. K., Song Y. H. and Johns A. T. (1993), Adaptive Three-phase Autoreclosure for Double-circuit Transmission Systems Using Neural Networks, *1993 IEE Advances in Power System Control, Operation and Management Conf.*, pp. 389-392.

Ahn S. P., Kim C. H., Aggarwal R. K. and Johns A. T. (2001), An Alternative Approach to Adaptive Single Pole Auto-Reclosing in High Voltage Transmission Systems Base on Variable Dead Time Control, *IEEE Trans. on Power Delivery*, **Vol. 16**, **No. 4**, pp. 676-686.

Balanathan R., Pahalawaththa N. C. and Udaya D. Annakkage (2002), Modelling induction motor loads for voltage stability analysis, *Electrical Power and Energy Systems*, **Vol. 24**, pp. 469-480.

Chaudhary A. K. S., Tam K. S. and Phadke A. G (1994), Protection System Representation in the Electromagnetic Transients Program, *IEEE Trans. on Power Delivery*, **Vol. 9**, **No. 2**, pp. 700-711.

Custsem T. V. and Vournas C. (1998), *Voltage Stability of Electric Power Systems*, pp. 93-129, Kluwer Academic Publishers.

Delfino B., Denegri G.B., Invernizzi M. and Morini A. (1996), Voltage Stability of Power System : Links between Static and Dynamic Approaches, *1996 Electrotechnical Conf.*, pp. 854-858.

Dube L. (1996), *MODELS RULE BOOK(New Version)*.

Funabashi T., Mizuma Y., Otoguro H. and Dube L. (1998), Fault locator simulation using MODELS language, *1998 IEEE Energy Management and Power Delivery Conf.*, **Vol. 2**, pp. 601-606.

Karlsson D. (2000), Voltage Stability-Modelling and System Protection Scheme, PSP 2000, pp. 37-42.

Kim C.H., Lee M.H., Aggarwal R. K. and Johns A. T. (2000), Educational Use of EMTP MODELS for the Study of a Distance Relaying Algorithm for Protecting Transmission Lines, *IEEE Trans. on Power Systems*, **Vol. 15, No. 1**, pp. 9-15.

Kim D. H., Ryu H. S., Lee J. G. and Moon Y. H. (2002), A Study on Voltage Collapse Mechanism using Equivalent Mechanical Model, *KIEE International Trans. on Power Engineering*, **Vol. 12A, No. 1**, pp. 6-14.

Kundur P. (1999), Voltage Stability Assessment Procedures and Guides, *IEEE/PES Power system Stability Subcommittee Special Publication*, pp. 3.1-3.18.

Martinez J. A. (1993), Educational Use of EMTP MODELS for the Study of Rotating Machine Transient, *IEEE Trans. on Power Systems*, **Vol. 8, No. 4**, pp. 1392-1399.

Morison G. K., Gao B. and Kundur P. (1993), Voltage Stability Analysis Using Static and Dynamic Approaches, *IEEE Trans. on Power Systems*, **Vol. 8, No. 3**, pp. 1159-1171.

Nagao T., Tanaka K. and Takenaka K. (1997), Development of Static Simulation Programs for Voltage Stability Studies of Bulk Power System, *IEEE Trans. on Power Systems*, **Vol. 12**, **No. 1**, pp. 273-281.

Peterson J. N. and Wall R. W. (1991), Interactive Relay Controlled Power System Modeling, *IEEE Trans. on Power Delivery*, **Vol. 6**, **No. 1**, pp. 96-102.

Wilson R. E. and Nordstorm J. M. (1993), EMTP Transient Modeling of a Distance Relay and a Comparison with EMTP Laboratory Testing, *IEEE Trans. on Power Delivery*, **Vol. 5**, **No. 3**, pp. 984-990.

ELSEVIER
IFAC
PUBLICATIONS
www.elsevier.com/locate/ifac

DGR-BASED INVESTMENT DECISION MODEL OF GENERATION EXPANSION PLANNING

Yu Cheng LiZi Zhang

Dept. of Electrical Engineering, North China Electric Power University, Beijing 102206,China

Abstract: In deregulated power markets, we must pay enough attention to generation expansion planning. This article applies degree of gray relationship theory to analyze investment decision of generation expansion planning. At last, according to proper practical case to show that DGR (degree of gray relationship) is really effective in analyzing investment decision of generation expansion planning. *Copyright © 2003 IFAC*

Keywords: generation expansion planning, investment decision model, degree of gray relationship.

1. INTRODUCTION

Due to the natural monopolization of electricity transportation and distribution, most countries introduce competition into electricity generation side and supply side. As we know, most countries' deregulated power market is developed from generation market. The crisis of California deregulated power market advise us that we must attach importance to generation expansion planning to avoid the lack of capacity during the construction of deregulated power market.

Under monopolization environment, Generation expansion planning pays much attention to technology than the market itself. But in competition environment, generation expansion planning should pay much attention to markets and utilize the economic signals produced by electricity price to improve the benefits of generation expansion planning. So investment decision of generation expansion planning should be regarded day by day.

2. PROBLEM

In deregulated power market, investment decision of generation expansion planning faces many new problems such as:

1) Independency of generation expansion planning. Traditional management pattern is vertical that means electricity generation, transmission, distribution and supply is integral. Relevant generation expansion planning and grid planning could work in line and share the information and drive for the common interests. In market environment, power generation, transmission, distribution and supply deregulated gradually and introduce competition so the interests of every segments have their own independency.

2) Goals of generation expansion planning is changed. In monopolization pattern, goals of generation expansion planning must be consistent with whole power system planning. The core goals of generation expansion planning is to make sure when, where, what kind of and how much capacity of power plant to construct within the planning period. Generation expansion planning also must meet the needs of the development of power load that is constrained by many restriction condition and technical economy targets of least expend of national economy. But in marketing environment each generation company take part in competition generation market independently. Interests of generation expansion planning are in common with generation investor. Generation expansion planning should still try to improve the power supply ability, security and reliability to meet the needs of power customers. But investors' core goal is to improve their profits by generation expansion planning. So the

frame of generation expansion planning must improve the profits of investors and also improve social benefits.

3) More investment risk. In the past, generation expansion planning is an integrate resource planning of electricity generation, transmission, distribution and supply and based on the "least expend of equal profits" rule. The planning could guarantee the return of generation equipments investment and operation costs. But in market environment, introduction of competition brings many uncertain factors. Profits of future investment are in deep relation with competition level of generation market and influenced by the fluctuation of electricity price. So it is very hard to make sure that investments could return. During the decision of generation expansion planning, paying more attention to the investment return is very important.

4) Difficulty of load forecast. The basic information of generation expansion planning that derived from long-term load forecasting is not enough. Traditional long-term load forecasting only focuses on the development of load itself and also involves the development of national economy. But it doesn't think of the response of end-users to the power market. In today's deregulated market environment, end-users have much more choices. When facing the fluctuation of electricity price, they will choose a proper power supplier to make expend least. So the relationship of load and electricity price is not only rigid but also elastic. This elastic relationship of load and electricity price will influence electricity generation and the operation and benefit of new planning capacity.

5) Importance of electricity price forecast. The function and characteristic of electricity price determine that electricity price is the relevancy of power enterprises and power end-users. Electricity price forecast is previsional and must forecast the development of electricity price. In investment decision process of generation expansion planning, unsuccessful electricity price forecast will make the whole investment scheme balance distortion. So electricity price forecast is very crucial to generation expansion planning. Generally speaking, electricity price forecast must think of such factors as: capacity cost, depreciation rate, fuel price, system load rate, interest rate, investment risk, exchange rate, technology and management level and electricity demand price elasticity.

Based on the analysis of context, generation expansion planning involves many factors. In the actual implement process, these factors are uncertain and fuzzy. Some rational index will exhibit some extent uncertainty and fuzzy. And also we must think of the psychological and physical condition of investors. So many uncertain factors make investors hard to make a successful decision. Thus, generation expansion planning investment decision system is a "gray system" that means information of the system is partly known and partly unknown.

3. DEGREE OF GRAY RELATIONSHIP (DGR)

Thinking of investment decision of generation expansion planning is a "gray system", this article uses degree of gray relationship theory (DGR) to set up the investment decision model of generation expansion planning. The basic concept of degree of gray relationship theory, see (Marco, *et al.*, 1996), is such:

$$X = \{x_i | i \in N, N = (1\Lambda\ m), m \geq 2\} \qquad (1)$$

In the above formula, X is a serial, if X has such characters as:

1) Data of X are approachable;
2) Data of X are proportional;
3) non-negative;

then X is called degree of gray relationship factors set.

Suppose that X is a degree of gray relationship factor set,

$$\begin{aligned} &X = \{x_i | i \in N, N = (1\Lambda\ m), m \geq 2\}; \\ &x_i = (x_i(1), \Lambda\ x_i(n)); \\ &x_i(h) \in x_i, h \in H, H = \{1,2\Lambda\ n\}, n \geq 3; \end{aligned} \qquad (2)$$

Make $x_0 \in X$ is reference row, $x_i \in X$ is compare row, $x_0\ (h)$ and $x_i\ (h)$ is the data of x_0 and x_i in the point h. Define that $R(x_0, x_i)$ is the degree of gray relationship of x_i to x_0, calculation formula is below:

$$R(x_0, x_i) = \frac{1}{n}\sum_{h=1}^{n} r(x_0(h), x_i(h)) \qquad (3)$$

In formula (3), $r(x_0\ (h),\ x_i(h))$ is the compare estimation of $x_0\ (h)$ to $x_i\ (h)$, defined as below:

$$r(x_0(h), x_i(h)) = \frac{\min\limits_{i\in N}\min\limits_{h\in H}|x_0(h) - x_i(h)| + \lambda \max\limits_{i\in N}\max\limits_{h\in H}|x_0(h) - x_i(h)|}{|x_0(h) - x_i(h)| + \lambda \max\limits_{i\in N}\max\limits_{h\in H}|x_0(h) - x_i(h)|} \qquad (4)$$

$r(x_0\ (h),\ x_i\ (h))$ satisfies such conditions as follows:

1) Criterion:

$$0 < r(x_0\ (h),\ x_i\ (h)) <= 1;$$

2) Integrity:

$$r(x_i,y_j) <> r(y_j,x_i),\quad x_i, y_i \in X;$$

3) Even symmetry:

$$r(x,y)=r(y,x),\ \text{if}\ X=\{x,y\};$$

4) Approachability:

$$|\ r(x_0\ (h)-x_i\ (h))| \propto r(x_0\ (h),\ x_i\ (h)).$$

From above we can get degree of gray relationship space that is the combination of distance space and point set topological space. In formula (4), $|\ r(x_0\ (h)-x_i(h))|$ is the estimation of distance, $\left[\min\limits_{i\in N}\min\limits_{h\in H}|x_0(h) - x_i(h)|, \max\limits_{i\in N}\max\limits_{h\in H}|x_0(h) - x_i(h)|\right]$ is the compare condition of x_i and x_0, it includes point set topological information. In this formula, λ which is a constant, $\lambda \in [0,1]$, is the differentiate coefficient of DGR, its function is to adjust compare condition When λ =0, no compare condition, when λ =1, compare condition keep unchanged. Usually we set λ =0.5.

4. INVESEMENT DECISION MODEL

In the process of investment decision, we must consider many indexes. These indexes associate and influence each other and weight of each index is different. The magnitude of each weight reflects the importance of the corresponding index. Each index has its own meanings and dimension, so it is very hard to make a compare among different indexes. Thus we should initialize these indexes first. In practical, it is crucial to identify weight of each index. Common methods include Delphi technique, AHP and relative entropy method etc.

4.1 Model construction based on DGR

Suppose that $S=\{ s_1,\ldots,s_m \}$ is the set of m decision-making objects. Every object is described by n index systems $P=\{ p_1,\ldots, p_n \}$ which have different meaning and dimension. Thus, we can obtain a $m \times n$ dimension decision indexes matrix. After standardization of this matrix, we obtain a new index matrix $[R]$:

$$[R]= \begin{array}{c} \\ s_1 \\ s_2 \\ \mathrm{M} \\ s_m \end{array} \begin{array}{c} \begin{array}{cccc} p_1 & p_2 & \Lambda & p_n \end{array} \\ \begin{bmatrix} x_{1,1} & x_{1,2} & \Lambda & x_{1,n} \\ x_{2,1} & x_{2,2} & \Lambda & x_{1,n} \\ \mathrm{M} & \mathrm{M} & \mathrm{M} & \mathrm{M} \\ x_{m,1} & x_{m,2} & \Lambda & x_{m,n} \end{bmatrix}_{m\times n} \end{array} \quad (5)$$

$x_{i,j}$ represents the *jth* index of the *i*th object. It is obvious that $[R]$ has the characteristic of DGR factor sequence.

$$r(p_i,\overline{p_i}) = \frac{1}{n-1}\sum_{j=1,j\neq i}^{n} r(p_i,p_j) \quad (6)$$

In the formula (6), $r(p_i,\overline{p_i})$ is the cluster DGR of p_i to other indexes, shown below:

$$r(p_i,p_j) = \frac{1}{m}\sum_{h=1}^{m} r[x_{h,i},p_{h,j}], i,j=1,\Lambda\ n \quad (7)$$

$r(p_i,\overline{p_i})$ reflects the influence of p_i to other indexes in some extent. If certain index has great influence to other indexes, we can draw the conclusion that the information contained in this index is great and vice versa. After standardization of these n DGRs, we could obtain every relative weights of each index. That is:

$$w_i = \frac{r(p_i,\overline{p_i})}{\sum_{i=1}^{n} r(p_i,\overline{p_i})}, i=1,\Lambda .n \quad (8)$$

In the format of matrix: $[W]=[w_i,\Lambda\ ,w_n]^T$, that means that $[W]$ is the weight matrix of index system. The construction of investment decision model is:

$$[Y]_{m\times 1} = [R]_{m\times n} \bullet [W]_{n\times 1} \quad (9)$$

$[Y]$ is the result matrix of investment decision.

4.2 Model Construction based on fuzzy DGR

Models of fuzzy DGR perform fuzzy reasoning considering the fuzzy characteristic of some indexes that can simulate the reasoning ability of human nature.

To construct a model, firstly we should define fuzzy cluster DGR $\bar{r}(p_i,\overline{p_i})$ as below:

$$\bar{r}(p_i,\overline{p_i}) = \frac{1}{n-1}\sum_{j=1,j\neq i}^{n} \bar{r}(p_i,p_j) \quad (10)$$

$$\bar{r}(p_i,p_j) = \frac{\sum_{h=1}^{m}(x_{h,i} \wedge x_{h,j})}{\sum_{h=1}^{m}(x_{h,i} \vee x_{h,j})} \quad (11)$$

"$\wedge$" and "$\vee$" is fuzzy less operator and greater operator.

According to the same reasons, $\bar{r}(p_i,\overline{p_i})$ also reflects the fuzzy influence of p_i to other indexes.

$$\overline{w_i} = \frac{\bar{r}(p_i,\overline{p_i})}{\sum_{i=1}^{n} \bar{r}(p_i,\overline{p_i})}, i=1,\Lambda\ ,n \quad (12)$$

$[W]=[\overline{w_i},\Lambda\ ,\overline{w_n}]^T$ is the matrix format of $[W]$, that means $[W]$ is the weight matrix of index systems. Investment decision model could be constructed as below:

$$[Y]_{m\times 1} = [R]_{m\times n} \bullet [W]_{n\times 1} \quad (13)$$

$[Y]$ is the result matrix after decision-making.

From what we have discussed above, we can obtain the DGR based investment decision algorithm as below:

1) Set up index system, then initialize and standardize of index matrix;
2) According to formula (7) and (11) to calculate DGR of any two indexes;
3) Calculate the cluster DGR $r(p_i,\overline{p_i})$ or $\bar{r}(p_i,\overline{p_i})$ of each index;
4) Calculate the relative weight of each index;

5) Perform decision-making process using formula (9) and formula (13).

5.MODEL APPLICATION

We choose investment decision of generation expansion planning as the analytical object. There are three generation expansion plans to implement which are expressed as s_1, s_2, s_3. Decision-maker will take into account three indexes, which are investment cost, operation cost and investment return. The first two indexes are cost-type index that should be the least and the last index is benefit-type index that should be the biggest.

Calculation of each index is shown below, see (Wang, 1990; Zhang and He, 1989):

1) Calculation of investment cost (thermal power plant).

$$Z = \sum_{j=1}^{N} \sum_{k=1}^{NG_j} C_{t,j,k} \bullet G_{t,j,k} \quad (14)$$

Z -- investment cost;
N -- numbers of power plants;
NG_j -- the unit numbers of the jth generating unit
$C_{t,j,k}$ -- the current value of the tth year the jth power plant the kth unit investment cost;
$G_{t,j,k}$ -- 0-1 variable, means whether the tth year the jth power plant the kth unit will be constructed ($G_{t,j,k}$=1, construct; $G_{t,j,k}$=0, not construct).

2) Calculation of operation cost (thermal power plant).

$$OC = \sum_{j=1}^{N} \sum_{k=1}^{NG_j} PC_{t,j,k} \bullet GC_{t,j,k} \bullet H_{t,j,k} \quad (15)$$

OC -- operation cost;
$PC_{t,j,k}$ -- the current value of the tth year the jth power plant the kth unit generation cost;
$GC_{t,j,k}$ -- the capacity of the tth year the jth power plant the kth unit.
$H_{t,j,k}$ -- the tth year the jth power plant the kth unit expectational generation hours.

3) Calculation of investment return rate.

$$F = \frac{(\sum_{j=1}^{N} \sum_{k=1}^{NG_j} P_t \bullet GC_{t,j,k} \bullet H_{t,j,k}) - OC - Z}{Z} \quad (16)$$

F -- investment return rate;
P_t -- electricity price forecasted of the tth planning year.

4) Calculation of reliability level

This article uses reserve capacity coefficient and reserve power coefficient to reflect reliability level.

$$RL = RL_c + RL_p \quad (17)$$

$$RL_c = \frac{\sum_{j=1}^{N} \sum_{k=1}^{NG_j} GC_{t,j,k}}{D_{\max}} - 1 \quad (18)$$

$$RL_p = \frac{\sum_{j=1}^{N} \sum_{k=1}^{NG_j} GC_{t,j,k} \bullet H_{t,j,k}}{PD_{\max}} - 1 \quad (19)$$

RL -- reliability level coefficient;
RL_c -- reserve capacity coefficient;
RL_p -- reserve power coefficient;
D_{max} -- the max load forecasted of the tth planning year;
PD_{max} -- the max power demand forecasted of the tth planning year.

According the formula above, we can obtain every index's numerical value of each scheme. As shown in table 1.

Table 1 Indexes' value for each scheme

Factor	Investmet cost (money)	Operation cost (money)	Investment return rate (%)	Reliability level (%)
Scheme s_1	1150	600	12%	18%
Scheme s_2	1000	500	10%	22%
Scheme s_3	1150	550	10%	25%

The calculation of index weight matrix (showed in table 2) is based on the decision of DGR:

$$[W]_{4\times1} = [0.318, 0.217, 0.344, 0.121]^T$$

Table 2 Decision results based on DGR

Scheme	S_1	s_2	s_3
[Y]	28.29%	37.06%	34.65%

The calculation of index weight matrix (Showed in table 3) is based on the decision of fuzzy DGR:

$$[W]_{4\times1} = [0.331, 0.225, 0.305, 0.139]^T$$

Table 3 Decision results based on fuzzy DGR

Scheme	S_1	s_2	s_3
[Y]	31.18%	35.74%	33.08%

From the results of these two decision methods, we can draw the consistent conclusion that the best scheme is s_2.

6. CONCLUSION

There are many uncertain and fuzzy factors in practical investment decision of generation expansion planning; these factors will influence the decision-makers' final decision results. Gray system theory is the right and feasible method to solve this problem. This article attempts to apply degree of gray relationship theory to construct models of investment decision of generation expansion planning. According to the detailed analysis, this article draw the conclusion that models based on DGR or fuzzy DGR is an effective way to provide successful investment decision for generation expansion planning.

REFERENCES

Marco Dorigo, Vittorio Maniezzc and Alberto Colorni(1996). Optimization by Acolony of Cooperating Agents. *Journal of IEEE Transactions on System*, **1**, 29-41.

Wang Xifan(1990). *Power System Planning*. Hydraulic Power Press, Beijing.

Zhang ben and He Dayu(1989).*Generation Expansion Planning*. Energy Press, Beijing.

ELSEVIER
IFAC
PUBLICATIONS
www.elsevier.com/locate/ifac

DECENTRALIZED SIMULATION OF LARGE-SCALE ELECTRICAL NETWORKS

S. M. Kouhsari and A. Kalantari

Department of Electrical Engineering
Amirkabir University of Technology, Tehran, Iran

Abstract: The paper reports the *Global Network Simulation (GNS)* project, which is a distributed computing approach for power system analysis. In this project we have provided facilities for distributed web-based simulation of large-scale electrical networks in a well-established power system simulator package, i.e. *PASHA (Power Apparatus & System Homological Analysis)*. Possibility of autonomous planning and operation of every individual utility of an interconnected network is the ultimate goal of the GNS. Reduction in simulation time, saving in hardware requirements, detailed and secure analysis, are the greatest gains of the GNS. The software has been successfully test on various networks up to 3000 buses. *Copyright © 2003 IFAC*

Key words: Diakoptics, distributed computing, Global Network Simulation (GNS), large-scale network, Internet, Large Change Sensitivity (LCS), piecewise method.

1. INTRODUCTION

Clean, inexpensive, transportable electricity is the indispensable commodity of modern civilization. But, the electric supply network is probably the largest and most complex man-made infrastructure in the world. It comprises a huge volume of different components, which are geographically dispersed, even in continental ranges. Therefore, many analyses are needed before any operations can be performed on the actual network. Nevertheless, analysis of such a bulk network by means of traditional centralized approaches is faced on serious challenges. Limitation in processing power is one of the biggest. In fact it is predicted that computational speeds a million times greater than today's high-end computers would be required (Fardanesh, 2002). Using the parallel processing super-computers may be a possible solution, but it is only affordable by large corporations moreover the application oriented and platform specific design of them, which is a hurdle to global accommodation. Nevertheless, even if the problem of processing power is to be solved, say through open design of the processing unit (Azevedo and Oliviera, 2001), there are other factors that make the centralized approaches useless for analysis of large-scale networks. The most important of them can be outlined as bellow:

× Lack of the sufficient knowledge around some parts of the network, say due to security considerations of some utilities or neighbored countries.

× Real-time control constraints due to wide geographical ranges covered by interconnected networks. For example to limit the signal transmission delays to *2 ms* round-trip, only an area with a maximum radius of approximately 150 miles can be under a central control umbrella (Fardanesh, 2002).

× Even if the above drawbacks are to be eliminated, a huge database, which consists of the network topology data and components values, must be updated in a regular time interval and also operators and system managers must analyze a huge volume of simulation results. This is an overwhelming work, especially when real-time operation, say for contingency analysis.

So, approximately, all of the centralized approaches employ a reduced version of the original network meanwhile its computational procedures. These reduced networks can be obtained using different network equivalencing techniques (Kato, 1994) from offline studies. But this process involved with degrees of approximation and doesn't provide a comprehensive view on different operational aspects of a given interconnected power system.

The GNS provides a proper solution to all of the above materials through employing a federative analysis approach. It utilizes the available computational resources of individual utilities control centers into a collaborative computational task to provide a possibility for *Detailed Analysis* of the interconnected networks. In such a scheme, every utility's control center is only responsible to modeling and analysis of its own local network, autonomously, while is coordinated with the rest of utilities through a system-wide coordinator.

In order to practical implementation of such an ideal scheme, there are need to special considerations in communication medium, mathematical algorithms, and programming styles. In the GNS project we combined a series of techniques consist of, *Diakoptics*, Large Change Sensitivity (LCS), socket and multi-thread programming, mixed-language programming, and etc, to implement the above decentralized strategy. The GNS project is completed now for piecewise solution of Load Flow and Fault analyses as the most common power system computations. Nevertheless the proposed technique in this project is completely flexible to extend to the other power system computations, without any restriction. In fact, this method is useable for solving of large physical systems, which can be represented as a set of linear (or linearized) algebraic equations, such that it is usual in the most actual engineering problems.

2. GNS AT A GLANCE

Briefly, GNS is *a collaborative distributed computing system over the Internet for piecewise simulation of large-scale interconnected networks.* The GNS operates in a combination of client/server and peer-to-peer computing models across web environment, as shown in Fig. 1. In such a scheme, every utility's control center (or factory, power-station, neighbored country, and etc.) operates, autonomously, while is coordinated with other involved utilities through a validated higher-level entity, i.e. a system-wide coordinator, for consistency of the overall solution. In this way, every utility is a black box from the view of the other utilities and system-wide coordinator is the unique common point among all of them. In fact, such a type of distributed simulation can be compared with a common whiteboard shared among several painters, which all of them create a unique portrait in a collaborative task, simultaneously.

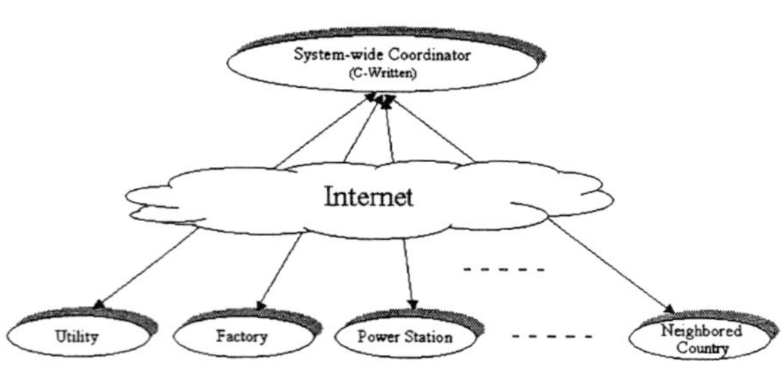

Fig. 1. GNS computational model

3. GNS FUNDAMENTALS

Different technical aspects of the GNS project are discussed here. The most important issues can be classified as below,

3. 1. Mathematical Basis

From the mathematical point of view, two key concepts i.e. *Diakoptics* and *Large Change Sensitivity (LCS)* constitute the basis of the GNS:

3. 1. 1. Diakoptics

Diakoptics is a general term, which is applied to piecewise solution of large physical systems (Happ, 1980; Brameller, *et al.*, 1969; Happ and Young, 1971). The basic idea of *diakoptics* is to solve a large system by tearing it apart into smaller subsystems, and then to combine and modify the solution of the original untorn problem. Schematically, this is shown in Fig. 2. for solving a power system in two and four pieces. Nevertheless, traditional implementations of *diakoptics* concept using tensor analysis, contour theory, and topological theory and also the utilized decomposition techniques, don't provide the GNS requirements in a convenience manner. They present degrees of complexities that are in contrast with the GNS goals i.e. minimum number of network-based interactions, minimum content of data transfer, and minimum local computational cost; by the way, accurate results must be preserved. So, we were developed a *diakoptical* algorithm based on Large Change Sensitivity (LCS) and bus-splitting (as decomposition technique) to satisfy the GNS requirements.

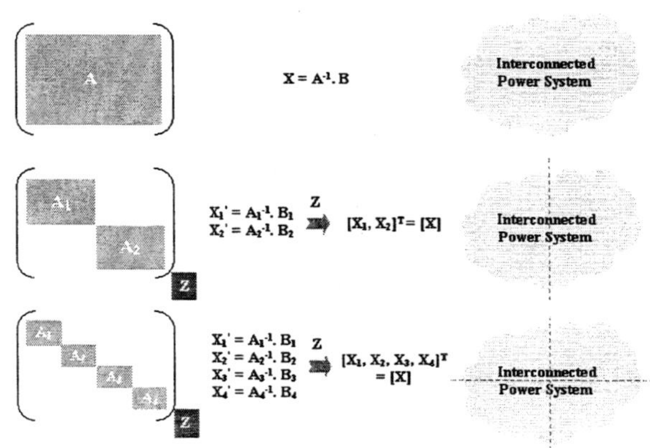

Fig. 2. Diakoptics solution procedure

3. 1. 2. Large Change Sensitivity (LCS)

Let a given network be solved for specified nominal values. The LCS makes it possible, without resolving the whole problem to obtain the exact solution of the network if some elements are changed arbitrarily (from zero to infinite). So, the LCS can be set forth as superiority in computer storage and computational time when solving of large and complex engineering problems. Mathematically, Consider the network equations (n-sized) with the elements being assigned their nominal values

$$\mathbf{T}_0\mathbf{X}_0 = \mathbf{W} \qquad (1)$$

where $\mathbf{T}_0$ is the network matrix, $\mathbf{X}_0$ is the network state vector, and $\mathbf{W}$ is the input excitation vector. Now, suppose *m* components of the network are subjected to variations that need not be small (α_i

$i=1...m$). The system modified by m component perturbations is described by:

$$\mathbf{TX} = \left(\mathbf{T}_0 + \mathbf{P}.\alpha\mathbf{Q}^t\right)\mathbf{X} = \mathbf{W} \qquad (2)$$

where **T** and **X** are the new network matrix and network state vector, respectively. Also **P** and **Q** are $n \times m$ matrices, which contain 0 and ±1 entries and indicate the location of the perturbed elements. The direct approach for solving (2) is factorization of the coefficient matrix followed by a series of forward/backward substitutions to find **X**. But, when the number of changed variables, *m*, is much smaller than the size of the system, *n*, the LCS approach offer a preference in computational cost. The LCS confines the required computations only on the changed elements instead of resolving the whole network. The following equation ensures the integrity of the results:

$$\mathbf{X} = \mathbf{X}_0 - \mathbf{T}_0^{-1}\mathbf{Pz} \qquad (3)$$

In the above equation **z** reflects the effect of perturbed components and is calculated through,

$$\mathbf{z} = \left(\alpha^{-1} + \mathbf{Q}^t\mathbf{T}_0^{-1}\mathbf{P}\right)^{-1}\mathbf{Q}^t\mathbf{X}_0 \qquad (4)$$

The reader can refer to (Vlach and Singhal, 1983) for detailed explanations of the LCS. Equations (3) and (4) are the key expressions, which were used to piecewise simulation of power systems in the GNS.

3. 2. Network Decomposition

The GNS uses a bus-splitting approach for network decomposition. This is done through insertion of fictitious Circuit Breakers (CBs) between the torn pieces of boundary buses (see Fig. 3.). Really, such condition is completely in consensus with the LCS rules for diakoptical solution of a large network. When the CBs are open the network is decomposed, mathematically. So, due to block diagonal form of the coefficient matrix in this manner (see Fig. 2.), every diagonal block can be solved separately and in parallel with others in a multi-computer environment, locally. Now, if the CBs are to be closed (i.e. occurrence a large change in these components), the overall interconnected network will be obtained. The solution of the recent state is equal to updating the previous solution of the individual subnetworks through LCS rules (i.e. Equations (3) and (4)) for large changes in the values of the CBs.

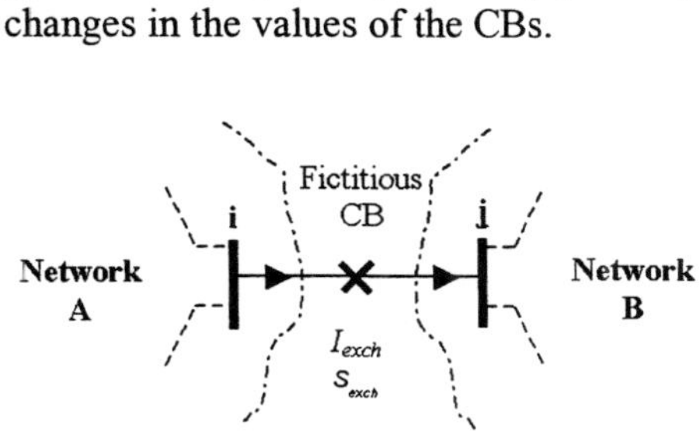

Fig. 3. Network decomposition strategy

In order to overcome the singularity and bad-conditioning problem in the subnetworks without reference node, a pair of symmetrical shunt elements (equal in value but opposite in sign) is added at the place of boundary buses, corresponding to Fig. 4. This provides a well-condition system matrix in those subnetworks without the need to employ temporary reference nodes for them. These symmetrical shunts neuter the effect of each other in the procedure of solution, implicitly.

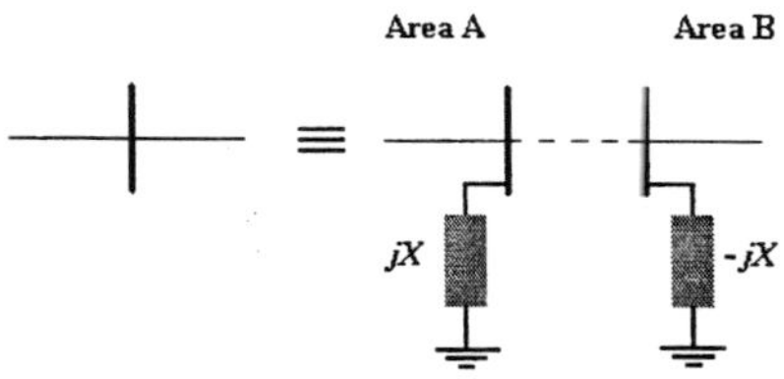

Fig. 4. Symmetrical shunts for well-conditioning the system matrix

3. 3. Communication Medium

Information technology (IT) at the beginning of 3rd millennium is the biggest revolution in human life for all eras. IT defines everything in a new way – the way technology is created, deployed, and the way we used it. Internet is set forth as the backbone of IT. Nevertheless, power engineering is one of the less IT affected areas among various branches of engineering. But, the GNS provides an opportunity to capture the Internet capabilities for distributed simulation of power systems. The Internet is set forth here as the communication medium that links all of the utilities control center computers (which are dispersed geographically) with a system-wide coordinator (see Fig. 1.). It also provides the possibilities to develop a various spectrum of Web-based services both for utilities and for their costumers.

4. LOAD FLOW CONSIDERATIONS

The GNS policy for Load Flow calculations can be outlined in three main steps,

1. Solving every individual subnetwork through traditional load flow procedure in corresponding utility,

$$\Delta\mathbf{V}_i^0 = \mathbf{J}_i^{-1}\Delta\mathbf{S}_i \quad (i = 1...s) \qquad (5)$$

where s is the number of subnetworks. Then the boundary bus values will be sent to the system-wide coordinator. Note that, this state is equivalent to opening the fictitious CBs or network decomposition.

2. Determination of the power exchange among adjacent subnetworks due to voltage difference in every related pair of boundary buses through (4) of LCS at system-wide coordinator,

$$\Delta\mathbf{S}_{exch} = \left(\sum_{i=1}^{s}\mathbf{C}_i^t\mathbf{J}_i^{-1}\mathbf{C}_i\right)^{-1}\mathbf{Q}^t\Delta\mathbf{V}^0 \qquad (6)$$

where C_i is the incidence matrix of the ith subnetwork, which determines the location of the fictitious CBs. Then the calculated power exchanges will be sent back to the utilities.

3. Taking the effect of the determined power exchanges in step 2 to update the obtained solution of step 1, corresponding to (3) of LCS,

$$\Delta \mathbf{V}_i = \Delta \mathbf{V}_i^0 - \mathbf{J}_i^{-1}\mathbf{C}_i \Delta \mathbf{S}_{exch} \quad (i = 1...s) \tag{7}$$

The above steps must be done repeatedly, as long as the desired power mismatch level in all subnetworks becomes satisfied. It may be noted here that the communication overhead for the above transactions is restricted to the values of boundary bus voltages and consequent power exchanges related to these buses. Also the computational task, which is involved with the system-wide coordinator is trivial and is limited to calculation of the fictitious CBs power flow.

It must be said here that the proposed method is a direct approach, which has formulated the problem, in depth. It leads to the same convergence characteristics with the original untorn algorithm. It provides the same results, iteration by iteration, provided synchronous communication. In contrast, there are iterative methods, which their convergence characteristic is questionable and is a function of the nature of network and the type of decomposition. For example, solution of a 888 bus system through iterative method presented by (Amano, *et al.*, 1996) converges in 12, 199, and 666 iterations when the system is decomposed into 1, 2, and 4 pieces, respectively. But it doesn't converge when the same network is decomposed into 8 pieces, even after 2000 iterations. Nevertheless, it can be solved through the proposed method in 12 iterations for different desired pieces. The reader can refer to (Kalantari, 2003; Kalantari, *et al.*, 2003) for detailed explanation of this section.

5. FAULT CONSIDERATIONS

The nodal Z_{bus} matrix constitutes the basis of fault studies (Stevenson, 1982; Glover and Sarma, 1987; Arrillaga, *et al.*, 1983). So, the GNS policy in this case, is to modify every subnetwork's nodal Z_{bus} matrix in such a way that reflects the effect of the whole interconnected network, exactly. For this purpose, it combines the LCS rules and fault equations in a Diakoptical framework, and finally provides the following key relation as the validated nodal Z_{bus} expression in every utility without any network equivalencing,

$$\mathbf{Z}_i = \mathbf{Z}_i^0 - \mathbf{Z}_i^0 \mathbf{C}_i \mathbf{K} \mathbf{C}_i^t \mathbf{Z}_i^0 \quad (i = 1\Lambda\ s) \tag{8}$$

where

$$\mathbf{K} = \left(\sum_{i=1}^{s} \mathbf{C}_i^t \mathbf{Z}_i^0 \mathbf{C}_i \right)^{-1} \tag{9}$$

$\boldsymbol{Z}_i^0$ and $\boldsymbol{Z}_i$ are, respectively, unmodified and modified nodal Z_{bus} matrix of the *i*th subnetwork. Using the $\boldsymbol{Z}_i$ every subnetwork is capable to do accurate fault studies, whether symmetrical or unsymmetrical, around its geographical boundaries, as offline. Of course, originally, **K** must be computed through the system-wide coordinator and this is the unique common term, which is required among all subnetworks. As can be seen there is only one transaction between the system-wide coordinator and subnetworks for determination of $\boldsymbol{Z}_i$. Nevertheless, in order to involve the effect of the loads and shunt elements in prefault voltage profile, instead of flat start, there is need to a piecewise nodal voltage equation that takes the loading effect of all subnetworks into consideration. It has the form of below for the *i*th subnetwork,

$$\mathbf{V}_i = \mathbf{V}_i^0 - \mathbf{Z}_i^0 \mathbf{C}_i \mathbf{I}_{exch} \quad (i = 1\Lambda\ s) \tag{10}$$

where

$$\mathbf{V}_i^0 = \mathbf{Z}_i^0 \mathbf{I}_i \quad (i = 1\Lambda\ s) \tag{11}$$

$$\mathbf{I}_{exch} = \mathbf{K}\mathbf{Q}^t \mathbf{V}^0 \tag{12}$$

Hence, originally, the network is solved in a decomposed manner through (11), then the network will be interconnected and the inter-subnetwork currents are determined through (12), and finally the initial subnetwork's solution is updated through (10). This is the exact prefault voltage of the *i*th subnetwork and can be used in conjunction with (8) for making a series of accurate fault studies. So, altogether, two transactions are needed for the GNS-fault module. The reader can refer to (Kalantari, 2003) for detailed explanations of this section.

6. GNS AND PASHA

PASHA is a well-established power system simulator package for approximately a quarter of century. It consists of a series of powerful computational routines for all needed calculations in power industry. But, it is primarily a desktop application, such as the other existing counterparts in the world. Nevertheless, the GNS provides a facility to capture the PASHA capabilities in large-scale simulation. At present, Load Flow and Fault modules of PASHA equipped with the GNS feature, but it can be freely developed to the other modules after an accurate inspection of them. So, using the GNS-enabled version of PASHA, a comprehensive set of distributed Web-based simulations is possible for an interconnected network, which is expanded to wide geographical ranges.

6. 1. Example

Consider the 126-bus network, consists of junctions, in Fig. 5. as a small template of an interconnected network. The technical data associated with this network can be found in (Taleshian, 2001). As can be seen in the recent figure and in order to piecewise solving of this network, it was decomposed into 4 pieces at 10 boundary buses. Every piece has been implemented on a specified computer located on computer network. Then, this collection has been subjected to distributed simulation process both for Load Flow and Fault calculation. The comparison of the piecewise solution with the traditional untorn counterpart is provided in Table. 1. only for boundary buses (due to space limitation). The reader can observe that there are no differences between the results of the two types of solutions, at all. By the way, the Load Flow solution procedure converged in 7 iterations for both states (with 0.001 MVA mismatch). Also the fault studies on this network shows that employing the equivalent networks instead of part B lead to 20% and 30% error

respectively, in the fault currents and relay tripping times (Taleshian, 2001) while using the proposed approach the true results have been obtained.
Separate of this moderate example, we have successfully test the software on various networks up to 3000 buses (IRAN network) in different pieces, even include a large steel mill plant represented down to 380 V level. In this regard we are able to simulate networks from 380 V up to 400 kV and more comprising a collection of different power components.

7. CONCLUSION

This paper was a report about the GNS project while its detailed technical documentation can be found in (Kalantari et al., 2003) and (Kalantari, 2003). Based on the GNS rules, every incorporated utility of an interconnected network, models, analysis, and operates the associated local network, autonomously. Strategically, this provides a reliable computational tool for *decentralized planning and operation* of power systems, as the future trend in power industry area (Fardanesh, 2002). From this point of view, it has several preferences as,

✓ Distribution of the total computational cost among the incorporated utilities. Really, ordinary desktop computers are set forth as the processing elements instead of a platform specific and application-oriented supercomputer.

✓ Possibility of detailed analysis of interconnected networks by developing the modeling process to the distribution and down to the customer level. So, the equivalent networks and their consequent errors will be eliminated and the network will be analyzed in a real sense.

✓ Integrated/decentralized operation and control with other corporate functions, such as accounting, customer services, and management information services. Also the effects of non-electrical factors such as weather and oil/gas industries can be involved in the procedure of power system operation, locally.

✓ Elimination of the real-time constraints by sectionalization of the control process.

✓ Security of the local network, which must be under supervision of local personnel, only (i.e. against neighbored utilities and countries). In fact, every utility seems as a black box from the view of the other utilities.

✓ Scaleable design such that new subnetworks can be added to the GNS forum, as plug & play, without any restriction and performance decrement.

✓ Adoption with tightly coupled parallel processing structures (i.e. supercomputers).

From the mathematical point of view a comprehensive exact infrastructure has been developed through *Diakoptics* and *Large Change Sensitivity (LCS)* concepts, which ensures the integrity of the results in a distributed computing environment. Distinctly, this method has been developed for Load Flow and Fault studies in the GNS project. But, due to comprehensiveness of the proposed method, it can be freely developed to other computational routines with some degrees of inspection. Practical implementation of the GNS is available on PASHA software.

Finally, it must be said that approximately, all of the future trends in power system planning and operation, are oriented to *decentralization*. Restructuring and privatization, distributed generation, and etc. are some of the outstanding templates. Under such circumstances, the GNS can be set forth as an appropriate decentralized computational tool.

REFERENCES

Amano M., Zečević A. I., and Šiljak D. D. (1996), "An improved block-parallel newton method via epsilon decomposition for load-flow calculations," *IEEE Trans. Power Systems*, vol. 11, no.3, pp. 1519-1527.

Arrillaga J., Arnold C. P., and Harker B. J. (1983), *Computer Modelling of Electrical Power Systems,* John Wiley & Sons.

Azevedo G. P. and Oliviera Filho A. L. (2001), "Control Centers with Open Architectures", *IEEE Computer Applications in Power*, pp. 27-32.

Brameller A., John M. N., and Scott M. R. (1969), *Practical Diakoptics for Electrical Networks*, by Chapman & Hall Ltd.

Fardanesh B. (2002), "Future Trends in Power System Control," *IEEE Computer Applications in Power*, vol. 15, no. 3, pp. 24-31.

Glover J. D. and Sarma M. (1987), *Power System Analysis and Design,* PWS Publishers.

Happ H. H. and Young C. C. (1971), "Tearing algorithms for large-scale network programs," *IEEE Trans. Power Apparatus and Systems*, vol. PAS-90, pp. 2639-2649.

Happ H. H. (1980), *Piecewise Methods and Applications to Power Systems,* by John Wiley & Sons, Inc.

Kalantari A. (2003), *Global Network Simulation,* MSc Dissertation, Dept. Elec. Eng., Amirkabir University of Technology, Tehran, Iran, July 2003.

Kalantari A., Kouhsari S. M., and Rastegar H. (2003), "Piecewise Fast Decoupled Load Flow Using Large Change Sensitivity," *in Proc. 2003 IEEE PES General Meeting*, Toronto, Canada, 13-17 July 2003.

Kato K. (1994), "External network modeling – recent practical approach," *Task Force - IEEE Trans. Power Systems,* vol. 9, no. 1, pp. 216-225.

Stevenson W. D. Jr. (1982), *Elements of Power System Analysis,* 4th edition, McGraw-Hill, Inc.

Taleshian M. (2001), *Assessment of over current and earth fault relay performance in interconnected networks,* MSc dissertation, Dept. Elec. Eng., Amirkabir University of Technology, Tehran, Iran, March 2001.

Vlach J. and Singhal K. (1983), *Computer Methods for Circuit Analysis and Design*, by Van Nostrand Reinhold Company Inc, New York.

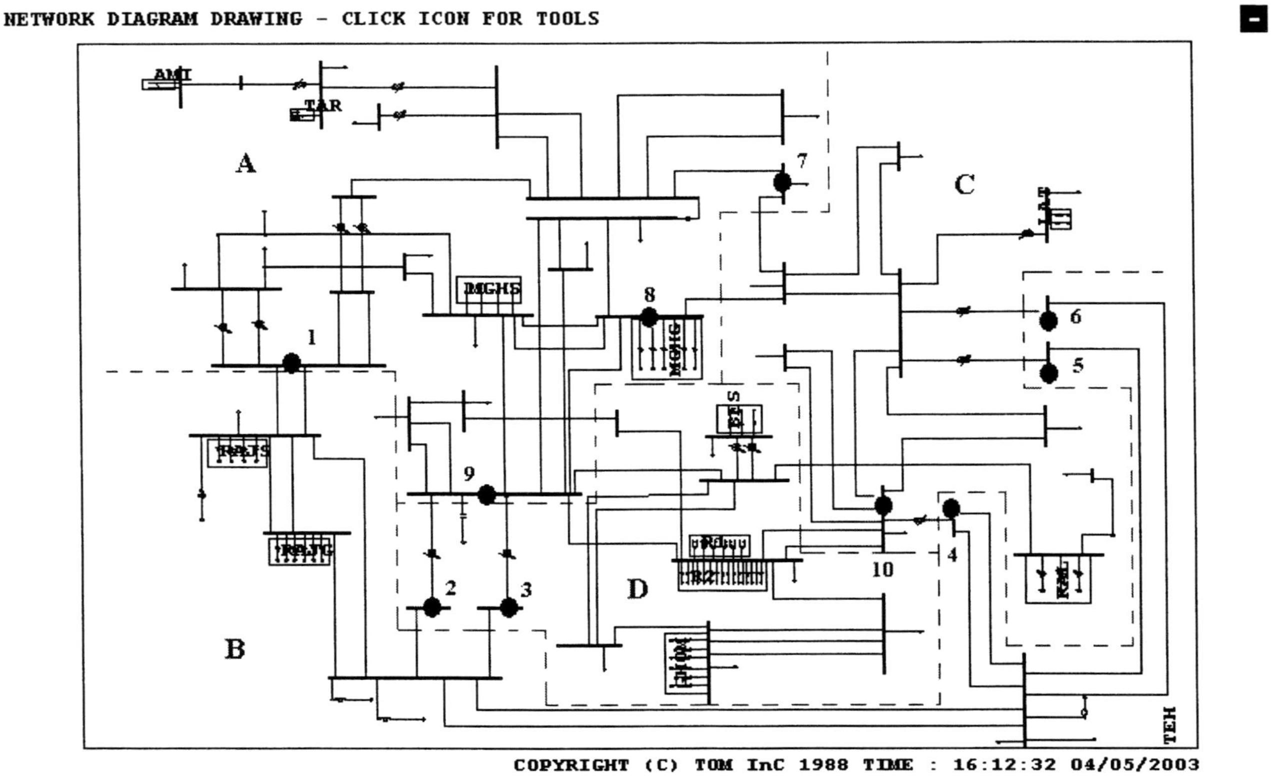

Fig. 5. 126-bus test network, which is decomposed into four pieces A, B, C, and D

Table 1 Test network solution - Both for Untorn and Torn states respectively, through traditional approaches and piecewise approach presented here

Bus	Load Flow (p.u./degree)				Fault (MVA)							
	Magnitude		Angle		LLL		LLG		LL		LG	
	Untorn	Torn	Untorn	Torn	Untorn	Torn	Untorn	Torn	Untorn	Torn	Untorn	Torn
1	1.028	1.028	-3.43	-3.43	9442	9442	9823	9823	8768	8768	10320	10320
2	1.018	1.018	-6.71	-6.71	6394	6394	6304	6304	5794	5794	6142	6142
3	1.018	1.018	-6.71	-6.71	6394	6394	6304	6304	5794	5794	6142	6142
4	1.012	1.012	-6.70	-6.70	6423	6423	6388	6388	5824	5824	6045	6045
5	1.007	1.007	-7.48	-7.48	5178	5178	5052	5052	4653	4653	4770	4770
6	1.007	1.007	-7.48	-7.48	5178	5178	5052	5052	4653	4653	4770	4770
7	0.978	0.978	-10.4	-10.4	7673	7673	7275	7275	7023	7023	6682	6682
8	1.003	1.003	-7.58	-7.58	10744	10744	11834	11834	10220	10220	13929	13929
9	0.995	0.995	-9.26	-9.26	11239	11239	11707	11707	10595	10595	12880	12880
10	0.987	0.987	-9.92	-9.92	8243	8243	8150	8150	7592	7592	8069	8069

ELSEVIER
IFAC
PUBLICATIONS
www.elsevier.com/locate/ifac

A STUDY ON WIND POWER GENERATION PLANNING AND CONTROL

***K.Watanabe A.Nishiyauchi Toshiya Ohtaka Shinichi Iwamoto**

Dept. of Electrical Engineering and Bioscience, Waseda University, Tokyo, Japan

Abstract: It is considered that wind power generation will be increased in the near future. In recent years, international measures against environmental problems have been paid attention to the deterioration of the earth environment. Thus many problems will be surfaced in the power systems. To solve these problems we propose a solution with regards to planning and operation of the wind power generation.

Keywords; system, wind power generation, voltage control, power system planning, analytic hierarchy process.

1. INTRODUCTION

In the power industry, thermal generators which have much CO2, NOx and SOx emissions are being reviewed and generation systems using natural energy are being introduced. Particularly, wind power generation has advanced technically, and is expected to grow in the future. However, the wind power is influenced by natural wind so the outputs fluctuate, creating some problems. When the wind power is connected to a power system, it is said that bus voltages fluctuate. Studies for controlling the voltage fluctuations are done widely. For example, SVCs and battery storage systems are considered for controlling the fluctuations. SVCs are electronic voltage control devices which can keep the voltage level and stability by controlling susceptances in the power system.

In this paper we consider a method of planning and operation of wind power generation. At first, we propose an optimal wind power allocation method which considers influences to the power systems. We consider several evaluation indices which can monitor the power systems stabilities, economics, and environment. These are not treated with the weights for the wind power allocation, and do not have the same units. Therefore, we take these indices hierarchically using ISM (Interpretive Structural Modeling). When we decide the optimal construction sites of wind power stations, we use AHP (Analytic Hierarchy Process). Thus it is impossible to compare the indices because of the difference of these units. Therefore, we use the concept of the membership function, and we can compare these indices in the same dimension. Second, concerning the operation, we propose a coordinated SVC control method which considers the effects of voltage fluctuations at other buses. SVCs which are located in wind power stations are to control the bus voltage fluctuations.

*K. Watanabe is presently with Sumitomo Corporation, Japan.

2. OPTIMAL PLANNING OF WIND POWER GENERATORS WITH CONSIDERTION OF POWER SYSTEM EVALUATIONS

If many wind power generators are connected to a power system, the power system is influenced by wind power outputs. When the outputs injected to the power systems are little, influences in the power systems are also little. However several outputs of wind powers will have a tremendous impact on the power systems. Consequently, we consider how we should construct wind power stations, and we consider optimal allocation of wind power stations pending on the influences to the power system. We assume that power storage devices such as NAS batteries are installed at the buses where wind power generators are connected. We use the AHP method to decide our objective, where we use a fuzzy theory (membership function) to compare to evaluation indices.

2.1 AHP [1] [2]

One must make decisions in their social lives. It is said that these are subjective judgments under complex or unclear situations. AHP is a decision making method often used under uncertain situations or existing several evaluation indices. This method is one of the decision methods which combine subjective judgments and system approaches. The process of solution is as follow.

Final objective → evaluation indices → alternative ideas

Subsequently a hierarchy model is made. The evaluation indices' importance (satisfaction) is obtained by looking up from the final objective. Next, the alternative ideas' importance is obtained by looking up evaluation indices, and finally these are converted to alternative ideas' evaluation looked up from the final objective. Thus, an optimal idea can be selected.

2.2 A Mathematical background of AHP

Weights - w_1、w_2、…、w_n to the elements in a level above the element-A_1、A_2、…、A_n are needed. At this time, if the importance (satisfaction) to a_j of a_i is a_{ij}, the pair comparison matrix of elements- A_1、A_2、…、A_n are represented by A $=[a_{ij}]$. If w_1、w_2、…、w_n are known, A=$[a_{ij}]$ is as follows.

$$A=[a_{ij}]=\begin{matrix} A_1 \\ A_2 \\ \vdots \\ A_n \end{matrix}\begin{bmatrix} w_1/w_1 & w_1/w_2 & \cdots & w_1/w_n \\ w_2/w_1 & w_2/w_2 & \cdots & w_2/w_n \\ \vdots & \vdots & \cdots & \vdots \\ w_n/w_1 & w_n/w_2 & \cdots & w_n/w_n \end{bmatrix} \quad (1)$$

($a_{ij}=w_i/w_j$、$a_{ji}=1/a_{ij}$、$w=[w_1 \quad w_2 \quad \cdots \quad w_n]^T$、$i\ j=1,2\cdots n$)

A vector n・w can be obtained to multiply pair comparison matrix A by the weight vector w. Namely,

$$A\cdot w=n\cdot w \quad (2)$$

This equation is considered as a eigen value ploblem.

$$(A-n\cdot I)\cdot w=0 \quad (3)$$

Here, n must be an eigen value of A when $w\neq 0$ holds. At this time, w is an eigen vector of A. Moreover, a rank of A is 1 so that $\lambda_i(i=1,2,\cdots n)$ are zeros but only one is non-zero. The sum of the diagonal elements is n. Thus, only one of λ_i which is not zero is λ_{max}, as below.

$$\lambda_i=0 \qquad \lambda_{max}=n(\lambda_{i\neq}\lambda_{max}) \quad (4)$$

Therefore, weight vector w to A_1、A_2、…、A_n become a normalized eigen vector ($\sum w_i=1$) to the maximum eigen value of A.

Actually, when problems are solved under the complicated situations, w is unknown and W' is needed, which is obtained by calculating the pair comparison matrixes selected by persons who make the decisions. These are expressed as follows.

$A'W'=\lambda_{max}W'$ (λ_{max} is a maximum eigen value) (5)

Therefore, W' is a normalized eigen vector to the maximum eigen value λ_{max} of A. Then the unknown W' is determined.

2.3 ISM [2]

A hierarchy is decided by the subjectivities of persons making decisions on AHP. Consequently, an optimal hierarchy model should be found more objectively by using mathematical models. In this case, a mathematical method ISM is often used. This method can clarify the problems to amend points of inconsistencies from the intuition people have and experiential judgments. The process is as follows.

1. Elements to be evaluated are determined.
2. Matrix E (to be shown in the example) is obtained to compare these indices and goals. If element i influences element j, the matrix elements are 1or else 0.
3. N is obtained from E + I (unit matrix).
   N* is obtained by N*=N×N×N×…×N.
4. Sets below are obtained by N*
   $R(t_i)=\{\ t_j \mid n_{ij}'=1\}$
   $A(t_i)=\{\ t_j \mid n_{ji}'=1\}$
   $R(t_i)$ is obtained by gathering the row element whose value is 1 when each column is checked. $A(t_i)$ is obtained by gathering the column element whose value is 1 when each row is checked.
5. Level decisions on the hierarchy of each element are as follows.
   $$R(t_i)\cap C(t_i)=R(t_i) \quad (6)$$
   If there are elements which satisfy Equation (6), the elements are level 1. Next, the level 1 elements are deleted. The elements which satisfy Equation (6) among the rest are level 2. Until all the elements are deleted, the process is repeated similarly.
6. From the relation between elements in each level and elements among neighboring levels, the hierarchy is obtained.

2.4 Simulation Example using the Proposed Method

In this paper we consider an optimal wind power generator allocation from the view point of system operators and power industries. At this time we use the AHP and ISM. We explain the proposed algorithm using the Ward & Hale 6 bus system, and we make the following assumptions.

1. Wind power outputs are constant (P=Q=0.2 [p.u.]).
2. Power storage systems like the NaS batteries are equipped at all wind power stations. Therefore、 outputs are controlled for these systems.

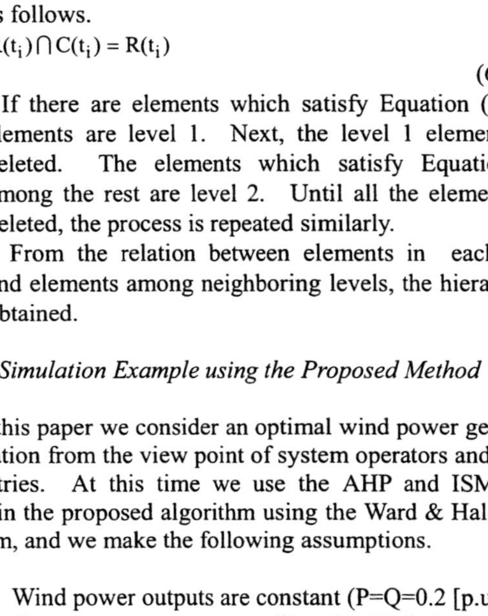

Fig. 1 Ward & Hale 6bus System

[STEP1]

Buses are selected for wind power stations placing (buses 3,4,5,6 are selected in this example). Optimal power flow calculations are carried out for 24 hours in consideration of load fluctuations with the interval of 1 hour. Thus, terminal conditions are obtained. In this paper, the objective function is fuel costs.

[STEP2]

Evaluation indices are calculated by using terminal conditions. These indices can evaluate the power system influences on the wind power station allocation. In this paper , as the evaluation indices we use voltage stability index Qloss, fuel cost, transmission congestion, generator output margin, and CO_2 outputs. Equations expressing these indexes are as follow.

- Qloss

  All reactive power losses / All active power supplies

- Fuel cost

  $\sum_{i \in NG}(a_i P_{Gi}^2 + b_i P_{Gi} + c_i)$

- Transmission congestion

  (Active power flow / maximum active power flow) × 100

- Generator outputs margin

  (Outputs margin of all generators / all loads of active power) × 100

- CO_2 outputs

Table 1 : Characteristics of generators [3]

	Units of CO_2 outputs	
	Variable g-C/kWh	Fixed g-C/kW
Coal thermal	222	3818
Combined LNG	97	3181
LNG thermal	121	3181
Oil thermal	172	3361

CO_2 outputs are calculated by using this Table.

[STEP3] [4]

These units are all differently, therefore these values are non-dimensionalized. Consequently, thus membership functions are defined. Boarder values for defining membership functions are calculated in consideration of terminal conditions. By using these values, the evaluation indices can be compared to each other.

[STEP4]

It is not advisable to evaluate these indexes on the same weight and to decide the weight subjectivity. These indices are then divided to consider the weight objectively using ISM. Each element is numbered as follows, 1. Wind generator stations construction, 2. Qloss, 3. Fuel cost, 4. Transmission congestion, 5. Generator output margin, 6. CO_2 outputs. Then, the matrix E is given as Table 2.

Table 2 Matrix E

	1	2	3	4	5	6
1	0	0	0	0	0	0
2	1	0	0	0	0	0
3	1	0	0	0	1	0
4	1	1	0	0	0	0
5	1	0	0	0	0	0
6	1	0	0	0	0	0

The hierarchy is obtained as in Table 3.

Table 3 Hierarchy

	1	2	5	6	3	4
1	1	0	0	0	0	0
2	1	1	0	0	0	0
5	1	0	1	0	0	0
6	1	0	0	1	0	0
3	0	0	1	0	1	0
4	0	1	0	0	0	1

[STEP5]

The hierarchy is a more objective method than the subjective hierarchy. AHP is used for the decision of intentions. The hierarchy in whole is shown in Fig.2.

[STEP6]

Table 4 shows the sum of the values of membership functions obtained by solving the optimal power flow calculation for 24 hours and their averages.

Weight vectors are made on the basis of Table 4. Weight vectors are shown in Table 5. Pair comparisons of the second level connecting the first level are as follows. At this time values of averages in Table 4 are used. Weights are obtained by elements i per elements j. The other weight vectors are obtained similarly.

Pair comparisons of the third level connecting the second level show that there is no pair so that the all weight vectors are 1. In this example, level exists up to the third level so that pair comparisons among the levels are carried out 2 times.

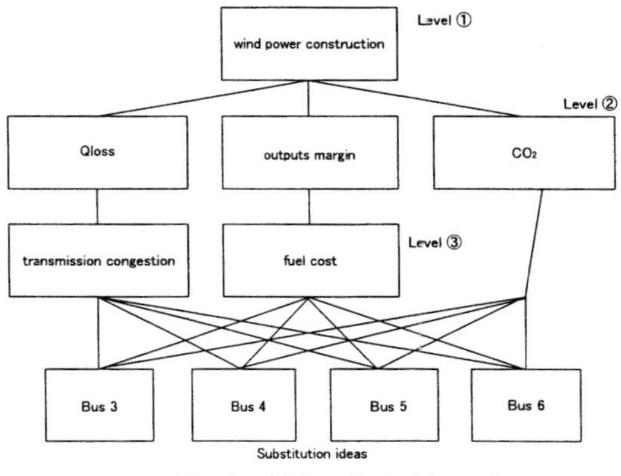

Fig. 2 The whole hierarchy

Table 4 Sum of values of membership functions and their averages

Candidate bus	Qloss	Fuel cost	Transmi-ssion congestio n	Generator outputs margin	CO_2
Bus3	23.6000	17.430 1	2.6689	24.0000	22.2408
Bus4	22.9880	17.311 7	3.1440	24.0000	21.9434
Bus5	22.8417	13.583 5	0.0000	24.0000	20.0528
Bus6	23.5430	17.370 9	1.8535	24.0000	21.9855
Average	23.2432	16.424 1	1.9166	24.0000	21.5556

Table 5 Pair comparison of the second level connected to the first level

	Element 2	Element 5	Element 6
Element 2	1.0000	0.9685	1.0783
Element 5	1.0326	1.0000	1.1134
Element 6	0.9274	0.8982	1.0000

At first, eigen values and eigen vectors of the weight matrix obtained by pair comparison of the alternative ideas and elements connecting their ideas need to be obtained, and the maximum eigen vectors to the maximum eigen values are obtained. Moreover, its eigen vectors are normalized.

Next, the normalized eigen vectors are obtained by pair comparisons among the levels. Initially, the maximum eigen values and eigen vectors to their values are needed. The values are weights of their elements.

Pair comparisons in the third level do not exist so that eigen values and vectors are 1. According to Fig.2 of ISM, Qloss is connected to Transmission congestion, and Generator outputs margin is connected to Fuel cost. Certainly to these connections, CO_2 is not connected to the other elements and directly connecting the substitution idea. Therefore, weights in the third level are revised as follows.

Real Weight = (weight of the element in the one level above the level where the element is connected) × (present level weight)

Hence, real weighs in the third level are as follow.

Real fuel cost weight =
Generator outputs margin weight × present Fuel cost weight

Real transmission congestion =
Qloss weight × present transmission congestion

This process is repeated until the element is connected to an alternative idea. Therefore the solution of the optimal wind power plant construction is as follows.

$$X = \begin{matrix} Bus3 \\ Bus4 \\ Bus5 \\ Bus6 \end{matrix} \begin{bmatrix} 0.2538 & 0.2653 & 0.3477 \\ 0.2473 & 0.2635 & 0.4096 \\ 0.2457 & 0.2068 & 0.0013 \\ 0.2532 & 0.2644 & 0.2415 \end{bmatrix} \begin{bmatrix} 0.3133 \\ 0.3488 \\ 0.3378 \end{bmatrix} = \begin{matrix} Bus3 \\ Bus4 \\ Bus5 \\ Bus6 \end{matrix} \begin{bmatrix} 0.2895 \\ 0.3077 \\ 0.1495 \\ 0.2531 \end{bmatrix}$$

In the obtained solution X, the maximum number is 0.3077. This is a value corresponding to the bus 4. Consequently, the location of the optimal construction is bus 4.

2.5 Comprehensive Evaluation

If wind power plants are constructed in the power system, it is possible to reduce the fuel costs and the CO_2 outputs. Therefore, it is essential to introduce the wind power plants for reducing CO_2 due to the electric power companies' contributions. In addition, decreasing the fuel costs is economical for electric power companies. However, it is impossible unless wind power costs are reduced. Therefore, wind power costs have to be cheaper than fuel costs. If wind power generators are prevailed, their costs will be reduced. It is important to advance the wind power techniques and to find the energy sources with the term ' wind ' for spreading wind powers. This is how to reduce the wind power costs. Moreover, it is possible that costs of generators like thermal systems increase because a market for exchanging a CO_2 output rights like securities may come in the future. Then wind power costs will be cheaper than fuel costs.

If many wind powers are connected to the power system, big influences occur in the power systems. Particularly, wind power generators which can generate much electric power are connected to the power system, and influences in the power systems need to be evaluated. Therefore, we consider a proposed method in this paper as one of the methods for solving these problems and lowering the barrier of introducing the wind powers.

3. COORDINATED SVC CONTROL OF WIND POWER GENERATORS CONNECTED TO POWER SYSTEMS [5]

3.1 Outline of the Proposed Method

We install SVCs in all the wind power plants, and do not install them at other buses. We then consider controlling voltage fluctuations using SVCs. In this method we consider controlling voltages not only for the buses with SVCs but also for the whole power system, and perform a coordinated control in consideration of the information of the other buses.

3.2 Method for determining SVC Susceptance

We consider a matrix equation for determining SVC susceptances.

$$\begin{bmatrix} \Delta V_1 \\ \vdots \\ \Delta V_n \end{bmatrix} = \begin{bmatrix} \frac{\partial V_1}{\partial B_1} & \cdots & \frac{\partial V_1}{\partial B_{nsvc}} \\ \vdots & \ddots & \vdots \\ \frac{\partial V_n}{\partial B_1} & \cdots & \frac{\partial V_n}{\partial B_{nsvc}} \end{bmatrix} \begin{bmatrix} \Delta B_1 \\ \vdots \\ \Delta B_{nsvc} \end{bmatrix} \quad (7)$$

Where ⊿Vi : voltage at bus i
⊿Bj : susceptance SVC j
n : total number of buses
nsvc : total number of SVCs

Equation(7) is represented as below.

$$\mathbf{\Delta V} = \mathbf{S} \cdot \mathbf{\Delta B} \quad (8)$$

where S : sensitivity matrix.

The equation for obtaining Δ B is given by Equation (9).

$$\begin{bmatrix} \Delta B_1 \\ \vdots \\ \Delta B_{nsvc} \end{bmatrix} = \begin{bmatrix} \frac{\partial V_1}{\partial B_1} & \cdots & \frac{\partial V_1}{\partial B_{nsvc}} \\ \vdots & \ddots & \vdots \\ \frac{\partial V_n}{\partial B_1} & \cdots & \frac{\partial V_n}{\partial B_{nsvc}} \end{bmatrix}^{-1} \begin{bmatrix} \Delta V_1 \\ \vdots \\ \Delta V_n \end{bmatrix} \quad (9)$$

If n is not equal to nsvc, the sensitivity matrix is not a square matrix, therefore we use the concept of the pseudo inverse. In general, SVC applications are considered only for diagonal elements of the sensitivity matrix. However, this method in this paper can consider the coordinated control by calculating the non-diagonal elements of the sensitivity matrix.

3.3 *Calculation Method of Sensitivity Matrix*

The sensitivity matrix cannot be calculated directly, therefore we explain how to calculate it analytically.

The changes of bus voltages Δ V to the changes SVC susceptance Δ B are represented as follows.

$$\mathbf{\Delta V} = \frac{\partial \mathbf{V}}{\partial \mathbf{X}}\mathbf{\Delta X} + \frac{\partial \mathbf{V}}{\partial \mathbf{B}}\mathbf{\Delta B} \quad (10)$$

Where X is the vector of bus voltages in the rectangular coordinates.

$$\mathbf{X} = (e_1, f_1, e_2, f_2, \cdots e_n, f_n) \quad (11)$$

The power flow equation of the rectangular coordinates is as follow.

$$\mathbf{G}(\mathbf{X},\mathbf{B}) = 0 \quad (12)$$

In Equation (12), if X and B change, we have

$$\mathbf{\Delta G} = \frac{\partial \mathbf{G}}{\partial \mathbf{X}}\mathbf{\Delta X} + \frac{\partial \mathbf{G}}{\partial \mathbf{B}}\mathbf{\Delta B} = 0 \quad (13)$$

solving for Δ X, we get

$$\mathbf{\Delta X} = -\left[\frac{\partial \mathbf{G}}{\partial \mathbf{X}}\right]^{-1} \frac{\partial \mathbf{G}}{\partial \mathbf{B}}\mathbf{\Delta B} \quad (14)$$

Then Equation (14) is substituted into (10), and we obtain

$$\mathbf{\Delta V} = -\frac{\partial \mathbf{V}}{\partial \mathbf{X}}\left[\frac{\partial \mathbf{G}}{\partial \mathbf{X}}\right]^{-1} \frac{\partial \mathbf{G}}{\partial \mathbf{B}}\mathbf{\Delta B} + \frac{\partial \mathbf{V}}{\partial \mathbf{B}}\mathbf{\Delta B} \quad (15)$$

The second term of Equation (15) is zero. Therefore, the sensitivity matrix is represented as follows.

$$\mathbf{S} \equiv -\frac{\partial \mathbf{V}}{\partial \mathbf{X}}\left[\frac{\partial \mathbf{G}}{\partial \mathbf{X}}\right]^{-1} \frac{\partial \mathbf{G}}{\partial \mathbf{B}} \quad (16)$$

The bus voltage deviation vector is as follows.

$$\begin{bmatrix} \Delta V_1 \\ \vdots \\ \Delta V_n \end{bmatrix} = \begin{bmatrix} V_1(t) - V_S \\ \vdots \\ V_n(t) - V_S \end{bmatrix} \quad (17)$$

where V(t) : a bus voltage at time t, Vs : specified bus voltage.

3.4 *Simulation*

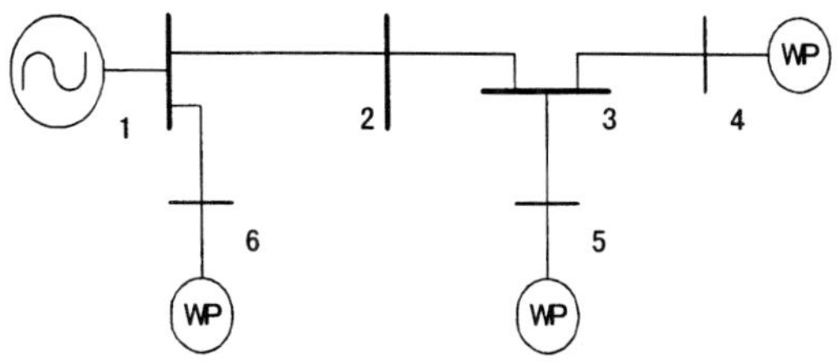

Fig. 3 Simulation Model

WP (bus number 4, 5, 6) represents the wind power generators, and bus number 1 is a large capacity generator. The bus 1 voltage is always kept at 1.00 [p.u.]. SVCs are installed at all the buses where wind power generators are installed.

Table 6 Impedance data

Transmission Line	Line Impedance	
	R (p.u.)	X (p.u.)
1-2	0.7743	1.3320
2-3	0.1795	0.3859
3-4	0.0858	3.0177
3-5	0.0000	3.7500
1-6	0.0000	1.8750

In this paper active powers and reactive powers of wind generators are given as a random number. This data is obtained by a repetition of power flow calculations and then specified values are given by random numbers. The slack bus Nò1 works for the sake of maintaining supply and demand balances.

At first, the initial values are calculated by a power flow calculations, and SVC susceptances are determined by the sensitivity matrix. SVC susceptances are added in the diagonal elements whose numbers equal an SVC installed bus number of the bus admittance matrix. A power flow calculation is carried out again using the updated bus admittance matrix, and new bus voltages are obtained. In this paper the specified bus voltages are 1.00 [p.u.].

3.5 *Simulation Results*

The simulation results at Bus2 and at Bus4 are shown below. These results are compared with non-control and coordinated control cases.

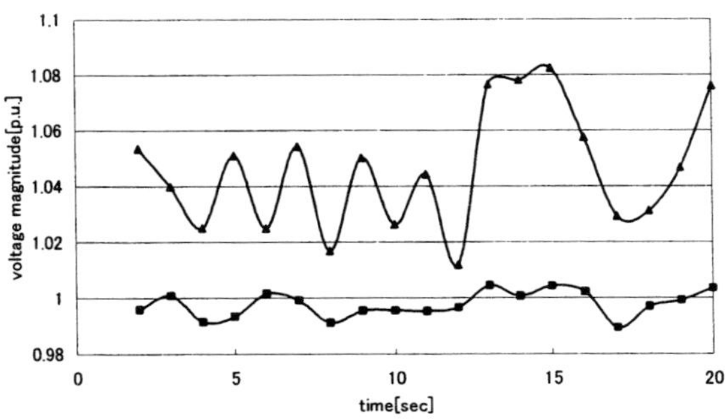

Fig. 4 Comparison of Bus-2 Voltage

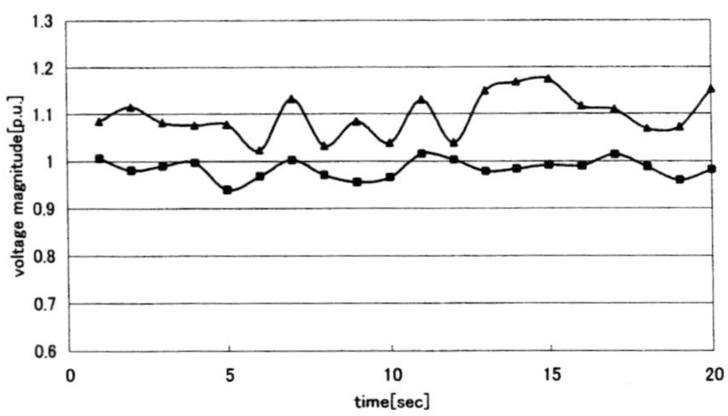

Fig. 5 Comparison of Bus-4 Voltage

3.6 *Remarks*

Voltage fluctuations were large for the non-control case. However, voltage fluctuations could be controlled by the proposed method, namely the coordinated control. The validity of the proposed method was verified.

Moreover, the best performance of this method is to control the voltage fluctuations at buses where SVCs are not installed. The reason is that the coordinated SVCs control obtains information of the other buses. It was possible to control the voltage fluctuations using the a minimum number of SVCs. Under the deregulation, minimum number of SVCs allocation may lead to cost reduction because SVCs are expensive.

4. CONCLUSIONS

The environment surrounded by the electric power industry is changing drastically. The structural reformation and the environmental problems are becoming big social problems. Renewable energy, such as wind power , is very important for the electric power industry, and it must be connected to the power systems in a large amount. However it is a risky power source because using natural energy that is connected to power systems can cause several problems. Voltage fluctuations are one of these problems, and in this paper we have proposed a method for controlling the voltage fluctuations. Finally, we have verified the validity by simulations.

Furthermore, it is possible to construct high capacity wind power generators if skills of the wind power generation are advanced. However, if large output quantities of wind powers are injected into the power system, influences to the power systems will be larger. Therefore, wind power plants should be constructed considering these influences. In this paper we have proposed a method using AHP for the optimal construction of wind power generators in consideration of power system evaluations. We have also considered the results of the simulations.

We consider the proposed approach as one of the methods for supplying the electric power stably even if wind power generators are connected to the power system in large quantities.

REFERENECES

[1] Kinoshita : 'A guide of easy decision of intention' (book) Kindaikagakusya, Japan

[2] Nakajima Takeda Ishii : 'Mathematics of Social science - a guide of fuzz theory' (book) Syoukabou, Japan

[3] Hamada, Katoh, Yokomizu, Okamoto, Suzuoki : 'Fuel-depend Pricing of IPP's Electricity under Limitation of CO_2 Emission in Electric Power Industry' Trans. of IEEJ Vol 119-B No12 1999

[4] Yokoyama and others : 'Liberalization of electric power and technical development' (book) Tokyo Denki University Press, Japan

[5] Watanabe, Ohaka, Iwamoto : 'Coordinated SVC Control for Wind Power Generators Connected to Power Systems' IEEJ PE-01-98 (in Japanese)

[6] David Milborrow : Understanding integration, Windpower Monthly, Vol.9,No.9,1993.

ELSEVIER
IFAC
PUBLICATIONS
www.elsevier.com/locate/ifac

STUDY OF UPFC OPERATION STRATEGIES IN KEPCO SYSTEM(I)-STEADY STATE CHARACTERISTICS

B. H. Chang, Y. S. Jeon, N. H. Kwak, J. B. Choo — KEPRI of KEPCO

X. K. Xu & B. P. Lam — PTI in USA

Abstract-- **This paper presents a study performed to investigate steady-state operational strategies of UPFC in the Jeollanam-Do system in Korea. The objective of the study was to determine the UPFC operating points in normal and contingency conditions. The study consists of developing load flow models to simulate different load levels with and without UPFC in the system, assessing the effectiveness of UPFC by contingency analysis, and introducing corrective actions by optimization for removing voltage problems caused by contingencies.**
The paper describes analytical tools, model and approach. It also includes analysis and discussion of the study results. The paper contributes to the area of transmission operational studies with FACTS applications. ***Copyright © 2003 IFAC***

Index Terms— **FACTS (Flexible AC Transmission System), UPFC (Unified Power Flow Controller), STATCOM (Static Synchronous Compensator), SSSC (Static Synchronous Series Compensator), Optimal Power Flow (OPF), Contingency Analysis, Reactive-Voltage (Q-V) Analysis**

I. INTRODUCTION

FIGURE 1 shows a simplified power system of the Jeollanam-Do region in Korea. Jeollanam-Do is divided into three areas including Gwangju, Mokpo and Yeosu. In the Gwangju area, Gwangju city is a major load center and there are six nuclear plants in Youngkwang generating about 6000 MW. In the Yeosu area, the Honam and Yeosu thermal plants generate about 1000 MW in total, and the Yeocheon industry zone is the major load while other zones consist mainly of rural loads. In the Mokpo area, most of the loads are small urban or rural consumers, and the Haenam HVDC Converter Station (bus 7440) operates to supply electric power to the Jeju Island (about 40% of the total island load) through the submarine DC cable.

If the 345kV double-circuit line from GJU to WHS or

B. H. Chang, Y.S. Jeon, N.H Kwak, J. B. Choo are Korea Electric Power Research Institute (KEPRI) of Korea Electric Power Corporation (KEPCO), Teajeon, Korea (e-mail: jang7@kepri.re.kr)
X. K. Xu and B. P. Lam are with Power Technologies, Inc. (PTI), Schenectady, NY 12305 U.S.A. (e-mail: xiaokang.xu@shawgrp.com).

from WHS to KJN is tripped, the Mokpo area would be electrically very remote from the generation sources in the Gwangju and Yeosu areas and would have to be supplied through 154kV lines. This would cause low bus voltages in the Mokpo area and some branch overloads in the Gwangju area.

For solving this problem, KEPCO (Korea Electric Power Corporation) had considered the construction of a 345kV line from KJN to GYG. But the plan was delayed due to increasing difficulties in obtaining new rights-of-way (ROWs). In the meantime, loads continue to grow rapidly in the Mokpo area, aggravating the low voltage problems. This prompted KEPCO to look into alternatives other than the line construction.

KEPRI (Korea Electric Power Research Institute), as the research center of KEPCO, has conducted many feasibility studies on FACTS (Flexible AC Transmission System) applications to the Korean power system. One study was about the application of FACTS to resolve the aforementioned problem in the Jeollanam-Do region [1][2]. The study results suggested the installation of a UPFC (Unified Power Flow Controller) that has a rating of ±40MVA each in the shunt and series parts each at the 154kV Kangjin Substation for regulating the voltage on bus 7420 and controlling the power flow on the Y-M interface shown in Fig. 1.

This paper describes a study performed to investigate the steady-state operational strategies of the UPFC in the Jeollanam-Do region. The objective was to determine the UPFC operating points under normal and contingency conditions. The study consists of developing load flow models to simulate different load levels with and without the UPFC in the system, assessing the relative effectiveness of the UPFC under contingency conditions, and applying corrective actions to remove voltage problems caused by the contingencies.

The paper describes analytical tools, model and approach. It also includes analysis and discussion of the study results. The paper contributes to the area of transmission operational studies with FACTS applications

II. Analytical Tools, Models And Approach

The analytical tools and models used in this study include:

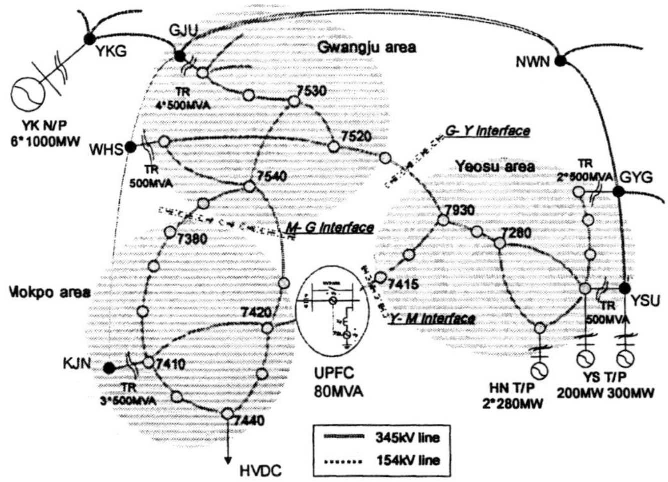

Fig. 1. Simplified Jeollanam-Do Power System

- Contingency Analysis
- Optimal Power Flow
- Reactive-Voltage (Q-V) Curve Analysis
- UPFC, Static Synchronous Series Compensator (SSSC) and Static Synchronous Compensator (STATCOM) Models

A. Contingency Analysis

Contingency analysis is based on an ac load flow model. This analysis identifies severe contingencies leading to thermal overloads, high or low voltages or voltage collapse under steady-state conditions. The contingency analysis is performed using a full AC Newton solution engine in a widely used commercial software package, PSS/E[2].

B. Optimal Power Flow (OPF)

The severe contingencies identified by contingency analysis are further analyzed using the OPF module in PSS/E [3]. This module uses an interior point algorithm based on the Newton and Kuhn-Tucker methods to solve a nonlinear problem consisting of an objective function and a combination of equality and inequality constraints [4]. In this application, the OPF is used to identify optimal locations and amount of reactive compensation required to eliminate the voltage limit violations.

C. Reactive-Voltage (Q-V) Curve Analysis

Fig. 2 is a typical Q-V that relates the reactive injection at a bus or a group of buses necessary to sustain a certain voltage. The minimum point ("Nose") on the curve (where dQ/dV=0) is the critical point, also referred to as the Voltage Collapse Point. The points on the curve to the left of the minimum represent unstable conditions, since lowering the voltage requires more reactive power injection. The points to the right of the minimum are stable. The intersection of the V-Q curve with the horizontal axis is the voltage at the bus without reactive compensation. The vertical distance between the horizontal axis and the critical point is the reactive power margin. If the minimum point of the V-Q curve is above the horizontal axis, the system is deficient in reactive power, and additional supply of reactive power is required to prevent a voltage collapse. If the critical point is below the horizontal axis, the system has a positive reactive margin.

Q-V curve analysis can be performed using another feature of PSS/E, IPLAN [5][6], to find the minimum reactive

---

[2] *PSS/E (Power System Simulator for Engineering) is the commercial tradename for PTI's transmission planning and operation software.*

compensation needed to maintain system voltages within the criteria. IPLAN is an environment independent programming language that can be used to control a host program, such as PSS/E. IPLAN can conduct two-way communication with the host. For example, IPLAN can provide instructions to PSS/E to sequentially decrease the controlled voltage, thereby developing Q-V curves. Upon detection of the critical "Nose" condition or an unconverged load flow condition, IPLAN then automatically terminates the calculations.

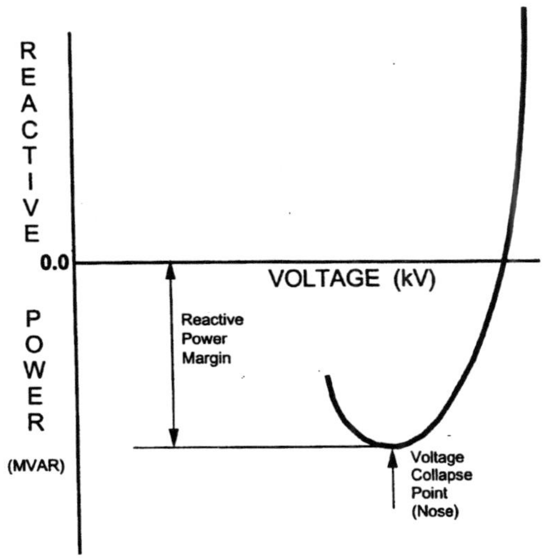

Fig. 2. Reactive Margin Measured by MVAr on Q-V Curve

D. UPFC, STATCOM and SSSC Models

Fig. 3 shows the block diagram of the steady state UPFC model implemented in PSS/E [7]. It has a series element, which is connected between two buses (*I, J*), and a shunt element, which is connected between the sending end bus (*I*) and the ground. One or both of these elements may be active, depending upon the type of device being represented. In typical operations, the series element maintains the desired active and reactive power flows (P_{DES} & Q_{DES} in the diagram) between the sending bus (*I*) and the terminal bus (*J*) and the shunt element maintains the desired sending end voltage magnitude (V_{SET}).

By properly selecting the data, different device types and control options can be represented. For example, a SSSC can be represented by setting the maximum bridge power transfer to zero and disabling the shunt component. A STATCOM can be represented by disabling the series component, while the reactive power output of the shunt element is varied to maintain the desired sending end bus voltage magnitude. In a UPFC, the current in the shunt connected bridge depends upon its var output and the amount of power transferred to the series connected element.

E. Study Approach

The following approach is adopted in the study:

1. Develop base power flow models representing peak, shoulder and light conditions, with and without UPFC, for a given year. These base models are prepared using OPF to minimize active and reactive losses and optimize UPFC operating points in normal conditions.

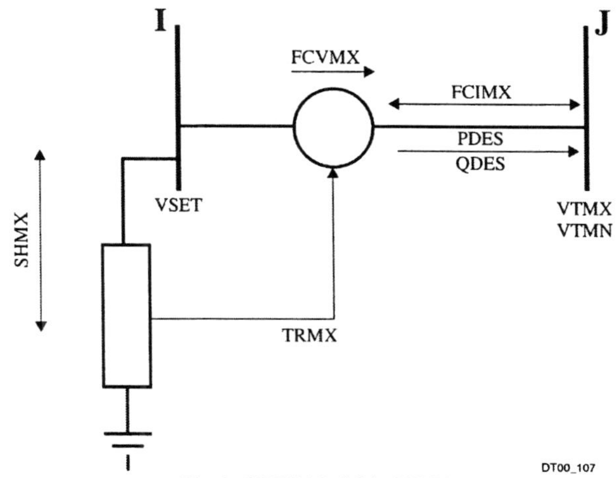

Fig. 3. UPFC Model in PSS/E

2. Perform contingency analysis to identify severe contingencies leading to thermal or voltage problems.
3. Find solutions to the above problems using OPF:
   (a) In system without UPFC, adjust transformer OLTC's (On Load Tap Changers) and add shunts (existing & future) to solve voltage problems. If necessary, use generation redispatch and load shed to solve overload problems.
   (b) In system with UPFC, adjust OLTC's and add shunts, adjust STATCOM to solve voltage problems. The OLTC's and shunt capacitor settings define preventive action (pre-contingency settings, if base case voltage criteria not violated). STATCOM var output is set to be near zero (or slightly negative) in pre-contingency to allow for dynamic response during contingencies. UPFC control will be corrective action (post-contingency adjustment).
   (c) In system with UPFC, after solving problems in (b), if STATCOM output is at maximum in post-contingency steady state, then consider the addition of more shunts to reserve STATCOM vars for dynamic response.
4. Based on (3), define pre-contingency and post-contingency settings of the STATCOM/UPFC for the three load levels as specified in (1).
5. Perform Q-V analysis on buses where voltages are observed to be lowest and/or shunt compensation is needed. Develop Q-V curves at these locations. Compare with and without STATCOM portion.

III. Case Studies and Results Analysis

The analytical tools, models and approach described in the previous section were applied to the KEPCO system with UPFC. The subsequent sections present system background, case studies and results.

A. Brief Description of KEPCO System

Power generation complexes in Korea are concentrated in coastal areas and loads are concentrated in inland metropolitan areas far away from generation sources. System peak demand in 2003 is expected to be about 47 GW and total generation capacity is about 52 GW. The KEPCO transmission system consists mainly of 765 kV, 345 kV and 154 kV networks. Fig. 4 shows the surrounding power system near Kangjin where the UPFC is installed. The shunt connection of UPFC is at the 154 kV Kangjin substation, and the series connection is on the double-circuit transmission line between Kangjin and Jangheung substations. Under normal conditions, the UPFC is meant to increase transmission line capacity and reduce system losses through coordinated control of power flow and reactive power compensation. During contingencies, the UPFC is used to relieve undervoltage and overload problems in the surrounding system, if possible.

The KEPCO system is modeled as seven areas in PSS/E. Area 6, GWANGJU, where UPFC is located, was defined as the Study Area. Other major voltage controlling equipment in the Study Area includes transformer on-load tap changers (OLTC's) and switchable shunts.

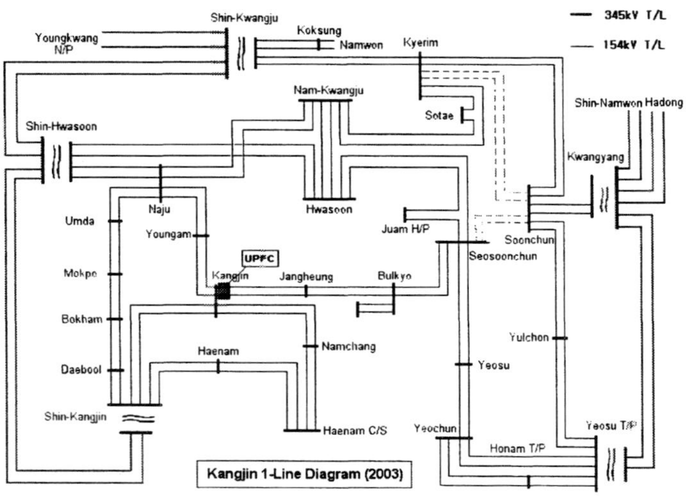

Fig. 4. Surrounding Power System near Kangjin

B. Power Flow Models Setup

The load flow scenarios were developed as follows:

- Peak case without UPFC
- Peak case with UPFC
- Shoulder case without UPFC
- Shoulder case with UPFC
- Light case without UPFC
- Light case with UPFC

In setting up these models, OPF solution was used to minimize active and reactive losses by controlling power flow on the series element of the UPFC and by allowing existing shunt capacitors and reactors to be switched. The reactive output of the shunt element of the UPFC was set to be near zero in pre-contingency to allow for response during contingencies. The UPFC operating points in the normal condition are listed in Table 1. Positive power flow indicates that direction is from Kangjin to Jangheung. Other parameters of the UPFC model in PSS/E used in this study are included in Appendix.

TABLE 1 UPFC OPERATING SETPOINTS IN NORMAL CONDITION

Loading	P_{DES} (MW)	Q_{DES} (MVAr)	V_{SET} (p.u.)	Q_{SHUNT} (MVAr)
Peak	-31.4	18.8	1.035	0.1
Shoulder	6.8	-9.2	1.016	0.0
Light	-6.8	-14.2	1.016	0.0

C. Contingency Analysis Results

In order to identify credible severe contingencies leading to thermal or voltage problems in the Study Area, contingency analysis was performed on all the base cases to evaluate the steady-state response of the system to outages of 345 kV and 154 kV double-circuit lines in the area. The monitored elements included all 154 kV and above branches and buses in the Study Area.

The KEPCO Operating Criteria were applied to the contingency analysis. The Criteria require that, under normal conditions, the power flow in any branch of 154 kV or above should not exceed Normal Ratings (Rate A). Under contingencies (loss of any parallel 345kV or 154 kV circuits), the power flow in any branch of 154 kV or above should not exceed Emergency Ratings (Rate B).

System voltages in normal and contingency conditions are to be controlled in a range in accordance with the Criteria, as shown in Table 2.

TABLE 2 VOLTAGE CRITERIA

Voltage Limits (for Both Normal and Contingency Conditions)		High (p.u.)	Low (p.u.)
345 kV	For All Loading Conditions	1.0435	0.9739
154 kV	Peak	1.0649	1.0129
	Shoulder	1.0454	0.9935
	Light	1.0389	0.9870

All the contingency cases were solved using the fixed sloped decoupled Newton-Raphson iterative algorithm with transformer taps, and switched shunts and DC tap controls disabled.

Contingency analysis shows that outages of the parallel circuits result in a few minor thermal overload problems but many voltage violations. Therefore, this study focused on the voltage performance of the Study Area. Table 3 shows statistics regarding the analysis results. There were about 60 double contingencies tested. Some of them resulted in multiple voltage violations. As expected, the system with the

UPFC operating shows fewer post-contingency voltage violations.

From the contingency results, five severe contingencies were selected for further study with corrective actions. These severe contingencies are:

(1) Loss of the YKG-GJU 345 kV double-circuit line

(2) Loss of the GJU-WHS 345 kV double-circuit line

(3) Loss of the WHS-KJN 345 kV double-circuit line

(4) Loss of a 154 kV double-circuit line (7610-7615/7616)

(5) Loss of a 154 kV double-circuit line (7610-7215)

TABLE 3 CONTINGENCY ANALYSIS STATISTICS

Case	No. of Voltage Violations		
	Peak	**Shoulder**	**Light**
W/ UPFC	113	105	47
W/O UPFC	184	133	61

D. Corrective Actions and OPF Results

For the severe contingencies, the following corrective actions were applied:

(a) In the cases without UPFC, existing OLTC's and switchable shunts were adjusted to relieve voltage problems.

(b) In the cases with UPFC, existing OLTC's and switchable shunts were adjusted to relieve voltage problems. In addition, the var output of the UPFC shunt part (STATCOM) also responded automatically to provide voltage support. The var outputs of STATCOM under the severe contingencies are shown in Table 3. It can be seen from the table that the outputs vary by contingency and loading level.

TABLE 3 STATCOM VAR OUTPUT DURING CONTINGENCY

Contingency	UPFC STATCOM Output (MVAr)		
	Peak	**Shoulder**	**Light**
Loss of the YKG-GJU 345 kV Double-Circuit Line	39.5	39.2	39.7
Loss of the GJU-WHS 345 kV Double-Circuit Line	34.5	39.2	40.1
Loss of the WHS-KJN 345 kV Double-Circuit Line	0.1	25.9	38.8
Loss of a 154 kV Double-Circuit Line (7610-7615/16)	40.5	40.0	19.4
Loss of a 154 kV Double-Circuit Line (7610-7215)	24.6	13.6	7.3

(c) The corrective actions in (a) and (b) would not completely remove the voltage violations caused by the severe contingencies. Therefore, OPF was used to find new corrective actions, including the location and amount of new switched shunts and the optimal control of existing OLTC's and switched shunts, summarized as below:

- In all base power flow cases, the voltage control ranges of the switched shunts at 154 kV buses were set at 1.02 to 1.05 pu during peak load. These limits are within the minimum and maximum voltage criteria at 154 kV buses. Using these limits allow the switchable shunts to respond to voltage violations during contingencies and initiate switching actions to correct the problems.
- The OLTC's of 345/154 kV transformers were set to control 154 kV voltages between 1.02 and 1.05 pu during peak load. Switchable shunts at the 154 kV buses of the transformers were set to control local voltage. This avoided control conflicts between the OLTC's and switchable shunts.
- After the above actions, the remaining low voltage violations under the severe 345 kV contingencies were eliminated by adding new shunt capacitors at buses 7675 (5x5.0 MVAR), 7695 (5x5.0 MVAR) and 7945 (3x5.0 MVAR) as determined by OPF.
- Similarly, the low voltage violations under the severe 154 kV contingencies were eliminated by adding new shunts calculated by OPF.

E. Q-V Curve Analysis Results

In order to verify the suggested shunt additions determined by OPF, Q-V curve analysis was performed using an IPLAN program to find the minimum reactive compensation needed to maintain system voltages within the Criteria.

In the study, Q-V curves were generated for the pre-contingency and post-contingency conditions for a set of selected severe contingencies and buses where voltages are worst and/or shunt compensation is needed. One case is selected for illustration here. The selected bus and corresponding critical contingency are:

- Monitored bus is 7590
- Contingency is the loss of a 154 kV double-circuit line (7160-7615/7676)

Fig. 5 shows the Q-V curves at the monitored bus in the peak load case with UPFC. The curve represents two conditions:

- Pre-Contingency (solid line in the figures)
- Post-Contingency (dashed line in the figures)

The curve also includes capacitor Q-V characteristics (10 MVAr, 50 MVAr and 100 MVAr) that represent the relationships between voltage and the reactive power produced by shunt capacitors.

It is observed from the figure that the minimum ("Nose")

points under pre-contingency and post-contingency conditions are below the horizontal axis. This indicates that the reactive margins in these conditions are positive and voltage collapse would not occur on the system. This further verifies the previous contingency analysis in which no voltage collapse was found.

However, it is noted that the intersection of the V-Q curve with the horizontal axis under the post-contingency condition is below 1.0 p.u. voltage, which is lower than the Voltage Criteria (1.0129~1.0649 p.u. for 154 kV buses during peak load operation). This indicates that additional shunt compensation is needed. The intersection between the set of capacitor Q-V curves and the system post-contingency Q-V curve establish the steady-state operating point. From the figure, to operate the monitored bus at 1.0129 p.u. voltage under this contingency, approximately 30-40 MVAr of shunt capacitors are required at this bus. This is close to the value identified by OPF.

IV. CONCLUSION

This study focused on the voltage issues in the Study Area. In general, UPFC improved the voltage performance of the Study Area. The study can draw the following conclusions on the operation strategies of UPFC with respect to the contingencies tested and the load flow scenarios studied:

1. UPFC should be set such that the var output of its shunt part (STATCOM) is near zero in normal conditions for all three load levels, thus providing maximum reactive reserves for contingencies.
2. The series element of the UPFC should be set to control flow such that area losses are minimized.
3. To maximize the effectiveness of UPFC, the voltage control ranges of OLTC's and switched shunts in the Study Area should be consistent with the Operating Criteria.
4. The control of 345/154 kV transformers and switchable shunts at 154 kV buses should be coordinated to avoid control conflicts.
5. Additional shunt compensation (approximately 100 MVAr determined by OPF) is needed to completely remove low voltages caused by the contingencies tested.

V. APPENDIX

Parameters of the UPFC model in PSS/E used in this study are set as follows:

- Maximum series voltage magnitude (VSMX)=0.0682
- Maximum series current magnitude (*IMX*)=586.8
- Maximum shunt current magnitude (*SHMX*)=40.0
- Maximum terminal voltage (*VTMX*)=1.05
- Minimum terminal voltage (*VTMN*)=0.95
- Maximum power transferable between the shunt and series connected bridges (*TRMX*)=40.0
- UPFC Rating: Series Part: ±40MVA; Shunt Part: ±40MVA

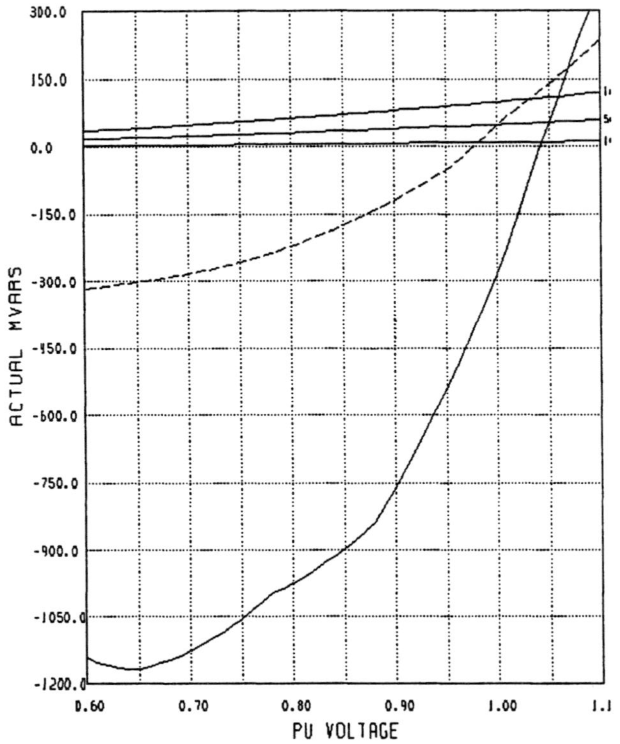

Fig. 5. Q-V Curve for Bus 7590 for Peak Case with UPFC

VI. ACKNOWLEDGMENT

The authors wish to thank Mr. James W. Feltes and Dr. Yuriy Kazachkov for their assistance and discussions during the study.

VII. REFERENCES

[1] Jin-Boo Choo and Byung-Hoon Chang, "Development of FACTS Operation Technology (Phase 1 : Decision of a Type and a Spec. Draft for the Installation of a Pilot Plant)", *KEPRI Technical Report*, April, 1999

[2] In-Sup, Kim, "long-term Forecast of KEPCO System Planning and Operation, in Korea", in *Proceedings of 1998 Power System Engineering Symposium*, KIEE, May, 1998

[3] "PSS/E Optimal Power Flow Manual – Version 29", Power Technologies, Inc., December 2002

[4] H. W. Kuhn and A. W. Tucker, "Nonlinear Programming," in *Proceedings of the Second Berkeley Symposium on Mathematical Statistics and Probability (J. Neyman, Editor)*, Berkeley, University of California Press, 1951

[5] J. V. Staron and A. R. Weekley "Expand Your Study Capability With IPLAN," *PTI Newsletter, Power Technology*, Issue No. 53, April 1988

[6] R. J. Koessler "Voltage Collapse Analysis Using IPLAN," *PTI Newsletter, Power Technology*, Issue No. 60, January 1990

[7] Xiaokang Xu, Baldwin P. Lam, Ricardo R. Austria, Wang Huaming, Fan Haihong, Yang Lin and Wang Xu, "Application of FACTS Technology for Increasing

Power Transfers of Jiangsu Transmission System," in *Proceedings of the CIGRE 2001 International Conference on Power Systems (CIGRE ICPS'2001)*, September 3-5, 2001, Wuhan, Hubei, China

VIII. BIOGRAPHIES

Byunghoon Chang received his B.S. and M.S. from Korea University, Korea. He received his Ph. D. from Korea University in 1999. He have worked at Korea Electric Power Research Institute as a researcher for 9 years. He is presently a Senior Member of Technical Staff of Power System Laboratory at Korea Electric Power Research Institute. His research interests include FACTS, power system stability and power quality.

Young-Soo Jeon(M'2002) received his B.S., from Sungkyuhnkwan University, Korea in 1983 and M.S., from Chungnam National University, Korea in 1999 respectively. Currently he is a senior member of technical staff in the Electric power system technology group at Korea Electric Power Research Institute, Korea. His research interests include FACTS and Power Quality.

NO-Hong Kwak received his B.S., M.S, Ph.D. degrees in Electrical Engineering from Yonsei university , Korea in 1978, 1980, and 1990 respectively. He worked at Korea Electric Power Corporation as a principal engineer in Korea Electric Power Research Institute(KEPRI). Currently he is a Group Leader at the Electric Power System Technology Group of Power System Laboratory in KEPRI. He is interested in area of operation and control of power system including FACTS, and Real time power system simulation.

Jin-Boo Choo (M'1994) received his B.S., M.S, Ph.D. degrees in Electrical Engineering from Seoul National University, Korea in 1977, 1987, and 1994 respectively. He worked at Korea Electric Power Corporation as a senior engineer in the Power System Operation Center. Currently he is a general manager and is in charge of the Power System Analysis Center at Korea Electric Power Research Institute, and also Real time power system analysis team leader of National Research Lab. at Ministry of Science & Technology, Korea. His research interests include power system operation and control, FACTS, and Real time power system simulation.

Xiaokang Xu (M'92, SM'00) is a graduate of The Southeast University and The Electric Power Research Institute (EPRI) in China and The University of Western Ontario in Canada from which he holds B.Sc., M.Sc. and Ph.D. degrees in Electrical Engineering. From 1984 to 1991 he was a research engineer for EPRI-China. During 1992 - 1993 he worked for Ontario Hydro of Canada as a planning engineer. From 1995 to 1998 he was with CAE Electronics Ltd., Canada as an EMS engineer. He joined Power Technologies, Inc. (PTI), USA in March 1998 and is presently an Executive Consultant. He is a Member of Sigma Xi.

Baldwin P. Lam is a BS and ME graduate of Rensselear Polytechnic Institute in Troy, New York. He has been with Power Technologies and its affliate since 1975. His main area of work includes load flow, stability and reliability studies of major power grids and industrial systems. He is presently an Associate Director of the Consulting Services Department. Mr. Lam is a Senior Member of IEEE.

ELSEVIER
IFAC
PUBLICATIONS
www.elsevier.com/locate/ifac

STUDY OF UPFC OPERATION STRATEGIES IN KEPCO SYSTEM(II)-DYNAMIC CHARACTERISTICS

B. H. Chang, S. Y. Kim, J. S. Yoon, J. B. Choo — KEPRI of KEPCO

Leonardo T.G. Lima & James W. Feltes — PTI in USA

Abstract*--Korea Electric Power Corporation (KEPCO) has installed an 80 MVA Unified Power Flow Controller (UPFC) at its 154kV Kang-Jin Substation in South Korea. The device, manufactured by Siemens & Hyusung, has been operational since Oct. 2002. The Korea Electric Power Research Institute (KEPRI)), a division of KEPCO was tasked to study the operation strategies that could be employed for the UPFC and surrounding reactive support devices to address problems of low voltages and overloads in the Mokpo & Gwangju areas for severe contingencies. A special concern involved potential delay in the installation of a new 345 kV transmission line from 2005 to beyond 2010. The studies were to determine if these problems could be eliminated by UPFC application. The studies included determining the UPFC operating point under the various conditions, investigations of the coordination between the UPFC and a HVDC line terminating in this area, and to design the supplementary damping controller for the UFPC. This paper summarizes the results of those studies, demonstrating the dynamic characteristics of the operation of this UPFC operation in the Korean power system. *Copyright © 2003 IFAC

***Index Terms*--Power system dynamic stability, Power electronics, power system control.**

I. INTRODUCTION

HE Figure 1 shows the local simplified power system of TJeollanam-Do in Korea. Jeollanam-Do is divided into three sub-areas include Gwangju, Mokpo and Yeosu area. In the Gwangju area, Gwangju city is the large load and the six Youngkwang nuclear plants generate about 6000MW. In Yeosu area, Honam thermal plant and Yeosu thermal plant generate totally about 1000MW and Yeocheon industry area is major load, other areas are rural loads. In Mokpo area, most of all load is small city or rural, and specially 7440 bus(Haenam converter Station of HVDC) operates to supply the electric power(about 40% of the total load in Jeju island) toward the Jeju island through the DC cable under the sea. If the 345kV line from GJU to WHS or from WHS to KJN could be opened, Mokpo area would be far from power source and would be supplied the power through the only 154kV line in the Kangju area and Yeosu area. And then there would be low voltage buses in the Mokpo area and

B.H. Chang, S.Y Kim, J.S. Yoon, J.B. Choo are Korea Electric Power Research Institute (KEPRI) of Korea Electric Power Corporation (KEPCO), Teajeon, Korea (e-mail: jang7@kepri.re.kr)
Leonardo T.G. Lima and James W. Feltes are with Power Technologies, Inc. (PTI), Schenectady, NY 12305 U.S.A. (e-mail: leonardo.lima@shawgrp.com).

overload branch in Kangju area. The voltage oscillation of 7440 bus, HVDC C/S would be happened.

For solving the problems, previously KEPCO had the construction plan of the 345kV line from KJN to GYG. But KEPCO should delay the plan due to increasing the difficulties in obtaining new right-of-ways (ROWs). However, the more loads increase in the Mokpo area, the severer problem can be happened. KEPCO needed alternatives other than the line construction.

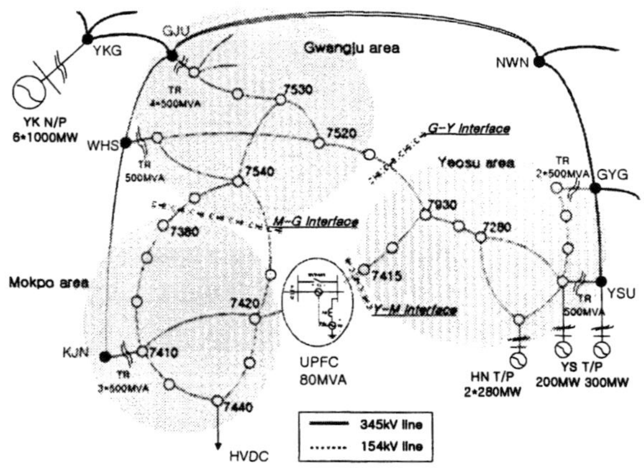

Figure 1. Simplified one-line of the Jeollanam-Do system

This paper investigates the dynamic problems and the operation effects of the KEPCO UPFC under the following disturbances.

o Jeollanam-Do, Mainland

- Severe contingency, 345kV WHS-KJN line 3-phase fault
- Large disturbance, 1 unit outage in YK N/P
- Small disturbance, 154kV Heanam bus(7440) – 2000[Mvar] on/off

o Jeju-Do, Island

- Severe contingency, 154kV JJU-DJJ line 3-phase fault
- Large disturbance, 1 unit outage in Jeju T/P
- Small disturbance, 154kV DJJ bus(130) –600[Mvar] on/off

The major dynamic problems under the disturbances is the oscillations of the ac bus voltage, the active power and dc voltage & dc current of HVDC.

KEPRI as the research center of KEPCO had implemented a lot of feasibility studies for FACTS application to the Korea power system. So KEPRI was requested the study for FACTS application to solve the problem in the Jeollanam-Do, Korea by KEPCO. Through the results of the study, the KEPRI suggested the site, the capability and the type of FACTS. The KEPCO UPFC which has ±40MVA inverter in shunt and in series each other, was installed at the 154kV Kangjin Substation for regulating the voltage of 7420 bus and for controlling the power flow of Y-M interface shown in the Figure 1.

II. MODELING

A. HVDC

The HVDC connected between the mainland and Jeju island in Korea has the following dynamic characteristics.

- Blocking : If the ac voltage of the rectifier terminal in HVDC goes down less than 0.6[pu], HVDC system blocks the current from the rectifier terminal bus.
- Unblocking : If the ac voltage of the rectifier terminal in HVDC recover more than 0.65[pu], HVDC system unblocks the current from the rectifier terminal bus.
- Bypass : If the dc voltage of the inverter terminal in HVDC goes down less than 108[kV], HVDC system bypasses the current from the inverter terminal bus.
- Unbypass : If the ac voltage of the inverter terminal in HVDC recovers more than 0.65[pu], HVDC system unbypass the current from the inverter terminal bus.
- Recovering Ramp : The recovering ramp of the dc voltage & the dc current in HVDC rectifier has the rate 0.2[pu/sec]. The ramp in inverter has the same rate.
- HVDC has the modulation control to maintain the frequency in Jeju island.

The CDC4 and SQBAUX in PSS/E models of PTI is used to represent the general dynamic characteristics and the modulation control of HVDC in Korea.

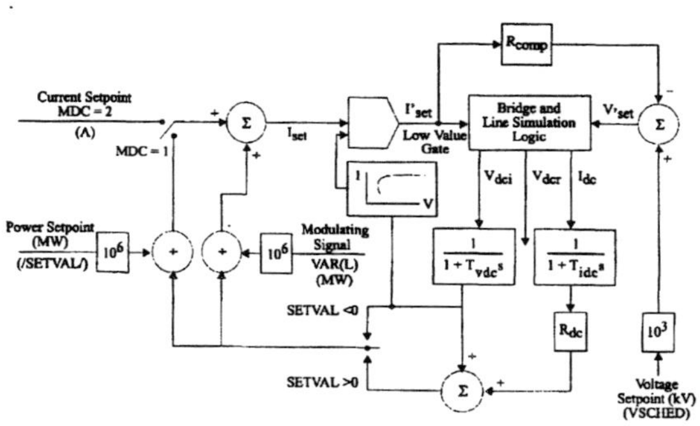

Figure 2. Block Diagram of CDC4

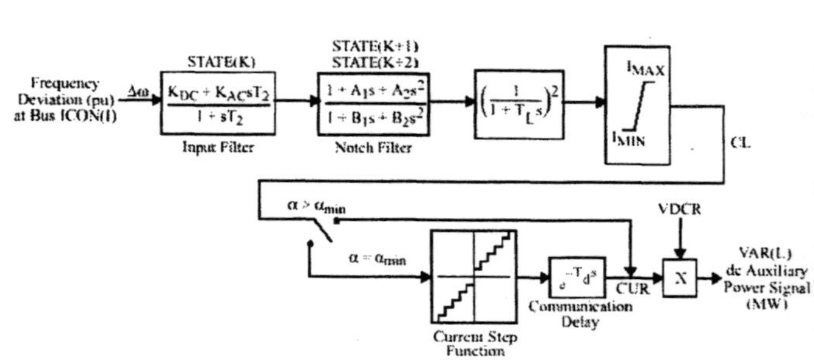

Figure 3. Block Diagram of SQBAUX

B. UPFC

The dynamic response of the UPFC to a disturbance is

very fast. Due to this fact no integrator is needed to model the UPFC and an algebraic model is sufficient for dynamics studies. An algebraic model in the PSS/E is a model that is called through CONET subroutine and contains no STATE. PQ control and Voltage insertion for series element are the two modes of operation that are implemented in this dynamics model.

A theoretical model for the UPFC consists of two voltage sources, one in series, corresponding to the series inverter and one in shunt corresponding to the shunt inverter, as shown in Figure. The two voltage sources are coupled by the real power flow Pdc which is drawn from the power system by the shunt inverter and injected into the power system by the series inverter. XP is the leakage reactance of shunt transformer.

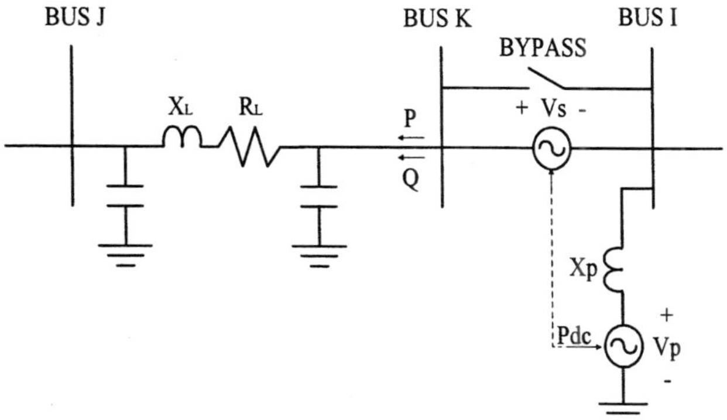

Figure 4. UPFC Dynamic Model

III. Operation Effects of UPFC

The interactions between HVDC and UPFC in Jeollanam-Do are investigated using the PSS/E models. The dynamic problems under the disturbances are investigated also. The following disturbances are simulated for investigation of the dynamic problems.

TABLE 1. CONTINGENCY CASES

Disturbance Type	**Main Land**	**Island**
Severe Line Fault	345kV Line Fault	154kV Line Fault
Large Disturbance	1000MVA Unit Off	70MVA Unit Off
Small Disturbance	-2000 MVar On/Off	-600 MVar On/Off

A. Jeolllanam-Do, mainland system

Three type disturbances are simulated in the local system of the Jeollanam-Do in which the UPFC and HVDC was installed. The first type is the 3-phase fault of 345kV non-looped line. The second one is the #3 unit outage of YK Nuclear Plant. Third one is the –2000[Mvar] reactor on/off instead of the various disturbances like the line-ground fault.

The operation effects of UPFC with respect to the dynamic problems at the disturbances are analyzed.

If the dynamic problems are transferred into Jeju island through the HVDC, the UPFC effects relating to the dynamics of HVDC are analyzed also.

1) Severe Line Fault

This is the severe contingency since it makes the long electric distance from voltage source in Kwangju area to load in Mokpo area. This results in significant weakening of the transmission line crossing the M-G interface.

a) Voltage dynamics

Figure 5 shows the voltage dynamics at the ac terminal bus of HVDC(7440) & UPFC(7420) with/without UPFC. The almost AC buses in Mokpo area have the voltage oscillations without UPFC operation, but the oscillations are mostly eliminated by UPFC operation. This figure 6 particularly shows the damping effect of the oscillations at the terminal buses of HVDC & UPFC during UPFC operation.

When the fault happens at 1.0[sec], HVDC acts the blocking of the rectifier since the terminal bus has the low voltage less than 0.6[pu], blocking reference voltage. The terminal bus of HVDC reaches the unblocking voltage 0.65[pu] at 5.1[sec], after the fault is cleared at 1.1[sec]. But the rectifier DC voltage of HVDC has the oscillation under the influence of the ac terminal bus of rectifier. Figure shows the dynamic relationship between AC and DC voltage of HVDC rectifier.

Figure 7 shows the damping effects of the dc voltage oscillations at the rectifier of HVDC after the during the UPFC operation.

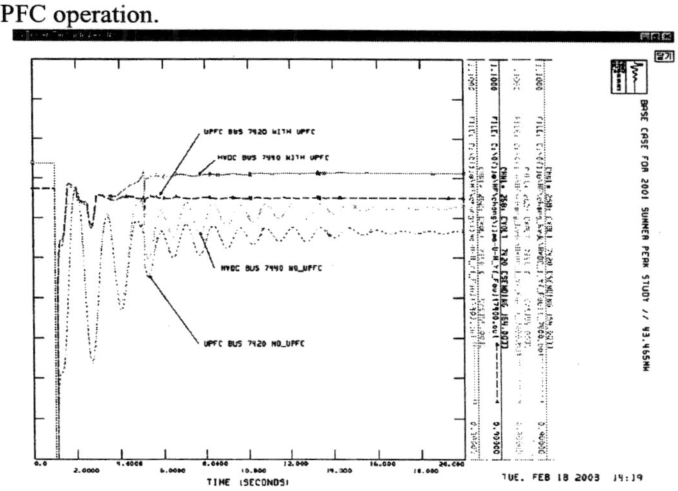

Figure 5. Effects of Voltage Control at the Severe Line Fault in Mainland

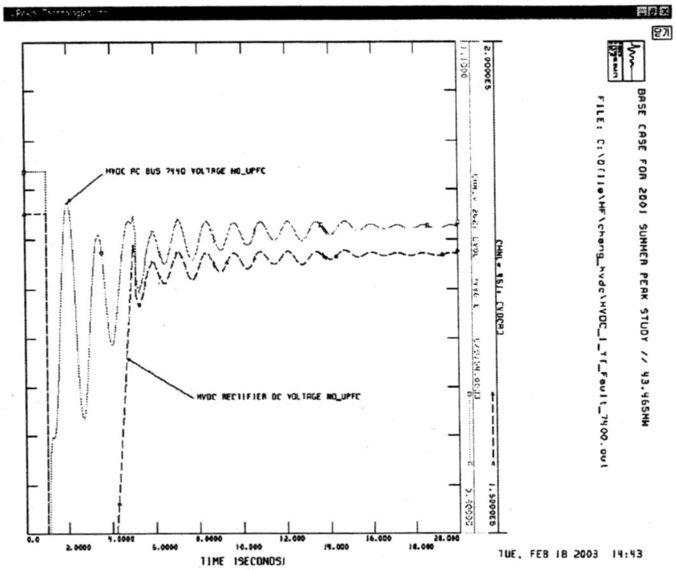

Figure 6. DC Voltage Characteristics of HVDC bus at Severe Line Fault in Mainland

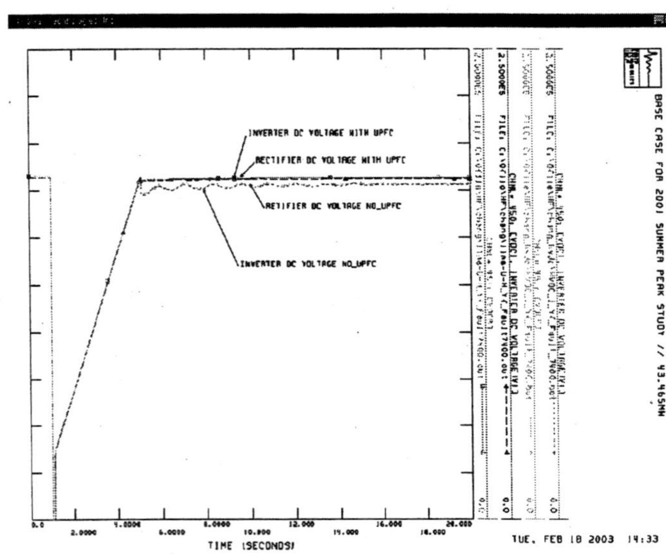

Figure 7. Effects of DC Voltage Control of HVDC bus at Severe Line Fault in Mainland

b) power flow dynamics

Figure 8 shows the active power dynamics flows into the HVDC at the 345kV line 3-phase fault. UPFC is operated for maintaining the desired P and Q of the line, but UPFC doesn't have the damping effect of the oscillation since post contingency. The reason of not damping the power oscillation near HVDC in Mokpo area is that the recovering action of HVDC after the blocking of the rectifier is more dominant than the constant power flow control action of UPFC. The another reason of not damping the power oscillation crossing M-Y interface is that th capability of UPFC is too small to handle the oscillation which happen in the whole Korea system.

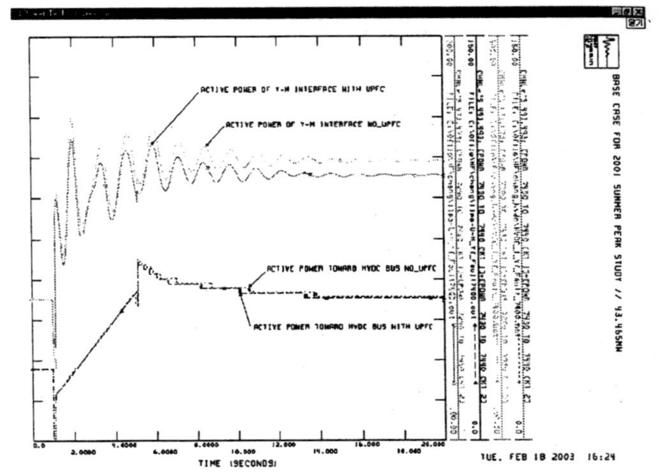

Figure 8. Effects of Power Flow Control at Severe Line Fault in Mainland

2) Large Disturbance

The large disturbance, #3 unit outage makes the large impact in YK nuclear plant complex. This result in significant dynamic problems like sustained oscillations in the whole Kepco system including Mokpo area.

a) Voltage dynamics

Figure 9 shows that the oscillations at the AC terminal buses of HVDC(7440) and UPFC(7420) is dramatically damped by UPFC operation.

Figure 10 shows the damping effects of the DC voltage oscillation in HVDC during UPFC operation.

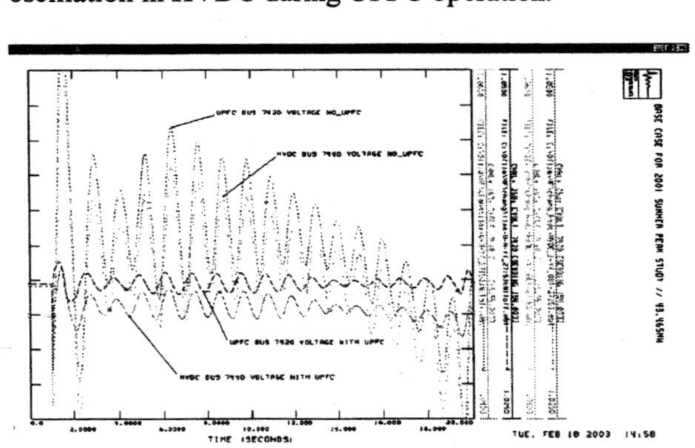

Figure 9. Effects of Voltage Control at the Large Disturbance in Mainland

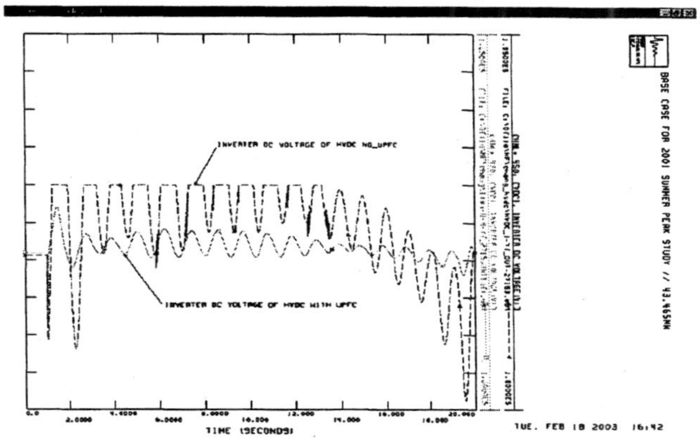

Figure 10. DC Voltage Characteristics of HVDC bus at Large Disturbance in Mainland

b) power flow dynamics

Figure 11 shows that UPFC is controlling the constant power flow of the line at the large disturbance.

Figure 12 shows that the power that flows into the AC terminal bus of HVDC is oscillating owing to the large disturbance. The power oscillation is damped by UPFC operation as shown the same figure.

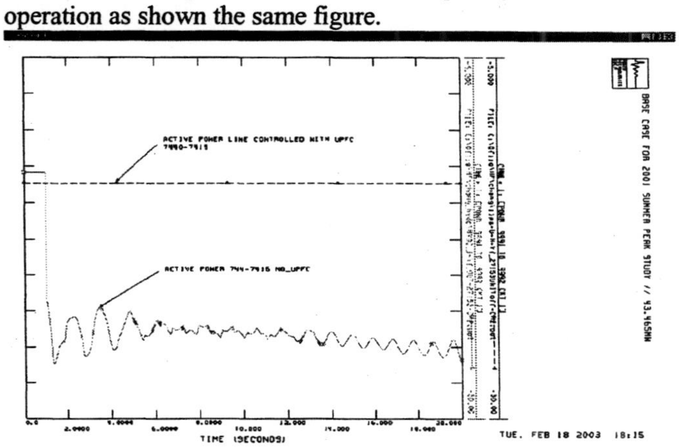

Figure 11. Effects of Power Flow Control at Large Disturbance in Mainland

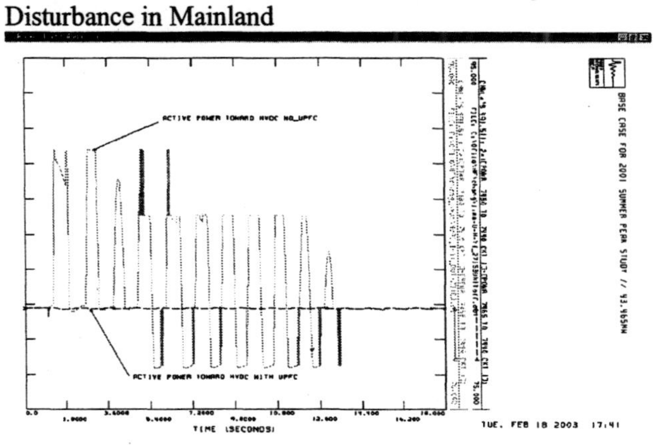

Figure 12. Effects of DC Power Flow Control of HVDC at Large Disturbance in Mainland

Figure 13 shows that the DC current and the firing angel of the HVDC is oscillating under the influence of the large disturbance. The oscillations in the HVDC are eliminated since the power oscillation which flow toward HVDC is damped by UPFC's constant PQ control.

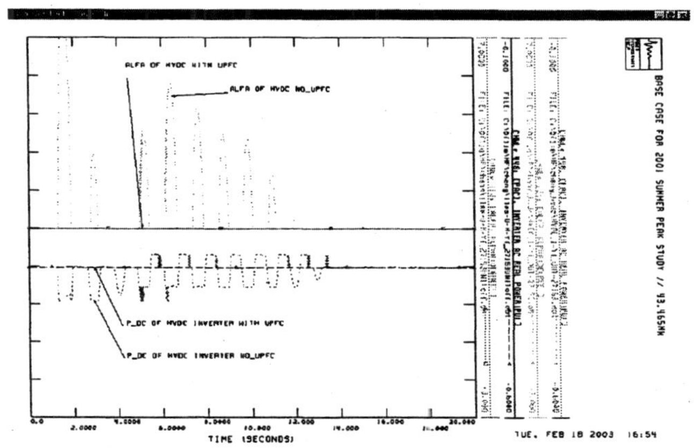

Figure 13. Effects of DC Current Control of HVDC at Large Disturbance in Mainland

The active power dynamic problems happen in Jeju island because HVDC which connected with the mainland makes the active power oscillation. Figure 14 shows that the power oscillation between HVDC and Jeju island. And the oscillation is damped by the effect of UPFC operation.

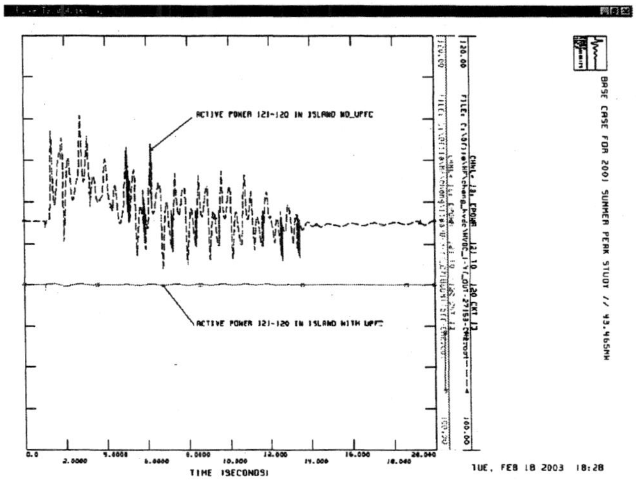

Figure 14. Effects of Power Flow Control of Island at Large Disturbance in Mainland

Figure 15 shows the electric power oscillation of the generators in Jeju Island. And the oscillation is damped by the effect of UPFC operation.

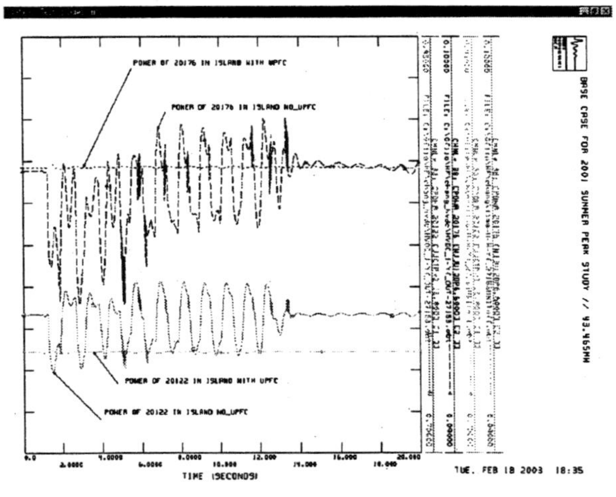

Figure 15. Power Oscillation Damping of Island Plant at Large Disturbance in Mainland

3) Small Disturbance

For the small disturbance, -2000[Mvar] reactor is connected instantly to 154kV bus nearby the rectifier ac bus of HVDC. This results the small oscillation of the voltage and power only in Mokpo area.

a) Voltage dynamics

Figure 16 shows that the oscillations at the AC terminal buses of HVDC(7440) and UPFC(7420) is well damped by UPFC operation.

The voltage oscillation of the terminal ac bus influences the dc voltage of HVDC as shown figure16.

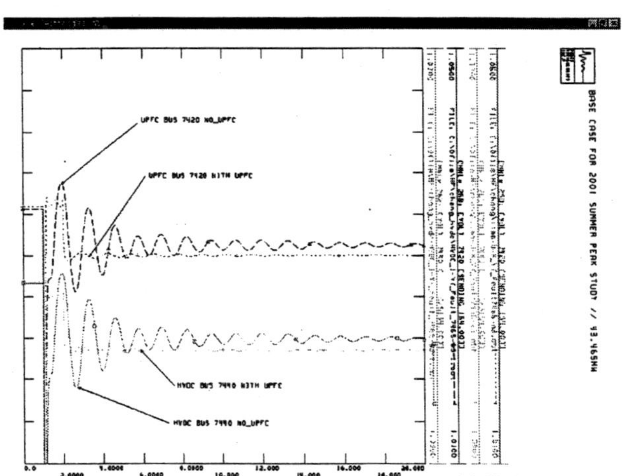

Figure 16. Effects of Voltage Control at the Small Disturbance in Mainland

b) power flow dynamics

Figure 17 shows that the small oscillation happens at the power flow toward HVDC. The oscillation is damped by UPFC operation, but the oscillation doesn't influence the dynamic actions of HVDC like Figure 18.

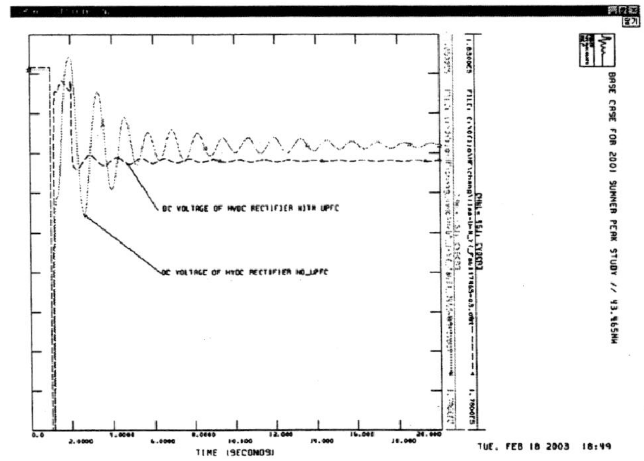

Figure 17. Effects of Power Flow Control at the Small Disturbance in Mainland

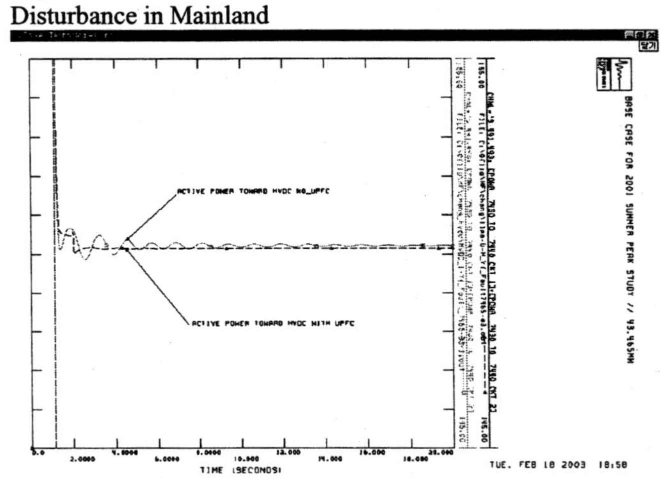

Figure 18. Effects of DC Power Flow Control of HVDC at the Small Disturbance in Mainland

B. Jeju-Do, island system

Three type disturbances are simulated in the Jeju-Do connected with mainland through the HVDC. The first type is the 3-phase fault of 154kV line which is one of two the line distributing the power from the inverter terminal of the HVDC to the load area in Jeju-Do. The second one is the #2 unit outage of Jeju Thermal Plant. Third one is the –600[Mvar] reactor on/off instead of the various disturbances like the line-ground fault.

The operation effects of UPFC with respect to the dynamic problems at the disturbances are analyzed when the dynamic problems is transferred from Jeju island to mainland through the HVDC.

1) Severe Line Fault

When the fault happens at 1.0[sec], HVDC acts the bypassing of the inverter since the terminal bus has the low voltage less than 108[dc V], bypassing reference voltage. The terminal inverter bus of HVDC reaches the unbypassing voltage 0.65[pu] at 5.1[sec], after the fault is cleared at 1.1[sec].

a) Voltage dynamics

Figure 19 shows the voltage dynamics at the dc current and AC & DC voltage of HVDC rectifier with/without UPFC. This shows the damping effects of the AC voltage oscillations at the rectifier bus of HVDC after DC voltage recovering to the nominal value by UPFC operation.

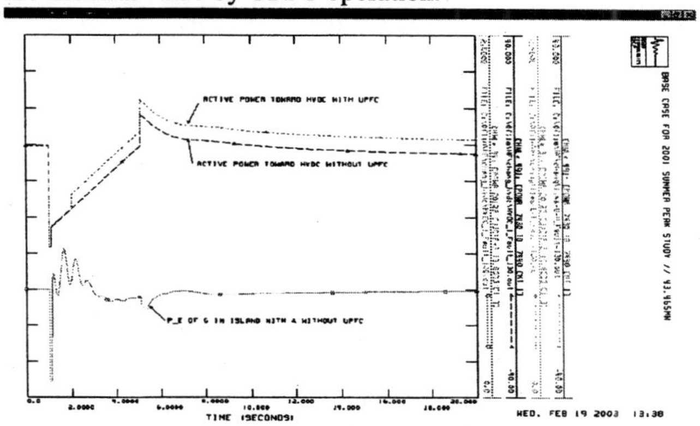

Figure 19. Effects of Voltage Control at the Severe Line Fault in Island

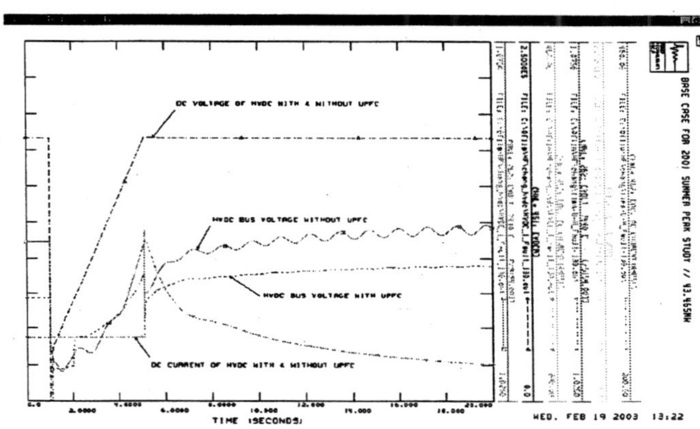

Figure 20. Effects of Power Flow Control at the Severe Line Fault in Island

b) power flow dynamics

Figure 20 shows the active power flows into the rectifier terminal bus of HVDC and shows the electric power of Jeju TP at the 154kV line 3-phase fault in Jeju island. This results the power oscillation in the Jeju TP, but there is no oscillation of active power flows into the rectifier terminal bus of HVDC because HVDC has the DC voltage recovering action while the Jeju TP has the power oscillation.

2) Large Disturbance

a) Dynamics of voltage and power flow

Figure 21 shows that the voltage at the AC terminal bus of HVDC(7440) and the power flow into the rectifier bus of HVDC. HVDC increase the power gently since HVDC has the modulation controller, SQBAUX model which responses slowly. So there is no dynamic oscillations.

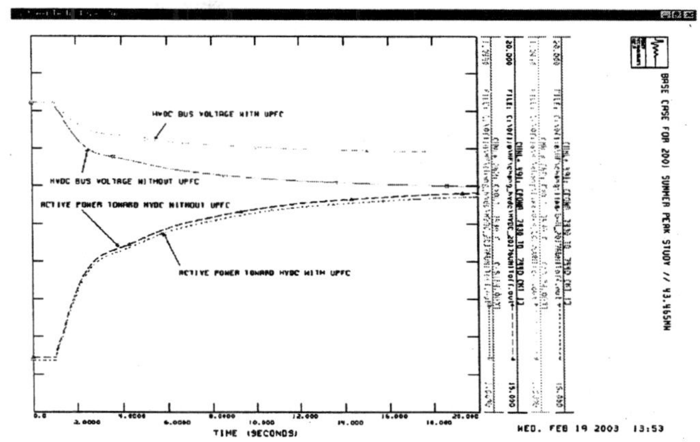

Figure 21. Effects of Voltage & Power Flow Control at the Large Disturbance in Island

3) Small Disturbance

a) Dynamics of voltage and power flow

Figure shows that the voltage at the AC terminal bus of HVDC(7440) and the power flow into the rectifier bus of HVDC. This results that the voltage at terminal bus is oscillating slightly because of the low voltage during the disturbance. But the oscillation is damped by UPFC operation. The oscillation of power flow doesn't happen in Jeollanam-Do.

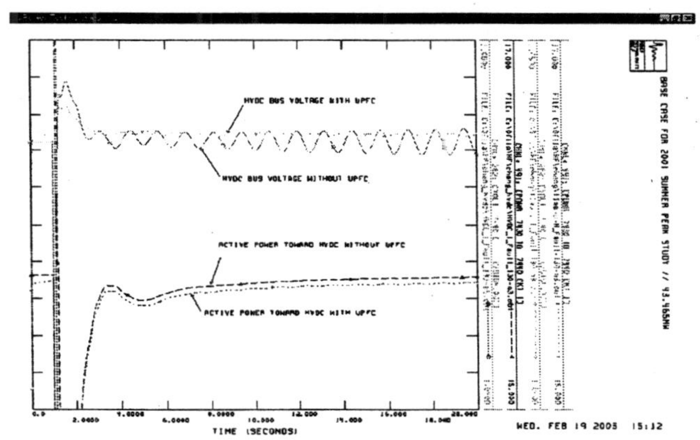

Figure 22. Effects of Voltage & Power Flow Control at the Small Disturbance in Island

IV. Control Strategy

A. Voltage Control

It is well known that shunt reactive power infection can be used be control bus voltage. The voltage compensation by 80MVA KEPCO UPFC influences the voltage dynamics nearby UPFC in Jeollanam-Do system of KOREA. Analyzing the results of the dynamic simulations to investigate the operation effects of UPFC to damp out the voltage oscillation that happens at the various disturbances as shown in table.

TABLE 2. Effects of Dynamic Voltage Control by UPFC Operation

voltage	Effects of UPFC operation

dynamics Types of Disturbance		Oscillation			Damping		
		Main	HVDC	Island	Main	HVDC	Island
Main	345kV Line Fault	O	X	X	O	-	-
	1000MVA Unit Off	O	O	O	O	O	O
	-2000MVar On/Off	O	X	X	O	-	-
Island	154kV Line Fault	O	O	O	O	O	O
	70MVA Unit Off	X	X	O	-	-	O
	-600 MVar On/Off	X	X	O	-	-	O

B. Power flow control

The real and reactive power flow in the line can be controlled independently using the series injected voltage. The series part of KEPCO UPFC has the two control modes. The one is the PQ control mode that maintains the constant active and reactive power desired by operator. The other is the V control mode which maintains the constant series injected voltage desired by operator. The V control mode is usually used, when the operator want the maximum series compensation of UPFC.

The series compensation by 80MVA KEPCO UPFC influences the power flow dynamics nearby KEPCO UPFC. The UPFC can damp out the power flow oscillation except the severe contingency, 345kV line fault in mainland as shown in table. It'd better change the control mode from PQ control mode to V control mode for the maximum series compensation, when the severe contingency happens.

TABLE 3. EFFECTS OF DYNAMIC POWER FLOW CONTROL BY UPFC OPERATION

Power dynamics Types of Disturbance		Effects of UPFC operation					
		Oscillation			Damping		
		Main	HVDC	Island	Main	HVDC	Island
Main	345kV Line Fault	O	X	X	X	-	-
	1000MVA Unit Off	O	O	O	O	O	O
	-2000MVar On/Off	O	X	X	O	-	-
Island	154kV Line Fault	X	X	O	-	-	-
	70MVA Unit Off	X	X	O	-	-	-
	-600 MVar On/Off	X	X	O	-	-	-

Figure 23 shows the simulated results of the series operation example of KEPCO UPFC. UPFC is operated on the PQ control mode until 1[sec] as the steady state. After the system has the large disturbance, YK NP 1unit outage at 1[sec], the dynamics of UPFC is simulated until 10[sec]. Finally the dynamics of UPFC is simulated until 20[sec] after the severe 345kV line fault happens at 10[sec] during 0.1[sec] and then fault clear at 10.1[sec].

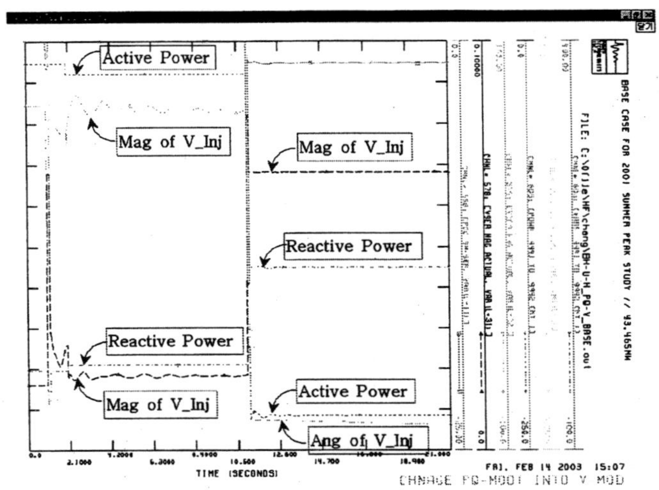

Figure 23. Dynamic Characteristics of UPFC though the changes of the operation mode

V. REFERENCES

[1] Jin-Boo Choo and Byung-Hoon Chang, "Development of FACTS Operation Technology (Phase 1 : Decision of a Type and a Spec. Draft for the Installation of a Pilot Plant)", *KEPRI Technical Report*, April, 1999.

[2] In-Sup, Kim, "long-term Forecast of KEPCO System Planning and Operation, in Korea", in *Proceedings of 1998 Power System Engineering Symposium*, KIEE, May, 1998.

[3] High Voltage Direct Current Handbook, EPRI TR?104166S, 1994.

[4] Schauder, C., et al, " AEP Unified Power Flow Controller Performance" , IEEE Transactions on Power Delivery, March, 1999.

[5] Gyugyi, et al. " The Unified Power Flow Controller: A New Approach to Power Transmission Control" , IEEE Transactions on Power Delivery, April, 1995.

[6] Gyugyi, L., Hingorani, N. " Understanding FACTS : Concepts and Technology of Flexible AC Transmission Systems" , IEEE press, 2000.

[7] Jin-Boo, Choo, et al, " Development of FACTS Operation Technology(Phase 1 : Decision of a Type and a Spec. Draft for the Installation of a Pilot Plant)" , KEPRI Technical Report, April, 1999.

[8] O. B. Nayak, A. M. Gole, D. G. Chapman, J. B. Davies, "Dynamic Performance of Static and Synchronous Compensators at HVDC Inverter Bus a very Weak AC

System", *IEEE Transactions on Power Systems*, Vol.9. No.3, pp.1350?1358, 1994.

[9] P. Kunder, *Power system stability an0d control*, Mcgraw?Hill, 1996.

VI. BIOGRAPHIES

Byunghoon Chang received his B.S. and M.S. from Korea University, Korea. He received his Ph. D. from Korea University in 1999. He have worked at Korea Electric Power Research Institute as a researcher for 9 years. He is presently a Senior Member of Technical Staff of Power System Laboratory at Korea Electric Power Research Institute. His research interests include FACTS, power system stability and power quality.

Jong Su Yoon received the B. S. and M. S. degrees in electrical engineering from Kyung Pook National University, Korea in 1993 and 1996 respectively. He is currently a senior member in the power system stabilization group at Korea Electric Power Research Institute. His research interests include FACTS and control theory application

Soo Yeol Kim received the B. S. and M. S. degrees in electrical engineering from the Chung-Ang University, Korea in 1990 and 1995 respectively. He worked in the field of generator excitation control for 6 years. He is currently an engineer in the power system stabilization group at Korea Electric Power Research Institute, Korea.. His research interests include FACTS and power quality.

Jin-Boo Choo (M'1994) received his B.S., M.S, Ph.D. degrees in Electrical Engineering from Seoul National University, Korea in 1977, 1987, and 1994 respectively. He worked at Korea Electric Power Corporation as a senior engineer in the Power System Operation Center. Currently he is a director and the group leader in the Power system stabilization group and Power system analysis group at Korea Electric Power Research Institute, and also Real time power system analysis team leader of National Research Lab. at Ministry of Science & Technology, Korea. His research interests include power system operation and control, FACTS, and Real time power system simulation.

Leonardo T.G. Lima received his B.Sc. degree in electrical engineering from the Federal University of Rio de Janeiro, Brazil, in 1986. He had worked in CEPEL and MAIN Engenharia S.A. an electirc consultant firm in Rio de Janeiro until he joined Power Technologies, Inc. (PTI), USA and he is presently an Executive Consultant.

James W. Feltes received his BSEE degree with honors from Iowa State University in 1979 and his MSEE degree from Union College in 1990. He joined PTI in 1979 and is currently an Associate Derector. He has participated in many studies involving planning, anlaysis and design of transmission and distribution systems. He is a Senior Member of the IEEE and is a registered professional engineer in the State of New York.

ELSEVIER
IFAC
PUBLICATIONS
www.elsevier.com/locate/ifac

FEEDER RECONFIGURATION USING CYCLIC BEST-FIRST SEARCH

Sang-Yule Choi[1)], Chang-Dea Yoon[2)], Myong-Chul Shin[3)]

[1)] *Dept. Of Digital Media Engineering, Anyang University, Anyang, 430 - 714, Korea*
[2)3)] *School of Information and Communication Engineering, Sungkyunkwan University, Suwon 440 - 746, Korea*

Abstract : Network reconfiguration in distribution systems is realized by changing the status of sectionalizing switches, and is usually done for loss reduction or load balancing in the system. This paper presents an effective heuristic based switching scheme to solve the distribution feeder loss reduction problem. The proposed algorithm consists of two parts. One is to set up a decision tree to represent the various switching operations available. Another is to apply a proposed technique called cyclic best first search. The proposed algorithm identifies the most effective the set of switch status configuration of distribution system for loss reduction. To demonstrate the validity of the proposed algorithm, numerical calculations are carried out the 32, 69 bus system models. *Copyright © 2003 IFAC*

Key words : Network Reconfiguration, Loss Reduction, Cyclic Best First Search

1. INTRODUCTION

Electric distribution networks maintain radial structure with normally closed sectionalizing switches along a feeder and normally open interfeeder tie switches for proper protection coordination. For every tie switch closed, another sectionalizing switch is opened. Under normal operating conditions, distribution feeders may be frequently reconfigured by changing the open/close state of each switch in order to reduce line losses or to avoid overloaded network branches. The resulting feeders must remain radial, without any violations of branches loading and voltage limits. Because of these requirements, the problem of finding network reconfiguration with minimum losses is a very complicated mixed-integer, non-linear optimization problem. Since there are a numerous number of switches in a practical distribution networks, a combinatorial analysis of switch options(2^m) could require prohibitively long computation time. Therefore, the problem appears to be best solved by heuristic search methods. Heuristic approaches do not guarantee optimal solutions, but they lead to sub-optimal solutions that are technically acceptable. Many heuristic algorithms dealing with feeder reconfiguration have been presented.

Cinvalar et al, 1988, proposed heuristics and computationally effective formulas for the loss change due to a switch exchange operation. Single loop optimization algorithm for determining the minimum loss configuration was presented J.Y. Fan et al 1996.. Baran et al, 1989, developed approximate power flow method for loss reduction and load balancing based upon considering branch exchange type switch operations. And Heuristic based switching indices were designed by W. M. Lin,1999, where it utilizes fuzzy notations for the loss reduction of distribution systems. Taylor et al. 1990, proposed a switch exchange type heuristic method to determine the network configuration for overloads, voltage problem, and for load balancing simultaneously. Its solution scheme sets up a decision tree which represents the various switching operations available, and a best first tree searching and heuristic rules are used to find feasible switching operations. Wu et al 1991, extended the method Taylor et al by developing explicit exhaustive method that solves the problem of overloads, phase current unbalance, service-restoration, and maintenance. This method taken is to set up a feasible switching options tree which represents possible switching options under constraint of radial structure. Evaluation functions and heuristic rules are used to find feasible switching options.

In this paper, the authors present a heuristic feeder reconfiguration algorithm based on an effective exhaustive search method. Its main steps have been implemented in two stages. First stage is to set up a sub-tree that was presented by Wu et al.1991. Second stage is to find feasible switching operations with a proposed new search technique called cyclic best-first search. This procedure favors solutions with a fewer number of switching operations where the switching sequences may be acceptable to the operators. Numerical calculations are carried out to show the effectiveness of the proposed algorithm.

2. FORMULATION OF A LOSS MINIMUM PROBLEM

In normal distribution operations, the object of the feeder reconfiguration algorithm is to minimize the total i^2r losses in the system by deciding the position of open tie switches under the current capacity, voltage drop limit and radial structure constraints. The loss-minimum problem can be formulated by W.M Lin 2000 as

Min $P_{loss}(S_V)$

With $P_{loss} = \sum_{i=1}^{N} R_i * |I_i|^2$

- Current capacity constraint

$I_i \le I_i^{max}$ i = 1 ~N

- Voltage drop constraint

$V_i^{min} \le V_i \le V_i^{max}$ i = 1 ~N

- Radial structure constraint

S_{close} = B-1

where

P_{loss} : total line losses of distribution feeders

S_V : the status vector of switches.

R_i : resistance of feeder section i

N : total numbers of feeder sections in the whole system.

I_i : Current magnitude in feeder section i

I_i^{max} : upper limit of feeder section current magnitude

V_i : Voltage magnitude of busbar i

V_i^{max} : upper limit of busbar voltage magnitude

V_i^{min} : lower limit of busbar voltage magnitude

S_{close} : total number of closed switches

B : total number of busbars

3. Solution algorithm

A feasible switching options tree was suggested in [6] in order to decrease searching space, and is called sub-tree in this paper. To find reasonable switching options effectively, the authors developed a cyclic best-first tree searching strategy. The whole solution procedure as follows:

3.1 Constructing the sub-tree

Sub-tree is a decision tree to represent the various switching options available under constraint of radial structure. Constructing the sub-tree methodology is the same that in the paper by Wu et al 1991 in that a tree is adopted in order to decrease searching space.

Under the constraint of the radial structure in the load transfer process, closing a normally open tie switch should follow the opening of a complementary normally closed sectionalizing switch. Therefore, If n tie switches are closed, then n sectionalizing switches has to be opened.

Fig. 1 shows a sample distribution network[3] consisting of three feeders with three normally opened tie switches and thirteen normally closed sectionalizing switches.

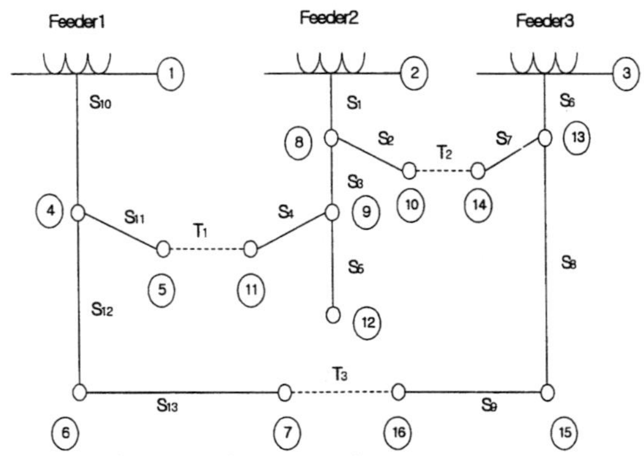

Fig 1. Three-feeder example system

If feeder 2 experiencing an overload, then the amount of overload on feeder 2 must be transferred to feeder 1 and/or 3 without creating an overload on either of these feeders. To transfer load at node 11 from feeder 2 to feeder 1, the notation (T_1, S_4) is used to denote the operation of closing switch T_1 and opening switch S_4, henceforth.

Feasible (close,open) switching options can be found by searching sectionalizing switches through the overloaded feeder.: When each tie switch of the overloaded feeder is closed, a complementary sectionalizing switch to be opened is found by searching from the tie switch, and moving upstream along the overloaded feeder to its source, the circuit breaker of the overloaded feeder.

Fig. 2 shows a searching path for finding feasible switching options when feeder 2 is overloaded.

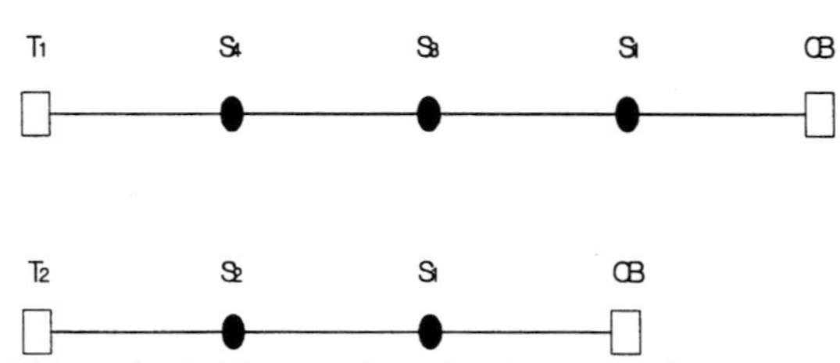

Figure 2. Main search paths for example system

If the amount of overload on feeder 2 be transferred to only feeder 1, then T_1 and either S_4, S_3 or S_1 constitute a switching pair. So feasible switching options are expressed as {(T_1,S_4), (T_1,S_3), (T_1,S_1)}. And one of switching options would be a solution for reliving the overload.

Similarly, the amount of overload on feeder 2 may be transferred to feeder 1 and 3 simultaneously by choosing one of following feasible switching options $\{(T_1,S_4), (T_2,S_2)\}$, $\{(T_1,S_4), (T_2,S_1)\}$, $\{(T_1,S_3), (T_2,S_2)\}$, $\{(T_1,S_3), (T_2,S_1)\}$, $\{(T_1,S_3), (T_2,S_2)\}$. But when T_1 and T_2 are used simultaneously, the switching option $\{(T_1,S_1), (T_2,S_1)\}$ is not a feasible one. Because it violates radial structure constraint.
If the results of these feasible options are examined, then the corresponding sub-tree of fig. 2 is obtained. In fig.2, both T_1 and T_2 are tie switches of the overload feeder 2 and dotted line represents switching option.pairs.

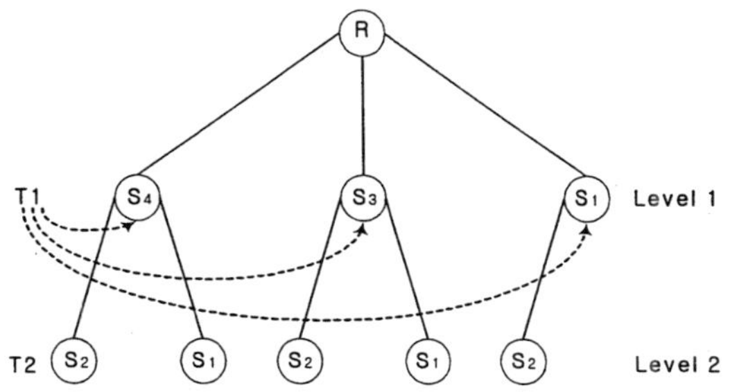

Fig. 2 Sub tree with two backup feeders

3.2 Cyclic best-first search

An exhaustive search will evaluate all feasible switching options of the above sub-tree and it guarantees optimal solution. But this is probably not realizable for large sub-tree because of heavy computation. On the other hand, by using heuristic search, time and effort can be saved by finding reasonable solution quickly. There are usually three heuristic search ways to find optimal(or near-optimal) switching pairs on the above sub-tree : depth-first, breadth-first, best-first.

The advantage of the best-first search usually, but not always, yields solution faster than any other heuristic search. But the problem is that it does not always give the optimal solution: unexplored path would have given an optimal solution. In this paper, the new methodology (so called cyclic best-first search) is presented. This methodology is based on best-first search. But, by using cyclic methodology, it can usually find more accurate solution than best-first search technique. In best-first search, node at each level in the above sub-tree is chosen for switching option with the most potential to lead to a loss minimization. Then that node is expended into its children nodes. Similarly, one of its children nodes is chosen for switching option with minimum loss. This procedure is continued until it finds a goal node. Consider fig. 3, where node ① is the start node and node ⑫ is a goal node. Node ① is expanded into its children node ②,③,④,⑤ Since the losses of node ② is less than other nodes, node ② is chosen for expansion. This is continued until a goal node has been found.

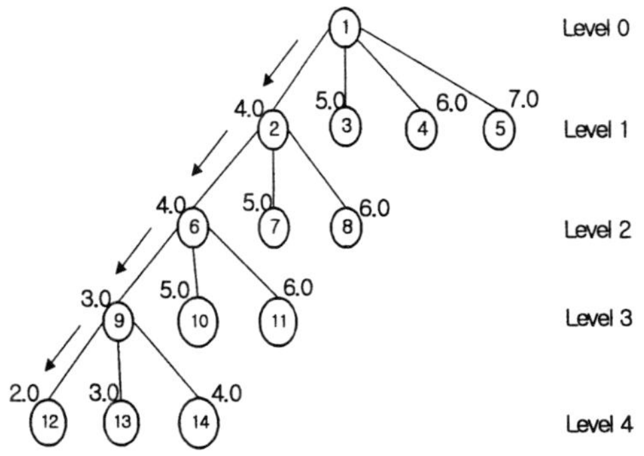

Figure 3. First Step of best-first search

In the end, nodes {①,②,⑥,⑨,⑫} are found by best-first search. And these nodes may be a solution switching pair set to minimize line losses. By always expanding the most likely node, it is possible to get to a goal node or a solution quickly. But this procedure achieves the trade-off between optimality and computational speed. In cyclic best-first search, it gives circulatory function for best-first search to find more proper switching pair set in the sub-tree. With circulatory reevaluating the unexplored nodes and path, an effective solution with minimum switching operations would be found. Although the search space of cyclic best-first search is slightly larger than that of best-first search, the computation difference is negligible. The cyclic best-first search process is as follows:
First step) A path from a start node to a goal node is selected by using best-first search.
In fig 3, a selected path is ①→ ②→ ⑥→ ⑨→ ⑫, a start node is ① and a goal node is ⑫. However, this path is usually (not always) not an optimal solution: Selected node ⑥ seems to be an optimal switching pair in level-2 on condition that nodes in lower level are not selected. But it is hard to say that node ⑥ is a optimal switching pair if nodes in lower level are selected by expansion Therefore, different(or more accurate) path and nodes may be found under different lower level conditions.
Second step) Constructing the reversed sub-tree. And a new path is selected by using best-first search.
Reversed sub-tree can be constructed by reversing the level of sub-tree that was used in first step: The level-4 of the sub-tree in first step becomes the level-0 of the reversed sub-tree, and the level-3 of the sub-tree in first step becomes the level-1 of the reversed sub-tree. Similarly, the levels of sub-tree in first step can define the rest levels of the reversed sub-tree. Therefore, a goal node in first step becomes the start node of the reversed sub-tree in second step. Fig. 4 shows second step of best-first search to find near-optimal path in a reversed sub-tree.

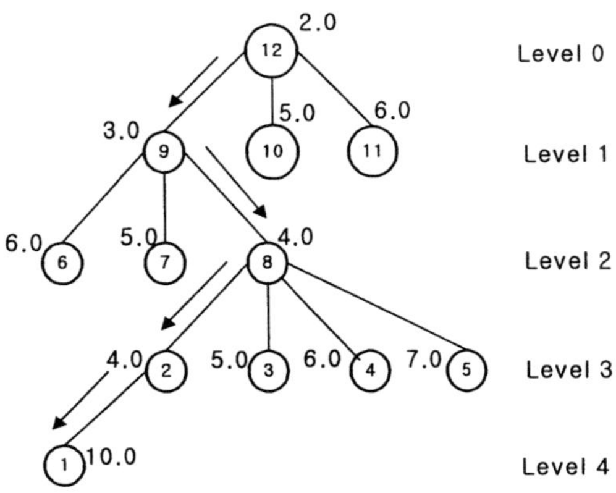

Figure 4. Second step of best-first search

In second step of best-first search, when nodes in each level are evaluated, it is assumed that nodes in the other lower levels are already chosen by first step of best-first search. therefore, node ⑨ is chosen assuming that nodes ⑥,②,① are already determined from first step. Similarly, Node ⑧ in level-2 is selected assuming that nodes ②,① are already determined from first step. This gradient measurement can determine nodes that appear to be closest to an optimal solution. Due to using near-optimal path from first step, new path from second step is more near-optimal solution than that from first step, After second step of best first search, a new path ⑫→ ⑨→ ⑧→ ②→ ① is selected. Comparing a new selected path with a previously selected path , a selected node ⑥ is changed into a node ⑧ from second step because of the other levels condition : In first step, the node ⑥ in level-2 is selected on condition that node ⑨ in level-3 and node ⑫ in level-4 are not selected by expansion. In second step, node ⑧ is choosing instead of node ⑥ on condition that node ⑫ in level-0 and node ⑨ in level-1 are already selected by expansion before evaluating nodes in level-2, and it is assumed that nodes ②,① are already determined from first step. Because two paths are different, more improved solution may be found if the way of second step is repeated to the next step.

Third step) Repeat the way of second step, until a newly selected path and a previously selected path are compared as the same. In third step, reversed sub-tree can be constructed by reversing the level of sub-tree that was used in second step. And the third step of best-first search is shown in fig.5.

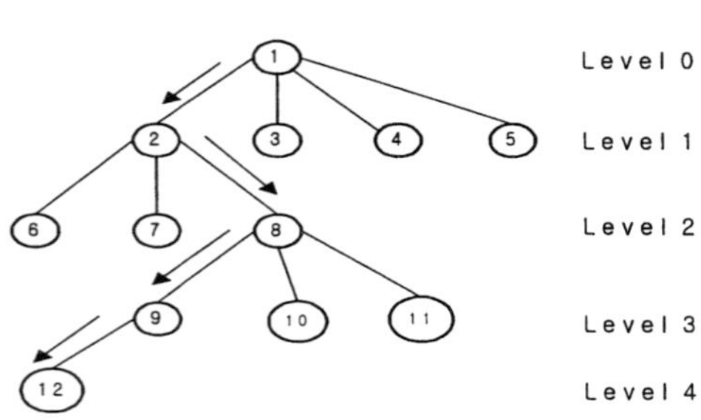

Fig 5. Third step of best-first search

New path is ①→②→⑧→⑨→⑫. This path and the path chosen in second step are compared as the same. Therefore this path is appears to be a proper solution. Fourth step) Terminate the iterative procedure, and nodes of newly selected path are converted to the complete switching pairs.

3.3 The Overall Solution Process.

The overall solution process for the proposed feeder reconfiguration algorithm is shown in fig 6.

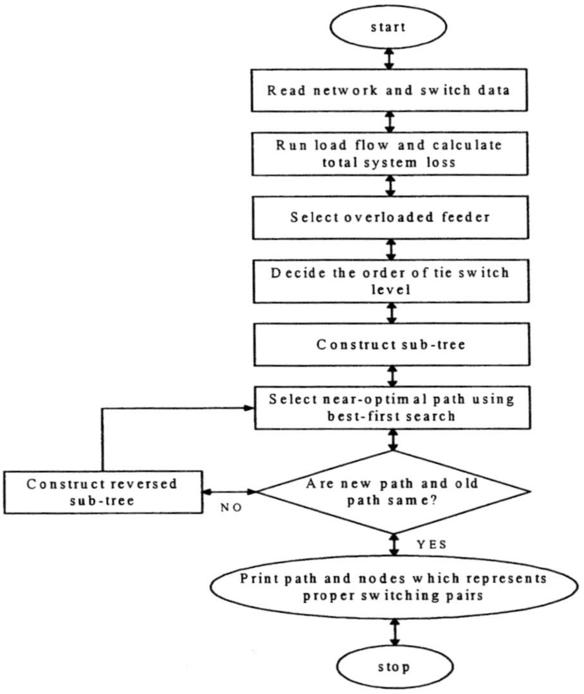

Fig. 6 Flowchart of the solution for the loss minimization algorithm

In Fig.6, when constructing the initial sub tree, the order of sub tree level is defined by tie switch having maximum voltage across (expecting that the largest voltage difference will cause maximum loss reduction when it selected).

5. Test Result

The distribution network for reconfiguration presented in M.E Baran, 1898 is used to demonstrate the validity and effectiveness of the proposed algorithm. The network consisting of two feeders with 32 busbars and 5 tie switches {T_{33}, T_{34},T_{35}, T_{36}, T_{37}}, as shown in Fig.7. The total load are 5084.26[kw] , 2457.32[kvar], and initial losses is 0.0203 p.u Table 1 shows initial feeder loadings and voltage difference across normally closed tie switches.

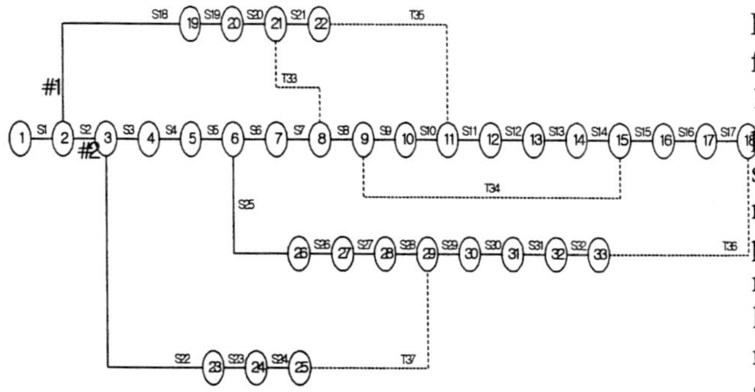

Fig. 7. Sample of distribution system with 32 bus

Table 1. Initial sample system data

Initial feeder loadings [KVA]		Voltage difference across the normally open tie switch [pu]	
		Tie switch number	Voltage difference across the tie switch
Feeder 1	Feeder 2	33	0.0401
		34	0.0193
		35	0.0653
417.4	514.9	36	0.0018
		37	0.0506

In table 1, feeder loadings of feeder 1 are 417.4[KVA] and those of feeder 2 are 514.9[KVA]. 1. The transferring loads from feeder 2 to relatively less heavy loaded feeder 1 may reduce the overall system losses because it relieves the level of loads on the feeder 2 and improves voltage profile along feeder 2. Therefore, each tie switch in the feeder 2 is used to construct initial sub-tree. The order of tie switch level for initial sub-tree may be selected depending upon voltage across: tie switch having maximum voltage across may be first level of sub-tree and the rest level are arranged by the same manor.

First step of best-first search for initial sub-tree is shown in fig.8. Initial sub-tree level is defined by T35, T37, T33, T34, T36 sequentially due to the different voltage across.

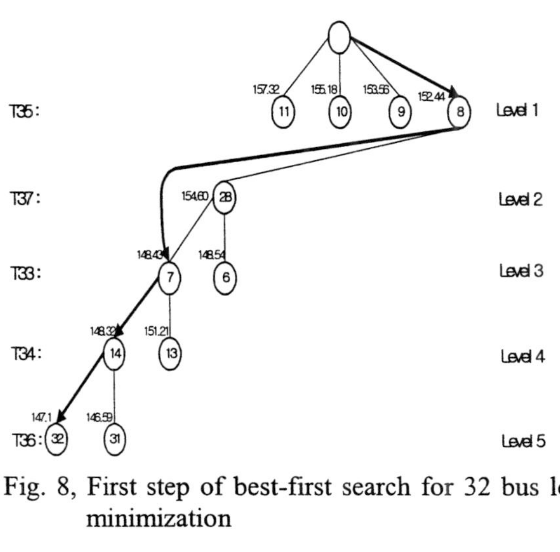

Fig. 8, First step of best-first search for 32 bus loss minimization

In Figure 8, an selected (close,open)switching pair for level 1 is (T35,S8) and system losses are 152.44[kw] for the switching operation. In the process of checking nodes of each level, if checked switching pair would increases system losses then the rest of unchecked nodes are ignored and searching proceeds to next level's nodes. And first checked node (T37, 28) in level 2 would increase system losses, therefore By pruning of the most unlikely nodes, this procedure make it possible to get a solution much faster even if it deep down in the tree. After first step of best-first search, selected (close, open) switching operations are (T35,8), (T33,7), (T34,14),(T36,32). This solution seems feasible but it is only locally optimal, because the first step of best-first search dose not examines all the possible nodes. It is possible that unexplored path would have presented more feasible solution. Thus, to find more feasible solution, reversed sub-tree is constructed by reversing the level of sub-tree that was used in first step and second step of best-first search is executed in figure 9.

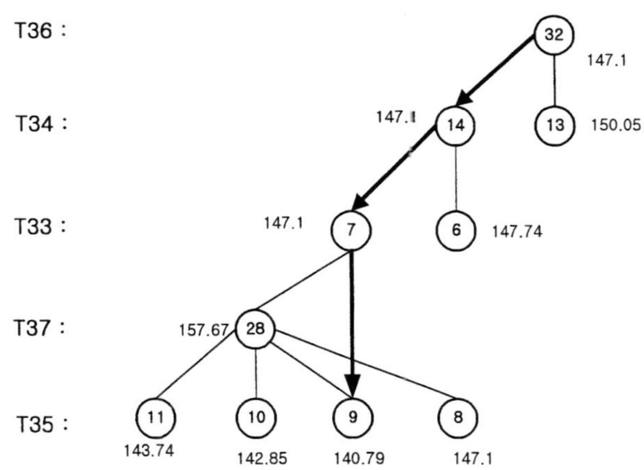

Fig. 9. Second step of best-first search for 32 bus loss minimization

In figure 9, first checked node (T37, 28) in level 4 would increase system losses, therefore the rest of unchecked nodes in level 4 are ignored and searching is proceeded to level 5. After second step of best first search, switch pairs (T35,9), (T33,7), (T34,14), (T36,32) are selected to minimize system losses. Comparing new switch pairs with those of first step of best-first search, switch pair (T35,8) is changed into (T35,9). The reason why newly selected switch pairs are different from those of previously selected is that selected switch pair (T35,8) is operated before evaluating switching pairs of lower levels in first step of best-first search. Therefore, (T35, 8) may not be an optimal switch pair if the rest of selected switch pairs are operated. Since newly selected switch pairs are different from those from first step of best-first search, solution wicth is obtained from second step of best-first search do not guarantee most feasible solution to minimize system losses. Third step of best-first search is proceeded in order to find more improved solution in figure 10.

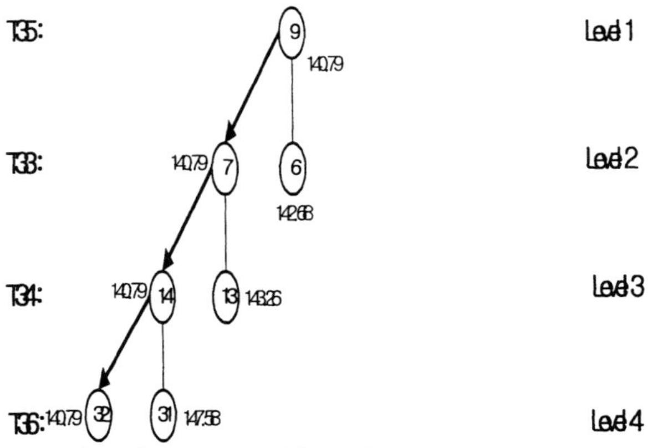

Fig. 10. Third step of best-first search for 32 bus loss minimization

As seen above figure [8, 9], close/open switch options using T37 are always increase system losses. At third step of best-first search, tie switch T37 is excluded from reversed sub tree to avoid unnecessary search. Selected switch pairs (T36,32), (T34,14), (T33,7), (T35,9) are the same that those of second step. It is indicate that the solution obtained from third step would be the complete solution in the sense that the further search would be the same After selected switching operations, the percentage of total loss reduction is 31[%]

4. Effectiveness comparison

. The three methods proposed by Goswami1992, the three methods proposed by Baran and Wu 1989and the method by W-M.Lin 2000 had used the same test system as shown in figure 7. Therefore, The effectiveness of proposed method was compared with these methods in table 2

Table 2 Summary of test results

Mesethod	Switching times		Initial open switches	Last Open Switches	Total loss reductions [%]
	M1	7			31.148
Goswami	M2	7		7, 9, 14, 32, 37	31.148
Et al	M3	4			31.148
	M1	5	33, 34, 35, 36, 37	11, 28, 31 ,33, 34	27.830
Baran					
Et al	M2	3		6, 11, 31 ,34, 37	23.826
	M3	3			23.826
W-M Lin Et al		5		7, 9, 14, 32, 37	31.148
Proposed method		5		7, 9, 14, 32, 37	31.148

Proposed method needs fewer switching times than Gowami[M1,M2] and it is more effective than Baran for loss reduction . Goswami [M1]and W-M Lin methods need more complicate numerical computation than proposed method.

5 Conclusion

A new heuristic algorithm has been presented in this paper for loss minimization of distribution networks. The proposed algorithm adopts the concept of sub-tree proposed by J.S Wu, 1991 and utilizes cyclic best-first search method developed by the authors. Cyclic best-first search is derived from best-first search, which gets a solution much faster even if it lies deep down in the tree. And it compensates best-first search for not obtaining the best solution every time by using revered sub tree. For a feeder reconfiguration algorithm, it is sufficient to know the relative change in the line loss due to close/open switching operations. And absolute accuracy is not essential. Therefore, proposed algorithm is suitable for such applications even though it may not guarantees the optimal solution.

REFERENCES

S. Cinavanlar, J. J. Grainger, H. Yin, S. S. H. Lee, "Distribution feeder reconfiguration for losses reduction", IEEE Trans. on Power Delivery, Vol. PWRD-3, 1988, pp.1217-1223, July 1988.

J-Y, Fan, L. Zhang, J.D. McDonald, "Distribution Network Reconfiguration Single Loop Optimization", IEEE Trans, Power Systems, Vol. 10, No. 3, pp. 1643-1647, August 1996.

M. E, Baran, F. F. Wu, "Network reconfiguration in distribution systems for loss reduction and load balancing", IEEE Trans. on Power Delivery, Vol. PWRD-4, 1989, pp. 1401-1407, April 1989.

W. M. Lin et al "An effective algorithm for distribution feeder loss reduction by switching operations", IEEE Trans on Power Delivery, pp. 597 - 602, 1999.

T. Taylor, D. Lubkeman, "Implementation of heuristic search strategies for distribution feeder reconfiguration", IEEE Trans. on Power Delivery, Vol. 5, No. 1, pp. 239 - 246. January 1990.

J. S. Wu. K. L. Tomsovic, C. S. Chen, "A heuristic search approach to feeder switching operations for overload, fault, unbalanced flow and maintenance", IEEE Trans. on Power Delivery, Vol. 6, No. 4, pp.1579 - 1585, October 1991.

S. K. Goswami, S. K. Basu, "A new algorithm for the reconfiguration of distribution feeders for loss minimization", IEEE Trans. on Power Delivery, Vol. 7, No. 3, pp.1484-1491, July 1992.

D. Shirmohammandi, H.W. Hong, "Reconfiguration of electrical distribution networks for resistive line losses reduction", IEEE Trans. On Power Delivery, 1989, Vol. 4, NO.2, April, 1989.

W-M.Lin, et al " Distribution feeder reconfiguration with refined genetic lgorithm" IEE Proc-Gener. Transm.Distrib,Vol. 147, No.6, November,2000

ELSEVIER
IFAC
PUBLICATIONS
www.elsevier.com/locate/ifac

AN APPLICATION OF GAIN SCHEDULING THEORY TO MULTI-MACHINE STABILIZATION CONTROL

Shinya Yamamoto Hiroshi Ohshita Toshiya Ohtaka Shinichi Iwamoto

Dept. of Electrical Engineering and Bioscience, Waseda University, Tokyo, Japan

Abstract: An application of gain scheduling theory to multi machine stabilization control is proposed in this paper. And, a design method of a generator control system based on the gain scheduling theory for multi-machine power systems is proposed. Subsequently a decentralized control is carried out which considers interaction between the generators directly, that is to say, we regard the non-diagonal blocks of the whole system matrix as a disturbance matrix. Moreover, we confirm that the multi input generator control system is effective in using the control system based on the H infinity control theory. *Copyright © 2003 IFAC*

Keywords: Multi-Machine Power System, LMI, Gain Scheduling Theory, H-infinity control

1. INTRODUCTION

PSS designs, which have the robustness for two or more operating points, have been investigated in recent year[1].

The conventional PSS (Power System Stabilizer) is generally designed for improving the damping by linearizing specify at a certain operating point. When the state of the system configuration and power flow in the power system change, neither the performance nor stability are theoretically guaranteed.

In this paper, a generator control system, which can cope with two or more operating points and has robustness in each operating point, is designed by using the gain scheduling theory based on LMI. This theory is applied to dynamic systems which are linearized and approximated by an affine parameter dependent model, and can adjust the controller to present operating conditions when the value of affine parameters can be measured in real time. For the system with a wide range of operating points, a high control performance is expected by the gain scheduling theory.

There are two fluctuation modes in a large-scale system. One is the local fluctuation which occurs in a system, and the other is the long cycle fluctuation which occurs between the systems. In recent years, for these two fluctuations, the former is controlled by a ΔP type PSS and the latter is controlled by a $\Delta P+\Delta\omega$ type PSS.

We design a ΔP type and a $\Delta P+\Delta\omega$ type generator control system by using the H infinity control theory based on LMI, and confirm the effectiveness of a $\Delta P+\Delta\omega$ typegenerator control system.

The gain scheduling theory and the H infinity control theory can be applied to multi-machine power systems by using a decentralized control[2][3]. We carry out a decentralized control which considers interaction between the generators directly, namely, we regard the non-diagonal blocks of the whole system matrix as a disturbance matrix[4].

In this paper, the scheduling parameters are generator constants $K_{1,ii}$-$K_{6,ii}$ because $K_{1,ii}$-$K_{6,ii}$ are determined by measurements at the present operating point. In addition, the effectiveness of the designed generator control system is confirmed and verified by running nonlinear simulations of a three-phase to ground fault for system condition changes.

*Shinya Yamamoto is presently with JR Tokai, Japan

2. GAIN SCHEDULING THEORY[5]

2.1 Gain Scheduling Theory

Gain scheduling is one of the practical control techniques to non-linear systems in which the dynamic characteristic changes along with the movement of the operating point. In short, one controller that guarantees the performance to the entire operating area is not designed. Multi controllers for operating points are designed. When the operating condition changes, the controller is switched. In the system with a wide range of operating points like a generator, a high control performance is expected by the gain scheduling using such a controller for the entire operating points.

2.2 Gain Scheduling Theory via LMI

The target of the control system design based on the gain scheduling theory is an LPV(Linear Parameter Varying) system shown below.

$$\begin{aligned} \dot{x} &= A_{cl}(\theta)x + B_{cl}(\theta)w \\ z &= C_{cl}(\theta)x + D_{cl}(\theta)w \end{aligned} \tag{1}$$

$A_{cl}(\theta), B_{cl}(\theta), C_{cl}(\theta)$ and $D_{cl}(\theta)$ depend affinely on θ. Moreover, the time varying parameter θ varies in a polytope $\Theta := \{\omega_1, \omega_2, \cdots, \omega_r\}$. As stated above, it is clear that the state-space matrices $A_{cl}(\theta), B_{cl}(\theta), C_{cl}(\theta)$ and $D_{cl}(\theta)$ vary in a polytope of matrices whose vertices are $\omega_1, \omega_2, \cdots, \omega_r$. In other words,

$$\begin{bmatrix} A_{cl}(\theta) & B_{cl}(\theta) \\ C_{cl}(\theta) & D_{cl}(\theta) \end{bmatrix} \in \begin{bmatrix} A_{cl}(\omega_i) & B_{cl}(\omega_i) \\ C_{cl}(\omega_i) & D_{cl}(\omega_i) \end{bmatrix}, i = 1, \cdots, r \tag{2}$$

Because of this property, this LPV system is called as a polytopic LPV system.

2.3 Quadratic H Iinfinity Performance

The LPV system (1) has a quadratic H infinity performance γ if there exists a single matrix $X_{cl}>0$ such that

$$\begin{bmatrix} A_{cl}^T(\theta)X_{cl} + X_{cl}A_{cl}(\theta) & X_{cl}B_{cl}(\theta) & C_{cl}^T(\theta) \\ B_{cl}^T(\theta)X_{cl} & -\gamma I & D_{cl}^T(\theta) \\ C_{cl}(\theta) & D_{cl}(\theta) & -\gamma I \end{bmatrix} < 0 \tag{3}$$

for all admissible values of the parameter vector θ. Then, the Lyapunov function $V(x) = x^T X_{cl} x$ establishes stability of the system, and the L_2 gain from the input to output is bounded by γ along all possible parameter trajectories θ.

A difficulty found with condition (3) is the infinite number of constraints it imposes. In the case of the polytopic LPV system, however, this condition can be reduced to a finite set of LMIs. Using convexity, it is easily shown that (3) will hold for all ($A_{cl}(\theta), B_{cl}(\theta), C_{cl}(\theta), D_{cl}(\theta)$) if it holds at the vertices $\omega_1, \omega_2, \cdots, \omega_r$.

Theorem 2.1

「Consider a polytopic LPV system described by (1) and (2). Then, the following statements are equivalent.

(i) This LPV system is stable with quadratic H infinity performance γ.

(ii) There exists a single matrix $X_{cl}>0$ such as (3) for all admissible values of the parameter vector θ.

(iii) There exists a single matrix $X_{cl}>0$ satisfying a set of LMIs as follows.

$$\begin{bmatrix} A_{cl}^T(\omega_i)X_{cl} + X_{cl}A_{cl}(\omega_i) & X_{cl}B_{cl}(\omega_i) & C_{cl}^T(\omega_i) \\ B_{cl}^T(\omega_i)X_{cl} & -\gamma I & D_{cl}^T(\omega_i) \\ C_{cl}(\omega_i) & D_{cl}(\omega_i) & -\gamma I \end{bmatrix} < 0 \quad i = 1, \cdots, r \tag{4}$$

2.4 Target Plant of Gain Scheduling and Scheduled Controller

The gain scheduling theory via LMI can be applied to the affine parameter depending system described by (5).

$$\begin{aligned} \dot{x} &= A(\theta)x + B_1(\theta)w + B_2u \\ z &= C_1(\theta)x + D_{11}(\theta)w + D_{12}u \\ y &= C_2x + D_{21}w + D_{22}u \end{aligned} \tag{5}$$

$$\theta(t) = (\theta_1(t), \cdots, \theta_n(t)) \quad \underline{\theta_i} \le \theta_i(t) \le \overline{\theta_i}$$

$A(\theta), B_1(\theta), C_1(\theta)$ and $D_{11}(\theta)$ are the affine functions of the parameter vector $\theta(t)$. For example, $A(\theta)$ is stated as follows.

$$A(\theta) = A_0 + \theta_1 A_1 + \cdots + \theta_n A_n \tag{6}$$

The assumption on the system (5) is as follows.

[1] $D_{22} = 0$

[2] B_2, C_2, D_{12}, D_{21} are parameter-independent.

[3]The pairs $(A(\theta), B_2)$ and $(A(\theta), C_2)$ are quadratically stabilizable and quadratically detectable for $\theta(t)$, respectively.

Assuming complete measurements of the parameter θ, the controller incorporating these measurements can be designed. The closed loop is shown in Fig.1 and the state-space matrices are formed as in (2).

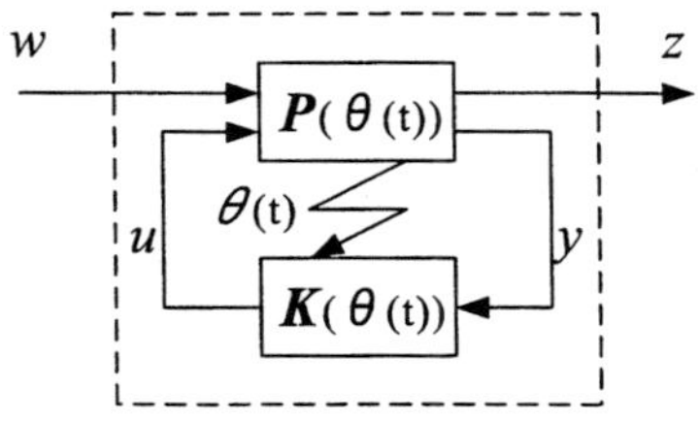

Fig.1. Closed loop system

$$\dot{x}_{cl} = A_{cl}(\theta)x_{cl} + B_{cl}(\theta)w$$
$$z = C_{cl}(\theta)x_{cl} + D_{cl}(\theta)w \tag{7}$$

We seek a LPV controller of the form

$$\dot{x}_k = A_k(\theta)x_k + B_k(\theta)y$$
$$u = C_k(\theta)x_k + D_k(\theta)y \tag{8}$$

that guarantees some quadratic H infinity performance γ for the closed loop system of Fig.1. Then the controller matrix is described as (9).

$$\Omega(\theta) = \begin{pmatrix} A_k(\theta) & B_k(\theta) \\ C_k(\theta) & D_k(\theta) \end{pmatrix} \tag{9}$$

This controller K ensures that

[1] The closed-loop system is quadratically stable over all admissible parameter trajectories $\theta(t)$.

[2] The L_2 gain from w to z in the worst case is bounded by γ for all admissible parameter trajectories $\theta(t)$.

Concretely, the gain scheduling is done in the following steps.

【Step1】 Solve a set of LMIs (4) and compute a symmetric matrix $X_{cl}>0$.

【Step2】 Compute the LTI vertex controller Ω_i at the vertices $\omega_i (i = 1,\cdots,r)$ from X_{cl}.

【Step3】 Define the LPV controller $\Omega(\theta)$ as an interpolation of the vertex controller Ω_i. Here, the interpolation coefficient α_i is based on the position of θ in the polytope Θ. That is,

$$\Theta = \left\{ \sum_{i=1}^{r} \alpha_i \omega_i : \alpha_i \geq 0, \sum_{i=1}^{r} \alpha_i = 1 \right\} \tag{10}$$

The LPV controller $\Omega(\theta)$ is given by the vertex controller Ω_i and the interpolation coefficient α_i as follows.

$$\begin{pmatrix} A_k(\theta) & B_k(\theta) \\ C_k(\theta) & D_k(\theta) \end{pmatrix} = \sum_{i=1}^{r} \alpha_i \Omega_i = \sum_{i=1}^{r} \alpha_i \begin{pmatrix} A_k(\omega_i) & B_k(\omega_i) \\ C_k(\omega_i) & D_k(\omega_i) \end{pmatrix} \tag{11}$$

【Step1】 and 【Step2】 are carried out off-line in advance. 【Step3】 is carried out whenever operating conditions vary.

3. Decentralized Control for Multi Machine Power Systems

In this paper, we regard a multi-machine power system as an aggregate of subsystems. Moreover a decentralized control, in which a controller of each individual subsystem is designed by considering other subsystem's influences, is considered. The disturbance matrix which represents mutual interference is incorporated into the plant. The ith subsystem of the multi-machine power system is stated as follows.

$$\dot{x}_i = A_{ii}x_i + B_{ii}u_i + \sum_{i \neq j} A_{ij}x_j \tag{12}$$

Here, the third term $\sum A_{ij}x_j$ is the disturbance matrix. We regard the non-diagonal blocks of the whole system matrix as the disturbance matrix, and consequently, we can consider interactions between the generators directly[4]. The state-space matrices of the ith subsystem is formed as follows.

$$\begin{bmatrix} \Delta\dot{\delta}_i \\ \Delta\dot{\omega}_i \\ \Delta\dot{E}'_{qi} \\ \Delta\dot{E}_{fdi} \end{bmatrix} = \begin{bmatrix} 0 & \omega_0 & 0 & 0 \\ -\frac{K_{1,ii}}{M_i} & -\frac{D_i}{M_i} & -\frac{K_{2,ii}}{M_i} & 0 \\ \frac{K_{4,ii}}{\tau'_{doi}} & 0 & -\frac{C_{3,ii}}{\tau'_{doi}} & \frac{1}{\tau'_{doi}} \\ -\frac{K_{Ai}}{T_{Ai}}K_{5,ii} & 0 & -\frac{K_{Ai}}{T_{Ai}}K_{6,ii} & -\frac{1}{T_{Ai}} \end{bmatrix} \begin{bmatrix} \Delta\delta_i \\ \Delta\omega_i \\ \Delta E'_{qi} \\ \Delta E_{fdi} \end{bmatrix}$$

$$+ \begin{bmatrix} 0 \\ 0 \\ 0 \\ \frac{K_{Ai}}{T_{Ai}} \end{bmatrix} u_i + \begin{bmatrix} 0 & 0 & 0 & 0 \\ \sum_{j\neq i}\frac{K_{1,ij}}{M_i} & 0 & \sum_{j\neq i}\frac{K_{2,ij}}{M_i} & 0 \\ -\sum_{j\neq i}\frac{K_{4,ij}}{\tau'_{doi}} & 0 & \sum_{j\neq i}\frac{C_{3,ij}}{\tau'_{doi}} & 0 \\ \sum_{j\neq i}\frac{K_{Ai}}{T_{Ai}}K_{5,ij} & 0 & -\sum_{j\neq i}\frac{K_{Ai}}{T_{Ai}}K_{6,ij} & 0 \end{bmatrix} \begin{bmatrix} \Delta\delta_{ave\_i} \\ \Delta\omega_{ave\_i} \\ \Delta E'_{qave\_i} \\ \Delta E_{fdave\_i} \end{bmatrix} \tag{13}$$

Equation (13) corresponds to (12). The multi-machine power system have N subsystems.

$$\Delta\delta_{ave\_i} = \frac{\sum_{j\neq i}\Delta\delta_j}{N-1}, \quad \Delta\omega_{ave\_i} = \frac{\sum_{j\neq i}\Delta\omega_j}{N-1}$$
$$\Delta E'_{qave\_i} = \frac{\sum_{j\neq i}\Delta E'_{qj}}{N-1}, \quad \Delta E_{fdave\_i} = \frac{\sum_{j\neq i}\Delta E_{fdj}}{N-1} \tag{14}$$

From (13), mutual interferences are represented by $\Delta\delta_j$, $\Delta E'_{qj}$ and generator constant K_{ij}. The change of the terminal voltage due to the fault is considered as the disturbance w_i, and therefore, the target subsystem of the controller design is formed as the following equation.

$$\begin{bmatrix} \Delta\dot{\delta}_i \\ \Delta\dot{\omega}_i \\ \Delta\dot{E}'_{qi} \\ \Delta\dot{E}_{fdi} \end{bmatrix} = \begin{bmatrix} 0 & \omega_0 & 0 & 0 \\ -\frac{K_{1,ii}}{M_i} & -\frac{D_i}{M_i} & -\frac{K_{2,ii}}{M_i} & 0 \\ \frac{K_{4,ii}}{\tau'_{doi}} & 0 & -\frac{C_{3,ii}}{\tau'_{doi}} & \frac{1}{\tau'_{doi}} \\ -\frac{K_{Ai}}{T_{Ai}}K_{5,ii} & 0 & -\frac{K_{Ai}}{T_{Ai}}K_{6,ii} & -\frac{1}{T_{Ai}} \end{bmatrix} \begin{bmatrix} \Delta\delta_i \\ \Delta\omega_i \\ \Delta E'_{qi} \\ \Delta E_{fdi} \end{bmatrix}$$

$$+\begin{bmatrix} 0 \\ 0 \\ 0 \\ \frac{K_{Ai}}{T_{Ai}} \end{bmatrix} u_i + \begin{bmatrix} 0 & 0 & 0 \\ \sum_{j\neq i}\frac{K_{1,ij}}{M_i} & 0 & \sum_{j\neq i}\frac{K_{2,ij}}{M_i} \\ -\sum_{j\neq i}\frac{K_{4,ij}}{\tau'_{doi}} & 0 & \sum_{j\neq i}\frac{C_{3,ij}}{\tau'_{doi}} \\ \sum_{j\neq i}\frac{K_{Ai}}{T_{Ai}}K_{5,ij} & \frac{K_{Ai}}{T_{Ai}} & -\sum_{j\neq i}\frac{K_{Ai}}{T_{Ai}}K_{6,ij} \end{bmatrix} \begin{bmatrix} \Delta\delta_{ave\_i} \\ w_i \\ \Delta E'_{qave\_i} \end{bmatrix} \quad (15)$$

4. DESIGNING GUIDE

4.1 Multi Input PSS

In this paper, a ΔP type and a $\Delta P+\Delta\omega$ type control systems are designed, and the validity of the $\Delta P+\Delta\omega$ type control system is considered.

There are two techniques of controlling fluctuation modes using PSS. One is a ΔP type control system controlling the local fluctuation mode (a cycle 0.8sec ~1.2sec). The other is a $\Delta P+\Delta\omega$ type-control system controlling the long cycle fluctuation mode (a cycle 2.0sec~4.0sec). In recent years, because flat and stable damping torque is obtained in the frequency band from the local fluctuation mode to the long cycle fluctuation mode, the multi input PSS ($\Delta P+\Delta\omega$ type) is put into practice.

4.2 Generator Control System

The ΔP type control system model of each generator is shown in Fig.2, and the $\Delta P+\Delta\omega$ type is shown in Fig.4.

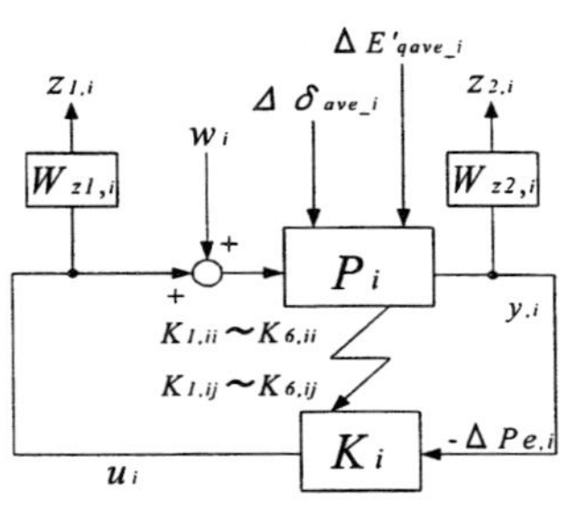

Fig.2 The control system model i (ΔP type)

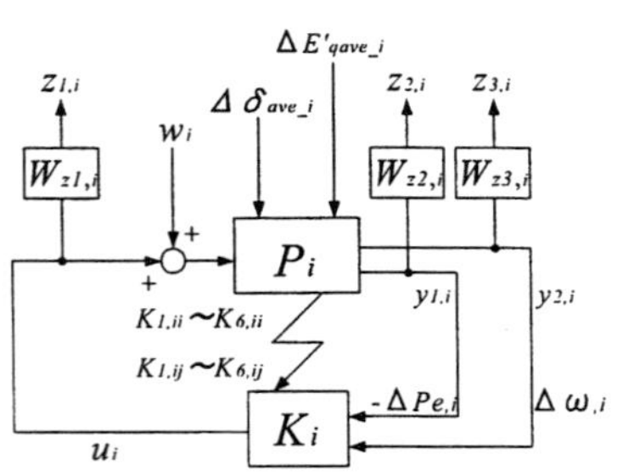

Fig.3. The control system model i ($\Delta P+\Delta\omega$ type)

We regard the change of the terminal voltage due to the fault as the disturbance w_i. The robust stability is designed from w_i to $z_{1,i}$, and the sensitivity decrease is designed from w_i to $z_{2,i}$ and from w_i to $z_{3,i}$($Wz_{1,i}$, $Wz_{2,i}$ and $Wz_{3,i}$ are weight functions). The input signal to the ΔP type control system is assumed to be $-\Delta Pe$, and the one to the $\Delta P+\Delta\omega$ type is assumed to be $-\Delta Pe$ and $\Delta\omega$. Since the generator constants $K_{1,ii}$~$K_{6,ii}$ and $K_{1,ij}$~$K_{6,ij}$ are derived from measurable variables at the present operating point in real time, the gain scheduling is done smoothly.

4.3 Control System Design Procedure

Because we want to consider stability in the entire multi-machine power system, the controller is designed in the following steps in this paper.

[Step1]
A generator control system is designed so that a closed-loop quadratic H infinity performance (the maximum value of the H infinity norm to the entire amplitude ranges of $K_{1,ii}$-$K_{6,ii}$) from w_i to z_i may become less than one for each generator.

[Step2]
The designed generator control system is built in, and it is checked to see if the H infinity norm in the entire multi-machine power system is less than one.

[Step3]
If Step 2 is satisfied, the design is completed. If not, return to Step 1, and the controller is designed again after the weight functions are changed.

Here, the weighting function $Wz_{2,i}$ concerning sensitivity decrease is set by checking the singular value plots from the disturbance w_i to $z_{2,i}$. The H infinity norm from the disturbance w_i to $z_{2,i}$ is bounded by one, which is described by Equation (16).

$$\left\| W_{z2,i}(I - P_i K_i)^{-1} P_i \right\|_\infty < 1 \quad (16)$$

from (16)

$$\left| \frac{P(j\omega)}{1 - P(j\omega)K(j\omega)} \right| < \left| W_{z2}(j\omega)^{-1} \right|, \quad \forall\omega \quad (17)$$

From (17), by the inverse function $W_{z2,I}^{-1}$, the singular value plots from the disturbance w_i to $z_{2,i}$ are suppressed. In the case of the $\Delta P+\Delta\omega$ type control system, the weighting function $Wz_{3,i}$ concerning sensitivity decrease is set in the same way as $Wz_{2,i}$.

5. SIMULATION

5.1 System Model for Design

A three-machine nine-bus system model shown in Fig.5 is used as a model for the proposed control design.

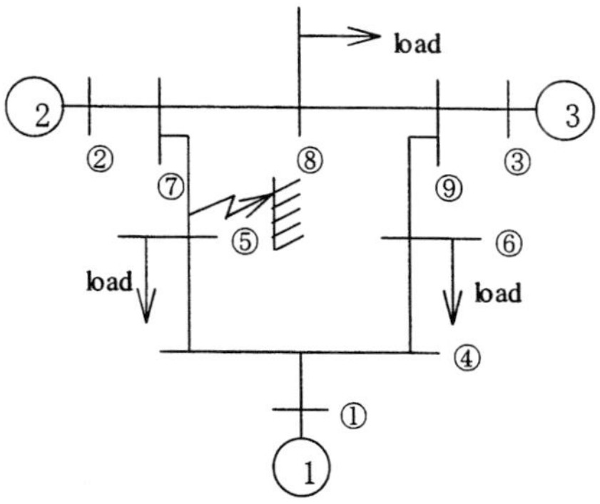

Fig.4 Three-machine nine-bus system[6]

Generator 2 and 3 are control targets, and designed controllers are installed to these two generators. In this model, we consider the change of operating conditions, in which the generator power output increases with load increase.

In this paper, in order to confirm the validity of the gain scheduling control, we compare the proposed method with the *H* infinity control.

The weight functions are described by Equation (18), and shown in Fig.6.

$$Wz_{1,i} = \frac{100s+1}{s+10000} \quad Wz_{2,i} = \frac{5.4s+2944}{s+90} \tag{18}$$

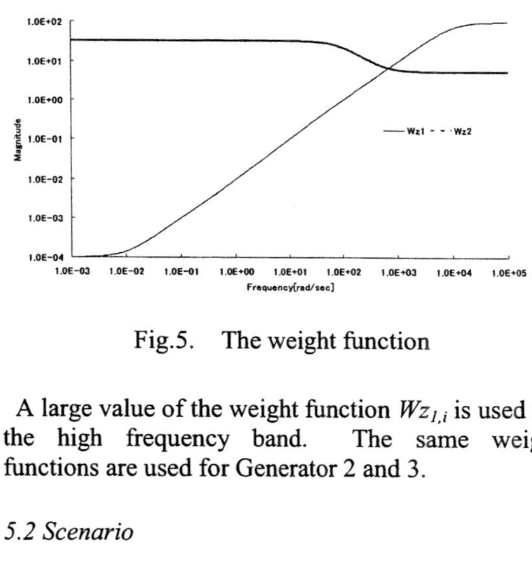

Fig.5. The weight function

A large value of the weight function $Wz_{1,i}$ is used for the high frequency band. The same weight functions are used for Generator 2 and 3.

5.2 Scenario

It is assumed that a three-phase to ground fault occurs near the bus 5 shown in Fig.4. The system has double circuit lines. The fault case is as follows : a fault occurs at 0.01 seconds, and one line of the double circuit is opened 0.07 seconds after the fault and re-closesd 0.6 seconds later.

6.3 Generator Control System via H Infinity Control

Simulation results of the ΔP type and the $\Delta P+\Delta\omega$ type control systems are shown in Fig.6,Fig.7. The simulations are done when the generator outputs are steady state and increased by 20%. Since the *H* infinity control has some degree of robustness, the controllers designed for steady state are applied when the generator outputs are increased by 20%.

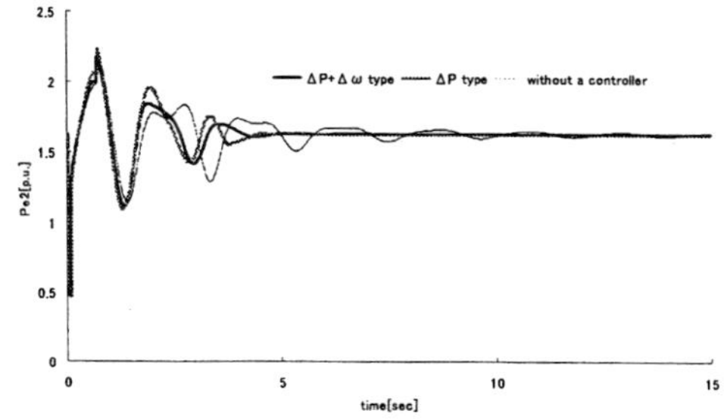

Fig.6. The active power for steady state (Gen2)

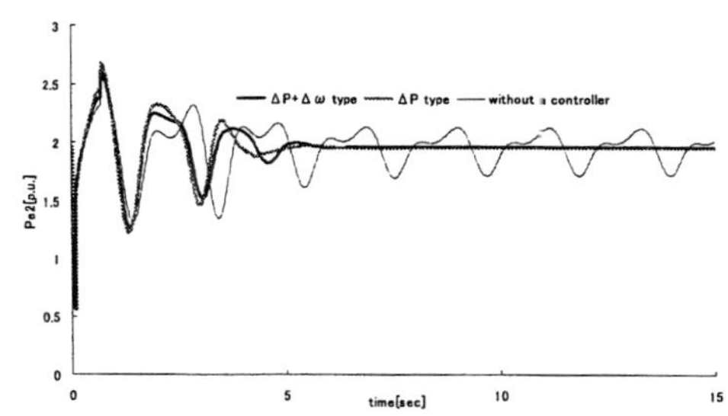

Fig.7. The active power when the generator outputs are increased by 20% (Gen2)

From Fig.6, Fig.7, we confirm that the *H* infinity control has some degree of robustness. When the generator outputs are steady state and increased by 20%, the fault response performance of the $\Delta P+\Delta\omega$ type control system is better than that of the ΔP type control system. The difference between the two is remarkably shown at 5 -15 sec when the fluctuation cycle is comparatively long. The $\Delta P+\Delta\omega$ type control system controls the local fluctuation mode and the long cycle fluctuation mode effectively.

5.4 Generator Control System via Gain Scheduling Control

Simulation results of the ΔP type *H* infinity control system and ΔP type gain scheduling control system are shown in Fig.8 to Fig.9. The simulations are done when the generator outputs are in a steady state and increased by 45%. Since the *H* infinity control has some degree of robustness, the controllers designed for a steady state are applied when the generator outputs are increased by 45%.

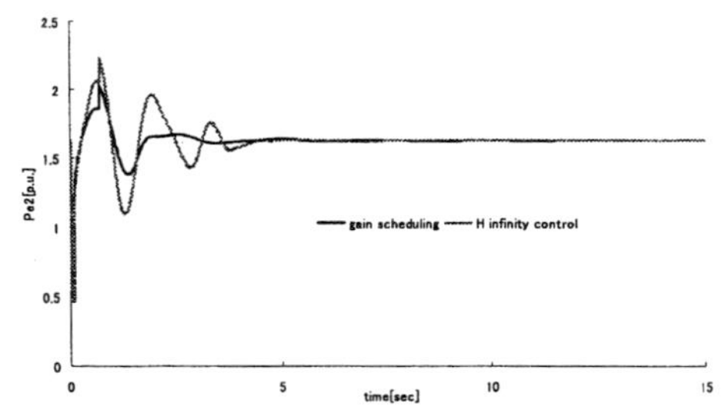

Fig.8. The active power for a steady state (Gen2)

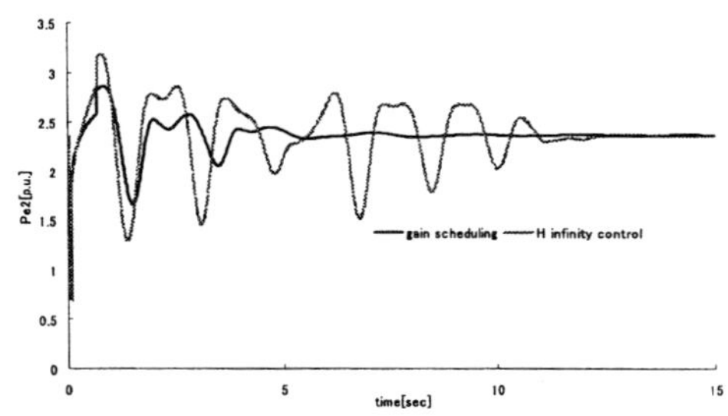

Fig.9. The active power when the generator outputs are increased by 45% (Gen2)

From Fig.8 to Fig.9, the H infinity control system designed for a steady state cannot suppress the fluctuation by the fault when the generator outputs are increased by 45%. Though the H infinity control has some degree of robustness, it cannot completely control that operating condition. In the case of the gain scheduling control, an excellent response can be confirmed when the generator outputs are in a steady state and increased by 45%. The generator control system is adaptable for the present operating state because of the gain scheduling. Here, to confirm this characteristics, we introduce the index J_i of generator i described by Equation (19).

$$J_i = \sum_{time=0\,sec}^{15\,sec} |\Delta Pe_i| \tag{19}$$

As an example, the change of index J_2(generator2) is shown in Fig.10.

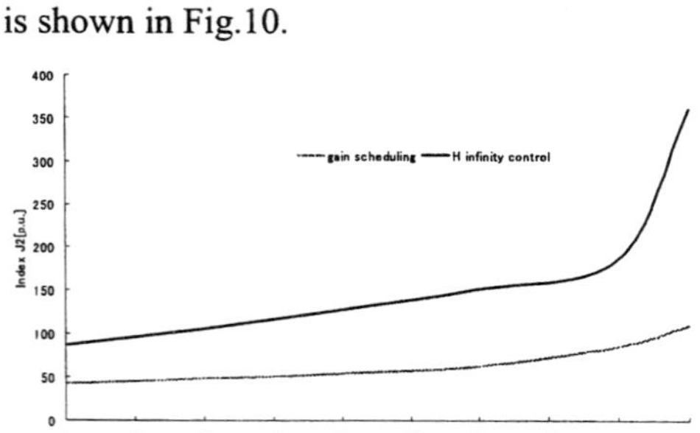

Fig.10. The change of index J_2 (Gen2)

From Fig.10, though the operating condition changes, the index J_2 of the gain scheduling control system changes little compared with that of the H infinity control system. We can confirm that the gain scheduling control system is adaptable for each operating state.

6. Conclusions

In this paper, the gain scheduling theory and the H infinity control theory based on LMI have been applied to a multi-machine power system, and generator control systems were designed using the proposed method. A decentralized control has been carried out which considers interactions between the generators directly, namely, we have regarded the non-diagonal blocks of the whole system matrix as a disturbance matrix.

In the case of the H infinity control system, it has been confirmed that the $\Delta P+\Delta\omega$ type control system showed an excellent response performance compared with the ΔP type control system when the generator outputs were in a steady state and increased by 20%. The difference between the two was remarkably shown at 5-15 sec when the fluctuation cycle was comparatively long. It has been confirmed that the $\Delta P+\Delta\omega$ type control system controlled the local fluctuation mode and the long cycle fluctuation mode effectively. Moreover, it has been confirmed that the H infinity control had some degree of robustness.

In the case of the gain scheduling control system, it has been confirmed that the gain scheduling control system has shown an excellent response at the operating condition where the H infinity control system could not cover. In addition, we introduced the index J_i of generator i, and showed that the index J_i of the gain scheduling control system changed little compared with that of the H infinity control system under changing the operating condition. We have confirmed that the gain scheduling control system was adaptable for each operating state.

REFERENCE

[1] S.Yamamoto, T.Ohtaka, S.Iwamoto:「A Design of Generator Control Systems Based on Gain Scheduling forMulti-Machine Power Systems」(in Japanese), IEEJ, PE-01-100, PSE-01-94,2001

[2]Y.Uriu, M.Ohashi, F.Koyanagi:「Stabilization for Multi-Machine Power System by Decentralized Generator Control Strategy Using H_∞ Control Theory」(in Japanese), Trans. of IEEJ, Vol.116-BNo.1, p7-13, 1996

[3]M.Makino, H.Ukai, S.Cui, M.Kobayashi, H.Kandoh「Stabilization of Multi-Machine Power Systems by Using Decentralized H_∞ Control」(in Japanese), Trans. of IEEJ, Vol.118-B No.1, p23-30, 1998

[4]S.Niioka, R.Yokoyama, G.Fujita, G.Shirai:「Decentralized Exciter Stabilizing Control for Multi-Machine Power Systems」(in Japanese), Trans. of IEEJ, Vol.120-B No.6, p808-814, 2000

[5] P.Apkarian, P.Gahinet, G.Becker:「Self-scheduled H_∞ Control of Linear Parameter-varying Systems: a Design Example」, Automatica, Vol.31, No.9, p1251-1261, 1995

[6] Y.Sekine:「Power System Transient Stability Analysis」(book in Japanese), Ohmusha, 1984

ELSEVIER
IFAC
PUBLICATIONS
www.elsevier.com/locate/ifac

VOLTAGE STABILITY CONTROL BASED ON VIPI INCLUDING A NAS BATTERY SYSTEM

Toshinori Esaka Arisa Takehara Toshiya Ohtaka Shinichi Iwamoto

Dept. of Electrical Engineering and Bioscience, Waseda University, Tokyo, Japan

Abstract: In recent years, the majority of voltage control schemes are local controls on each bus. However there are a few on-line voltage stability control schemes that consider the whole system. Hence, we propose on-line voltage stability control schemes in addition to conventional local controls (Tanaka, *et al.*, 1998; Lobato, *et al.*, 2000; Feng, *et al.*, 1999; Wang, *et al.*, 1998; Tran-Quoc, *et al.*, 2000). In this paper, we regard an NAS battery system as control equipment and we propose a voltage stability preventive control method based on the VIPI using the NAS battery system. Simulations have been run to verify the effectiveness of the proposed method. The results showed that the proposed methods coped well with severe faults. *Copyright © 2003 IFAC*

Keywords; NAS battery system, Voltage stability control, VIPI, Preventive control,

1. INTRODUCTION

Recently, as power system loads are located further away from power plants and are more unevenly distributed, reactive power losses tend to increase due to the heavier power flow with high load growth and long distance transmission. The power demand differences, between the daytime and the night-time or between summer and winter, worsen the load factor. Furthermore, the Power Products and Suppliers (PPS) have been participating in the power generation market since the deregulation of the electric power industry in Japan. There is a possibility that their generators might affect voltage stability. Therefore, concerns about voltage instability phenomena such as a great voltage drop and a voltage collapse are being raised. Thus, an effective voltage control scheme is becoming more indispensable (Ilic, *et al.*, 1995).

A control index is needed when the preventive control and the emergency preventive control are implemented. Most of the proposed indices for the evaluation of voltage stability vary peculiarly near the critical point or scarcely vary for heavy loaded conditions. Compared with them, the voltage stability index VIPI has some considerable characteristics. The value decreases linearly with an increase of loads and becomes zero near the critical point. It is not a local bus index but an entire system index. The VIPI makes it easier to grasp a margin to the critical point than other indices (Hashimoto, *et al.*, 1988). Therefore, we have adapted VIPI as a voltage stability index.

Some energy storage systems have been proposed, such as energy storage batteries, Superconducting Magnetic Energy Storage devices (SMES devices), flywheel devices and compressed air storage devices. Above all, from a technological point of view, the NAS battery system is superior to the others and most practical at present. They can control active and reactive power outputs promptly and individually. In this point, the NAS battery system is more flexible and practical than the other devices. It has been proposed that the NAS battery system should be applied to voltage reactive power control in power systems (Ohtaka and Iwamoto, 2000).

It is adequate to convert the MW margin standard in the whole system to set the reference value of VIPI. If we find the fault for which the value of VIPI becomes below the reference while on-line contingency analysis is being carried out, we initiate the preventive control against the severest case. We also carry out Optimal Power Flow (OPF) calculations. However, if the entire equipment is already used or the operation amount reaches its limits or the OPF has no solution, we

*Arisa Takehara is presently with CRIEPE.Japan

consider load curtailment as an emergency preventive control in a conventional method. It has become more indispensable to investigate load shedding schemes as corrective control (Begonic, *et al.*, 1995; Moors, *et al.*, 2000). In this paper, we can take the NAS battery system and carry out the preventive control again before we consider load curtailment. When we perform load curtailment, we can scarcely adjust the ratio of active power output to reactive power output. As a result, we can consider more effective voltage stability control.

Simulations have been run using the 11 bus system to verify the effectiveness of the proposed methods. The results showed that the proposed methods when compared with the conventional methods, coped with more severe faults owing to the NAS battery system.

2. VOLTAGE STABILITY INDEX VIPI

Power flow equation has several solutions. The number of these solutions decreases as a power system becomes heavy loaded. It is known that two closely located solutions exist near the critical point. As they approach the critical point, these two solutions become closer, and at the critical point they merge. Focusing on this point, VIPI has been proposed (Tamura *et al.*, 1988).

VIPI is a scalar index for evaluating the voltage stability margin by using the angle between the critical vector and the specified value vector.

In rectangular coordinates, the power flow equation is represented as follows:

$$\boldsymbol{y}_s = \boldsymbol{y}(\boldsymbol{x}) \quad \text{.............................(1)}$$

where

$\boldsymbol{y}_s$: specified value

$$\boldsymbol{x} = (e_1, f_1, e_2, f_2, \cdots, e_n, f_n)^T$$

e_i : real part of bus voltage

f_i : imaginary part of bus voltage

Here, we assume that two kinds of voltage vectors $\boldsymbol{x}$ (operational solution) and $\boldsymbol{x}^*$ (ficticious solution) satisfy the same specified value vector. Using critical vector $\boldsymbol{a}$ and deflection vector $\boldsymbol{b}$, $\boldsymbol{x}$ and $\boldsymbol{x}^*$ can be expressed as follows:

$$\boldsymbol{x} = \boldsymbol{a} + \boldsymbol{b}, \quad \boldsymbol{x}^* = \boldsymbol{a} - \boldsymbol{b} \quad \text{..................(2)}$$

From these, $\boldsymbol{a}$ and $\boldsymbol{b}$ are expressed as follows:

$$\boldsymbol{a} = \frac{\boldsymbol{x} + \boldsymbol{x}^*}{2}, \quad \boldsymbol{b} = \frac{\boldsymbol{x} - \boldsymbol{x}^*}{2} \quad \text{...............(3)}$$

Because VIPI is the angle between the specified value vector $\boldsymbol{y}_s$ and the critical vector $\boldsymbol{y}(\boldsymbol{a})$, VIPI is defined as follows (see Fig.1).

$$\text{VIPI} = \theta = \cos^{-1} \frac{\boldsymbol{y}_s^T \boldsymbol{y}(\boldsymbol{a})}{\|\boldsymbol{y}_s\| \cdot \|\boldsymbol{y}(\boldsymbol{a})\|} \quad \text{...........(4)}$$

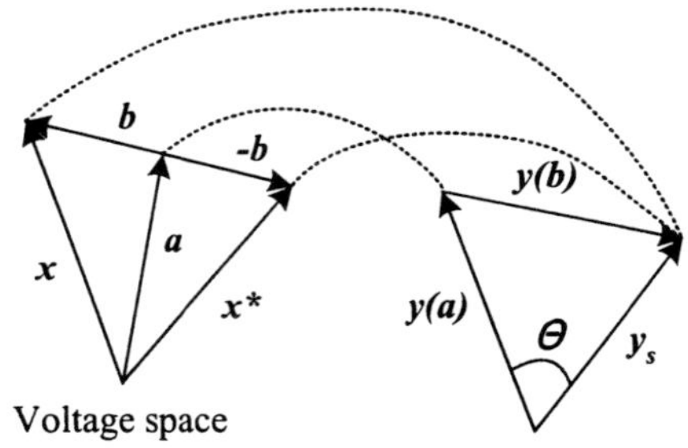

Fig.1 Concept of VIPI

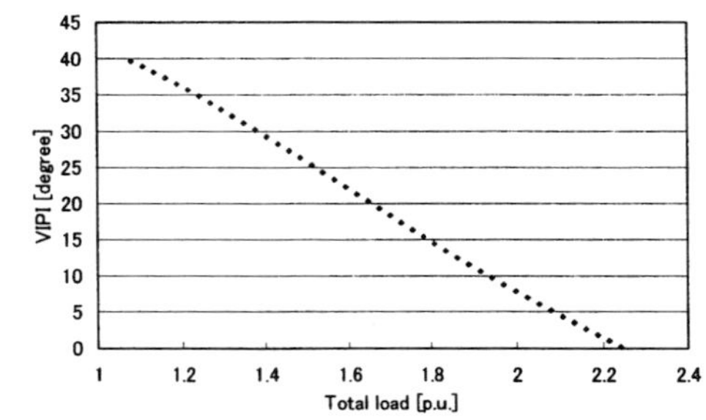

Fig.2 Relationship between VIPI and total load

VIPI has the following characteristics.

- Fig.2 shows that the value decreases linearly with an increase of loads and becomes zero near the critical point.
- It is not a local bus index but an entire system index.

Oue to these characteristics, the VIPI makes it easy to grasp the margin to the critical point and the VIPI is suited for on-line system monitoring and control near the critical point. Therefore, we have adapted VIPI as a voltage stability index.

3. NAS BATTERY SYSTEM

An NAS battery is made of fused sodium as a cathode-active material, fused sulfur and sodium polysulfide as an anode-active material, and utilizes beta alumina, which conducts sodium ions selectively as an electrolyte (Okuno,1996; Kodama and Tanaka,1999; Standing Special Committee of Chemical and Electrical Energy Transformation, 1980; The Institute of Applied Energy, 1992).

NAS battery systems have the following features.

- They can control active and reactive power outputs promptly and individually.
- They are capable of discharging the power approximately from two to five times as much as the rated capacity if the capacity of an AC-DC converter device is sufficient to do so.
- They can be placed compactly in such a small space as distribution systems and power demand areas. The size of the energy density is approximately three times that of a lead battery.
- They have high efficiency for both charging and discharging but do not have self-discharging

characteristics, i.e. Therefore, they can store electricity effectively.

- They can withstand more than 2,250 times of charging or discharging. They have a notable lifetime with 15-year durability.
- A clean battery having a completely closed structure ensures safety and no maintenance for operation.

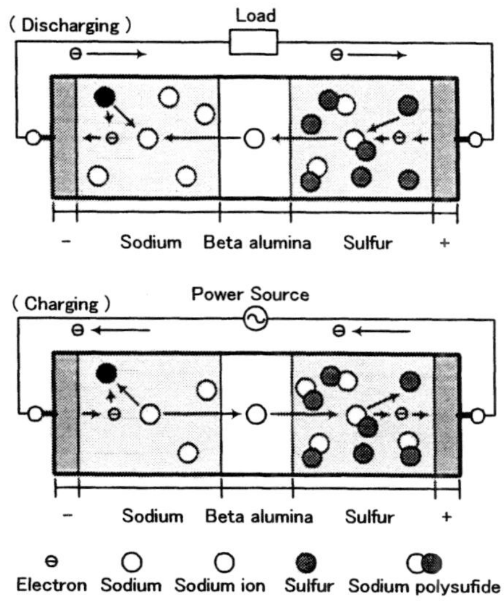

Fig.3 Mechanism Chart of Charging and Discharging

4. PREVENTIVE CONTROL AND EMERGENCY PREVENTIVE CONTROL

The concept of preventive control is proposed to avoid voltage instability phenomena beforehand by improving the voltage stability. However in case the preventive control cannot cope with severe faults, emergency preventive control is carried out to curtail loads.

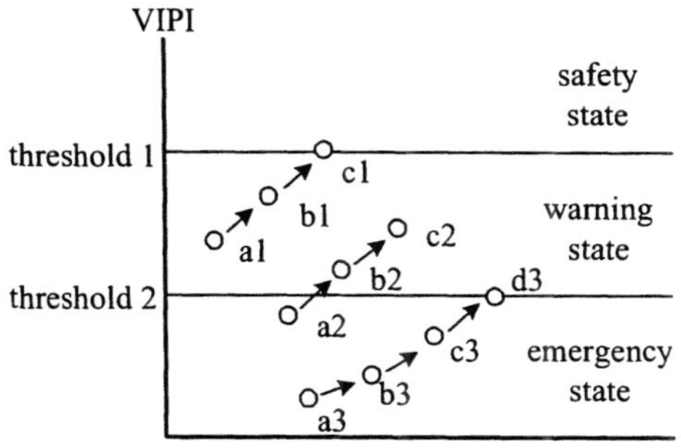

Fig.4 Concept of preventive control and emergency preventive control

One needs a reference value in order to determine when one should perform and finish the preventive control. There is considerable validity in converting the MW margin standard in a whole system into the value of VIPI. Thus, we the authors determine "threshold 1" and "threshold 2" and divide the power system state into three states; safety, warning and emergency states.

For example, if one finds that the operating point (a1) has moved to (b1) after a fault has occurred, we carry out preventive control to move the post fault operating point to a safe state (c1). When a post fault operating point (b2) can be moved only to (c2) by preventive control, such state is defined as a warning state. Lastly in case that the operating point after preventive control (c3) is still in the emergency state, we carry out emergency preventive control to move the operating point to the warning state (d3).

5. PROPOSED METHODS

5.1 PREVENTIVE CONTROL 1

Next, we will look at the conventional method. For further details of preventive control 1, see (Takehara, *et al.*, 2002; Takehara ,*et al.*, 2000; Esaka, *et al.*, 2001).

Formulation of Optimal Power Flow (OPF) is as follows.

If the value of VIPI is not above threshold 1 after contingency analysis is implemented, we implement preventive control 1.

C1=assumed fault condition
C2=operating condition

- Objective function

$$Minimize \quad f = (VIPI - threshold1)^2 \quad \ldots\ldots\ldots\ldots\ldots (5)$$

- Equality constraints

Power flow equation with operational solution at each bus (C1 and C2) and with fictitious solution at each bus (C1)

- Inequality constraints

Limits of generator outputs (C1 and C2), voltages at buses (C1 and C2), line flows (C1 and C2), SC and ratios of a transformer.

5.2 PREVENTIVE CONTROL 2

Here, we talk about the proposed method. The improvement is that we use the NAS battery system as control equipment. If the value of VIPI is not above threshold 1 after preventive control 1 is implemented or the solution is not found, we implement preventive control 2. In this case, all SCs are used.

Formulation of Optimal Power Flow (OPF) is as follows.

- Objective function

$$Minimize \quad f = (VIPI - threshold1)^2 \quad \ldots\ldots\ldots\ldots\ldots (6)$$

- Equality constraints

Power flow equation with operational solution at each bus (C1 and C2) and a fictitious solution at each bus (C1).

- Inequality constraints

Limits of generator outputs (C1 and C2), voltages at buses (C1 and C2), line flows (C1 and C2), ratios of a transformer and the NAS battery's output.

5.3 EMERGENCY PREVENTIVE CONTROL

Here is another proposed method. The improvement is that we use the NAS battery system as control equipment (Concerning the conventional method, see [19]-[21]). If the value of VIPI is not above threshold 2 after preventive control 2 is implemented, we implement emergency preventive control. VIPI sensitivity for a change of real power and reactive power at each load bus is calculated in order to select effective loads to curtail (Nanba, *et al.*, 1998).

$$\text{VIPI sensitivity}^{[22]}: \frac{d\text{VIPI}}{dP_L} + \frac{d\text{VIPI}}{dQ_L}\tan\phi \;\cdots\cdots\cdots(7)$$

where

$\tan\phi$ is calculated by power factor $\cos\phi$.

In this case, all SCs are used.

Formulation of Optimal Power Flow (OPF) is as follows.

- Objective function

  Minimize f = Total amount of load curtailment...(8)
- Equality constraints

  Power flow equation with an operational solution at each bus (C1 and C2) and with a fictitious solution at each bus (C1).
- Inequality constraints

  Limits of generator outputs (C1 and C2), voltages at buses (C1 and C2), line flows (C1 and C2), ratios of a transformer, the NAS battery's output and the amount of load curtailment at a selected bus.

6. ALGORISM

The algorithm is as follows.

Step1: Determine the value of VIPI thresholds.

Step2: Carry out on-line contingency analysis every few minutes or ten minutes. If the value of VIPI becomes below threshold 1, we regard that fault as a target.

Step3: Initiate preventive control 1.

Step4: If the value of VIPI is above threshold 1, the preventive control 1 finishes successfully. If it is between threshold 1 and threshold 2, the operating point is in a warning state. Keep a constant observation. If it is below threshold 2 or the OPF has no solution, initiate preventive control 2.

Step5: In a similar fashion to Step4, the following matter can be explained. If the value of VIPI is above threshold 1, preventive control 2 finishes. If it is between threshold 1 and threshold 2, keep a constant observation. If it is below threshold 2 or the OPF has no solution, proceed to Step6.

Step6: Calculate sensitivity of VIPI for each load bus. Take the load bus having the biggest value of VIPI sensitivity.

Step7: Initiate emergency preventive control.

Step8: If the value of VIPI is above threshold 1, the emergency preventive control has finished successfully. If it is below threshold 1, take the load bus having the next biggest value of VIPI sensitivity and go back to Step7.

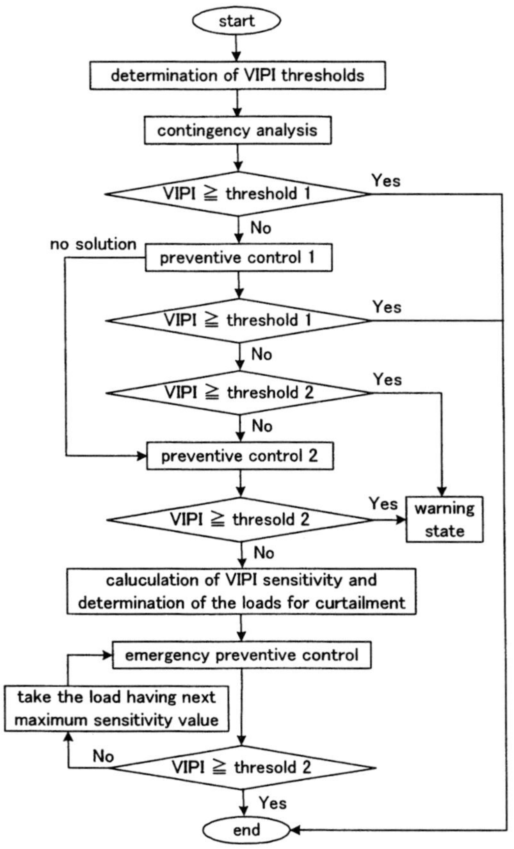

Fig.5 Flow chart

7. SIMULATIONS

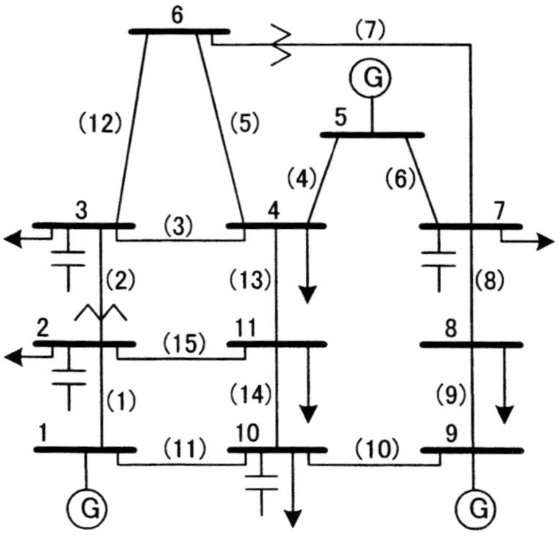

Fig.6 11-bus system

Two simulations were carried out in appropriate system conditions using an 11-bus system (Klos and Kerner, 1975) shown in Fig.5. Some voltage controllers were added to this system for the purpose of this paper.

Assumptions are as follows:

- VIPI threshold 1 is 10 degrees and threshold 2 is 9 degrees.
- As a contingency, we assume one transmission line has opened in a double transmission. In general, there are many contingency accidents. Therefore, in this simulation, we set ten monitored contingencies.
- SCs are set at bus 2, 3, 7 and 10. As one bank is 0.1j, 3 banks are set at each bus.
- In the initial condition, the amount of SC is 0 and LTC is set at 1.0 for all devices.
- The NAS battery is set at bus 2 and the capacity is approximately one-twentieth as much as the total amount of the generator capacity of the bus system.

7.1 SYSTEM CONDITION 1

A result of a contingency analysis is shown in Table.1 and a result of the amount of NAS battery is shown in Table2. For the fault of opening a number 1 transmission line for double circuit transmission lines, the value of VIPI was below threshold 1. Thus, we singled this fault out as the target of preventive control.

Table1 contingency analysis (system condition 1)

Transmission line opening	Initial condition	Preventive control 2
operating	18.742	19.767
1	**4.949**	**9.127**
2	14.629	14.648
3	16.300	17.622
4	13.078	13.826
5	18.699	20.064
6	18.437	19.730
7	17.221	18.465
8	17.856	19.264
9	16.607	17.911
10	18.520	19.961
11	18.212	20.306
12	17.189	18.587
13	15.700	16.804
14	17.214	18.966
15	18.374	19.929

Conventional preventive control 1 resulted in no solution because it could not satisfy the inequality constraints of the voltage limits at the bus. Hence, as proposed preventive control 2 was implemented. It satisfied all the inequality constraints and the value of VIPI was beyond threshold 2. The operating point moved from the emergency state to warning state, which indicated avoidance of a severe fault. However, a constant observation is needed.

The result clearly shows that preventive control 2 with an NAS battery is capable of coping with more severe faults in comparison with conventional preventive control 1.

Table2 the amount of NAS battery system (system condition 1)

P [p.u.]	Q [p.u.]	S [p.u.]
1.8247	2.3813	3.0000

7.2 SYSTEM CONDITION 2

The result of contingency analysis is shown in Table3; a result of VIPI sensitivity for a load bus is shown in Table4, a result of the amount of NAS battery is shown in Table5 and a result of the amount of load curtailment is shown in Table6.

Owing to the result of contingency analysis, we singled this fault out as the target of preventive control. This system condition is heavier loaded than condition 1. Therefore, preventive control 1 resulted in no solution. Subsequently, preventive control 2 was implemented. As a result, the value of VIPI increased to 7.291. However, it was still in an emergency state below threshold 1. This situation leaded us to the consideration of emergency-preventive control.

The problem, which we have to consider before emergency-preventive control, is the choice of a load bus to curtail. Therefore, we calculated VIPI sensitivity for the load bus. As the Table1 indicates, the most effective load bus to curtail is the No.3 bus. This method minimizes the total amount of curtailment.

Table3 contingency analysis (system condition 2)

Transmission line opening	Initial condition	Preventive control 2	Emergency preventive control
operating	16.706	18.172	20.368
1	**3.368**	**7.291**	**9.009**
2	13.249	12.995	15.166
3	15.029	16.046	18.058
4	11.727	12.243	14.080
5	17.244	18.183	20.625
6	16.969	17.843	20.257
7	15.813	16.713	18.976
8	16.288	17.332	19.790
9	14.910	15.856	18.289
10	17.021	18.085	20.547
11	16.334	18.420	20.766
12	15.828	16.881	19.075
13	14.053	14.882	17.173
14	15.672	17.094	19.546
15	16.901	18.079	20.469

After emergency preventive control, the value of VIPI increased to 9.009 and it altered its state from the emergency to the warning state. The result shows that the emergency preventive control with an NAS battery is capable of coping with more severe faults in comparison with conventional emergency preventive control. Furthermore, we can safely say that the proposed method can decrease the total amount of load curtailment because the NAS battery system can control active power outputs. This is the notable difference from the conventional control equipment such as a Static Var Compensator (SVC) and Static Synchronous Compensator (STATCOM).

Table4 VIPI sensitivity for each load bus (system condition 2)

load bus	VIPI sensitivity	load bus	VIPI sensitivity
2	0.0348	8	0.0143
3	**0.0381**	10	0.0103
4	0.0288	11	0.0280
7	0.0179		

Table5 The amount of NAS battery (system condition 2)

NAS battery's output	Preventive control 2	Emergency Preventive control
P [p.u.]	1.207	1.487
Q [p.u.]	2.746	2.605
S [p.u.]	3.000	3.000

Table6 The amount of load curtailment (system condition 2)

bus	the amount of curtailment	
	[p.u.]	[%]
3	0.9990	8.680

8. CONCLUSIONS

In recent years, concerns about voltage instability phenomena, such as a great voltage drop and a voltage collapse, have been raised. In this situation, the importance of voltage stability control schemes cannot be overemphasized. Hence, we proposed an on-line voltage stability control schemes.

We focused on the NAS battery system features and regarded the NAS battery system as control equipment. The difference between conventional control equipment and the NAS battery system is that the latter cannot only control reactive power output but also active power output. In this point, the NAS battery system is more flexible and practical than the other devices.

Simulations have been run using the 11 bus system to verify the effectiveness of the proposed method. The results showed that the proposed methods i.e., the preventive control 2 and the emergency preventive control, which compared with the conventional methods, coped with more severe faults owing to the NAS battery system. Curtailment of load bus and output of active power by the NAS battery system have the same effect. We can safely say that proposed method could decrease the total amount of load curtailment due to the active power outputs of the NAS battery system. Thus, the effectiveness of the proposed methods is verified.

REFERENCE

Y. Tanaka, S. Iwamoto (1998). A Preventive Voltage Stability Control Using Optimal Power Flow and VIPI, *IEEJ*, PE-98-167

E. Lobato, L. Rouco, M. I. Navarrete, R. Casanova, J. G Castillejo, G Lopez (2000). An Integrated Tool for Analysis of Power System Constraints in the Spanish Electricity Market, *Proceedings of IEEE PES Summer Meeting*, pp.1627-1633.

Z. Feng, V. Ajjarapu, D. J. Maratukulam (1999). A Comprehensive Approach for Preventive and Corrective Control to Mitigate Voltage Collapse, PE-290PRS.

X. Wang, GC. Ejebe, J. Tong, J.G Waight (1998). Preventive/Corrective Control for Voltage Stability Using Direct Interior Point Method", *IEEE Trans. on Power Systems*, Vol.13, No.3.

T. Tran-Quoc, Ch. Praing, R. Feuillet, J. C. Sabonnadiere, U. La-Van, C. Nguyen-Duc (2000). Improvement of Voltage Stability on The Vietnam Power System, *Proceedings of IEEE PES Winter Meeting*

M. Ilic, X. Liu, G Leung, M. Athans, C. Vialas, P. Pruvot (1995). Improved Secondary and New Tertialy Voltage Control, *IEEE Trans. on Power Systems*, Vo. 10, No.4

J. Hashimoto, N. Yorino, Y. Tamura (1988). On the Continuous Monitoring of Voltage Stability Margin in Electric Power Systems, *The Trans. of IEEJ*, Vol. 108-B, No. 2

T.Ohtaka, S.Iwamoto (2000). Optimal Allocation and PQ Output Ratio of NAS Batteries for Improving Voltage Stability in Distribution Systems, *Transaction of IEEJ*, Vol.120-B, No.10, pp1254-1262,

M. Begonic, D. Fulton, M. R. Gonzalez, J. Goossens, E. A. Guro, R. W. Haas, C. F. Henville, G Manchur, G L. Michel, R. C. Pastore, J. Postforoosh, G L. Schmitt, J. B. Williams, K. Zimmerman (1995). SUMMARY OF SYSTEM PROTECTION AND VOLTAGE STABILITY", *IEEE Trans. on Power Delivery*, Vol.10, No.2

C. Moors, D. Lefebvre, T. Van Cutsem (2000). Design of Load Shedding Schemes against Voltage Instability, *Proceedings of IEEE PES Winter Meeting*

Y. Tamura, K. Sakamoto, Y. Tayama (1988). Voltage Instability Proximity Index (VIPI) based on multiple load flow solution in Ill-conditioned power system, *Proceedings of the 27th IEEE Conference on Decision and Control, Texas, USA*

A.Okuno (1996). Energy Storage NAS Battery – Present Status of Technical Development of NAS Batteries for Load Leveling, *The New Material of Japan Industrial Publishing*, Vol.7, No.8

E.Kodama, K.Tanaka (1999). The Newest Battery Technology 5 – Energy Storage Battery", *Journal of IEEJ*, Vol.119, No.7

Standing Special Committee of Chemical and Electrical Energy Transformation (1980). Application of Secondary Batteries for Power Systems, *Technical Report of IEEJ*, Vol.103,No.2,

New Energy and Industrial Technology Development Organization, The Institute of Applied Energy (1992). *Study on Prevalence of New Battery Power Storage System (II)*, NEDO-p9158,

A.Takehara, Y. Tanaka, S.Iwamoto (2002). Voltage Stability Preventive and Emergency Preventive Controls Using VIPI Sensitivity, *The Trans. of IEEJ* Vol.1 122-B

A.Takehara, T.Ohtaka. S.Iwamoto, (2000). Voltage Stability Control Based on VIPI Sensitivity Considering Contingency Analysis, *Power Engineering and Power System Engineering Symposium of IEEJ*, PE-00-140, PSE-00-144.

T..Esaka, T.Ohtaka, S.Iwamoto (2002). Voltage Stability Emergency preventive Control Considering Priority Order of Curtailment Loads, *Proceedings of the Twelfth Annual Conference of Power and Energy Society of IEEJ*, vol.A.pp.496-497

M. Nanba, Y. Huang, T. Kai, S. Iwamoto (1998). Studies on VIPI based control methods for improving voltage stability, *Electrical Power & Energy Systems*, Vol.20, No.2, pp.141-146

A.Klos and A.Kerner (1975). The Uniqueness of Load-Flow Solution" Proc. PSCC V, 3.1/8 Cambridge

ELSEVIER
IFAC
PUBLICATIONS
www.elsevier.com/locate/ifac

A GENERATOR POWER TRACING METHOD FOR LINE POWER FLOWS

M. Suwa* **E. Masaki** **T. Ohtaka** **S. Iwamoto**

Dept. of Electrical Engineering and Bioscience, Waseda University, Tokyo, Japan

Abstract: "Tracing Power Flow of Electricity" is the concept that has become a focus of attention recently, as the deregulation or the liberalization of electric power markets proceeds. In this paper, we focus on line flows and static capacitors in particular, and we study how to deal with the problem. Although a number of methods, such as "Tracing Method by Power Flow Proportion", have already been published, reactive power contribution has not been studied much thus far. However, applying 'Tracing Power Flow of Electricity' to ancillary service, for example compensation of reactive power, is very important. Simulations are run using a sample system and the tracing results are shown. *Copyright © 2003 IFAC*

Keywords: ancillary service, deregulated power market, line flow, power flow calculation, reactive power, static capacitor, tracing power flow of electricity

1. INTRODUCTION

Recently, the reduction of electricity bill is expected by the introduction of a free competitive market to the electric power market, while deregulation and liberalization are pushed forward. In this deregulated environment, IPP (Independent Power Producer) and PPS (Power Producer and Supplier) supply electric power using the existing transmission lines. Thus the opinion that the state of the power system has to be made clear has arisen for the purpose that the wheeling rate has to been decided impartially. Therefore the concept of tracing power flows from generators means how the power flows arrive at loads through many transmission lines. This is the concept of "Tracing Power Flow of Electricity".

A number of tracing methods already have been published, and they are classified in three ways. The first tracing method is the "Tracing Method by Power Flow Proportion" [1][2]. In this method, contributions are calculated on the assumption that the effects of some power sources in the power flow flowing out from a node are proportional to the ratio of that in the power flow which flows in the node. The second is the "DC Flow Tracing Method" [2]. In this method, contributions are considered by simplifying the state of the power system using DC Flow Method. The third is the "Tracing Method Using Complex Impedances". In this method, a more detailed power system becomes possible by using complex impedances [3].

Ancillary services, for example compensation of reactive power, are given as one of application of the tracing methods. Static capacitors(SCs) are generally treated as passive devices and considered in susceptance units. When they are connected to a power system, however, they are considered in MVAR units. Therefore, the problem of how to treat SCs occurs as impedance loads or power sources. In traditional papers, only active power contributions have been considered and thus the

*M. Suwa is presently with Tohoku Electric Power Company, Japan.

reactive power problem, containing treatment of SCs, has not been studied so far. Hence, a tracing method is formulated for SCs using "Tracing Method Using Complex Impedances". Then, we run computer simulations using a sample system and show the tracing results.

2. CONVENTIONAL METHODS [1][2]

TMPFP (Tracing Method by Power Flow Proportion) is one of the most famous tracing methods that have already been published, and it is very easy to understand its concept. In this method, contributions are calculated on the assumption that the effects of some power sources in the power flow, which flows out from a node, are in proportion to the ratio of that in the power flow which flows in the node. Fig. 1 shows the process of this method.

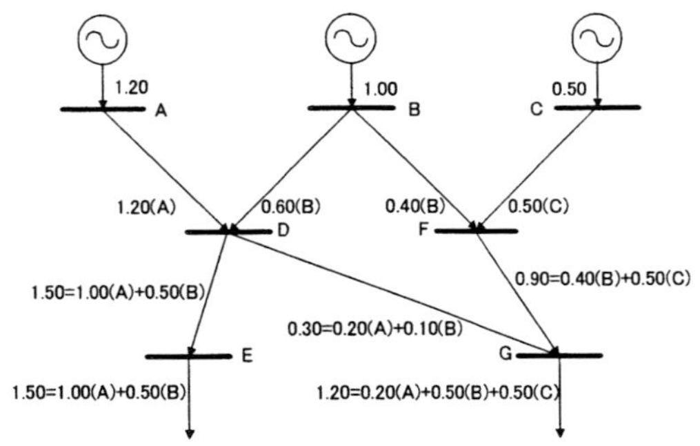

Fig. 1. Concept of TMPFP

In Fig. 1 there are power sources at node A, B, and C. If a node is connected to a power source node directly, the effect of each power source in the node is proportional to the line flows from each power source node. For example, the power flow from node A (1.20), and that from node B (0.60) flow into node D. Thus, the ratio of the effect of each power source in the power flow from node D is 2 to 1. If a power flow arrives at a branch node through some nodes, the total of the power sources' effects is compared to the effect of the power source in the power flow flowing out from the node. For example, the power flow from node D and that from F flow into node G. 0.20 of the power flow from node D is supplied by the power source A and 0.10 by source B. Also, 0.40 of the power flow from node F is supplied by source B and 0.50 by source C. So the total powers from each power source are 0.20, 0.50 and 0.50. The ratio of the effects of each power source in the power flow from node G is 2 to 5 to 5.

The effect of some power source at line *i-k* is described as Eq. (1).

$$flow_{ik}^{g} = \frac{\sum_{m} flow_{mi}^{g}}{\sum_{m} flow_{mi}} \times flow_{ik} \qquad (1)$$

where $flow_{ik}$ is the power flow from node i to node k, and $flow_{ik}^{g}$ is the effect of the power source g.

Fig. 2 shows the process of TMPFP, and thus the contributions are calculated as below with Eq. (1). The detail of this process is explained in reference [4] and [5].

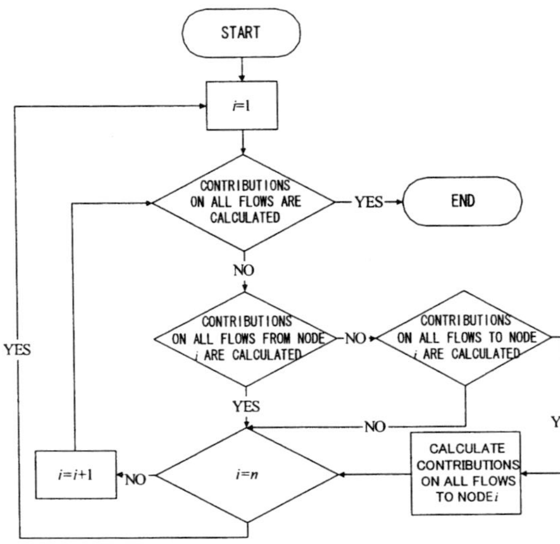

Fig. 2. Flow chart of TMPFP

CIDCFM (Complex Impedance DC Flow Method) proposed in [2] is typical TMUCI (Tracing Method Using Complex Impedances). In this method, generators are compared to current sources, and the contributions are calculated by applying the principle of superposition, using bus voltages and current flows when only one current source is connected. In CIDCFM, a load, generally treated as a constant-power load in power flow calculations, is treated as a constant-impedance load. Then by regarding a generator to current source, we can consider a complex DC network as in Fig. 3. In this network, a current equation described as a matrix equation and divided into a generator side and a loads side is as follows,

$$\begin{bmatrix} \dot{\mathbf{I}}_G \\ \mathbf{0} \end{bmatrix} = \begin{bmatrix} \dot{\mathbf{Y}}_{GG} & \dot{\mathbf{Y}}_{GL} \\ \dot{\mathbf{Y}}_{LG} & \dot{\mathbf{Y}}'_{LL} \end{bmatrix} \begin{bmatrix} \dot{\mathbf{V}}_G \\ \dot{\mathbf{V}}_L \end{bmatrix} \qquad (2)$$

where G means generator nodes, L means load nodes.

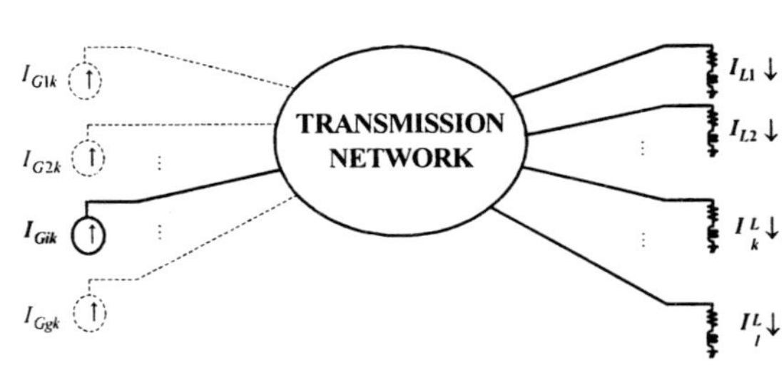

Fig. 3. Concept of CIDCFM

Next, we calculate voltages in this network connected by only one current source as in Eq. (3). Then the state of currents from each current source is calculated as in Eq. (4).

$$\begin{bmatrix} \dot{I}_1 \\ 0 \\ \vdots \\ 0 \\ \hline \mathbf{0} \end{bmatrix} = \left[\begin{array}{cccc|c} & & & & \dot{\mathbf{Y}}_{1L} \\ & & & & \vdots \\ & \dot{\mathbf{Y}}_{GG} & & & \vdots \\ & & & & \dot{\mathbf{Y}}_{gL} \\ \hline \dot{\mathbf{Y}}_{L1} & \cdots & \cdots & \dot{\mathbf{Y}}_{Lg} & \dot{\mathbf{Y}}'_{LL} \end{array}\right] \begin{bmatrix} \dot{V}_1^1 \\ \vdots \\ \vdots \\ \dot{V}_g^1 \\ \hline \dot{\mathbf{V}}_L^1 \end{bmatrix}$$

$$\vdots \tag{3}$$

$$\begin{bmatrix} 0 \\ \vdots \\ 0 \\ \dot{I}_g \\ \hline \mathbf{0} \end{bmatrix} = \left[\begin{array}{cccc|c} & & & & \dot{\mathbf{Y}}_{1L} \\ & & & & \vdots \\ & \dot{\mathbf{Y}}_{GG} & & & \vdots \\ & & & & \dot{\mathbf{Y}}_{gL} \\ \hline \dot{\mathbf{Y}}_{L1} & \cdots & \cdots & \dot{\mathbf{Y}}_{Lg} & \dot{\mathbf{Y}}'_{LL} \end{array}\right] \begin{bmatrix} \dot{V}_1^g \\ \vdots \\ \vdots \\ \dot{V}_g^g \\ \hline \dot{\mathbf{V}}_L^g \end{bmatrix}$$

$$\dot{\mathbf{I}}^1 = \dot{\mathbf{Y}}\dot{\mathbf{V}}^1, \cdots, \dot{\mathbf{I}}^g = \dot{\mathbf{Y}}\dot{\mathbf{V}}^g \tag{4}$$

Therefore, contributions are calculated as in Eq. (5) with bus voltages and the state of current coming from each current source.

$$\dot{W}_i^k = \dot{V}_i \times \dot{I}_i^{k*} \tag{5}$$

3. PROPOSED METHOD

In CIMM (Complex Impedance Matrix Method), the power flow equation is described as a matrix equation divided into a generator side and a load side. Then by transforming the equation matrix, contributions can be calculated. The process of CIMM is described in the following.

The power flow equation at node i is Eq. (6). Therefore, the equations on the whole power system are described as a matrix equation as follows,

$$P_i + jQ_i = \dot{V}_i \times \{\sum_k (\dot{Y}_{ik} \times \dot{V}_k)\}^* \tag{6}$$

$$\begin{aligned} \dot{\mathbf{W}} &= \mathbf{P} + j\mathbf{Q} \\ &= \begin{bmatrix} \dot{V}_1\dot{Y}_{11}^* & \cdots & \dot{V}_1\dot{Y}_{1n}^* \\ \vdots & \ddots & \vdots \\ \dot{V}_n\dot{Y}_{n1}^* & \cdots & \dot{V}_n\dot{Y}_{nn}^* \end{bmatrix} \begin{bmatrix} \dot{V}_1^* \\ \vdots \\ \dot{V}_n^* \end{bmatrix} \\ &= \begin{bmatrix} \dot{A}_{11} & \cdots & \dot{A}_{1n} \\ \vdots & \ddots & \vdots \\ \dot{A}_{n1} & \cdots & \dot{A}_{nn} \end{bmatrix} \begin{bmatrix} \dot{V}_1^* \\ \vdots \\ \dot{V}_n^* \end{bmatrix} = \mathbf{AV}^* \end{aligned} \tag{7}$$

Dividing Eq. (7) into a generator side and a load side, Eq.. (8) and (9) are obtained. In addition, the current equation on the load side is described as Eq. (10) by regarding constant-power loads as constant-impedance loads.

$$\dot{\mathbf{W}}_G = \dot{\mathbf{A}}_{GG}\dot{\mathbf{V}}_G^* + \dot{\mathbf{A}}_{GL}\dot{\mathbf{V}}_L^* \tag{8}$$

$$\dot{\mathbf{W}}_L = \dot{\mathbf{A}}_{LG}\dot{\mathbf{V}}_G^* + \dot{\mathbf{A}}_{LL}\dot{\mathbf{V}}_L^* \tag{9}$$

$$\mathbf{0} = \dot{\mathbf{Y}}_{LG}^*\dot{\mathbf{V}}_G^* + \dot{\mathbf{Y}}_{LL}'^*\dot{\mathbf{V}}_L^* \tag{10}$$

Thus, the following equations are given by arranging Eq. (10) for $\mathbf{V}_L^*$ and substituting it into Eq. (8) and (9).

$$\begin{aligned} \dot{\mathbf{W}}_G &= \dot{\mathbf{A}}_{GG}\dot{\mathbf{V}}_G^* - \dot{\mathbf{A}}_{GL}\dot{\mathbf{Y}}_{LL}'^{*\,-1}\dot{\mathbf{Y}}_{LG}^*\dot{\mathbf{V}}_G^* \\ &= (\dot{\mathbf{A}}_{GG} - \dot{\mathbf{A}}_{GL}\dot{\mathbf{Y}}_{LL}'^{*\,-1}\dot{\mathbf{Y}}_{LG}^*)\dot{\mathbf{V}}_G^* \end{aligned} \tag{11}$$

$$\begin{aligned} \dot{\mathbf{W}}_L &= \dot{\mathbf{A}}_{LG}\dot{\mathbf{V}}_G^* - \dot{\mathbf{A}}_{LL}\dot{\mathbf{Y}}_{LL}'^{*\,-1}\dot{\mathbf{Y}}_{LG}^*\dot{\mathbf{V}}_G^* \\ &= (\dot{\mathbf{A}}_{LG} - \dot{\mathbf{A}}_{LL}\dot{\mathbf{Y}}_{LL}'^{*\,-1}\dot{\mathbf{Y}}_{LG}^*)\dot{\mathbf{V}}_G^* \end{aligned} \tag{12}$$

In addition, the relationship between the generators and loads as in Eq. (13) is obtained by arranging Eq. (11) for $\mathbf{V}_G^*$ and substituting it into Eq. (12).

$$\begin{aligned} \dot{\mathbf{W}}_L &= (\dot{\mathbf{A}}_{LG} - \dot{\mathbf{A}}_{LL}\dot{\mathbf{Y}}_{LL}'^{*\,-1}\dot{\mathbf{Y}}_{LG}^*) \\ &\quad \times (\dot{\mathbf{A}}_{GG} - \dot{\mathbf{A}}_{GL}\dot{\mathbf{Y}}_{LL}'^{*\,-1}\dot{\mathbf{Y}}_{LG}^*)^{-1}\dot{\mathbf{W}}_G \\ &= \dot{\mathbf{K}}\dot{\mathbf{W}}_G \end{aligned} \tag{13}$$

Thus, the contribution of generator k at node i is described as in Eq. (14). Therefore, the active power contributions and the reactive power contributions are calculated as in Eq. (15) and (16).

$$\dot{W}_{Li}^k = \dot{K}_{ik}\dot{W}_{Gk} \tag{14}$$

$$P_{Li}^k = \mathrm{Re}\{\dot{W}_{Li}^k\} \tag{15}$$

$$Q_{Li}^k = \mathrm{Im}\{\dot{W}_{Li}^k\} \tag{16}$$

By observing Eq. (13) the relationship between $\dot{W}_L$ and $\dot{W}_G$ is clearly seen as one equation. This is the primary feature of the proposed method. Because the previous method and proposed method use the impedance loads, the obtained results of Eq. (5) and Eq. (14) become identified. In the proposed method, we extend the concept of Eq. (13) to the relationship between the generator outputs and line flows.

Considering the contributions to line flows, the line flow equations between node i and k are described as in Eq. (17) and (18). Also by describing them as a matrix equation, Eq. (19) is obtained.

$$\dot{W}_{ik}^{lf} = \dot{V}_i \frac{\dot{V}_i - \dot{V}_k}{r_{ik} + jx_{ik}} \tag{17}$$

$$\dot{W}_{ki}^{lf} = \dot{V}_k \frac{\dot{V}_i - \dot{V}_k}{r_{ik} + jx_{ik}} \tag{18}$$

$$\begin{aligned} \dot{\mathbf{W}}^{lf} &= \begin{bmatrix} \dot{W}_{13}^{lf} & \dot{W}_{31}^{lf} & \cdots & \dot{W}_{ik}^{lf} & \dot{W}_{ki}^{lf} & \cdots \end{bmatrix}^T \\ &= \begin{bmatrix} \dot{V}_1\dot{y}_{13}^* & 0 & -\dot{V}_1\dot{y}_{31}^* & 0 & \cdots & \cdots & \cdots & \cdots & 0 \\ \dot{V}_3\dot{y}_{31}^* & 0 & -\dot{V}_3\dot{y}_{13}^* & 0 & \cdots & \cdots & \cdots & \cdots & 0 \\ \vdots & \vdots & \vdots & \vdots & \vdots & \vdots & \vdots & \vdots & \vdots \\ 0 & \cdots & \cdots & 0 & \dot{V}_i\dot{y}_{ik}^* & 0 & -\dot{V}_i\dot{y}_{ki}^* & \mathbf{0} & 0 \\ 0 & \cdots & \cdots & 0 & \dot{V}_k\dot{y}_{ki}^* & 0 & -\dot{V}_k\dot{y}_{ik}^* & \mathbf{0} & 0 \\ \vdots & \vdots & \vdots & \vdots & \vdots & \vdots & \vdots & \vdots & \vdots \end{bmatrix} \dot{\mathbf{V}}^* \\ &= \dot{\mathbf{A}}^{lf}\dot{\mathbf{V}}* = \dot{\mathbf{A}}_G^{lf}\dot{\mathbf{V}}_G^* + \dot{\mathbf{A}}_L^{lf}\dot{\mathbf{V}}_L^* \end{aligned} \tag{19}$$

Then, similarly to the case above on calculating the contributions on loads, Eq. (20) is obtained by arranging Eq. (10) for $\mathbf{V}_L^*$ and substituting it into Eq. (19).

$$
\begin{aligned}
\dot{\mathbf{W}}^{lf} &= \dot{\mathbf{A}}_G^{lf}\dot{\mathbf{V}}_G^* - \dot{\mathbf{A}}_L^{lf}\dot{\mathbf{Y}}_{LL}'^{*\,-1}\dot{\mathbf{Y}}_{LG}^*\dot{\mathbf{V}}_G^* \\
&= (\dot{\mathbf{A}}_G^{lf} - \dot{\mathbf{A}}_L^{lf}\dot{\mathbf{Y}}_{LL}'^{*\,-1}\dot{\mathbf{Y}}_{LG}^*)\dot{\mathbf{V}}_G^*
\end{aligned} \tag{20}
$$

In addition, the relationship between generators and line flows is obtained as in Eq. (21) by arranging Eq. (11) for $\mathbf{V}_G^*$ and substituting it for Eq. (20) and the contributions of line flows are calculated as in Eq. (22), (23) and (24).

$$
\begin{aligned}
\dot{\mathbf{W}}^{lf} &= (\dot{\mathbf{A}}_G^{lf} - \dot{\mathbf{A}}_L^{lf}\dot{\mathbf{Y}}_{LL}'^{*\,-1}\dot{\mathbf{Y}}_{LG}^*)\ldots \\
&\quad (\dot{\mathbf{A}}_{GG} - \dot{\mathbf{A}}_{GL}\dot{\mathbf{Y}}_{LL}'^{*\,-1}\dot{\mathbf{Y}}_{LG}^*)^{-1}\dot{\mathbf{W}}_G \\
&= \dot{\mathbf{K}}^{lf}\dot{\mathbf{W}}_G
\end{aligned} \tag{21}
$$

$$\dot{W}_{im}^{lfk} = \dot{K}_{im}^{lfk}\dot{W}_{Gk} \tag{22}$$

$$P_{im}^{lfk} = \mathrm{Re}\{\dot{W}_{im}^{lfk}\} \tag{23}$$

$$Q_{im}^{lfk} = \mathrm{Im}\{\dot{W}_{im}^{lfk}\} \tag{24}$$

Where, $\dot{W}_{im}^{lfk}$ is the line flow contribution of generator k, $\dot{K}_{im}^{lfk}$ is the relationship coefficient between the line flow (i-m) and generator k, P_{im}^{lfk} is the active power contribution, and Q_{im}^{lfk} is the reactive power contribution.

4. TREATMENT OF SC ON THE METHOD USING COMPLEX IMPEDANCE

Though active power contributions have already been considered, reactive power contributions have not been studied much so far in the previous references because the concept of power flow tracing is devised for deciding the wheeling rate. However, if we consider applying the tracing method to ancillary services, reactive power contribution has to be considered. Thus, a problem of SC-contribution occurs.

SCs are generally treated as passive devices in power flow calculations, and the reactive power is calculated as the calculation result. However connecting SC into a power system for voltage control, the unit of SC is not a susceptance unit but a MVAR unit. Therefore SCs are considered as reactive power sources.
Though the susceptance of SC is inserted in diagonal elements of the bus admittance matrix in power flow calculation, SC has to be treated as a reactive power source. Therefore, by dividing the bus admittance matrix into transmission line elements and SC elements, it is described as in Eq. (25).

$$
\dot{\mathbf{Y}} = \dot{\mathbf{Y}}^{line} + \dot{\mathbf{Y}}_{SC}
$$

$$
= \begin{bmatrix} \dot{y}_{11}^{line} & & \dot{y}_{1i}^{line} & \\ & \ddots & & \\ \dot{y}_{i1}^{line} & & \dot{y}_{ii}^{line} & \\ & & & \dot{y}_{nn}^{line} \end{bmatrix} + \begin{bmatrix} 0 & & 0 & \cdots & 0 \\ 0 & \dot{y}_{SCi} & & \ddots & \vdots \\ \vdots & & \ddots & & 0 \\ 0 & \cdots & & 0 & \dot{y}_{SCj} \end{bmatrix} \tag{25}
$$

Where, $\mathbf{Y}_{line}$ is the transmission line elements of the bus admittance matrix, $\mathbf{Y}_{SC}$ is SC element of the bus admittance elements. All SC elements at the node without SC are zero.

Next, the power flow is calculated as the value added to the value of SC's reactive power with transmission line elements of the bus admittance matrix. Thus, the calculated power flow value is described as in Eq. (26). In Eq. (26), Q_{SC} is equal to zero at all nodes without SC as similarly to $\mathbf{Y}_{SC}$.

$$
\dot{\mathbf{W}}' = \dot{\mathbf{W}} + \dot{\mathbf{W}}_{SC}
$$

$$
= \dot{\mathbf{W}} + j\mathbf{Q}_{SC} = \begin{bmatrix} P_1 + jQ_1 \\ \vdots \\ P_i + jQ_i \\ \vdots \\ P_n + jQ_n \end{bmatrix} + j\begin{bmatrix} 0 \\ 0 \\ Q_{SCi} \\ Q_{SCk} \\ 0 \end{bmatrix} \tag{26}
$$

Consequently, the contributions are calculated by treating generators and SCs as power sources individually. In this way, the concept of tracing electricity is described in Fig. 4(a) and (b).

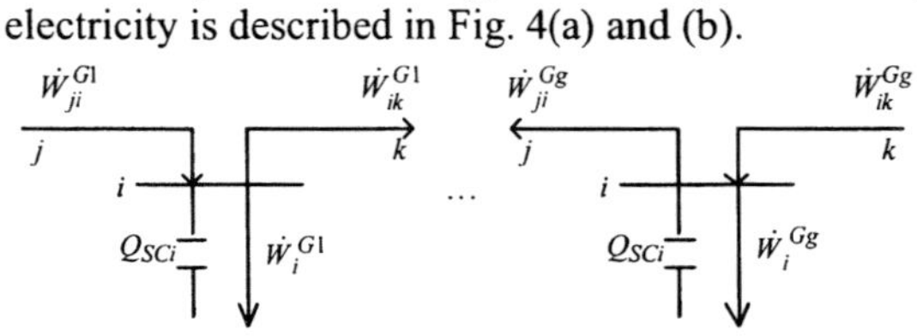

Fig. 4(a). Contribution of each generator that treats SC as a power source

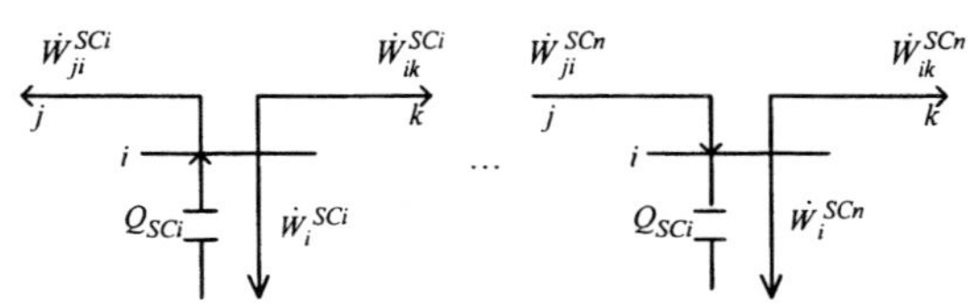

Fig. 4(b). Contribution of each SC that treats SC as a power source

We consider the case where SCs are treated as power sources in CIMM. In CIMM, the power flow equation is divided into generator side and a load side as in Eq. (27).

$$
\begin{bmatrix} \dot{\mathbf{W}}_G \\ \dot{\mathbf{W}}_L \end{bmatrix} = \begin{bmatrix} \dot{\mathbf{A}}_{GG} & \dot{\mathbf{A}}_{GL} \\ \dot{\mathbf{A}}_{LG} & \dot{\mathbf{A}}_{LL} \end{bmatrix} \begin{bmatrix} \dot{\mathbf{V}}_G \\ \dot{\mathbf{V}}_L \end{bmatrix}^* \tag{27}
$$

However, treating SCs as power sources, because SCs are connected at both the generator side and the load side, dividing the matrix into a SC and load sides is impossible. Therefore to resolve the problem, calculating contributions is proposed by the following process.
By preparing the bus admittance matrix that treats loads as constant-impedance loads and calculating power flows, the power flow equation is described as in Eq. (28). As a result, the calculated power flow becomes the summation of power flows and SCs flow, thus equaling zero at all nodes where no power source is connected.

$$\dot{\mathbf{W}}_{(G,SC)} = \dot{\mathbf{W}}_G + j\dot{\mathbf{Q}}_{SC}$$

$$= \begin{bmatrix} \dot{V}_1\dot{y}'^*_{11} & \cdots & \dot{V}_1\dot{y}^*_{1i} & \cdots & \dot{V}_1\dot{y}^*_{1n} \\ \vdots & \ddots & & & \vdots \\ \dot{V}_i\dot{y}^*_{i1} & & \dot{V}_i\dot{y}'^*_{ii} & & \dot{V}_i\dot{y}^*_{in} \\ \vdots & & & \ddots & \vdots \\ \dot{V}_n\dot{y}^*_{n1} & \cdots & \dot{V}_n\dot{y}i & \cdots & \dot{V}_n\dot{y}'^*_{nn} \end{bmatrix} \dot{\mathbf{V}}^*$$

$$= \dot{\mathbf{A}}_{(G,SC)}\dot{\mathbf{V}}^* \tag{28}$$

By arranging Eq. (28) for the bus voltage vector and substituting it into Eq. (19), the relationship between generators and line flows is calculated as Eq. (29).

$$\dot{\mathbf{W}}^{lf} = \dot{\mathbf{A}}^{lf}\dot{\mathbf{A}}^{-1}_{(G,SC)}\dot{\mathbf{W}}_{(G,SC)} = \dot{\mathbf{K}}^{lf}_{(G,SC)}\dot{\mathbf{W}}_{(G,SC)} \tag{29}$$

At that point, the contribution of each power source containing SCs is calculated with $\dot{\mathbf{K}}^{lf}_{(G,SC)}$ and vectors of power flows and SC flows described as in Eq. (30) and (31).

$$\dot{W}^{Gk}_{im} = \dot{K}^{lfk}_{(G,SC)im} \times \dot{W}_{Gk} \tag{30}$$

$$\dot{W}^{SCk}_{im} = \dot{K}^{lfk}_{(G,SC)im} \times jQ_{SCk} \tag{31}$$

The total of the line flows from node i equals the power flow at node i. Hence, the total contribution to line flow from node i is compared to the contribution to the load at node i. If power sources and loads are connected at the same node, in the case where i equals g in Eq. (32) and (33), by decreasing the power flow of the power source, the contribution of each power source is calculated.

$$\dot{W}^{Gg}_{Li} = \sum_{k}^{n} \dot{W}^{Gg}_{ik} - \dot{W}^{Gg}_{i} \tag{32}$$

$$\dot{W}^{SCg}_{Li} = \sum_{k}^{n} \dot{W}^{SCg}_{ik} - jQ^{SCg}_{i} \tag{33}$$

5. SIMULATION

In this chapter, the tracing results using a model system is shown. The state of the system is shown in Table 1. A simulation is run for the case, treating SC as power sources.

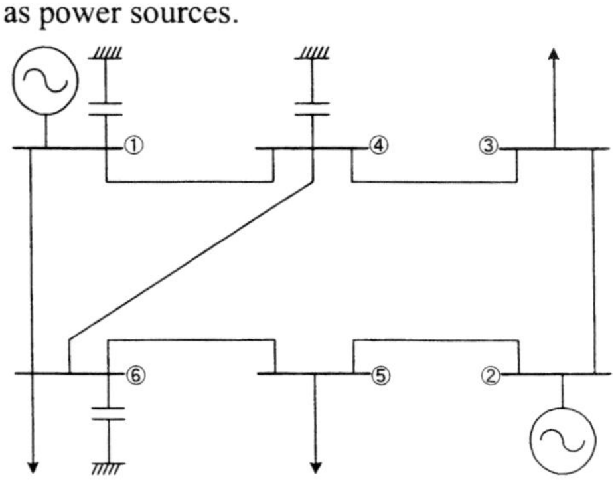

Fig. 5. Ward & Hale 6-Bus System[6]

Table 1(a) State of node flow

NODE	P[p.u.]	Q[p.u.]
1	0.9522	0.4355
2	0.5000	0.1854
3	-0.5500	-0.1300
4	0.0000	0.0000
5	-0.3000	-0.1800
6	-0.5000	-0.0500

Table 1(b) State of SC flow

SC	Q[p.u.]
NODE 1	0.0374
NODE 4	0.0253
NODE 6	0.0296

Table 1(c) State of line flow

LINE	P[p.u.]	Q[p.u.]
1->6	0.4431	0.2025
6->1	-0.4166	-0.0910
1->4	0.5091	0.2704
4->1	-0.4850	-0.1588
4->6	0.0892	0.0046
6->4	-0.0883	-0.0009
6->5	0.0049	0.0715
5->6	-0.0049	-0.0698
2->5	0.3282	0.1854
5->2	-0.2951	-0.1102
2->3	0.1718	0.0001
3->2	-0.1541	0.0255
4->3	0.3959	0.1795
3->4	-0.3959	-0.1555

Table 2 shows the tracing results of the simulation where SCs are treated as power sources. In Table 2, the contributions of SCs become negative quantities, similarly to the contributions of generators. In addition, though the contributions of the reactive powers are 2-5 times larger than the ones of the active powers, it is shown that the latter have some values. Also by observing the contribution of SC at NODE 6, we see that, the per-load contribution to NODE 6 with the SC is largest of all the other nodes. Therefore, it is clear that the contributions of SCs match the electric distance.

Table 2(a) Contribution of each generator to loads

	GENERATOR ①		GENERATOR ②	
NODE	P[p.u.]	Q[p.u.]	P[p.u.]	Q[p.u.]
3	-0.3685	-0.0644	-0.1772	-0.0308
5	-0.1894	-0.1148	-0.1155	-0.0445
6	-0.3273	-0.0075	-0.1636	-0.0125

Table 2(b) Contribution of each SC to loads

	SC at NODE 1		SC at NODE 4		SC at NODE 6	
NODE	P[p.u.]	Q[p.u.]	P[p.u.]	Q[p.u.]	P[p.u.]	Q[p.u.]
3	-0.0034	-0.0129	-0.0016	-0.0106	0.0006	-0.0113
5	0.0009	-0.0079	0.0019	-0.0055	0.0021	-0.0073
6	-0.0046	-0.0107	-0.0015	-0.0084	-0.0030	-0.0109

Table 2(c) Contribution of each generator to line flows

LINE	GENERATOR ①		GENERATOR ②	
	P[*p.u.*]	*Q*[*p.u.*]	*P*[*p.u.*]	*Q*[*p.u.*]
1->6	0.4358	0.1939	0.0068	-0.0089
6->1	-0.4087	-0.0850	-0.0042	0.0089
1->4	0.5165	0.2416	-0.0068	0.0089
4->1	-0.4871	-0.1327	0.0046	-0.0088
4->6	0.0780	0.0202	0.0098	-0.0192
6->4	-0.0779	-0.0167	-0.0089	0.0194
6->5	0.1594	0.0942	-0.1506	-0.0407
5->6	-0.1557	-0.0916	0.1470	0.0395
2->5	0.0355	0.0337	0.2946	0.1483
5->2	-0.0337	-0.0232	-0.2625	-0.0840
2->3	-0.0355	-0.0337	0.2054	0.0371
3->2	0.0369	0.0250	-0.1898	-0.0027
4->3	0.4090	0.1125	-0.0144	0.0280
3->4	-0.4054	-0.0894	0.0126	-0.0281

Table 2(c) Contribution of each SC to line flows

LINE	SC at NODE 1		SC at NODE 4		SC at NODE 6	
	P[*p.u.*]	*Q*[*p.u.*]	*P*[*p.u.*]	*Q*[*p.u.*]	*P*[*p.u.*]	*Q*[*p.u.*]
1->6	0.0002	0.0170	-0.0010	0.0042	0.0013	-0.0036
6->1	-0.0033	-0.0145	0.0001	-0.0037	-0.0005	0.0034
1->4	-0.0002	0.0204	0.0010	-0.0042	-0.0013	0.0036
4->1	-0.0029	-0.0178	-0.0002	0.0038	0.0006	-0.0034
4->6	0.0005	0.0028	-0.0004	0.0082	0.0013	-0.0074
6->4	-0.0006	-0.0028	0.0001	-0.0082	-0.0009	0.0073
6->5	-0.0007	0.0066	-0.0017	0.0035	-0.0016	0.0081
5->6	0.0007	-0.0064	0.0016	-0.0034	0.0015	-0.0079
2->5	-0.0006	0.0017	-0.0007	0.0025	-0.0006	-0.0008
5->2	0.0003	-0.0014	0.0003	-0.0021	0.0006	0.0006
2->3	0.0006	-0.0017	0.0007	-0.0025	0.0006	0.0008
3->2	-0.0003	0.0016	-0.0003	0.0023	-0.0007	-0.0006
4->3	0.0024	0.0149	0.0007	0.0133	-0.0019	0.0108
3->4	-0.0031	-0.0145	-0.0013	-0.0130	0.0013	-0.0106

6. CONCLUSION

In this paper, the relationship between the power generators and AC line flows identified by using constant-impedance loads and the treatment of SCs has been considered. In the proposed method, SCs are treated as power sources and the relationship between SCs and loads or line flows becomes clear. In such a manner, the contributions of SCs at load nodes or branches become clearly understood. Therefore, applications of 'Tracing Power Flow of Electricity' to ancillary services, for example compensation of reactive power, become possible in the future.

REFERENCES

[1] J. Bialek: "Tracing the flow of electricity" IEE Proc. Transm. Distrib. , Vol. 14, No. 4, July 1996

[2] K. Tanaka, K, Okada, N. Uchida, M. Nagata: "Method of Load Electricity for costing Supply Service –Approach based on Electrical Distance from Load to Generations- (in Japanese)" CREEPI Komae Research Laboratory Rep. NO. T98040, May 1999

[3] M. Suwa, T. Ohtaka, S. Iwamoto: "Tracing Power Flow of Electricity and its Application to Power Systems (in Japanese)" IEEJ PE-01-174, PSE-01-168, October 2001

[4] Daniel Kirschen, Ron Allan, Goran Strbac: "Contributions of Individual Generators to Loads and Flows" IEEE Transactions on Power Systems, Vol. 12, No. 1, 1997

[5] Felix F. Wu, Yinxin Ni, Ping Wei "Power Transfer Allocation for Open Access Using Graph Theory –Fundamentals and Applications in Systems Without Loopflow" IEEE Transactions on Power Systems, Vol. 15, No. 3, August 2000

[6] J. B. Ward, H. W. Hale: "Digital Computer Solution of Power Flow Problem" AIEE Transactions on Power Apparatus and Systems, PAS-75, (1956)

ELSEVIER
IFAC
PUBLICATIONS
www.elsevier.com/locate/ifac

OPTIMAL PLACEMENT OF FACTS FROM DYNAMIC PERFORMANCE VIEWPOINT

Motoki Yanagitani* Ryousuke Shizawa Toshiya Ohtaka Shinichi Iwamoto

Dept. of Electrical Engineering and Bioscience, Waseda University, Tokyo, Japan

Abstract: UPFC, which is one of the FACTS devices, can control transmission line reactances and phase angles. In this paper we propose a method for determining the optimal allocation of UPFC devices. In this method, for a multi-machine power system first through numerical simulations severe fault conditions are found. Then a UPFC device is allocated to different buses and behaviors of the control inputs under the severe fault conditions are studied. The less control effort for the system equipped with UPFC means the most efficient UPFC allocation. The effectiveness of the proposed method is shown using a four-machine eleven bus power system model. *Copyright © 2003 IFAC*

Keywords; power system , FACTS , UPFC , transient stability , optimal placement

1.INTRODUCTION

Recently, because of the difficulty of constructing new transmission lines, long-distance and large-capacity transmissions have been unavoidable. This may cause transient/voltage stability problems and loop flow problems that are not desirable in mesh systems. In 1988, Electric Power Research Institute (EPRI) of the United States, proposed a novel concept of a transmission system called FACTS (Flexible AC Transmission Systems) as one of the solutions to such problems. FACTS introduces the modern power electronics technology into the AC transmission system, and regulates bus voltages, line reactances and phase angles of the transmission systems fast and flexibly.
It can greatly increase transmission power and load flow control capability, enhance power system stability, and damp power oscillations.

* Motoki Yanagitani is presently with RECRUIT Co, Ltd , Japan.

So far, a variety of FACTS devices have been researched. UPFC (Unified Power Flow Controller) is one of the excellent FACTS devices. It can control both the power transmission line reactances and the phase angles in high-speed, and it is field-tested in an electric power system of the United States. Originally, FACTS devices have been developed to control the power flow fast and flexibly. However, using the FACTS devices only for power flow control does not make the best use of the high speed of the FACTS devices to its maximum usage. Therefore, a lot of studies, which used the FACTS devices for transient stability came out.

Moreover, the optimal placement problem can come out when FACTS devices are installed in an electric power system. In most of the latest studies about the optimal placement of FACTS devices, aims of the optimal placement are the overload elimination, steady state stability and voltage stability. These are optimal placement examples where the merit of FACTS devices which has the high-speed control performance is not employed fully.

It is desired to tackle the optimal placement problem using FACTS devices for power system stabilization control. In this paper, it considers applying FACTS devices to system stabilization first. UPFC is applied to system stabilization as typical FACTS device. As the procedure, first, the equivalent model of the FACTS devices which is needed in the case of the transient stability analysis is created, and a linearized model of a power system is obtained based on it. The optimal control theory (LQR) is used as the control technique. Next, we propose an optimal placement method of FACTS devices for transient stability. This is the optimal placement method considering stability aspect and the financial aspect (efficiency aspect) at the time of using FACTS devices for system stabilization.

In the actual situation, no effective index about transient stability exists. Then, in this paper, the amount of changes of the generator phase differences and the amount of changes of the control operation variables of FACTS devices at the time of carrying out system stabilization are used as the index. When there are few amounts of control operations or few amounts of changes of generator phase differences at the time of using FACTS devices for system stabilization, it will be judged that it is an excellent installation location for economical aspect (efficiency aspect). As a procedure, first, FACTS apparatus is installed in all the places in the system, and all the faults considered in each of the installation place are examined. The optimal location place is selected to measure the amount of control operations of the FACTS devices and the amount of changes of generator phase differences to all the faults.

2. SYSTEM STABILIZATION CONTROL USING UPFC

2.1 UPFC (Unified Power Flow Controller)

UPFC is one of the FACTS devices which can control the power transmission line reactances and phase angles in the transmission electric power component in high speed. Fig. 1 shows the outline of UPFC. UPFC has a parallel transformer, a series transformer, and a self-excited inverter of the GTO (Gate Turn Off Thyristor) control.

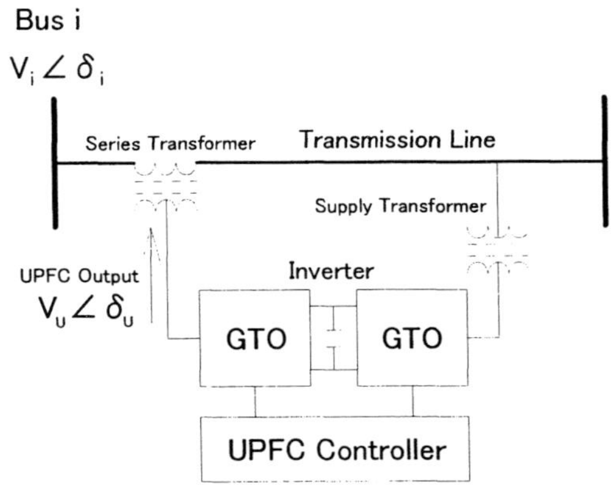

Fig.1 Outline of UPFC

UPFC is a FACTS device which can apply the voltage $V_U \angle \delta_U$ with an arbitrary magnitude and a phase shift angle for the transmission voltage to control the secondary voltage of the series transformer using the voltage generated by the self-excited inverter. Moreover, UPFC applied voltage $V_U \angle \delta_U$ is controlled with the power converter. However, UPFC controller receives the limitation according to the converter characteristics. In this paper, UPFC is assumed to impress the voltage $V_U \angle \delta_U$ with arbitrary magnitude and phase shift angle on the bus voltage at bus *i*.

2.2 Formulation of power system with UPFC

First, the differential equation of the generator *i* is given by (1)-(8).

$$\dot{\delta}_{(i)} = \omega_0(\omega_{(i)} - 1) \quad (1)$$

$$\dot{\omega}_{(i)} = \frac{P_{m(i)} - P_{e(i)} - D_{(i)}(\omega_{(i)} - 1)}{M_{(i)}} \quad (2)$$

$$\dot{E}'_{q(i)} = \frac{E_{fd(i)} - E'_{q(i)} - (X_{d(i)} - X'_{d(i)})I_{d(i)}}{\tau'_{do(i)}} \quad (3)$$

$$\dot{E}_{fd(i)} = -\frac{E_{fd(i)} - E_{fd0(i)} + K_{a(i)}(V_{t(i)} - V_{t0(i)})}{\tau_{a(i)}} \quad (4)$$

$$V_{d(i)} = X_{q(i)} I_{q(i)} \quad (5) \qquad V_{q(i)} = E'_{q(i)} - X'_{d(i)} I_{d(i)} \quad (6)$$

$$V_{t(i)} = \sqrt{V_{d(i)}^2 + V_{q(i)}^2} \quad (7) \qquad P_{e(i)} = V_{d(i)} I_{d(i)} + V_{q(i)} I_{q(i)} \quad (8)$$

$\delta_{(i)}$: phase angle [rad] $\omega_{(i)}$: generator speed [pu]

$E'_{q(i)}$: internal voltage [pu] $E_{fd(i)}$: field voltage [pu]

$I_{d(i)}, I_{q(i)}$: generator current [pu]

$V_{d(i)}, V_{q(i)}$: terminal voltage [pu]

$V_{t(i)}$: magnitude of terminal voltage [pu]

$P_{e(i)}$: active power [pu]

ω_0 : nominal generator speed [rad/s]

$P_{m(i)}$: mechanical input [pu]

$M_{(i)}$: inertia constant [sec]

$D_{(i)}$: damping [pu]

$X_{d(i)}$, $X_{q(i)}$: synchronous reactance [pu]

$X'_{d(i)}$: transient reactance [pu]

$\tau_{do(i)}$: time constant of rotator [sec]

$V_{t0(i)}$: initial value of $V_{t(i)}$ [pu]

$E_{fd0(i)}$: initial value of $E_{ed(i)}$ [pu]

$K_{a(i)}$: AVR gain [pu]

$\tau_{a(i)}$:time constant of AVR [sec]

From (1)-(8), it is understood that $\delta_{(i)}$, $\omega_{(i)}$, $E'_{q(i)}$, and $E_{fd(i)}$ are determined depending on $I_{d(i)}$ and $I_{q(i)}$. Then, the generator current considering UPFC effect is derived.

Generally, the terminal voltage of the n-machine electric power system is given by (9).

$$\bar{V} = e^{j\delta}E'_q - jX'_d\bar{I} + (X_q - X'_d)e^{-j(\frac{\pi}{2}-\delta)}I_q \quad (9)$$

$\bar{V}$: terminal voltage (row vector, phasor)

$\bar{I}$: generator current (row vector, phasor)

E'_q, I_q : row vector, phasor

$e^{j\delta}$, X'_d, $(X_q - X'_d)$, $e^{j(\pi/2-\delta)}$: diagonal matrix

We assume that UPFC is installed near a generator p. The UPFC output $V_U \angle \delta_U$ is impressed on the terminal voltage of the generator p. Therefore, the terminal voltage considering UPFC effect changes from (9) into (10).

$$\bar{V} + \bar{V}_U = e^{j\delta}E'_q - jX'_d\bar{I} + (X_q - X'_d)e^{-j(\frac{\pi}{2}-\delta)}I_q + \bar{V}_U \quad (10)$$

$$\bar{V}_U \equiv \begin{bmatrix} 0 & \cdots & 0 & \underset{(i=p)}{V_U e^{j\delta_U}} & 0 & \cdots & 0 \end{bmatrix}^T$$

The generator current considering UPFC effect is given by (11).

$$\bar{I} = \bar{Y}_R(\bar{V} + \bar{V}_U) \quad (11)$$

$\bar{Y}_R$: reduced admittance matrix

When (10) is substituted into (11) and $\bar{V}$ is eliminated, (12) is obtained.

$$\bar{I} = \bar{Y}e^{j\delta}E'_q + \bar{Y}(X_q - X'_d)e^{j(\frac{\pi}{2}-\delta)}I_q + \bar{Y}\bar{V}_U \quad (12)$$

$$\bar{Y} \equiv \left(\bar{Y}_R^{-1} + jX'_d\right)^{-1}$$

From (12), generator i current is given by (13).

$$\bar{I}_{(i)} = \sum_{\substack{k=1 \\ (k\neq p)}}^{n} \bar{Y}_{(i,k)}\left\{e^{j\delta_{(k)}}E'_{q(k)} + (X_{q(k)} - X'_{d(k)})e^{-j(\frac{\pi}{2}-\delta_{(k)})}I_{q(k)}\right\}$$

$$+\bar{Y}_{(i,p)}\left\{e^{j\delta_{(p)}}E'_{q(p)} + (X_{q(p)} - X'_{d(p)})e^{-j(\frac{\pi}{2}-\delta_{(p)})}I_{q(p)} + \bar{V}_{U(i)}\right\} \quad (13)$$

(14), (15) are obtained when (13) is changed in the d-q coordinates and decomposed into the d and q axis elements.

$$I_{d(i)} = \mathrm{Re}\left(\bar{I}_{(i)}e^{j(\frac{\pi}{2}-\delta_{(i)})}\right)$$

$$= \sum_{\substack{k=1 \\ (k\neq p)}}^{n} Y_{(i,k)}\left[-S_{(i,k)}E'_{q(k)} + (X_{q(k)} - X'_{d(k)})C_{(i,k)}I_{q(k)}\right]$$

$$+Y_{(i,p)}\left[-S_{(i,p)}E'_{q(p)} + (X_{q(p)} - X'_{d(p)})C_{(i,p)}I_{q(p)} + V_{U(i)}S_{U(i)}\right] \quad (14)$$

$$I_{q(i)} = \mathrm{Im}\left(\bar{I}_{(i)}e^{j(\frac{\pi}{2}-\delta_{(i)})}\right)$$

$$= \sum_{\substack{k=1 \\ (k\neq p)}}^{n} Y_{(i,k)}\left[C_{(i,k)}E'_{q(k)} + (X_{q(k)} - X'_{d(k)})S_{(i,k)}I_{q(k)}\right]$$

$$+Y_{(i,p)}\left[C_{(i,p)}E'_{q(p)} + (X_{q(p)} - X'_{d(p)})S_{(i,p)}I_{q(p)} + V_{U(i)}C_{U(i)}\right] \quad (15)$$

$$\bar{Y}_{(i,k)} \equiv Y_{(i,k)}e^{j\phi_{(i,k)}}$$

$$S_{(i,k)} \equiv \sin\left(-\delta_{(i)} + \delta_{(k)} + \phi_{(i,k)}\right) \quad C_{(i,k)} \equiv \cos\left(-\delta_{(i)} + \delta_{(k)} + \phi_{(i,k)}\right)$$

$$S_{U(i)} \equiv \sin\left(-\delta_{(i)} + \phi_{(i,p)} + \delta_{u(i)}\right) \quad C_{U(i)} \equiv \cos\left(-\delta_{(i)} + \phi_{(i,p)} + \delta_{u(i)}\right)$$

From (14) and (15), $I_{d(i)}$ and $I_{q(i)}$ considering UPFC effect are determined.

By solving equation (1)~(8), (14), and (15), the dynamics of the generator can be obtained. However, the initial values of phase angles, angular velocities, internal voltages, and field voltages must be obtained by power flow calculations. In addition, we have to give UPFC imputs during the simulation period.

2.3 UPFC output calculation method

This chapter shows a method of calculating the UPFC output $V_U \angle \delta_U$. In this paper, the discrete-time linear control theory is applied to the control of the UPFC output $V_U \angle \delta_U$. From (14) and (15), the UPFC output magnitude V_U has strong correlation with Generator p current $I_{d(p)}$, $I_{q(p)}$. And, UPFC output phase shift angle δ_U is strong correlation in Generator p phase angle $\delta_{(p)}$. Then, V_U is calculated by ΔI_{dU}, ΔI_{qU} (infinitesimal deviation of $I_{d(p)}$, $I_{q(p)}$), and δ_U is calculated by $\Delta\delta_U$ (infinite simal deviation of $\delta_{(p)}$).

Thus, the continuous-time linear system equation for the UPFC control system design has been obtained. When (1)-(8) is approximated linearly (taking an infinitesimal deviation), the continuous-time linear system equation is obtained shown in (16).

$$\begin{cases} \dot{x}(t) = A_0 x(t) + B_0 u(t) \\ y(t) = C_0 x(t) + D_0 u(t) \end{cases} \quad (16)$$

$$x(t) = \begin{bmatrix} \cdots & \Delta\delta_{(i)} & \Delta\omega_{(i)} & \Delta E'_{q(i)} & \Delta E_{fd(i)} & \cdots \end{bmatrix}^T$$

$$u(t) = \begin{bmatrix} \Delta I_{dU} & \Delta I_{qU} & \Delta\delta_U \end{bmatrix}^T$$

$$y(t) = \begin{bmatrix} \cdots & \Delta\omega_{(i)} & \Delta E_{fd(i)} & \Delta P_{e(i)} & \Delta V_{t(i)} & \Delta I_{t(i)} & \cdots \end{bmatrix}^T$$

$I_{t(i)}$: magnitude of generator current [pu]

A_0 , B_0 , C_0 , D_0 : matrix determined by $x(0), u(0), y(0)$

Eq. (16) can be converted into a discrete-time linear system equation (17), and the digital control theory shown in chapter 4 can be applied. Then, the UPFC output $V_U \angle \delta_U$ is determined. Concretely, the UPFC output $V_U \angle \delta_U$ is obtained from (18), (19).

$$\begin{cases} x[i+1] = A x[i] + B u[i] \\ y[i] = C x[i] + D u[i] \end{cases} \quad (17)$$

$$V_U \equiv \sqrt{\left(V_{d0(p)} + X_{q(p)}\Delta I_{qU}\right)^2 + \left(V_{q0(p)} - X'_{d(p)}\Delta I_{dU}\right)^2} - V_{t0} \quad (18)$$

$$\delta_U \equiv \Delta \delta_U \quad (19)$$

$V_{d0(p)}$, $V_{q0(p)}$, V_{t0} : initial value

3. OPTIMAL PLACEMENT OF UPFC FOR TRANSIENT STABILITY

In this paper, we use two indices for optimal placement of UPFC in consideration of transient stability. They are the amount of changes of the phase differences of the generators and the amount of changes of the UPFC outputs at the time of stabilizing the system using UPFC. These two different indices can select the optimal location based on different purposes, namely, the stability aspect and the efficiency aspect (financial aspect). How to find each index is shown below.

3.1 Amount of changes of phase differences

The amount of changes of the generator phase differences is calculated at the time of stabilizing the system using UPFC in the n-machine power system.

That is, if the simulation during t second is performed, the integrated value of the generator phase differences for t seconds is calculated for all generators. Let maximum among $\int_0^t \delta_{1\sim n} dt$ be an index for this fault.

In this paper, we define this index as

$$S_{ge} = MAX\left(\int_0^t \delta_{1\sim n} dt\right) \quad (20)$$

3.2 Amount of changes of UPFC outputs

The same as Chapter Ⅲ. i , the amount of changes of the UPFC outputs is calculated at the time of stabilizing the system using UPFC.

That is, if the simulation is performed for t seconds, let the integrated values of the UPFC outputs for t seconds be an index. However, there are two control operation variables (V_U and $\angle \delta_U$) as described in Chapter Ⅱ.

In order to take into consideration both these two operation variables、in this paper , we define this index as

$$S_{UPFC} = \left(\int_0^t \angle\delta_U dt\right) + R \times \left(\int_0^t V_U dt\right) \quad (21)$$

where R is a suitable coefficient

$$\left(\int_0^t \angle\delta_U dt\right) = R \times \left(\int_0^t V_U dt\right) \quad (22)$$

3.3 Optimal placement procedure

The features of each index is summarized in Table 1.

Table 1 The features of each index

index	Features		
S_{ge}	Select the place where amount ofchanges of generator phase	→	amount of the place which is excellent for stability aspect
S_{UPFC}	Select the place where amount of control operations of UPFC is smallest	→	the place which is excellent for efficiency aspect (financial a spect)

＊It is a severe fault when each index is large, and it is mild fault when each index is small.

An optimal location selection procedure is shown based on this index.

where generator end → 1 ～ i ～ I
fault point → 1 ～ j ～ J

Selection procedure

Step 1. UPFC is installed at the generator end i.

Step 2. A 3- phase to ground fault is caused at each bus near the end j. 1 in 2 circuits is opened, and re-closed after 1 sec. The value of each index during that time is calculated.

Step 3. step 1.～step 2. is repeated J times.

Step 4 maximum among each J index is defined as $MAX(S_{ge}(i,j))$ $MAX(S_{UPFC}(i,j))$

Step 5. Step 1.～Step 4.is repeated I times.

Step 6. minimum among each I index is defined as $MIN(MAX(S_{ge}(i,j)))$ $MIN(MAX(S_{UPFC}(i,j)))$

step7. let the installation place i $MIN(MAX(S_{ge}(i,j)))$ $MIN(MAX(S_{UPFC}(i,j)))$ be the optimal allocation.

4. SIMULATION

4.1 System stabilization simulation

In this chapter, a 4-machine 11-bus power system model with UPFC is used which is shown in Fig. 2. It is assumed that the UPFC output is impressed on bus 4 (terminal voltage of Generator 1).

As shown in the figure, fault points are ten places of A - J points. In each case a three-phase to ground fault of one line in double circuits transmission lines occurs. The faulted transmission line is opened after 100 [ms], and reclosed after 1 [sec].

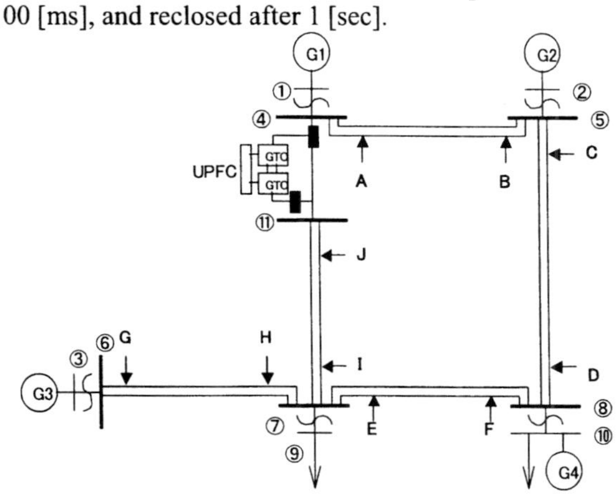

Fig.2 4-machine 11-bus power system model with UPFC

4.2 The optimal placement of UPFC in consideration of transient stability

4.2.1 When index S_{ge} is used

Index values for each fault are indicated in Table 2.

Table 2 index S_{ge} for each fault (rad)

	installed at generator 1 end	installed at generator 2 end	installed at generator 3 end	installed at generator 4 end
A	21.2083	22.1371	21.6837	21.7314
B	20.9981	22.0701	21.6249	21.6544
C	**24.6314**	23.5549	24.3132	24.1200
D	24.5678	23.4141	**24.3555**	24.3465
E	20.3493	23.4928	21.5691	21.4972
F	20.3648	24.1168	21.5729	21.5088
G	20.8041	24.8649	22.4332	22.6239
H	20.8376	25.2153	22.7620	22.8856
I	23.7183	26.1179	23.7509	**25.1390**
J	24.1707	**26.7798**	23.8763	24.1253

From Table 2. we can select the severest fault point in each generator end. The result is shown in Table 3.

Table 3 the severest fault point and index S_{ge} in case UPFC is installed in each generator end.

	installed at generator 1 end	installed at generator 2 end	installed at generator 3 end	installed at generator 4 end
Severest fault	C	J	D	I
S_{ge}	24.6314	26.7798	**24.3555**	25.1390

The installed point which have the smallest index S_{ge} in Table 3 is the best point for stability aspect. Therefore we can conclude that the optimal location of UPFC from the viewpoint of stability aspect is **generator 3 end.**

4.2.2 When index S_{UPFC} is used

Index values for each fault are indicated in Table 4.

Table 4 index S_{UPFC} for each fault(R=1.3266)

	installed at generator 1 end	installed at generator 2 end	installed at generator 3 end	installed at generator 4 end
A	1.2311	1.4286	1.9663	1.8959
B	1.1284	1.3883	1.7929	1.7583
C	2.0477	1.5338	1.4478	1.7627
D	2.1958	1.5843	1.7708	2.6710
E	0.8101	2.6543	1.3563	1.4401
F	0.8028	2.8989	1.3284	1.4125
G	2.0905	2.9959	2.0322	2.0635
H	2.1031	3.0361	2.3842	2.3020
I	1.6562	2.9999	1.6458	2.2825
J	1.8821	3.0794	1.7596	2.2461

From Table 4 we can select the severest fault point in each generator end. The result is shown in Table 5. The installed point which has the smallest index S_{UPFC} in Table 5 is the best point for efficiency aspect (financial aspect). Therefore we can conclude that the optimal locatin of UPFC from the viewpoint of efficiency aspect (financial aspect) is **generator 1 end.**

Table 5 Severest fault point and index S_{UPFC} in case UPFC is installed in each generator end.

	installed at generator 1 end	installed at generator 2 end	installed at generator 3 end	installed at generator 4 end
Severest fault	D	H	H	D
S_{UPFC}	**2.1958**	3.0361	2.3842	2.6710

4.2.3 When both S_{ge} and S_{UPFC} are used.

Table 3 and 5 are collectively shown in Table 6.

Table 6 Severest fault point and index S_{ge} S_{UPFC} in case UPFC is installed in each generator end.

	installed at generator 1 end	installed at generator 2 end	installed at generator 3 end	installed at generator 4 end
S_{ge}	**24.6314**	26.7798	**24.3555**	25.1390
S_{UPFC}	**2.1958**	3.0361	**2.3842**	2.6710

From Table 6 , index S_{ge} in the case UPFC is installed at **generator 1 end** is small to the second, and index S_{UPFC} in the case UPFC is installed at **generator 3 end** is small to the second.

Installing at generator 1 end is good to second for index S_{ge} , that is, it is good to second in consideration of stability. At the same time,it is best for index S_{UPFC} , that is, it is best in consideration of efficiency side (finansial side). Installing at generator 3 end is best for index S_{ge} , that is, it is best in consideration of efficiency side (finansial side). At the same time,it is good to second for index S_{UPFC} , that is, it is good to second in consideration of stability. From the above If it is installed at the generator end 1 or 3, it is good for both index S_{ge} and S_{UPFC} . That is, the differences is whether the priority is given to efficiency aspect (financial aspect) or stability aspect.

5. CONCLUSION

In this paper, an optimal placement method of UPFC in consideration of transient stability has been verified. When choosing the optimal location based on different indices, naturally different results are obtained. Therefore If a diffrent index in consideration of the index other one than the one used in this paper, for example, steady state stability, voltage stability, etc. is used, a different result will come out. However, it is shown, even if it takes two or more indices into consideration, the installation place which was excellent to two or more indices can be selected by changing the priority.

If the purpose is narrowed down to one, the optimal location can be selected easily. However, various elements are entangled intricately in an actual power system. Therefore, in case the optimal location is selected, it is thought that elements, such as the stability aspect and the financial aspect, become entangled. It is thought that the proposed method is having a useful approach to solve the electric power system planing problem containing FACTS devices.

REFERENCES

O.Kunitomo, H.Satomura, S.Iwamoto (1998).
A Study on Robust control using FACTS devices (in Japanese), Trans. of IEEJ, pp.288-289, 1998

M.Yanagitani, A. Sanai Sabzevary, T.Ohtaka, S.Iwamoto (2001).
A Study on Optimal Placement of UPFC in Power System (in Japanese), IEEJ, PE-01-28, PSE-01-22, 2001

A.Kurita (1993).
FACTS vision proposed by EPRI –Fundamental Surve and Numerical Study- (in Japanese), CRIEPI survey report, T92020, pp.1-36, 1993

Y.Sekine (1984).
Power System Transient Stability Analysis (in Japanese), Ohmusha

T.Kouno, T.Nkajima, A.Yokoyama (1999).
Power System Damping Enhancement using Unified Power Flow Controller(UPFC) (in Japanese), Trans. of IEEJ, vol119-B No.3, pp.344-353, 1999

ELSEVIER
IFAC
PUBLICATIONS
www.elsevier.com/locate/ifac

USE OF CRITICAL CLEARING TIME FOR TRANSIENT STABILITY PREVENTIVE CONTROL

Yoshinori Kato Kaoru Yoda Toshiya Ohtaka Shinichi Iwamoto

Dept. of Electrical Engineering and Bioscience, Waseda University, Tokyo, Japan

Abstract: Deregulation of electric power industries has begun in many countries and power system operations have become more difficult. Therefore, power system operators should consider not only economic load dispatch but also on-line stability aspects. In this paper, we pay attention to one of the most important transient stability indices, Critical Clearing Time (CCT), and propose a new transient stability preventive control method using linear relationships between CCTs and generator rotor angles. A preventive control to achieve a more stable operating point is carried out by generator output rescheduling, which are determined using the relationships between CCTs and generator rotor angles. *Copyright © 2003 IFAC*

Keywords: Power System, Transient Stability, Preventive Control, Critical Clearing Time, generator rotor angle

1. INTRODUCTION

The transient stability monitoring, some methods based on the single machine equivalent (SIME) (Ernst, *et al.*, 2001), the transient energy function (Vaahedi, *et al.*, 1997) and the coherency of generators (Li and Bose, *et al.*, 1998) have been proposed. Several methods have been tested using actual power systems and put into actual practice. As the transient stability preventive control, some methods based on (non)linear optimization (La Scala, *et al.*, 1998), post contingency information (Kuo and Bose, 1995) and the transient energy function (Fouad and Jianzhong, 1993) have been proposed.

In this paper, a new transient stability preventive control method is proposed, which is needed when some severe contingencies are found. A severe contingency means that the critical clearing time (CCT) is smaller than the actual operating time of circuit breakers in the power system. As the preventive control action, generator output rescheduling is used in this paper. The purpose of the preventive control is to make all CCTs longer than the circuit breaker operating times. The method proposed in this paper utilizes a relationship between CCT and the generator rotor angle, which is almost linear. (Kato and Iwamoto, 2002) This linear relationship can be found by a One Machine to Infinite Bus (OMIB) system analysis and a simulation study for a multimachine power system in this paper. The generator rotor angles in a power system are influenced by all generators and loads, therefore, to use the relationship means that the method can observe the whole power system, and all contingencies. To speed up the calculation time to determine generator output rescheduling, the generator rotor angles are estimated by a method using the Jacobian matrix calculated during the power flow computation. The estimation method is introduced in this paper.

The validity of the proposed method is shown by a simulation study using a Japanese standard power system model, called the IEEJ (IEE of Japan) EAST 10 machine system.

*Yoshinori Kato is presently with Tokyo Electric Power Co., Inc., Japan.

2. POWER SYSTEM MODEL

In this paper, we use a simplified generator model, which is called a classical model, and the center of angle (COA) quantities. COAs in the rotor angles and the angular velocities of generators are defined at each time. The new variables correspond to the deflection from the COA reference frame. The swing equations in the COA reference frame are shown below, where the damping coefficient is neglected.

$$\dot{\theta}_i = \hat{\omega}_i \tag{1}$$

$$\dot{\hat{\omega}}_i = \frac{\omega_0}{M_i}\left[Pm_i - Pe_i - \frac{M_i}{M_T}P_{COA}\right] \tag{2}$$

where

$$P_{COA} = \sum_{i=1}^{N}(Pm_i - Pe_i) \tag{3}$$

$$M_T = \sum_{i=1}^{N} M_i \tag{4}$$

θ_i : rotor angle in the COA reference frame

$\hat{\omega}_i$: angular velocity in the COA reference frame

M_i : inertia constant

Pm_i : mechanical power input

Pe_i : electrical power output

N: total number of generators

3. RELATIONSHIP BETWEEN CCT AND GENERATOR ROTOR ANGLE

3.1 One Machine to Infinite Bus System

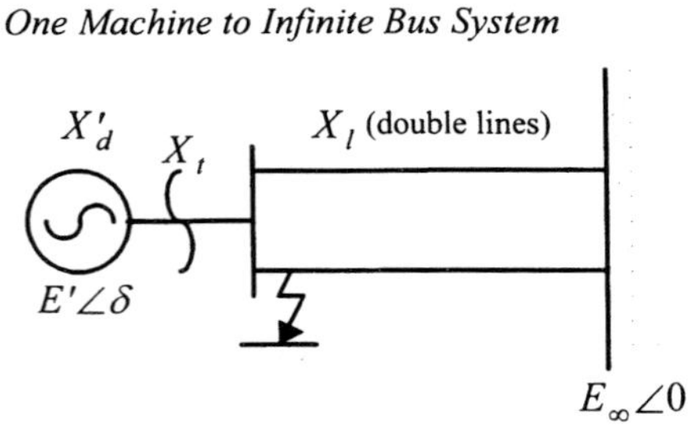

Fig. 1 One Machine to Infinite Bus (OMIB) system

The proposed transient stability preventive control uses a relationship between CCT and the generator rotor angle at the faulted bus. The relationship is useful because it is almost linear.

In this section, the relationship is found using a One Machine to Infinite Bus (OMIB) system with the double circuit lines shown in Fig. 1. The generator model used here is the classical model (see section III). The resistance of the circuit line (G) and also the control systems (for example, AVR and Governor) are neglected.

A three phase to ground fault at the bus near the generator is considered here. The fault occurs and is cleared by tripping one circuit of a double circuit lines. From the equal area criterion, the rotor angle at the fault clearing time δ_c is obtained below.

$$\delta_c = \cos^{-1}\left(\frac{P_0(\delta_u - \delta_0) + Pe^c_{\max}\cos\delta_u}{Pe^c_{\max}}\right) \tag{5}$$

where

δ_0 : rotor angle before fault

δ_u : rotor angle at an unstable equilibrium point

P_0 : generator output before fault

$Pe^c_{\max}$: maximum generator output after fault clearing

The swing equation in the OMIB system is shown below.

$$\frac{d^2\delta}{dt^2} = \frac{\omega_0}{M}(Pm - Pe_{\max}\sin\delta) \tag{6}$$

where $Pe_{\max} = \dfrac{E' \cdot E_\infty}{X_t + X_l + Xd'}$

E' : voltage behind the transient reactance

E_∞ : voltage of infinite bus

X_t : reactance of transformer

X_l : reactance of double circuit lines

Xd' : transient reactance

The value of CCT can be obtained (Sumida and Iwamoto, 1998) from the analysis of the following swing equation.

$$CCT = \sqrt{\frac{2M}{\omega_0 P_0}(\delta_c - \delta_0)} \tag{7}$$

Substituting (5) into (7), when the value of P_0 is changed, the values of CCT and δ_0 are also changed. Therefore, the relationship between CCT and δ_0 is obtained as in Fig. 2. From Fig. 2, we can observe that the relationship is virtually linear.

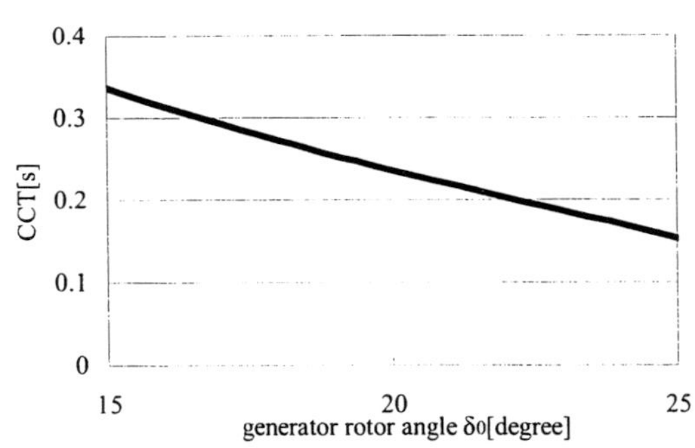

Fig. 2 Relationship between CCT and generator rotor angle (one machine to infinite bus system)

3.2 Multimachine System

In this section, the relationship between CCTs and the generator rotor angles is shown by a simulation study using a multimachine power system. The sample system is the IEEJ EAST 10 machine system (IEEJ Study Committee, 1999) which is shown in Fig. 3. The contingencies are three phase to ground faults occurred at buses near all 10 generators, that is, the faults at buses 11 - 20.

It is assumed that the fault-cleared system condition is the same as the pre-fault system. The initial condition of this simulation is a 75% load condition of the original data. The simulations are carried out increasing the loads 5% step by step of the original data. The relationships between CCTs and generator rotor angles are shown in Fig. 4. Here, the generator rotor angles are in the COA reference frame. In Fig. 4, the relationships are almost linear. However, the values of the generator rotor angles cannot determine the values of the CCTs. We can see how large the CCT changes are, when the generator rotor angles are changed.

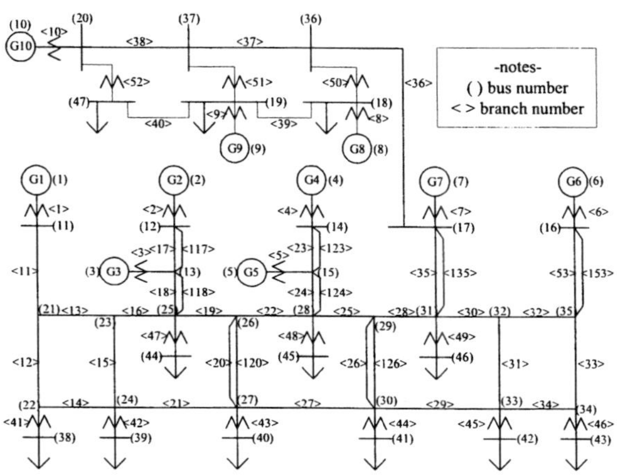

Fig. 3 IEEJ EAST 10 machine system

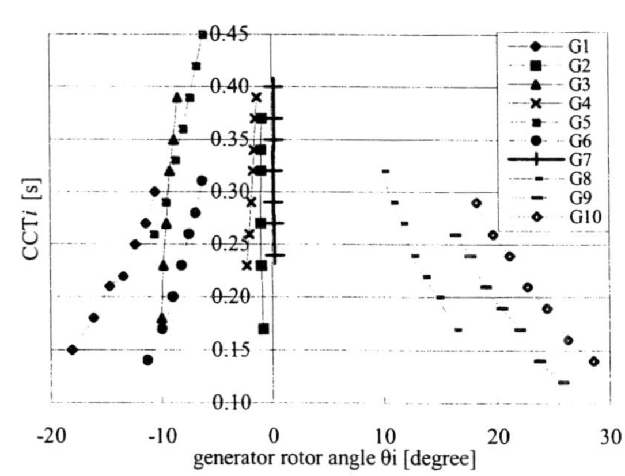

Fig. 4 Relationships between CCTs and generator rotor angles (IEEJ EAST 10 machine system)

4. GENERATOR ROTOR ANGLE ESTIMATION

In this section, a generator rotor angle estimation method is introduced using the Jacobian matrix in the power flow calculation.

Because a transient stability preventive control method using the relationships between CCTs and generator rotor angles is proposed, it is important to estimate the generator rotor angle changes when the preventive control is carried out. A generator rotor angle estimation method using the Jacobian matrix is introduced here. The inverse of the Jacobian matrix can approximately estimate voltages of buses connected to generators and generator rotor angles.

The Jacobian matrix used in the power flow calculation is shown below. Here, the bus voltage $V_i\angle\varphi_i$ is represented using the polar coordinates.

$$\mathbf{J} = \begin{bmatrix} & & \cdots\cdots\cdots & & \\ \cdots \dfrac{\partial P_{gi}}{\partial \varphi_j} & \cdots & \dfrac{\partial P_{gi}}{\partial \varphi_k} & \dfrac{\partial P_{gi}}{\partial V_m} & \cdots \\ & & \cdots\cdots\cdots & & \\ \cdots \dfrac{\partial P_{lr}}{\partial \varphi_j} & \cdots & \dfrac{\partial P_{lr}}{\partial \varphi_k} & \dfrac{\partial P_{lr}}{\partial V_m} & \cdots \\ \cdots \dfrac{\partial Q_{ls}}{\partial \varphi_j} & \cdots & \dfrac{\partial Q_{ls}}{\partial \varphi_k} & \dfrac{\partial Q_{ls}}{\partial V_m} & \cdots \\ & & \cdots\cdots\cdots & & \end{bmatrix} \quad (8)$$

In (8), the subscript g and l mean the generator bus and the load bus, respectively. The bus voltage changes by the generator output change ΔP_{gi}, the real power change of load ΔP_{lj} and the reactive power change of load ΔQ_{lj} are shown as below.

$$\begin{bmatrix} \vdots \\ \Delta\varphi_i \\ \vdots \\ \Delta\varphi_j \\ \Delta V_j \\ \vdots \end{bmatrix} = \mathbf{J}^{-1} \cdot \begin{bmatrix} \vdots \\ \Delta P_{gi} \\ \vdots \\ \Delta P_{lj} \\ \Delta Q_{lj} \\ \vdots \end{bmatrix} \quad (9)$$

From the calculation in (9), the voltage changes of all buses are obtained. Next, the values of real power Pi and reactive power Qi of all buses are calculated from the estimated bus voltages. Therefore, the generator bus current $\bar{I}_{gi}$ is calculated as in (10).

$$\bar{I}_{gi} = \frac{P_{gi} - jQ_{gi}}{\bar{E}^*_{gi}} \quad (10)$$

where $\bar{E}^*_{gi}$:conjugate of i-th generator bus voltage

The voltage $\bar{E}'_{gi}$ behind the transient reactance x'_{di} can be calculated as in (11).

$$\bar{E}'_{gi} = \bar{E}_{gi} + jx'_{di}\bar{I}_{gi} \quad (11)$$

Finally, the estimated generator rotor angle δ_i is obtained as below. (El-Abiad, 1968)

$$\delta_i = \tan^{-1}\left(\frac{\mathrm{Im}(\overline{E}'_{gi})}{\mathrm{Re}(\overline{E}'_{gi})}\right) \qquad (12)$$

The rotor angels in the COA reference frame can be calculated by the calculation results of (12).

5. PROPOSED PREVENTIVE CONTROL METHOD

In this section, a transient stability preventive control method by generator output rescheduling is proposed using the linear relationships between CCTs and generator rotor angles shown in section IV. Considering that one generator rotor angle is influenced by all generators and loads in the power system, the use of the relationships means that we can observe the influence of the preventive control action on all contingencies in the power system. It also means that the proposed method can treat contingencies (not a contingency) and stabilize them simultaneously. The outline of the proposed method is as follows.

(1) As the on-line transient stability monitoring, the values of CCTs are calculated for the present system (load) condition and also for the condition 5 or 10 minutes later. It is assumed that the contingencies are the three phase to ground faults occurring at the buses near the generators, because these faults are the most severe cases for transient rotor angle stability. (The times of 5 or 10 minutes are thought of as the target time to carry out transient stability monitoring. If the generator output rescheduling is used for preventive control, the time is necessary to reschedule generator outputs.)
(2) If the CCTs are larger than the predefined values which are larger than the actual operating times of circuit breakers in the power system, no preventive control is necessary. If one or more contingencies have smaller CCTs than the predefined values, preventive control is necessary.
(3) The relationships between CCTs and generator rotor angles are made from the result of (1), since the relationships are linear.
(4) From the expressions of CCTs and generator rotor angles confirmed in (3), the rotor angles are estimated to have larger CCTs than the predefined values. To satisfy the rotor angles, preventive control action (generator output rescheduling) is determined. Here, the generator rotor angle estimation method introduced in section V is used to shorten the calculation time.

When the predefined value is 0.12[s], the proposed method is shown graphically using Fig. 5 (Case1: CCT < 0.12[s]) and Fig. 6 (Ca.se2: CCT > 0.12[s]).

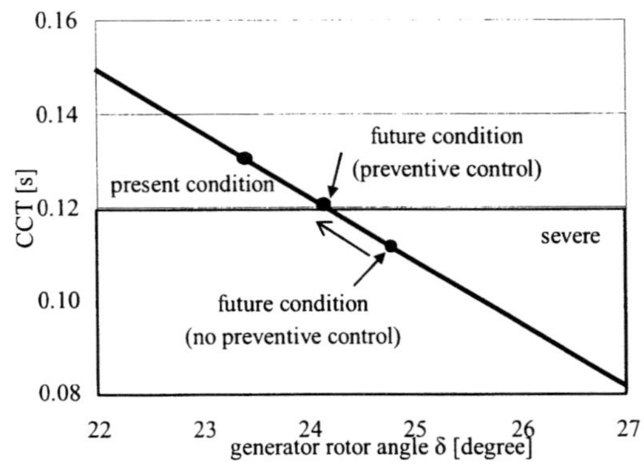

Fig. 5 CCT < 0.12[s]

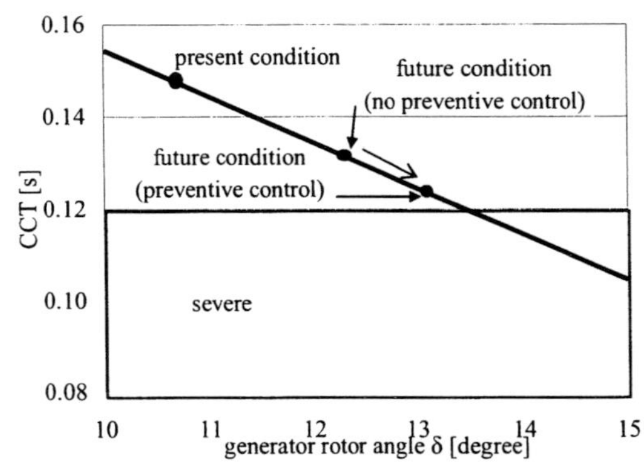

Fig. 6 CCT > 0.12[s]

Case1: The CCT of the contingency is smaller than the predefined value

When CCT of the three phase to ground fault at the bus near the generator is smaller than the predefined value, the generator rotor angle is estimated to make the CCT longer than the predefined value as in Fig. 5. The generator rotor angle is satisfied with the generator output rescheduling.

Case2: The CCT of the contingency is larger than the predefined value

When CCT of the three phase to ground fault at the bus near the generator is larger than the predefined value, the generator rotor angle is estimated not to shorten the CCT more than the predefined value as in Fig. 6.

The generator rotor angles which satisfy the above two conditions are regarded to have the ability of preventive control, and thus the preventive control cannot be uniquely determined. We propose a minimization formulation which is demonstrated in the next section.

6. SIMULATION STUDY

In this simulation study, the predefined target value of CCT is 0.20[s]. For the simulation study, the IEEJ 10 machine system (Fig. 3) is used. All generator outputs can be changed. The contingencies are the three phase to ground faults occurring at the generator buses, which are buses 11 - 20. It is assumed that the fault-cleared system condition is the same as the pre-fault system.

Table 1 and 2 show the generator output *Pgi*, generator terminal voltage *Vgi*, generator rotor angle in the COA reference frame θi and CCT*i* at the present and future conditions, respectively. Here, it is assumed that the future load condition is increased by about 5% from the present load condition.

Table 1 Present condition

gener-	present condition			
ator	*Pgi* [pu]	*Vgi* [pu]	θi [degree]	CCT*i* [s]
1	6.300	1.007	-13.442	0.23
2	9.900	1.005	-1.008	0.27
3	4.849	1.020	-9.535	0.27
4	9.900	1.005	-1.715	0.32
5	5.220	1.020	-8.016	0.36
6	9.900	1.005	-8.191	0.23
7	9.900	1.020	0.137	0.32
8	6.300	1.020	12.598	0.24
9	6.300	1.020	20.388	0.19
10	4.500	1.015	22.671	0.21

Table 2 Future condition

gener-	future condition			
ator	*Pgi* [pu]	*Vgi* [pu]	θi [degree]	CCT*i* [s]
1	6.650	1.007	-14.649	0.21
2	10.450	1.005	-1.016	0.23
3	5.221	1.020	-9.813	0.23
4	10.450	1.005	-1.872	0.29
5	5.510	1.020	-8.739	0.33
6	10.450	1.005	-8.970	0.20
7	10.450	1.020	0.133	0.29
8	6.650	1.020	13.623	0.22
9	6.650	1.020	21.928	0.17
10	4.750	1.015	24.357	0.19

In Table 1 and 2, the CCTs of the contingencies near the generators G9 and G10 are smaller than the predefined target value (=0.20[s]). Therefore, the generator outputs assumed for the future load condition are not good for the stable operation, and preventive control should be considered. From Table 1 and 2, the relationships between CCTs and generator rotor angles can be obtained using the linear expressions below.

$$CCT_1=0.0166\,\theta_1+0.4527 \quad CCT_2=4.9444\,\theta_2+5.2548$$
$$CCT_3=0.1439\,\theta_3+1.6423 \quad CCT_4=0.1907\,\theta_4+0.6470$$
$$CCT_5=0.0415\,\theta_5+0.6930 \quad CCT_6=0.0385\,\theta_6+0.5452$$
$$CCT_7=6.9444\,\theta_7-0.6303 \quad CCT_8=-0.0195\,\theta_8+0.4856$$
$$CCT_9=-0.0130\,\theta_9+0.4548 \quad CCT_{10}=-0.0119\,\theta_{10}+0.4788$$

For example, the expression of CCT_1 is obtained as follows.

$$CCT_1=\frac{0.21-0.23}{-14.649-(-13.442)}\{\theta_1-(-13.442)\}+0.23$$

From these expressions, the generator rotor angles which have larger CCTs than the predefined value (=0.20[s]) can be estimated. Therefore, to satisfy the generator rotor angle with desired CCTs, preventive control action is determined. However, considering the error of the calculations, the target value of CCT is set as 0.22[s], a little longer than 0.20[s]. Then, when the generator rotor angles are as follows, the new CCTs are expected to become larger than the predefined values.

$$\theta_1>-14.0457,\ \theta_2>-1.0182,\ \theta_3>-9.8828,$$
$$\theta_4>-2.2392\ \theta_5>-11.3872,\ \theta_6>-8.4504,\ \theta_7>0.1224$$
$$\theta_8<13.6234,\ \theta_9<18.0782,\ \theta_{10}<21.8276 \qquad (13)$$

Here, θ_1 is calculated as follows, for example,

$$0.22 < 0.0166\theta_1 + 0.4527$$
$$\therefore \theta_1 > \frac{0.22-0.4527}{0.0166} = -14.0457$$

At the future condition, to satisfy the generator rotor angles above, the generator outputs are estimated. The generator rotor angle estimation method using the Jacobian matrix is used for this calculation. The Jacobian is used for the future condition and the preventive control actions are the changes of the generator outputs from the assumed future condition. That means the following expression is used for the estimation.

$$\begin{bmatrix} \vdots \\ \Delta\varphi_i \\ \vdots \\ \Delta\varphi_j \\ \Delta V_j \\ \vdots \end{bmatrix} = \mathbf{J}^{-1}\cdot \begin{bmatrix} \vdots \\ \Delta P_{gi} \\ \vdots \\ 0 \\ 0 \\ \vdots \end{bmatrix} \qquad (14)$$

Since the Jacobian is for the future condition, the changes of the loads are not in this expression, that is, $\Delta P_{lj}=0$ and $\Delta Q_{lj}=0$. When the preventive control action is determined to keep the assumed future condition, the generator output changes are minimized. Moreover, the generator terminal voltage constrains and generator output change constrains are considered. Therefore, the preventive control is determined under the following formulation. Here, the symbols with superscript *f* represent the values assumed for the future condition.

$$Minimize \sum_{generators} |\Delta Pgi| \qquad (15)$$

$$Subject\ to\ |\Delta Pgi| \le |\Delta Pgli| \qquad (16)$$

$$rotor\ angle\ constrains\ such\ as\ in\ (13)$$

$$where \quad Vi = Vi^f + \Delta Vi$$
$$\varphi_i = \varphi_i^f + \Delta\varphi_i$$
$$Pi = Pi(V,\varphi)$$
$$Qi = Qi(V,\varphi)$$

$$\theta_i = \theta_i(P, Q, V, \varphi)$$

(ΔVi and $\Delta\varphi_i$ are calculated from (14))

Table 3 shows the preventive control results for the future load condition. ΔPgi is the generator output change (the change from the assumed future generator output).

The CCTs in Table 3 are longer than the predefined value (=0.20[s]). (They do not achieve 0.22[s]) When the preventive control is not used for the assumed future load condition, the operation becomes severe, however when the preventive control is applied, the operation becomes stable with desired CCTs.

The validity of the proposed method has been shown by the example above.

Table 3 Preventive control results

generator	ΔPgi [pu]	Pgi [pu]	CCT [s]
1	-0.150	6.500	0.22
2	0.066	10.516	0.22
3	0.600	5.821	0.21
4	0.000	10.450	0.28
5	0.000	5.510	0.32
6	0.000	10.450	0.20
7	0.122	10.572	0.29
8	0.000	6.650	0.22
9	-0.164	6.487	0.20
10	-0.500	4.250	0.22

7. CONCLUSIONS

The proposed method assumes the near future system condition considering the frequency of stability monitoring and the time to reschedule generator outputs. In the proposed method, the linear relationships between CCTs and the generator rotor angles, and the generator rotor angle estimation method using the Jacobian matrix are utilized. In the simulation study, the proposed method can achieve the desired values of CCTs. The proposed method can control the CCTs of all contingencies in the power system. The target value of the CCT can be changed for each contingency. CCT, one of the most important indices for transient stability, can be controlled by the proposed preventive control by generator output rescheduling. Moreover, not only the severe contingencies become stable, but also the stable contingencies do not become unstable but keep some stability standard. Therefore, this can be thought of as a reliable method. Power system operators can keep balance of transient stability of the whole power system.

At present, it might be difficult to calculate CCTs at on-line transient stability monitoring. However, with the development of computers and methods to calculate CCTs, it will be possible to operate power systems with transient stability preventive control observing some important transient stability index such as CCT.

REFERENCES

Damien Ernst, Daniel Ruiz-Vega, Mania Pavella, Peter M. Hirsch, Dejan Sobajic : "*A Unified Approach to Transient Stability Contingency Filtering, Ranking and Assessment*", IEEE Transactions on Power Systems, Vol. 16, No. 3, August 2001

Ebrahim Vaahedi, Yakout Mansour, E. K. Tse : "*A General Purpose Method for On-line Dynamic Security Assessment*", PE-577-PWRS-0-05-1997

Wenping Li, Anjan Bose : "*A Coherency Based Rescheduling Method for Dynamic Security*", IEEE Transactions on Power Systems, Vol. 13, No. 3, August 1998

M. La Scala, M. Trovato, C. Antonelli : "*ON-LINE DYNAMIC PREVENTIVE CONTROL : AN ALGORITHM FOR TRANSIENT SECURITY DISPATCH*", IEEE Transactions on Power System, Vol. 13, No. 2, May 1998

ADing-Heh Kuo, Anjan Bose : "*A GENERATION RESCHEDULING METHOD TO INCREASE THE DYNAMIC SECURITY OF POWER SYSTEMS*", IEEE Transactions on Power System, Vol. 10, No. 2, February 1995

A. A. Fouad and Tong Jianzhong : "*STABILITY CONSTRAINED OPTIMAL RESCHEDULING OF GENERATION*", IEEE Transactions on Power Systems, Vol. 8, No. 1, February 1993

Yoshionori Katoh, Shinichi Iwamoto: "*Transient Stability Preventive Control For Stable Operating Condition With Desired CCT*", IEEE Transactions on Power Systems, Vol. 17, No. 4, November 2002

Yoshihisa Sumida, Shinichi Iwamoto: "*Fast Stability/Transmission Limit Assessment Using Critical Clearing Time*", PROCEEDINGS of the 30th North American Power Symposium pp. 69-75, October 1998

IEEJ Study Committee, "*Power System Standard Models (in Japanese)*", IEEJ Technical Report N0.754, 1999 (http://www.iee.or.jp/pes)

Stagg, El-Abiad : *Computer Methods in Power System Analysis*, McGraw-Hill Kogakusha, Ltd., 1968

ELSEVIER
IFAC
PUBLICATIONS
www.elsevier.com/locate/ifac

FORECASTING SHORT-TERM SYSTEM MARGINAL PRICE USING A NEURAL NETWORKS

Jeong-Kyu Lee Min-Soo Kim Jong-Bae Park Joong-Rin Shin

Konkuk University, Seoul, Korea

Abstract: This paper presents an application of the Artificial Neural Network (ANN) technique to forecast the short-term System Marginal Price (SMP). Recently, the deregulation in power industries has impacted on the power system operational problems. The energy market bidding strategies among market participants are highly dependent on the short-term price. Therefore, short-term SMP forecasting is a very important issue to market participants to maximize their profits, and to market operator who may wish to operate the electricity market in a stable sense.
The proposed neural network based forecasting scheme is composed of three layers. In this process, firstly input data are appropriately set up to reflect market conditions. And secondly λ, the coefficient of activation function is suitably adjusted in order to give a proper signal to a neuron considering the inversion of output neuron's signal. Also, selection rule is made to improve the efficiency of the proposed techniques.
The proposed techniques are trained, validated and tested with historical data from Korea Power Exchange (KPX). *Copyright © 2003 IFAC*

Keywords: Electric power systems, Forecasts, Neural networks, Backpropagation

1. INTRODUCTION

The electric power industry in many countries all over the world has been undergoing deregulation, privatisation and restructuring through the introduction of competition. The core objectives of competitive electricity industry are to reduce electricity price, to improve the electricity supply service and to give a right of choice to consumers.

In Korea, Korea Power Exchange (KPX) has been made a start at April 2001, and the generation sector is divided into six companies. Currently, Cost Based Pool (CBP) electricity market is operated. In the future it will be introduced the wholesale and retail competitive market through a division of distribution sector.

The objective of power system planning and operation in vertically integrated industry was to minimize production and operation cost, but the objective of the competitive electricity market is to maximize profits or returns of market participants. Also, power system planning. In the competitive electricity market, is established based on the price signal. It is highly determined by the long-term forecast of the electricity price in deciding on the transmission and generation expansion planning. The energy trading levels between market participants, however, is highly dependent on the short-term price forecast, see (Martini, *et al.*, 2001; Gao, *et al.*, 2000; Szkuta, *et al.*, 1999)

Indeed, the accuracy of short-term market price forecasting is connected directly with the profits of market participants. Therefore, short-term market price forecasting must be performed precisely to give effective information to market participants who would to maximize their profits.

From now on, it will be introduced Two Way Bidding Pool (TWBP) in Korea. In TWBP market, the electricity suppliers and purchasers submit their bidding data to KPX and then, the KPX decides the Market Clearing Price (MCP) using submitted data. In this process, all of electricity suppliers and purchasers need a price forecasting data at each time period to maximize their profits.

There are several techniques for forecasting the SMP. Artificial neural networks (ANN) are used to forecast SMP in Victorian Power System by B. R Szkuta *et al.* (Szkuta *et al.*, 1999). In that paper, it has been performed the analysis of factors that have impact to SMP. Dynamic fuzzy system has been applied by Liu Honghie, *et al.*, to California Power Market (Honghie, *et al.* 2002). Finally, Time series models using dynamic regression and transfer function approaches to overcome the drawback of the linear regression models have been applied by F. J. Nogales *et al.* to Spain and California Markets (Nogales, *et al.* 2002).

In this paper, a new technique to forecast SMP using time and day information approaches based on Back-propagation Neural Network is proposed. To improve the efficiency of proposed technique. An additional selection rule is proposed. The proposed techniques are tested on KPX data, and the results shows that the proposed approach is a promising way.

2. FORMULATION

SMP is determined by taking into account the characteristics of generators and system conditions. Conceptually, market price is the offer price of the highest offer price that is satisfied the balance of supply and demand (highest successful bid). Also, all generators that have facilities and capacities exceed specific level should sell only through electricity markets in Korea competitive electricity market.

In general, the factors such as SMP and priority list are determined by the Independent System Operator (ISO) or Power Exchange (PX) according to the bidding data. And there are several factors that have influence on SMP such as market demand, forced outage rates, capacity of generators, I/O characteristics of generators and conditions of generators etc. Furthermore, SMP is influenced by market structure, correlation of each markets and competitor's bidding strategies too. Therefore, SMP can be represented as follows;

$$SMP = f(MD, FOR_i, C_i, IO_i, OP_i) \quad \forall i \in N \qquad (1)$$

where,

MD : market demand

FOR_i : forced outage rate of ith generator

C_i : capacity of ith generator

IO_i : I/O characteristics of ith generator

OP_i : offer price priority list of ith generator

N : all participated generators

3. NEURAL NETWORKS

Many researchers have been studying about human and human's brain because they would like to realize for computing, judging and recognizing like a human. The recent resurgence of interest in neural networks has its roots in the recognition that the brain performs computations in a different manner than do conventional digital computers. Computers are extremely fast and precise at executing sequences of instructions that have been formulated for them. A human information processing system is composed of neurons switching at speeds about a million times slower than computer gates. Yet, humans are more efficient than computers at computationally complex tasks such as speech understanding. Moreover, not only human, but even animals, can process visual information better than the fastest computers. Especially, in power systems the neural networks have already been used to solve problems such as load forecasting, component and system fault diagnosis, security assessment, unit commitment, etc. (Zurade, 1992; Fausett, 1994).

In this paper, the backpropagation neural networks are applied among the several neural network techniques. During feed-forward, each input unit receives an input signal and broadcasts this signal to the each of the hidden units. Each hidden unit then computes its activation and sends its signal to each output unit. Each output unit computes its activation to form the response of the net for the given input pattern. This activation is compared with target value and then weights are adjusted to reduce difference between the activation and target value. Fig. 1 represents a basic structure of backpropagation neural networks.

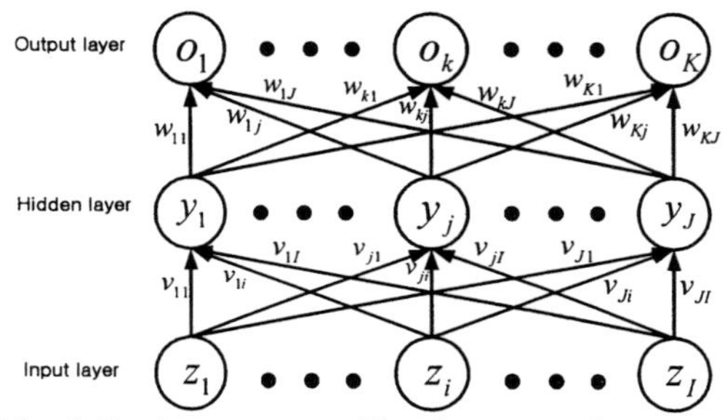

Fig. 1 Basic structure of backpropagation neural networks

where,

z_i : input unit ($i = 1,\Lambda, I$)

y_j : hidden unit ($j = 1,\Lambda, J$)

o_k : output unit ($k = 1,\Lambda\ K$)

v_{ji} : weight between input layer and hidden layer

w_{kj} : weight between hidden layer and output layer

Also, the objective function of neural networks can be formulated to minimize an error (err) as follows;

$$\min\ err = \sum_{p=1}^{P}\sum_{k=1}^{K}\frac{1}{2}(d_{pk} - o_{pk})^2 \qquad (2)$$

where,

P : number of pattern

K : number of output unit

d_{pk} : target value of kth output unit at pth pattern

o_{pk} : calculated value of kth output unit at pth pattern

Neural networks adjust weights during learning process to minimize the objective function. And then, we can get optimal weights. In this paper, the bipolar sigmoid activation function is adopted as follows;

$$f(x)=\frac{2}{1+\exp(-\lambda x)}-1 \qquad (3)$$

where,
λ : coefficient of activation function
x : input value of neural networks

The slope of activation function varies according to λ. Gauss-Newton Method is applied to adjust λ in accordance with input data of each forecasting time period. And momentum is used to increase a learning velocity. In backpropagation neural networks with momentum, the weight change is in a direction that is a combination of the current gradient and the previous gradient. This is a modification of gradient descent whose advantages arise chiefly when some training data are very different from the majority of the data (Fausett, 1994).

4. APPLICATION OF NEURAL NETWORKS

SMP forecasting using the neural networks is impacted by the activation function, the number of hidden layer and the number of units in each layer. Also, even if it is used a same input data and activation function, it may be led to very different results according to the manner of organizing the input data pattern. Therefore, it is very important to search an appropriate manner of organizing the input data pattern for improving the efficiency of algorithm.

In this paper, it is decided the slope of activation function by taking into account input/output characteristics of learning data. And SMP is forecasted with time and day axes information for each time period. To select more accurate results, it is proposed the selection rule.

4.1 Approximation

Generally, SMP is impacted by several factors as mentioned above. However, currently published data are demand forecasting data and past SMP data for each time period. Hence, SMP can be approximately formulated as follows;

$$SMP' = F(MD) \qquad (4)$$

Although, equation (4) does not contain all factors to represent SMP function, SMP is very strongly related with market demand, so it can be approximated like that.

4.2 Time axis approach

In time axis approach, 4 weeks and 4 times data are used, represented in Fig. 2.

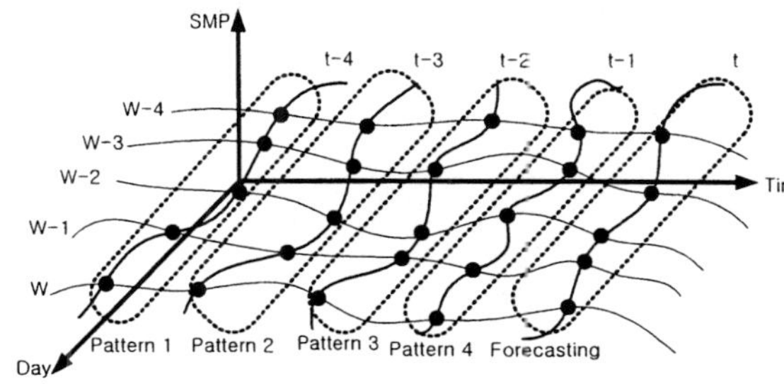

Fig. 2 Time axis approach

4.2.1 Learning process of time axis approach

In learning process, theobjective function value of neural networks for four patterns is minimized to get optimal weights. The optimal weights for $SMP_W^T(t)$ are expressed as follows;

$$\begin{aligned}
&weight_T^* = \\
&\{input: weighted\ avg.of\ SMP_{W-3}(t-i)\cdot SMP_{W-2}(t-i)\cdot SMP_{W-1}(t-i),\\
&\quad SMP_{W-4}(t-i), SMP_{W-3}(t-i), SMP_{W-2}(t-i), SMP_{W-1}(t-i),\\
&\quad Load_{W-4}(t-i), Load_{W-3}(t-i), Load_{W-2}(t-i), Load_{W-1}(t-i),\\
&\quad Load_W(t-i)\\
&output: SMP_W(t-i)\}
\end{aligned} \qquad (5)$$

where,
t : time period to forecast
i : patterns (4, 3, 2, 1)
W : week to forecast

4.2.2 Forecasting process of time axis approach

In forecasting process, $SMP_W^T(t)$ is computed using optimal weights obtained through learning process, i.e. $weight_T^*$, and input data. Following equation expresses this process.

$$\begin{aligned}
&SMP_W^T(t) = \\
&\{weighted\ avg.of\ SMP_{W-3}(t)\cdot SMP_{W-2}(t)\cdot SMP_{W-1}(t),\\
&SMP_{W-4}(t), SMP_{W-3}(t), SMP_{W-2}(t), SMP_{W-1}(t),\\
&Load_{W-4}(t), Load_{W-3}(t), Load_{W-2}(t), Load_{W-1}(t),\\
&Load_W(t)\}
\end{aligned} \qquad (6)$$

4.3 Day axis approach

In day axis approach, 4 weeks and 3 times data are used. The data and patterns are represented in Fig. 3.

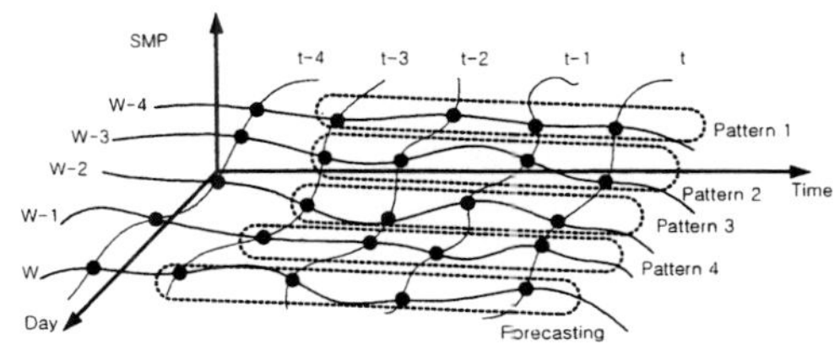

Fig. 3 Day axis approach

4.3.1 Learning process of day axis approach

In learning process, objective function value of neural networks for four patterns is minimized to get optimal weights. The optimal weights for $SMP_W^D(t)$ are expressed as follows;

$$\begin{aligned}
&weight_D^* = \\
&\{input : weighted\ ave.of\ SMP_{W-i}(t-3)\cdot SMP_{W-i}(t-2)\cdot SMP_{W-i}(t-1), \\
&\quad SMP_{W-i}(t-3), SMP_{W-i}(t-2), SMP_{W-i}(t-1), \\
&\quad Load_{W-i}(t-3), Load_{W-i}(t-2), Load_{W-i}(t-1), \\
&\quad Load_{W-i}(t) \\
&output : SMP_{W-i}(t)\}
\end{aligned} \tag{7}$$

where,
t : time period to forecast
i : patterns (4, 3, 2, 1)
W : week to forecast

4.3.2 Forecasting process of day axis approach

In forecasting process, $SMP_W^D(t)$ is computed using optimal weights obtained through learning process, i.e. $weight_W^D$, and input data. Following equation expresses this process.

$$\begin{aligned}
&SMP_W^D(t) = \\
&\{weighted\ ave.of\ SMP_W(t-3)\cdot SMP_W(t-2)\cdot SMP_W(t-1), \\
&SMP_W(t-3), SMP_W(t-2), SMP_W(t-1), \\
&Load_W(t-3), Load_W(t-2), Load_W(t-1), \\
&Load_W(t)\}
\end{aligned} \tag{8}$$

4.4 Selection rule for time and day axes approaches

The selection rule means that how to select better approach out of time and day axes approaches. Fig. 4 represents the points used in the selection rule.

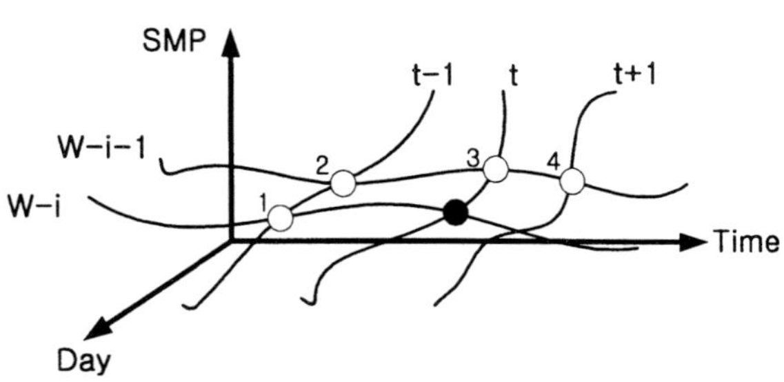

Fig. 4 Points to be used in the selection rule

The variables to be used in the selection rule are calculated for 4 weeks, and are formulated as follows;

Point 1

$$\begin{aligned}
corr_1^{axis}(t) &= \rho_t\{SMP_{axis,1}^{W-i}(t-1), SMP_{axis}^{W-i}(t)\} \\
corr_1^{axis} &= \frac{1}{T}\sum_{t=1}^{T} corr_1^{axis}(t)
\end{aligned} \tag{9}$$

Point 2

$$\begin{aligned}
corr_2^{axis}(i) &= \rho_i\{SMP_{axis,2}^{W-i-1}(t-1), SMP_{axis}^{W-i}(t)\} \\
corr_1^{axis} &= \frac{1}{I}\sum_{i=1}^{I} corr_1^{axis}(i)
\end{aligned} \tag{10}$$

Point 3

$$\begin{aligned}
corr_3^{axis}(i) &= \rho_i\{SMP_{axis,3}^{W-i-1}(t), SMP_{axis}^{W-i}(t)\} \\
corr_3^{axis} &= \frac{1}{I}\sum_{i=1}^{I} corr_3^{axis}(i)
\end{aligned} \tag{11}$$

Point 4

$$\begin{aligned}
corr_4^{axis}(i) &= \rho_i\{SMP_{axis,4}^{W-i-1}(t+1), SMP_{axis}^{W-i}(t)\} \\
corr_3^{axis} &= \frac{1}{I}\sum_{i=1}^{I} corr_4^{axis}(i)
\end{aligned} \tag{12}$$

$$\%corr_i^{actual} = \frac{corr_i^{actual}}{\sum_{i=1}^{I} corr_i^{actual}} \times 100\% \tag{13}$$

$$r_i^T = corr_i^T \times \%corr_i^{actual} \tag{14}$$

$$r_i^D = corr_i^D \times \%corr_i^{actual} \tag{15}$$

where,
$corr_i^{actual}$: average value of correlation coefficients between actual SMPs at i point
$\%corr_i^{actual}$: percentage value at i point
$corr_i^T$: average value of correlation coefficients between time axis approach results for i point
r_i^T : effective rate for i point in time axis
$corr_i^D$: average value of correlation coefficients between day axis approach results for i point
r_i^D : effective rate of i point in day axis

The values of r_i^T and r_i^D are used as a index in the selection rule to select better approach. Fig. 5 represents the flowchart of selection rule.

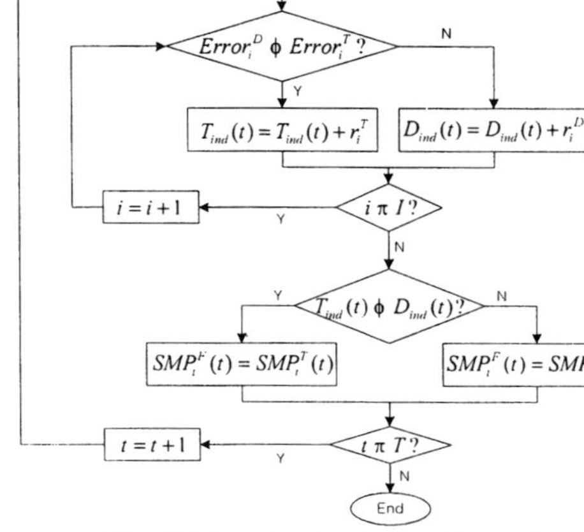

Fig. 5 Flowchart of selection rule

5. CASE STUDY

The actual time data that was used for testing was derived from the Korea Power Exchange. Case studies are implemented for weekdays during April 2002.

The SMP of holiday changes very extremely, because the range of demand is placed between 25,000MW and 32,000MW and there exist different fuel type units. In that region, although demand changes very small, SMP may be changes very large. Therefore, case studies are simulated on weekdays only. Fig. 6 represents the hourly curve for demand and SMP.

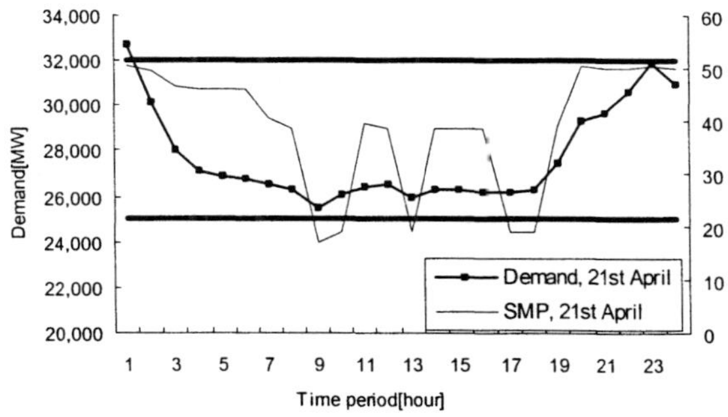

Fig. 6 Demand and SMP at 21st April

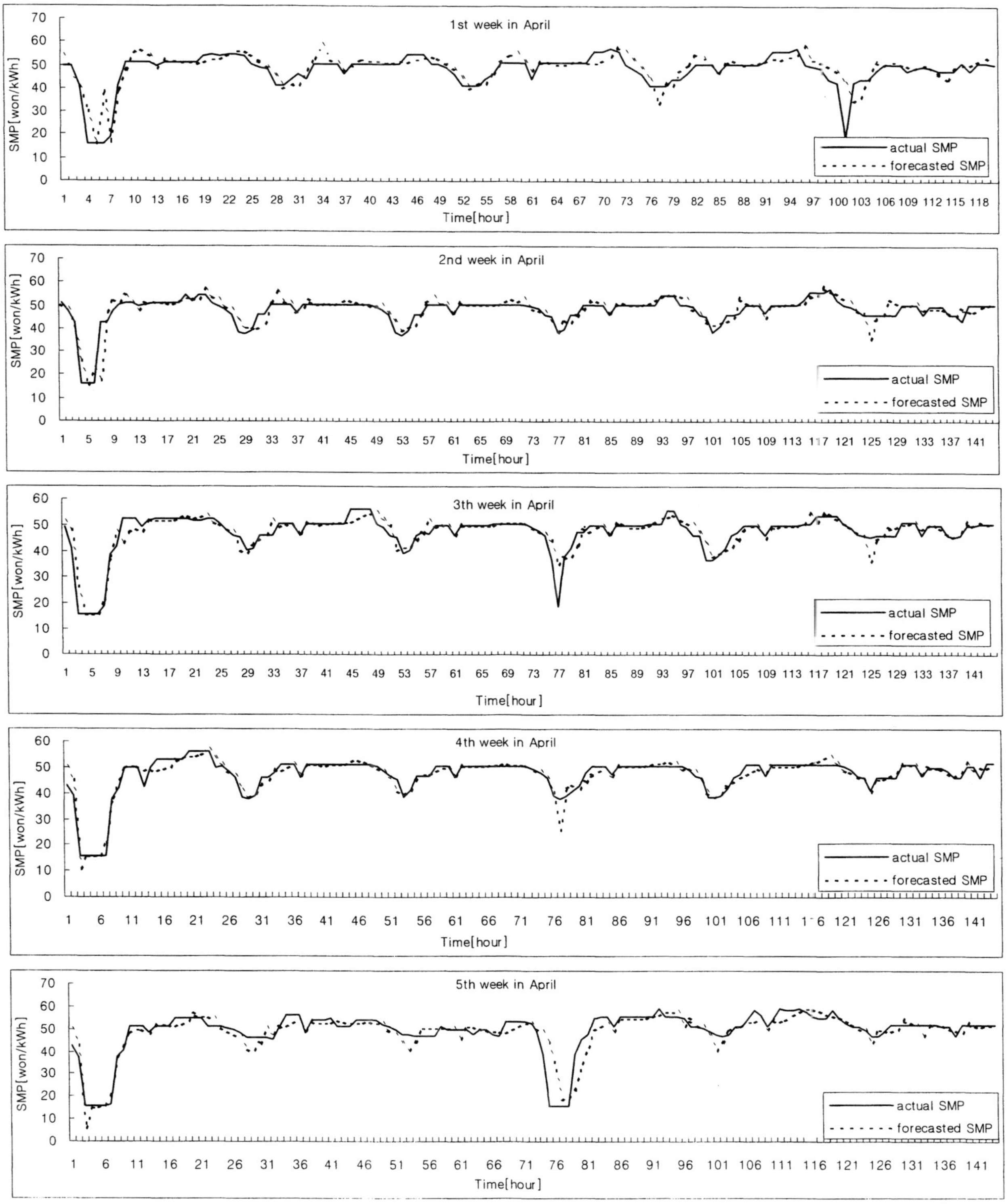

Fig. 7 Actual and forecasted SMP for April 2002

Table 1 The results during April 2002

Day	Monday			Tuesday			Wednesday			Thursday			Friday			Saturday			Average		
Date	1/4/2002			2/4/2002			3/4/2002			4/4/2002			5/4/2002			6/4/2002					
Axis	Time	Day	Slt.	Time	Day	Slt.	Time	Day	Slt.	Time	Day	Slt.	Time	Day	Slt.	Time	Day	Slt.	Time	Day	Slt.
Err(%)	19.60	12.70	12.42	3.75	5.62	4.75	3.89	4.28	3.92	4.47	5.32	5.07				11.55	11.09	10.55	8.64	7.80	7.34
Date	8/4/2002			9/4/2002			10/4/2002			11/4/2002			12/4/2002			13/4/2002					
Axis	Time	Day	Slt.	Time	Day	Slt.	Time	Day	Slt.	Time	Day	Slt.	Time	Day	Slt.	Time	Day	Slt.	Time	Day	Slt.
Err(%)	13.31	9.95	8.37	3.48	4.40	3.37	3.03	4.44	2.89	2.98	4.06	2.97	3.16	4.18	2.82	4.07	4.48	4.45	5.00	5.25	4.15
Date	15/4/2002			16/4/2002			17/4/2002			18/4/2002			19/4/2002			20/4/2002					
Axis	Time	Day	Slt.	Time	Day	Slt.	Time	Day	Slt.	Time	Day	Slt.	Time	Day	Slt.	Time	Day	Slt.	Time	Day	Slt.
Err(%)	16.65	6.06	6.97	2.65	2.89	2.82	2.59	3.34	2.65	6.65	6.65	6.63	3.83	4.17	3.13	2.78	3.41	2.22	5.86	4.42	4.07
Date	22/4/2002			23/4/2002			24/4/2002			25/4/2002			26/4/2002			27/4/2002					
Axis	Time	Day	Slt.	Time	Day	Slt.	Time	Day	Slt.	Time	Day	Slt.	Time	Day	Slt.	Time	Day	Slt.	Time	Day	Slt.
Err(%)	13.33	6.49	6.80	3.41	1.74	1.96	2.05	1.84	1.67	3.47	3.13	3.88	3.29	2.18	2.27	2.31	2.85	2.07	4.64	3.04	3.11
Date	29/4/2002			30/4/2002			1/5/2002			2/5/2002			3/5/2002			4/5/2002					
Axis	Time	Day	Slt.	Time	Day	Slt.	Time	Day	Slt.	Time	Day	Slt.	Time	Day	Slt.	Time	Day	Slt.	Time	Day	Slt.
Err(%)	12.06	7.10	7.22	3.71	4.02	4.39	3.60	4.71	4.71	24.30	20.23	19.95	3.67	4.89	4.98	2.48	3.63	2.89	8.30	7.43	7.36

Although it is a weekday, the characteristics of the day after holiday are different from other weekdays because of maintenance scheduling of generators and demand curve etc. As a result of that, the error of the day after holiday is higher than the error of others. Fig. 7 represents the results of the proposed method, and that are compared with the actual values.

Table 1 and table 2 represent the errors of daily and monthly respectively. From table 1, the average error of SMP of the day after holiday is 10.33% and the average error of SMP of weekdays is 3.6%. Furthermore, in the situation that the demand is placed between 25,000MW and 32,000MW at daybreak of weekdays, the error is higher than the error in the daytime.

Table 2 The average errors for April 2002

	Time	Day	Comb.
avg. err(%)	6.49	5.59	5.20

From table 2, the average error of time and day axes is 6.49% and 5.59% respectively. And the average error using selection rule is 5.20%. That is, the average error using selection rule is lower than the average error of time and day axes approaches.

6. CONCLUSION

In this paper, two forecasting methods and one selection rule has been proposed. The proposed methods are time and day axes approaches, and to improve the efficiency of the proposed methods selection rule is developed. The proposed methods are simulated using backpropagation neural networks. And to adjust λ it is applied to the Gauss-Newton method in accordance with input data of each forecasting time period. Also, correlation coefficients are calculated to apply to selection rule.

From the results, the average error using selection rule is about 5% and daily error is varying from 1.67% to 19.95%, but the errors except the day after holiday almost maintains 3% therefore proposed method may can applied short-term SMP forecasting.

Forecasting errors of the day after holidays are relatively high because of the approximation of SMP function and the shortage of input data. That is, SMP is impacted by several factors but we assume that the SMP is the function of demand only.

In the future, the precise method should be developed to reflect the characteristics of the day after holiday, and the analysis of input data has to be followed.

Acknowledgments

This work was financially supported by MOCIE through IERC program.

REFERENCES

Fausett, Laurene (1994). *Fundamentals of Neural Netweorks*, Prentice Hall.

Gao Feng, Xi-Ren Cao and Alex Papalexopoulos (2000). Forecasting power Market Clearing Price and Quantity Using a Neural Network. *Power Engineering Society Summer Meeting*, IEEE Vol. 4, pp 2183-2188

Hongjie, Liu, Qang Xiufeng, Zhang Weicun and Xu Guohua (2002). Market Clearing Price Forecasting Based on Dynamic Fuzzy System. *Power System Technology*, pp 890-896

Martini, A., P. Pelacchi, L Pellegrini, M.V. Cazzol A, Garzillo and M. Innorta (2001). A Simulation Tool for Short Term Electricity Market. *22nd IEEE Power Engineering Society International Conference on*, pp 112-117

Nogales, F.J., Contreras J., Conejo A.J. and Espinola R. (2002). Forecasting Next-day Electricity Prices by Time Series Models. Power Systems, IEEE Transaction on, Vol. 17 Issue 2, pp 342-348

Szkuta, B.R., L.A. Sanabria and T.S. Dillon (1999). Electricity Price Short-term Forecasting Using Artificial Neural Networks, *IEEE Transactions on Power System, Vol. 14, No. 3*, pp 851-857

Zurade, Jacek M., (1992). *Introduction to Artificial Neural Systems*, West Publishing Company.

ELSEVIER
IFAC
PUBLICATIONS
www.elsevier.com/locate/ifac

A STUDY OF CHANNEL MODELING FOR BROADBAND POWER LINE COMMUNICATION

Young-Kwan Choi*, Jeong-Hoon Lee, Jae-Sang Cha***Jong-Wan Seo*, Myung-Chul Shin***

**sungkyunkwan Univ., **hynix, ***seokyeong Univ.*

Abstract: Power lines are directly exposed to noise emitted from various electrical equipment, home appliances and office equipment. The transmission characteristics of power lines are also changed frequently by when and where load is connected to them. The existing narrow band channel models have too many drawbacks to improve the poor environment for power line communication (PLC) and communication error correction capability. Therefore a modeling technique for PLC, which would be used in the newly approved broadband frequency, needs to be developed. - A task which requires a variety of empirical studies based on remodeling of PLC.
Therefore using a mixture noise modeling techniques,this thesis analyzed power line channel model which includes factors like load change and random noise. To validate its results, the paper calculated bit error rates by applying Code Division Multiple Access (CDMA). *Copyright © 2003 IFAC*

Keywords: power line communication (PLC), a mixture noise model, power line channel model

1.INTRODUCTION

With mush research on and development of Power Line Communication (PLC) being actively conducted recently, the Korean government plans to extend the frequency allowance range of transmission channels for PLC from the current 450 kHz narrowband to 30 MHz broadband. Now, high-speed PLC technologies to transmit lots of data are being actively developed. The key to developing and using the communication system technology is the analysis of power line channels at 30 MHz (pending broadband approval)
Predicting the characteristics of power lines is difficult. Power lines are directly connected to and affected by load, of which types are vary and shows irregular characteristics depending on the time. Power lines' characteristics are changed according to their lengths, types and installment methods. The structure of power greed also changes the characteristics of high frequency, thereby making it hard to exactly predict the characteristics of power lines.
Recently, many models on transmission wires such as cable have been developed from various perspectives. But because of their short history, little work has been done on modeling and theorizing power line channels. Therefore, creating and thoroughly analyzing a formula for power line channels are essential to getting basic data for an accurate performance analysis of a PLC system.

2.NOISE COMPONENTS FOR BROADBAND PLC

Significant characteristic-determining factors of transmission channels are noise, signal attenuation, impedance mismatching, etc. Types and characteristics of these factors are as below.

2.1 Noise

Power lines are originally designed for power supply, not for communication. Thus they are vulnerable to communication-hindering noise. To maintain stable communication performance of PLC, measures need to be taken against noise.

Types of noise There are three kinds of noise generally occurring in power lines.

1) noise synchronized with 60Hz frequency
2) white noise
All existing communication channels on earth have background noise. The noise is created by heat noise, which occurs when free electrons are heated and start moving randomly. Another cause of the noise is powered electronic appliances such as a motor for an electric drill. When the brush switches to maintain the motor's speed steadily, it takes on traits similar to white noise. General background noise in various frequencies is expressed as follows.

$$P_{back}(f) = N_0/2 \quad \text{watts/hertz} \quad (-\infty<f<\infty) \text{ ----- (1)}$$

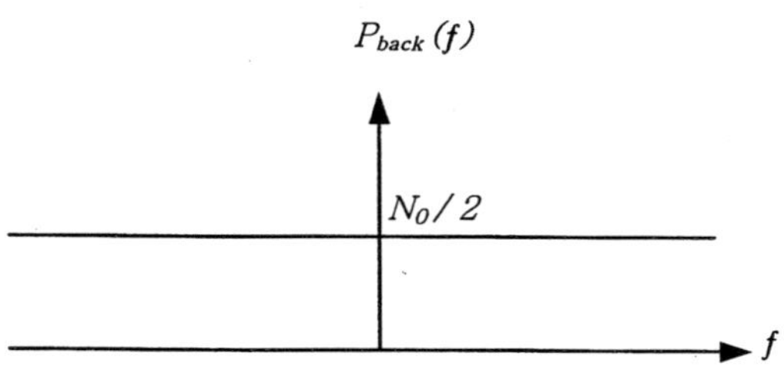

Fig.1 white noise

The figure 1, which is actually measured date, indicates that background noise is -50dBmV and N_0 is as follows.

$$20\log u_{reff}/1mV = -50dBmV \text{ ------------------------ (2)}$$

$$u_{reff} = 10^{-(50/20)} \cdot 1\text{mV} = 3.16\mu V \text{ -------------------- (3)}$$

$$N_0 = (u_{reff})^2 = 9.9\times10^{(-8)}\ V^2/Hz \text{ ------------------ (4)}$$

3) Periodic and random impulse noise
Impulse noise is divided into periodic noise emitted from some electronic appliances and random impulse noise caused by things like switching. A prime example of a device causing periodic impulse is TV. A TV creates impulse noise by being periodically synchronized with power frequencies.
When you switch an electronic appliance on or off, sparks are transmitted to power lines, triggering random impulse noise. The noise is irregular and it is hard to predict how many times they would occur.

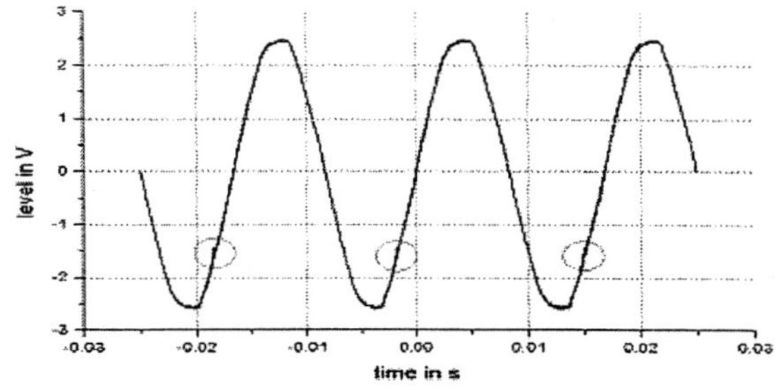

Fig.2 periodic impulse noise

Periodic impulse noise was measured as shown in the figure 3.7. It indicates that periodic impulse noise occurred once in every 16.67[ms], which is one cycle of 60[Hz]. It lasted for 0.1[ms] and was 0.2[V].

In the figure 2 and 3, impulse noise is expressed as circles. The figure 3, an enlargement of the figure 2, well demonstrates periodic impulse. (The voltage marked in the figure 2 and 3 is 1/00 of actual voltage.)

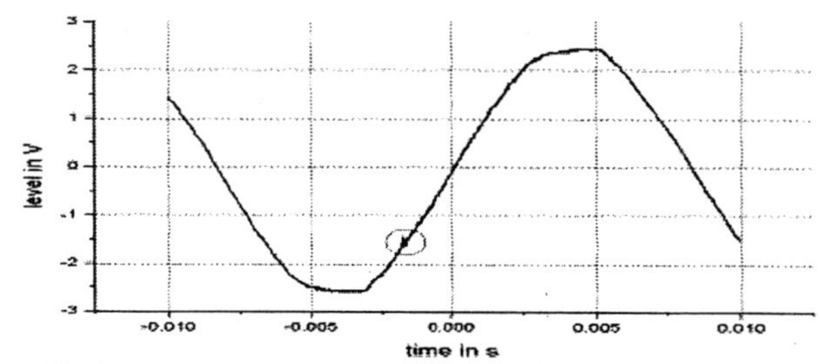

Fig.3 an enlargement of the periodic impulse noise

Random impulse noise mostly lasted for 0.1[ms] and its voltage was different depending on load. The figure 4 shows a random impulse noise wave, an impulse that occurs when a fluorescent lamp is turned on or off. (The voltage marked in the figure 4 is 1/00 of actual voltage.)
It lasted for 0.1[ms] and was 160[V]

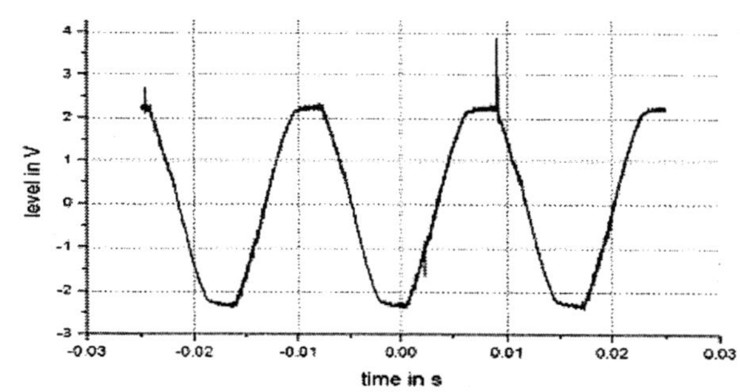

Fig.4 random impulse noise

The impulse created by turning on and off an electric fan was 200[V] and usually lasted for 0.1[ms], sometimes even 0.4[ms]. When a hair dryer was in use, the impulse was 200[V] and lasted for 0.1[ms]. There are also burst noise, of which sources are not yet clear.

Characteristics of Noise Since the noise affecting PLC coexists with AC power, a high proportion of the nose is spike noise created by changes in an electronic device's load. In case of outdoor power lines, lots of high frequency noise could occur.

2.2 Attenuation

Wired transmission channels including a power line have diverse attenuation coefficients depending on a medium's characteristics. The distance that a transmission signal reaches varies according to the attenuation coefficient of a channel. Let's assume that a PLC system with a cellular network is designed, its attenuation coefficient would be closely related to the system's cell coverage.

2.3 Impedance Mismatching

Impedance matching is a critical factor in transmitting signals from a transmitter to a receiver and facilitating lowering of frequencies. In particular, when a switch built in load is turned on or off, the momentary change in impedance causes impedance mismatching. Since the mismatching is closely related to error rates in data transfer, it undermines systems performance.

3. THE ANALYSIS OF A PLC SYSTEM'S BIT ERROR RATE UNDER THE ENVIRONMENT WITH WHITE NOISE AND IMPULSE NOISE

On the basic assumption that the problems of attenuation and impedance matching are resolved, this paper analyzed every factor of a formula for a PLC channel model mainly focusing on white noise and impulse noise. They are major determining factors of bit error rate, a measurement of a communication system's basic function.

The paper applied mixture noise modeling technique to PLC channels. The technique was suggested by Miller and Thomas and could be widely used in terms of the characteristics of power lines and noise-occurring area. The method proposed by the model expresses total noise in the form of $n(t) = n_B(t) + n_I(t)$. The formula is produced by the combination of AWGN (Additive White Gaussian Noise) element (n_B) and Impulse Noise element (n_I).

The formula (5) is the first order probability density function of the mixture noise model.

$$f_n(x) = (1-\varepsilon)\, f_B(x) + \varepsilon f_I(x) \quad \text{-------------------- (5)}$$

In this formula, f_B is AWGN's probability density function, of which power spectrum density is $N_0/2$. Also f_I is a probability density function that has higher variance than AWGN. Its probability is $0 \leq \varepsilon \leq 1$.

The ratio of f_B 's variance (σ_I^2) to f_I 's variance (σ_B^2) is the formula (6)

$$\gamma^2 = \sigma_I^2 / \sigma_B^2 \quad \text{------------------------------------ (6)}$$

Here, γ^2 is a parameter chonsen for this experiment, which indicates the quantity of an impulse. Although the variance of an impulse noise factor here increases as shown in the formula (6), it is considered Raplacian distribution similar to Gaussian distribution.
The total noise variance is given as a formula like this.

$$\sigma^2 = N_0/2[(1-\varepsilon)+\varepsilon\gamma^2] \quad \text{------------------------------ (7)}$$

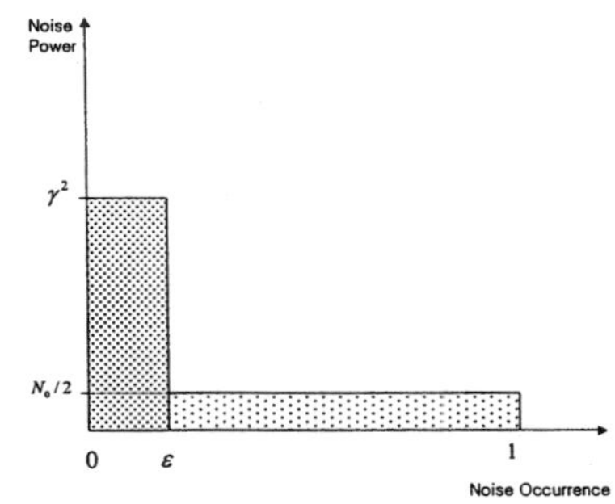

(a)

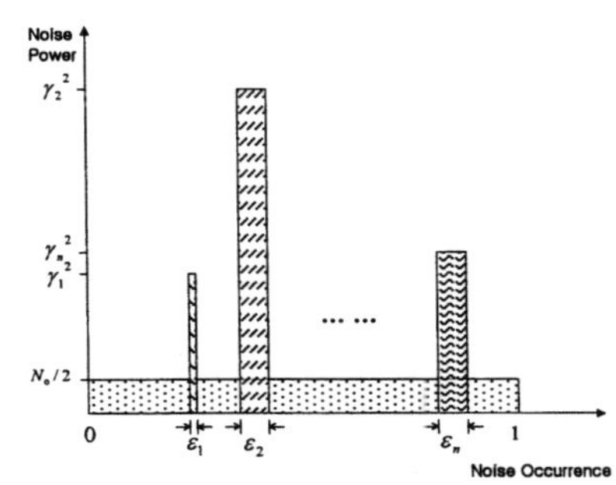

(b)

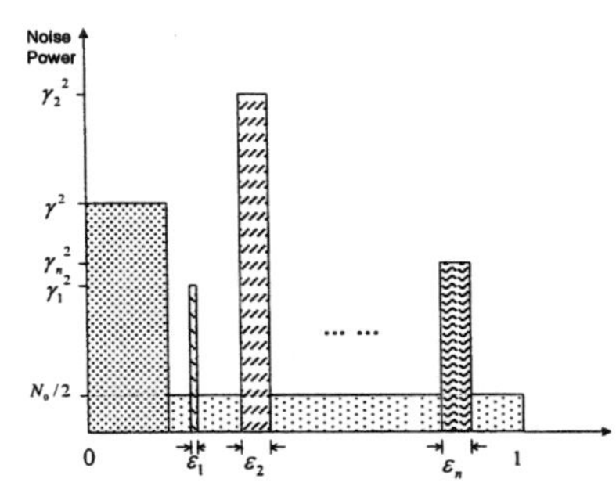

(c)

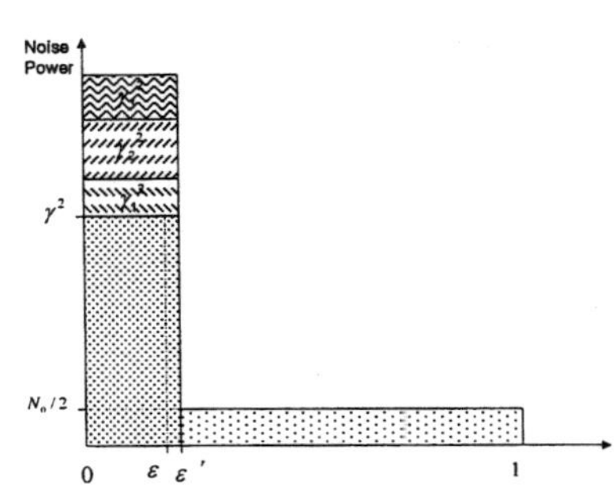

(d)

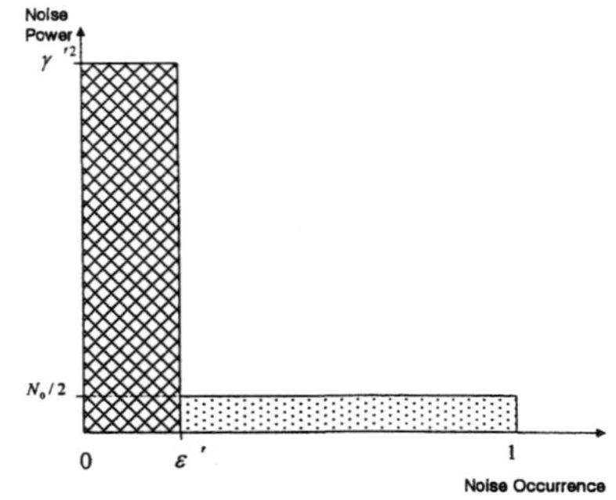

(e)

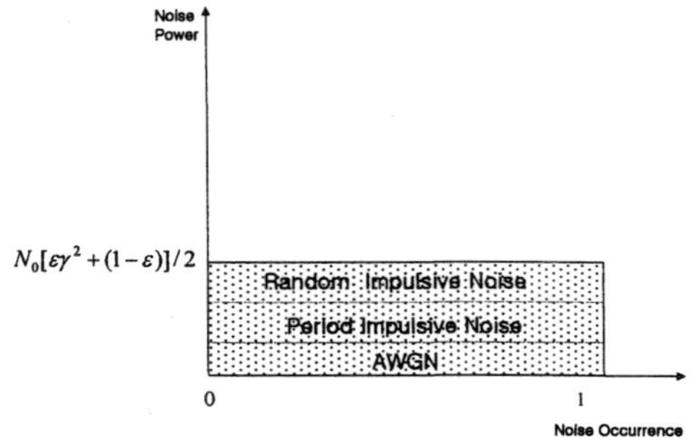

(f)

Fig.5 noise

The figure 5 (a) indicates periodic noise. The figure 5 (b) indicates random noise, which is a unique characteristic of a power line channel. It is hard to predict how many times, how strong (γ^2), how long(ε) an impulse would occur. The figure 5 (c) is the combination of periodic and non-periodic noise. It shows the kinds of noise existing in a power line channel. When $\varepsilon' = \varepsilon_1 + \varepsilon_2 + \varepsilon_3 + \cdots + \varepsilon_n$, the sum of duration periods, and $\gamma'^2 = \gamma^2 + \gamma_1^2 + \gamma_2^2 + \cdots + \gamma_n^2$, the sum of noise power are combined, they are expressed again like the figure 5 (d). In other words, ε and γ^2 in figure 5 (a) increased to ε' and γ'^2 respectively, thereby being expressed again as in figure 5 (d).

Periodic impulse noise lasted for 0.1[ms] and was 0.2[V]. It occurred every 16.67[ms], which is one cycle of 60[Hz]. Most random impulse noise lasted for 0.1[ms], sometimes even 0.4[ms]. Its voltage differed depending on load, but most home appliances were about 160[V] or 200[V].

From these measured data, we can analogize that the occurrence of random noise has little impact on noise duration period (ε) but significantly affect noise power (γ^2). Therefore, as shown in the figure 5 (e), the traits of periodic and random noise can be expressed in the form of increased noise duration and power of periodic noise.

The total noise power of the impulse noise model is the whole spreading area, as demonstrated in the figure 5 (f). It is average noise power as shown in the figure 5 (e), where periodic and random noise uniformly spread.

The greatest merit of this model is the parameter, which is calculated from physically measured quantity, not chosen under a certain circumstance. Different theses have applied different average incidence rates and average duration areas of a impulse noise wave as well as ratios of impulse noise waves to AWGN power (γ^2).

Considering the fact that impulse noise lasts for 0.1[ms] in 16.67[ms] once it occurs. this paper adopted as the sums of impulse duration hours(ε'), 0.2/16.67 ≒ 0.01 (when one periodic noise and one random noise occur in one cycle) and 1.6/16.67 ≒ 0.1 (when one periodic noise and 15 random noise occur during one cycle). It also applied 10~10,000 as the sum of noise power (γ'^2) given the fact that impulse noise power is higher than that of white noise by 10dB and sometimes 40dB.

Based on this impulse noise model, the performance analysis of 16QAM CDMA's BER could be conducted.

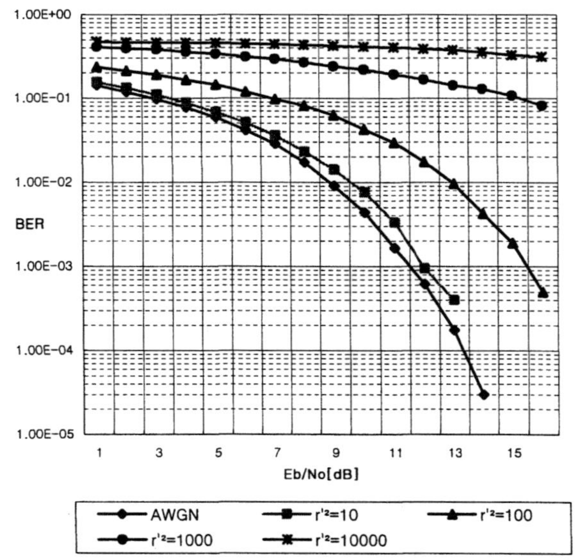

6.1 CDMA's BER for $\varepsilon' = 0.01$

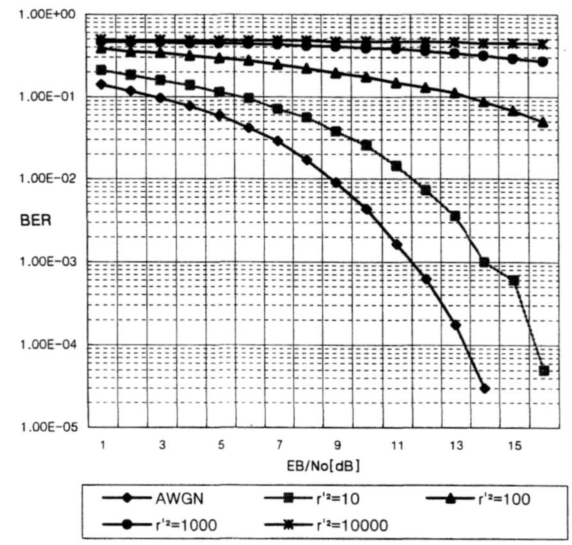

6.2 CDMA's BER for $\varepsilon' = 0.001$

As shown in the figure 6.1 and 6.2, when noise power (γ'^2) was 10,000, in other words, when impulse power was higher than that of white noise by more than 40dB, communication was almost impossible. Neither was it when noise power (γ'^2) was 1,000.

As the two figures show, not only noise power (γ'^2) but also impulse noise duration (ε') has significant impact on the performance of BER.

4.CONCLUSION

In this paper, we measured periodic and random impulses experimentally. In observing periodic and random impulses, the paper used the sum of their noise power and the sum of their duration hours to produce a new form of periodic impulse noise shown in figure (5) and applied the result to a mixture noise model. To confirm the validity of the result, the study used the mixture noise modeling technique to establish a PLC channel model and again applied the result to CDMA to calculate bit error rates.

[reference]

B. Aazhang and H. V. Poor, "Performance of DS/SSMA comunications in impulsive channels-part I: Linear correlation receivers," *IEEE Trans. Comm.*, vol. COM-35. no. 11, pp. 1179~1183, 1987.

J. H. Miller and J. B. Thomas, "The detection of signals in impulse noise modeled as a mixture process," *IEEE Trans. Comm.,* vol. COM-24, pp. 559~563, 1976.

R.M. Vines, et al. "Noise on residential power distribution circuits", *IEEE Trans. Electromag. Compat.,* vol. 26, pp.161~168, Nov. 1984

R. M. Vines, H. J. Trusell, L. J. Gale, "Noise on Residential Power Distribution Circuits", *IEEE Trans. on electromagnatic Compatibility*, vol. EMC-26, No.4, 1984

S. Kapoor, V. P. Telang, J. D. Mills, and M. G. Makhija, "Mitigating impulse noise over HFC network for telephony-over-cable", Proc. *GLOBECOM'95*, vol. 2, pp. 967~971, 1995.

ELSEVIER
IFAC
PUBLICATIONS
www.elsevier.com/locate/ifac

THE INFLUENCE OF CONDUCTOR SAG ON SPATIAL DISTRIBUTION OF TRANSMISSION LINE MAGNETIC FIELD

Babak Porkar Kumleh[1], M.H.Varahram [1], Siyamak .Porkar Kumleh [2]

[1] *Electrical Engineering Dept. Sharif University of Technology, Iran*
[2] *Electrical Engineering Dept. Amir Kabir University of Technology, Iran*

Abstract: Common methods for the calculation of electric fields created by power transmission lines assume straight lines parallel to a flat ground and the influence of the sag due to the line weight is neglected. In this paper the magnetic fields of a catenary 3-Phase line that has a change in direction was calculated and discussed. It is believed that this work has not been published before. *Copyright © 2003 IFAC.*

Keywords: Magnetic Fields, Transmission Lines, Sag Conductor.

1. INTRODUCTION

The calculation and measurement of fields generated by high voltage power transmission lines has been a major concern in the past years. Of the possible health implications and environmental impact caused by electric and magnetic fields of high voltage electric power lines, the most difficult issue to face will often be public perceptions of the possible health hazard, specifically an increased risk of contracting cancer, caused by the magnetic field of the power line(Web sites; Olsen, 1983; EPRI, 1982) Also utilities are challenged by the need to expand transmission system capacity to meet growing customer requirements. The acquisition of new transmission Right-of-Way (ROW) is very difficult, expensive and time consuming due to land use public concerns. Standardization of right-of-way distances for power lines sometimes include consideration of expected magnetic fields (Mamishev, 1996) Other research has centered upon the design of power line proximity detectors, instrumentation and measurements equipment, and new line configuration. The ability to calculate magnetic fields levels that corrective with measurement data is important in all of these investigation (Mamishev, 1996). Several publications have been made to calculate and measure the electric and magnetic fields created by power transmission lines. Most assume lines parallel to a flat ground, and the sag due to the line weight is neglected (EPRI, 1982). One publication (Mamishev, 1996) calculated the magnetic field of a catenary line in state of single phase. In this paper presented method in (Mamishev, 1996), was extended not only to 3-Phase transmission lines but also to those that have a change in direction.

2. CATENARY GEOMETRY

Figure 1 depicts the basic catenary geometry for a single-conductor line, where H is the maximum height on the extremes of the line where it is anchored, h is the minimum height at midspan (hence the sag is S=H-h) and L is the span length between adjacent towers.
A concise analytical method for calculation of magnetic field in the direct vicinity of power lines

has been developed and implemented for three-dimensional field modeling in (Mamishev, 1996). In (Mamishev, 1996) the expression for the total field has been presented:

$$\hat{B}_0 = \frac{\mu_0}{4\pi}\sum_{i=1}^{M}\sum_{k=-N}^{N}\int_{-L/2}^{L/2}\left(J_x a_x + J_y a_y + J_z a_z\right)dx \tag{1}$$

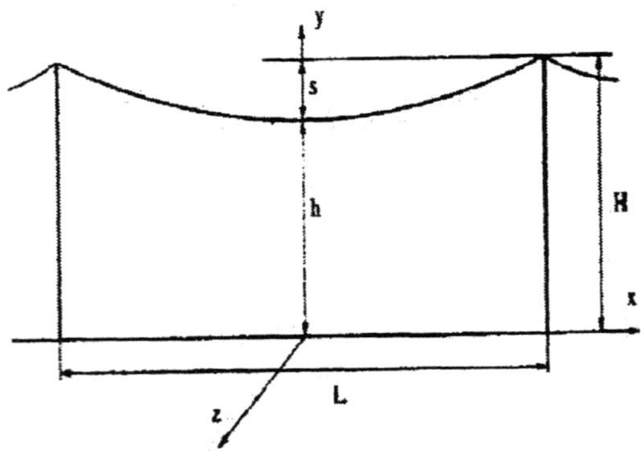

Fig. 1. Basic catenary geometry

Where the parameter M in (1) represents the number of individual conductors on the support structures. Also the number of spans to the right and to left the generic one is represented by N in (1). And

$$J_x = \frac{\hat{I}_i z_0}{4\pi}\left[\frac{\sinh(a_i x)}{d_i} - \frac{\sinh(a_i x)}{d_i'}\right] \tag{2}$$

$$J_y = -\frac{\hat{I}_i z_0}{4\pi}\left[\frac{1}{d_i} - \frac{1}{d_i'}\right] \tag{3}$$

$$J_z = -\frac{\hat{I}_i}{4\pi}\left[\frac{\frac{\cosh(a_i x)}{a_i} - (y_0 - y_{di})}{d_i} - \frac{(x - x_0 + kL)\sin(a_i x)}{d_i} - \frac{\frac{\cosh(a_i x)}{a_i} - (y_0 - 2y_{di} - \alpha)}{d_i'} + \frac{(x - x_0 + kL)\sin(a_i x)}{d_i'}\right] \tag{4}$$

$$d_i = \left[\left((x - x_0 + kL)^2 + z_0^2\right) + \left(\frac{\cosh(a_i x)}{a_i} - y_0 - y_{di}\right)^2\right]^{\frac{3}{2}} \tag{5}$$

$$d_i' = \left[\left((x - x_0 + kL)^2 + z_0^2\right) + \left(\frac{\cosh(a_i x)}{a_i} - (y_0 - 2y_{di} - \alpha)\right)^2\right]^{\frac{3}{2}} \tag{6}$$

Also the complex depth α of each element of image current can be found as given in (Olsen, 1983; Wait, 1969)

$$\alpha = \sqrt{2}\delta_g e^{-j\frac{\pi}{4}} \tag{7}$$

The parameter δ_g in (7) is the skin depth of the earth represented by

$$\delta_g = 503\sqrt{\frac{\rho_g}{f}}$$

Where ρ_g is the resistivity of earth, and f frequency of the source current in Hz.

3. EXAMPLES OF CALCULATIONS

A computer program is developed based on mentioned equations. In proof of the correctness of the developed computer program, example in (Mamishev, 1996) is solved again by the developed computer program. The distance L between the poles is 100m, the sag is 2m, and the height of suspension is 13m. A current of 100A is assumed to by flowing through the conductor. The effect of six neighboring spans (N=3) is taken into account. Figures 3, 4 and 5 present spatial distribution of mentioned power line magnetic field.

These profiles are similar to the ones in (Mamishev, 1996). From Figures 3 and 4, it is evident that with increasing sag conductor, magnetic field especially in midspan will increase.

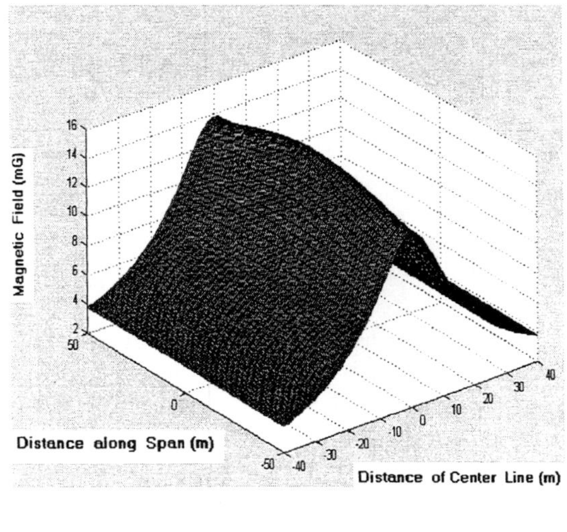

Fig. 2. Field distribution on the ground level under the power line (H=13m, h=11m, Sag =2m).

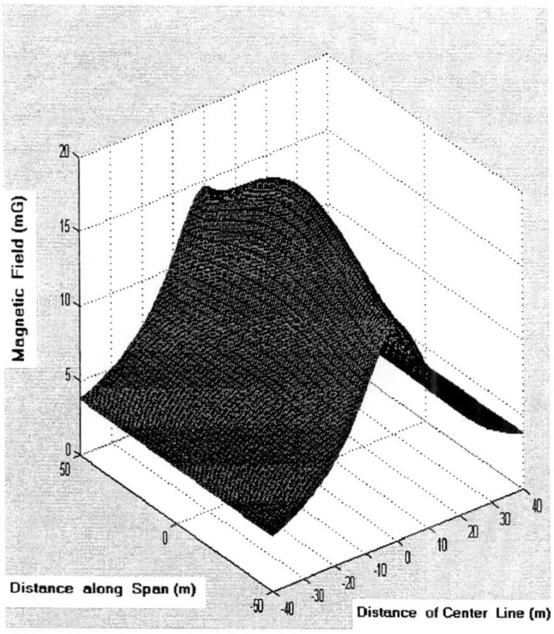

Fig. 3. Field distribution on the ground level under the power line (H=13m, h=9m, Sag = 4m).

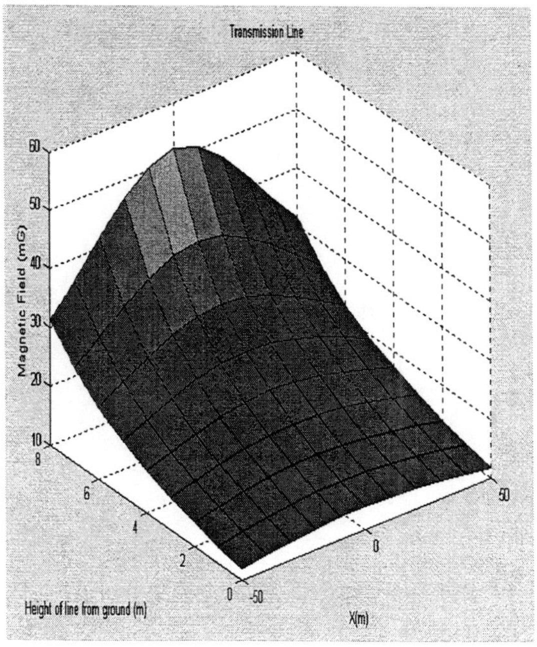

Fig. 4. Distribution of magnetic field, below the catenary.

4. THE EFFECT OF NEIGHBORING SPANS

In this section, we evaluate the effect of neighboring spans on spatial distribution of mentioned power line magnetic field.

The parameter N in (1) represents the number of spans to the right and the left from the generic one. One important question will be that how many is the parameter N? Fig. 5 presents the difference of the mentioned power line magnetic field between two status, with considering N=2 and N=3. in other that profile 5 presents : (magnetic field with N=3) - (magnetic field with N=2). From this plot, it is evident that:

In midspan, the spans of the right and the left from the generic one are far from points and the their effects on spatial distribution in these points is negligible, but with approach the towers, the their influences will be considerable, and in these points the amount of the difference increased. Therefore, selection the parameter N depends on the amount of the difference in the location towers. Figure 6, 7 present the spatial distribution of the differences from (*N=3&4, N=4,5*).

These plots illustrate that $N=3$ is the best options. Because with $N=4$, the most increasing of magnetic field will be smaller than *0.006 mG* in location of towers. Also with selection $N\rangle 3$, the execution time of calculation will increase.

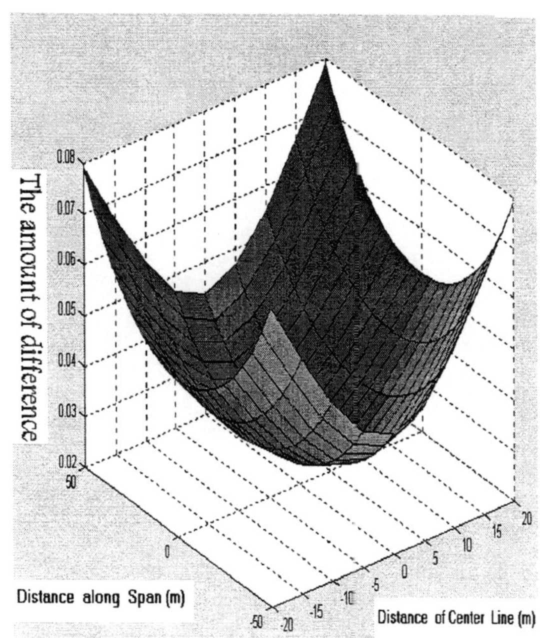

Fig. 5. the spatial distribution of the power line magnetic field difference with N=2 and N=3.

5. SPATIAL DISTRIBUTION OF 3-PHASE POWER LINES MAGNETIC FIELD THAT HAVE A CHANGE IN DIRECTION

Figures 8 and 9 illustrates the spatial distribution on 3-phase power lines magnetic field that have a change in direction.

The specifications of the 3-phase transmission lines are:

Flat configuration, the phase distance is 20m, the distance L between the towers is 50m, the sag is 2m, the suspension is 10m, the change of direction is 10.8 deg., the resistivity of earth is 100 Ω, the frequency of the source current is 60Hz, the phase current is 100 A and N=3.

Also the mentioned software was developed for calculation spatial distribution of the magnetic field of real transmission lines. In this status, the height of the conductors in the location of towers is different from a common surface. Figure 10 presents the mentioned status.

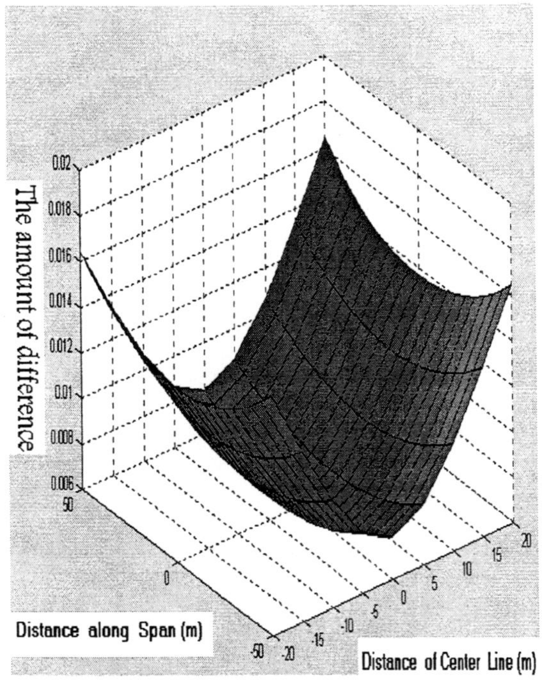

Fig. 6. the spatial distribution of the power line magnetic field difference with N=3 and N=4.

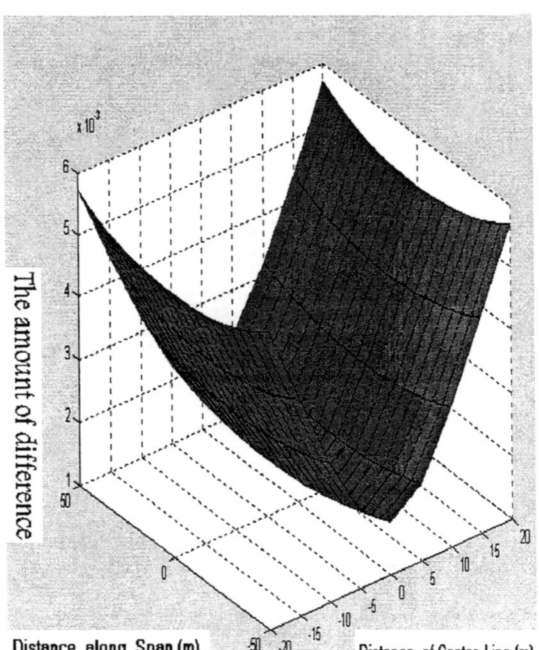

Fig. 7. the spatial distribution of the power line magnetic field difference with N=4 and N=5.

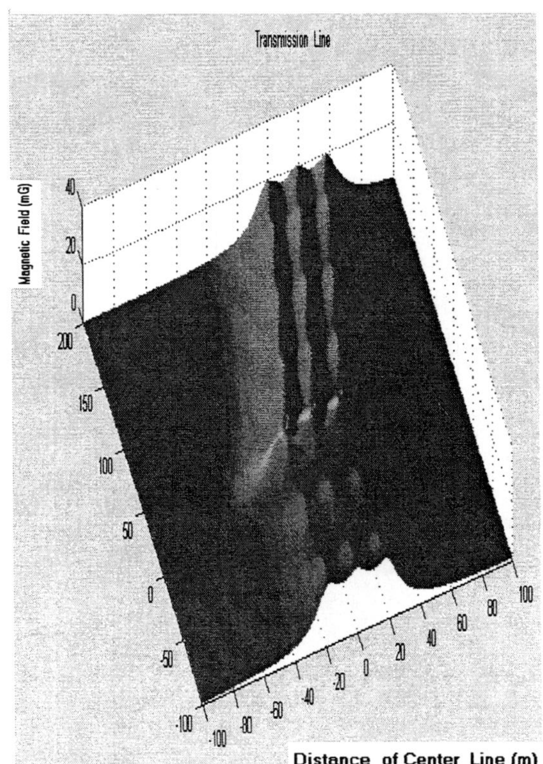

Fig. 8. The spatial distribution of a 3-phase power lines magnetic field that has a change in direction.

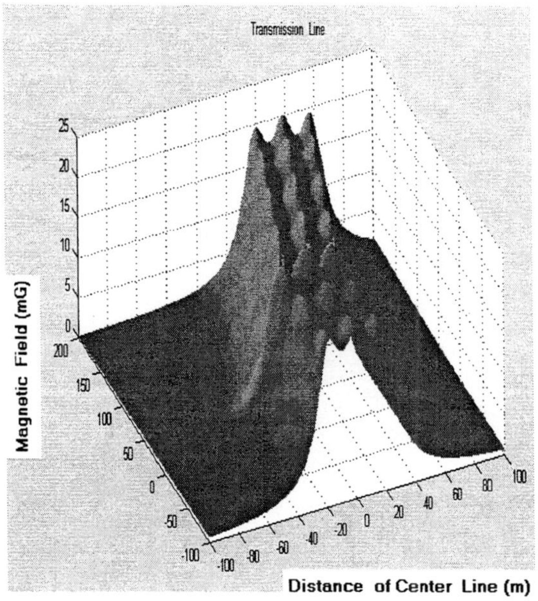

Fig. 9. The spatial distribution of a 3-phase power lines magnetic field that has a change in direction.

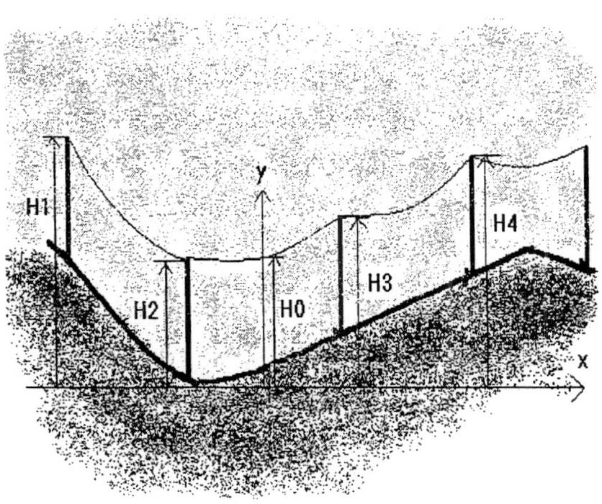

Fig. 10. The layout of a real transmission line.

5. CONCLUSIONS

In this paper, calculation of magnetic fields in the direct vicinity of 3-Phase power lines that have a change in direction has been developed and implemented for three-dimensioned field modeling. The method takes into account the sag of the conductors of power lines. Also the effect neighboring spans on spatial distribution of power line magnetic field was evaluated and discussed.

6. REFERENCES

http://www.esaa.com.org.au
http://www.nrpb.org.uk
http://www.electricity.org.uk
http://www.iee.org.uk
http://www.enertech.net
http://www.niehs.nih.gov/emfrapid

Darveniza, M., D. Mackerras (2002). Electric and Magnetic Field Effects From Power Lines: Review of Possible Health Effects. *Lightning and Transient Protection PTY. LTD.1*

EPRI, (1982). *Transmission Line Reference Book, 345 kV and Above*, Second Edition, Report EL-2500, EPRI, Palo Alto, CA.

Mamishev, A. V., R. D. Nevels, B. D. Russel, (1996). Effects of Conductor Sag on Spatial Distribution of Power Line Magnetic Field. *IEEE Trans. On Power Delivery*, Vol . 11, no. 3, July, pp. 1571-1576.

Olsen, R. G., T. A. Pankaskie (1983). On the Exact, Carson and Image Theories for Wires at or Above the Earth's Interface. *IEEE Transaction on Power Apparatus and System* , Vol.102, no.3, pp.769-774.

Wait, J. R., K. P. Spies (1969). On the Image Representation of the Quasi-Static Fields of a Line Current Source Above the Ground. *Canadian Journal of Physics*, Vol.47, pp.2731-2733.

Copyright © IFAC Power Plants and Power Systems Control,
Seoul, Korea, 2003

ELSEVIER
IFAC
PUBLICATIONS
www.elsevier.com/locate/ifac

MITIGATION OF AC VOLTAGE OCCURRING IN PIPELINE LOCATED CLOSED TO ELECTRIC TRANSMISSION LINES WITH DESIGN OF JOINT RIGHT OF WAY

Babak Porkar Kumleh[1], Mohammad Hadi Varahram[1], Siyamak Porkar Kumleh[2]

[1] *Electrical Engineering Dept. Sharif University of Technology, Iran*
[2] *Electrical Engineering Dept. Amir Kabir University of Technology, Iran*

Abstract: In joint–use corridors where both pipelines and AC electric transmission lines are present, a portion of the energy contained in the electromagnetic field surrounding the electric transmission lines is captured by each pipeline, resulting in induced AC voltages. In this paper a joint power line/pipeline Right of Way is designed. Suitable single and double circuit configuration of overhead transmission line or suitable distance of buried pipelines from center electric power lines for mitigation of buried pipeline voltage due to AC Inductive Coupling is introduced and discussed. *Copyright © 2003 IFAC.*

Keywords: Mitigation, Induced AC Voltage on buried pipeline, suitable single and double circuit configuration of transmission line.

1. INTRODUCTION

A pipeline which shares a common corridor whit AC transmission lines becomes energized by the magnetic and electric fields surrounding the power system in the air and soil. This AC interference can result in a electrical shock hazard for people touching the pipeline or metallic structures connected to the pipeline or simply standing nearby. Furthermore, damage to the pipeline coating, insulating flanges, rectifiers or even direct damage to the pipeline wall itself can occur. In recent years, study of interference effects from AC power lines in pipelines, railways, communication lines and other such structures and mitigation AC voltage induced in pipelines has resulted in numerous research reports, papers and standards (Dabkowski, 1979a; Dabkowski, 1979b; Dabkowski, 1979c; Dabkowski, 1979d; Sarfi, 1993; Christoforidis 2001; Dawalibi, 1989; Dawalibi, 1990; Southey, 1994). The Iranian National Gas Company was also concerned about the reports of electric shock which occurred on people working on gas pipelines and also, the customers. There was also concern on about the possible effects of induced AC Voltage the performance of cathodic protection systems employ for protection of their steel pipelines against corrosion. In this paper a method will be introduced for determination of electromagnetic induced AC voltage in gas pipeline under overhead transmission lines and based on it, suitable single and double circuit configuration of electric power lines or suitable distance of buried pipeline from center overhead line for mitigation on buried pipeline is introduced and discussed.

2. DETERMINATION OF INDUCED VOLTAGE ON BURIED GAS PIPELINES

Generally there are two possible mechanisms for voltage induction on buried gas pipeline placed under transmission lines:

a) Electrostatic induction (Capacitive Coupling)
b) Electromagnetic induction (Inductive Coupling)

2.1 Electrostatic Induced Voltage

The presence of a special distribution of electrical charge in space produced the distribution of an electric field proportional to the value of those charges. Consequently, if a conductive element is placed in this field, it will be in a potential with respect to the ground in the basis of its location around the area of specific equipotential line. However, in the case of insulated buried pipelines which are laid down at ground level, the pipeline potential is approximately equal to the potential of the nearby ground, since they are located on the same equipotential line and their difference is almost zero. Gas utilities are not concerned about this kind of induction.

2.2 Electromagnetic Induced Voltage

AC interference in a pipeline sharing a corridor with a power line consists of a inductive component and a conductive component. The inductive component is present during both normal load conditions and fault conditions on the power line, and is induced in the pipeline by the magnetic field generated by the power line.

Conductive interference typically arises when a power line structure injects a large magnetic current into the earth during a single – phase-to- ground fault and therefore raises soil potentials in the vicinity of structure. If a pipeline is located near such a faulted structure, then the earth around the pipeline will be at a relatively high potential with respect to the pipeline potential, which will typically remain relatively low, especially if the pipeline coating has a high resistance. The difference in potential between the pipeline metal and the earth surface above the pipeline is the touch voltage to which a person would be subjected when standing near the pipeline and touching an exposed metallic appurtenance of the pipeline.

The induced voltage due to inductive component on pipe g in figure 1, under a single phase line, carrying a current of I , whit a return path through the ground, can be determinated based on Carson formula(Kulman, 1967) :

$$V_g = I_a \frac{f}{60}\left(0.0254 + j0.2794 \log \frac{D_{eg}}{D_{ag}}\right) \quad \left(V/_{km}\right) \qquad (1)$$

Where D_{eg} is the equivalent depth of the return path under the conductor " a" as follows:

$$D_{eg} = 660\sqrt{\frac{\rho}{f}}$$

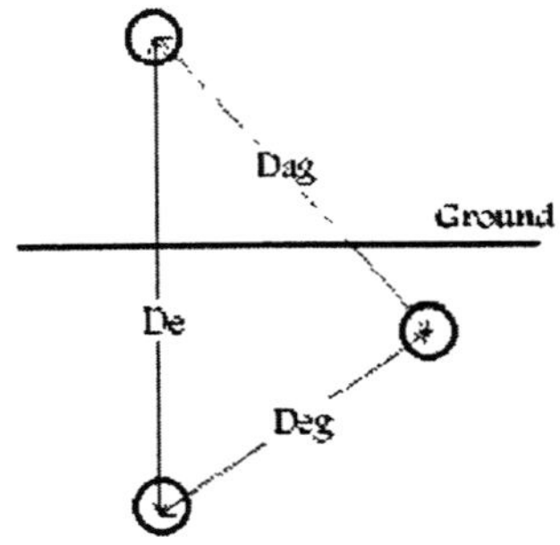

Fig. 1. Arrangement of a single phase line and a pipeline

Where ρ is the specific resistivity of the earth. Consequently, for a three- phase line the induced voltage would be:

$$V_g = 0.286\frac{f}{60} + j0.2794\frac{f}{60}\left[I_a \log\frac{D_{eg}}{D_{ag}} + I_b \log\frac{D_{eg}}{D_{bg}} + I_c \log\frac{D_{eg}}{D_{cg}}\right] \qquad (2)$$

3. DESIGN OF JOINT RIGHT OF WAY FOR SINGLE TRANSMISSION LINES AND BUIED GAS PIPELINE

For calculation induced AC voltage due to single and double circuit electric power line on buried gas pipeline, a computer program is developed based on equation (2). Figure 2 shows different single circuit configuration and specification of transmission lines (FLAT, VERTICAL, DELTA). Figure 3 presents a plot of the induced voltage on buried pipeline that placed in depth 1m and with different of center of overhead lines.

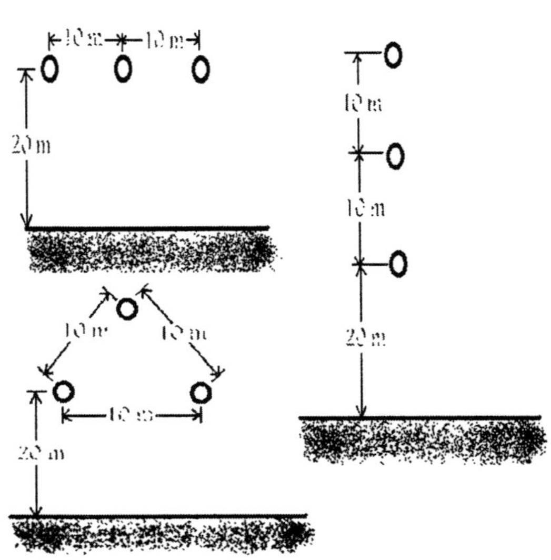

Fig. 2: Different single circuit configuration and specification of transmission lines (FLAT, VERTICAL, DELTA).

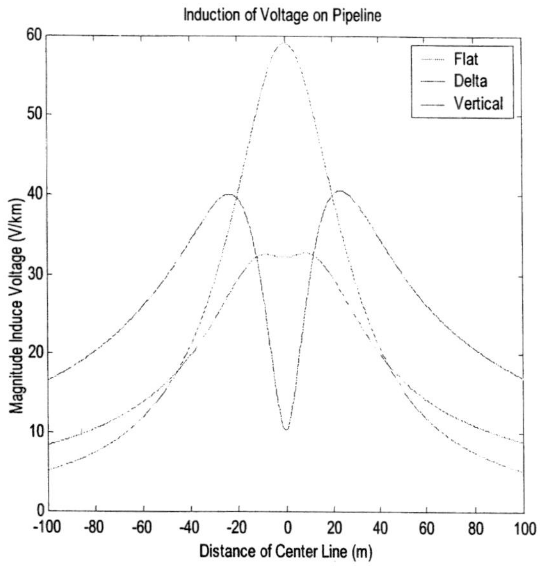

Fig. 3 Plot of the induced voltage on buried pipeline

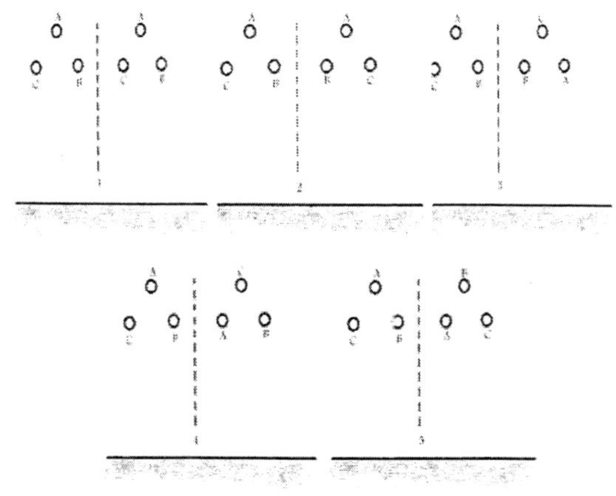

Fig. 4. Different double circuit configuration and specification of transmission lines

From this plot it is evident that:

1- In Right of Way, minimum induced voltage is due to Flat configuration and Vertical configuration will induced maximum voltage on buried gas pipelines.

2- Out of Right of Way, maximum induced voltage on buried pipelines is due to Flat configuration and Vertical configuration will minimum induced voltage.

3- From 1 and 2 is deduced that selection suitable configuration with minimum induced voltage on buried pipeline is dependent on distance gas pipeline from transmission line.

4- From 1,2 and 3 is deduced that one can determine suitable distance of buried pipe from transmission line set up before.

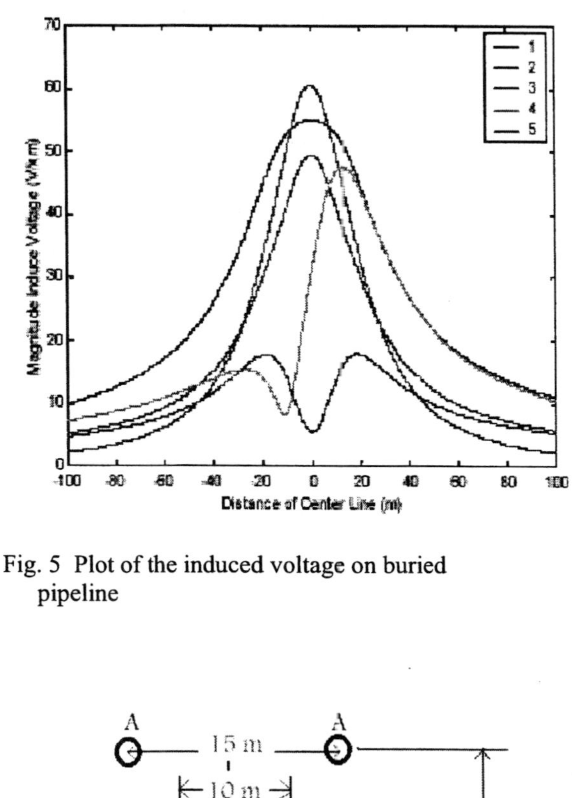

Fig. 5 Plot of the induced voltage on buried pipeline

4. DESIGN OF JOINT RIGHT OF WAY FOR DOUBLE TRANSMISSION LINES AND BUIED GAS PIPELINE

Figure 4 and 6 shows different double circuit configurations and specifications of transmission lines. Figure 5 and 7 presents a plot of the induced voltage on buried pipeline that placed in depth 1m and with different of center of overhead lines.

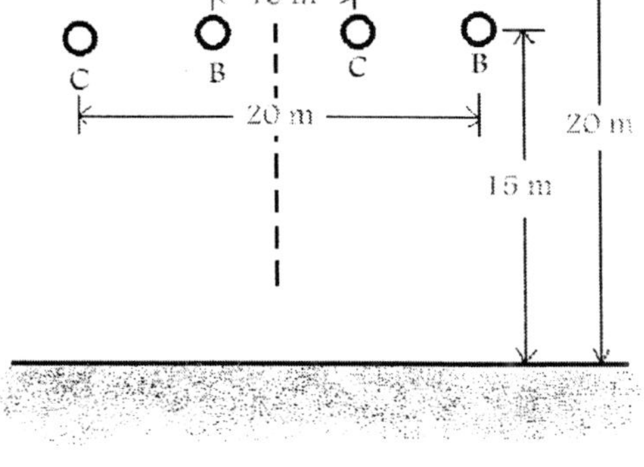

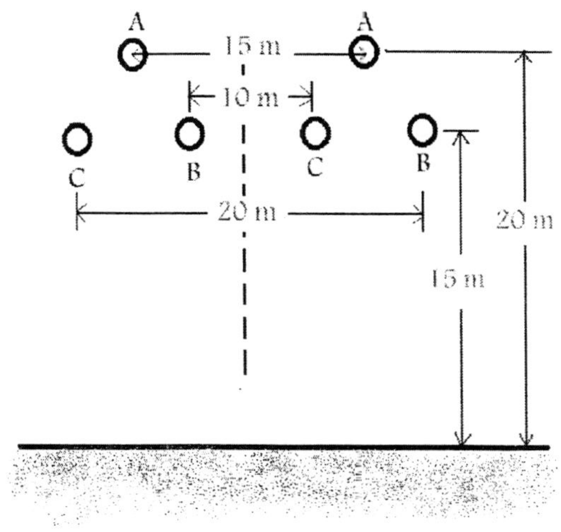

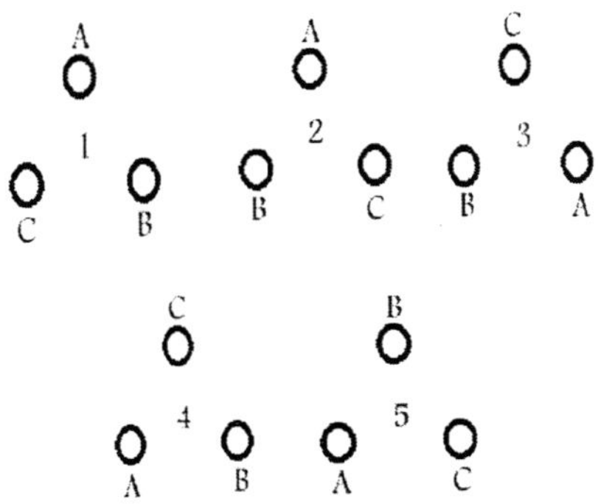

Fig. 6 Different double circuit configuration and specification of transmission lines

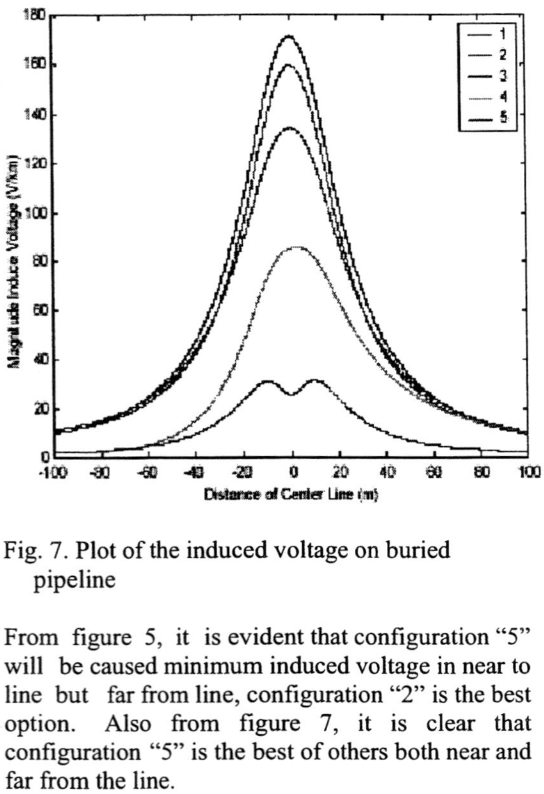

Fig. 7. Plot of the induced voltage on buried pipeline

From figure 5, it is evident that configuration "5" will be caused minimum induced voltage in near to line but far from line, configuration "2" is the best option. Also from figure 7, it is clear that configuration "5" is the best of others both near and far from the line.

5. CONCLUSIONS

In this paper Suitable single and double circuit configuration of overhead transmission line or suitable distances of buried pipelines from center electric power lines for mitigation of buried pipeline voltage due to AC Inductive Coupling is introduced and discussed. Selection of the best options of line configurations is dependent on distance of buried gas pipeline from center of transmission line.

6. REFERENCES

Dabkowski, J., A. Taflove (1979a). Mitigation of Buried Pipeline Voltage die to 60 HZ AC Inductive Coupling Part I–Design of Join Rights-of-Way. *IEEE Trans. On Power Apparatus and system*, Vol.PAS-98, no.5, pp.1806-1813.

Dabkowski, J., A. Taflove (1979b). Mitigation of Buried Pipeline Voltage die to 60 HZ AC Inductive Coupling Part II – Pipeline Grounding Methods *IEEE Trans. On Power Apparatus and system*, Vol.PAS- 98, no.5, pp.1814-1821.

Dabkowski, J., A. Taflove (1979c). Prediction Method for Buried Pipeline Voltage Due to 60 Hz AC Inductive Coupling, Part I–Analysis. *IEEE Trans. On Power Apparatus and system*, Vol.PAS- 98, no.3, pp.780-787 .

Dabkowski, J., A. Taflove (1979d). Prediction Method for Buried Pipeline Voltage Due to 60 Hz AC Inductive Coupling, Part II– Field Test Vertification. *IEEE Trans. On Power Apparatus and system*, Vol.PAS-98, no.3, pp.788-794. Sarfi,

R.J., M. M. A. Salama, C. Gebotys, A. Y. Chikhani (1993). Optimal Design of Cathodic Protection Schemes:a Power Engineering Applications. *Electrical and Computer Engineering,. Canadian Conference,* vol.2, pp.664-667.

Christoforidis, G. C., P.S.Dokopoulos, K.E.Psannis (2001), Induced Voltage and Currents on Gas Pipelines with Imperfect Coating Due to Faults in a Nearby Transmission Line.*IEEE Porto Power Tech Conference,* 10th-13th September, Porto, Portugal.

Dawalibi, F.P., Y.Li, J.Ma, Safety of Pipelines in Close Proximity to Transmission Lines, [on web site: www.sestech.com].

Tranen, J.D., G. L. Wilson (1971), Electrostatically Induced Voltages and Current on Conducting Objects Under EHV Transmissions, *IEEE Trans.on PowerApparatus and System* , Vol . PAS- 90, no.2, pp.768-776.

Dawalibi, F.P., R. D. Southey (1989). Analysis of Electrical Interference From Power Lines to Gas Pipeline Part I: Computation Method. *IEEE Trans. On Power Delivery*, Vol .4, no.3, pp.1840-1846.

Dawalibi, F.P., R. D. Southey (1990). Analysis of Electrical Interference From Power Lines to Gas Pipeline Part I:Parametric Analysis. *IEEE Trans. On Power Delivery*, Vol .5, no.1, pp.415-421.

Southey, R. D., F. P. Dawalibi, W.Vukonich (1994). Recent Advances in the Mitigation of Voltage of AC Voltages Occurring in Pipeline Located Closed to Electric Transmission Lines. *IEEE Trans.on Power Delivery*,Vol.9, no.2, pp.1090-1097.

Kulman, F.E. (1967). Effect of alternating current on corrosion of steel gas pipes. *American* Gas Association.

ELSEVIER
IFAC
PUBLICATIONS
www.elsevier.com/locate/ifac

A SERIES ACTIVE POWER FILTER FOR HARMONICS AND REACTIVE POWER COMPENSATION

Jin-Sun Kim, Young-Seok Kim

Research Center for Next-generation High Voltage and Power Technology, Dept. of Elec., Inha University 253, Yong Hyun-Dong, Nam-Gu, inchon, 402-752, Korea

Abstract: This paper suggests a control algorithm of 3-phase 3-wire series active power filter. This suggested algorithm can compensate source harmonics and reactive power in 3-phase 3-wire power distribution systems. This control algorithm extracts a compensation voltage reference from performance function without phase transformation and multiplying harmonic current value by gain which is presented in generalized p-q theory. Therefore, this control algorithm is simpler than any other conventional control algorithms; the performance of the series active power filter is improved because the delay in compensation voltage reference calculation is shortened. 3-phase 3-wire series active power filters are manufactured and experiments are carried out to verify the effectiveness of suggested control algorithm. *Copyright © 2003 IFAC*

Keywords: Harmonic currents, Reactive power, Series active power filter, Passive filter, THD.

1. INTRODUCTION

As the society becomes industrialized and globalized, the loads using the power conversion systems such as SMPSs, UPSs, power suppliers of computer systems are widely propagated recently. Most of these loads use switch mode type power converters which draw excessive harmonic currents. Owing to harmonic problems, the engineering capacity of power system equipments such as generators, power transmission lines, transformers etc, should be large and source voltage can be distorted and the power equipments such as electric machines, cable, and power capacitors for improving power factor may be heated. Therefore, the life-time shortage and insulation breakdown may result from these harmonics. The malfunction of electronic equipments like measuring equipments, the occurrence of EMI and the interference of communication systems etc, can be also made by harmonics. Moreover, the abnormal vibration for motors can be occurred and the iron and copper losses are increasing. In order to solve these harmonic problems, kinds of passive filter and hybrid type active power filters have been proposed. The passive filters have been traditionally used to absorb harmonics which are generated by non-linear loads because of their low cost to install and high performance efficiency. However, they have several drawbacks. Generally, many passive filters should be installed to eliminate all harmonics which may be presented in the source stage widely because the passive filter can eliminate only one harmonic current that tunes to the frequency of the passive filter. Therefore, volume and cost are increased when all harmonics, which widely presented in the source, should be eliminated. Also, the hazard of resonance with source impedance is always existed [4].

To solve the preceding problems of the passive filter, active power filters have been studied and developed in recent years. According to the construction shape, the parallel active power filter which is suitable for compensating of harmonic current sources and series active power filter which is suitable for compensating of harmonic voltage sources are have been developed [1,4,6].

Between these active power filters, the parallel active power filter is selected for industrial application and studies on it have been started in the late of seventies to overcome the defects of the passive filter. In contrast, this parallel active power filter is suitable for harmonic current source compensation method. However, the power electronic equipments such as UPSs, electronic ballasts and power supplies of computer system, etc., which have a large capacitor on the dc side of rectifier circuit are more increased recently. Because these equipments behave as a harmonic voltage source, the parallel active power filter which is only suitable for compensating harmonic current can not be applied to the harmonic voltage source loads of computers, UPSs and so on. Accordingly, it is more and more increasing that the development of series active power filter is an important matter for the compensation of harmonic voltage source [7].

In order to defeat these drawbacks, the control algorithm for 3-phase 3-wire series active power filter to compensate the harmonics and reactive power simultaneously with the series active power filter is proposed in this paper. In relation with the type of loads, the active power filter can be applied to the loads consisting of the harmonic voltage source consists of the parallel connection of a resistor and a capacitor in the dc-link of three-phase diode rectifier, and the harmonic current source which consists of the series connection of a resistor and an inductor in the dc-link of three-phase diode rectifier.

The proposed control algorithm directly extracts compensation voltage references without multiplying the gain and phase transformation. Therefore, the calculation of the compensation voltage reference will turn out to be simpler than other control algorithms. In order to verify the effectiveness of the proposed control algorithm, the series active power filters is manufactured and the experiments are executed with the loads consisting harmonic voltage source and current source in 3-phase 3wire power distribution system [3].

2. PRINCIPLE OF COMPENSATION

In this paper, the control algorithm of using performance function for a 3-phase 3-wire system as shown Fig. 1 is presented. Instead of reactive power which is delivered by loads, the power of each phase being generated from active power filter is considered as an instantaneous reactive power

It is noted that the active power filter does not generate the instantaneous active power because an amount of power delivered from a certain line of the power system into the active power filter is instantly fed to the other lines of the power system.

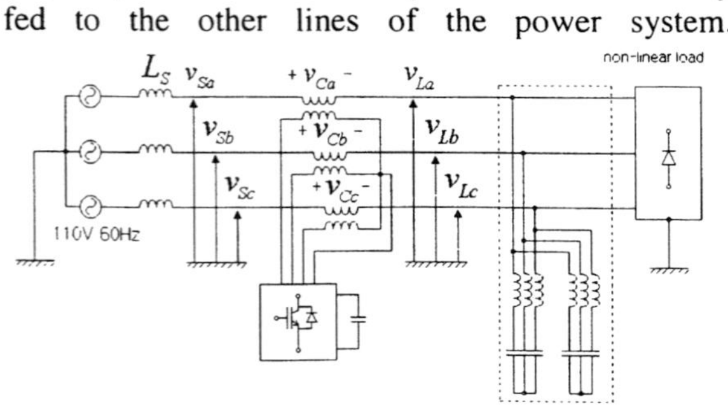

Fig 1. Three-phase circuit structure

Therefore, the instantaneous reactive power is shown in equation (1)

$$q_k = v_{Ck} \cdot i_{Sk} \quad (k = a, b, c) \tag{1}$$

In equation (1), k represents the phase rotations of phase a, b, c and v_{ck} means the compensation voltage to be generated by active power filter and i_{sk} specify the per-phase source currents.

Since the active filter does not generate the instantaneous active power, the sum of reactive power is zero. Consequently, the equation (1) can be represented as follows.

$$v_{Ca}i_{Sa} + v_{Cb}i_{Sb} + v_{Cc}i_{Sc} = 0 \tag{2}$$

Because there is no neutral line in the 3-phase 3wire-wire system, the zero sequence does not reflected on the neutral line. Following equation (3) satisfies the term of the sum for compensation voltage to be zero.

$$v_{Ca} + v_{Cb} + v_{Cc} = 0 \tag{3}$$

The compensation voltage of the series active power filter can be determined when the value of following equation satisfying the terms of equation (4) with the constraint of equation (2) and (3) minimizes the performance function "L".

$$L = (v_{La} - v_{Ca})^2 + (v_{Lb} - v_{Cb})^2 + (v_{Lc} - v_{Cc})^2 \tag{4}$$

In equation (4), v_{La}, v_{Lb}, v_{Lc} represents the per-phase voltage. Since each term has the form of square for a difference between a load voltage and a compensation voltage, if the performance function has the minimum value, the component of source voltage harmonics is also to be the minimum values. The compensation voltage of series active power filter for making the performance function to be minimum value can be derived from differential equation in constraint (5) with the constraints of equation (2), (3),

$$\frac{dL}{dv_{Ca}} = 0 \qquad \frac{dL}{dv_{Cb}} = 0 \qquad \frac{dL}{dv_{Cc}} = 0 \tag{5}$$

The proposed compensation voltage which can be derived from equation (2), (3), (5) as follows

$$\begin{aligned} v_{Ca}^* &= \frac{\sqrt{3}(i_{Sb} - i_{Sc}) \cdot q}{2(i_{Sa}^2 + i_{Sb}^2 + i_{Sc}^2 - i_{Sa}i_{Sb} - i_{Sb}i_{Sc} - i_{Sc}i_{Sa})} \\ v_{Cb}^* &= \frac{\sqrt{3}(i_{Sc} - i_{Sa}) \cdot q}{2(i_{Sa}^2 + i_{Sb}^2 + i_{Sc}^2 - i_{Sa}i_{Sb} - i_{Sb}i_{Sc} - i_{Sc}i_{Sa})} \\ v_{Cc}^* &= \frac{\sqrt{3}(i_{Sa} - i_{Sb}) \cdot q}{2(i_{Sa}^2 + i_{Sb}^2 + i_{Sc}^2 - i_{Sa}i_{Sb} - i_{Sb}i_{Sc} - i_{Sc}i_{Sa})} \end{aligned} \tag{6}$$

where,

$$q = \frac{1}{\sqrt{3}}((i_{Sb} - i_{Sc})v_{La} + (i_{Sc} - i_{Sa})v_{Lb} + (i_{Sa} - i_{Sb})v_{Lc}) \tag{7}$$

Since the constraint of equation (6) is imposed on equation (2), the equation of specifying the reactive power "q" is the same equation as it is presented for reactive power in the general p-q theory. This relationship can be certified by the substituting equation (9), (10) into equation (8) in p-q theory.

$$q = v_{L\alpha} i_{S\beta} - v_{L\beta} i_{S\alpha} \tag{8}$$

$$v_{L\alpha} = \sqrt{\frac{2}{3}}\left(v_{La} - \frac{1}{2}v_{Lb} - \frac{1}{2}v_{Lc}\right)$$
$$v_{L\beta} = \sqrt{\frac{2}{3}}\left(\frac{\sqrt{3}}{2}v_{Lb} - \frac{\sqrt{3}}{2}v_{Lc}\right) \tag{9}$$

$$i_{S\alpha} = \sqrt{\frac{2}{3}}\left(i_{Sa} - \frac{1}{2}i_{Sb} - \frac{1}{2}i_{Sc}\right)$$
$$i_{S\beta} = \sqrt{\frac{2}{3}}\left(\frac{\sqrt{3}}{2}i_{Sb} - \frac{\sqrt{3}}{2}i_{Sc}\right) \tag{10}$$

$$q = \frac{1}{\sqrt{3}}\left(v_{La}i_{Sb} + v_{Lb}i_{Sc} + v_{Lc}i_{Sa} - v_{La}i_{Sc} - v_{Lb}i_{Sa} - v_{Lc}i_{Sb}\right)$$
$$= \frac{1}{\sqrt{3}}\left((i_{Sb} - i_{Sc})v_{La} + (i_{Sc} - i_{Sa})v_{Lb} + (i_{Sa} - i_{Sb})\right) \tag{11}$$

Consequently, the instantaneous reactive power which is involved in the compensation voltage reference can be divided into ac and dc components. If it is necessary to compensate only the harmonic components, the ac component of instantaneous reactive power can be used by substituting the instantaneous reactive power "q" of equation (6) into $\tilde{q}$ as following equation (12).

$$v_{Ca}^{*} = \frac{\sqrt{3}(i_{Sb} - i_{Sc}) \cdot \tilde{q}}{2(i_{Sa}^{2} + i_{Sb}^{2} + i_{Sc}^{2} - i_{Sa}i_{Sb} - i_{Sb}i_{Sc} - i_{Sc}i_{Sa})}$$
$$v_{Cb}^{*} = \frac{\sqrt{3}(i_{Sc} - i_{Sa}) \cdot \tilde{q}}{2(i_{Sa}^{2} + i_{Sb}^{2} + i_{Sc}^{2} - i_{Sa}i_{Sb} - i_{Sb}i_{Sc} - i_{Sc}i_{Sa})} \tag{12}$$
$$v_{Cc}^{*} = \frac{\sqrt{3}(i_{Sa} - i_{Sb}) \cdot \tilde{q}}{2(i_{Sa}^{2} + i_{Sb}^{2} + i_{Sc}^{2} - i_{Sa}i_{Sb} - i_{Sb}i_{Sc} - i_{Sc}i_{Sa})}$$

The $\tilde{q}$ can be obtained by subtracting the dc component which can be achieved by the extraction of the dc component from the calculated instantaneous reactive power with a low-pass filter [2]. If the compensation voltage is calculated by this control method, we can extract the compensating voltage of the series active power filter without phase transformations and multiplying harmonic current value by gain. Therefore, the calculating time is short and the control method is simple compared with conventional methods.

In Fig. 2, $i_{saf}, i_{sbf}, i_{scf}$ are the fundamental components of 3-phase source currents which are obtained by low-pass filtering. A PI controller is added to the voltage compensation references for an inverter dc-link voltage control. The inverter dc-link voltage can be controlled to follow the inverter dc-link voltage reference V_{dc-inv}^{*} through PI controller

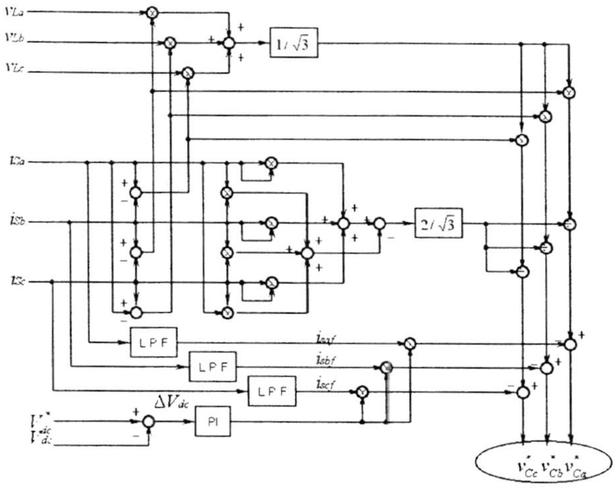

Fig. 2. Block-diagram of the control algorithm

3. SYSTEM CONFIGURATION

In Fig. 3, the 3-phase 3-wire series active power filter and the combined system of the parallel passive filter and the series active power filter which are used in experiments are represented. In the 3-phase 3-wire series active power filter, the harmonic voltage source consists by the parallel connection of a resister and a capacitor in the dc-link of 3-phase diode rectifier, whereas the harmonic current source consists by the series connection of a resistor and an inductor in the dc-link of 3-phase diode rectifier. The rms voltage, 110[V], is applied in the source and the inductance of source inductor Ls is $0.1[mH]$.

$30[\Omega] - 2400[uF]$ RC load is used in the harmonic voltage source circuit, while $30[\Omega] - 15[mH]$ RL load is used in the harmonic current source circuit.

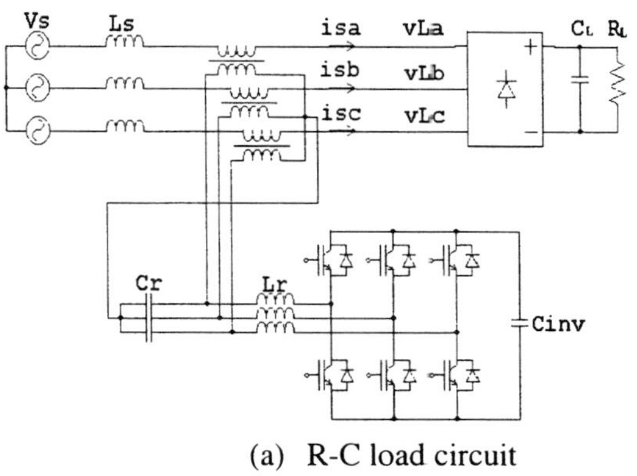

(a) R-C load circuit

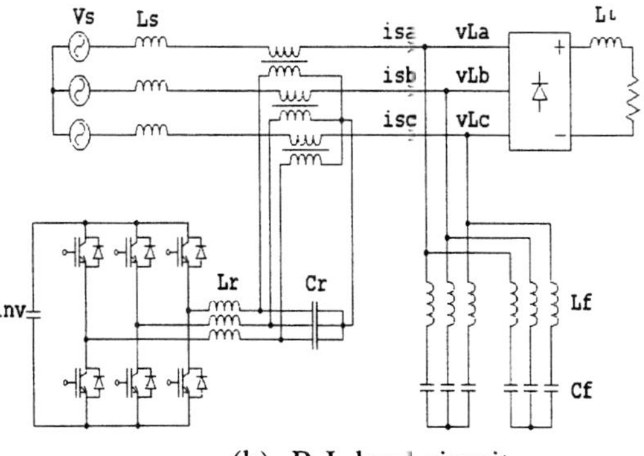

(b) R-L load circuit

Fig. 3. 3-phase 3-wire series active power filter Circuits

The capacitance of the inverter dc-link capacitor is $2350[uF]$. The LC filter to reduce the switching ripple of inverter switching frequency is composed of $4[mH]$ inductors and $0.5[uF]$ capacitors. The 5th passive filter is made up using the $2[mH]$ inductor and $140[uF]$ capacitor, while the 7th passive filter is made up using the $2[mH]$ inductor and $70[uF]$ capacitor.

4. EXPERIMENTAL RESULTS

4. 1. Experimental Results for Harmonic voltage source

The typical waveforms with the load to be considered as a harmonic voltage source are shown in Fig. 4. Fig. 4-(a) shows the source current and the source voltage waveforms with power factor about 0.925 lagging, whereas Fig, 4-(b) and Fig. 4-(c) shows the source current and source voltage spectra. The source current and source voltage THD are 29.8% and 10.5%, respectively. From these waveforms, it is observed that source voltages and currents are distorted by harmonic voltage source. Fig. 5 and Fig.6 show the transient waveforms and the steady state waveforms when compensated by using the series active power filter.

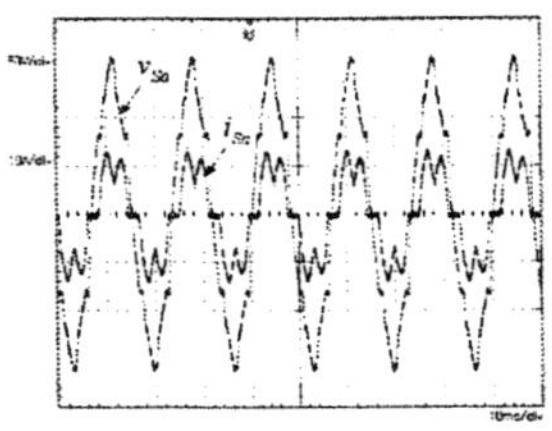

(a) Source current and voltage waveforms

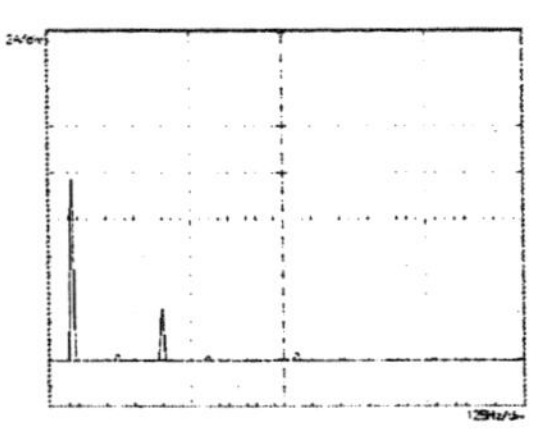

(b) Source current spectrum

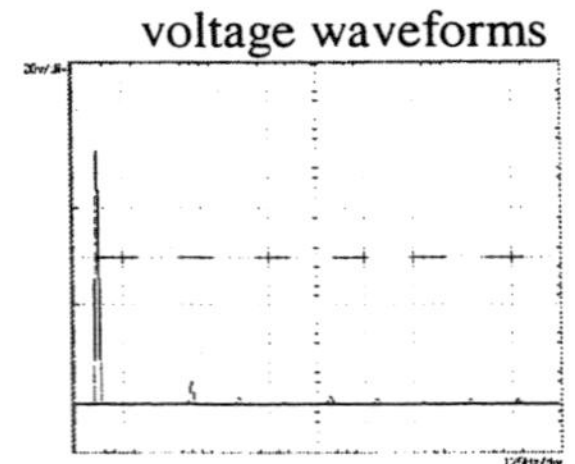

(c) Source voltage spectrum

Fig. 4. Experimental waveforms for the harmonic current source without compensation

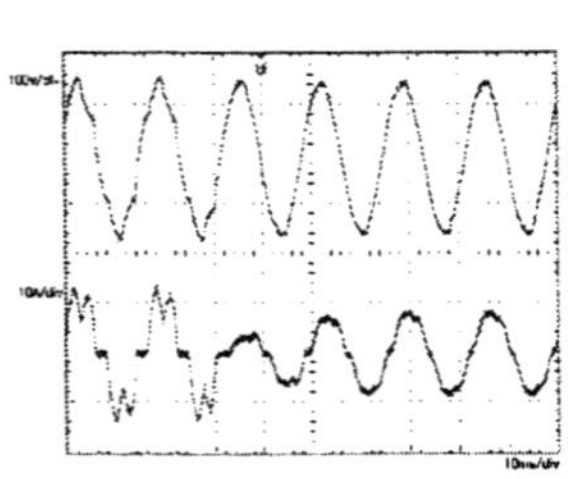

(a) Source current and voltage waveforms

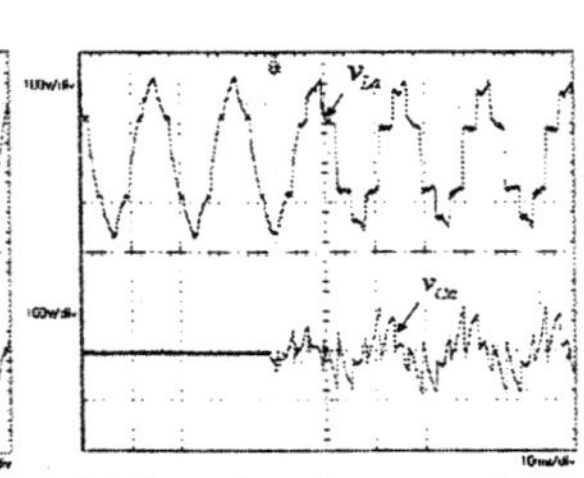

(b) Load voltage and compensating voltage waveforms

Fig. 5. Transient states experimental waveforms for the harmonic voltage source

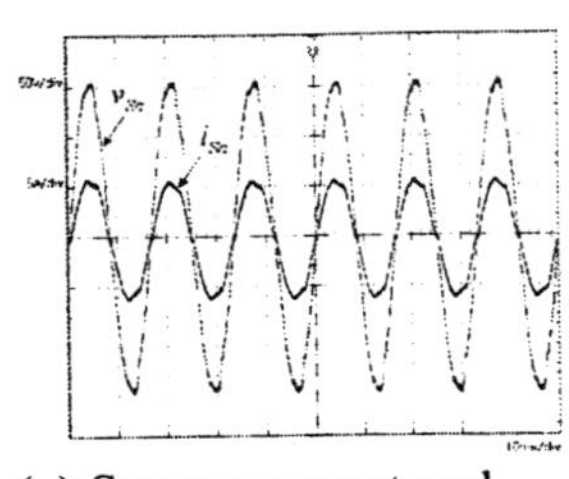

(a) Source current and voltage waveforms

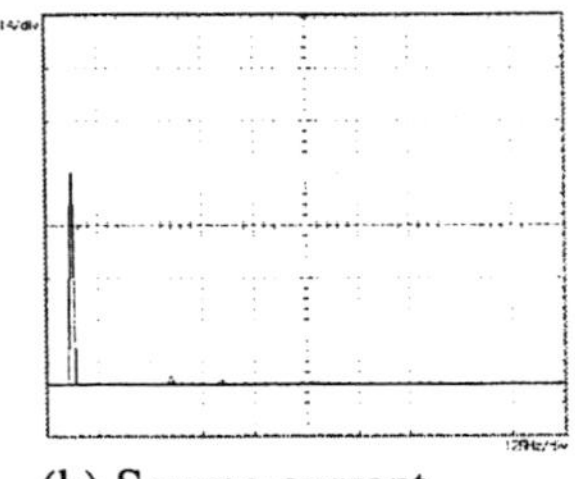

(b) Source current spectrum

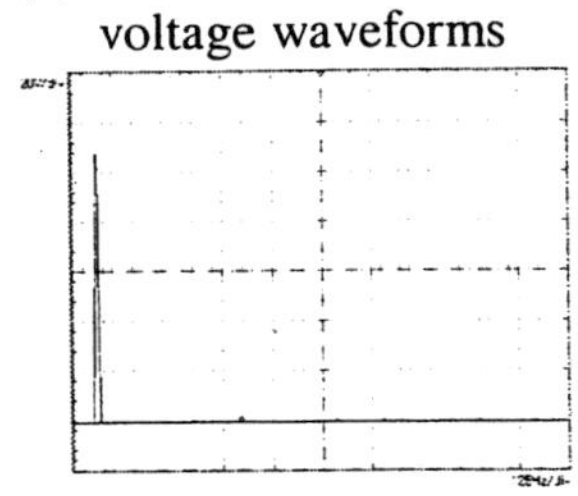

(c) Source voltage spectrum

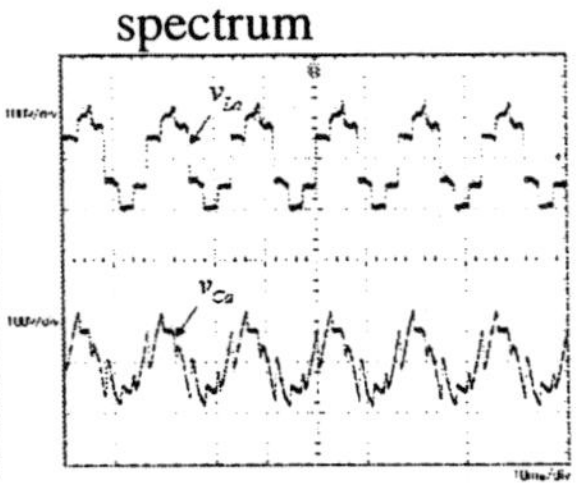

(d) Load voltage and compensating voltage

Fig. 6. Experimental waveforms for the harmonic voltage source when compensated with the series active power filter

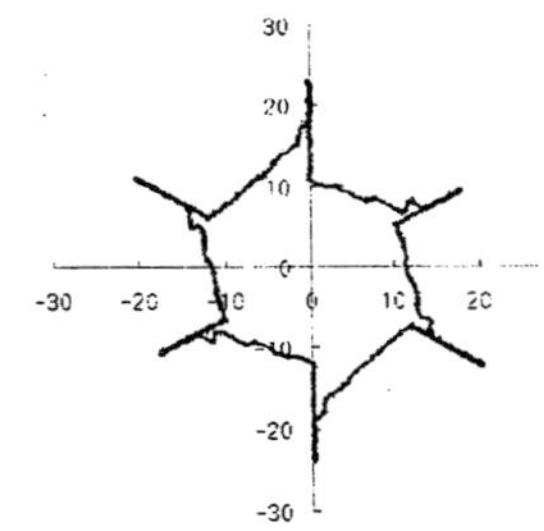

(a) before compensation

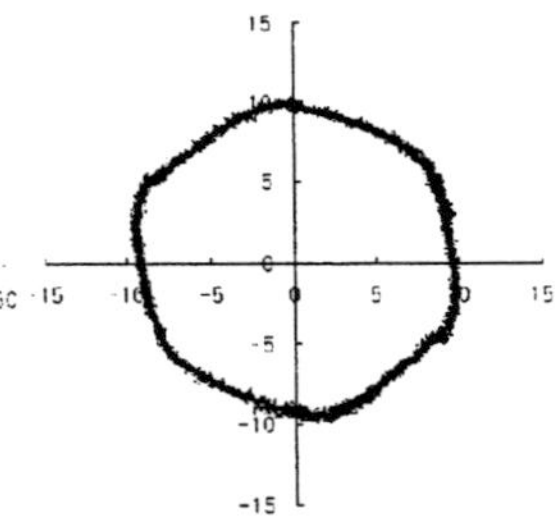

(b) after compensation

Fig. 7. Three-phase to two-phase vector transformation waveforms for harmonic voltage source

Fig. 5-(a) shows that harmonic compensation for the source current is response slowly within 5-6 cycle, but the voltage compensation performed rapidly. Fig. 5-(b) shows that the reactive and harmonic components are compensating. Fig. 6-(a) shows the source current and source voltage waveforms. Fig. 6-(b) and Fig. 6-(c) show the source current and source voltage spectra, respectively.

The source current and source output voltage THD are 3.8% and 2.27%, respectively. The power factor is measured as 0.99 lagging. Fig. 6-(d) shows the load voltage and compensation voltage as an active power filter output.

Fig. 7 shows the currents in two-phase coordinates for the harmonic voltage source. In Fig. 7, the harmonic compensation effect can be observed as clearly. The waveform after compensation constitutes a near circle-like orbit. But the source current THD after compensation is about 4%, so the waveform of Fig. 7-(b) is not a perfect circle.

4. 2. Experimental Results for the Harmonic Current Source

Fig. 8 shows the typical waveforms in the case of the harmonic current source. Fig. 8-(a) shows the source current and source voltage waveforms at a power factor 0,95 lagging. Fig. 8-(b) and Fig. 8-(c) show the current and voltage spectra, respectively. The source current and source voltage THD are 22.9% and 9.5%, respectively. As shown in Fig, 8, the source output voltage and source current waveforms are distorted by the harmonic current source as well as the harmonic voltage source.

These waveforms, which are shown in Fig. 9, are the compensated waveforms by using only the parallel passive filter. Fig. 9-(a) shows the source current and source voltage waveforms with a power factor of 0.6 leading, whereas Fig. 9-(b) and Fig. 9-(c) illustrate the source current and source voltage spectra, respectively. The source current and source voltage THD are 6.3% and 1.7%, respectively. The amplitude of the source current is increased, and the power factor is leading because of the effectiveness of the parallel passive filter.

From Fig. 9-(b), it is known that 5th and 7th harmonic components of the source current are almost cancelled by the parallel passive filter. But, the source current THD is over 5%, as usual. Moreover, the power factor of a source stage is leading. The leading power factor is also a reason of malfunction in load.

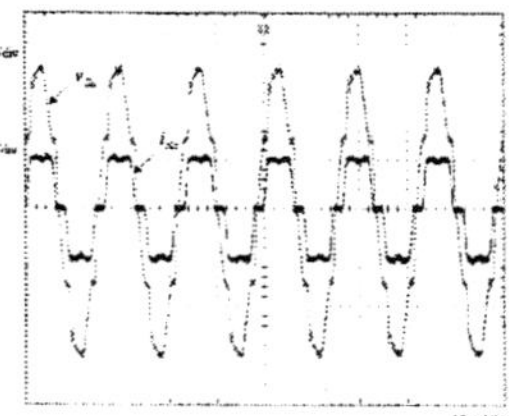

(a) Source current and voltage waveforms

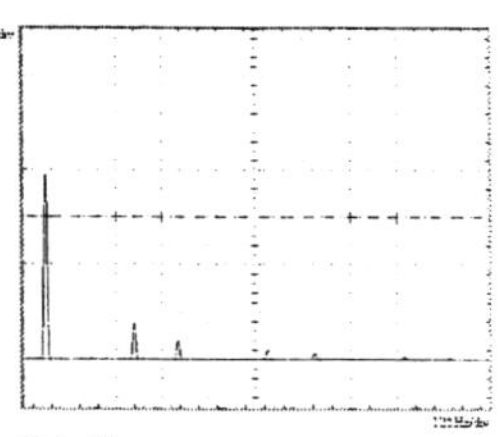

(b) Source current spectrum

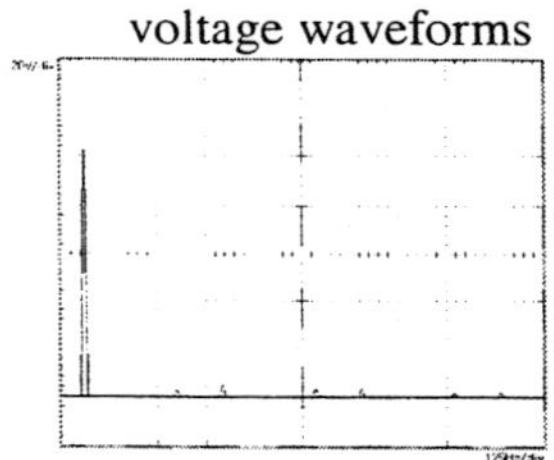

(b) Source voltage spectrum

Fig. 8. Experimental waveforms for the harmonic current source without compensation

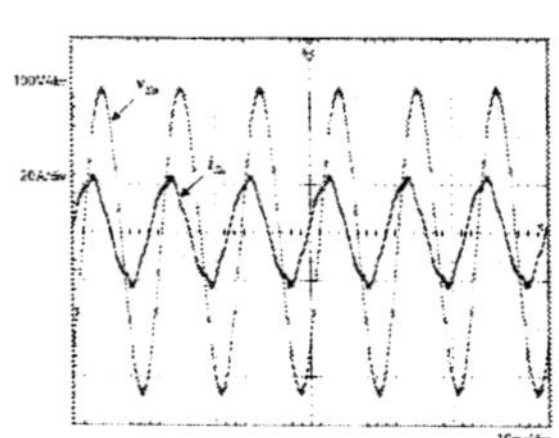

(a) Source current and voltage waveforms

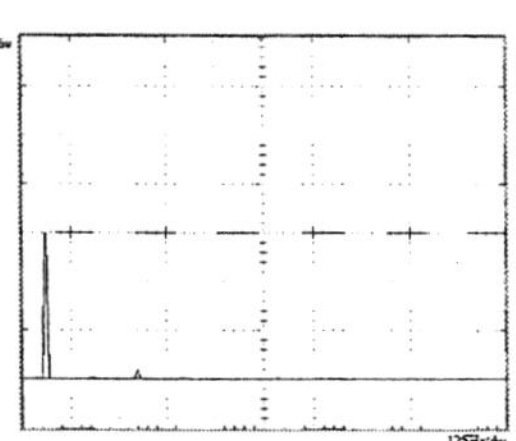

(b) Source current spectrum

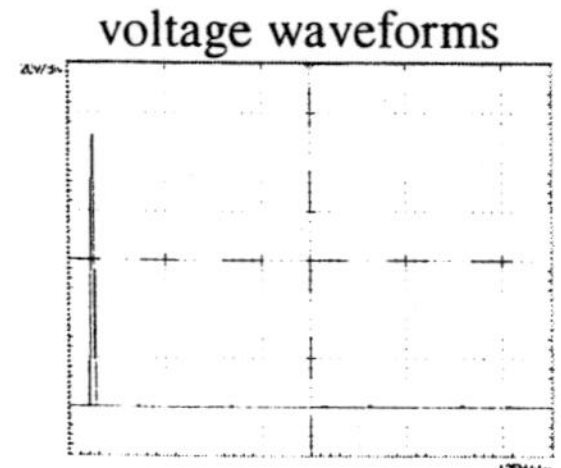

(c) Source voltage spectrum

Fig. 9. Experimental waveforms for the harmonic current source when compensated with the parallel passive filter

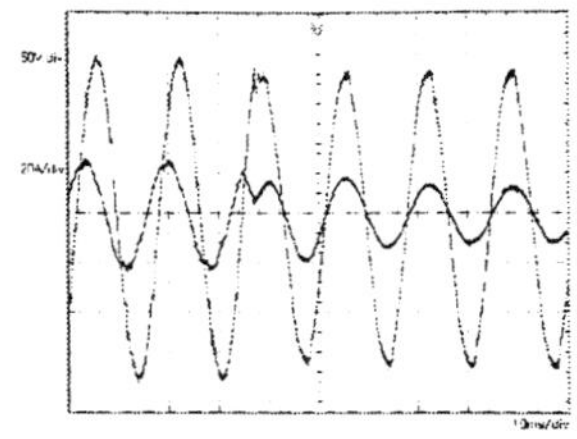

(a) source current and voltage waveforms

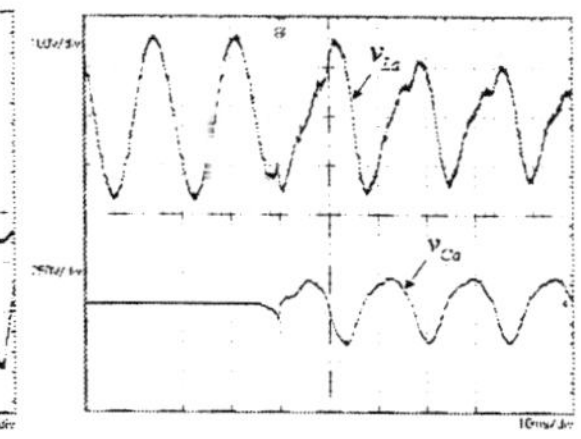

(b) load voltage and compensating waveforms

Fig. 10. Transient states experimental waveforms for the harmonic current source when compensated with the combined system of the parallel passive filter and the series active power filter

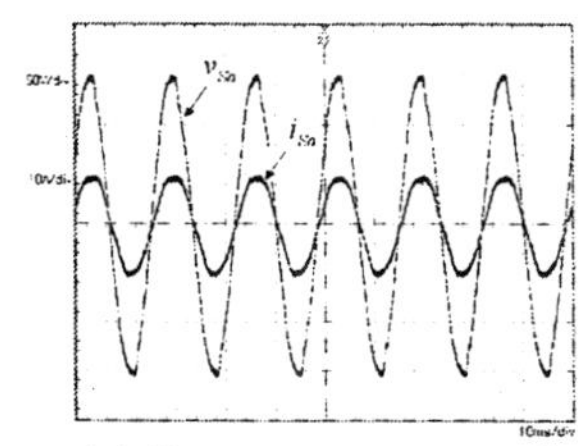

(a) Source current and voltage waveforms

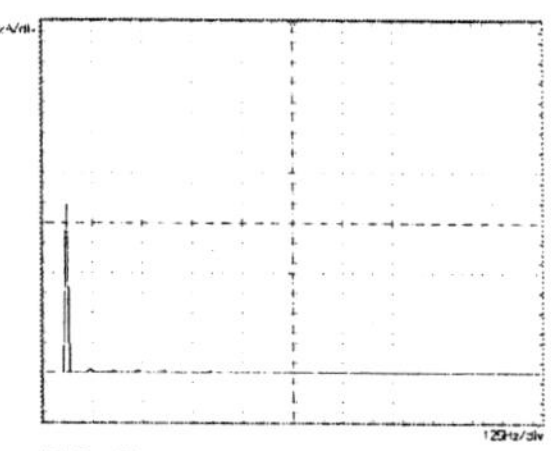

(b) Source current spectrum

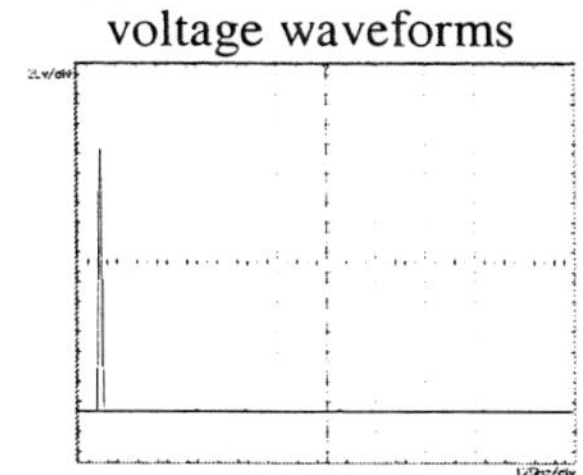

(b) source voltage spectrum

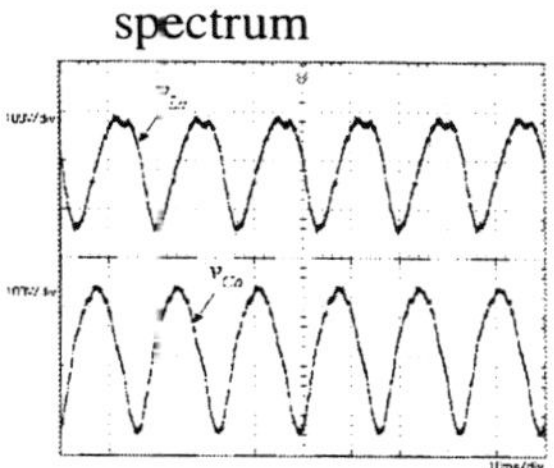

(d) load voltage and compensating waveforms

Fig. 11. Experimental waveforms for the harmonic current source when compensated with the combined system of the parallel passive filter and the series active power filter

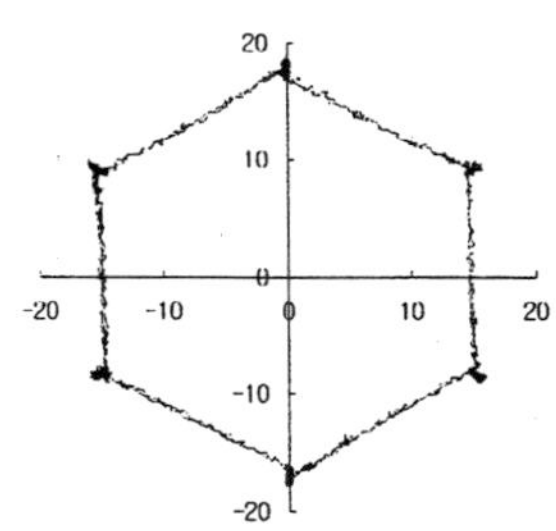

(a) Without compensation

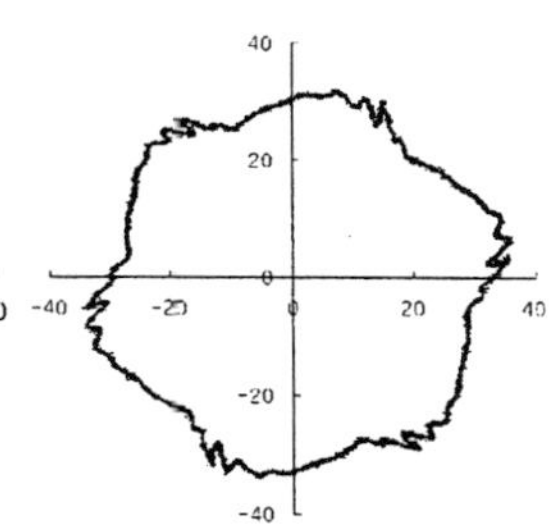

(b) Compensation with the passive filters

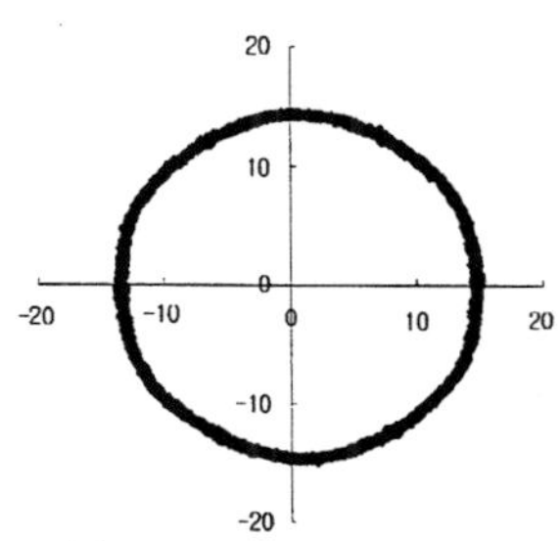

(c) compensation with the passive and active filters

Fig. 12. Three-phase to two-phase vector transformation waveforms for source currents

Fig. 10 and Fig. 11 shows the transient waveforms and the steady state waveforms of the phase which are compensated by the combined system of the parallel passive filter and the series active power filter, respectively. As shown in Fig. 10-(a), the phase shift and harmonic compensation is performed within a half-cycle. Fig. 10-(b) shows that the reactive components for improving power factor are mainly compensated.

Fig. 10-(b) shows that the VA rating of the active power filter can be reduced, because the passive filter compensate the 5th and 7th harmonic components and only the other harmonics are compensated by series active filter. However, Fig. 5-(b) shows that active power filter simultaneously compensates harmonics and reactive power.

Fig. 11-(a) shows the source current and source voltage waveforms. Fig. 11-(b) and Fig. 11-(c) shows the source current and source voltage spectra, respectively. The source current THD is 3.3%, and the source voltage THD is 1.6%. Power factor is almost unity. Fig. 11-(d) shows the load voltage and the compensating voltage as an active filter output. These experimental results show that the series active power filter complements the defects of the parallel passive filter and cooperates in harmonic compensation. Moreover, the series active power filter can compensate the leading power factor as well as the source current harmonics.

Fig. 12 shows the source current waveform from three-phase to two-phase transformation. Fig. 12-(a) is the waveform before compensation, while Fig. 12-(b) is the one after compensation only the parallel passive filter applied. Finally, Fig 12-(c) is the waveform after compensation by the combined system of the parallel passive filter and the series active power filter. From Fig. 12, the harmonic compensation effect is clearly observed. Comparing these three waveforms, the better harmonics compensate, the more smooth the waveforms become and the closer the circles get. Because the harmonic components of the source current are almost zero, when they are compensated by the combined system of the parallel passive filter and the series active power filter, the waveform of Fig. 12-(c) forms almost a circle.

As the result of these experiments, the THD of both cases meets the IEEE std. 519. But it turns out that it is more effective when using the combined system of the parallel passive filter and the series active power filter. This is drawn from the fact that the parallel passive filter and the series active power filter are complementary and cooperate with each other.

5. CONCLUSION

This paper suggests a control algorithm which can compensate both the harmonic current source and the harmonic voltage source. 3-phase 3-wire series active power filter and the combined system of the parallel passive filter as well as the series active power filter are manufactured. Then, experiments are carried out to verify the effectiveness of the proposed control algorithm. This proposed control algorithm extracts the voltage compensation reference, directly. Therefore, the calculation time is shorter than other control algorithms of the series active power filters. Furthermore, the difficulty on finding the voltage reference gain is disappeared. The experiments on the harmonic current source and the harmonic voltage source are executed. The THD of source currents is lower than 5%, which meets the regulations of IEEE std. 519 and the power factor is almost unity after compensation.

These experimental results verify the effectiveness of the control algorithm and it is expected that the series active power filter contributes to resolve power quality problems, such as the harmonics and the power factor degradation.

ACKNOWLEDGEMENTS

This work was financially supported by MOCIE through EIRC program.

REFERENCES

[1] F. Z. Peng, H. Akagi and A. Nabae (1990). *A New Approach to Harmonic Compensation in Power Systems-A Combined System of Shunt Passive and Series Active Filters,* IEEE Transactions on Industry Applications, Vol. 26, NO. 6, pp 983-990.

[2] Fang Zheng Peng, Ott, G.W. and Adams, D.J. (1997). *Harmonic and reactive power compensation based on the generalized instantaneous reactive power theory for 3-phase 4-wire systems,* Power Electronics Specialists Conference, 1997. PESC '97 Record, 28th Annual IEEE , Vol. 2 , pp 1089 –1095.

[3] Furuhashi T., Okuma S. and Uchikawa Y. (1990). *A study on the theory of instantaneous reactive power* Industrial Electronics, IEEE Transactions on , Volume: 37 Issue: 1 , pp 86 -90T.

[4] H. Akagi, Y. Kanazawa and A. Nabae (1984). *Instantaneous Reactive Power Compensators Comprising Switching Devices Without Energy Storage Components*, IEEE Transactions on Industry Applications, Vol. IA-20, NO.3, pp 625-630.

[5] J. Montano, and P. Salmeron (2000). *Different Approaches For Compensation In Three-Phase Four-Wire Circuits With Active Power Filters,* Mediterranean Electrotechnical Conference, Vol. III , 0-7803-6290-X, pp 920-923.

[6] Qun Wang, Weizheng Yao, Jinjun Liu, and Zhaoan Wang (1999). *Voltage Type Harmonic Source and Series Active Power Filter Adopting New Control Approach*, The 25th Annual Conference of the IEEE , Vol. 2, pp 843 -848.

[7] Wang Qun, Yao Weizheng, Liu Jinjun, and Wang Zhaoan (1999). *A Control Approach for Detecting Source Current and Series Active Power Filter*, IEEE International Conference on Power Electronics and Drive System, pp910-914.

ELSEVIER
IFAC
PUBLICATIONS
www.elsevier.com/locate/ifac

DESIGN OF WEB BASED MONITORING AND CONTROL SYSTEM FOR PHOTOVOLTAIC SYSTEM

Byung-Bog Kang[1], Jung-Yeol Lim[2], Jeong-Phil Yoon[1], Se-Jun Park[1], In-Su Cha[1]

1) Dept. of Electrical & Electronic Eng. Dongshin Univ.
2) School of Computer & Electronic Eng. Nambu Univ

Abstract: Digital environment that is represented to internet is displacing business way of industry and business achievement way with the fast speed being giving great change on life whole, improve existence business process utilizing internet and Web connection technology, information superhighway to tradition industrialist manufacture and e-transformation's propulsion that wish to maximize productivity and administration efficiency is spread vigorously. In this paper, we wish to accomplish generation equipment's heighten stability and believability through remote monitoring and control of PV system. This paper describes the design of the monitoring system for sensing the monitoring data and indirect controlling of the PV system. Most of the conventional monitoring system depend on the special hardware and software. Basic design goal of monitoring system is to provide the convenience for the user and the portability for the system. In order for the system to fulfill its requirements, it was designed using Labview GUI facility based on the Windows 2000 environment of IBM PC compatible and Add-on card based on the TCP/IP protocol. Advantage of the monitoring system are a personnel expenses curtailment effect, free of the place restriction and unmanned system of the generation plants, etc.. *Copyright © 2003 IFAC*

Keywords: Photovoltaic, Irradiation, Remote Monitoring, Web

I. INTRODUCTION

In view of the limitation of the fossil fuel and the upward tendency of energy demand due to the raise of the living standard, the alternative energy must be developed without delay especially in Korea because of the poorness of natural resources and dependency on the thermal power generation.

The development of the solar energy is necessary since the future alternative energies that have no pollution and no limitation are restricted.

Photovoltaic generation have an advantage of unlimited and unpolluted amount of energy resource. Since there is such an advantage in these energies, they are being studied and developed consistently.

In this paper, we wish to accomplish generation equipment's heighten stability and believability through remote monitoring and control of PV system. This paper describes the design of the monitoring system for sensing the monitoring data and indirect controlling of the PV system.

The string monitoring is connected directly to the inverter via an RS485 bus and allows for current measurement and monitoring of each string in our PV system.

By connecting a radiation sensor and a temperature sensor directly to the field point and acquisition card it is possible to acquire not only electrical data but also the corresponding meteorological values measured. An web server and a LAN was installed inside Field server. This allows to monitor PV system via the Internet – the most convenient of possibilities to communicate.

II. SYSTEM STRUCTURE

Basis structure of the system is same with figure 1. It is consisted of the PV system become target of the monitoring, the monitoring system that can bring and control to their state. Figure 1 Main structure of Monitoring system. The PV system is 10kW utility interactive type.

The monitoring system is consisted of the Add-on card that translate information and transducer measurement voltage, current, temperature etc.. Software for monitoring designed using LABVIEW. In the add-on card, bring information that is measured in PV system through serial communication or pass established information by administrator.

Field server is consisted of HTTP server and system that collects and store and analyze monitoring data. PV system is used 3 PV array(17 Serial × 4 Parallel)and its total power are about 10KVA. The specification of PV module is as follow.

- Maximum Power : 50 [W]
- Maximum Voltage : 17.39 [V]
- Maximum Current : 1.94 [A]
- Efficiency : 11.7 [%]

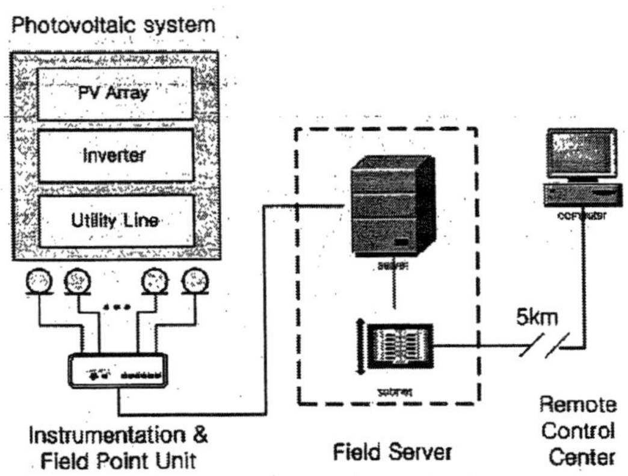

Figure 1. Basic structure of monitoring system

Figure 2 shows apparatus of inverter system. It is used the utility interactive type inverter with utility line. The specification of utility interactive type inverter is as follow.

- Input Voltage : DC 252 ~ 400 [V]
- Output Voltage : AC 220 [V]
- Output Power : 10 [kVA]
- Output Frequency : 60 [Hz]
- Type : PWM, 3ϕ
- Maximum Efficiency : 93.9 [%]

III. WEB BASED MONITORING SOLUTIONS

In order for the system to fulfill its requirements, it was designed using Labview GUI facility based on the Windows 2000 environment of IBM PC compatible and add-on card based on the TCP/IP protocol. The transmission control protocol (TCP) is intended for use as a highly reliable host-to-host protocol between hosts in packet-switched computer communication networks, and in interconnected systems of such networks. The internet datagram provides a means for addressing source and destination TCPs in different networks.

The internet protocol also carries information on the precedence, security classification and compartmentation of the TCP segments, so this information can be communicated end-to-end across multiple networks. To read data from various sensors or data measuring device, remote data processing module need that serve to convert analog signal to digital signal [Standard : RS485 : 9600. N. 8. 1 (Half Duplex)]. Figure 3 shows the measurement unit that is included Transducer, Transformer and Field Point, etc.. It measures data(voltage, current) value of the PV system and send to the Field Server.

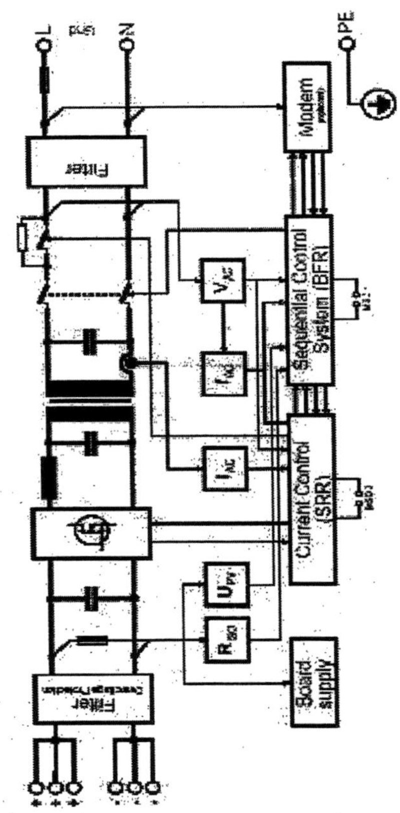

(a) Block diagram

(b) Actuality picture

Figure 2. Apparatus of inverter system

Figure 4 shows the program flow chart by Labview. And it means that control method of the monitoring element.

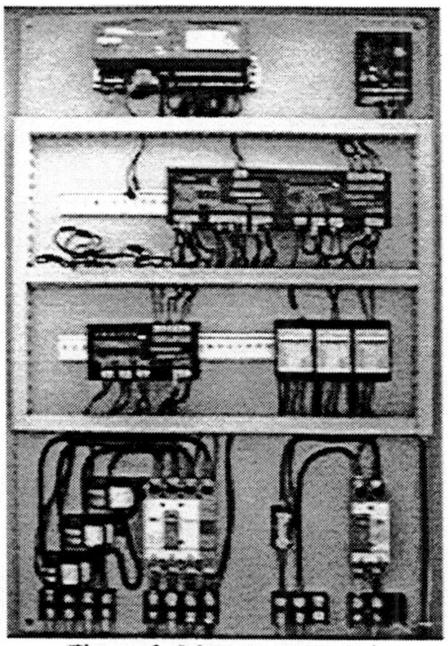

Figure 3. Measurement unit

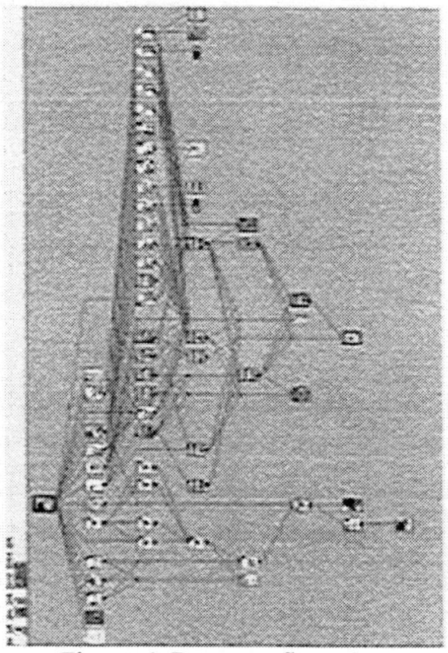

Figure 4. Program flow chart

IV. SIMULATIONS AND EXPERIMENTS

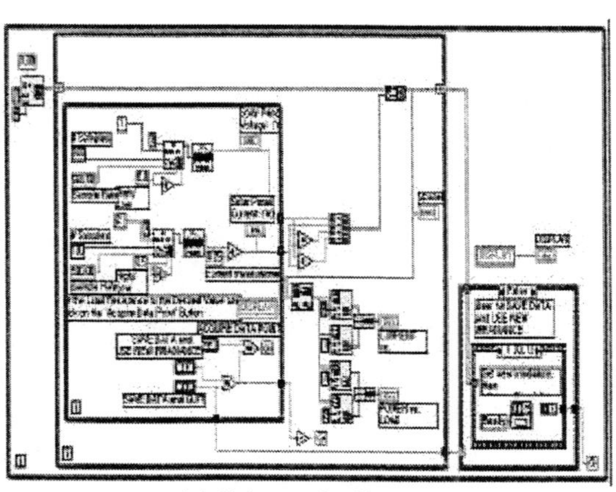

(a) Schematic diagram

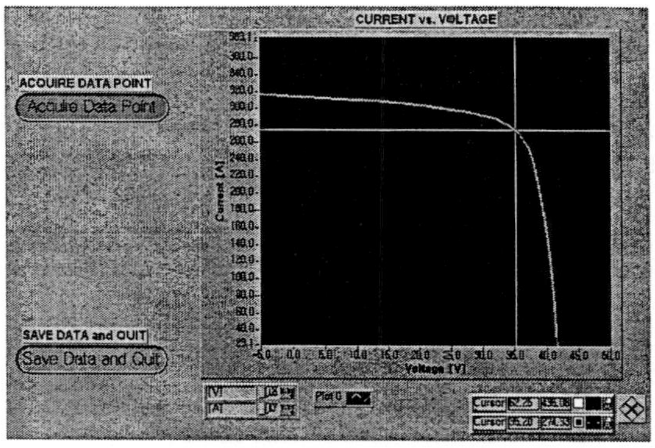

(b) Output characteristic

Figure 5. Characteristic of PV array

Figure 5 shows the schematic diagram and output characteristic of PV array.

Maximum power point formed in the voltage 274V and current 35.2[A].

Figure 6 is schematic diagram and output waveform of inverter that is used in this paper.

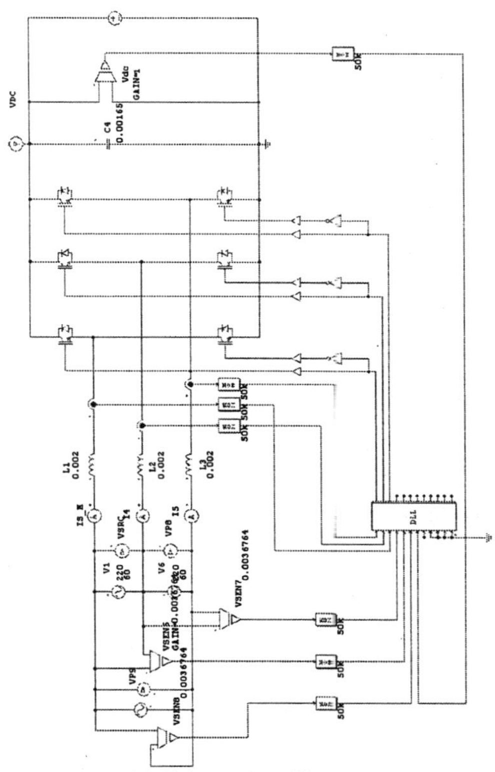

(a) Schematic of inverter

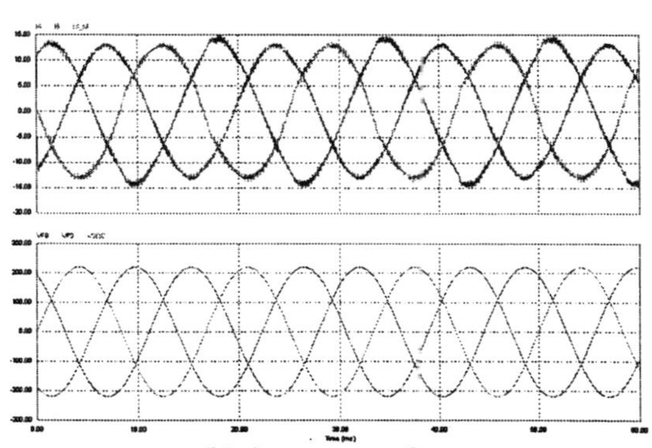

(b) Output waveform

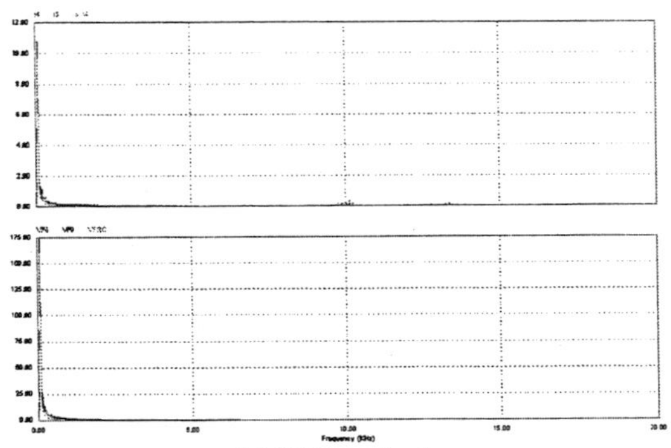

(c) FFT analysis

Figure 6. Inverter simulation

Figure 7 shows the main screen that is displayed power, voltage(DC, AC), current (DC, AC), irradiation and temperature (outside, solar cell surface). When compared remote control center data receiving speed with field server, remote control center is indicated delay time about 10 [sec] than field server. Because of the data translation speed of field point and field server is faster than communication speed of field and remote control center. In additional, because of the field server and remote control center are far away about 5km.

Also, it can control the PV system through this panel. Figure 8 shows the trend screen that is displayed power, voltage, irradiation and temperature.

It is displayed graphic of data.

Figure 9 shows output characteristic of PV system and this waveforms can be plotted by accumulated data of data acquisition system. Where, x axis is converted by minute.

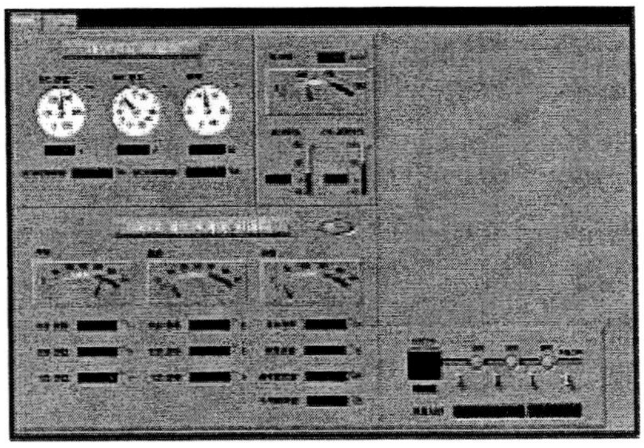

(a) Field Server

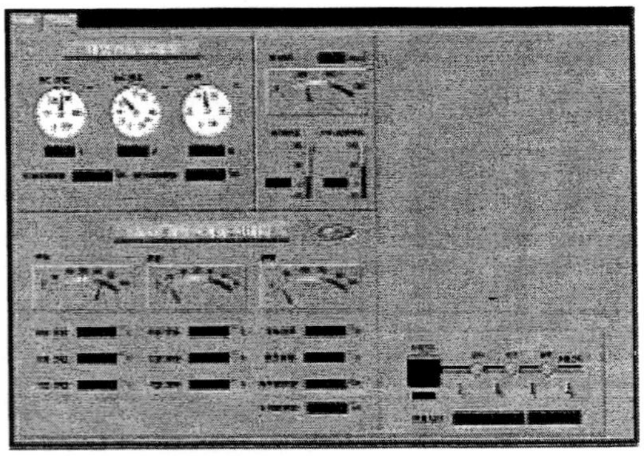

(b) Remote Control Center

Figure 7. Monitoring system

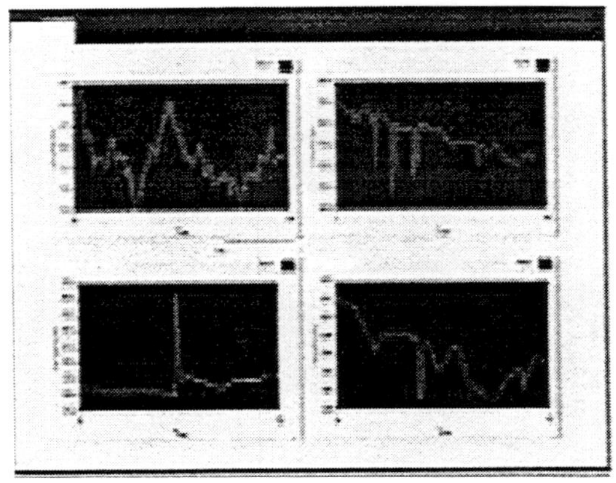

Figure 8. Trend Screen

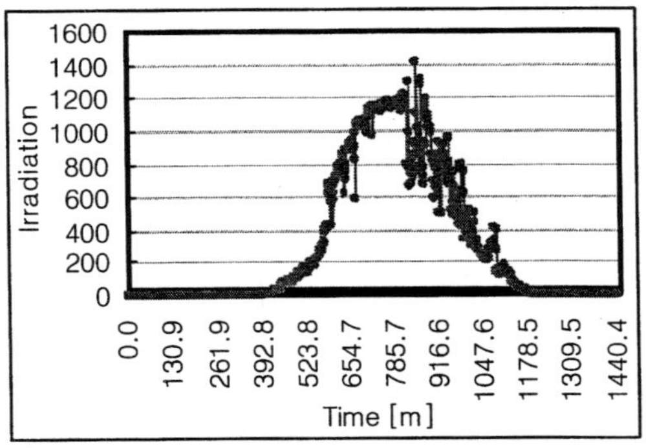

(a) Irradiation characteristics

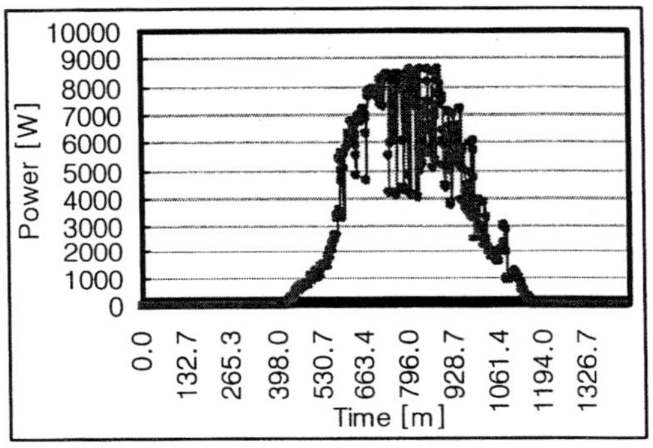

(b) Power waveforms

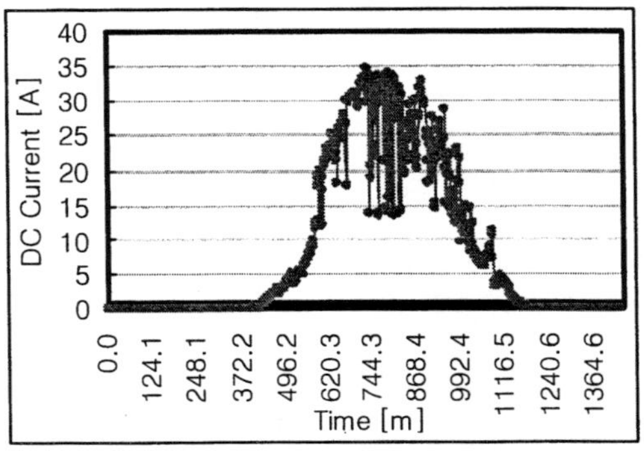

(c) Current waveforms

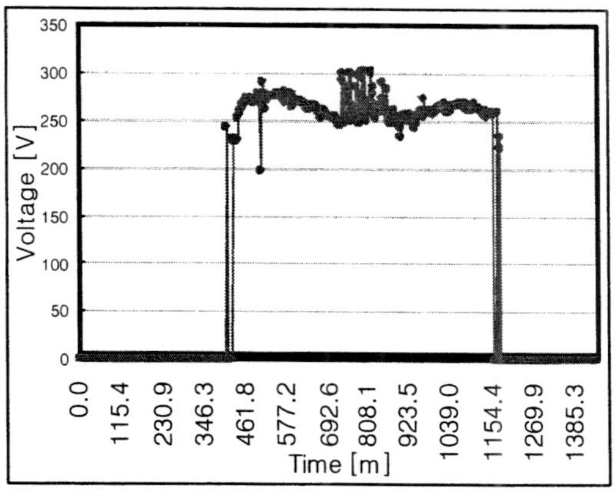

(d) Voltage waveforms

Figure 9. Output characteristic

V. CONCLUSIONS

In this paper configured and analyzed remote monitoring system for manless and independent operation of 10kW photovoltaic. Through this, we wished to minimize stability, reliance, time, spatial restriction. To supplement problem of locality, compatibility, extensity, we embodied Web-based monitoring and control system. We composed hardware for monitoring system, design and drove considering user interface for MGDC that can do monitoring and control with this. But, it will necessary research about error etc. that application of web technology occur problem of security by network protocol and emergency situation of equipments.

In addition to we wish to compose photovoltaic that independent operation strengthen considering voltage variation and problem of system using construed data through monitoring system influenced to supply of electric power system.

ACKNOWLEDGMENTS

This work was supported by grant Gwangju·Jeonnam Technopark Support Center In Korea "The development for MPPT Control System of Utility-Type 10kW Photovoltaic that is considered Locality Characteristics."

REFERENCES

M.Benghanem, and A. Maafi(1997), "Data acquisition system ofr photovoltaic systems performance monitoring ", IEEE Instrumentation and Measurement Technology Conference, Ottawa, Canada, pp. 19-21

Phuong T. Huynh(1999), "Design and Analysis of a Regulated Peak-Power Tracking System", IEEE Trans. On Aerospace & Electronic Systems, Vol. 35, No. 1, pp 84-91

T. Takebayashi, H. Nakata, M. Eguchi, and H. Kodama(1997), "New current Feed Back Control Method for Solar Energy Inverter using Digital Signal Processor", PCC-Nagaoka

Victorio Arcidiacono, Sando Corsi, Luciano Lambri, "Maximum Power Point Tracker for Photovoltaic Power Plants", IEEE. pp. 507-512, 1982.

Z. Salameh, D. Taylor, "Step-Up Maximum Power Point Tracker for Photovoltaic Arrays", Solar Energy, Vol. 44, No. 1, pp. 57-61, 1990.

ELSEVIER
IFAC
PUBLICATIONS
www.elsevier.com/locate/ifac

EFFECT OF FAULT CURRENT LIMITER IN AC/DC HYBRID POWER SYSTEM

T.Kamiya, H.Yamada, K.Yukita, Y.Goto, K.Ichiyanagi
Aichi Institute of Technology

T.Matsumura
Nagoya University

Abstract—this paper, the case in which current limiter is introduced into power transmission system of DC interconnection is assumed. By viewing from the responsibility of DC transmission system in AC line fault, the current limiter introduction effect is examined. The effect by the current limiter introduction is reported, because in the place, the stable driving of the DC system was able to be maintained and be able to be confirmed with the control of fault current. As the result, the stable operation of the DC system was able to be maintained with the suppression of fault current, and the effect by the current limiter introduction was able to be confirmed. *Copyright © 2003 IFAC*

Keywords: HVDC transmission lines, power-system, Power transmission, Voltage stability, Series resistance

1. INTRODUCTION

With the increasing demand for electric power, the electric power system enlarges scale, and becomes complicated. In such situation, it consisted in order to introduce the direct current power transmission as an electrical power transmission facility which was excellent in economical efficiency and reliability at large capacity in conventional ac transmission (Y.Sekine, et al.,1995) (T. Machida, et al.,1999) (IEEJ, et al., 1999). The application to low voltage system and power distribution system would be recently examined in the DC power transmission system of small capacity. On the other hand, for current suppression in the transmission line fault, the introduction of the current limiter is examined, and there are mainly the many report on the application effect by AC system (D. Iioka et al.,2002)(J. Zou et al.,2002). In this paper, the case in which current limiter is introduced into power transmission system of DC interconnection is assumed. By viewing from the responsibility of DC transmission system in AC line fault, the current limiter introduction effect is examined. For AC line fault of DC interconnection transmission line, the computer simulation was carried out using P S CAD/EMTDC. The effect by the current limiter introduction is reported, because in the place, the stable driving of the DC system was able to be maintained and be able to be confirmed with the control of fault current.

As the result, the stable operation of the DC system was able to be maintained with the suppression of fault current, and the effect by the current limiter introduction was able to be confirmed.

2. AC/DC HYBRID TRANSMISSION MODEL

It consisted in order to introduce the direct current power transmission in conventional ac transmission. The application to low-pressure system and power distribution system would be recently examined in the direct current power transmission system of small capacity (M.Saisho et al.,2002). As direct current power transmission system linked in the AC system, latitudinal model shown in Fig. 1 is taken up. In the figure, the DC transmission system is assumed as the capacity of 120MW and transmission distance of 100km. And the DC system consists of the rectifier with constant current control equipment and the inverter with constant margin angle. The transformers for the commutation are connected in the each AC sides of the rectifier and the converter and they are linked in the AC system. The constants of the model are shown in the Table 1.

Generally, the converter for the direct current power transmission can suppress the over current by the DC reactor with the large inductance value and the operation of protection circuit. In the Fig. 1, the over current in the case of DC line fault can be sufficiently suppressed by the operation of constant current control of the rectifier and the DC reactor. Therefore, the current limiter is supposed to be inserted in the AC line only. The fault of the DC line is not made to be an object, and only AC line fault is made to be an object, introduction effect of the current limiter is examined.

The two AC transmission lines are connected with the inverter. The current limiter has respectively been installed in each AC line in sending and receiving terminals. As a current limiter model, resistive type is assumed, and the basic configuration is shown in Fig. 2. The resistance increase of current limiter is shown in Fig. 3. The broken line in the figure shows the resistance increase characteristics in case of super conductive type current limiter (Y.Gotou et al., 1999). In this report, current-limiting resistor is supposed to be arising in the middle in current-limiting start and completion time.

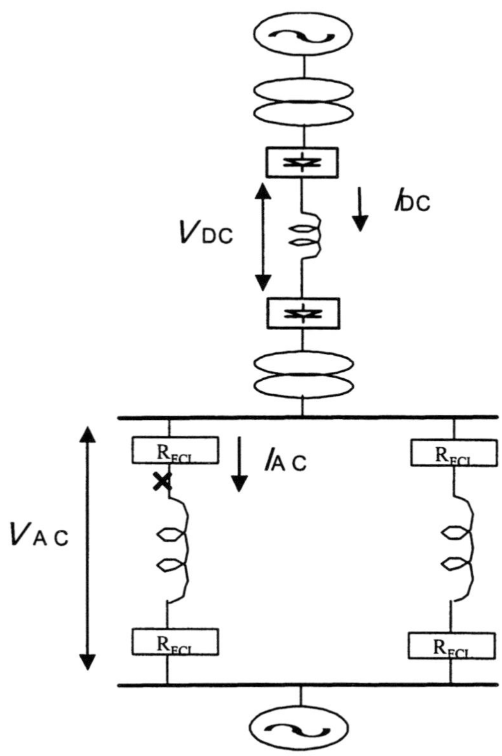

Fig.1 DC interconnection power transmission model

Table 1. Model Parameters

DC system	Rated power	120 MW
	Rated voltage	25 kV
	Rated current	4.5 kA
	Line inductance	16 mH
	Line resistance	1.45 Ω
	Inductance of filter	250 mH
AC system	Rated power	144 MVA
	Rated voltage	200 kV
	Rated current	0.24 kA
	Line inductance	80 mH
	Line resistance	2 Ω

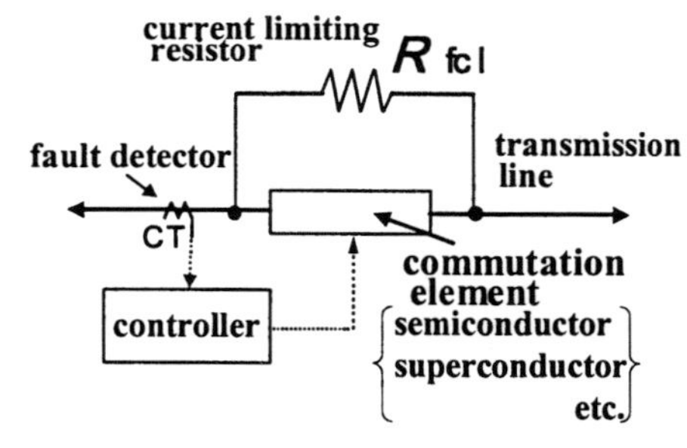

Fig.2 Resistive type fault current model

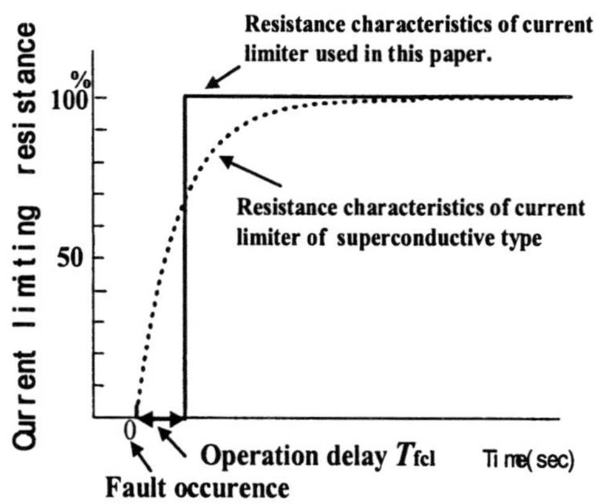

Fig.3 Resistance characteristics of current limiter

3. SIMULATION RESULT IN CASE OF AC LINE FAULT

Using the model system shown in Fig. 1, the fault point is supposed to occur only at the AC line. Based on time sequence of Fig. 4, the simulation was carried out. In this case, the fault line is opened after 70ms of permanent fault occurrence. The PSCAD/EMTDC was used for the calculation. As the resistance for current-limiting and the operation delay time, R_{fcl}=200 Ohm and T_{fcl} =4msec were respectively used. The sending terminal as fault point of the AC line The simulation was carried out on the case in which sending terminal (point P shown in Fig.1) of the AC line was assumed as a fault point. The results in case of with and without current limiter are respectively shown in Fig. 5 and Fig. 6.

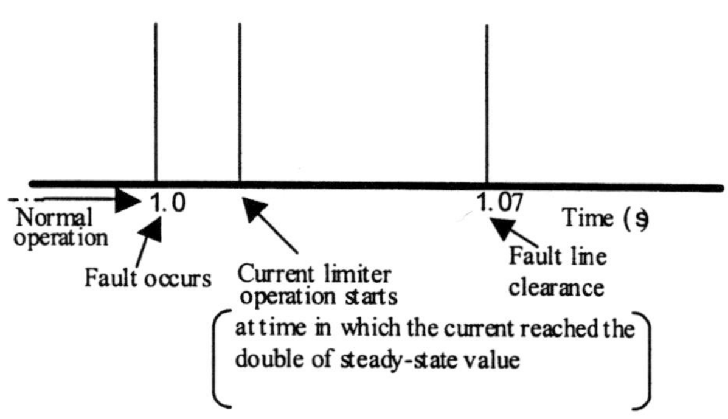

Fig.4 Time sequence of simulation

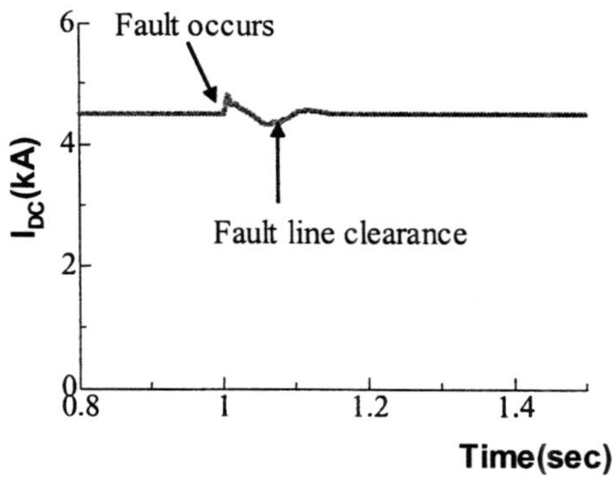

(a)DC current wave form

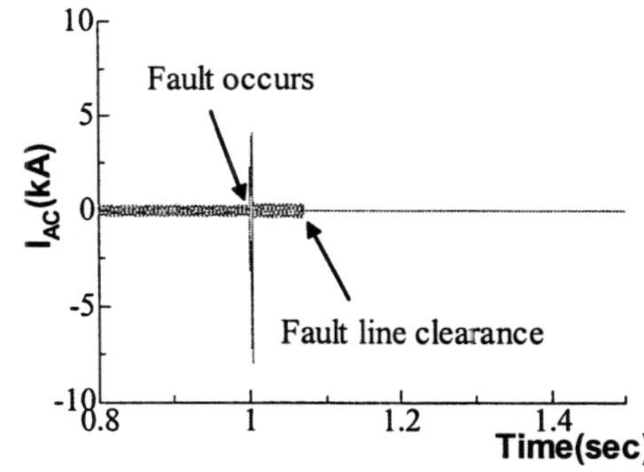

(b)AC current wave form

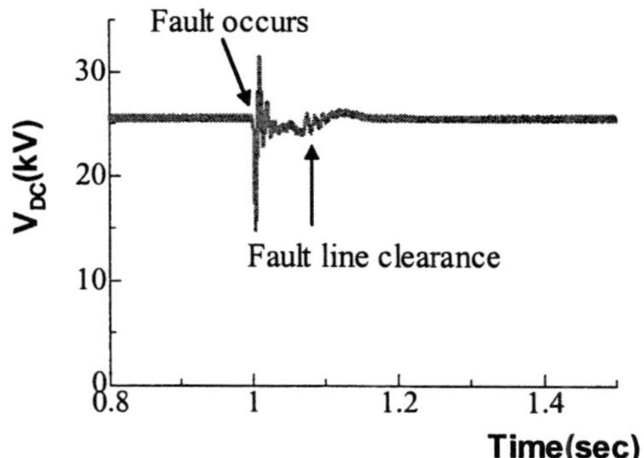

(b)DC voltage wave form

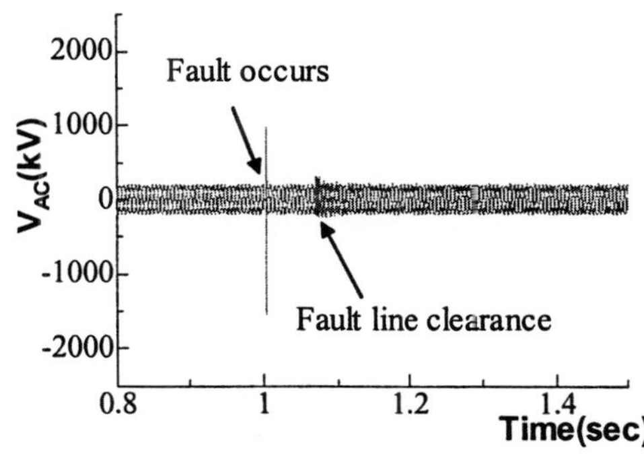

(b)AC voltage wave form

Fig.5 DC system response in AC transmission line fault with fault current limiter in case of fault point P

Fig.7 Fault line voltage-current waveform in AC transmission line fault with fault current limiter in case of fault point P

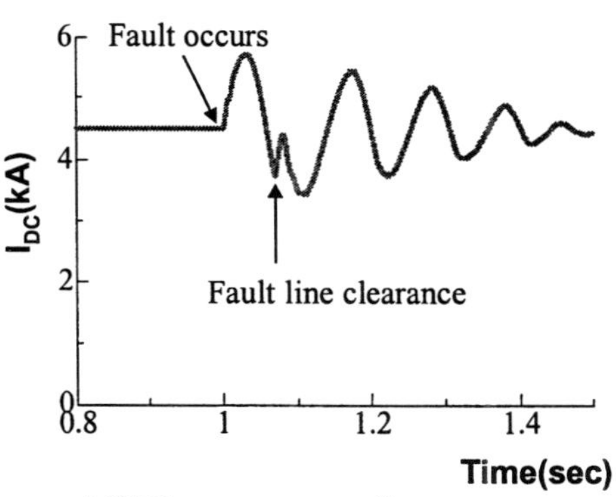

(a)DC current wave form

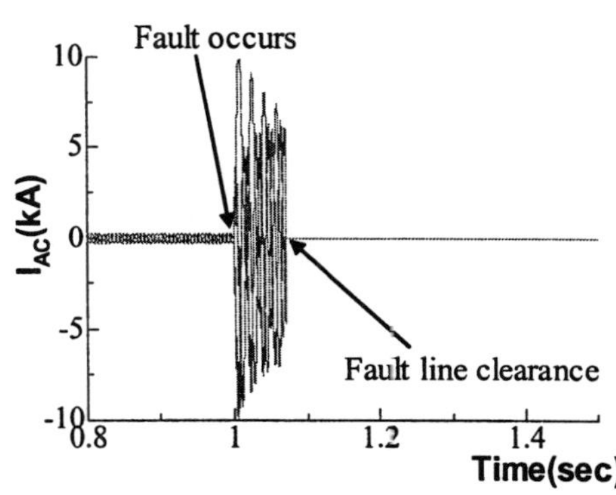

(a)AC current wave form

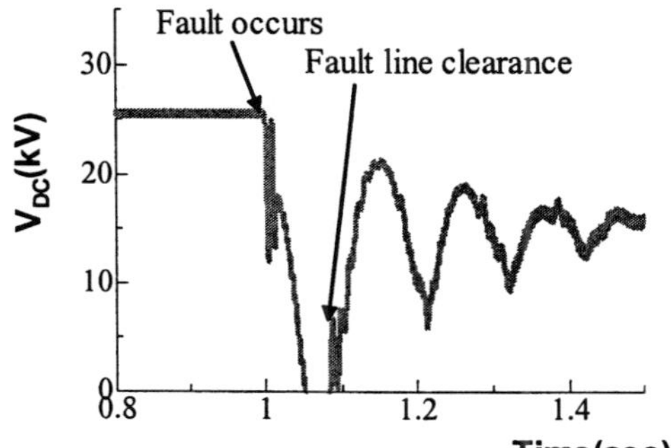

(b)DC voltage wave form

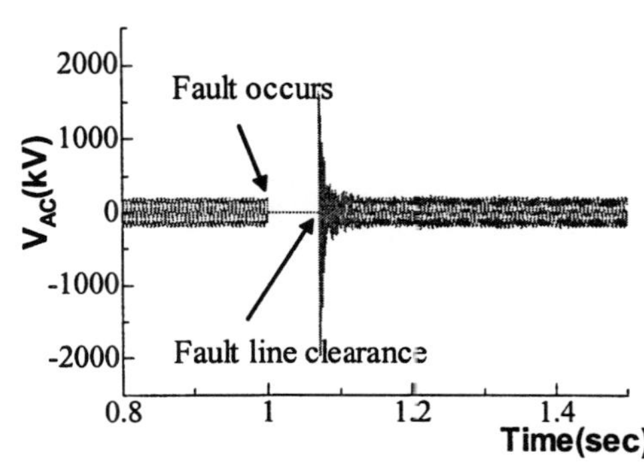

(b)AC voltage wave form

Fig.6 DC system response in AC transmission line fault without fault current limiter in case of fault point P

Fig.8 Fault line voltage-current waveform in AC transmission line fault without fault current limiter in case of fault point P

In both figures, the current and voltage of the DC line are respectively shown at (a) and (b). According to figure (a), the voltage and current of the DC system greatly change in the case without the current limiter. According to figure (b), by the current limiter introduction, it is proven that the fluctuation of voltage and current of the DC system becomes slight.

The fault current waveforms in case of with and without current limiter are respectively shown in Fig. 7 and Fig. 8. By the current limiter introduction, it is proven that the fault current of the AC line is suppressed from 10 kA to 5 kA.

In case of the DC interconnection power transmission system used in this report, the above result shows that the fault current in the AC line fault can be suppressed by current limiter operation installed in AC transmission line. In addition, it was confirmed that the stable driving in the direct current system could be maintained.

4. EXAMINATION OF CURRENT- LIMITING RESISTOR VALUE R_{fcl}. SIMULATION RESULT IN CASE OF AC LINE FAULT

The value of current-limiting resistor R_{fcl} inserted in the AC transmission lines was made to change from zero to 800 Ohm. The simulation was carried out as is mentioned above. The value of current-limiting resistor, the peak of the DC line and AC line current were respectively examined. Those results are shown in Fig. 9 and Fig. 10. In the both figures, the peak current shows almost constant value within being bigger than R_{fcl}=100 Ohm.

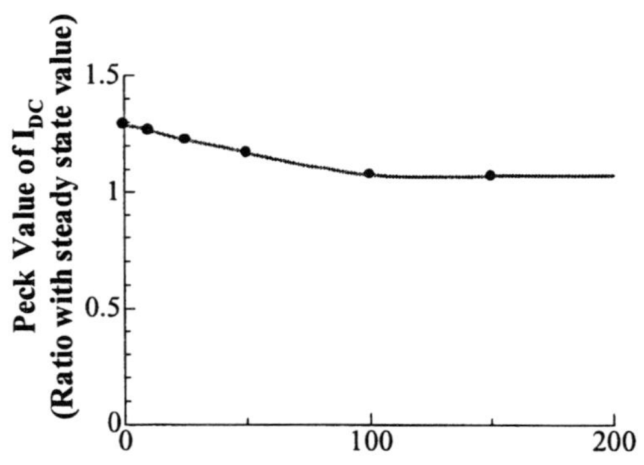

Fig.9 Characteristics of fault current limiter R_{fcl} resistance vs peak current of DC line

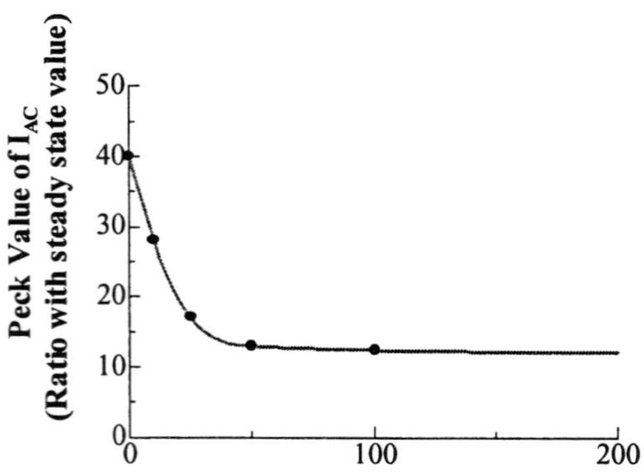

Fig.10 Characteristics of fault current limiter R_{fcl} resistance vs peak current of AC fault line

V. EXAMINATION IN OPERATION DELAY TIME OF CURRENT LIMITER

The operation delay time T_{fcl} of current limiter after the fault current detection was made to change from three to ten msec, and for AC line fault, the similar simulation was carried out. T_{fcl} and direct current line current peak, fault current peak were similarly examined as is mentioned above. Those results are shown in Fig. 11 and Fig. 12. In the both figures, the peaks of AC and DC current shows almost constant value within being less than T_{fcl} =5 msec. In viewing from the stable operation of DC power transmission, it can be confirmed that current limiter operation delay time T_{fcl} =3-5 msec is suitable.

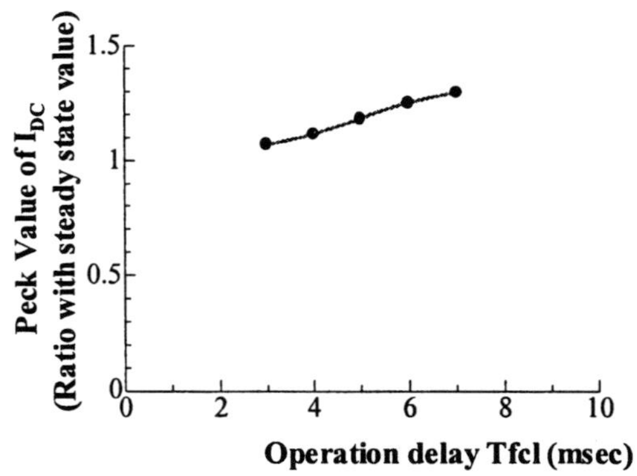

Fig.11 Characteristics of operation delay *T*fcl of fault current limiter vs peak current of DC line

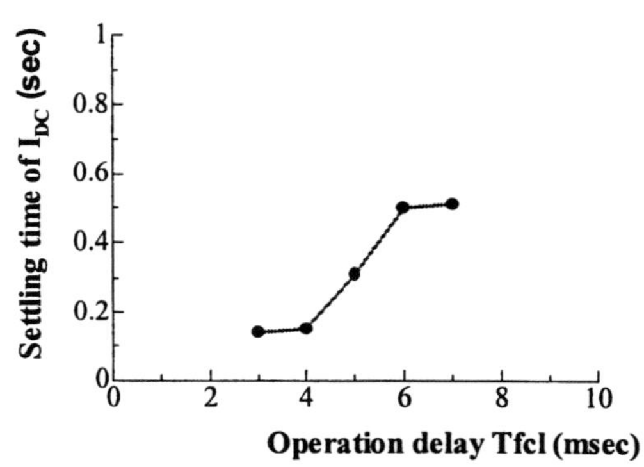

Fig.12 Characteristics of operation delay T_{fcl} of fault current limiter vs peak current of AC fault line

5. CONCLUSION

With the liberalization of the electric power market, the electric power supply according to the quality of electric power became possible. By the popularization of distributed generation, the new electricity distribution system is proposed. In such environment, the responsibility of DC system by AC line fault was examined on the power transmission system that direct current power transmission is linked with the AC system.

From simulation result, it could be confirmed that the stable operation of the DC system could be maintained by the introduction of current limiter with the control of fault current. In the future, the examination will be carried out by the similar method by taking direct current power transmission system using self-excitation type DC converter into consideration.

6. REFERENCE

S.Tanaka, T.Maeda, H.Konishi, H.Aizawa "Control and Protection Scheme of HVDC System with Self-commutated Converter", Proceedings of the Tenth Annual Conference of Power & Energy Society. IEE of Japan, Vol.A, No.244,pp.598-599 (1999) (in Japanese)

D.Iioka,H.Shimizu,Y.Yokomizu,M.Goto, T.Matsumura and I. Kano :"Suppressing Effect of Fault Current Limiter on Shaft Torsional Torque of Synchronous Generatorin Consumer System"IEEE/PES Transmission and Distribution Conference and Exhibition 2002,Asia Pacific, po24-4, (2002)

M.Saisho T.Ise K.Tsuji:"DC Loop Type Quality Control Center – One Possible Configuration of Power Factor Correctors", National Convention Record, IEE Japan, Vol.6, No.6-023, pp39-40 (2002) (in Japanese)

J. Zou, J. Chen, E.Dong:"Fault Current Limiter with Fast-Closing Switch", IEEE Power Engineering Review, pp40-41 June (2002)

T. Machida:"DC Transmission Engineering", Tokyo Denki University Press (1999) (in Japanese).

Y.Gotou T.Niimi K.Yukita K.Mizuno K.Ichiyanagi Y.H.GuoY Yokomizu and T.Matumura "Experimental Studies on Effect of Superconducting Fault Current Limiters on Improvement of Power System Stability", T.IEEJapan,Vol.119-,No.11,pp1241-1248 (1999) (in Japanese)

Y.Sekine, S.Ito, et al. :KII Channel HVDC Link between SIKOKU and KANSAI Electric Power Companies by Submarine Cables, CIGRE Symposium Tokyo,1995.5.

ELSEVIER
IFAC
PUBLICATIONS
www.elsevier.com/locate/ifac

RENEWABLE ENERGY IN DEREGULATED POWER MARKET

Richard Nieh, Loi Lei Lai, N Rajkumar
Energy Systems Group, City University, London, UK

Yuping Lu, Guoqing Tang
Institute of Power System and its Automation, Southeast University, Nanjing, P.R.China

Abstract: This paper discusses briefly what are renewable energies and how renewable energy could be included successfully in the deregulated power market. Problems have been identified and some solutions have also been proposed to reduce the risk. Due to the lead time and large investment required to implement a large generation, distributed generators will be the future trend under the new restructured environment. It points out that system integration is an important factor to enhance the use of renewables. *Copyright © 2003 IFAC*

Keywords: Renewables, risk management, environment issues, distributed generation

1. INTRODUCTION

Renewable energy refers to natural energy resources that are inexhaustible: wind, solar, geothermal, biomass, and hydropower. However, for the purpose of qualifying for benefits under the RPS (Renewables Portfolio Standard), "renewable energy" can be defined in the policy process to include only those renewable resources that the nation wishes to particularly encourage. For example, in *U.S*, they would be wind, solar, geothermal and biomass because those resources and technologies that currently produce an insignificant amount of total electricity generations. However, they have great potential to provide a significant share of the nation's power in the next decade and beyond. Generally, the small-hydro energies should be included (Nancy Rader, 1997)

2. THE ADVANTAGES AND IMPORTANCE OF THE RENEWABLE ENERGY

The main advantages can be discussed in three ways: Environmental advantages, Strategic advantages and Social economic advantages.

2.1 Environmental advantages

First of all, the most significant effect that the power generation can produce is air pollutions. Energy produce from fossil fuels (petroleum, gas, coal) produce CO_2 emissions that has caused more air pollution than those produced by vehicles and other kind of air pollution producers. On the other hand, renewable energies do not produce CO_2 emissions and other atmosphere polluting gases, which make them extremely environmental friendly, especially in term of air contamination.

Secondly, nuclear energy and fossil fuels produce wastes, which are threats to the environment. Especially, the infamous nuclear waste costs money, time and labour to clean. It is extremely dangerous to living beings. Conversely, the renewable energies do not generate wastes hard to handle.

Finally, nuclear and fossil fuel supply is limited. The oil crisis that has been talked about since 1970 is getting worse. On the contrary, renewable energies are inexhaustible.

2.2 Strategic advantages

The renewable energies are indigenous. The resources are everywhere and easy to find, even though it is not easy to collect them efficiently and still depend on the climatic and geographic conditions of the country. They still have much better availability than fossil fuels or nuclear power, which are limited to a small number of countries.

Subsequently, the fossil fuels increase the energy import in the European Union while the renewable energies avoid external dependency. Thus the renewable are more consistent in terms of economy reason and availability.

2.3 Social economic advantages

The traditional energies create fewer jobs in comparison to the business that they produced and the renewable energies create five times more jobs than the conventional energies. For example, it is very difficult to ask a weather predictor to work in a fire-power station but the same job can be extremely important for a wind-power station.

Additionally, renewable energies contribute decisively in the inner-territorialize balance because they usually are installed in rural areas. On the other hand, the traditional energy sources are usually located near highly developed areas. Finally, the technology of the renewable energies can be developed in countries which need them while the great part of the technology used for traditional energies is usually imported from other countries.
(web page: www.appa.es/dch/eolic_en.htm).

3. POTENTIAL PROBLEM OF THE RENEWABLE ENERGIES AND THEIR SOLUTIONS

Even though renewable energies have countless advantages over traditional power generation, they still have a few problems since it is still at developing stages. Essentially, they can be sorted into three major parts: technological problem, economical problem and political problems.

3.1 What are renewable technologies and what are their problems?

The technologies of renewable energies are very diverse because there are so many different kinds of renewable energies. The term renewable energies indicate that they are inexhaustible. They are natural power from the planet and the main concern is how we can fully utilise the technologies available.

Renewable technologies include:

- Passive solar design strategies, which combine basic energy efficiency practices with building designs that maximize natural heating, cooling, and day lighting
- Solar thermal systems for hot-water heating and ventilation air preheating
- Solar cells (photovoltaic or PV) for electricity generation
- Wind systems for water pumping and electricity generation
- Geothermal technologies for water and space heating
- Bio fuels (ethanol and methanol).

There seems to be many different kinds of technologies but they all actually share one problem: they are all very young. The wind turbines that were developed 20 years ago seem like prehistoric monument. During the last 20 years, renewable energy technologies have developed considerably and there is no technical reason anymore not to switch to renewables. However, the development compared to the conventional energies is still very much young.

In addition, new technical challenges will have to be met for an energy system that is based completely on renewable sources. Energy from wind or solar irradiation for instance is produced when the wind blows or when the sun shines. This is not always the moment that energy is needed or asked for.

3.2 Economical and political perspectives of the renewable energies

In general, renewable energy is still a bit more expensive than energy from conventional sources. This is partly due to the fact that the costs of the use of conventional sources (like damage to human and monuments caused by acid rain) are not included in their price. On top of that, in many countries, conventional sources like coal or nuclear energy profit from large subsidies. If these differences are taken into account the cost of renewable energy is already very close to that of conventional sources. All it needs is a little extra provided by special subsidies or by voluntary customers of green energy (web page: www.worldwidegreen.com/index.asp?page=remewables-extended.htm&menumenure.htm).

Subsequently, the economical environment is hugely affected by the political decisions. For example, the subsidies from the government can tremendously influence the pressures on renewable energies. The subsidies can reduce the cost of the energy but also reduce the competition under a deregulated market, which can cause the loss of profit to the company. The reason that happens is the energy shortage may happen if the government does not subsidise the company.

3.3 Unravelling uncertainty

There is little or no dissent that the required increase in distributed generation can only happen with significant focus on distribution network – the question is – what is it to focus on?

The sum of the total existing embedded energy generation [Gas Turbine(GT) /Diesel/ Other/ Renewables/ Combined heat and power(CHP)] is about 5 GW in 2002. However, the potential growth of the generation is predicted as over 20 GW in total energy generation by the year 2010.

The plan is through re-configuring, re-engineering and re-in-forcing from 2002 to 2010. The embedded generation can change from self-sufficient passive to integrated active.

There is a need to develop technical and commercial solutions in harmony with each other – rather than independently. This brings additional complexity. Considering the least cost forward, technical options and commercial acceptability, the future business environment may cause increasing fluidity and responsiveness and changing in technical and business risks, incentives and legal obligations.

In the business uncertainty point of view, for all the stakeholders, the most challenging aspect is the level of uncertainty associated with the issues - in this case it is double headed – 1. What is the exact problem? 2. What will result if we solve it?

The Distributed Generation Co-ordination Group (DGCG) and its Technical Steering Group (TSG) has been set up to ensure representatives from the key stakeholders are brought together to address a few issues. The key stakeholders are Generators, suppliers and regulator, such as OFGEM. In the process, TSG need to consider:1. Co-ordination of work-streams. 2. Programme of projects. 3. Technical and commercial issues. Ofgem will need to consider: 1. Internal Ofgem DG project, 2. Links to price control review project, 3.Statutory roles and consultation.

Under the direction of the DGCG, the TSG has a suite of work-streams that are targeted at unravelling uncertainty that attends the many network related issues. They are, for example, status & Projections, standardisation of solutions, short-term network solutions. Questions that arise are: 1. what are the possible attendant risks? 2. What are we trying to cater for? 3. What is active management? 4. How challenging is connecting millions of domestic micro generators? 5. How could we DG for network security? (Unravelling uncertainty. Distributed Generation Co-operation Group. IEE London).

3.4 Distribution and local generation

There are a few points that can be discussed about distribution and local generation. First, there is a duty to connect within reasonable time but sometimes need for better processes. Subsequently, the generation unit should get the whole benefit of the cost reduction: 1. Connection costs. 2. Charges for additional costs.

Network reinforcement is due to demands on transmission capacity and the need for increased short circuit current is due to flicker. Balancing costs due to forecast difficulties may be substantial (Unravelling uncertainty. Distributed Generation Co-operation Group, IEE London).

3.5 Large-scale wind generation

Additional future demands participate in voltage regulation (demands on reactive power generation) and in secondary frequency control (10-15 minutes). In addition, disturbance endurance for frequency variations and dynamic stability needs to be improved excessively.

Also the large scale wind generation has problem of slow voltage variation, over tones, that means the generation output is not pure sine wave and the extra cost is needed to rectify it.

More problems should be solved while Research & Development programs work together with generator producers. Ongoing discussion will be necessary with generation investors (Unravelling uncertainty. Distributed Generation Co-operation Group. IEE London).

3.6 A need to re-think

Last ten years is a long time in electricity, the energy is transferring from coal to gas to the latest nuclear power. The obligation of the electricity and its appeal to the market as a firm tool are in a time of uncertainty. Both of environmental problems of traditional power generation and less economic opportunities need to be noticed.

The policy framework should address all three objectives of sustainable development – economic, environmental and social – as well as energy security. Climate change objectives must largely be achieved through the energy system. Where energy policy decisions involve trade-offs between environmental and other objectives will tend to take preference (PIU energy review executive summary, February 2002).

In terms of environmental issues, the regulatory policies will need to be promoted and not to be inhibited. A fundamental review of the financing, management and regulation of electricity network (like the national grid) is necessary in order to encourage renewable energy sources.

The electricity grid needs to be transformed to accommodate much greater number of small-scale, local generators, many using renewable energy sources. Major part of these plants will have intermittent or unpredictable output. The electricity distribution system will need to undergo major changes to cope with these developments.

However, the subject is broad. We will have to focus on distributed generation, NETA (new electricity trading arrangements), strategic redesign and the role of the government. For all of them, they are inter-linked issues.

3.7 Incentives of renewables, CHP and distribution networks

Any incentive mechanism must be effective, predictable, simple and consistent. Also they have to be encouraging economic and efficient behaviour and avoid creating perverse or conflicting incentives. Subsequently, it is better to be co-ordinated with existing incentives and mechanisms. Finally, it should not be conflicting with longer-term development of distribution networks, as they become more like Transmission systems.

3.7.1 Incentives for connecting & for utilising distributed generation (DG)

The distribution generation has shallow connection charge, so there will be more appropriate return. Also, there will be higher revenue if more capacity is connected. Similar to connecting DG, there will be higher revenue for more generation supplied. In addition, while utilising DG, there will be a small loss of energy because the mechanism of the systems already exist or even enhanced.

If we consider both connecting and utilising the DG, the designated power zones can promote a proactive network development. They will also be considered to give a special treatment as renewable energies.

How will the longer term look? At the moment, the renewable energy, CHP network is passive. Their roles are the asset owner and RPI-X price, which are, controlled "traditionally". However, in the future, the network is becoming increasingly active and becoming energy "transportors". There are incentives for efficient & secure transportation but the potential problems still can arise: Are there any management of constraints? Secure access rights for generators? Is there role for storage.

(http://www.ofgem.gov.uk/renewables/sept2002conf/ofgem_pres.pdf)

3.8 Small generation

How small is small. The size of the generators are relative, one regulator's "small" can be another's "fairly large". Renewables includes a wide range of size including anticipated multi-hundred MW offshore projects. In fact there are debates as to whether small generators are relevant to a large, but intermittent generator. Renewables represent a large percentage of the freely traded electricity – compared to that which is sold via contracts for differences, etc. this is the price of aggregation.

Is it the end of large-scale distributed generation? Thinking about 2010 and beyond, not 1990, the developing world is unlikely ever to have large systems but is likely to have smaller, distributed networks. We can learn as well as teach (and export) smart networks and net-metering. We are not advocating wholesale re-building of the system, rather the need to modify the design rules in order to achieve the objectives.

3.9 Risk Management

The risk management issues for distributed generation are largely dependent upon the electricity market structure and regulatory environment in which the generation is sited. The United Kingdom has recently undergone a dramatic restructuring of the electricity market with the introduction of New Electricity Trading Arrangements (NETA) in March 2001. The structure of NETA, and in particular the pricing policy in the Balancing Mechanism, has introduced a number of risk management issues for distributed generation, which is now exposed to significantly different risks from those experienced under the Pool. In order to operate successfully under NETA it is important that participants understand the risks to which they are now exposed and implement strategies to manage these risks effectively.

3.9.1 Imbalance Risk

The significant risk for distributed generation is the two-price policy of the balancing mechanism (BM). If a generator delivers less energy than it has contracted for in a settlement period then it must pay the system buying price (SBP) for the shortfall. This is the weighted-average of offers accepted in the period. If it over generates in a period then it receives the, normally lower, system sell price (SSP) for the excess. This is the weighted-average of bids accepted in the period. The mechanism was designed to encourage market participants to contract in the long-term markets and power exchanges ahead of gate-closure. In the first few months of NETA, SBP has been extremely volatile and so the earnings risk if a generator does not deliver its contract position is high. Effective risk management strategies are therefore important for all market participants, including distributed generation. It has been said that NETA rewards reliable plant. However it would be more accurate to say that NETA rewards predictable plant, as the contract position for the generator must be submitted at gate closure, which is currently 1 hour ahead of the delivery period. A large number of the distributed generators in the UK are CCGTs, which have a relatively predictable output. It might therefore be expected that the earnings risk for these generators due to imbalance will be low. However, the volatility of SBP is such that a forced outage occurring during a price spike could have a catastrophic effect on the earnings of an independent generator. As environmental concerns become more important, there is an increasing drive to encourage more renewable generation to connect to the distribution networks. The latest measure is an obligation on supply companies to purchase a proportion of their power from renewable sources. A large proportion of this renewable generation will consist of wind turbines, which inherently have an unpredictable output. It is therefore not possible to reliably match the contract position that is submitted at gate closure to the delivered energy and there will be significant imbalance risk. It is particularly important to have effective risk management solutions for renewable generators 'under NETA, if the UK government targets for generation from renewable sources are to be met (C McClay and P Litchev, 2002). Comparisons between several Risk Management Solutions, namely, Licensed Supplier, Aggregation Agent and Independent Trading have been made in (C McClay and P Litchev, 2002).

3.10 Weather risk solutions

Weather insurance, in terms of protection against the effects that catastrophic events such as hurricanes, flood, windstorm etc can have on property, has been around for many years. But the use of risk management products to protect a company's profitability against non-catastrophic weather conditions is still a relatively new concept. Since the first weather derivative transaction between two energy companies, Eron and Koch in 1997/98, there have been considerable developments in respect of the use and proliferations of both weather derivatives and non-catastrophic weather insurance. The types of weather indices on which derivatives quotes can be obtained and insurance policies purchased are increasing; and the products are becoming increasingly available internationally.

The energy industry is currently the largest user of non-catastrophic weather risk management products. The reasons for the participation of the energy market are:

1. The extent to which weather impacted not only on the business of the energy companies but also the operational risks experienced by their main clients.
   There is a strong relationship between weather and residential consumption of power. A slight variation from the normal weather conditions has a significant impact on electricity supply and demand projections.
2. Weather risk management products easily added to the suite of risk mitigation products that energy risk management companies could offer their clients
3. Many energy trading companies were comfortable trading weather derivatives off the back of an understanding of their other commodity trading operations.

Weather has a significant impact on the energy industry. The energy industry will always play an important role in the market. While weather impacts many sectors, few are impacted to the same degree as the energy sector. The potential application of weather risk management products will continue to grow as the energy industry grows.

3.11 The deeper consideration

There are technical challenges and high costs as all other new technologies. However, the technical problem can be overcome and whatever energy solution is taken, there will be costs. Therefore, the system will be changing in any case. How do we influence that inevitable process?

Three main areas: 1. "Grid" including connection charging. 2. Price, including new additional costs. 3. Planning. They are all institutional, not technical. The main progress is on "Green panel" and the challenge of working within Ofgem's statutory duty. The "future offshore" is imminent and need consultation, which will probably include strategic transmission consideration. The Ofgem's preliminary consultation on Embedded Generation: price controls, incentives and connection charging. It is a very clear that rapid implementation is necessary.

Increasing levels of wind and other types of renewables are being introduced to the UK power system. The intermittent output of wind turbines raises new issues of finding the economic balance between unpredictable output and an increased range of possible operating conditions.

There is an increasing need for co-ordination between distributed generators regardless of their placement, either at the distribution level or at the customer level. The steady state and dynamic interaction between generating devices often occur simultaneously. Basic system problems include load sharing and tracking among distributed sources, reactive power flow, power factor control, voltage profiles, protection, dynamic stability and system security. There is a need to assess the effect of different mixes of distributed sources in terms of their spatial distribution and expected output fluctuations. Optimisation techniques for resources allocation should take into account the cost of incorporating distributed generation into power system, the cost of outages and the cost of maintenance subject to their capacity, mode of operation, local grid stability, power quality and reliability criteria. New controllers for individual unit or several units in close proximity and their co-ordination with the overall system operation and control strategy are required.

2002 is UK wind's most successful year with 206MW of consented wind farms installed. However, only 95 MW will be used for this year. From this fact we can see that a change in renewable energy management and system integration is exceptionally essential. In 2001, Germany installed 2640MW.

4. CONCLUSION

If we review our energy system in the last few decades, there are problem with our energy production. The conventional power generations are not environmental friendly and are going to be used up. They are cheaper than the green energy, however if we consider the cost to clean up the waste or pollution, they are not as cheap as they look.

The renewable energies are inexhaustible and environmental friendly and they are the future of the power market. Just give them time and they will become cheaper and will replace the traditional power generation. At the moment we have to spot the problems that the renewable energy have, and figure ways to resolve them. For the benefit of our future generations, governments and regulators must involve the process of pushing for green energy.

5. REFERENCES

Nancy Rader, Policy Advisor American Wind Energy Association September 1997, the Mechanics of a Renewables Portfolio Standard Applied at the Federal Level. **http://www.awea.org/policy/rpsmechfed.html**

Asociacion de Productors de Energias Renovables – APPA. Small-hydro. http://www.appa.es/dch/min_en.htm

World Wide Green http://www.worldwidegreen.com/index.asp?page=renewables-extended.htm&menumenure.htm

Unravelling Uncertainty. Distributed Generation Co-ordination Group. IEE Savoy Place.

http://www.ofgem.gov.uk/renewables/sept2002conf/ofgem_pres.pdf

PIU energy review executive summary, February 2002.

C McClay and P Litchev, 'Risk management issues for distributed generation', 2002 IEEE Power Engineering Society, Winter Meeting Jan 2002, New York.

AUTHOR INDEX